New England Wildlife

New England Wildlife

Habitat, Natural History, and Distribution

Richard M. DeGraaf and Mariko Yamasaki

University Press of New England HANOVER AND LONDON

Published by University Press of New England
One Court Street, Lebanon, NH 03766
www.upne.com

Printed in the United States of America 5 4 3 2

Library of Congress Cataloging-in-Publication Data
DeGraaf, Richard M.
New England wildlife : habitat, natural history, and distribution / by Richard M. DeGraaf, Mariko Yamasaki.
p. cm.
ISBN 0-87451-957-8 (pbk. : alk. paper)
1. Zoology—New England. 2. Zoogeography—New England. I. Yamasaki, Mariko, 1952- II. Title.
QL157.N48 D4 2000
591.974—dc21 99-86702

To

Dolores E. DeGraaf

and to

Gary A. Getchell

for their help and forbearance

Contents

Preface

New Englanders have been interested in wildlife since the time of earliest settlement. However, until now there has been no complete compilation of the natural history, distribution, and habitat relationships of the region's amphibians, reptiles, birds, and mammals. There are excellent atlases for breeding birds and preliminary atlases for reptiles and amphibians, but no publication provides detailed information for all taxa. This volume is a compilation of all relevant scientific literature, including much that is seldom available to anyone but biologists.

This book is a complete revision of R. M. DeGraaf and D. D. Rudis' *New England Wildlife: Habitat, Natural History, and Distribution* (1986, USDA Forest Service, Northeastern Forest Experiment Station, Broomall, Pa., 491 pp.). That volume was the first reference to provide managers of public lands a current compilation of the natural histories of all the species whose habitats are potentially impacted by land management. It was reprinted four times and was widely used by the public and in university courses, as well as by land managers, because of its scope and level of detail. Still, there were many gaps in our knowledge of the natural histories and distributions of many species, especially the amphibians and reptiles. Even with these limitations, the original volume was extensively used in planning wildlife habitat management by state and federal agencies in New England, in identifying environmental impacts from development, and in risk assessments by ecotoxicologists, among others.

These uses and popular appeal necessitated a revision of the original, for which data and information gathering ceased in 1983. In the intervening 15 years, research on most species, especially on amphibians, reptiles, and mammals, has increased dramatically. The scope of recent work has expanded beyond describing species biology and distributions to include their habitat relationships. Land management agencies, conservation organizations, and the public have moved beyond setting aside habitat for "wildlife"; they are concerned with maintaining or improving habitat for diverse arrays of species on public and private lands through time.

The information in this volume is still incomplete. More information is presented for some species than for others. The distributions and life histories of some are still poorly known. We hope that this book stimulates research to fill in missing information on species life histories, habitat relationships, and distribution. The most critically needed data are minimum patch size requirements; they are essential for enlightened natural resources planning yet are unknown for virtually all species.

As much as the wealth of new specific information, wildlife population changes have necessitated this revision. Rapid changes in land use over much of New England have had profound effects on wildlife; grassland species and young-forest species are declining over much of the region, while mature-forest species are increasing. Agricultural lands in many parts of New England no longer gradually revert to forest. Now many are built upon directly, and the habitats of species that occupy old fields and brushy thickets do not occur. Knowing the habitat associations of species can help predict their population trends as the landscape continues to change.

This book is in two parts: the natural history accounts, which include an illustration and a distribution map for each species, and the species/habitat matrices, which present relationships between species and forested and nonforest upland and wetland habitats. We include the life histories and habitat associations of 338 inland species that breed, winter, or reside in New England, arranged in phylogenetic order by taxonomic class. We do not provide accounts for extirpated species except for wolf and mountain lion. Individual wolves have been recently reported in Maine; habitat and prey resources are present, and dispersal from southeastern Canada may occur or reintroductions may be made in the future. The few reliable mountain lion reports are of individuals that appear to have been released. Both species arouse great public interest; we include accounts to help clarify the likelihood of their establishment in the future. Information is provided on each species' range, distribution and status in New England, habitat, special habitat requirements, age/size at sexual maturity, breeding period, number of eggs or young, home range or territory size, and food habits, and there are comments on related aspects of species' status and life history, as well as historical information. We have not included "conservation and management" sections for each species. Endangered species already have specific management plans in place; for most others, land management affects whole groups of species—the effects of a given land management practice cannot be restricted to one or even a few species. Management of vegetation structure to enhance wildlife diversity is provided in *New England Wildlife: Management of Forested Habitats* (R. M. DeGraaf, M. Yamasaki, W. B. Leak, J. W. Lanier, 1992, U.S. Forest Service, Radnor, Pa., 271 pp.).

We compiled these accounts from the available scientific literature, expert reviews, and original field research. Distribution maps for each species have been compiled from numerous sources and subjected to extensive, critical expert review and represent the best estimates currently available of species' distributions at the end of the twentieth century. Approximate continuous range in New England is shown, which may include only scattered areas where a species is found or which may include areas where a species has not been found but is presumed to occur where its required habitat conditions are present.

In Part II the matrix of species/habitat relationships use a forest-cover type classification based upon *Forest Cover Types of the United States and Canada* (F. H. Eyre, editor, 1980, Society of American Foresters, Washington, D.C., 148 pp.).

The matrix of upland and wetland nonforest habitat relationships contains the full range of such habitats in New England that relate to wildlife needs. The upland habitat types are based on descriptions in *Land Resources in New Hampshire* (U.S. Dept. of Agriculture, Soil Conservation Service, 1982, Durham, N.H., 43 pp.). The wetland habitat types are based on *Classification of Wetlands and Deepwater Habitats of the United States* (L. M. Cowardin, V. Carter, F. C. Golet, and E. T. LaRoe, 1979, U.S. Dept. of the Interior, Washington, D.C., 103 pp.).

We do not provide life history accounts for strictly coastal species, those that only migrate through the region, or species of the coastal and offshore waters of New England. These are listed in the Appendices.

Application of Information

We present this information for considering the potential responses of amphibians, reptiles, birds, and mammals to habitat change in New England. We emphasize the word *potential,* for there is no substitute for sound field work and judgment when assessing the potential impacts of a specific project or management action. The habitat associations described in this publication represent a set of hypotheses that need to be tested through additional research. This information provides insights into which species likely inhabit a given site and which do not. But it is most useful for land management and project planning; species occurrence can be more accurately predicted with larger units of land. Larger areas contain more special habitat features, more habitat interspersion and successional stages of forest growth, and hence more species diversity than do smaller areas. Conversely, smaller areas need more detailed field work and biological experience to compile accurate species lists because species occurrence can not be as accurately predicted.

Regional land-use changes and local habitat changes necessitate periodic updates of species occurrence and habitat distribution information. We urge readers to identify the species applicable to their area of interest or responsibility. We provide a generalized New England distribution map for each species, but breeding bird and other atlases provide detailed local information. We have provided a local occurrence column on the matrices in Part II for users to check species that apply to their area of concern.

Accuracy of Information

This information is not a substitute for professional field work, nor for thoroughly checking each site proposed for management. At the very least, managers need field information on the special habitat features present or lacking on each site proposed for management.

This volume lists the species that potentially occur in a given habitat. More are listed than will likely occur—the smaller the site, the fewer the species that will actually occur there. Factors in addition to habitat features, such as territoriality, diseases, and irruptive movements, among others, affect a species' occurrence in a given area. This effect diminishes with increasing size of area under consideration. Still, several site visits will be required to determine where a given species actually occurs on a given site.

No information is included on habitat size. A clue to help determine whether a given species will likely occur, assuming its special habitat requirements are present, can be found by comparing its territory or home-range size with that of the area under consideration and its landscape context. No detailed information is provided here on how many of a given species will occur in a given area. Merely dividing the project area by the territory/home-range area of a species is not recommended because not all parts of a habitat patch will be occupied, and numbers will be overestimated. We have provided sample densities when such information was available. Note the localities when consulting these entries.

We hope that the habitat associations provided here encourage avid naturalists and professionals alike to think about the opportunities to improve wildlife habitat conditions in the increasingly human-influenced New England landscape. Providing a broad range of habitats for New England's wildlife requires active land management, not further reductions in vegetation management across the landscape. It is time for thoughtful discussions about the types of habitats most needed in the region and how different types of landownerships can provide them over time. The future of New England's wildlife depends on the choices made.

Acknowledgments

We gratefully acknowledge those who contributed to this effort: James E. Cardoza, Todd K. Fuller, William M. Healy, and Matthew J. Kelty critically reviewed the introduction; Scott Jackson, Joan Milam, Christine Costello, Michael Medeiros, Timothy Stone, Richard Martin, and Thomas Maier assisted with literature reviews for the species accounts; illustrations of amphibians and reptiles by Abigail Rorer originally appeared in *Amphibians and Reptiles of New England* by Richard M. DeGraaf and Deborah D. Rudis (1983, University of Massachusetts Press); illustrations of birds are by Charles Joslin, Nancy Haver, and Andrew Magee; illustrations of mammals are by Roslyn A. Alexander, D.V.M.; all except black-tailed jackrabbit and gray wolf are reprinted from *Mammals of Ontario* by Anne Innis Dagg (Otter Press, Waterloo, Ontario, 1974). The maps of species distribution were prepared by Linda Cahillane and greatly benefited from the expertise provided by Wayne Petersen, Randall B. Boone, David E. Capen, William Krohn, Eric Orff, Gary Donovan, Scott Jackson, Bradford Blodgett, Mark Ferguson, HW Heusmann, Patrick O. Corr, Edward G. Robinson, William Crenshaw, Charles Allin, and Paul Merola. For expertise on various species and groups we thank Rachel Stevens, Chrissie Henner, Linda Thomasma, James E. Cardoza, John Litvaitis, Theodore Walski, Kristine Bontaites, and William Staats. Special status designations were provided or reviewed by Mark McCollough, John Kanter, Christopher Raithel, and Ellen Snyder. We also thank personnel of the New Hampshire Fish and Game Department, the Maine Department of Inland Fisheries and Wildlife, the Vermont Agency of Natural Resources—Department of Fish and Wildlife, the Massachusetts Division of Fisheries and Wildlife, and the New Hampshire Natural Heritage Inventory.

We thank William B. Leak for sharing his forestry expertise and John W. Lanier for encouraging the compilation of this information.

We acknowledge those who reviewed various sections in the original publication: T. J. Andrews, Terri E. Graham, Michael W. Klemens, James D. Lazell, Margaret M. Stewart, and Thomas F. Tyning reviewed the amphibian and reptile accounts; Bradford Blodgett and Chandler S. Robbins reviewed the bird accounts; and Wendell E. Dodge, Edward N. Francq, Gordon L. Kirkland, Thomas H. Kunz, Harvey R. Smith, and Dana P. Snyder reviewed the mammal accounts.

Sarah Lupis and Beth Dziokonski kept our respective map and text files organized, Mary A. Sheremeta diligently typed the manuscript, and Todd K. Fuller facilitated its publication.

Finally, we thank Blackwell Science, Inc., for permission to reprint fig. 4 from J. A. Litvaitis, "Response of early successional vertebrates to historic changes in land use," *Conservation Biology* 7 (1993):866–873.

New England Wildlife

Introduction

Recorded observations of New England's wildlife date almost from the earliest English settlement and give some glimpses of the species the colonists encountered. Some of the early accounts have been regarded as little more than promotional material for the colonies and the wildlife descriptions mere menus to allay fears of starvation. For example, Reverend Higginson in 1630 listed several birds, essentially those that could be eaten, among them partridge "so bigge they could hardly fly," "turkies . . . bigger, sweeter, fatter than those in England," and pigeons, geese, and ducks. Higginson, at odds with the Church of England, could hardly return to preach in England, a fact that may help explain his glowing accounts (Emerson 1976).

The focus of the earliest accounts was on edible, useful, or pest species; those not found in England, such as the hummingbird, were likely included for their novelty. But, unlike other seventeenth-century writers, who included anecdotal natural history observations in works with religious or political intent, William Wood (1634) focused on the land and vegetation and its human and animal inhabitants and provided the first comprehensive record of New England's natural resources at the beginning of European settlement.[1]

In Wood's record of New England's fauna, species that provide food and fur or that might depredate livestock (including the "ounce," a wildcat or bobcat) are prominent. But his rhyming verse includes keen, albeit anthropomorphized, observations "Of Beasts that Live on the Land":

> The kingly lion and strong-armed bear,
> The large-limbed mooses, with tripping deer,
> Quill-darting porcupines, and raccoons be
> Castled in the hollow of an aged tree . . .
>
> The grim-faced ounce, and
> ravenous howling wolf,
> whose meager paunch sucks
> like a swallowing gulf.

Birds are similarly addressed, but amphibians and reptiles, except for rattlesnakes, are scarcely noted. A "long black snake" that glides through the woods swiftly (almost certainly a racer), two frogs (one clearly the spring peeper, the other likely the bullfrog), a "tree-climbing toad" (the gray treefrog), and "the tortoise sought by the Indian squaw" complete the list.

In all, Wood includes more than 50 recognizable species, 20 mammals and more than 30 birds, plus some with curious, unfamiliar names: "morning-mounting lark," "eel-murthering hearn," and "drowsy madge" active in evening. *The Oxford English Dictionary* sheds light on these antiquated terms: The "morning-mounting lark" refers to the behavior of the European skylark, which makes high, hovering courtship flights in early morning (hence "morning-mounting"). While skylarks are not a North American species, the horned lark is, and it makes similar courtship flights. Horned larks have bred in Massachusetts, and may have been present in the early 1600s, although Josselyn (1672:12) states that there are no larks. The "hearn" was the name applied to the grey heron of Europe; "eel-murthering hearn" could apply to any heron-like bird, but the grey heron looks like our great blue heron. In English country folk customs, some birds have English nicknames: Jenny Wren and Robin (diminutive of Robert) Redbreast, for example. Madge, diminutive of Margaret, is the name often applied to the barn owl; "drowsy madge" is an owl.

Also curious are the missing species. Wood specifically states that he did not observe moles but omits presumably common woodchucks, bats, and rodents such as white-footed mice, which surely would have infested stored grain or barns. Among birds, Wood states that there are no magpies, jackdaws, cuckoos, jays, or sparrows; blue jays would presumably be obvious, along with also-omitted red-winged blackbirds, pileated and other woodpeckers, and bluebirds, among others. It is difficult to imagine such an observer not noticing wrens or song sparrows in spring or summer, but these species may not have adapted to human habitation in earliest settlement times. We can forgive his omitting the robin; it was in all likelihood then a highly secretive forest bird, as are woodland robins today. Wood's omissions, of course, do not mean that these species were not present; writers of the period typically noted species that were dangerous, useful, or peculiar. The rest may have made no impression, or may not have been recognized for what they were. For example, Wood may not have recognized the blue jay as a jay, probably being familiar with the pinkish-brown European jay. Likewise, the robin may have been overlooked because Wood was familiar with the little English robin, a different bird entirely. Josselyn (1672:13) also states that there are no robins, but refers to "singing Birds are Thrushes with red-breasts, which will be very fat and are good meat."

John Josselyn in *An Account of Two Voyages to New England* ([1675] 1833; republished by University Press of New England, 1988) expanded upon Wood's accounts and included a fascinating account of "The golden or yellow-hammer, a Bird about the biggness of a Thrush, that is all over red as bloud." This seems to refer to the cardinal, which is considered a relative newcomer to New England, becoming well established after 1940. Was it present earlier in the patchy field-forest presettlement landscape of southern New England? The European yellowhammer is indeed yellowish and not red; perhaps it was the closest species in size and especially call (the yellowhammer emits a metallic chip, as does the cardinal) to convey Josselyn's intent.[2]

Josselyn's account of the "troculus" (our chimney swift) also is interesting because it illustrates the European superstition of fate being represented by some birds (storks, for example) and because it includes a keen observation of the structure of the bird's tail feathers, in which the feather rachis extends beyond the vane to form a sharp point serving to prop the bird in the chimney. The chimney-looker (a kind of fire safety officer of the period, whose job it was to make sure that chimneys were clean and in good repair for use amid the closely spaced, thatched or wood-roofed houses) would have had good opportunity to view the birds up close. Josselyn reports that "the points of the feathers are sharp, which they stick into sides of the chimney . . . the nests are of a glewy substance. They produce 4–5 young, and when they depart, never fail to throw down one of their young birds into the room by way of gratitude . . . against the ruin of the family, these birds will suddenly forsake the house, and come no more."

It wasn't until Alexander Wilson and John James Audubon in the nineteenth century that substantial increases in knowledge of New England's fauna occurred. Not until the advent of economic ornithology in the late nineteenth and early twentieth centuries was New England's birdlife adequately described in the works of William Brewster, Edward Howe Forbush, and Arthur Cleveland Bent.[3]

Although field guides have included identification and approximate ranges since the early 1950s (e.g., Burt and Grossenheider 1952), the first reliable comprehensive account of New England's mammals was compiled by Godin in 1977,[4] and the first account of amphibians and reptiles by DeGraaf and Rudis in 1983. For a region so long settled, the natural histories and current distributions of its fauna are only recently fairly well known. And the distributions of most amphibians are still rather poorly known.

Today there are approximately 338 regularly occurring inland wildlife species in New England. We say "approximately" because New England's geography, topography, physiography, and land-use history, among other factors, influence the region's fauna. All species' populations are always in flux, responding to conditions that either favor or work against individuals' survival.

For example, many northern species, including mink frog, merlin, Wilson's warbler, northern bog lemming, and moose, find their southern limits of distribution in New England. Others meet their northern limits here, including slimy salamander, five-lined skink, timber rattlesnake, acadian flycatcher, opossum, and least shrew.

Their distributions reflect New England's latitudinal and elevational gradients (fig. 1), which together provide an extremely wide array of habitats, ranging from boreal forests in the north to temperate coastal forests in southern Connecticut, and from alpine tundra on the highest peaks to tidal marshes.

New England's geography and physiography set the stage for a diverse regional fauna. Against this background, human actions and uses of the land in large measure deter-

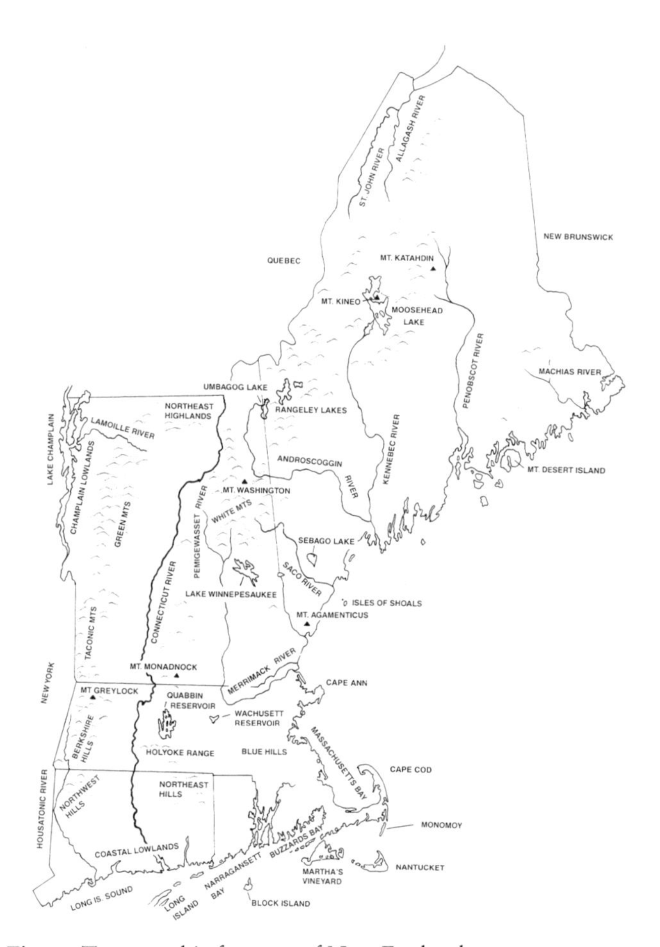

Fig. 1. Topographic features of New England.

mine which species thrive and which decline over time. Other factors are at work as well: Land clearing on the Central and South American wintering grounds affects many Neotropical migratory bird populations in New England; perhaps recent climatic trends also influence our region's fauna. The abundance and distribution of wildlife are in constant flux and reflect the interplay of these and other factors. We cannot know which species occupied New England 2,500 or even 250 years ago. The total number of species may not be much different from the number present today. There have been several extinctions and extirpations, invasions and introductions. Most species have likely had very different distributions through time. In 50 or 100 years, both the species present and their distributions will be different.

Connecticut River Valley, Northfield, Massachusetts. Photo: Richard M. DeGraaf.

New England

Extending from the Canadian border south to Long Island Sound, New England is the northeasternmost part of the United States. The six states that comprise the region—Maine, New Hampshire, Vermont, Massachusetts, Connecticut, and Rhode Island—together cover 63,012 square miles. Five of the six states are coastal; New England's coastline is more than 6,100 miles long. Vermont is the only inland state; still, much of its western border is defined by 120-mile-long Lake Champlain. New England is a region of high mountains, valleys, rolling plateaus, lakes, rivers, and seacoasts, an extremely varied landscape with a diverse fauna and a unique land-use history.

Geology and Landform

Portions of the New England landscape are among the oldest on earth. The pre-Cambrian ancestors of the Berkshire Hills and Green Mountains emerged from an ancient inland sea 1 billion years ago (B. F. Thompson 1977). Eroded by water and cracked by freezing and thawing, the rock that was to become New England slowly folded into north-south ridges and thrust upward into the now worn down Berkshire and Green Mountains that extend from western Connecticut into Vermont. To the west and separated by a long valley, the Taconic Mountains literally are thought to have slid off the top of the Green Mountains during intense crustal folding (Jorgensen 1977:32), resulting in older rock lying over younger (Johnson 1998:13). The White Mountains and the Mahoosics of New Hampshire and Maine are younger—150 to 500 million years old—and were formed by folding, faulting, and an upward infusion of magma that thrust the land into a mountainous bulge.

As the mountains were rising, running water was eroding them. Sediments washed from above accumulated in

Salt and brackish marsh, Narragansett, Rhode Island. Plate 20, *National Wetland Classification* (Cowardin et al. 1979–1985 and 1997 reprinting). Photo: Frank C. Golet.

Tuckerman's Ravine, White Mountains, New Hampshire. White Mountain National Forest photo.

Presidential Range, White Mountains, New Hampshire. White Mountain National Forest photo.

Agricultural landscape, Tunbridge, Vermont. Photo: Vermont Department of Tourism and Marketing.

the central lowland, the ancient precursor of the Connecticut Valley of Massachusetts and Connecticut. Dinosaur tracks are numerous in the shales formed from ancient lake sediments (Jorgensen 1977:54–58).

In virtually all of New England the bedrock lies within about 20 feet of, and so shapes, the surface. Almost all of the elevations over 200 feet are knobs and ridges of bedrock, and most of the lakes and river valleys lie in basins and channels in the bedrock. Only in the Cape Cod region is the bedrock so deeply buried that it influences the landscape only by its absence (Jorgensen 1977:15). Today, most of the New England landscape is composed of isolated mountains and even-topped hills, the products of erosion since major mountain-forming events ceased more than 200 million years ago.

Northeast Hills, Litchfield, Connecticut. Photo: Amy Ziffer.

All of New England has been glaciated. A mile-thick sheet of ice covered the region until 12,000 years ago. Boulder trains fanning south and southeast from parent ledges and numerous drumlins indicate the direction of the ice movement. Cirques, notably Tuckerman's Ravine in the White Mountains and the Knife Edge and Great Basin on Mt. Katahdin, were carved by mountain glaciers pulling out the rock. Broad U-shaped valleys were scoured out by glaciers; good examples exist in Crawford Notch and Franconia Notch. Glacial erratics, huge boulders moved far from their parent rock, litter the landscape. Cape Cod, Nantucket, and Martha's Vineyard are glacial terminal moraines and outwash plains formed at the southernmost glacial extent.

Besides shaping the land, glaciation scraped off the soil, mixed it with rocks, and redeposited it as glacial till in a patchy manner. Thus there is a wide range of soil texture, depth, acidity, and fertility in New England. Later glacial melting produced outwash plains, sites that were subject to frequent fire before settlement.

New England is well watered with rivers, lakes, and innumerable ponds, bogs, and swamps. The Housatonic River in western Massachusetts and Connecticut and the 410-mile-long Connecticut River flow southward into Long Island Sound. The Androscoggin, Kennebec, and Penobscot Rivers flow southward into the Gulf of Maine. The Lamoille River flows west into Lake Champlain in northern Vermont. Only the Green Mountains are without lakes, probably because the glaciers moved south along the north-south folds of the mountains. Elsewhere, lakes are numerous, especially in northern Maine, as a result of glaciation. Large glacial lakes once existed, their locations

Industrial timberlands, northern Maine. Photo by J. M. Hagan.

marked by flat topography, fine sediments, and dunes. Ancient Lake Hitchcock flooded the Connecticut Valley when a glacial readvance plugged the valley at Rocky Hill, Connecticut; the lake extended 160 miles north to Lyme, New Hampshire. After its outlet stream cut the dam about 10,000 years ago, the lake drained. Winds blew the dry sediments, and dunes 50 feet high and 2 miles long were formed at Amherst and Chicopee, Massachusetts. Other large glacial lakes occurred at Keene and Concord, New Hampshire, and Lake Vermont filled the Champlain Valley (Jorgensen 1977:126–131).

Vegetation Responses to Glaciation

The pattern and composition of New England's vegetation have developed since the retreat of the Wisconsin glaciation 12,000 years ago. The evidence from pollen records indicates that, when the ice sheet melted, both deciduous and coniferous tree species moved northward independently of one another. Throughout this period—the Holocene—the data indicate that forest communities have been, in all likelihood, rather chance combinations of species that were rarely stable for more than 2,000 or 3,000 years (see Davis 1981 for review). Tree species that are now mapped as forest types (e.g., Braun 1950) have been shown to have followed quite different migration routes. Beech, for example, lagged 4,000 years behind hickory when those species migrated into Ohio deciduous forests. In Connecticut, only 700 to 800 km (450 to 500 miles) to the northeast, beech preceded hickory by 3,000 years (Davis 1976).

Davis summarized the migration rates of trees during the Holocene in eastern North America and concluded that the speed with which forests change in response to climate is exceeded by the speed of climate change, even though the rate at which trees have extended their ranges northward is quite rapid—about 300 meters per year on average. Movement rates vary among tree species: 100 m/yr for chestnut, 200 m/yr for balsam fir, maple, and beech, 350 m/yr for oaks, and up to 400 m/yr for pines. Considering the differences in seed dispersal rates between conifers and hardwoods, between heavy-seeded and light-seeded species, and between wind-dispersed and animal-dispersed species, it is clear that forest composition has been very different through time since the melting of the last glacier.

The course of forest change in New England is fairly well understood. The pollen record shows that spruce forests thrived in southern New England as the climate warmed, beginning about 12,000 years ago. Spruces, especially white spruce, were among the first trees to appear after the ice melted, probably accompanied by balsam poplar, black ash, and hophornbeam. White spruce, the most abundant tree species at that time, suddenly and synchronously became rare over a wide region from Minnesota to Nova Scotia (Watts 1983). Spruce was replaced by a more diverse forest of conifers, including jack pine and red pine, balsam fir, and white pine, and hardwoods including paper birch, elm, and oaks. The decline of spruce was so rapid that paleo-Indians, probably hunters of woodland caribou, may have witnessed the forest change (Pielou 1991).

Forest Regions and Climate

The ever-changing forests of New England have been classified into regions by various authors (Braun 1950, Hawley and Hawes 1912, Kuchler 1964, Shantz and Zon 1924, Westveld et al. 1956) on the basis of presettlement forest conditions, original forest vegetation, or potential natural vegetation. Gradual transitions and distinct boundaries exist between these regions, depending on physiography, climate, bedrock mineralogy, topography, and soils.

Six forest regions based on the work of Kuchler (1964) and Braun (1950) are described here (fig. 2). Characteristic tree species in each region form the basis for the regional names. Certain physiographic features and climatic conditions are characteristic of each of these forest regions (fig. 3). Many of the 11 major forest-cover types found in New England forests and woodlands (see Part II) consistently occur throughout these forest regions. Forest cover types are important to forest management because timber management guides are written by cover type, and forest management *is* wildlife habitat management in heavily forested New England.

- *Spruce-Fir Forest Region.* Red spruce and balsam fir are the major tree species in the spruce-fir forest region.

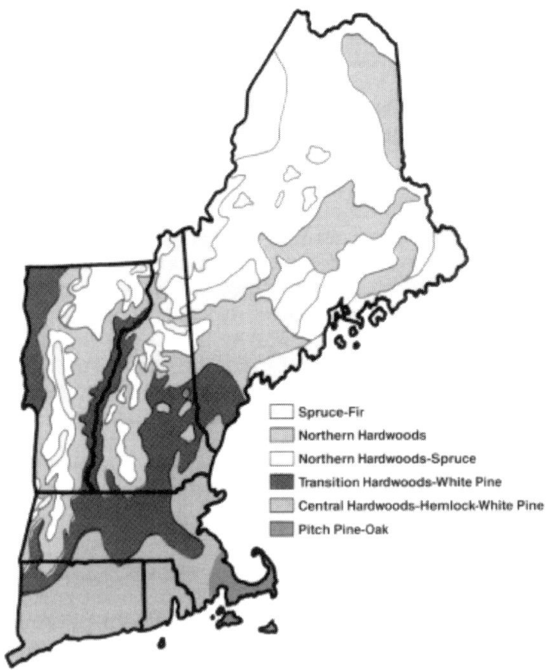

Fig. 2. Forest regions of New England (After Braun [1950] and Küchler [1964]).

Black spruce can be found in the colder, poorly drained to very poorly drained uplands, in addition to swamps and bogs. White spruce occurs frequently on abandoned agricultural fields. Other conifer associates of the spruce-fir forest region include tamarack and hemlock. Most characteristic of the hardwood component in the spruce-fir forest region are red maple, paper birch, quaking and bigtooth aspen, and mountain ash.

The spruce-fir forest region occurs in the coolest sections of New England. It is predominant above elevations of 500 ft (Bormann et al. 1970) in the central, northern, and easterly coastal counties of Maine, and above elevations of 2,600 ft in the White and Green Mountains of New Hampshire and Vermont (Leak and Graber 1974). Krummholz is characteristic at high elevations. Average date of last frost is June 1; average date of first frost is September 15, with an average of 90 to 120 frost-free days. Mean annual total snowfall ranges from 96 to more than 160 in (Kingsley 1985, Lull 1968).

- *Northern Hardwoods–Spruce Forest Region.* The major tree species in the northern hardwoods-spruce forest region are beech, white and yellow birches, sugar maple, and associated conifers on the better drained soils. Hemlock is found at lower elevations in the mountains of northern New England; red spruce and balsam fir replace hemlock at higher elevations. Imperfectly to poorly drained soils, mountain tops, and northerly exposures are characterized by spruce and fir stands. Hardwood stands with a component of hemlock and white pine characterize southerly exposures and the southern portion of this forest region. White pine is an early successional species that can be found on abandoned agricultural areas in river valleys and outwash plains.

  Characteristic of lower elevations, this forest region occurs below 2,800 ft in the White Mountains in northern and central New Hampshire (Leak and Graber 1974) and Green Mountains in southern Vermont. Average date of last frost is June 1; average date of first frost is September 15, with an average of 90 to 120 frost-free days. Mean annual total snowfall ranges from 80 to 120 in (Kingsley 1985, Lull 1968).
- *Northern Hardwoods Forest Region.* The northern hardwoods forest region includes beech, sugar maple, and yellow birch. Hemlock, balsam fir, basswood, black cherry, and white ash are also components to a lesser extent. Quaking and bigtooth aspens and paper birch are early successional species across the range of sites in this forest region. Red maple, American elm, black ash, hemlock, white cedar, and spruce are characteristic of wetter site conditions.

  The northern hardwoods forest region occurs across the higher elevations of the New England upland, in the Aroostook area and upper Penobscot, Kennebec, and Androscoggin river basins in Maine, portions of the Berkshires in western Massachusetts, and throughout the Champlain Valley in western Vermont. Elevations range from 500 to 2,600 ft. Average date of last frost is

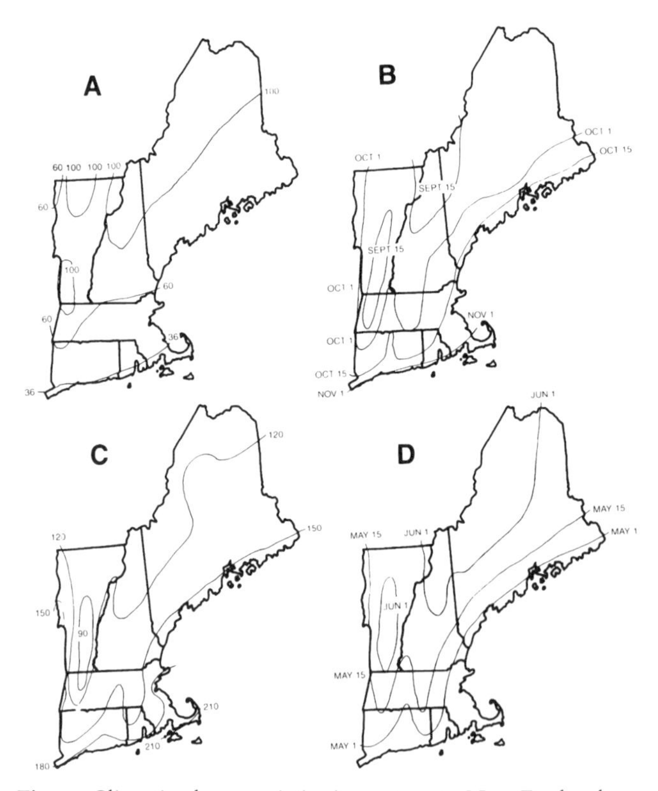

Fig. 3. Climatic characteristics important to New England forests. **A:** mean annual snowfall (inches); **B:** mean date of first frost; **C:** mean number of frost-free days; and **D:** mean date of last frost.

May 15; average date of first frost is October 1, with an average of 120 to 150 frost-free days. Mean annual total snowfall ranges from 80 to 96 in (Kingsley 1985, Lull 1968).

- *Transition Hardwoods–White Pine Forest Region.* This is a transition region. Northern hardwoods (yellow and paper birches, beech, and sugar and red maple) are the major species, with the oaks and hickories found on the warmer and drier site conditions of this region. Hemlock occurs on the cooler sites; white pine is characteristic of the well-drained, sandy sites. Red maple, black ash, and American elm can be found on the imperfectly drained to very poorly drained sites of this region.

  The transition hardwoods–white pine forest region occurs to 1,500 ft throughout the lower elevations (Bormann et al. 1970) of the New England upland in north-central Massachusetts and southern New Hampshire, reaching northward through the Champlain and Connecticut river valleys. It also occurs in southwestern Maine and the lower reaches of the Androscoggin, Kennebec, and Penobscot river basins. Average date of last frost is May 15; average date of first frost is October 1, with an average of 120 to 150 frost-free days. Mean annual total snowfall ranges from 48 to 80 in (Kingsley 1985, Lull 1968).

- *Central Hardwoods–Hemlock–White Pine Forest Region.* This region is a mixture of northern and central hardwoods. Formerly, American chestnut was the dominant tree species of this region. Now, however, red, black, and white oaks, hickories, gray, yellow, and black birches, and beech are the major species. The primary conifers in the region are white pine and hemlock. Red maple occurs on all sites and forms nearly pure stands on the wetter sites. Pitch pine can be found on sandy outwash sites.

  The central hardwoods–hemlock–white pine forest region occurs across the southern portions of the New England upland and seaboard; throughout Massachusetts, Connecticut, and Rhode Island; and along the southwestern Maine and New Hampshire coast. Elevations range from sea level to 1,000 ft. Average date of last frost is May 1; average date of first frost is October 15, with an average of 150 to 180 frost-free days. Mean annual total snowfall ranges from 32 to 64 in (Kingsley 1985; Lull 1968).

- *Pitch Pine–Oak Forest Region.* The pitch pine–oak forest region is composed of pitch pine, scrub oak, and scarlet and black oaks. Repeated burning creates favorable conditions for pitch pine and scrub oak, especially on sandy, well-drained soils. Lowbush blueberry and other ericaceous shrubs also abound on these burned areas.

  The pitch pine-oak forest region of New England essentially lies on the northeastern extension of the coastal plain formation, represented by Cape Cod, and ranges in elevation from sea level to 150 ft. Much of the Cape is underlain with poorly stratified sheets of sand, gravel, and clay. The Montague Plain in western Massachusetts is an inland example of the vegetation characteristic of this region. An isolated example in southern Maine exists because of severe forest fires in the late 1940s. Average date of last frost is May 1; average date of first frost is November 1, with an average of 180 to 210 frost-free days. Mean annual total snowfall is less than 32 in (Kingsley 1985; Lull 1968).

Forest Disturbance

Eastern forests have been changing gradually since the end of the last glaciation, in response to climate change. Oaks, beech, hickories, maples, birches, and chestnut, among others, were common and extended their distributions northward at varying rates during the 10,000 years, creating a shifting pattern of forest-cover types on the landscape (Davis 1976). Five major types of disturbance have altered New England's forests: windthrow, fire, exotic pests and pathogens, agriculture, and logging. In addition, native insects, disease, ice storms, drought, flood, landslide, and avalanche have caused minor, but occasionally major, disturbances. For example, the spruce budworm, a native insect, periodically damages millions of acres of spruce-fir forest in northern New England and eastern Canada; recorded outbreaks date back to the 1700s. Also, an ice storm in January 1998 caused severe forest damage on more than 12 million acres in Vermont, New Hampshire, and Maine (Irland 1998).

Catastrophic disturbances from windthrow and fire occur at intervals of about 1,150 and 800 years, respectively, in New England (Lorimer 1977). However, localized disturbances—sometimes severe—occur at much shorter intervals. Major hurricanes and windstorms occurred several times during the twentieth century; the last severe hurricane occurred in 1938, when several billion board feet of timber were blown down, from Rhode Island to central New Hampshire (Curtis 1943). The effect of that hurricane was great because of the structure and composition of forest—200 years of occupancy and subsequent abandonment produced a high proportion of pine, which suffered far greater damage than hardwoods (Foster and Boose 1992). Storms of similar magnitude occurred in 1635 and 1815 (Channing 1939) and are estimated to occur, on average, at 150-year intervals. Although wind has a dramatic effect on forest overstories, it has little impact upon successional trends and overall species composition because of the presence of a shade-tolerant understory and the aggressive nature of New England pioneer species. Wind has minimal effects upon soils in New England except for some churning and mounding from uprooted trees.

Intense forest fires generally occur on dry sites at high or low elevations, and less frequently on mid-slopes. Fire-site soils tend to be outwash sands and gravels, fractured

Hurricane damage, 1938, Bartlett, New Hampshire. Photo: U.S. Forest Service.

or loose rock, or shallow over bedrock. Generally, these sites support mixed-wood or softwood types such as white pine, oak-pine, pitch pine, or spruce-fir. The impact of fire on forest conditions is more severe than that of wind. Pitch pine barrens occur on repeatedly burned areas that often were originally in white pine or oak-pine. Most bare-rock mountain tops below 3,800 ft in New England are the result of fire, which destroyed organic matter and allowed the thin soils to wash away. Productivity of outwash and shallow-bedrock sites seems to be reduced by fire because of the destruction of organic matter. However, even very large, intense burns revegetate with surprising rapidity; the 7,600-acre Beddington Burn in Maine was 97 percent stocked with tree species and other vegetation 5 years after the 1952 burn (Mills 1961). Low-intensity ground fires have maintained open habitats in Maine for at least the past 900 years (Winne 1988).

In addition to natural fire, Indians in southern New England burned the forest periodically to drive game for hunting, clear fields for planting, and open the forest for traveling (Day 1953, Cronon 1983); many such fires burned until extinguished by rain (Morton [1632] 1883). The use and frequency of fire varied; burning was not always extensive (Russell 1983). Native Americans in southern New England used fire to cultivate the land and open the forest more than those in northern New England. Fire exclusion throughout much of the twentieth century has reduced the occurrence of open habitats in much of New England.

Only a few tracts in New England remain unlogged. Much of the logging (as opposed to agricultural land clearing in the 1700s) took place in the mid to late 1800s when the best softwoods in mixed-wood stands were high-graded and the softwood stands were clear-cut. Until recently, the only hardwood stands that were heavily cut were those along railroads and those cleared for agriculture in the past. Logging was similar to windthrow in that it did not have an impact on successional pathways or soils. However, intense fires fueled by slash sometimes followed, especially in softwood stands on dry sites. Much of the paper birch in the White Mountains originated after such fires early in this century.

More than half of New England was cultivated or grazed in the past, though much of the area has reverted to forest. The effects of agriculture are dramatic: loss of nutrients, changes in soil profiles, and major cover-type conversions from hardwoods to spruce-fir in the north and from hardwoods to old-field white pine farther south (Bormann 1982).

The introduction of exotic pests and insects has caused rapid and irreversible changes in New England's forests. For example, the range of American chestnut had been expanding slowly across North America for about 8,000 years (Davis 1976). Within 50 years of its introduction in 1904, the chestnut blight fungus had eliminated chestnut as a dominant forest species throughout its range (Brewer 1995). American elm and American beech have been greatly reduced as canopy trees as the result of introduced insects and pathogens (Bey 1990, Tubbs and Houston 1990).

Acorns and other hardwood seeds represent the most valuable and energy-rich plant food available in the dormant season (Robbins 1993), and supported vast flocks of passenger pigeons, once the most abundant bird in the world (Bucher 1992). Oak increased in importance during the twentieth century as American chestnut declined; oaks have also declined subsequent to the onset of fire protection (Abrams et al. 1995) and defoliation by the gypsy moth, introduced into Massachusetts in 1869 (Forbush and Fernald 1896).

Among all sources of forest disturbance, agriculture has had the greatest impact on the forested landscape in New England because it caused major changes in cover types and soils over a wide range of sites. Although fire lowered the productivity of dry-site softwood stands, which reverted to earlier successional stages, fires did not cause major shifts from hardwood to softwood successional paths. Windthrow and logging continue to maintain diversity by initiating earlier successional stages but have little impact on the forest ecosystem. Introduced pests and pathogens have had important effects on species composition within forests.

Legacies and Lagtimes: Land-Use History and Wildlife

Three major trends in New England's wildlife are apparent in the last several decades of the twentieth century: Forest species are increasing, grassland/shrubland species are declining, and many southern birds, especially coastal nesting species, have been spreading northward into New England. A few species have extended their ranges southward in the twentieth century. For example, ravens have again been nesting quite far south in recent years, after being extirpated long ago, and even moose are now found in Connecticut. Other species have undergone great changes in abundance and distribution throughout the period for which reliable records exist, but the trends of increasing forest species and declining open-country species are the most dramatic and most clearly understood because they have their roots in our past uses of the land.

Pre-European Conditions

Before European intrusion, substantial parts of southern New England were quite open due to the presence of native prairies, Indian agricultural clearing and fuelwood cutting, and periodic hurricanes. Throughout the region, abundant beaver meadows and periodic fire on dry sites imparted a shifting mosaic of open habitats to the forested landscape. In the period 1500–1000 BP, Indians south of the Kennebec River in Maine shifted from food gathering to food production and storage (Likens 1972). Indians raised corn, beans, squash, pumpkins, and tobacco as part of their annual subsistence cycles (Mood 1937, Russell 1961). Fields were, on average, used for 8 to 10 years (Wood [1634] 1977:35) until the soil fertility declined and new fields were needed. This shifting agriculture, in combination with the presence of native prairies, made for a fairly open landscape in southern New England along the coast and along the major rivers of the region. In the forested interior, abundant beaver also imparted a substantial component of open habitats in all stages of succession as they created and abandoned flowages.

Corn cultivation resulted in major alterations of the landscape (Whitney 1994:100–107). Indians possessed the technology for clearing large areas of a heavily wooded landscape for corn planting; girdling and fire together provided an efficient mechanism for clearing the forest (Loskiel 1794:1–55). New England Indians were knowledgeable about the use of fire (Patterson and Sassaman 1988); Indian burning resulted in a mosaic of fields and forests in all stages of succession in the vicinities of villages and old village sites.

Indians of southern New England generally burned the uncleared forests in spring or fall to facilitate travel and drive game (Russell 1983). This annual burning produced open, park-like woodlands with rich herbaceous layers. In addition, the forest was burned to open land for cultivation (Trumbull 1797, Parker 1910), to improve the forage for game (Dwight 1823 Vol. 4:50), and to enhance the production of strawberries, blueberries, and other fruiting shrubs (Bromley 1935). Periodic low-intensity ground fires tended to eliminate white pine (*Pinus strobus*) (Patterson and Backman 1988) and enhance mast-producing species such as oaks (Brown 1960).

The resulting abundance of deer, turkey, rabbits, and "partridge" (whether ruffed grouse, bobwhite quail, or heath hen is uncertain—all were present) was noted by the earliest English colonists (Wood [1634] 1977; Josslyn [1675] 1833; Higginson 1630). This abundance of game was produced by Indian agriculture, fuelwood cutting and periodic intentional burning of woodlands, and hurricanes, which together created a patchy landscape of fields and forests in various stages of succession. Nonagricultural Indians in northern New England did not create such an open landscape; populations were lower, and travel was by canoe rather than overland. The net result was a heavily forested landscape north of the corn-planting (at least 120 frost-free days) zone.

Prior to the Colonial period, much of the northeastern coastal forest of the United States from southern Maine to Virginia had a considerable amount and variety of open habitats. The dominant habitat of the region was forest, but there is ample evidence that extensive grasslands and oak openings were common in eastern North America both along the coast and inland before European settlement (e.g., Askins 1997).

The heath hen occupied scattered grasslands, native prairies, and blueberry barrens from Virginia north at least to Massachusetts and possibly to Maine (Palmer 1949) and on the larger offshore islands. The openness of the landscape is confirmed by the abundance of the heath hen in the mid seventeenth century: it was then "so common in the area around Boston that laborers and servants stipulated that employers not serve it more than a few times in a week" (Winthrop [1636] in Nuttall 1832:67–68).

Where agricultural Indians had lived, the forests that had reclaimed their old clearings by the early eighteenth century probably did not contain large, old trees. Indian populations dwindled soon after contact with Europeans, due to persecution and disease[5]; early settlers who penetrated interior southern New England likely encountered a landscape that was much more heavily forested than that which had existed a century earlier. For example, in 1642, John Winthrop (in Hosmer 1959) described "many thousands of acres of rich meadow" along the Saco River in present-day Maine at the time of the second European expedition to the White Mountains.

European Settlement Period

Now mostly forested, the New England landscape has undergone dramatic changes over the last 350 years. Land was cleared for agriculture, slowly until the 1750s and then at an increased pace until, between 1800 and 1860, 75 percent of the arable land in southern and central New England was in pasture and farm crops. One hundred years later, New England was again mostly forested—the result of an era of land abandonment that began soon after the opening of rich farmlands in Ohio via the Erie Canal in 1825. There were few takers for the abandoned farms so arduously brought under cultivation. Except for the miles of stone walls and old cellar holes in the woods, there is little to indicate that industrious farmers once inhabited the land.

Around 1910 the cutting of the white pine that had seeded into impoverished tilled land and dry pastures constituted the last major land clearing in southern and central New England. Once cut, most sites grew up to hardwoods, which in the 1920s and 1930s produced ruffed grouse populations that sustained market shooting, dramatic testimony to the wildlife population levels that can be attained when ideal habitat conditions exist over the entire landscape.

Today, about 65 percent of southern New England and more than 90 percent of northern New England is forested. Each year, except on the industrial timberlands in Maine, the age and extent of the forest increase (Brooks and Birch 1988).

Wildlife Responses to Landscape Change

Most of the faunal changes in the past 200 to 300 years were due to the vast scale and intensity of habitat change that swept New England during settlement and to subsequent land abandonment and reforestation.

No forest mammal became extinct during the settlement period in New England, although the gray wolf, mountain lion, elk,[6] and caribou, among others, were extirpated. The sea mink, known only from fragmentary skeletal remains, is now considered a subspecies of mink; it is believed to have become extinct by about 1900 (Manville 1966, Cardoza in press). White-tailed deer and black bear have been among the most utilized large mammal species in New England forests, and their abundances serve as excellent indicators of change during the past 5,000 years. White-tailed deer and black bear were the most common food animals recorded in Indian middens, approximately 4,300 years ago in southern Massachusetts and Rhode Island (Loskiel 1794:65, Waters 1962). Deer were taken in autumn and bear in winter during hibernation (Godman 1826:124).

White-tailed deer densities in disturbed forests of the northeastern United States immediately prior to the Colonial period were estimated at between 3.8 and 5.8 per square kilometer (Mattfeld 1984). In contrast, undisturbed northern hardwood forests were estimated to have supported 1 deer/km^2 (Cooperrider 1974). Deer were exploited for food and hides more or less continuously as settlement progressed. White-tailed deer in southern New England were reduced to very low levels soon after settlement. Hunting of deer was first prohibited in 1646 in the town of Portsmouth, Rhode Island, and harvest seasons were set in 1698 in Connecticut (Allen 1929) and 1694 in Massachusetts (J. E. Cardoza pers. commun.). Hides were shipped to England and Europe from the period of earliest settlement; for example, Quebec exported 132,271 deerskins in 1786 alone (Schoen et al. 1981).

Former agricultural land reverting to forest, North Stonington, Connecticut. Photo: Amy Ziffer.

In Maine, which was settled both differently and much later than southern New England and the Canadian Maritimes, white-tailed deer populations have changed dramatically over relatively short time periods in response to the logging and land clearing. From early accounts, deer were common primarily along the coast and up the major river valleys. Interior Maine supported few deer. Browse was scarce under the closed canopy, and wolves and deep winter snows limited the inland herd. For the first 140 years, settlement in Maine progressed very slowly due to strife with the Indians. In 1742 the known Colonial population of Maine was 12,000, mostly confined to the coast west of Penobscot Bay (Smith 1949). After the final peace treaty with the Indians in 1760, settlement advanced inland along the Saco, Androscoggin, Kennebec, and Penobscot rivers (Fobes 1944), following the logging of select pines for ship masts. Deer increases also occurred in north-central New Hampshire after the early logging of pine (Silver 1957). As logging advanced north and westward, deer abundance in the interior followed. However, as human settlement increased along the Maine coast, extensive clearing for agriculture, fuelwood cutting, logging for the West Indies cooperage trade, and the pasturing of sheep and cattle marked the temporary end of coastal deer abundance. The town of Wells, York County, which had annually elected deer and moose reeves,[7] or overseers, from the early 1700s, discontinued the position around 1800 because of the absence of deer and moose (Remick 1911).

From 1820 to 1840, the cutting of pine spread northward, and pine was essentially eliminated by 1861 (Wood 1935). Though deer were still uncommon north and west of Mt. Katahdin by 1850, deer were so abundant in east-central Maine that market hunting was profitable (see Banasiak 1961). Huge herds built up by the time of the Civil War, until the regenerating forest grew out of reach of the deer, and a massive die-off occurred during 1865–1870.

A second wave of logging for spruce after the Civil War again created favorable habitat for deer in eastern and central Maine; deer abundance reached levels not attained before or since, with marked browse lines clearly evident on cedar along pond shores by about 1900. In agricultural southern Maine, however, deer did not reappear until 1890. The effects of periodic unregulated logging on deer are dramatically revealed in the settlement history of Maine.

Presettlement bear density in the primeval deciduous forest in the East has been estimated at between 19/100 km^2 (Shelford 1963) and 8/100 km^2 (Seton 1929 Vol. 2 pt. 1:129). Black bears reached their lowest population between 1860 and 1880 but were apparently never extirpated completely from Massachusetts (Cardoza 1976). Bear density in western Massachusetts in 1993 was about 30/100 km^2 and the state population about 975 to 1,175; in 1998 the black bear population was estimated to be 1,750 to 1,800 animals (Fuller 1993).

Extirpations and Extinctions

Among mammals, only the elk, caribou, wolf, and mountain lion have not recolonized New England since extirpation. Elk were extirpated soon after settlement and caribou by 1910. Caribou were extirpated from northern Maine and New Hampshire by 1910, with the last sighted on Mt. Katahdin in 1908 (Lorenz 1917). Reintroductions at the Corbin Preserve (New Hampshire) in 1890 and in Maine, the most recent in the late 1980s, quickly failed. Wolves and mountain lions were extirpated by the mid nineteenth century. Numerous sightings of mountain lions have been reported in New England during the past 50 years, indeed throughout this century (Leach 1909), although sightings are useless as a basis for describing mountain lion abundance or distribution (VanDyke and Brocke 1987). Wolves have recently been observed in Maine. Human-caused mortality—as indexed by road density and thus human access—seems to limit wolf distribution and numbers: Wolf survival is low when road densities exceed 0.6 to 0.7 km/km^2 (Thiel 1985, Mech et al. 1988, Mech 1989, Fuller et al. 1992). Mountain lions in good habitat in the western United States occur where road densities are less than 0.4 km/km^2 (Van Dyke 1983, Van Dyke et al. 1986). Road densities exceed these thresholds throughout most of New England. Both species may now be replaced ecologically, to some extent, by eastern coyotes,[8] which now occur widely throughout northeastern North America (Moore and Parker 1992).

The historical evidence associated with the well-known bird extinctions in eastern North America illustrates the composite factors that generated these events. For example, human persecution figured prominently in the extinctions of the great auk (Montevecchi and Kirk 1996) and Labrador duck (Chilton 1997). Simultaneous taking in great numbers and rapid clearing of habitat led to the extinction of the passenger pigeon (Bucher 1992). The heath hen was extirpated by agriculture on the mainland by 1869 but persisted on Martha's Vineyard well into the twentieth century (Gross 1928). The small remaining population was subject to extreme predation pressure by goshawks and to low productivity. A wildfire reduced the spring population by more than 80 percent—to levels below minimum viability—in 1916. A subsequent ban on fire allowed forests to replace the open heaths, and the heath hen—a lone male—was last seen in 1932 (Gross 1932).

Local and regional extirpations of large predators, fur bearers, and game are well documented as the landscape was settled. Wolves were persecuted from earliest settlement and were gone from New England by about 1850 (Goodwin 1935). The last reported Connecticut mountain lion was killed in 1767 (Adams 1896). The last New England specimens were shot in Vermont in 1881 (Titcomb 1901), in New Hampshire in 1885 or 1887 (Jackson 1922, Preble 1942:28), and in Maine in 1891 (Goodwin 1936). Beaver were extirpated in southern New England soon after settlement. The intensity of exploitation is indicated by the records of John Pynchon, a fur trader in Springfield, Massachusetts. He shipped 8,992 beaver pelts from the Connecticut River drainage between 1652 and 1657 (Judd 1857).

Many game species reached extremely low population levels in New England and throughout the Northeast around the turn of the century, following the so-called "era of exploitation," 1850–1900. The second half of the nineteenth century was a period of intense market hunting of deer, waterfowl, passenger pigeons, and shorebirds as well as gamebirds, and also plume hunting of herons. It was widely believed at the time that most species were about to disappear anyway, so they might as well be taken sooner rather than later. Laws reflecting society's desire to protect and manage wildlife were largely established in the twentieth century, and recovery from persecution has been rapid and dramatic. Nowhere else in the world was so large an area cleared and rapidly abandoned as in New England, yet there were relatively few extinctions, and virtually all extirpated wildlife species later recolonized the region.

Introductions and Reintroductions

Many exotic species have been introduced into New England, and several extirpated native species have been reintroduced. Among the earliest intentionally introduced species was the ring-necked pheasant, which was released in several parts of New England in the Colonial Period. Red foxes were also introduced between 1650 and 1750 (Churcher 1959), as were rock doves or domestic pigeons. House mice and black rats are believed to have arrived about the time of first settlement or soon after; the Norway rat arrived about 1775 (Godin 1977). The black rat has not been detected in New England since the 1930s (Parker 1939). House sparrows and starlings spread into New England from releases in the New York City area in 1852 and 1890, respectively (Kieran 1959:320). The mudpuppy, a large neotenic salamander native to the Mississippi and Great Lakes drainages including Lake Champlain, was introduced into a tributary of the lower Connecticut River in this century (Warfel 1936). The red-eared slider, the once-common green hatchling pet turtle, exists in the wild in a few places in southern New England but is not known to breed successfully.

Introduced big-game animals that persisted for a time in the wild include European boar in New Hampshire and fallow deer in Massachusetts. Boar were introduced from Germany in 1899 by Austin Corbin to stock his preserve in Sullivan County, New Hampshire. An unknown number escaped after the 1938 hurricane and a small population existed in the area of the Corbin Preserve through the 1950s; the last reported ones killed from that population were shot in 1961 (Godin 1977), but individuals are still (rarely) taken in the area. Fallow deer were introduced from a park in Worcester to Nantucket (date unknown) and to Martha's Vineyard in 1932 and 1938 for hunting (Shaw and McLaughlin 1951). By the 1949 opening of deer hunting, there were an estimated 150 fallow deer on Martha's Vineyard. They persisted in low numbers until about 1980. Black-tailed jackrabbits were introduced to Nantucket from 1925 to the 1940s and are well established there.

Reintroductions of formerly present native species are for the most part more recent than exotic introductions, although white-tailed deer were released as early as the late nineteenth century in Vermont and elsewhere, having been rare or absent for a century or more. Beaver were successfully reintroduced in Lenox, Massachusetts, in 1932 (Pell 1944), having been extirpated in the early 1800s. After 1932 beaver also naturally reinvaded their former Massachusetts range from New York (Shaw 1948). Wild turkeys were successfully reestablished from trapped-and-transplanted birds from New York in the 1970s and now are widespread throughout and beyond their former range. Peregrine falcons, extirpated from most of their eastern North American breeding range by 1960, because of eggshell thinning from accumulation of DDT and other persistent pesticides (Hickey 1969), have been successfully reestablished through hacking of young from captive-reared adults in the 1970s and 1980s.

Bobwhite quail are present on Cape Cod and the Islands and elsewhere in eastern and southern New England, but they are not the native bird. The large, pale New England phenotype,[9] a bird of open forests that probably benefited from Indian burns, was greatly reduced by overshooting and mortality in heavy winter snows by about 1860 (Griscom and Snyder 1955) and was swamped by many reintroductions from southern stocks.

Other introductions of species extirpated from the region have occurred. Among these are marten, reintroduced to Vermont in the past decade from wild-trapped individuals in New York and Maine. Snowshoe hares were imported to Massachusetts for hound hunting largely from New Brunswick until 1990; some private beagle clubs continue the practice (J. E. Cardoza pers. commun.).

Recent Changes in Wildlife Distributions

The current trends in New England wildlife populations, notably the increases in forest species and the declines of open-habitat species, underscore both the dynamic nature of the landscape and the long-term effects of past human activities. Within the past several decades a number of species have undergone great changes in abundance and distribution in New England. The state rare and endangered species monitoring programs and atlas projects, especially for birds, have documented many changes.

Grassland and shrubland species, especially birds, are the most rapidly declining species in New England (Askins 1993). Mature-forest birds in eastern North America have been shown to be quite tolerant of patchy disturbance within extensive areas of forest (Webb et al. 1977, Maurer et al. 1981, DeGraaf 1991), but grassland and shrubland birds are specialists that quickly disappear from a site as habitat patches become too small or as succession proceeds (Askins 1993). Bobolinks, for example, abandon grassy hayfields when field area drops below 25 acres (Bollinger and Gavin 1992). Grasshopper sparrows occur only in grassland that is interspersed with bare ground (Smith 1963, Whitmore 1981), and yellow-breasted chats inhabit brushy old fields only until they begin to be invaded by overtopping trees (Shugart and James 1973, C. F. Thompson 1977). The decline of agriculture and reversion of forest have essentially eliminated grassland birds from most of the New England landscape (fig. 4), although they were common half a century ago (Askins 1993, Bagg and Eliot 1937). Resident forest birds have increased in abundance in extensive woodlands; the pileated woodpecker, which requires large trees (at least 20 inches in diameter) for nest and roost cavities, has significantly increased in abundance over the last 25 years (Bull 1974, Norse 1985). Some Neotropical migratory birds, especially species that are primary-forest specialists on the wintering grounds, such as wood thrush,

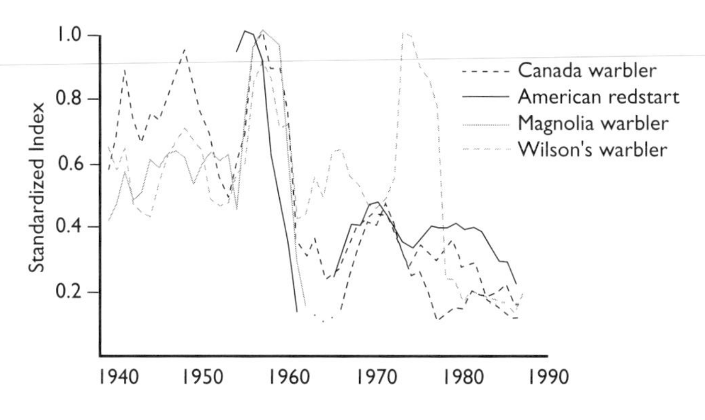

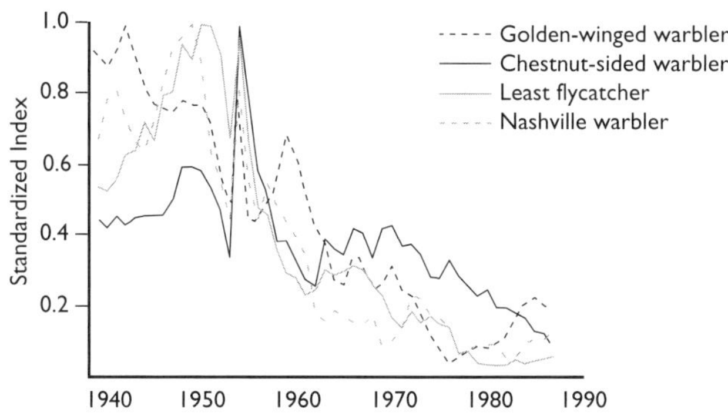

Fig. 4. Annual indices of some migrating land birds that nest in the forests of New England. Samples (number observed/hour) were obtained in eastern Massachusetts during spring migrations (Hill and Hagan 1991). Indices were standardized by dividing the highest annual observation rate for a particular species into each annual observation rate for that species. (After Litvaitis [1993]. Reprinted by permission of Blackwell Scientific Publications, Inc.)

are declining in some parts of their breeding range (Rappole et al. 1992), largely as the result of events on the wintering grounds or during migration (DeGraaf and Rappole 1995, Terborgh 1989, Hagan and Johnston 1992).

During the last three decades, 41 Massachusetts breeding birds have increased significantly in abundance (Veit and Petersen 1993:26). They are:

Common loon	Northern goshawk
Double-crested cormorant	Peregrine falcon
Great blue heron	Wild turkey
Great egret	American oystercatcher
Snowy egret	Willet
Little blue heron	Wilson's phalarope
Cattle egret	Herring gull
Glossy ibis	Great black-backed gull
Mute swan	Mourning dove
Canada goose	Barn owl
Gadwall	Red-bellied woodpecker
Turkey vulture	Acadian flycatcher
Osprey	Willow flycatcher
Bald eagle	Fish crow
Common raven	Worm-eating warbler
Tufted titmouse	Northern cardinal
Carolina wren	Brown-headed cowbird
Blue-gray gnatcatcher	Orchard oriole
Northern mockingbird	House finch
Blue-winged warbler	Evening grosbeak
Cerulean warbler	

Some of these increases are due to direct human efforts to restore extirpated species. For example, wild turkeys have been reestablished throughout New England through releases of wild-trapped birds from other states. Peregrine falcons and bald eagles are now well established through hacking programs, and ospreys and barn owls have responded to the placement of nesting platform and nest boxes. House finches spread rapidly throughout New England in the 1970s to early 1980s from caged birds released in New York City in the 1940s (Bull 1974).

Winter bird feeding has probably allowed northern cardinals and tufted titmouse to colonize New England by reducing mortality in severe winter weather.

Less certain are the causes of the spread of southern coastal waders and shorebirds northward along the New England coast: Great egret, snowy and cattle egrets, little blue heron, glossy ibis, American oystercatcher, and willet are examples. Oystercatchers and willets formerly nested in New England and may be recovering from shorebird hunting, which continued until the mid 1920s. Northern incursions of herons and egrets during summer have occurred for a long time (Cottrell 1949); Veit and Petersen (1993:27) suggested that their range extension may be due to general population increases of these species. Such increases may also be due to recovery from plume hunting early in this century. Increases of herring gull and great black-backed gull populations are generally understood to have occurred largely in response to open landfills and disposal of commercial fishing wastes near shore and island nesting sites (Drury 1973, 1974).

Populations of southern forest birds have also recently increased in Massachusetts and elsewhere in New England: Carolina wren, red-bellied woodpecker, blue-gray gnatcatcher, worm-eating warbler, and orchard oriole, as well as the aforementioned cardinal and titmouse, are examples.

The mockingbird presents an interesting case. A sedentary fruit-eater, it apparently expanded its range northward from the Southeast largely following the planting of multiflora rose in highway median strips after World War II and its subsequent spread (Stiles 1982). Now it thrives in suburban lots landscaped with fruiting shrubs and in more rural places in multiflora rose thickets.

Global warming has frequently been proposed as a possible cause of species range changes, but several cases suggest that a single cause may not be responsible. In the preceding list of increasing species, for example, several northern species have extended their ranges southward: common loon, common raven, and evening grosbeak.

Veit and Petersen (1993:28) listed 24 bird species that have declined significantly in Massachusetts since 1955:

Pied-billed grebe
American bittern
American black duck
Northern harrier
Cooper's hawk
Red-shouldered hawk
King rail
Common moorhen
Piping plover
Upland sandpiper
Roseate tern
Common tern
Arctic tern
Short-eared owl
Whip-poor-will
Olive-sided flycatcher
Eastern bluebird
Gray-cheeked (now Bicknell's) thrush
Golden-winged warbler
Northern parula
Vesper sparrow
Grasshopper sparrow
Bobolink
Eastern meadowlark

Marshland, coastal, and early-successional birds dominate the list. Wetland habitats have declined due to development and pollution, and grasslands have given way to development and succession. Terns have declined in the face of growing gull populations. Some forest species have been declining even as forests have reclaimed more of the New England landscape: Cooper's hawk, red-shouldered hawk, olive-sided flycatcher, and northern parula. Habitat destruction on the wintering grounds may be a factor in some species' declines, especially the whip-poor-will, olive-sided flycatcher, gray-cheeked thrush, golden-winged warbler, and northern parula (Veit and Petersen 1993:29).

In tracking species population changes it is difficult to separate the effects of land-cover change, that is, increasing forest and decreasing grassland/shrubland, from external "effects" such as global warming or destruction of wintering ground habitat in the case of Neotropical migrants. Some species seem to respond to short-term climate fluctuations—Carolina wrens periodically nest in southern New England during years of mild winters but are pushed southward after severe winters.

Among mammals there are some dramatic changes in abundance and distribution. Population increases are more obvious than decreases, and perhaps no changes are more obvious than those of moose and black bear. In the past 30 years moose have extended their range from northern Maine to Massachusetts, and stragglers occur as far south as Connecticut, as they did at the time of settlement (Morton 1637, Adams 1896). In the past 20 years, black bear have been increasing throughout their range and are now commonly seen in suburban areas. Coyotes now occupy all of New England, after first arriving in northern Maine in the 1930s and Vermont in the 1940s; coyotes first appeared in other New England states in the 1950s (Parker 1995:20) and in southern New England in 1957 (Pringle 1960), a century or more after wolf extirpation. Fishers have increased their range southward into Connecticut, largely in response to the return of the forest but also to introductions in 1989 to 1990 (J. E. Cardoza pers. commun.).

The New England cottontail has become much reduced in distribution with the decline of old field and brushy habitats, and now has a quite spotty distribution in much of the region (Litvaitis and Villafuerte 1996).

Conclusions

An examination of reliable records shows that species population changes are nothing new. There is also no baseline from which to measure species changes, only changes in the landscape itself. For example, Goodhue's (1877) "The Birds of Webster and Adjoining Towns" described bird abundances typical of southern New Hampshire shortly after the peak of agriculture. Vesper sparrows were listed as the most abundant sparrow and the chestnut-sided the most abundant warbler, a clear indication of recent land abandonment. Absent from the area were wood thrush and mourning dove. The wood thrush was first reported in summer in New Hampshire in 1894 near Mt. Moosilauke (Elkins 1957). Jeremy Belknap's 1792 *History of New Hampshire* listed "killdee" (killdeer) and "turtle dove" (mourning dove) as typical of open land, yet both were absent from Goodhue's list 85 years later. Urbanization results in stark changes. In Cambridge, Massachusetts, for example, there were 26 nesting bird species in the period from 1860 to 1873, and 9 from 1960 to 1964 (Walcott 1974).

Over the 350 years of settlement and early exploitation of wildlife, only three species, all birds, have been rendered extinct: Labrador duck, great auk, and passenger pigeon. The heath hen was a subspecies, which is also extinct. Among mammals, only the sea mink and eastern elk, both subspecies, are extinct. Changes in reptile and amphibian populations are not as well known. Some species are probably greatly reduced in range. Rattlesnakes, for example, were once found in Maine (Wood [1634] 1977), and five-lined skinks, now found in southwestern Connecticut and the West Haven area of Vermont, were formerly found in Barre, Massachusetts (Storer 1840:19), and New Bedford (Allen 1870:260), Massachusetts.

In sum, the effects of landscape change have exerted lasting effects on more species than did direct exploitation. A century and a half after the peak of land clearing, the effects of forest regrowth are still occurring, as revealed by declining open-country species and increasing forest species (fig. 5).

All species populations are in constant flux, now responding for the most part to human activities. Those activities that fundamentally alter land cover have far-reaching and long-term effects on wildlife. Given current trends in increasing age and extent of forests, commercial forestry has the best potential for maintaining habitat for early-successional forest birds, as well as for white-tailed deer. Upland sandpipers and grasshopper sparrows essen-

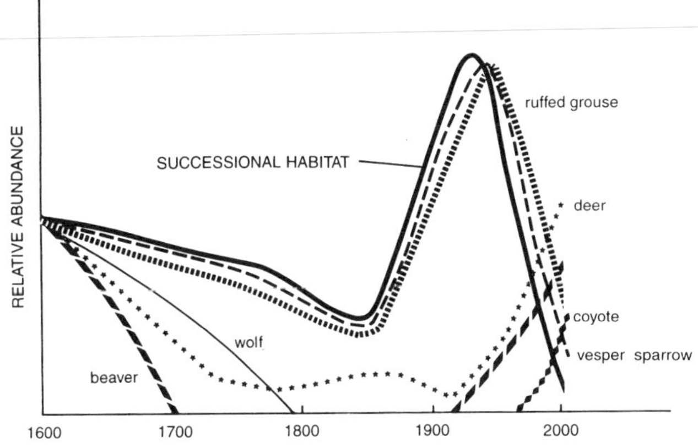

Fig. 5. Schematic depiction of historical changes in representative wildlife species and successional habitat in New England. Several factors are involved in the changes indicated. Wolf, beaver, and deer were persecuted from earliest settlement. Vesper sparrow and ruffed grouse responded to vegetation change with settlement and subsequent land abandonment. Deer and beaver recovered due to protection/reintroduction in the early twentieth century, while coyote colonized the region independently. Note that the abundance of early successional habitat declined with Indian retreats and declines shortly after initial contact. Modified from Bickford and Dymon (1990).

tially are found only at large airports now. Other grassland and old-field species will likely continue to decline in New England unless conservation measures are undertaken to provide early-successional habitat for them.

The New England landscape is not tending toward some ideal pre-Colonial condition—there is none. There were native prairies in southern New England, natural and Indian-set fires periodically burned large areas, and beaver flowages were extensive and imparted a substantial open character to much of the presettlement forest. Today those prairies are long since developed, fire is strictly controlled, and the extent and maturity of the forest probably exceed conditions that existed at any time previously.

As subsistence agriculture gave way to modern society, people's relationships to the land and wildlife changed. Consumption and persecution of wildlife have greatly diminished; modern society largely values wildlife for its aesthetic, ecologic, and scientific attributes. These modern values are not found in the early literature. Utilitarian values prevailed in our early history: deer were used for meat and leather, beaver for fur, and wolves were vermin. Some conflicts between people and wildlife remain; deer and bear populations are increasing in suburban areas, and their reduction is loudly debated. But the main threats to many species today are habitat availability and benign neglect. Many early-successional species are quietly disappearing from New England as forests grow in extent and age. Acceptance of such changes as "natural" will reduce populations of early-successional rabbits and grouse as surely as snares or guns.

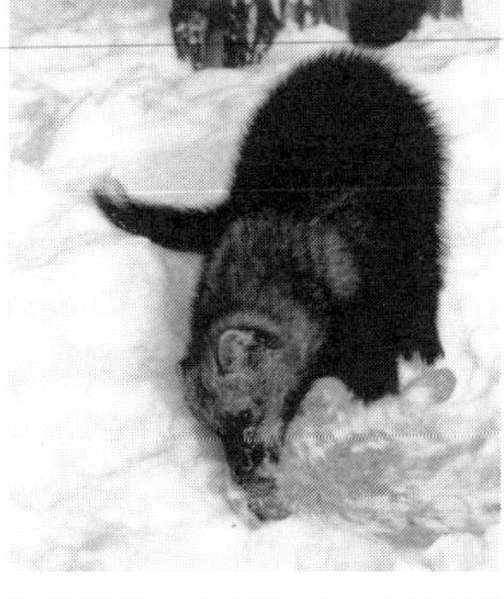

Species respond to landscape change and human action. Top row: (a) Fishers have increased with the return of the New England forest; (b) chestnut-sided warblers are birds of young forest, and respond to even-aged forest management. Middle row: (c) New England cottontails have declined with the loss of early-successional brushy habitat; (d) eastern meadowlarks have greatly declined with the loss of hayfields and grasslands. Bottom row: Both the (e) wild turkey and (f) bald eagle have benefited from direct human intervention, turkeys by trap and transplant after they were extirpated in the mid-1800s and bald eagles by rearing young at hacking sites after pesticides had reduced their numbers. Photos (a), (e), and (f): Ted Levin. Photos (b), and (d): Paul J. Fusco. Photo (c): John Litvaitis.

No landscape condition is intrinsically ideal or "natural"; we shape the land and our actions dictate which species thrive and which decline. All will not survive if we do nothing. It will take active vegetation management to maintain the diversity of wildlife in New England into the future.

Notes

1. John Brereton, a member of the Bartholomew Gosnold's 1602 voyage, listed at length the "commodities" of southeastern Massachusetts; Brereton's account with the list of species is in G. P. Winship, *Sailors' Narratives of Voyages along the New England Coast, 1524–1624* (Boston: Houghton-Mifflin, 1905), 292 pp.

2. The cardinal did occur casually in southern New England during that time period. See W. B. O. Peabody, Report on the Ornithology, in D. H. Storer and W. B. O. Peabody, *Reports on the Ichthyology, Herpetology, and Ornithology of Massachusetts* (Boston: Dutton and Wentworth, 1839), 426 pp. See also B. G. Bedall, Range extension of the cardinal and other birds in the northeastern United States, *Wilson Bulletin* 75 (1963):140–158.

3. There were many volumes, booklets, and pamphlets published from 1840 on. The species accounts contain historical information from the major early works describing the birds of the New England states.

4. The first annotated list of mammals of New England was by G. M. Allen, *Fauna of New England. 3. Terrestrial and Marine Mammals of Massachusetts and Other New England States.* (Brockton, Mass.: privately published, 1904), 151 pp.

5. Regarding the decline of northeastern Indians, see, among others, H. F. Dobyns, *Their Numbers Became Thinned: Native American Population Dynamics in Eastern North America* (Knoxville: University of Tennessee Press, 1983), 378 pp., and B. G. Trigger (ed.) *Handbook of North American Indians,* Vol. 15, *Northeast* (Washington, D.C.: Smithsonian Institution, 1978), 924 pp.

6. The eastern elk is believed by some to be extinct, although it is difficult to ascertain because there are few specimens from early times for comparative purposes. See L. D. Bryant and C. Maser, Classification and distribution, pp. 1–59, in J. W. Thomas and D. E. Toweill (eds.), *Elk of North America: Ecology and Management* (Harrisburg, Pa.: Stackpole Books, 1982), 698 pp.

7. In Massachusetts, deer reeves were elected in most towns by law around 1698; such elections continued until deer were essentially gone. The last Massachusetts deer reeve elected was in Lancaster in 1830 (J. E. Cardoza, Mass. Division of Fisheries and Wildlife, pers. commun.).

8. Coyotes may have developed a unique niche in Maine, making demands on resources that are different from those of wolves. See V. B. Richens and R. D. Hugie, Distribution, taxonomic status, and characteristics of coyotes in Maine. *J. Wildl. Manage.* 38 (1974):447–454.

9. Phillips (1915) first described the New England bobwhite as *Colinus virginianus marilandicus* after its alleged "extinction" in the 1870s or earlier. Ripley (1960) suggested that his investigations supported the existence and identity of this subspecies. However, neither the American Ornithologists Check-list (5th ed., 1957, the last to list subspecific names) nor Rosene (1969) recognize this subspecies. Thus, the putative *marilandicus* is likely synonymous with *virginianus* and probably represents only a phenotype. The literature does not support the "extinction" of a "New England" subspecies of bobwhite, although local phenotypes may have been swamped by numerous introductions. See J. C. Phillips, The old New England bobwhite, *Auk* 32 (1915):204–207; T. H. Ripley, Weights of Massachusetts quail and comparisons with other geographic samples for taxonomic significance, *Auk* 77 (1960):445–447; and W. Rosene, *The Bobwhite Quail: Its Life and Management* (New Brunswick, N.J.: Rutgers University Press, 1969), 418 pp.

Literature Cited

Abrams, M. D.; Orwig, D. A.; Demeo, T. E. 1995. Dendrochronological analysis of successional dynamics for a presettlement-origin white pine-mixed oak forest in the southern Appalachians, USA. *J. Ecology* 83:123–133.

Adams, S. W. 1896. *The native and wild mammals of Connecticut.* Hartford, Conn.: Case, Lockwood & Brainard. 16 pp.

Allen, G. M. 1929. History of the Virginia deer in New England. *Proc. New England Game Conf.* (1929):19–38.

Allen, J. A. 1870. Notes on Massachusetts reptiles and batrachians. *Proc. Boston Soc. Nat. Hist.* 13:260–263.

Askins, R. A. 1993. Population trends in grassland, shrubland, and forest birds in eastern North America. *Curr. Ornithol.* 11:1–34.

Askins, R. A. 1997. History of grasslands in the northeastern United States: Implications for bird conservation. Pages 119–136. In: Vickery, P. D.; Widdie, P. W. (editors). *Grasslands of Northeastern North America.* Lincoln: Massachusetts Audubon Society. 297 pp.

Bagg, A. C.; Eliot, S. A., Jr. 1937. *Birds of the Connecticut Valley in Massachusetts.* Northampton, Mass.: The Hampshire Bookshop. 813 pp.

Banasiak, C. F. 1961. *Deer in Maine.* Game Division Bull. No. 6. Augusta: Maine Dept. of Inland Fisheries and Game. 163 pp.

Belknap, J. 1792. *The history of New Hampshire.* Boston: Bradford and Read.

Bey, C. F. 1990. American Elm. Pages 801–807. In: Burns, R. M.; Honkala, B. H. (techn. coordinators). *Silvics of North America,* Vol. 2, *Hardwoods.* Ag. Handbook 654. Washington, D.C.: U.S. Department of Agriculture, Forest Service. 877 pp.

Bickford, W. E.; Dymon, V. J. 1990. *An atlas of Massachusetts river systems. Environmental designs for the future.* Amherst: University of Massachusetts Press.

Bollinger, E. K.; Gavin, T. A. 1992. Eastern bobolink populations: Ecology and conservation in an agricultural landscape. Pages 497–506. In: Hagan, J. M.; Johnston, D. W. (editors). *Ecology and conservation of Neotropical migrant landbirds.* Washington, D.C.: Smithsonian Institution Press.

Bormann, R. E. 1982. Agricultural disturbance and forest recovery at Mt. Cilley. Ph.D. dissertation, Yale University, New Haven, Conn. 244 pp.

Bormann, F. H.; Siccama, T. G.; Likens, G. E.; Whittaker, R. H. 1970. The Hubbard Brook ecosystem study: Composition and dynamics of the tree stratum. *Ecol. Monogr.* 40:373–388.

Braun, E. L. 1950. *Deciduous forests of eastern North America.* New York: The Hafner Press. 596 pp.

Brewer, L. G. 1995. Ecology of survival and recovery from blight in American chestnut trees (*Castanea dentata* [Marsh.] Borkh.) in Michigan. *Bull. Torrey Botanical Club* 122: 40–57.

Bromley, S. W. 1935. The original forest types of southern New England. *Ecol. Monogr.* 5:61–89.

Brooks, R. T.; Birch, T. W. 1988. Changes in New England forests and forest owners: Implications for wildlife habitat resources and management. *Trans. N. Am. Wildl. Nat. Resource Conf.* 53:78–87.

Brown, J. H. Jr. 1960. The role of fire in altering the species composition of forests in Rhode Island. *Ecology* 41:310–316.

Bucher, E. H. 1992. The causes of extinction of the passenger pigeon. *Curr. Ornithol.* 9:1–36.

Bull, J. 1974. *Birds of New York State.* Garden City, N.Y.: Doubleday/Natural History Press. 655 pp.

Burt, W. H.; Grossenheider, R. P. 1952. *A field guide to the mammals.* The Peterson Field Guide Series, 3rd edition. Boston: Houghton Mifflin. 289 pp.

Cardoza, J. E. 1976. *The history and status of the black bear in Massachusetts and adjacent New England states.* Research Bull. 18. Westborough: Massachusetts Division of Fish and Wildlife. 113 pp.

Cardoza, J. E. Sea mink. In: French, T. W.; Cardoza, J. E. (editors). *Rare vertebrates of Massachusetts.* Lincoln: Massachusetts Audubon Society. In press.

Channing, W. 1939. *New England hurricanes 1635, 1815, 1938.* Boston: Walter Channing.

Chilton, G. 1997. Labrador duck (*Camptor hynchus labradorensis*). In: Poole, A.; Gill, F. (editors). *The birds of North America.* No. 307. Philadelphia: Academy of Natural Science; Washington, D.C.: American Ornithologists' Union.

Churcher, C. S. 1959. The specific-status of the New World red fox. *J. Mammal.* 40:513–520.

Cooperrider, A. Y. 1974. Computer simulation of the interaction of a deer population with northern forest vegetation. Ph.D. dissertation, SUNY College of Environmental Science and Forestry, Syracuse, N.Y. 220 pp.

Cottrell, G. W., Jr. 1949. The southern heron flight of 1948. *Bull. Mass. Audubon Soc.* 33:101–104, 155–160.

Cronon, W. 1983. *Changes in the land: Indians, colonists, and the ecology of New England.* New York: Hill and Wang. 241 pp.

Curtis, J. D. 1943. Some observations on wind damage. *J. Forestry* 41:877–882.

Davis, M. B. 1976. Pleistocene biogeography of temperate deciduous forests. *Geosci. Man* 13:13–26.

Davis, M. B. 1981. Quaternary history and the stability of forest communities. Pages 132–153. In: West, D. C.; Shugart, H. H.; Botkin, D. B. (editors). *Forest Succession.* New York: Springer-Verlag.

Day, G. M. 1953. The Indian as an ecological factor in the northeastern forest. *Ecology* 34:329–346.

DeGraaf, R. M. 1991. Breeding bird assemblages in managed northern hardwood forests in New England. Pages 153–171. In: Rodiek, J. E.; Bolen, E. G. (editors). *Wildlife and habitats in managed landscapes.* Washington, D.C.: Island Press.

DeGraaf, R. M.; Rappole, J. H. 1995. *Neotropical migratory birds: Natural history, distribution, and population change.* Ithaca, N.Y.: Cornell University Press. 676 pp.

DeGraaf, R. M.; Rudis, D. D. 1983. *Amphibians and reptiles of New England: Habitats and natural history.* Amherst: University of Massachusetts Press. 85 pp.

Drury, W. H. 1973. Population changes in New England seabirds. *Bird-banding* 44:267–313.

Drury, W. H. 1974. Population changes in New England seabirds (continued). *Bird-banding* 45:1–15.

Dwight, T. 1823. *Travels in New England and New York.* London: Printed for W. Baynes and Son and Ogle, Duncan and Co. 4 Vols.

Elkins, K. C. 1957. The wood thrush in New Hampshire. *N.H. Bird News* 10:33–36.

Emerson, E. (editor). 1976. *Letters from New England: The Massachusetts Bay Colony; 1629–1638.* Amherst: University of Massachusetts Press. 263 pp.

Fobes, C. B. 1944. Path of the settlement and distribution of population of Maine. *Econ. Geogr.* 20:65–69.

Forbush, E. H.; Fernald, C. H. 1896. *The gypsy moth.* Boston: Massachusetts State Board of Agriculture. 495 pp.

Foster, D. R.; Boose, E. R. 1992. Patterns of forest damage resulting from catastrophic wind in central New England, USA. *J. Ecol.* 80:79–98.

Fuller, D. P. 1993. *Black bear population dynamics in western Massachusetts.* M.S. thesis, University of Massachusetts, Amherst. 136 pp.

Fuller, T. K.; Berg, W. E.; Radde, G. L.; Lenarz, M. S.; Joselyn, G. B. 1992. A history and current estimate of wolf distribution and numbers in Minnesota. *Wildl. Soc. Bull.* 20:42–55.

Godin, A. J. 1977. *Wild mammals of New England.* Baltimore, Md.: Johns Hopkins University Press. 304 pp.

Godman, J. D. 1826. *American natural history.* Vol.1. Philadelphia: H. Carey and I. Lea. 362 pp.

Goodhue, C. F. 1877. The birds of Webster and adjoining towns. *Forest Streams* 8:33–34, 49, 96, 113, 146.

Goodwin, G. G. 1935. *The mammals of Connecticut.* Bull. 53. Hartford: Connecticut State Geological and Natural History Survey. 221 pp.

Goodwin, G. G. 1936. Big game animals in the northeastern United States. *J. Mammal.* 17:48–50.

Griscom, L.; Snyder, D. E. 1955. *The birds of Massachusetts: An annotated and revised check list.* Salem, Mass.: Peabody Museum. 295 pp.

Gross, A. O. 1928. The heath Hen. *Mem. Boston Soc. Nat. Hist.* 6:491–588.

Gross, A. O. 1932. Heath hen. Pages 264–280. In: Bent, A. C. (editor). *Life histories of North American gallinaceous birds.* U.S. Natl. Mus. Bull. 162, Washington, D.C.: Smithsonian Institution.

Hagan, J. M. III; Johnston, D. W. (editors). 1992. *Ecology and conservation of Neotropical migrant landbirds.* Washington, D.C.: Smithsonian Institution. 609 pp.

Hawley, R. C.; Hawes, A. F. 1912. *Forestry in New England.* New York: John Wiley and Sons. 479 pp.

Hickey, J. J. (editor). 1969. *Peregrine falcon populations: Their biology and decline.* Madison: University of Wisconsin Press. 596 pp.

Higginson, F. 1630. New England's plantation. *Mass. Hist. Soc. Proc.* 62(1929):311.

Hill, N. P.; Hagan, J. M. III. 1991. Population trends of some northeastern North American landbirds: A half-century of data. *Wilson Bull.* 103:165–182.

Hosmer, J. K. 1959. *Winthrop's journal, "History of New England," 1630–1649.* New York: Barnes & Noble.

Irland, L. C. (coordinator) 1998. Ice storm 1998 and the forests of the Northeast. A preliminary asssessment. *J. Forestry* 96:32–40.

Jackson, C. F. 1922. Notes on New Hampshire mammals. *J. Mammal.* 3:13–15.

Johnson, C. W. 1998. *The nature of Vermont.* Hanover, N.H.: University Press of New England. 354 pp.

Jorgensen, N. 1977. *A guide to New England's landscape.* Chester, Conn.: Pequot Press. 256 pp.

Josselyn, J. [1672] 1865. *New England's rarities discovered.* Boston: W. Veazie. 169 pp.

Josselyn, J. [1675] 1833. An account of two voyages to New England. *Mass. Hist. Soc. Collection, 3rd series, 3.* 273 pp.

Judd, S. 1857. The fur trade on Connecticut River in the seventeenth century. *New England Hist. General Reg. N. S.* 1:217–219.

Kieran, J. 1959. *A natural history of New York City.* Boston: Houghton Mifflin. 428 pp.

Kingsley, N. P. 1985. *A forester's atlas of the Northeast.* Gen. Tech. Rep. NE-95. Broomall, Pa.: U.S. Department of Agriculture, Forest Service, Northeastern Forest Experiment Station. 96 pp.
Kuchler, A. W. 1964. *The potential natural vegetation of the conterminous United States.* Spec. Publ. No. 36. New York: American Geographic Society, 154 pp.
Leach, N. P. 1909. Panthers in New England. *Forest Stream* 73:412.
Leak, W. B.; Graber, R. E. 1974. *Forest vegetation related to elevation in the White Mountains of New Hampshire.* Res. Pap. NE-299. Upper Darby, Pa.: U.S. Department of Agriculture, Forest Service, Northeastern Forest Experiment Station. 7 pp.
Likens, G. E. 1972. Mirror Lake: Its past, present, and future? *Appalachia* 39:23–41.
Litvaitis, J. A. 1993. Response of early-successional vertebrates to historic changes in land use. *Conserv. Biol.* 7:866–873.
Litvaitis, J. A.; Villafuerte, R. 1996. Factors affecting the persistence of New England cottontail metapopulations: The role of habitat management. *Wildl. Soc. Bull.* 24:686–693.
Lorenz, A. 1917. Notes on the Hepaticae of Mt. Katahdin. *Brydogest* 20:41–46.
Lorimer, C. G. 1977. The presettlement forest and natural disturbance cycle of northeastern Maine. *Ecology* 58:139–148.
Loskiel, G. H. 1794. *History of the Mission of the United Brethren among the Indians of North America.* London: Printed for the Brethren Society for the Furtherance of the Gospel. 233 pp.
Lull, H. W. 1968. *A forest atlas of the Northeast.* Upper Darby, Pa.: U.S. Department of Agriculture, Forest Service, Northeastern Forest Experiment Station. 46 pp.
Manville, R. H. 1966. The extinct sea mink, with taxonomic notes. *Proc. U.S. Natl. Mus.* 122(3584):1–12.
Mattfeld, G. F. 1984. Northeastern hardwood and spruce-fir forests. Pages 305–330. In: Halls, L. K. (editor). *White-tailed deer ecology and management.* Harrisburg, Pa.: Stackpole. 870 pp.
Maurer, B. A.; McArthur, L. B.; Whitmore, R. C. 1981. Effects of logging on guild structure of a forest bird community in West Virginia. *Am. Birds* 35:11–13.
Mech, L. D. 1989. Wolf population survival in an area of high road density. *Am. Midl. Naturalist* 121:387–389.
Mech, L. D.; Fritts, S. H.; Radde, G. L.; Paul, W. J. 1988. Wolf distribution and road density in Minnesota. *Wildl. Soc. Bull.* 16:85–87.
Mills, B. B. 1961. *The Beddington Burn stages a comeback.* Res. Note 117. Upper Darby, Pa.: U.S. Department of Agriculture, Forest Service, Northeastern Forest Experiment Station. 6 pp.
Montevecchi, W. A.; Kirk, D. A. 1996. Great auk (*Pinquinus impennis*). In: Poole, A.; Gill, F. (editors). *The birds of North America.* No. 260. Philadelphia: Academy of Natural Sciences; Washington, D.C.: American Ornithologists' Union.
Mood, F. 1937. John Winthrop, Jr. on Indian corn. *New Engl. Q.* 10:128–129.
Moore, G. C.; Parker, G. R. 1992. Colonization by the eastern coyote (*Canis latrans*). Pages 23–27. In: Boer, A. H. (editor). *Ecology and management of the eastern coyote.* Fredericton, N.B.: Wildlife Research Unit, University of New Brunswick. 194 pp.
Morton, T. [1632] 1883. New English Canaan. In: Admans, L. F. (editor). *Publications of the Prince Society, XIV.* Boston. 177 pp.
Morton, T. [1637]. *New English Canaan or New Canaan.* Amsterdam: J. F. Stam. 188 pp.
Norse, W. J. 1985. Pileated woodpecker. Pages 168–169. In: Laughlin, S. B.; Kibbe, D. P. (editors). *The atlas of breeding birds of Vermont.* Hanover, N.H.: University Press of New England. 456 pp.
Nuttall, T. 1832. *Manual of the ornithology of the United States and Canada: The land birds.* Cambridge, Mass.: Hilliard and Brown. 683 pp.
Palmer, R. S. 1949. Maine birds. *Bull. Mus. Comp. Zool.* 102:1–656.
Parker, A. C. 1910. *Iroquois uses of maize and other food plants.* Bull. 144. Albany: N.Y. State Museum. 119 pp.
Parker, G. 1995. *Eastern coyote: The story of its success.* Halifax, N.S.: Nimbus. 254 pp.
Parker, H. C. 1939. A preliminary list of the mammals of Worcester County. *Proc. Boston Soc. Nat. Hist.* 41:403–415.
Patterson, W. A. III; Sassaman, K. E. 1988. Indian fires in the prehistory of New England. Pages 107–135. In: Nicholas, G. P. (editor). *Holocene human ecology in northeastern North America.* New York: Plenum.
Patterson, W. A. III; Backman, A. E. 1988. Fire and disease history of forests. Pages 603–632. In: Huntley, R.; Webb, T. III (editors). *Vegetation history.* Dordrecht: Kluwer.
Pell, K. W. 1944. Story of a beaver colony. *Bull. Mass. Audubon Soc.* 27:262–266.
Pielou, E. C. 1991. *After the Ice Age: The return of life to glaciated North America.* Chicago: University of Chicago Press.
Preble, N. A. 1942. The mammals of New Hampshire. Master's thesis, Graduate School, Ohio State University, Columbus.
Pringle, L. P. 1960. Notes on coyotes in southern New England. *J. Mammal.* 41:278.
Rappole, J. H.; Morton, E. S.; Ramos, M. A. 1992. Density, philopatry, and population estimates for songbird migrants wintering in Veracruz. Pages 337–344. In: Hagan, J. M. III; Johnston, D. W. (editors). *Ecology and conservation of Neotropical migrant landbirds.* Washington, D.C.: Smithsonian Institution Press.
Remick, D. 1911. *History of Kennebunk from its earliest settlement to 1890.* Portland, Me.: Lakeside Press. 542 pp.
Robbins, C. T. 1993. *Wildlife feeding and nutrition.* San Diego: Academic Press. 352 pp.
Russell, E. W. B. 1983. Indian-set fires in the forests of the northeastern United States. *Ecology* 64:78–88.
Russell, H. S. 1961. New England Indian agriculture. *Bull. Mass. Archaeol. Soc.* 22:58–61.
Schoen, J. W.; Wallmo, O. C.; Kirchhoff, M. D. 1981. Wildlife-forest relationships: Is a re-evaluation of old growth necessary? *Trans. N. Am. Wildl. Nat. Res. Conf.* 46:531–544.
Seton, E. T. 1929. *Lives of game animals.* Garden City, N.Y.: Doubleday, Doran and Co. 4 Vols.
Shantz, H. L.; Zon, R. 1924. Natural vegetation. In: *Atlas of American agriculture.* Washington, D.C.: U.S. Department of Agriculture. 29 pp.
Shaw, S. P. 1948. *The beaver in Massachusetts.* Bull. 11. Massachusetts Division of Fish and Game Reserves. 48 pp.
Shaw, S. P.; McLaughlin, C. L. 1951. *The management of white-tailed deer in Massachusetts.* Bull. 13, Mass. Div. of Fish and Game Res. 59 pp.
Shelford, V. E. 1963. *The ecology of North America.* Urbana: University of Illinois Press. 610 pp.

Shugart, H. H., Jr.; James, D. 1973. Ecological succession of breeding bird populations in northwestern Arkansas. *Auk* 90:62–77.
Silver, H. 1957. *A history of New Hampshire game and furbearers.* Concord: New Hampshire Fish and Game Dept. 466 pp.
Smith, M. L. 1949. *A history of Maine, from wilderness to statehood.* Portland, Me.: Falmouth Publishing House. 348 pp.
Smith, R. L. 1963. Some ecological notes on the grasshopper sparrow. *Wilson Bull.* 75:159–165.
Stiles, E. W. 1982. Expansions of mockingbird and multiflora rare in the northeastern United States and Canada. *Am. Birds* 36:358–364.
Storer, D. H. 1840. A report on the reptiles of Massachusetts. *Boston J. Nat. Hist.* 3:1–64.
Terborgh, J. W. 1989. *Where have all the birds gone?* Princeton, N.J.: Princeton University Press.
Thiel, R. P. 1985. The relationship between road densities and wolf suitability in Wisconsin. *Am. Midl. Naturalist* 113:404–407.
Thompson, B. F. 1977. *The changing face of New England.* Boston: Houghton Mifflin. 188 pp.
Thompson, C. F. 1977. Experimental removal and replacement of territorial male yellow-breasted chats. *Auk* 94:107–113.
Titcomb, J. W. 1901. Animal life in Vermont. *The Vermonter* 5:2.
Trumbull, B. 1797. *A complete history of Connecticut.* Hartford, Conn.: Hudson and Goodwin. 587 pp.
Tubbs, C. H.; Houston, D. R. 1990. American beech. Pages 325–332. In: Burns, R. M.; Honkala, B. H. (tech. coordinators). *Silvics of North America,* Vol. 2. *Hardwoods.* Ag. Handbook 654. Washington, D.C.: U.S. Department of Agriculture, Forest Service. 877 pp.
Van Dyke, F. G. 1983. A western study of cougar track surveys and environmental disturbances affecting cougars related to the status of the eastern cougar *Felis concolor couguar*. Ph.D. dissertation, State University of New York, Syracuse. 244 pp.
Van Dyke, F. G.; Brocke, R. H. 1987. Sighting and track reports as indices of mountain lion presence. *Wildl. Soc. Bull.* 15:251–256.
Van Dyke, F. G.; Brocke, R. H.; Shaw, H. G. 1986. Use of road track counts as indices of mountain lion presence. *J. Wildl. Manage.* 50:102–109.
Veit, R. R.; Petersen, W. R. 1993. *Birds of Massachusetts.* Lincoln: Massachusetts Audubon Society. 514 pp.
Walcott, C. F. 1974. Changes in birdlife in Cambridge, Massachusetts from 1860 to 1964. *Auk* 91:151–160.
Waters, J. H. 1962. Some animals used as food by successive cultural groups in New England. *Bull. Archaeol. Soc. Conn.* 31:32–45.
Warfel, H. E. 1936. Notes on the occurrence of *Necturus maculosus* (Raffinesque) in Massachusetts. *Copeia* 1936:237.
Watts, W. A. 1983. Vegetation history of the eastern United States. In: Porter, S. C. (editor). *Late Quaternary environments of the United States,* Vol. 1, *The Late Pleistocene.* Pages 294–310. Minneapolis: University of Minnesota Press.
Webb, W. L.; Behrend, D. F.; Saisorn, B. 1977. Effect of logging on songbird populations in a northern hardwood forest. *Wildl. Monogr.* 55:6–36.
Westveld, M.; Ashman, R. I.; Baldwin, H. I.; Holdsworth, R. P.; Johnson, R. S.; Lambert, J. H.; Lutz, H. J.; Swain, L. C.; Standish, M. 1956. Natural forest vegetation zones of New England. *J. Forestry* 54:332–338.
Whitmore, R. C. 1981. Structural characteristics of grasshopper sparrow habitat. *J. Wildl. Manage.* 45:811–814.
Whitney, G. G. 1994. *From coastal wilderness to fruited plain: Temperate North America 1500 to the present.* Cambridge, U.K.: Cambridge University Press.
Winne, J. C. 1988. *History of vegetation and fire on the Pineo Ridge blueberry barrens in Washington County, Maine.* Unpublished M.S. thesis, University of Maine, Orono.
Winship, G. P. (editor). 1905. *Sailor's narratives of voyages along the New England coast, 1524–1624.* Boston: Houghlin-Mifflin. 292 pp.
Winthrop, J., Jr. [1636] 1863. Letter to John Winthrop, Sr. (April 7, 1636). *Mass. Hist. Soc. Collection, 4th ser.* 6:514.
Wood, R. G. 1935. *A history of lumbering in Maine, 1820–1861.* University of Maine Studies, 2nd series, No. 33. 267 pp.
Wood, W. [1634] 1977. *New England's prospect.* Vaughan, A. T. (editor). Amherst: University of Massachusetts Press. 132 pp.

Part I

Natural Histories of New England's Wildlife Species

Species Accounts

Natural history accounts are given for 338 currently occurring inland species. It is difficult to state clearly and objectively which species should be included and which excluded. Some inland species breed or winter in New England infrequently; others are very rare or apparently recently extirpated. Our objective is to provide information that is useful in land management; we present detailed information for those species that predictably breed or winter in the region. Because New England has an extensive coastline, which extends from the warm waters of Long Island Sound to the cold north Atlantic water off Maine, even the coastal wildlife is rich and varied, ranging from the southerly diamondback terrapin to arctic seabirds and seals. These coastal species and those that neither breed nor winter in New England but only migrate through the region are included in Appendix B.

In each species account, the distribution in New England is included in text and in map form, as well as habitat and life history information. Distribution maps are, in many cases, difficult to prepare. The ranges of many species are changing, and those of others poorly known. The boundaries of breeding or wintering ranges cannot be accurately described by neat shading on a map. That is how ranges are shown here, but in reality the range indicated, especially near the boundaries, is a zone of "islands" or patches of suitable habitat. Except for truly ubiquitous species, the same principle holds across the indicated range. The maps herein are intended as a guide to probable breeding or wintering occurrence where suitable habitat exists. The zones indicated will likely shift for many species as forests mature, open habitats decline, and as southern birds extend their ranges northward into New England and retreat in response to severe winter weather.

Naturalized introduced species, including mudpuppy (*Necturus maculosus*), mute swan (*Cygnus olor*), gray partridge (*Perdix perdix*), ring-necked pheasant (*Phasianus colchicus*), rock dove (*Columba livia*), European starling (*Sturnus vulgaris*), house sparrow (*Passer domesticus*), Norway rat (*Rattus norvegicus*), and house mouse (*Mus musculus*), are included. Marine turtles, pelagic birds, and accidental species are omitted, as are the introduced but nonbreeding red-eared slider (*Trachemys scripta elegans*), the European rabbit (*Oryctolagus cuniculus*), which occurs only on certain islands in Boston Harbor, and feral cats (*Felis catus*).

Amphibians and Reptiles

This section provides a compilation of natural histories, distributions, and habitat associations for the 23 amphibian and 29 reptile species and subspecies occurring in New England. The current distributions of several species including the Jefferson salamander (*Ambystoma jeffersonianum*), blue-spotted salamander (*Ambystoma laterale*), four-toed salamander (*Hemidactylium scutatum*), eastern spadefoot (*Scaphiopus holbrookii*), eastern hognose snake (*Heterodon platyrhinos*), eastern worm snake (*Carphophis a. amoenus*), and black rat snake (*Elaphe o. obsoleta*) are not well known in New England, and the maps need to be periodically updated. Nomenclature follows *Standard Common and Current Scientific Names for North American Amphibians and Reptiles,* 4th edition (Collins 1997). Electronically compiled updates by Collins and others can be found at the following website addresses: <http://eagle. cc. ukans.edu/~cnaar/CNAARNAChecklist. html> and <http://www. embl-heidelberg. de/~uetz/LivingReptiles. html>.

We have included the mudpuppy (*Necturus maculosus*), introduced into southern parts of the region. We have omitted the Allegheny or mountain dusky salamander (*Desmognathus ochrophaeus*), known from one unconfirmed juvenile specimen in central Vermont (Lazell 1976), and western chorus frog (*Pseudacris triseriata*), which has not been reported for a decade in Vermont. We have omitted the eastern mud turtle (*Kinosternon s. subrubrum*) because individuals in Connecticut are believed to have been released and no breeding populations are known to exist. The northern diamondback terrapin (*Malaclemys t. terrapin*) is omitted because it is strictly coastal. We omitted the red-eared slider (*Trachemys scripta elegans*), which was introduced by the release of pets and survives in the wild but does not reproduce successfully (Klemens 1993). Rare inland amphibians and marine reptiles often found along the Atlantic coast are listed in Appendices B and C.

Species are listed in phylogenetic order. Measurement units here are as reported in the original work. When the original work used English units, metric equivalents have been supplied. Variations in development and hatching times for a species may be attributed to genetic and environmental factors. The species accounts are as complete as the literature permits, but many gaps still exist in our knowledge of amphibian and reptile ecology and distribution.

Literature Cited

Collins, J. T. 1997. *Standard common and current scientific names for North American amphibians and reptiles.* 4th ed. Society for the Study of Amphibians and Reptiles Herpetological Circular. 25:1–40.

Klemens, M. W. 1993. The amphibians and reptiles of Connecticut and adjacent regions. *State Geol. Nat. Hist. Surv. Conn. Bull.* 112. 318 pp.

Lazell, J. D., Jr. 1976. Geographic distribution: *Desmognathus orchophaeus. SSAR Herp. Review* 7:122.

Sources used to prepare maps of species distributions in addition to expert reviews include:

The amphibians and reptiles of Maine. 1992. Hunter, M. L., Jr.; Albright, J.; Arbuckle, J. (editors). Maine Agr. Exper. Sta. Bull. 838. Orono: University of Maine. 188 pp.

Amphibians and Reptiles of New England. 1983. DeGraaf, R. M.; Rudis, D. D. Amherst: University of Massachusetts Press. 85 pp.

Amphibians and reptiles of Connecticut and adjacent regions. 1993. Klemens, M. W. Bull. 112. Hartford: State Geological and Natural History Survey of Connecticut. 318 pp.

Salamanders of the United States and Canada. 1998. Petranka, J. W. Washington, D.C.: Smithsonian Institution Press. 587 pp.

Amphibian and Reptile Species

The following lists references on amphibian and reptiles for further reading on life histories and distribution.

Citations by Region

Eastern North America

Carr, A. 1995. *Handbook of turtles: The turtles of the United States, Canada, and Baja California.* 9th ed. Ithaca, N.Y.: Cornell University Press. 542 pp.
Dickerson, M. C. 1969. *The frog book.* New York: Dover Publications. 253 pp.
Ernst, C. H.; Barbour, R. W. 1972. *Turtles of the United States.* Lexington: University Press of Kentucky. 347 pp.
Ernst, C. H.; Barbour, R. W. 1989. *Snakes of eastern North America.* Fairfax, Va.: George Mason University Press. 282 pp.
Petranka, J. W. 1998. *Salamanders of the United States and Canada.* Washington, D.C.: Smithsonian Institution Press. 587 pp.
Wright, A. H.; Wright, A. A. 1949. *Handbook of frogs, amphibians and toads of the United States and Canada.* 3rd ed. Ithaca, N.Y.: Comstock. 640 pp.

Northeastern United States

DeGraaf, R. M.; Rudis, D. D. 1983. *Amphibians and reptiles of New England.* Amherst: University of Massachusetts Press. 85 pp.

Hunter, M. L., Jr.; Albright, J.; Arbuckle, J. (editors). 1992. *The amphibians and reptiles of Maine.* Bull. 838, Maine Agricultural Experiment Station. Orono: University of Maine. 188 pp.

Hunter, M. L., Jr.; Calhoun, A. J. K.; McCollough, M. (editors). 1999. *Maine amphibians and reptiles.* Orono: University of Maine Press. 252 pp.

Klemens, M. W. 1993. *Amphibians and reptiles of Connecticut and adjacent regions.* Bull. 112. Hartford: State Geological and Nat. History Survey of Connecticut. 318 pp.

Lazell, J. D., Jr. 1976. *This broken archipelago: Cape Cod and the Islands, amphibians and reptiles.* New York: Demeter Press. 260 pp.

Mudpuppy

(*Necturus maculosus*)

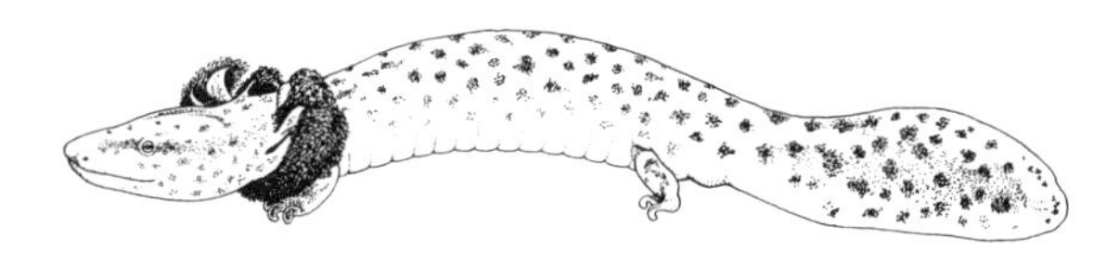

Range: St. Lawrence River w. to se. Manitoba, s. to e. Kansas and n. Alabama and through central Pennsylvania to New York and the Champlain Valley. Absent from the Adirondacks. Introduced in parts of New England.

Distribution in New England: The Lake Champlain drainage, the Connecticut River drainage, Scituate (R.I.) Reservoir drainage, Kennebec County, Me., and other areas where introduced.

Status in New England: Uncommon.

Habitat: Entirely aquatic. Clear (preferred [Bishop 1926]) or muddy waters of lakes, rivers, ditches, and large streams. Often found in submerged log piles around the bases of bridge pilings in larger rivers and around obstructions and bottom debris in streams (Shoop and Gunning 1967). One individual found at 90 ft (27.4 m) in Lake Michigan (Behler and King 1979:283). Absent from small ponds, creeks, marshes in Indiana (Minton 1972).

Special Habitat Requirements: Usually in running water; also in lakes, reservoirs, canals, ditches. Presence of suitable hiding cover may be a critical factor in habitat use (Harris 1959).

Age/Size at Sexual Maturity: At 5 yr and at 8 in (20 cm) total length; adults measure 20–49 cm (average 30 cm) total length (Bishop 1941:43). Retains external gills as an adult.

Breeding Period: Autumn (Bishop 1947:42).

Egg Deposition: May and June of the year following mating. Reproduces in flowing water (Oliver 1955:211). Prefers water depths of at least 3 ft (0.9 m) and bottoms with weeds and rocks to provide nesting cover. Nest sites are near riffles, often under large rock slabs in water depths of 6 to 8 in (15 to 20 cm) in New York (Stewart 1961:68). May prefer areas where water surface exposed to sunlight most of the day (Bishop 1926). Males abound on nest sites before females begin laying eggs (Bishop 1941).

No. Eggs/Mass: 18 to 180 eggs (average 60 to 100) in water beneath objects, attached singly by stalks (Bishop 1941:26).

Time to Hatching: 38 to 63 days, female guards eggs; hatchlings have prominent yolk sacs and measure 21–25mm total length (Bishop 1941:27).

Home Range/Movement: Displacement of individuals in Louisiana suggests homing ability; occupy restricted areas throughout the year (Shoop and Gunning 1967). Mean distance between captures, female 298 m, male 204 m.

Food Habits: In New York, aquatic insects were 30 percent of the diet by weight, particularly nymphs and larval forms, crustaceans 33 percent, small fish 13 percent, also mollusks, spawn, other amphibians, worms, leeches, and plants (Hamilton 1932). Identified vertebrate prey include over 15 species of fish, 5 species of salamanders, and tadpoles (Harris 1959). Most food captured at night along the bottom.

Comments: The mudpuppy is chiefly nocturnal, bottom dwelling, and active through the winter, when it moves to deeper water. This species was first found in the Connecticut River in Massachusetts in 1931, perhaps released laboratory specimens (Warfel 1936). The Maine population also originated from released individuals (Mairs 1992); however, the Rhode Island population origin is unknown but is presumed to be introduced (Vinegar and Friedman 1967).

Marbled Salamander

(*Ambystoma opacum*)

Range: New Hampshire and central Massachusetts, w. to central Pennsylvania and s. Illinois, s. to Missouri, e. Texas and Georgia.

Distribution in New England: Throughout Connecticut and Rhode Island, and Massachusetts e. of the Connecticut River, and in w. Massachusetts in the Berkshire Hills. One specimen from w. Vermont and two specimens from s. New Hampshire (Taylor 1993).

Status in New England: Uncommon.

Habitat: Floodplain forests with oxbows and cut off stream channels usually abound with adults of this species (Petranka 1998:88).Well-drained sandy or gravelly soils of mixed deciduous woodlands, and trap rock slopes (Klemens 1993:46), especially oak-maple and oak-hickory (Minton 1972:46). During breeding season, found in palustrine wetlands and low areas around ponds, and quiet streams (Seibert 1989) with forested upland present within 200 m (Whitlock et al. 1994). Outside the breeding period, inhabits somewhat drier areas than other species of *Ambystoma* (Bishop 1941). During the summer usually found under logs and rocks. Found at high elevation (>335 m) in Connecticut (Klemens 1993:46, Babbitt 1937). Larvae usually found in temporary water throughout the winter (Anderson 1967b). Probably hibernates in deep burrows.

Special Habitat Requirements: Temporary ponds, vernal pools, or fishless swamps in wooded areas for breeding. Seasonally ephemeral pools that fill in fall (Petranka 1989).

Age/Size at Sexual Maturity: 15 to 18 months.

Breeding Period: During the fall, adults migrate to breeding areas (September in northern parts of range). Adults in northern populations tend to breed earlier than those in southern populations—a pattern that is the reverse of that in other *Ambystoma* species in the eastern U.S. (Anderson and Williamson 1973).

Egg Deposition: September to early October in northern parts of range (Bishop 1941:138). Temperature taken at the nest sites in both New Jersey and South Carolina ranged from 11 to 15° C (52–59° F) (Anderson and Williamson 1973).

No. Eggs/Mass: 50 to 232 (average 100) eggs laid singly in shallow depressions beneath surface materials (Bishop 1941:142) or leaf litter (Petranka 1990). This is one of only two *Ambystoma* species that mate and oviposit on land (Petranka 1998:89). Eggs laid in dry beds of temporary ponds and streams or on land, or at the edge of ponds or swamps, where they will be washed into the water to hatch.

Time to Hatching: Embryos develop to the hatching stage within 9 to 15 days after oviposition but do not hatch until flooded (Kaplan and Crump 1978). Female forms a nest site and may brood eggs (Oliver 1955:234).

Eggs Hatch: Usually in fall or early winter within a few hours or 1–2 days after being submerged, but without rain will hatch in spring. Anderson (1972) found a wide range of temperature tolerance, 3–14° C (37–57° F), for egg development. Hatchlings usually measure 10–14 mm total length (Kaplan 1980); more advanced embryos may be 19 mm in total length at hatching (Bishop 1941).

Larval Period: Larvae overwinter with little growth until spring, and transform to terrestrial form in late May to June (Noble and Brady 1933). A higher temperature and abundant food supply will hasten metamorphosis (Stewart 1956b). The larval period was 135 days in New Jersey (Hassinger et al. 1970).

Home Range/Movement: Adults migrate an average of 194 m from breeding sites to summer range in Indiana (Williams 1973, in Semlitsch 1980b:320).

Food Habits: Arthropods, including adults and larval insects and crustaceans. Also takes earthworms and mollusks. Marbled salamander larvae eat small aquatic insects, crustaceans, other amphibian larvae and other small invertebrates and are cannibalistic (Minton 1972:47). Larvae rise in the water column to feed.

Comments: Terrestrial and nocturnal, often uses runways of other animals or tunnels through loose soil. Larvae are aquatic and primarily nocturnal. Marbled salamanders are undoubtedly preyed upon by owls, raccoons, skunks, snakes, and other woodland predators; when molested, adults secrete copious amounts of milky secretions from the tail that repel predators (Petranka 1998:96).

Jefferson Salamander

(*Ambystoma jeffersonianum* and associated hybrid forms including *A. platineum*)

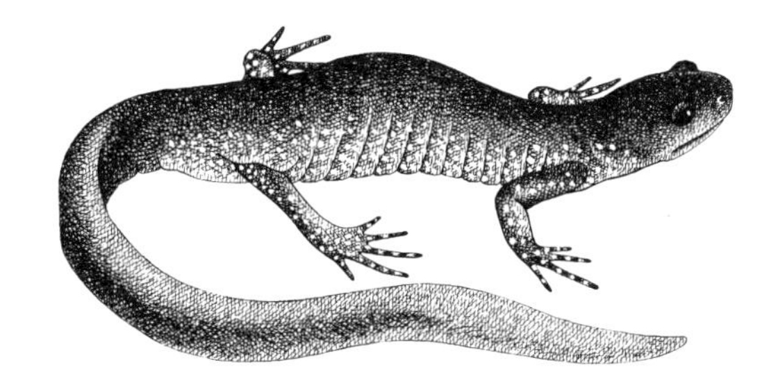

Range: Western New England to w.-central Indiana, s. to central Kentucky to w. Virginia and n. New Jersey. *Ambystoma platineum* occurs with *A. jeffersonianum*, pri marily n. of Wisconsin glacial moraine where ranges of *A. jeffersonianum* and *A. laterale* meet or overlap.

Distribution in New England: West of the Connecticut

River in Vermont, Massachusetts, and Connecticut, and e. of the Connecticut River in n. Franklin County, Massachusetts, and adjacent New Hampshire (Klemens 1993:25; French and Master 1986).

Status in New England: Locally common to rare.

Habitat: Adults are terrestrial, found primarily in undisturbed, shady deciduous (but also mixed) forest, commonly in steep rocky areas with rotted logs and heavy duff layers (Klemens 1993:29). Hides beneath leaf litter, in small mammal burrows, under stones or in decomposing logs and stumps. Found in upland hardwood forests on glaciated limestone areas northwest of the Great Swamp in New Jersey (Anderson and Giacosie 1967), in shale ravines in Connecticut (Babbitt 1937), and Taconic Uplift areas of sw. New England (Klemens 1993:25). Hibernates on land in winter months, usually near breeding sites in palustrine wetlands, sometimes within rotten logs (Blanchard 1933b).

Special Habitat Requirements: Requires temporary (vernal) ponds with deep leaf litter (Cook 1983) in upland deciduous forests with some portion of the wetland bordered by forest and isolated from urbanization or disturbance (Whitlock et al. 1994). Cleared strips create a barrier for dispersal (Pough and Wilson 1976). Will use older (>40 yr) man-made ponds for breeding (Williams 1973, Cortwright 1988). Not found in ponds with pH <4.62 (Freda and Dunson 1986).

Age/Size at Sexual Maturity: Females at 21 months (Bishop 1941:102), snout to vent length 70 to 75 mm in males, 75 to 80 mm in females (Minton 1954). Juveniles probably enter the breeding population at 2 to 3 yr of age (Wilson 1976, cited in Thompson et al. 1980:119). The smallest mature males and females examined by Uzzell (1967a) measured 62 and 76 mm SVL, respectively (Petranka 1998:62).

Breeding Period: February to April, migrates to ponds and vernal pools for spawning (Brandon 1961). Breeds earlier than *A. maculatum* in central Pennsylvania (Gatz 1971) and Massachusetts (Jackson 1990).

Egg Deposition: February to April, often in ice-covered pools. Will tolerate pH of 4 to 8, with best hatching success at pH range 5 to 6 (Pough and Wilson 1976). Egg mortality exceeded 60 percent in pools more acid than pH 5 in Tompkin's County, N.Y. (Pough 1976). Isolated upland pools bordered by shrubs and surrounded by forest were primary breeding sites in Maryland (Thompson et al. 1980).

No. Eggs/Mass: About 30 eggs/mass; 107 to 286 eggs/female (Oliver 1955:234). Laid in small masses ranging from 6 to 80 eggs/mass: typically 20–50 eggs/mass, in cylindrical masses attached to twigs or vegetation under water surface. Rarely if ever attached to pond bottom debris (Uzzell 1967c:49.1).

Time to Hatching: 13 to 45 days (Bishop 1941:135, Oliver 1955:234).

Larval Period: 56 to 125 days (Bishop 1941:99). Found overwintering in Nova Scotia (Bleakney 1952). Larvae feed voraciously on zooplankton soon after hatching (Petranka 1998:61).

Home Range/Movement: Adults migrated an average of 252 m from breeding ponds to summer range in Indiana. Newly metamorphosized individuals moved an average 92 m from the ponds (Williams 1973, cited in Semlitsch 1980b:320). In hardwood forest of northern Kentucky, adults moved an average of 250 m from ponds in a series of 6 to 8 moves in 45 days (Douglas and Monroe 1981). Jefferson salamanders have been recorded as far as 624 m from a breeding pond (Williams 1973).

Food Habits: Small invertebrates, including worms, millipedes, spiders, insects, and aquatic crustaceans. Feeds on most animal life that it can capture.

Comments: The Jefferson salamander is a rare and elusive member of the family of mole salamanders, so named because they spend much of their lives underground. The Jefferson is closely related to the more common spotted salamander (*A. maculatum*), and, like spotteds, travels from wintering sites in the upland forest to breeding pools. Jefferson salamanders breed earlier than spotteds, often traveling over snow and ice to get to the breeding pool. The Jefferson salamander breeds in isolated pairs, unlike spotteds, which court in large groups called congresses. Jefferson salamander populations are heavily skewed toward females, ranging from 4:1 to as much as 20 females per male in a given pond. This imbalance is due to the presence of unisexual biotypes, various hybrids between Jefferson and blue-spotted salamanders that persist in the wild from interbreeding after glacial retreat (see Petranka 1998:122–129). Hybrids may possess two (diploid), three (triploid), or four (tetraploid) sets of chromosomes. One common hybrid contains two sets of chromosomes from the Jefferson and one from the blue-spotted salamander, and is sometimes called the silvery salamander, but is not a true species. See Jackson (1994 and Petranka 1998) for reviews. The genetics of the Jefferson salamander and its hybrids are undergoing revision; readers are urged to seek the most current information. Before 1964, almost all references to specimens in the *Ambystoma jeffersonianum* complex (including *A. jeffersonianum, A. laterale, A. tremblayi,* and *A.*

platineum) were reported as *A. jeffersonianum* (Uzzell 1964). Many papers have since dealt with the genetics and taxonomy of this complex. This ongoing taxonomic revision has resulted in many apparently erroneous locality records. *Ambystoma jeffersonianum* is currently believed not to occur east of the Connecticut River Valley south of Montague, Massachusetts, and all museum specimens from this area identified as *A. jeffersonianum* have been found to refer to the diploid blue-spotted salamander (*A. laterale*) or the triploid hybrid "Tremblay's salamander" (formerly *A. tremblayi*) (Thomas French pers. commun.).

Blue-spotted Salamander

(*Ambystoma laterale* and associated hybrid forms including *A. tremblayi*)

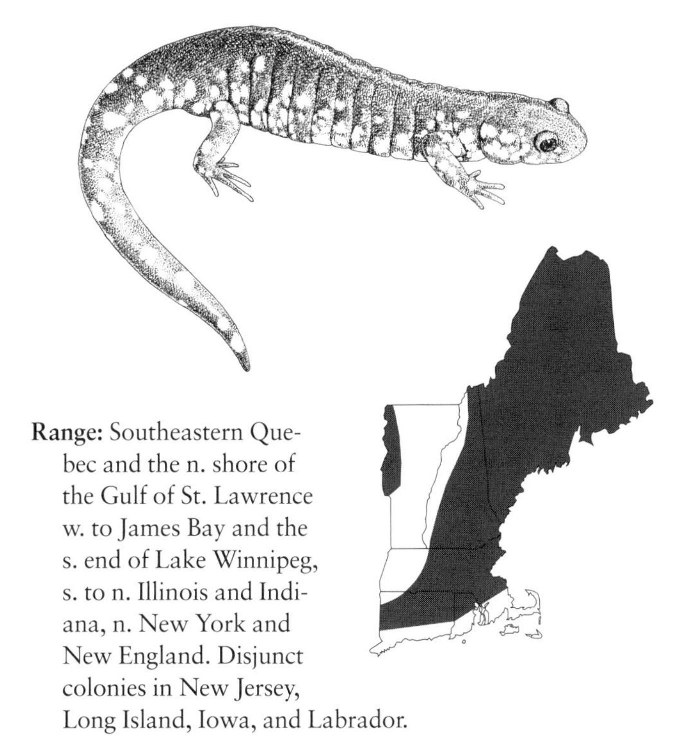

Range: Southeastern Quebec and the n. shore of the Gulf of St. Lawrence w. to James Bay and the s. end of Lake Winnipeg, s. to n. Illinois and Indiana, n. New York and New England. Disjunct colonies in New Jersey, Long Island, Iowa, and Labrador.

Distribution in New England: In New England, e. and central Massachusetts, se. New Hampshire, Maine, and Lake Champlain lowlands of Vermont, with scattered populations throughout sw. New England. Not reported from Cape Cod (Klemens 1993:33).

Status in New England: Widely distributed, but uncommon in local sites; threatened in southern portions of range.

Habitat: Adults occur primarily in deciduous forests with loamy soils (Downs 1989), wooded swamps (Minton 1954, Nyman et al. 1988). Occasionally in overgrown wet pastures. Sometimes occurs where soil is sandy, and may be found under logs or other forest debris. Occurs in a wide range of elevations in w. Connecticut and along the Connecticut river floodplain (Klemens 1993:33).

Special Habitat Requirements: Wooded swamps, ponds, marshes, ditches or semipermanent water for breeding; relatively open or forested aquatic sites (Downs 1989).

Age/Size at Sexual Maturity: About 2 years (Wilbur 1977). Snout to vent length of 47 to 55 mm in Indiana (Minton 1954); total length 7.5 to 13 cm (Petranka 1998:63).

Breeding Period: During early spring rains when night temperatures are above freezing.

Egg Deposition: March to early April. Eggs laid singly or in small, poorly defined masses on the bottoms of temporary shallow forest ponds, roadside drainage ditches, temporary pasture ponds, kettle holes (Landre 1980), or attached to litter or in bottom detritus (Stille 1954), and twigs (Uzzell 1976b:48.1).

No. Eggs/Mass: *Ambystoma laterale:* 2 to 10 eggs per mass (Landre 1980); often laid singly (Uzzell 1967b); 199 to 247 eggs/female (Uzzell 1964).

Time to Hatching: About 1 month (Smith 1961:28). Survival of embryos 100% at pH >4.0 (Mirick 1985).

Larval Period: Typically 2 to 3 months (Petranka 1998:66). *Ambystoma laterale:* extending to late June or mid-August (Smith 1961:28).

Home Range/Movement: Unreported.

Food Habits: Larvae: Daphnia, Amphipoda (Wilbur 1972). Adults: Arthropods, annelids, and centipedes; slugs most common (Gilhen 1974).

Comments: "Tremblay's salamander" is a triploid of hybrid origin from *A. laterale* and *A. jeffersonianum*; it receives two sets of chromosomes from *A. laterale* and one set from *A. jeffersonianum* and is not considered a true species. All hybrid forms, including the "silvery salamander" (see Jefferson salamander), are females. In order to reproduce, hybrid females must collect spermatophores from either Jefferson or blue-spotted males. The males' genes are not passed on—or at least not beyond one generation. The sperm serves to stimulate the triploid egg to begin division; if none of the male's genetic material is incorporated into the developing embryos, genetic clones of the adult female are produced. If the male's genes are incorporated into the developing daughters, the male's genes are normally lost when the daughters develop eggs. Her offspring then carry only the original triploid set of chromosomes and are identical to their "grandmother." Occasionally the male sperm is incorporated into this generation also. In this case, the "granddaughters" will be genetically diverse, although these male genes will not be passed on either.

See Jackson (1994) for review. The genetics of blue-spotted salamanders and their various hybrids are undergoing revision; readers are advised to seek the most current information. See Petranka 1998:122–129.

Spotted Salamander

(*Ambystoma maculatum*)

Range: Nova Scotia and the Gaspé Peninsula w. to s. Ontario, s. through Wisconsin, s. to s. Illinois (excluding prairie regions), e. Kansas and Texas, and throughout the eastern United States except Florida, the Delmarva Peninsula, and s. New Jersey.

Distribution in New England: Throughout the region. Reportedly on Martha's Vineyard, but not Nantucket (Scott Jackson pers. commun.).

Status in New England: Common though populations declining, possibly due to acid precipitation.

Habitat: Adults fossorial; found in moist woods, streambanks, beneath stones, logs, boards. Prefers deciduous or mixed woods in or adjoining floodplains, on rocky hillsides and shallow woodland ponds or marshy pools that hold water through the summer for breeding (Thompson and Gates 1982). Usually does not inhabit ponds containing fish (Anderson 1967a). Terrestrial hibernator. In summer often wanders far from breeding ponds. Found in low oak-hickory forests with creeks and nearby swamps in Illinois (Cagle 1942, cited in Smith 1961:30). Have been found in the pitch pine-scrub oak community of the Albany Pine Bush (Stewart and Rossi 1981), dense oak forests in Rhode Island, and streamside meadows in peatland (Stockwell and Hunter 1989). Breeds in man-made ponds >50 yr old (Cortwright 1986).

Special Habitat Requirements: Mesic woods with fish-free permanent, semipermanent, or ephemeral water for breeding.

Age/Size at Sexual Maturity: During second year. Males may mature 1 yr earlier than females (Wacasey 1961).

Breeding Period: March to mid April. Mass breeding migrations occur in this species; individuals enter and leave breeding ponds using the same track each year, and exhibit fidelity to breeding ponds (Shoop 1965, 1974). Individuals may not breed in consecutive years (Husting 1965).

Egg Deposition: 1 to 6 days after first appearance of adults at ponds (Bishop 1941:114). Eggs tolerate pH range of 6 to 10 with best hatching success pH 7 to 9 (Pough and Wilson 1976).

No. Eggs/Mass: 100 to 200 eggs, average of 125, laid in large masses of jelly, sometimes milky, attached to stems about 15 cm (6 in) under water. Each female lays 1 to 10 masses (average 2 or 3) of eggs (Wright and Allen 1909). Woodward (1982) reported that females breeding in permanent ponds produced smaller, more numerous eggs than females using temporary ponds.

Time to Hatching: 31 to 54 days (Bishop 1947:145). In a cold (≤10° C, 50° F) spring-fed pond, eggs developed in 60 days in Rhode Island (Whitford and Vinegar 1966); Shoop (1974) reported 8 to 14 days.

Larval Period: 61 to 110 days, and as short as 15 to 60 days (Shoop 1974); found overwintering in Nova Scotia (Bleakney 1952) and Rhode Island (Whitford and Vinegar 1966). Transforms July to September. High embryonic mortality occurred in temporary pools with pH below 6.0 in New York (Pough 1976).

Home Range/Movement: Individuals have been found up to a quarter of a mile (400 m) from the nearest breeding site in North Carolina (Gordon 1968). Will travel 91.2 to 182.4 m (300 to 600 ft) from woods to ponds in open meadows in New York (M. Stewart pers. commun.). Adults often leave the pond near the same point where they entered (Douglas and Monroe 1981) and tend to arrive in the same order from year to year (Stenhouse 1985). Some adults return to the same burrows from which they migrated (Downs 1989). Individuals were found to use subterranean rodent burrows as retreats; tagged salamanders that were monitored were found within a 300-cm^2 area of these burrows. Displaced adults moved up to 500 m to return to breeding ponds in Massachusetts (Shoop 1968). Average migration of 150 m from breeding ponds in Kentucky, 6- to 220-m range in thick oak-hickory forest. Linear migration was unaffected by the presence or absence of vegetation or change in the topography (Douglas and Monroe 1981).

Food Habits: Adults consume earthworms, snails, slugs, insects, spiders, particularly larval and adult beetles (Wacasey 1961). Small larvae stage eat zooplankton, larger larvae also eat chironomids, chaborids, and isopods (Nyman 1991). Cannibalism by larvae occurs under crowded conditions.

Comments: Nocturnal; travel only on ground surface for migrations to and from breeding pools. Rainfall, snowmelt, or high humidity coupled with air temperature of 10° C (50° F) or more is necessary for migrations to breeding pools.

Red-spotted Newt

(Notophthalmus v. viridescens)

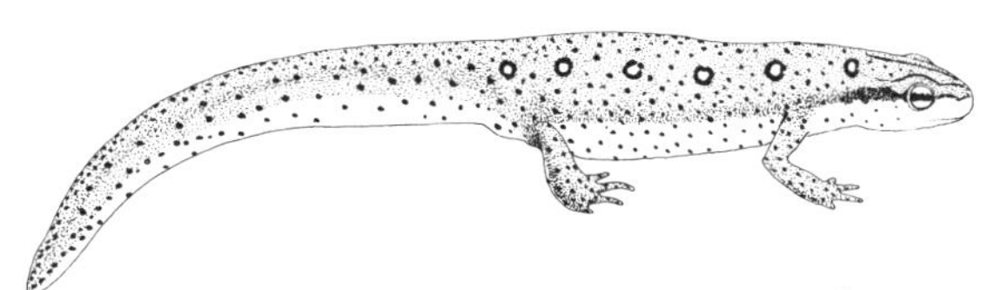

Range: Nova Scotia and the Gaspé Peninsula w. to the n. shore of Lake Superior and e. Michigan, s. to central Alabama and n.-central Georgia. Absent along coast from se. South Carolina s.

Distribution in New England: Throughout the region except for outermost Cape Cod.

Status in New England: Common.

Habitat: Adults found in ponds, particularly water with abundant submerged vegetation, and in weedy areas of lakes, marshes, ditches, backwaters, and pools of shallow slow-moving streams or other unpolluted shallow or semipermanent water. Terrestrial juveniles (efts) live in moist areas on land, typically under damp leaves, under brush piles or logs and stumps, usually in wooded habitats. More common in areas of higher elevation in Connecticut (Klemens 1993:89); from sea level to an elevation of 1.6 km on Mt. Marcy in the Adirondacks (M. Stewart, personal observation). Moist beech-maple-hemlock woods in New York (Hurlbert 1969), and oak-pine woods in Massachusetts (Healy 1974). May be seen moving about on wet days in spring and summer. Efts hibernate on land, burrowing under logs and debris, but most adults remain active all winter underwater in pond bottoms or in streams. During winter months often found semiactive in groups of 20 to 40 (Morgan and Grierson 1932). Newts abundant in Maine peatlands containing or near permanent water (Stockwell and Hunter 1989). Efts move up to 800 m from breeding pond (Healy 1975), density 300 efts/ha (Mass.).

Special Habitat Requirements: Water with aquatic vegetation for the adult newt.

Age/Size at Sexual Maturity: 2 to 8 yr (Healy 1974). Size at sexual maturity varies depending on life history, geographic race, and geographic locale. On average, northern coastal adults are smaller than inland populations. Adult males as small as 51 and 64 mm total length have been found in two northern coastal regions (Noble 1926, 1929). In contrast, the smallest adult from a w. North Carolina population was 80 mm in total length (Chadwick 1944). Aquatic juveniles feed almost year-round and mature in 2 yr. The eft feeds only during rainy summer periods and requires more time to reach maturity. In New York, efts were more abundant in old growth versus disturbed forest stands and least abundant in conifer plantations (Pough et al. 1987).

Breeding Period: Spring (April to June), fall (August to October), sometimes November to December (Hurlbert 1969). Characteristically breed in lakes, ponds, and swamps (Hurlbert 1970).

Egg Deposition: Late April to June (Petranka 1998:458).

No. Eggs/Mass: 200 to 375 eggs (Bishop 1941:64), laid in water, attached singly to the leaves of aquatic plants.

Time To Hatching: 3 to 5 weeks (Logier 1952:64), temperature dependent.

Larval Period: About 2 months in New York (Bishop 1941). Postlarval migration from aquatic to terrestrial habitat occurs from summer through late fall during diurnal rainfall in New York (Hurlbert 1970). Red efts are exceedingly abundant in the Northeast (Evans 1947, Petranka 1998:459).

Home Range/Movement: Approximately 270 m^2 for red efts (juveniles) in an oak-pine woodland in w. Massachusetts; maximum daily movement was 13 m (Healy 1974). Average movement along the edge of a small pond in Pennsylvania was 10.1 ft (3.1 m) for females, and 11.2 ft (3.4 m) for males; most individuals remained within 5 ft (1.5 m) of shore (Bellis 1968). Harris (1981) reported that all movement was random for 323 males in a Virginia pond and so considered males to be nonterritorial.

Food Habits: Both larvae and adults (Hamilton 1932) are generalist carnivores (Burton 1977): insects and their larvae, particularly mayfly, caddisfly, midge and mosquito larvae (Ries and Bellis 1966), springtails (MacNamara 1977); tadpoles, frog eggs, worms, leeches, small mollusks and crustaceans, spiders, mites, occasionally small minnows (Hamilton 1932), salamander eggs are also a major food item (T. Tyning pers. commun.). Also ingest molted skin. Snails are an important food source for the red eft (Burton 1976). Cannibalism on their own larvae provides an important component of the diet in July and August (Burton 1977).

Comments: The red-spotted newt is the second most widely distributed salamander in North America, occurring from boreal forests to subtropical forests (Petranka 1998:452). Red-spotted newts mate in ponds and streams. The red eft remains on land for 2 to 7 yr, feeding in the forest leaf litter; most remain on land 4 to 5 yr, then return to the water where they transform to aquatic adults (Healy 1974). Neotenic individuals have been found on the Coastal Plain in Massachusetts and in New York (Bishop 1941:73–75). Some individual populations omit the terrestrial eft stage. Skin secretions of red efts are highly toxic—about 10 times more toxic than those of adults (Brodie 1968).

Northern Dusky Salamander

(*Desmognathus fuscus*)

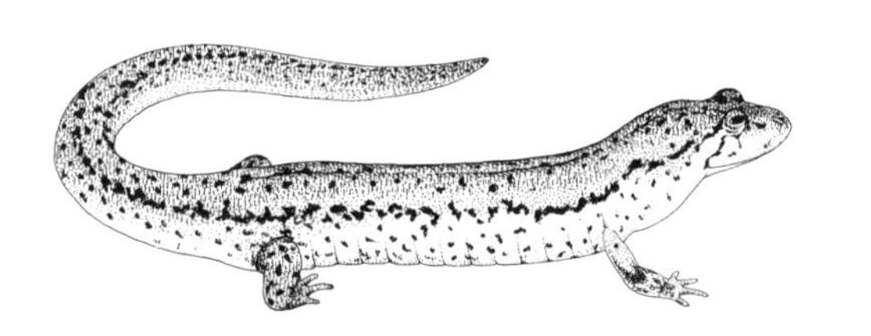

Range: Southern New Brunswick and s. Quebec west to se. Indiana and central Kentucky and s. to the Carolinas; throughout the Northeast excluding s. New Jersey.

Distribution in New England: Throughout the region except for Cape Cod and perhaps northernmost Maine. Apparent spotty distribution in e. Maine (Markowsky 1992).

Status in New England: Common.

Habitat: Adults occur in woodlands at the margins of cool running water—inhabits clear rocky streams in springy banks, seepage areas and beds of semidry brooks (Wyman 1988); in forested seeps and headwater streams (Whitlock et al. 1994), under the cover of wet leaves, moss, rock piles, other debris, or in burrows in the soil. Ventures from streamside only during wet weather. Occurs from sea level to high elevations >275 m (Klemens 1993:52). Moves under logs and rocks in deeper water to hibernate in September. May remain active throughout the winter in stream bottoms or deep in unfrozen soil (Ashton and Ashton 1978).

Special Habitat Requirements: Permanent or intermittent streams or seeps in woodlands.

Age/Size at Sexual Maturity: Variable: about 3 yr (Dunn 1926:92), most males at 3.5 yr, females deposit first eggs at 5 yr (Organ 1961). Some males mature at 2 yr, females at 3 yr (Danstedt 1975). Body size at maturity varies among populations (Tilley 1968). Adults range from 6 to 14 cm total length (Petrank 1998:174).

Breeding Period: Breeding occurs in either late spring or fall (Bishop 1941:212–213). Females may breed biennially (Organ 1961). Breeds in ponds or streams.

Egg Deposition: June to September in Connecticut (Babbitt 1937). Female guards the eggs in damp hollows beneath large rocks (Krysik 1980), under loose bark of logs, or between wet leaf litter layers and in moss close to the water's edge. Larvae move to water where development continues. Clutches found less than 50 cm from the edge of streams and springs or in seepage areas (Krysik 1980).

No. Eggs/Mass: 8 to 28 stalked eggs in compact clusters, average 17 (Bishop 1941:314).

Time to Hatching: 7 to 8 weeks in Massachusetts (Wilder 1917), 5 to 8 weeks in New York (Bishop 1941:318), about 10 weeks in Connecticut (Babbitt 1937:16).

Larval Period: 7 to 10 months, usually transform in June (Wilder 1913:295). From 9 to 12 months in Maryland (Danstedt 1975). Survive in wide range of pH values; highest density of larvae pH 5.0 to 6.0 (Gore 1983).

Home Range/Movement: Less than 10 ft (3 m) along a stream in a wooded ravine in Pennsylvania (Barthalmus and Bellis 1969). Average range of 1.4 m^2 in a gravel-bottom stream in Ohio (Ashton and Ashton 1978). Average about 150 ft^2 (14 m^2) along a stream in Kentucky, maximum movement of 100 ft (30.5 m) as open water dried up (Barbour 1971:57). Average weekly movement less than 0.5 m (Ashton 1975). In an intermittent mountain stream, average for 5 individuals was 48 m^2, daily movements less than 2 m (Barbour et al. 1969b).

Food Habits: Small aquatic and terrestrial invertebrates, insects— 96 percent of prey by weight (Burton 1976), grubs, worms, crustaceans, spiders, and occasionally mollusks; sometimes larvae of own species. Nocturnal feeder, also active on cloudy or rainy days.

Comments: Larval stage is aquatic; adults are riparian. Healy (1974) found efts most active on the forest floor when temperatures were above 13° C (55° F) and substrate was moist.

Northern Redback Salamander

(*Plethodon cinereus*)

Range: Nova Scotia w. to s. Ontario and e. Minnesota, s. in scattered colonies to Missouri, in the Smoky Mountains, in s. Tennessee and e. to Cape Hatteras.

Distribution in New England: Occurs throughout the region from sea level to at least 1,050 m elevation in the White Mountains.

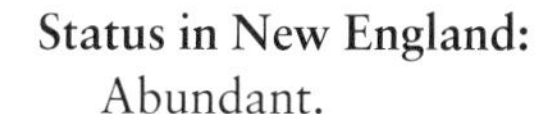

Status in New England: Abundant.

Habitat: Entirely terrestrial. Forest litter habitats in mixed deciduous or coniferous woods, inhabiting interiors of decaying logs and stumps, also found underneath stones, moist leaf litter, and bark. Most abundant in well-drained forested habitats (Petranka 1998:338); wet areas and extremely moist bottomland avoided. Enters xeric, sandy habitats where moist microhabitats exist (Klemens 1993:74). Hibernates down to 15 in (38 cm) soil depth (Oliver 1955:121) in deep leaf litter, or in rock crevices. May be active during mild winter weather (Minton 1972:67). In Indiana, individuals were found active in an ant mound throughout the winter (Caldwell 1975). Found hibernating at 30 to 36 in (76.2 to 91.4 cm) depth in decaying root systems of dead white oaks in se. Massachusetts (Hoff 1977). Has been found hibernating in aquatic situations in Maryland (Cooper 1956).

Special Habitat Requirements: Logs, stumps, rocks, woody debris piles.

Age/Size at Sexual Maturity: Generally during second year (Oliver 1955:277), but female usually reproduces in third year (Burger 1935). Males at 42.0 mm snout to vent length, females 44.8 mm snout to vent length in Michigan (Werner 1971).

Breeding Period: Biannual cycle, spring and late fall (October through December) in Maryland (Sayler 1966).

Egg Deposition: June to July in Massachusetts (Lynn and Dent 1941).

No. Eggs/Mass: 1 to 14 pale eggs, average 8 (Lynn and Dent 1941) in small grape-like clusters attached to roof of small natural chamber or crevice, laid in and under rotted logs and stumps. Reproduce annually in Connecticut (Lotter 1978). Both adults may brood the eggs (Friet 1995), or the female may guard the eggs (Highton and Savage 1961).

Time to Hatching: 30 to 60 days (Oliver 1955:234), extending to 84 days in Maine (Banasiak 1974). Hatch in August to September. Larval stage is completed within egg.

Home Range/Movement: Home range is small owing to restricted horizontal movement (Taub 1961). Movement of less than 1 ft (30.5 cm) for 14 individuals in hardwood forest habitat in New Jersey; individuals usually found under the same object where initially captured (Taub 1961). Home ranges of 13 m^2 for females, about 24 m^2 for males, were determined in a northern hardwood forest in Michigan (Kleeberger and Werner 1982).

Food Habits: Small insects and their larvae, earthworms, snails, slugs, spiders, sowbugs, millipedes, mites (Surface 1913:95). Mites were the most important food, accounting for 65 percent of the prey items in a New Hampshire study (Burton 1976), insects 73 percent by weight in a New York study (Jameson 1944). During rainy summer nights, found on leaf litter presumably foraging for food (Burton and Likens 1975). Often climbs tree trunks and shrubs in search of food, particularly during wet nights. Occasionally cannibalistic.

Comments: Three distinct color phases occur: redback, leadback, and erythristic. In Connecticut, the redback morph occurs almost exclusively in cold upland areas; in areas of more moderate climate and elevation, both redback and leadback morphs occur (Klemens 1993:74). All records of erythristic individuals occur north of 41° and south of 47° latitude (Tilley et al. 1982). The redback salamander is the most abundant terrestrial vertebrate in New England and accounts for the greatest amount of vertebrate biomass in the Hubbard Brook Experimental Forest in New Hampshire (Burton and Likens 1975). In northern New England, the typical redback and the leadback phases are both common. The leadback reportedly is most common in coniferous forests, especially near the coast in Maine (Witham 1992); the redback is more common in deciduous forest. The erythristic or red phase, not found in Maine, replaces the leadback in deciduous boglands in other parts of its range, and these two color phases rarely occur together (Tilley et al. 1982).

After clear-cut harvesting, numbers are reduced; recovers to preharvest levels in about 50 yr in New England northern hardwood forest (DeGraaf and Yamasaki 1992). In New York, populations in 60-yr-old second growth forest were similar to those in adjacent old growth (Pough et al. 1987).

Northern Slimy Salamander

(*Plethodon glutinosus*)

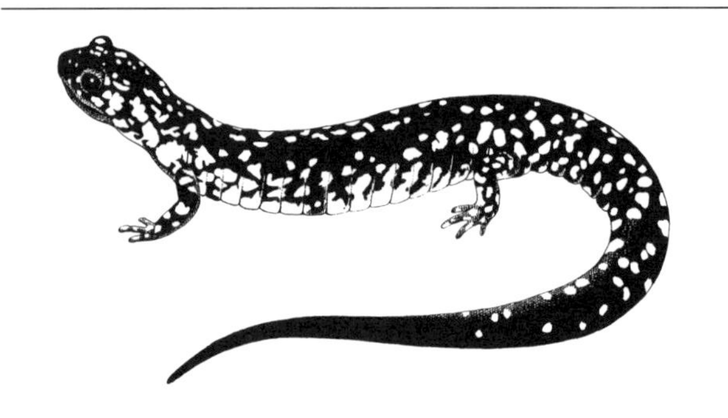

Range: Extreme w. Connecticut w. through central New York to e. Oklahoma, Arkansas, s. to Louisiana and Florida. Scattered colonies in s. New Hampshire and Texas.

Distribution in New England: Extreme w. Connecticut and perhaps sw. Massachusetts (Klemens 1993:79). In New Hampshire, two specimens found near Rindge in 1962; otherwise unreported (Taylor 1993).

Status in New England: Uncommon to rare.

Habitat: Moist wooded hillsides and ravines. Terrestrial, found underneath moist humus, manure piles, in crevices in rock, shale banks, and under logs in woodland areas. Bishop (1941:718) found the species most abundant in banks along highways and woodland openings. Has been found in second-growth oak-hickory forests and steep hemlock slopes of ravines in the Helderberg Mountains, New York (M. Stewart pers. commun.), and to an elevation of 1,768 m in Great Smoky Mountain National Park (Powders and Tietjen 1974). Also in mature mixed deciduous forests (Semlitsch 1980a). Hibernates underground from November to March or April.

Special Habitat Requirements: Rock outcroppings, logs in wooded areas.

Age/Size at Sexual Maturity: Females mature at about 4 yr and lay eggs in the fifth year, males at 4 yr (Highton 1962). Snout to vent length is 59 to 74 mm in females, 53 to 70 mm in males (Highton 1962).

Breeding Period: Autumn and spring (Highton 1956).

Egg Deposition: Probable biennial oviposition occurring in late spring or early summer in northern populations (Highton 1962). Eggs laid within rock crevice or rotted logs (Smith 1961:58), also found in caves (Bishop 1941:224).

No. Eggs/Mass: 13 to 34 eggs, average 16 to 17 (Highton 1962). Eggs aggregated in a thin envelopment. Fecundity increases with body size (Semlitsch 1980a). Females guard their eggs and often coil about them (Petranka 1998:358).

Time to Hatching: Probably in late summer; entire larval period spent within egg. Embryonic period lasts 2 to 3 months; hatching varies with latitude (Highton 1956).

Home Range/Movement: Twenty-two individuals in n. Florida were recaptured at or within 4 ft (1.2 m) of the original capture point (Highton 1956). Adult home ranges are less than 9 m diameter; immatures' range is less than 6 m in diameter, in oak-hickory forest with thick leaf litter in North Carolina. Mean movement distances were 17.5 m for males, 14.3 m females, and 4.2 m juveniles. Probably capable of movements more than 90 m beyond home-range area (Wells and Wells 1976).

Food Habits: Euryphagic, consuming a wide variety of prey (Powders and Tietjen 1974): Mostly insects, also sowbugs, worms, centipedes, spiders, slugs, and snails (Hamilton 1932). Ants and beetles were the most abundant food items in a Virginia study, accounting for 58 percent of the total weight of food (Davidson 1956). Availability of prey probably governs feeding habits.

Comments: Nocturnal, may be active during some rainy days. During hot, dry spells found deep underground or under logs in dense aggregations (Wells and Wells 1976). The slimy salamander is so named because it produces copious amounts of adhesive skin secretions from the tail when handled roughly (Petranka

1998:354). These secretions function to deter predators (Brodie et al. 1979) and when dry are difficult to remove from human skin (Petranka 1998:354).

Four-toed Salamander

(*Hemidactylium scutatum*)

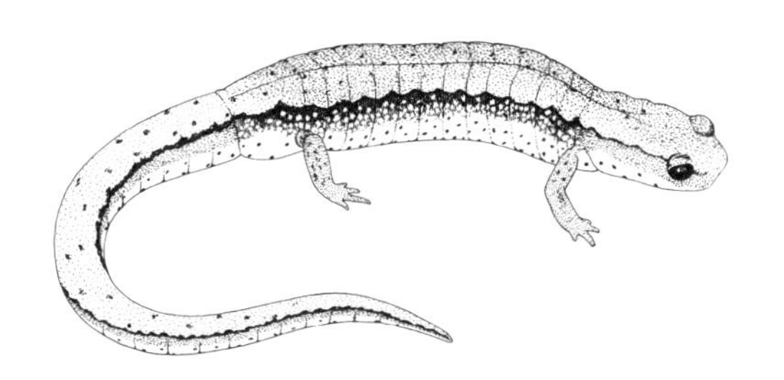

Range: Nova Scotia w. to s. Ontario and Wisconsin, s. to Alabama and Georgia. Absent from most of n. New England. Scattered disjunct populations occur in the e. United States.

Distribution in New England: Throughout Connecticut and Rhode Island, uncommon in w. Massachusetts uplands. In Maine, Mt. Desert Island and along Penobscot River.

Status in New England: Uncommon to rare.

Habitat: Shaded or open and wet woodlands, preferably with sphagnum moss; shallow woodland pools; tamarack bogs. Hides in moss, in moist decaying wood, under stones or wet leaves. Prefers an acidic environment. Found in beech/maple, yellow birch/maple and other hardwood forests, found less often in coniferous woods (Neill 1963:2.1). In mixed forests in New York (Bishop 1941:190). Larval stage is aquatic, found in pools and quiet streams with an abundance of moss. Typically hibernates in decaying root systems of trees. Aggregations may appear during hibernation with rotted wood or leaf litter (Blanchard 1933b).

Special Habitat Requirements: Acidic wet woodlands with sphagnum mats (Breitenbach 1982) in Connecticut. Prefers sandy acid woods adjacent to red maple swamps in Connecticut (Klemens 1993:68); adjacent to conifer ous woods in Michigan (Blanchard 1923).

Age/Size at Sexual Maturity: About 2½ yr (Barbour 1971:74).

Breeding Period: Late summer and autumn, peak in fall.

Egg Deposition: March to April or May (Blanchard 1934, Barbour 1971:73). Nests located next to and just above water. Females construct crude nest cavities either below or several centimeters within moss mats (Bishop 1941). When ovipositing, the female usually turns upside down and deposits her eggs attached singly to rootlets, moss strands, or other substrates in a loose cluster within the cavity. Several minutes are required to deposit a single egg (Bishop 1941).

No. Eggs/Mass: 19 to 50 eggs (Dunn 1926:200, 202), average 50 in New York (Bishop 1941:183). Communal nesting may occur with up to 800 eggs laid per nest. One to four females will remain with eggs (Wood 1953).

Time to Hatching: 38 to 60 days (Blanchard 1934).

Larval Period: 6 weeks (Blanchard 1923); as long as 18 weeks; variation in larval development depends upon pond conditions (Bishop 1941:186).

Home Range/Movement: Unreported.

Food Habits/Preferences: Adult: small invertebrates, including insects, spiders, and earthworms. Also springtails, ground beetles, ants, snails, true bugs. Larvae: zooplankton.

Comments: A nocturnal and secretive species, difficult to locate. When molested, often coils tightly, tucks head beneath the tail, and remains immobile (Brodie 1977).

Northern Spring Salamander

(*Gyrinophilus p. porphyriticus*)

Range: Through the Appalachian Mountains from w.-central Maine and extreme se. Quebec, s. to e. Ohio and central Alabama, Pennsylvania, and n. New Jersey. Absent from the Coastal Plain.

Distribution in New England: Vermont, New Hampshire, s.-central Maine (Conant and Collins 1991), nw. Rhode Island (Lazell and Raithel 1986); in Connecticut and Massachusetts the distribution is divided into e. and w.

portions by the Connecticut River Valley; in e. Massachusetts and Connecticut, rare and localized, primarily subterranean in springs and wells.

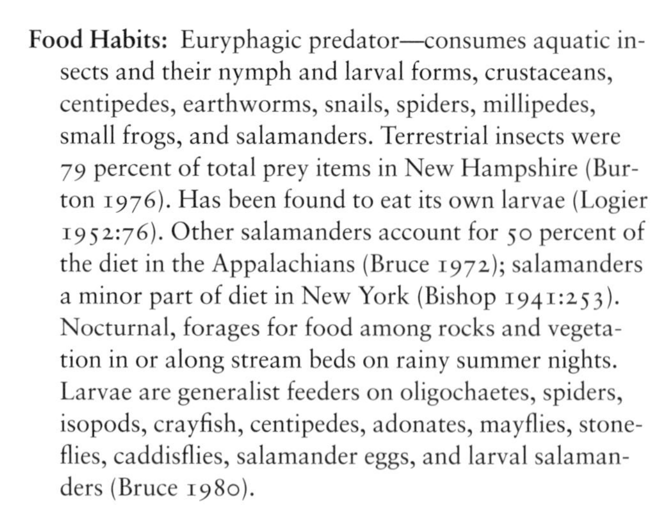

Status in New England: Uncommon to rare, except in Vermont and nw. Berkshire County, Mass., where common.

Habitat: Adults found in, but not restricted to (hardwood) forested areas with clear, cold water, springs, mountain streams, creeks, boggy areas. Also in depressions under stones or other cover adjacent to water. Usually occurs at higher elevations in spruce/fir forests, typically in moist situations, in underground water courses and limestone caves, beech/maple/hemlock forests, in shale ravine streams in Tompkins and Albany Counties, N.Y. (M. Stewart pers. commun.). Have been found in hillside meadow streams, swamps, and lake margins. Not found in suburban, urban, or disturbed areas (Whitlock et al. 1994).

Special Habitat Requirements: Cold (<12° C) streams, seeps, or springs containing large flat rocks or rock crevices. In winter, also in wet soil near water where it remains somewhat active in burrows.

Age/Size at Sexual Maturity: 4 to 5 yr (New York), at total length of about 5½ in (14 cm) in New York (Bishop 1941:370).

Breeding Period: Mid October to winter months (Bruce 1972). Annual reproduction cycle (Bruce 1969).

Egg Deposition: April to summer (most) and into the fall (Bruce 1972), female guards eggs (Organ 1961). Most females probably lay their eggs in deep underground recesses in streams or seeps as evidenced by the fact that so few nests have been found (Petranka 1998:283).

No. Eggs/Mass: 9 to 63 (Bruce 1972), 44 to 132 eggs in New York (Bishop 1941:247), 44 to 66 in Virginia (Organ 1961). Eggs laid in running water under logs and stones, usually in monolayers, sometimes attached singly, to the undersides of rocks or other objects (Petranka 1998:283).

Time to Hatching: Fall (Organ 1961). Hatch late summer, early fall. The young from one clutch may remain near the nest site for several months after hatching (Bruce 1980).

Larval Period: Variable larval period, average of about 4 yr. Metamorphosis occurs in late spring, summer (Bruce 1980). Larvae are aquatic.

Home Range/Movement: Unreported.

Food Habits: Euryphagic predator—consumes aquatic insects and their nymph and larval forms, crustaceans, centipedes, earthworms, snails, spiders, millipedes, small frogs, and salamanders. Terrestrial insects were 79 percent of total prey items in New Hampshire (Burton 1976). Has been found to eat its own larvae (Logier 1952:76). Other salamanders account for 50 percent of the diet in the Appalachians (Bruce 1972); salamanders a minor part of diet in New York (Bishop 1941:253). Nocturnal, forages for food among rocks and vegetation in or along stream beds on rainy summer nights. Larvae are generalist feeders on oligochaetes, spiders, isopods, crayfish, centipedes, adonates, mayflies, stoneflies, caddisflies, salamander eggs, and larval salamanders (Bruce 1980).

Comments: Formerly called the purple salamander. Adults are most frequently found beneath surface or subsurface objects in or near springs or seeps, or on roads on rainy nights (Petranka 1998:285). Spring salamanders commonly eat other salamanders (Bishop 1941) but this is likely more frequent in the southern Appalachians where the surface density of salamanders is very high (Bruce 1972); northern adult populations feed mostly on invertebrates (Surface 1913, Bishop 1941, Petranka 1998:286). Spring salamanders are sometimes eaten by water snakes and garter snakes (Uhler et al. 1939), but their noxious skin secretions repel shrews (Brodie et al. 1979).

Northern Two-lined Salamander

(*Eurycea bislineata*)

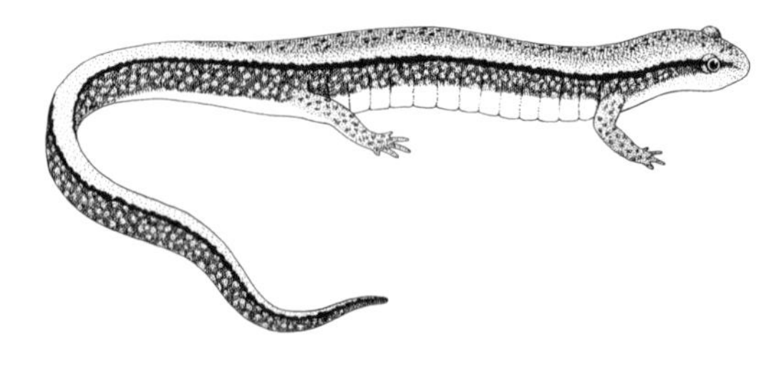

Range: Gaspé Peninsula, Quebec, and e. Ontario sw. through Ohio to e. Illinois, s. to extreme ne. Mississippi and e. to Virginia.

Distribution in New England: Throughout the region except Cape Cod.

Status in New England: Common to abundant.

Habitat: Deciduous floodplain bottoms to moist deciduous or mixed forest at high elevations (Behler and King

1979:321). Along rocky lakeshores, brooks and streams, boggy areas near spring or seeps. Adults found under objects at water's edge in moist soil or in coarse sand and gravel at stream bottoms or edges, leaf litter, and crayfish burrows (Ashton and Ashton 1978). In wet woodlands or pastures. During wet or humid weather will move into moist woods more than 100 m from water courses (D. Rudis, personal observation). Soils where *E. bislineata* occur have higher pH than soils where *E. bislineata* do not occur (Wyman 1988).

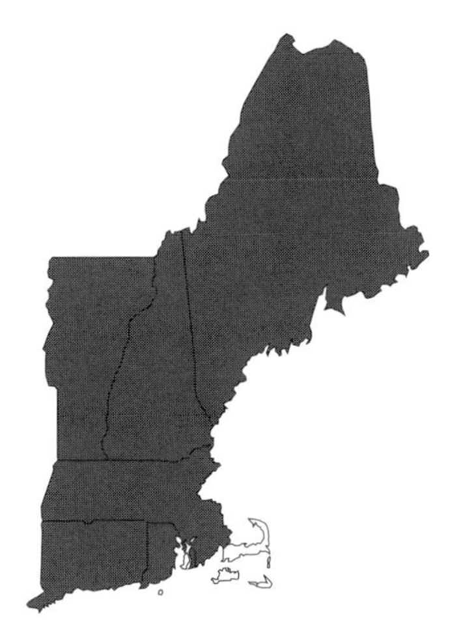

Hibernates under water, or remains active in feeding aggregations in springs and cold-flowing streams in New York (Stewart 1956a) and adjacent unfrozen soil (Ashton and Ashton 1978).

Special Habitat Requirements: Alkaline sandy or silty streams with gravel and flat rocks or cobble inclusions for breeding.

Age/Size at Sexual Maturity: The majority (New York) mature during the second fall after metamorphosis (Stewart 1956a).

Breeding Period: September through May, based on the seasonal development of male secondary sexual characteristics and the presence of sperm caps in the cloacae of females (Sever 1979). Breeds in streams.

Egg Deposition: May to early June in Massachusetts (Johnson and Goldberg 1975).

No. Eggs/Mass: 12 to 36 eggs, average of 18 eggs in Massachusetts (Wilder 1924). Eggs deposited singly in monolayers attached to bottoms of stones or logs in running water (Baumann and Huels 1982). Several females may use the same stone as a nest site, one female remains with eggs until hatching (Bishop 1941).

Time to Hatching: 4 to 10 weeks, depending on local stream temperature (Bishop 1941).

Larval Period: 2 or 3 yr, aquatic (Wilder 1924).

Home Range/Movement: Average area less than 14 m^2 for 20 monitored individuals along a stream in Ohio (Ashton and Ashton 1978). Territories were aggressively defended in an artificial environment (Grant 1955).

Food Habits: Adults eat insects, particularly beetles, beetle larvae, mayflies, stonefly nymphs, and dipterans; also spiders, mites, millipedes, sowbugs, and earthworms (Hamilton 1932). Most prey are of terrestrial origin (Burton 1976). Larvae forage on stream bottoms for chironomid larvae, copepods, fly pupae, and stonefly nymphs (Caldwell and Houtcooper 1973).

Comments: Will travel in the open during wet or rainy nights, rarely during rainy days. Adults are extremely agile and when disturbed often escape through a series of rapid jumps. Two-lined salamanders are eaten by a variety of predators: eastern screech-owls, garter snakes, ringneck snakes (Uhler et al. 1939). Rainbow trout and two-lined salamanders are mutual predators: trout feed on adult salamanders, and salamanders feed on trout fry and embryos (Mathews 1982). Two-lined salamanders are distasteful to shrews (Brodie et al. 1979).

Eastern Spadefoot

(Scaphiopus holbrookii)

Range: Southeastern Massachusetts w. to New York and se. Missouri, s. to e. Louisiana and Florida. Not found in the higher elevations of the Appalachians or the Everglades in Florida.

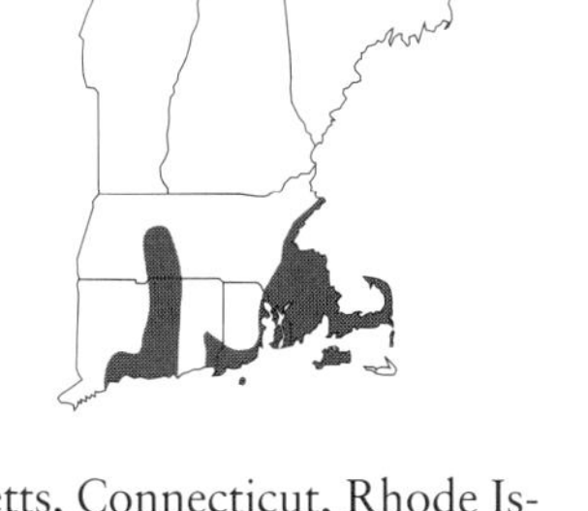

Distribution in New England: Cape Cod, Martha's Vineyard, Nantucket (Lazell 1976), Connecticut River Valley, e. (to Cape Ann) and se. Massachusetts, Connecticut, Rhode Island.

Status in New England: Uncommon to rare.

Habitat: In sandy or loose soils in sparse shrub growth or open forest areas. Terrestrial and subterranean, only enters water to breed, usually in temporary rain pools. Prefers forest areas with sparse leaf litter (Pearson

1955). In farmland areas in Connecticut River Valley, Massachusetts, and pitch pine–scrub oak dunes in New York (Stewart and Rossi 1981). Colonies occur along floodplains of major rivers. Emerge in spring from hibernation when soil moisture is sufficient.

Special Habitat Requirements: Loose sandy soils without thick leaf litter, temporary pools for breeding.

Age/Size at Sexual Maturity: During second year after metamorphosis, males at 15 months, females at 19 months (Pearson 1955).

Breeding Period and Egg Deposition: Usually April or May, extending into August; breeding is initiated by a heavy rainfall (Gosner and Black 1955). Breeds in congregations of many individuals if population is high. Usually a one-night phenomenon.

No. Eggs/Mass: 1,000 to 2,500 eggs in masses of 6 to 110 in irregular bands among plants in temporary water. Eggs are very adhesive.

Time to Hatching: 5 to 15 days (Oliver 1955:236).

Tadpoles: Late broods transform in 16 to 20 days (Gosner and Black 1955), 48 to 63 days for early broods (Driver 1936).

Home Range/Movement: Mean home range about 10 m^2 (108 square feet) in n. Florida, for 90 percent of captures average home range was about 6.2 m^2 (67 ft^2); occupy one or several underground burrows within home range (Pearson 1955). Maximum dispersal distances of 9.8 m (32 ft); individuals were recaptured in the same home ranges after 5 yr (Pearson 1955).

Food Habits: Flies, spiders, crickets, caterpillars, true bugs, other ground-dwelling arthropods, earthworms, and snails. Moths are eaten when they can be caught (Bragg 1956:36). Tadpoles are planktonic feeders for the first few days (Richmond 1947), later becoming carnivorous and sometimes even cannibalistic.

Comments: Nocturnal; peaks of activity occur just after sundown and before sunrise. Fossorial; individuals have remained in burrows an average of 9.5 days at a time, emerging to feed (Pearson 1955). Can remain underground for weeks or months during dry periods, to depths of 3 to 7 ft (1 to 2 m) (Ball 1933, cited in Babbitt 1937:20). As evidence of the spadefoot's secretive and nocturnal habits, there was a total of 16 reported sightings from 1811 to 1936 in the ne. part of its range (Ball 1936, cited in Bragg 1956).

Eastern American Toad

(Bufo a. americanus)

Range: Nova Scotia and the Gaspé Peninsula w. through central Ontario to Lake Winnipeg, s. to e. Kansas, central Indiana, central Alabama, and central North Carolina.

Distribution in New England: Throughout New England, Martha's Vineyard, Cape Cod e. to Dennis; introduced on Cuttyhunk Island, Mass.

Status in New England: Common.

Habitat: Found in almost any habitat: gardens, woods, yards with cover, damp soil and a food supply. Sea level to mountain elevations. Usually in moist upland woods.

Special Habitat Requirements: Needs shallow water for breeding. Hibernates in burrows underground to 12 in (30.5 cm) deep (Oliver 1955:122) from October to late March or April.

Age/Size at Sexual Maturity: 3 to 4 yr (Dickerson 1969:72), 2 to 3 yrs (Hamilton 1934).

Breeding Period and Egg Deposition: Early April to July, peak in late April in the Northeast. Travels to breeding ponds at night in large numbers (Maynard 1934).

No. Eggs/Mass: 4,000 to 12,000 eggs (Dickerson 1969:67). Laid in long curling strings amid aquatic vegetation.

Time to Hatching: About 3 to 12 (average 4) days.

Tadpoles: 5 to 10 weeks.

Home Range/Movement: Exhibits homing behavior by returning to breeding sites; 264 individuals used the same site annually in Ontario (Oldham 1966). Newly metamorphosed toads showed celestial orientation when leaving ponds; as most movement is nocturnal, course determination is probably during daylight hours (Dole 1973).

Food Habits: Terrestrial arthropods, including insects, sowbugs, spiders, centipedes, and millipedes. Slugs and earthworms are other invertebrate foods. Some vegetable matter is taken accidentally. Food taken determined by availability (Hamilton 1954). Feeds from twilight through the evening hours.

Comments: Most active during evening hours. May bask but will seek cover during the heat of the day. Calls and breeds during the day at the peak of breeding season.

Fowler's Toad

(*Bufo fowleri*)

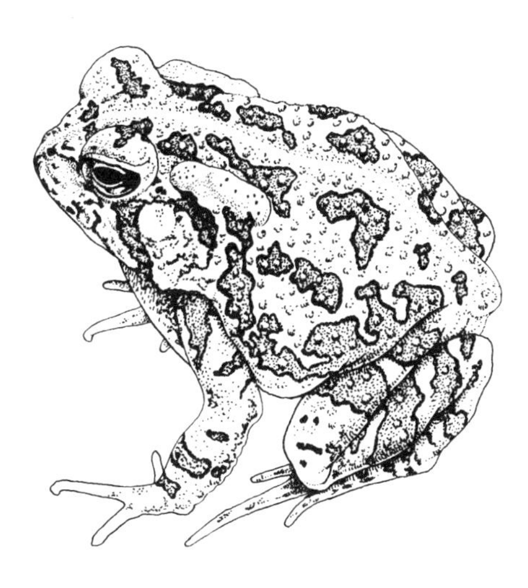

Range: Southern New England w. to central Pennsylvania, the n. shore of Lake Erie and e. shore of Lake Michigan, s. to Missouri, e. Oklahoma, Texas, central Georgia, and South Carolina.

Distribution in New England: Connecticut, Rhode Island, and Massachusetts w. to Berkshire County (where not yet confirmed), found in Connecticut River Valley to White River Junction, Vt. (Klemens 1993:101); one record from Hanover, N.H. (Taylor 1993). Primarily coastal plain from se. New Hampshire to Cape Cod and Narrasamset Bay and sand and alluvial deposits of central Connecticut lowlands (Klemens 1993:101).

Status in New England: Locally common.

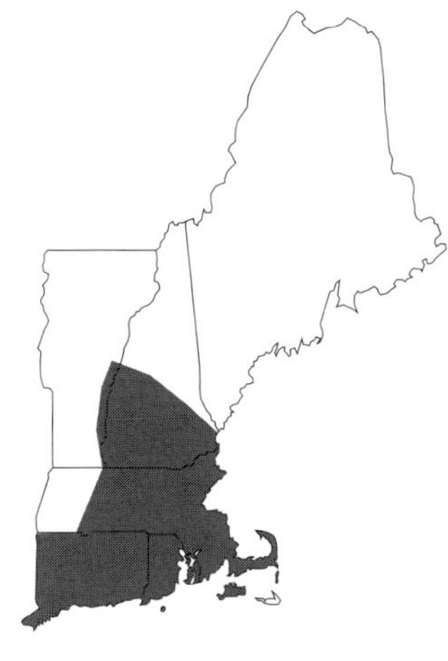

Habitat: Prefers areas with sand and alluvial deposits—shorelines, river valleys, beaches, and roadside areas. Usually found in lowland areas, but frequently in open pine and oak forests, gardens, lawns and fields; also found in small marshy ponds. Hibernates in burrows in well-drained sandy soils to 3 ft (0.9 m) deep from early fall to late spring (May in Connecticut).

Special Habitat Requirements: Sandy alluvial soils, shallow water (not dominated by *Sphagnum* spp.) for breeding.

Age/Size at Sexual Maturity: Probably breeds during third year (Stille 1952).

Breeding Period and Egg Deposition: From mid May extending to mid August (2 to 4 weeks later than *Bufo a. americanus*). Shallow water (with aquatic-emergent vegetation) of pools, lake margins, ditches, and so on, necessary for breeding.

No. Eggs/Mass: Up to 8,000 eggs laid in long strings in submerged sticks or aquatic vegetation (Wright and Wright 1949:212); sometimes laid in double strings (Klemens 1993).

Time to Hatching: About 1 week.

Tadpoles: 40 to 60 days, usually transform midsummer.

Home Range/Movement: Average distances between captures ranged from 22 to 32 m during a 3-yr period on a golf course in Connecticut (Clarke 1974a). Night movements of 200 to 700 ft (61 to 213 m) or more to reach water's edge (Lake Michigan). Toads usually found within 100 ft (30.5 m) of previous capture point (Stille 1952).

Food Habits: Chiefly ground-dwelling insects, particularly ants and beetles (Clarke 1974b); also consumes earthworms, spiders, snails, and slugs.

Comments: During evening hours may move to edge of water to replenish body moisture (Stille 1952). May be active during the day, but typically crepuscular (Minton 1972:95). Activity periods vary with populations, mostly nocturnal in Connecticut (Clarke 1974a). Hybridizes with *Bufo americanus* where breeding seasons overlap (Klemens 1993:101). Found at elevations to 1,150 feet (351 m) at Wanzer Mountain, Conn. (Klemens 1993:101).

Northern Spring Peeper

(*Pseudacris c. crucifer*)

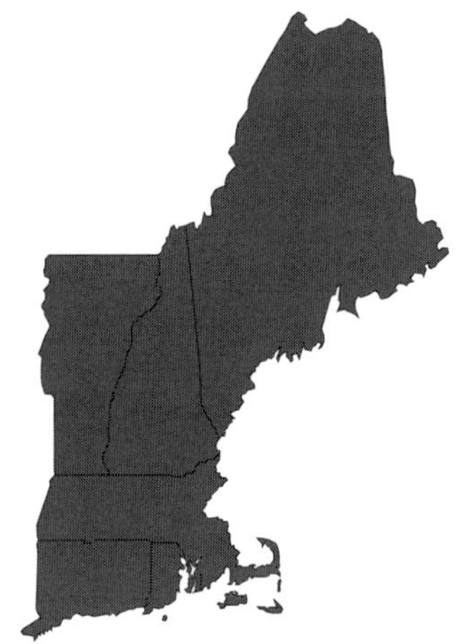

Range: Nova Scotia, the Gaspé Peninsula and Quebec w. to the s. tip of James Bay w. through Ontario to Lake Winnipeg, s. to e. Texas and the e. United States except Florida and s. Georgia.

Distribution in New England: Throughout the region. Widespread and ubiquitous, occurring from sea level barrier beach ponds to high elevation lakes (e.g., 2,000 ft, 610 m) in the Berkshires in w. Massachusetts.

Status in New England: Common to abundant.

Habitat: Marshy or wet woods, second-growth deciduous woodlots, sphagnum bogs, nonwooded lowlands, near ponds and swamps. Also fields, meadows, sandy coastal and pine barren habitats, and coniferous forests. Found on the ground or burrowed into the soil. Breeds in permanent or temporary water, usually woodland ponds with aquatic debris. Found in cool moist woods after breeding (M. Stewart pers. observ.). Hibernates on land during late November to January or early spring, under moss and leaves.

Special Habitat Requirements: Wetlands dominated by emergent vegetation for breeding.

Age/Size at Sexual Maturity: Early in second year at about 20 mm snout to vent length (Delzell 1958).

Breeding Period and Egg Deposition: Early March to June (in the north).

No. Eggs/Mass: 800 to 1,800 eggs (Wright 1914:16). Laid singly near the bottom of shallow weedy ponds, attached to submerged plants (Oliver 1955:236).

Time to Hatching: 6 to 12 days.

Tadpoles: 90 to 100 days (Wright 1914:42). Usually transform during July (Wright and Wright 1949:314).

Home Range/Movement: In se. Michigan, home-range diameters ranged from 4 to 18 ft (1.2 to 5.5 m), established around forest debris and vegetation; average daily travel was 20 to 130 ft (6.1 to 39.6 m), reported by Delzell (1958).

Food Habits: Small nonaquatic insects: primarily ants, flying bugs, beetles, flies, springtails, and spiders; also mites, ticks, and small snails. Foods taken probably reflect availability, catchability, and size rather than preferences (Oplinger 1967).

Comments: Young frogs are terrestrial in first year (Delzell 1958). May move long distances from breeding areas in summer and fall; single calls heard from woods, shrubby openings, far from water.

Gray Treefrog

(*Hyla versicolor*)

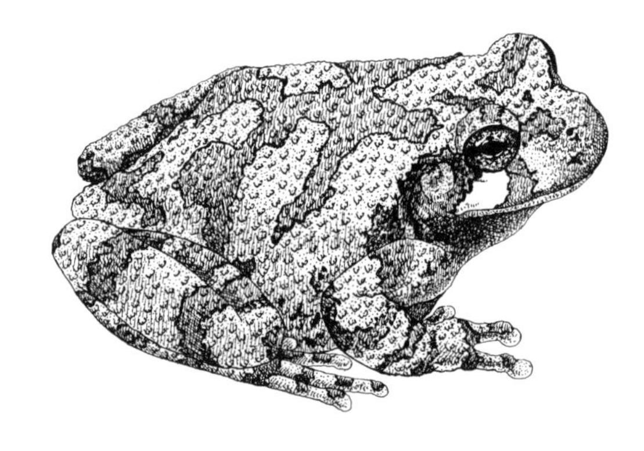

Range: Eastern United States and se. Canada from s. Maine w. to Manitoba and s. through central Texas and the Gulf states to central Florida.

Distribution in New England: Throughout the region except outermost Cape Cod and n. and easternmost Maine (Davis 1992).

Status in New England: Common.

Habitat: Forested areas with small trees, shrubs and shrubs near or in shallow water. Often found on moss or lichen on bark of old trees. Will breed in temporary pools or permanent water, swamps, bogs, ponds, weedy lakes, and roadside ditches; breeding sites are extremely variable. Commonly inhabit moist areas in hollow trees, under loose bark, or in rotted logs during summer months (Smith 1961:93). Hibernate under tree roots, under leaves (Babbitt 1937).

Special Habitat Requirements: Aquatic sites for breeding.

Age/Size at Sexual Maturity: Breeds at 3 yr (Palmer 1949:455).

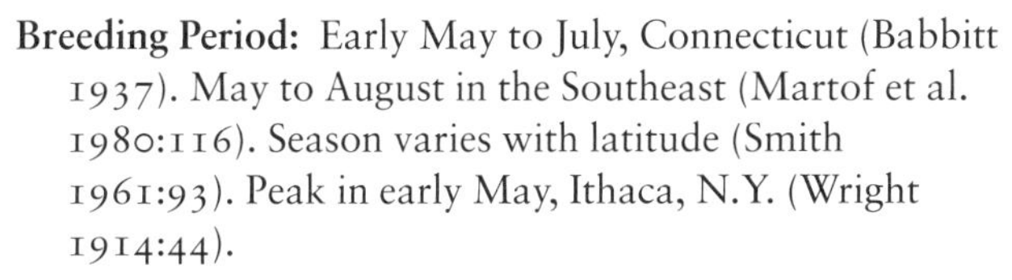

Breeding Period: Early May to July, Connecticut (Babbitt 1937). May to August in the Southeast (Martof et al. 1980:116). Season varies with latitude (Smith 1961:93). Peak in early May, Ithaca, N.Y. (Wright 1914:44).

Egg Deposition: Generally 20 to 35 days between first appearance and first eggs (Wright 1914:47). Loosely attached to vegetation on the surface of shallow water (Martof et al. 1980:116).

No. Eggs/Mass: Total of 1,800 to 2,000 eggs (Wright 1914:49). Packets of 10 to 40 eggs (Martof et al. 1980:116), or 4 to 25 eggs (Smith 1961:93).

Time to Hatching: 4 to 5 days (Babbitt 1937).

Tadpoles: 50 to 60 days, shorter period in warmer areas. Transform late in June to September.

Home Range/Movement: Unreported.

Food Habits: Small insects, spiders, plant lice, mites, and snails. Forages in vegetation and on the ground (Martof et al. 1980:116).

Comments: Most active during evening hours when vocal both during and out of breeding season. Rarely found outside of breeding period. Able to change color from gray to green. Young are emerald green. Single calls heard occasionally in summer during humid days, often before a storm. *Hyla versicolor* is a tetraploid species with 48 chromosomes (Martof et al. 1980:115).

Bullfrog

(*Rana catesbeiana*)

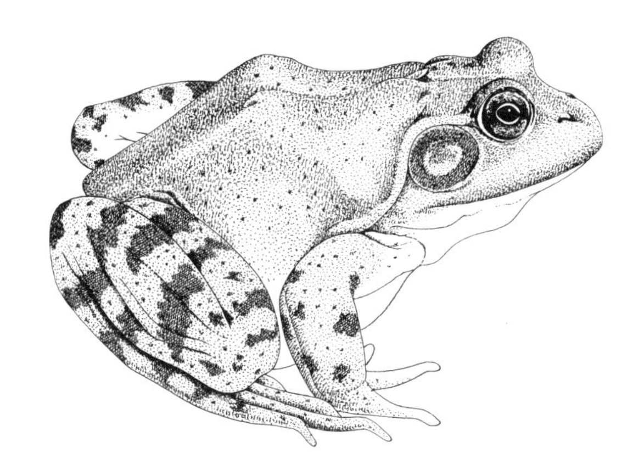

Range: Nova Scotia w. to Wisconsin, s. through the Great Plains to e. Colorado, Texas, and ne. Mexico; throughout the e. United States, except s. Florida and parts of n. Maine. Introduced in California and British Columbia.

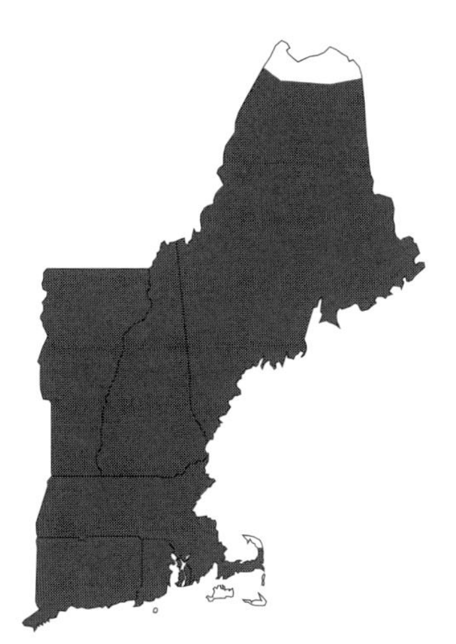

Distribution in New England: Throughout the region except for northernmost Maine.

Status in New England: Common, but formerly more abundant.

Habitat: Near shorelines of large bodies of water with emergent vegetation, lakes, river oxbows. Highly aquatic. Tend to remain in same pools for the summer months if water level is stable (Raney 1940). Will occupy floating logs far from shore. Breed close to shore in areas sheltered by shrubs (Raney 1940). Hibernates under water in mud and leaves about mid October, emerges late February to March, May in New York (Wright 1914:78).

Special Habitat Requirements: Deep permanent water and emergent vegetation for successful breeding. Adult frogs will also use semipermanently or seasonally flooded habitats in the growing season.

Age/Size at Sexual Maturity: In fourth or fifth year.

Breeding Period and Egg Deposition: Late May to July (in the North), peak in July.

No. Eggs/Mass: 12,000 to 20,000 eggs (Wright 1914:82). Eggs laid singly in a large jelly envelope that floats as a thin film in lakes, quiet streams, and ponds.

Time to Hatching: 5 to 20 days (Oliver 1955:237). Often 4 days or less (Wright 1914:83).

Tadpoles: For 2 (usually) to 3 winters. Transform to adult form in 10- to 20-day period from June to August.

Home Range/Movement: Average distance traveled in summer, 200 to 300 ft (61 to 91 m) in a woodland lake and pond in New York (Raney 1940, Ingram and Raney 1943). Evening movement of 200 to 700 ft (61 to 213 m) to water in Michigan (Stille 1952). Home range of 131 bullfrogs in an Ontario pond had an average mean activity radius of 8.6 ft (2.6 m) with minimum and maximum movements of 2.0 ft (0.6 m) and 37.1 ft (11.3 m), respectively (Currie and Bellis 1969). Males defend territories during breeding season. In a Michigan study (Emlen 1968), the average distance between males within a chorus was 17.8 ft (5.4 m), implying an average minimum territorial radius of approximately 9 ft (2.7 m).

Food Habits: Adults feed upon any available small prey: fish, other frogs, salamanders, newts, young turtles, snakes, small birds, mice, crayfish, insects, snails, and spiders. Also cannibalistic. Feeds among water weeds; an indiscriminate and aggressive predator. Tadpoles are primarily vegetarian but also scavenge dead animals, primarily fish.

Comments: The bullfrog has become rare in many areas, presumably due to harvest and to toxic effects of pesticides and other pollutants (M. Stewart pers. commun.). Tolerates a wide range of water acidity and alkalinity (Dale et al. 1985).

Green Frog

(Rana clamitans melanota)

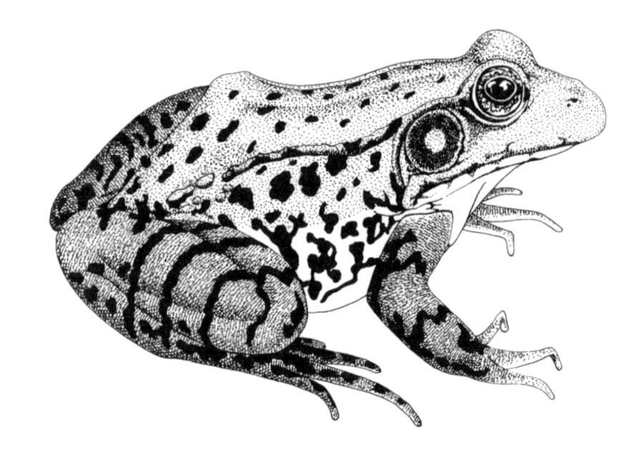

Range: Nova Scotia w. through Quebec and s. Ontario to central Minnesota, s. to e. Oklahoma and e. to n. Georgia and South Carolina. Absent from central Illinois.

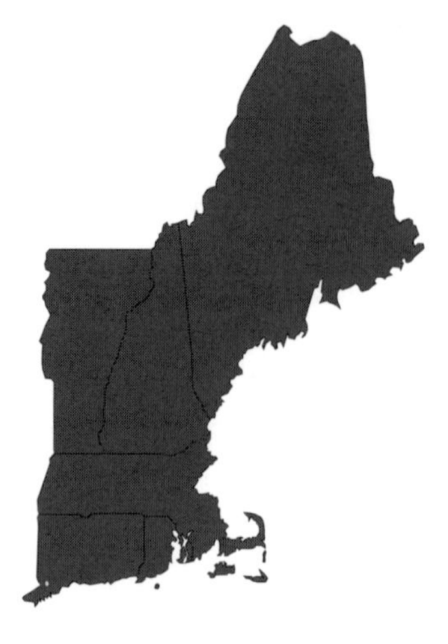

Distribution in New England: Throughout the region.

Status in New England: Common.

Habitat: Riparian, inhabiting margins of a wide variety of shallow permanent or semipermanent freshwater habitats and shores and banks of lakes and ponds, creeks, woodland streams, limestone quarry pools, springs, vernal pools, moist woodlands near water, fens, bogs, tidal creeks, and mill ponds. Seldom more than a few meters from the water. Young often found in semipermanent water. Hibernates underground or underwater from October until March, usually within its home range (Martof 1953b). May be active on warm winter days.

Special Habitat Requirements: Riparian areas.

Age/Size at Sexual Maturity: Males sexually active the season following metamorphosis when 60 to 65 mm long; females mature during the second or third year when 65 to 75 mm long (Martof 1956). Some females reached maturity at 90 mm at Cranberry Lake, N.Y. (M. Stewart pers. commun.). Some may not breed until the second year after transformation (Wells 1977).

Breeding Period and Egg Deposition: April to August, peak in mid May, varies with locality. Some females may lay two clutches (Wells 1976). Emerge from hibernation in early spring but do not breed until mid May in Connecticut (Babbitt 1937).

No. Eggs/Mass: 3,500 to 4,000 eggs (Wright 1914:16), to 5,000 (Pope 1944). Eggs deposited in floating masses of jelly attached to underwater twigs and stems in permanent water.

Time to Hatching: 3 to 6 days (Babbitt 1937).

Tadpoles: 1 to 2 yr. Less than 1 yr in s. parts of range. May transform in same season eggs are laid (Martof 1956).

Home Range/Movement: Ranged from 20 m^2 to 200 m^2 with an average of 61 m^2 in southern Michigan near a stream and lake; daily movements were less than 10 m for 80 percent of the 824 individuals recaptured (Martof 1953b). During breeding season, males maintained 2 to 3 m distance between each other (Martof 1953a). Territory size depends on cover density, 1 to 1.5 m between males in areas of dense cover. Territories with diameters of 4 to 6 m defended in open areas in New York (Wells 1977).

Food Habits: Terrestrial feeders among shoreline vegetation. Insects and their larvae, worms, small fish, crayfish and other crustaceans, newts, spiders, small frogs, and mollusks are taken. Beetles, flies, grasshoppers, and caterpillars constituted over 60 percent of food items (Hamilton 1948). Terrestrial beetles are the most important food item (Stewart and Sandison 1972). Tadpoles are herbivorous.

Comments: Found in or at edge of water during daylight hours; evening hours spent along the banks feeding or in water defending territories (Wells 1977).

Mink Frog

(*Rana septentrionalis*)

Range: Nova Scotia, n. New England and New York w. to n. Wisconsin and Minnesota, n. through Ontario to James Bay and n. Quebec and Labrador.

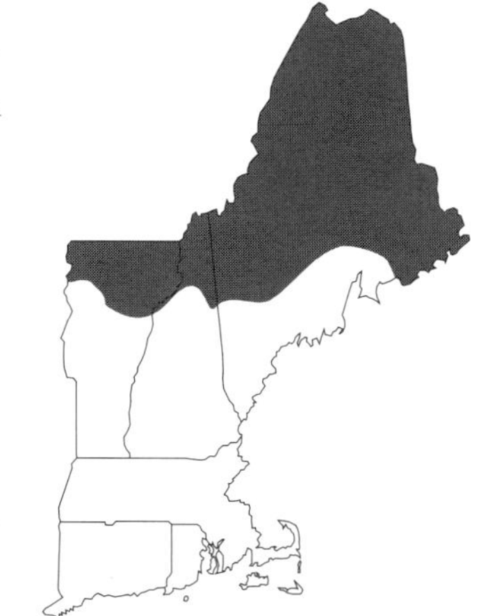

Distribution in New England: North of 43° N latitude in Maine, New Hampshire, and Vermont (excluding Champlain Valley).

Status in New England: Only in extreme n. areas, locally common to rare.

Habitat: Northern lakes and ponds, cold, oxygen-rich springs, inlets where cold streams enter ponds and stream edges. Prefers shallow water with abundant lily pads and pickerel weed. Sometimes found in northern bogs.

Special Habitat Requirements: Breeds and hibernates only in permanent waters with lily pads.

Age/Size at Sexual Maturity: Males 1 yr after metamorphosis, females 1 to 2 yr after metamorphosis (Hedeen 1972).

Breeding Period and Egg Deposition: June to early August (Hedeen 1972), peak in July (Wright and Wright 1949:535).

No. Eggs/Mass: One individual laid 509 eggs (Hedeen 1972). Eggs laid in globular jelly-like masses attached to submergent aquatic vegetation such as stems of spatterdock (*Nuphar*) then drop to bottom where they develop.

Time to Hatching: Unreported.

Tadpoles: For 1 to 2 yr. Transform during July and August.

Home Range/Movement: Unreported.

Food Habits: Adults feed from lily pads on animal matter, including adult insects and larvae, particularly aphids and chrysomelids (Kramek 1972, 1976), also minnows, millipedes, leeches, snails, spiders; plant material taken inadvertently. Most prey taken from the water surface—usually opportunistic feeders, but can be selective (Kramek 1972). Diet is a reflection of prey species availability. Tadpoles feed primarily on algae (Hedeen 1970).

Comments: Very similar to *R. clamitans melanota* in appearance and habits. Adults produce a musky scent, especially when handled roughly (Conant 1975:342). Competition from green frogs and bullfrogs may be an important factor in habitat selection in the Northeast. In ponds treated with rotenone in the Adirondacks, the anuran community of green, mink, and bullfrogs probably requires 10 to 15 years to recover to pretreatment levels (Stewart 1975).

Wood Frog

(*Rana sylvatica*)

Range: Atlantic Provinces and n. Quebec w. to Alaska (northern limit is along treeline), s. to North Dakota, the Great Lakes states, and the Appalachians to

Tennessee and extreme n. Georgia.

Distribution in New England: Throughout the region but apparently absent from outermost Cape Cod, Martha's Vineyard, and Nantucket.

Status in New England: Widespread and ubiquitous, from sea level to high elevations.

Habitat: Terrestrial; in mesic woods, often far from water during the summer months as woodland ponds dry up; xeric woods with moist microhabitats (Klemens 1993:140). Prefers wooded areas with small ponds for breeding (Heatwole 1961). Found in boreal conifer forests, swamps, and upland hardwood forests to elevations of 1,158 m (Trapido and Clausen 1938). Found in bogs and traprock slopes in Connecticut (Klemens 1993:140). Found in peatlands in Maine (Stockwell and Hunter 1989). Hibernates under moist forest floor debris or flooded meadows (Klemens 1993:140) from October to late March. Embryos and larvae showed limited tolerance to water with a high humic content in a Minnesota peat bog (Karns 1980). Wetland types: Palustrine, as defined by Cowardin et al. (1979).

Special Habitat Requirements: Prefers temporary woodland pools, backwaters of slow-moving streams.

Age/Size at Sexual Maturity: Males at 2 yr, females at 3 yr (Bellis 1961).

Breeding Period and Egg Deposition: March to July at temperatures of about 10° C (50° F) (Smith 1956:113). Moore (1939) found maximum temperature tolerance of 24° C (75° F) for egg development. Often breeds before ice is off the water (Martof 1970:86.2). Egg-laying usually completed within 4 to 6 days (Herreid and Kinney 1967).

No. Eggs/Mass: 2,000 to 3,000 eggs (Wright 1914:16), 1,019 average in Massachusetts (Possardt 1974). Dark globular egg masses attached to submerged twigs or free on the bottom.

Time to Hatching: 10 to 30 days, average 20 (Oliver 1955:236), temperature dependent.

Tadpoles: 6 to 15 weeks (Minton 1972:132). May overwinter in n. Canada.

Home Range/Movement: Average home-range size for 453 individuals in a Minnesota peat bog was 77.2 yd² (65.5 m²), range 3.5 to 440.5 yd² (2.9 to 368.3 m²). Distance between captures averaged 12.3 yd (11.2 m)

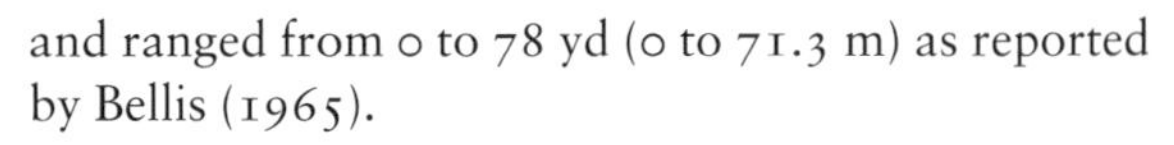

and ranged from 0 to 78 yd (0 to 71.3 m) as reported by Bellis (1965).

Food Habits: Insects; particularly beetles, flies and hymenopterans (Moore and Strickland 1955), also spiders, snails, slugs, and annelids.

Comments: Wood frogs range farther north than any other cold-blooded tetrapod in North America. Breeds before all other ranids in the Northeast. Adults have been observed migrating across surface ice toward other chorusing wood frogs (T. J. Andrews pers. observ.). Brush piles, grassy hummocks, and other terrestrial objects used as cover rather than utilizing aquatic escape (Marshall and Buell 1955).

Northern Leopard Frog

(*Rana pipiens*)

Range: Nova Scotia, Labrador w. to se. British Columbia, s. to e. Oregon, Washington and California, n. Arizona and New Mexico, Ohio, n. New York and New England.

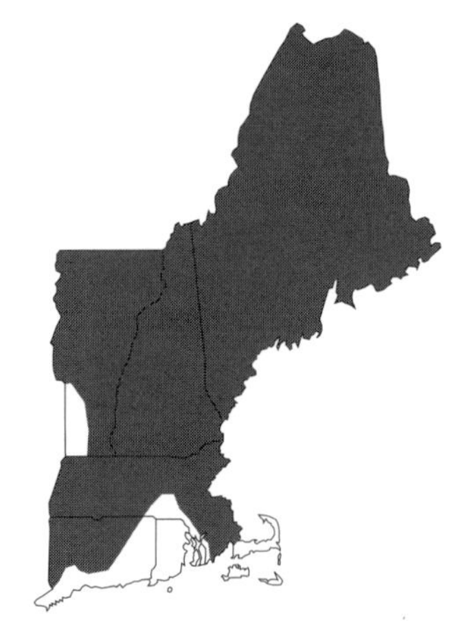

Distribution in New England: Accounts vary; in Connecticut, localized in Housatonic and Connecticut River drainages and absent from much of e. Massachusetts (Klemens 1993:134). Throughout New Hampshire (Taylor 1993) and Maine (Hinshaw 1992).

Status in New England: Common; spotty distribution in s. part of range, very uncommon in parts of formerly occupied range.

Habitat: Commonly found in wet open meadows and fields and wet woods during summer months, including river floodplains, in Connecticut (Klemens 1993:134). Breeds in ponds, marshes, slow shallow streams, and weedy lake shores. Usually hibernates from October or November to March; hibernates under water or in caves (Rand 1950). Sometimes emerges in early February (Smith 1956:110) and during warm days in winter (Zenisek 1964).

Special Habitat Requirements: Wet meadows.

Age/Size at Sexual Maturity: At 3 yr of age in Michigan (Force 1933).

Breeding Period and Egg Deposition: March to May, congregates to breed (Wright and Wright 1949:482).

No. Eggs/Mass: Up to 6,000 eggs laid in compact oblong masses in shallow water, sometimes attached to twigs.

Time to Hatching: 13 to 20 days (Wright 1914:58).

Tadpoles: 9 to 12 weeks, transform July and August in Connecticut (Klemens 1993:134) and August to September in New Hampshire (Taylor 1993). Overwinter as tadpoles in Nova Scotia (Bleakney 1952).

Home Range/Movement: Daily travel within home range reported to be usually less than 5 to 10 m in wet pasture and marsh (Dole 1965). Average nightly movement during rainy periods was 36 m in Michigan (Dole 1968). Occasional long-range movement, often exceeding 100 m during rainy nights (Dole 1965).

Food Habits: Insects, particularly beetles, lepidopteran larvae, wasps, bugs, crickets, grasshoppers, and ants; also takes sowbugs, spiders, small crayfish, snails, and myriopods. Almost 99 percent of foot items were insects and spiders (Drake 1914). Occasional records of having taken small birds and snakes. Food species taken correlates with peaks in insect prey abundance (Linzey 1967).

Comments: During dry summer days frogs may sit in "forms," small clearings made in wet soil within their home range (Dole 1965). Generally found at low elevations, such as below 350 m in Connecticut (Klemens 1993:134).

Pickerel Frog

(Rana palustris)

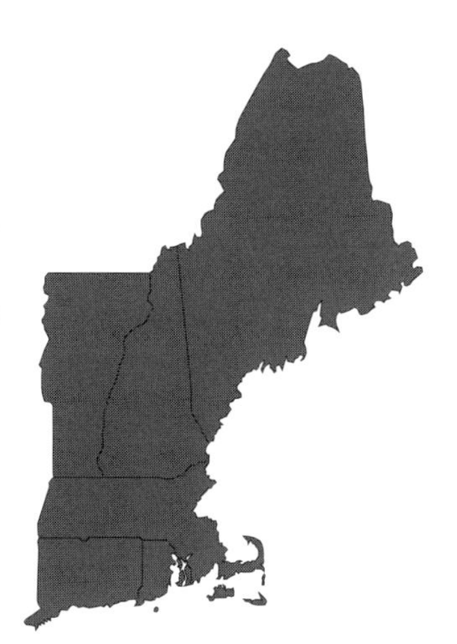

Range: Nova Scotia and the Gaspé Peninsula w. through se. Ontario to Wisconsin s. to e. Texas and South Carolina. Absent from central Illinois, nw. Ohio and parts of the South.

Distribution in New England: Throughout the region, occurring from sea level to over 1,700 feet (518 m) (Klemens 1993:130).

Status in New England: Locally common. Only the green frog (*R. pipiens*) is more abundant and widespread in New England (Klemens 1993:130).

Habitat: A wide variety of moist habitats: colder waters of lakes, ponds, clear streams, springs, meadows, sphagnum bogs, limestone quarry pools. In Massachusetts, fairly ubiquitous along streams and shores of permanent ponds and lakes. In summer found in wet pastures, fields, rocky ravines, or woodlands, often at a distance from water. Prefers water with thick vegetation at edges for cover. Hibernates in mud at bottom of ponds or in ravines under stones from October to March. Some individuals found wintering in caves in Indiana (Rand 1950).

Special Habitat Requirements: Shallow, clear water of bogs and woodland ponds for breeding.

Age/Size at Sexual Maturity: Females: 51–79 mm body length; males 41–58 mm (Klemens 1993:132).

Breeding Period and Egg Deposition: April to May.

No. Eggs/Mass: 2,000 to 3,000 eggs (Wright 1914:67). Eggs laid in firm globular masses attached to submerged plants and branches (Wright and Wright 1949).

Time to Hatching: 11 to 21 days (Wright 1914:67).

Tadpoles: 80 to 100 days. Transform July to September.

Home Range/Movement: Unreported.

Food Habits: In adults, 95 percent of food items were terrestrial arthropods (Smith 1956:108). Snails, small crayfish, aquatic amphipods and isopods are also eaten.

Comments: Diurnal; may be crepuscular during hot weather. Sensitive to pollution and changes in water quality. Skin secretions may be toxic to other amphibians confined with pickerel frogs.

Common Snapping Turtle

(*Chelydra s. serpentina*)

Range: Across the e. United States and s. Canada w. to the Rocky Mountains, s. to the Gulf of Mexico and into Central America.

Distribution in New England: Throughout Connecticut, Rhode Island, and Massachusetts. Not reported from nw. Maine, n. New Hampshire, and ne. Vermont.

Status in New England: Common.

Habitat: Bottom dweller in any permanent and many semipermanent bodies of fresh or brackish or even salt water; occasionally in temporary water. Marshes, swamps, bogs, pools, lakes, streams, rivers, frequently in areas with soft muddy banks or bottoms. Formerly thought to prefer permanent water. Almost entirely aquatic, but will travel overland. Nests in well-drained, open areas of sandy loam, silt, or sandy gravel within

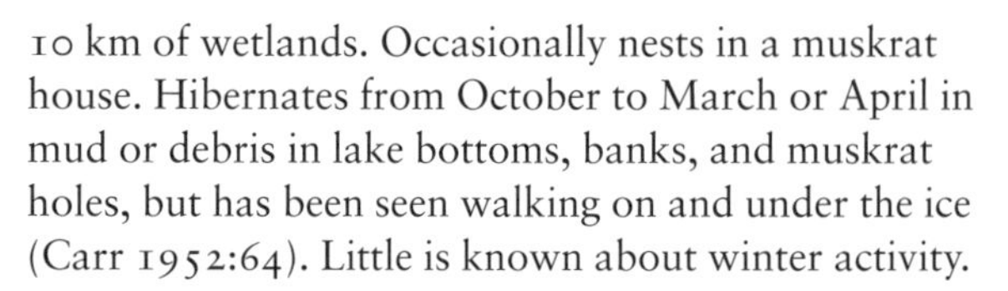

10 km of wetlands. Occasionally nests in a muskrat house. Hibernates from October to March or April in mud or debris in lake bottoms, banks, and muskrat holes, but has been seen walking on and under the ice (Carr 1952:64). Little is known about winter activity.

Special Habitat Requirements: Aquatic habitat; muddy bottomed wetlands preferred.

Age/Size at Sexual Maturity: Carapace length of 10 in (25.4 cm) reported by Hammer (1969).

Breeding Period: Mating occurs from late April to November, sperm may remain viable in females for several years. In s. New England, nesting occurs late May to June (Klemens 1993:152).

Egg Deposition: Mid June. Nests made in soil of banks or in muskrat houses. Also on lawns, driveways, fields, sometimes far from water.

Clutch Size: 11 to 83 eggs; females may lay two clutches per year in southern portions of range. Typically 20 to 30 eggs per clutch (Cahn 1937, cited in Conant 1938:128).

Incubation Period: 55 to 125 days (Hammer 1969), typically 80 to 91 days, depending on environmental conditions.

Eggs Hatch: Late August to early October, may overwinter in nest until spring in n. portions of range. Nests often destroyed by mammalian predators.

Home Range/Movement: Average distance traveled by 107 individuals was 0.69 mile (1.1 km), with most movement within the same marsh in South Dakota (Hammer 1969). In a New York marsh, movement of 100 m was the average for 85 individuals; home ranges from 3 to 9 ha (Kiviat 1980). Established range in Pennsylvania 4.5 acres (1.8 ha), reported by Ernst (1968). Quite migratory. Females exhibit strong nesting site fidelity and will travel more than 0.5 km overland through forest and uneven terrain between water bodies in Ontario. Maximum round-trip distance of 16 km between home range and nesting site (Obbard and Brooks 1980).

Food Habits: Snapping turtles are omnivorous; animal matter accounts for 54 percent of prey items including fish (40 percent), crayfish, aquatic invertebrates, reptiles, birds, mammals; plant material 37 percent (Alexander 1943). Primary fish species in diet included suckers, bullheads, sunfish, and perch in Connecticut (Alexander 1943). May occasionally take young waterfowl; not destructive to natural populations of fish or waterfowl. Scavenges for any food readily available.

Comments: Tolerant of pollution. High levels of persistent organochlorine contaminants found in the tissues of Hudson River specimens without apparent ill effects (Stone et al. 1980).

Spotted Turtle

(*Clemmys guttata*)

Range: Southern Maine to s. Quebec w. to Lake Michigan, s. to e. Virginia and n. Florida.

Distribution in New England: From s. Maine, s. New Hampshire, extreme se. Vermont s. to all but extreme nw. Massachusetts and throughout Connecticut and Rhode Island.

Status in New England: Uncommon to locally common. In Connecticut, widespread below 700 ft (213 m) elevation, becomes increasingly scarce up to 1,100 ft (335 m). Elevation appears to be a limiting factor in distribution (Klemens 1993:166).

Habitat: Inhabits unpolluted, small shallow bodies of water such as brooks, emergent marshes, wet sedge meadows, fresh water bogs, brackish tidal marshes, woodland streams, ditches, and forested wetlands surrounded by vegetation. Prefers areas with aquatic vegetation. Frequently hides in mud and detritus at bottom. Basks along water's edge on brush piles in water, and on logs or vegetation clumps (Ernst 1976). Occasionally found in cranberry bogs. Heavy grazing or mowing of marsh vegetation detrimental (Minton 1972). In Rhode Island, found in salt marshes and small bogs or ponds with adjacent dry upland oak-pine forest (C. Raithel, personal communication). In Massachusetts, found in small, isolated vernal pools located in mixed deciduous-coniferous forest, and spends up to 93 days (average 37 days) aestivating in upland forests from 13 to 412 m (average 178 m) from permanent wetland edges (Milam 1997). Females move over land from 75 to 312 m (average 182 m) to nest in upland fields (Milam 1997). Hibernates in water under tree root balls in vernal pools and in forested, scrub-shrub, and emergent wetlands (Carroll 1991, Graham 1995, Milam 1997), or in the muddy bottoms of shallow, freshwater habitat (Ernst et al. 1994).

Special Habitat Requirements: Unpolluted shallow (0.5 m) water and vernal pools adjacent to upland forest (Milam 1997); well-drained loamy or sandy soils within 0.5 km.

Age/Size at Sexual Maturity: Males about 83.4 mm plastron length, females about 80.8 mm plastron length in Pennsylvania (Ernst and Barbour 1972:73).

Breeding Period: March to June, peak usually in May.

Egg Deposition: June to early July. Eggs usually laid in well-drained soil of pastures or fields or in tussocks (Klemens 1993:167).

Clutch Size: 1 or 2 to 8 eggs (Adler 1961, Glowa 1992), average 3 to 5.

Incubation Period: 70 to 83 days.

Eggs Hatch: Late August (Ernst and Barbour 1972:74) to September (Finneran 1948). Overwintering in nest may occur.

Home Range/Movement: For 11 adults in a Pennsylvania marsh, range averaged 1.3 acres (0.5 ha) and movement less than 0.5 mile (0.8 km) (Ernst 1970). For 26 adults in central Massachusetts, home range averaged 8.8 acres (3.5 ha), and movement averaged 313 m (Milam 1997).

Food Habits: Omnivorous. Eats aquatic insects, tadpoles, amphibian eggs, vegetable matter, crustaceans, mollusks, spiders, and earthworms; occasionally takes frogs, small fish, and carrion.

Comments: A strongly diurnal species.

Bog Turtle

(*Clemmys muhlenbergii*)

Range: Scattered colonies from sw. New England w. through New York, s. to ne. Maryland, s. Virginia, w. North Carolina and Georgia.

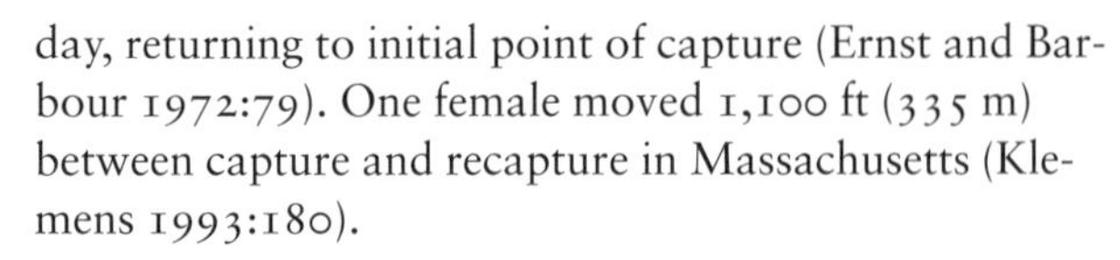

Distribution in New England: Discontinuous from s. Berkshire County, Mass., s. through w. Connecticut. Coincident with the distribution of calcareous wetlands in the region (Klemens 1993:177–8).

Status in New England: Threatened. See Appendix A.

Habitat: Unpolluted open sphagnum bogs or wet meadows; sluggish clear meadow streams with muddy or mucky bottoms (Zappalorti et al. 1979). Frequents shallow meandering waterways in swamps and wet meadows. In Connecticut, associated with open canopy and calcareous wetlands (Klemens 1993:178). Hibernates mid autumn to late March or April. Hibernaculum is in a subterranean rivulet or seepage area with continually flowing water in New Jersey (Zappalorti and Farrell 1980) and Massachusetts (Whitlock unpub. data). Commonly basks in spring and early summer on sedge grass tussocks or in open shallow pools.

Special Habitat Requirements: Open calcareous wet meadows and fens, characterized by continuous flow of water through the surface and occurring at an elevation of 500–700 ft (152–213 m), and bordered by shrubs and red-maple swamps.

Age/Size at Sexual Maturity: At 5 yr and plastron length of 75 mm (Barton and Price 1955). From 6 to 8 yr, at plastral length of 70 mm in Pennsylvania (Ernst 1977). Minimum 10 yr (Klemens 1993:181).

Breeding Period: Late May to June.

Egg Deposition: June to July, often in tussocks or on top of sphagnum in open, sunny areas (Zappalorti et al. 1979; Whitlock, unpub. data).

Clutch Size: 2 to 5, typically 2 to 3 in New Jersey (Zappalorti et al. 1979). 2 to 6 in (mean = 3.4 in), Massachusetts and Connecticut (Whitlock unpub.).

Incubation Period: 7 to 8 weeks (Nemuras 1969), ranges from 48 to 101 days (Whitlock, unpub.).

Eggs Hatch: July to early September (Ernst and Barbour 1972:77–78). In Massachusetts, September to October (Whitlock unpub.).

Home Range/Movement: Average range was 1.28 ha for 19 individuals in Lancaster County, Pa. (Ernst 1977). Ranging from 0.008 to 0.943 ha, traveling through wet runs (Barton 1957, cited in Ernst 1977:246). Average movement was 12 m between recaptures for a male; when displaced, the same individual moved 0.4 km in 1 day, returning to initial point of capture (Ernst and Barbour 1972:79). One female moved 1,100 ft (335 m) between capture and recapture in Massachusetts (Klemens 1993:180).

Food Habits: Omnivorous. Eats berries (20 percent), insects (80 percent) (Surface 1908:158), also slugs, earthworms, crayfish, frogs, snakes, nestling birds, seeds of pondweeds and sedges, snails, carrion; availability determines food consumption (Barton and Price 1955). Forages on land and under water.

Comments: The bog (formerly Muhlenberg's) turtle is the smallest turtle in New England. May aestivate during dry summer months (Ernst and Barbour 1972:77). Seldom active during the hottest part of the day (Zappalorti and Farrell 1980). Bog turtles disperse among patches of suitable habitat in the face of changing vegetation within a stable wetland complex, and beaver, deer, or cattle may be important in maintaining open wetlands for this species (Kiviat 1978). Succession of open wetlands to forest after beaver removal upon settlement likely led to bog turtle declines long ago (Klemens 1993:178).

Overcollection of this species is a problem, and locality information should be reported with discretion to prevent exploitation. Formerly abundant; population decreases related to wetland drainage and filling, succession, and collection for the pet trade.

Wood Turtle

(*Clemmys insculpta*)

Range: Nova Scotia w. through the Great Lakes region to e. Minnesota; in the East extending range s. to extreme n. Virginia.

Distribution in New England: Throughout the region except Cape Cod, Martha's Vineyard, and Nantucket; absent or local in coastal or pine barrens habitats (Klemens 1993:172).

Status in New England: Once common, population declining due to habitat loss.

Habitat: Frequents slow-moving meandering streams with sandy bottoms and overhanging riparian vegetation. Basks during morning hours along banks of streams. Disperses from water sources during summer months to meadows, woods, and roadsides.

Special Habitat Requirements: Wooded banks of sandy-bottom streams with adjacent meadows; open sandy nesting areas.

Age/Size at Sexual Maturity: Seems to vary geographically and between individuals. In New Jersey, specimens at 165 mm carapace length, aged between 7 and 8 yr were thought to be sexually mature (Harding and Bloomer 1979). About 10 yr and 160 mm carapace length in Michigan (Harding 1977). Well into second decade of life (about 14 yr) (Farrell and Graham 1991, Klemens 1993:173).

Breeding Period: March, May, October (Ernst and Barbour 1972:82), when stream temperature reaches about 15° C (59° F) (Farrell and Zappalorti 1979). Mating occurs in shallow water.

Egg Deposition: May to June. Eggs laid in prepared depressions in open areas with sandy soils or gravel, not necessarily near water.

Clutch Size: 4 to 12 eggs (Carr 1952:122), averages 8 to 9 (Farrell and Zappalorti 1979), 5 to 18 in Michigan (Harding 1977).

Incubation Period: 77 days (Allen 1955); 58 to 69 days in laboratory (Farrell and Zappalorti 1980).

Eggs Hatch: August to October. Late hatchlings may overwinter in the nest in northern parts of range (Klemens 1993:174).

Home Range/Movement: One male moved an average of 90 m for three recaptures; one female was found 15 m from initial capture point (Ernst and Barbour 1972:83). Exhibited fidelity to a particular stream or brook in New Jersey (Farrell and Zappalorti 1979), and Pennsylvania (Strang 1983); mean home range was 447 m² for 10 individuals in lowland forest.

Food Habits: Omnivorous. Eats young vegetation, grass, moss, mushrooms, berries, insects and their larvae, worms, slugs, snails (Surface 1908:161–162); also carrion, tadpoles, frogs, and fish. Feeds in water or on land.

Comments: The wood turtle is the largest member of the genus *Clemmys*. Formerly thought to be one of the most terrestrial turtles, it is actually found as frequently in water as on land. Lives in large groups or colonies in New Jersey (Farrell and Zappalorti 1979). Diurnal. Development of wooded river banks and widespread commercial collection are factors contributing to population decline. Not tolerant of pollution; young not often encountered. Unlike most other New England turtles, the sex of wood turtles is determined by their genes, not by the temperature at which the eggs are incubated (Taylor 1993).

Eastern Box Turtle

(*Terrapene c. carolina*)

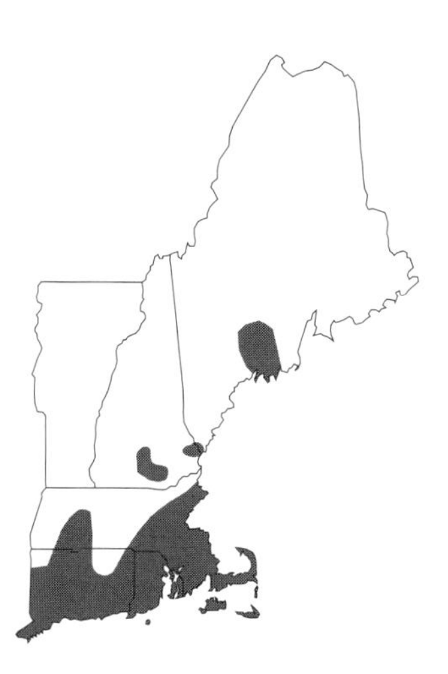

Range: Southwestern Maine w. through the Thousand Islands region of New York to the Mississippi River, central Illinois and s. to n. Florida.

Distribution in New England: Southeastern Maine, se. New Hampshire, most of e. Massachusetts including Cape Cod, Martha's Vineyard, and Nantucket, the Connecticut River Valley. Throughout Rhode Island and all but nw. Connecticut.

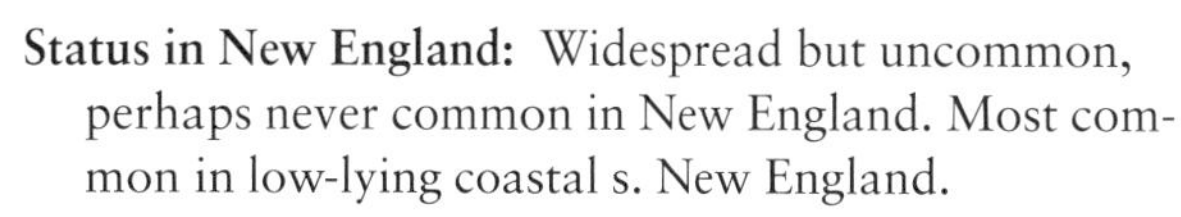

Status in New England: Widespread but uncommon, perhaps never common in New England. Most common in low-lying coastal s. New England.

Habitat: Woodlands, field edges, thickets, marshes, bogs, stream banks; typically found in well-drained forest bottomland (Stickel 1950). In Connecticut, restricted to low-lying, especially coastal, habitats and rarely found above 700 ft (213 m) elevation. Young semiaquatic. Has been observed swimming in slow-moving streams and ponds. Found chiefly in open deciduous forests.

Also found on mountain slopes in Massachusetts (T. Tyning pers. commun.). Does not aestivate in southern New England, but during hot dry weather may rest in mud or water or burrow under logs or decaying vegetation for extended periods. When not active, rests in brush piles and thickets. Hibernates on land from depths of several inches to 2 ft (0.6 m) below surface in loose soil, decaying vegetation, mud, or in stream banks from late fall to April.

Special Habitat Requirements: Old fields, powerline clearings, ecotones with sandy soils favored, seldom far from water (Klemens 1993:192).

Age/Size at Sexual Maturity: 4 to 5 yr in Kentucky (Ernst and Barbour 1972:43), 5 to 10 yr in Indiana (Minton 1972:165). Adults average 149 and 150 m straight-line carapace length for females and males, respectively (Klemens 1993:196).

Breeding Period: Mates on land (Connecticut) after emerging from hibernation in April, sometimes continuing to September 1. Females may lay viable eggs for up to 4 years after mating (Ewing 1943).

Egg Deposition: June to July in New England. Females often seen crossing roads in Massachusetts during the nesting season (T. Graham pers. commun.).

Clutch Size: 3 to 9 eggs, average 4 to 5.

Incubation Period: 87 to 89 days (Allard 1935, cited in Carr 1952:146).

Eggs Hatch: August to September, hatchlings may overwinter in nest.

Home Range/Movement: From 150 to 750 ft (45.7 to 228.4 m); 12 individuals averaged movement of 390 ft (118.8 m) on Long Island (Breder 1927). For 62 individuals in mixed woodlands and open habitat on Long Island, average range was less than 750 ft (228.4 m) as reported by Nichols (1939). Stickel (1950) reported average diameter of 350 ft (106.6 m) in Maryland. One individual was found within 0.25 mile (0.4 km) from point of release 60 yr previously (Allen 1868, cited in Babcock 1919:412). Very sedentary. Maintains same home range for many years, occasionally leaves normal home range for random wandering or egg laying (Stickel 1950). Homing instinct displayed by 45 out of 60 turtles (Nichols 1939).

Food Habits: Younger individuals are chiefly carnivorous, older individuals more herbivorous. Food items include animals such as earthworms, slugs, snails, insects and their larvae, particularly grasshoppers, moths and beetles; crayfish, frogs, toads, snakes, and carrion; vegetable matter such as leaves, grass, berries, fruits and fungi.

Comments: Terrestrial and diurnal. Digs into leaf litter toward end of day. Bisection of habitat by roads can reduce or destroy populations. The reversion of much agricultural land to woodland may be a beneficial change to populations (Klemens 1993:197). Estimated age at full growth is 20 yr. May live 60 to 80 yr (Nichols 1939). Some individuals may live more than 100 yr (Graham and Hutchinson 1969). Sexual dimorphism is marked. Male box turtles have a bright red iris, a brightly colored head, a concave rear plastral lobe, and a distinctly flared carapace. Females have a yellow or brown iris, a relatively unmarked head, a convex rear plastral lobe, and a weakly flared carapace (Klemens 1993:193).

Map Turtle

(*Graptemys geographica*)

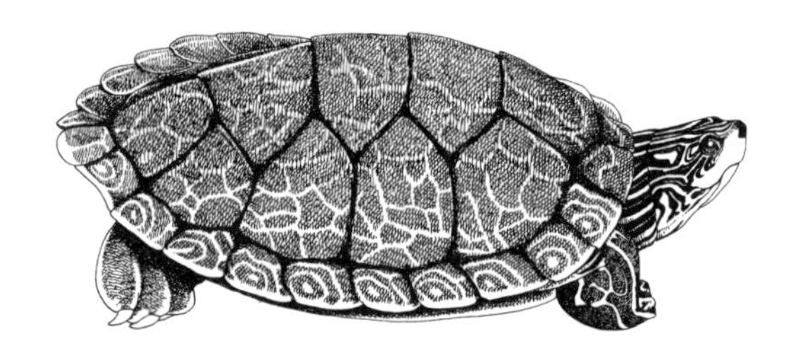

Range: Lake Champlain to the Great Lakes w. to the Mississippi drainage to e. Minnesota, s. to Louisiana and nw. Georgia. Along Susquehanna drainage. Introduced to Delaware River. Nests as far south as Poughkeepsie, Dutchess County, N.Y.

Distribution in New England: Along Lake Champlain and associated rivers in Vermont.

Status in New England: Uncommon and of limited distribution in New England; primarily a midwestern species.

Habitat: Aquatic, inhabiting slow-flowing rivers and lakes (Pluto and Bellis 1988). Prefers large bodies of water with soft bottoms and aquatic vegetation. Hibernates in mud of shallow water from late fall to early spring. May be active on or under ice. Gregariously basks on logs or rocks or along beaches and grassy shores. In Michigan, found in riffles of pebble-bottom streams that have interspersed, deeper, muddier pools (Klemens 1993:151).

Move from shallow bays to nesting areas and reenter bays to overwinter in Quebec (Gordon 1980).

Special Habitat Requirements: Shallow water bodies with muddy or soft bottom substrate; freedom from disturbance, basking sites, and deep areas (>2 m) within the same system.

Age/Size at Sexual Maturity: Females at 7.5 in (190.5 mm) and larger (Newman 1906, cited in Pope 1939:169).

Breeding Period: April and autumn (Ernst and Barbour 1972:110).

Egg Deposition: May to July, peak mid June. Nesting season begins in mid June in Quebec and averages 2 weeks in duration (Gordon 1980). Nests made in soft sand or soil close to shore but above high water level.

Clutch Size: 10 to 16 eggs (Cahn 1937), typically 12 to 14 eggs. More than one clutch may be laid.

Eggs Hatch: Late August to early September (Carr 1952:199); some may overwinter in the nest.

Home Range/Movement: Will move up to 6 km in water (Pluto and Bellis 1986).

Food Habits: Aquatic feeders on mollusks—snails and clams are the primary prey; other small mollusks, crayfish, vegetable matter, fish, insects, and carrion occasionally eaten (Carr 1952:199).

Comments: Map turtles bask communally, are easily disturbed, and move away from areas of high recreational use (Gordon and MacCulloch 1980).

Plymouth Redbelly Turtle

(Pseudemys rubriventris bangsi)

Range: Plymouth County, Mass. Recently, skeletal remains and a shell found in Ipswich, Essex County, Mass. (Graham 1982).

Status in New England: Endangered (federal list).

Habitat: Ponds of different sizes in Plymouth County. Frequents shallow coves (Graham 1971a).

Special Habitat Requirements: Muddy-bottom shallows with abundant aquatic vegetation, especially milfoil (*Myriophyllum*) and bladderwort (*Utricularia*) (Graham 1980).

Age/Size at Sexual Maturity: Probably not reached during first 9 yr (Graham 1971a). Average life span estimated at 40 to 55 yr (Graham 1980).

Breeding Period: Probably early spring and fall (T. Graham pers. commun.).

Egg Deposition: Mid June to early July. Prefer to nest in disturbed sites (T. Graham pers. commun.).

Clutch Size: Range 12 to 17 eggs, average 14.5 (T. Graham pers. commun.).

Eggs Hatch: Probably September, fall (T. Graham pers. commun.), July if they overwinter. Average hatching time of 75 days for 17 eggs incubated in a laboratory at 29° C (84° F) (Graham 1971b). If hatchlings overwinter, emerge during the following July.

Home Range/Movement: Unknown but wanders on land, especially during fall and late spring. Found 0.5 to 2.0 miles (0.8 to 3.2 km) from water on occasion. Significance of wandering unknown (T. Graham pers. commun.).

Food Habits: Primarily herbivorous, feeding mainly on milfoil, also feeds on bladderwort (Graham 1980) and arrowhead (*Sagittaria*) (Graham 1971a). Dietary shift to crayfish in fourth season (Graham 1971a).

Comments: Discovered in Plymouth, Mass., in 1869 (Lucas 1916). Population estimate about 200 to 300 in Plymouth County (T. Graham pers. commun.). Basks during early morning hours on elevated sites or in water by floating or resting on weed mats (Graham 1980).

Painted Turtle

(*Chrysemys picta*) includes Eastern Painted Turtle (*C. p. picta*) and Midland Painted Turtle (*C. p. marginata*)

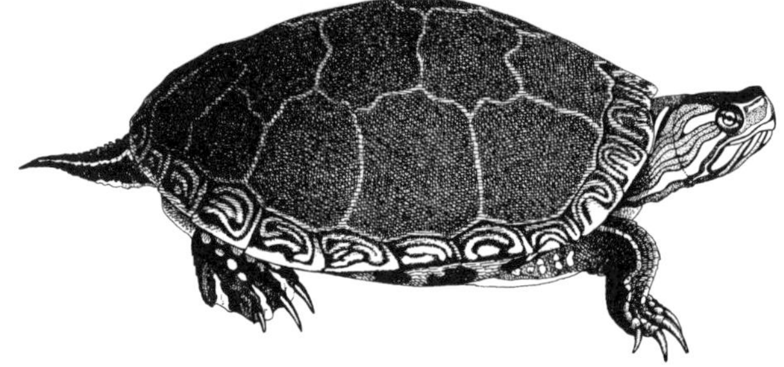

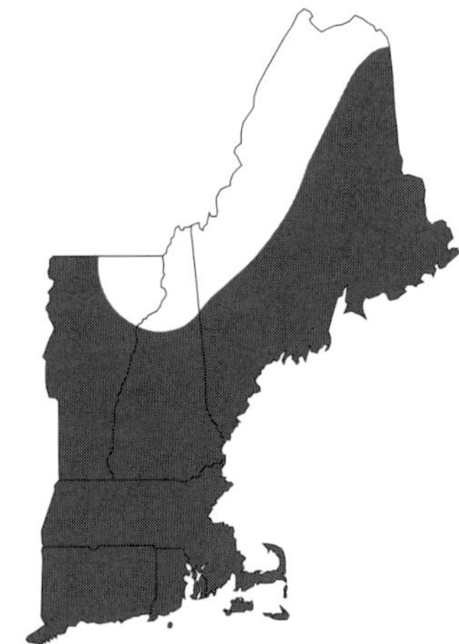

Range: Nova Scotia w. to ne. New York, s. to Cape Hatteras and inland to e. Alabama. In the Northeast the ranges of the midland and eastern painted turtles intergrade.

Distribution in New England: Throughout the region.

Status in New England: Common, often abundant.

Habitat: Quiet, muddy-bottom shallow ponds, marshes, woodland pools, rivers, lake shores, wet meadows, bogs, and slow-moving streams. Sometimes in brackish tidal waters, salt marshes (Pope 1939:183). Stagnant and polluted waters are sometimes inhabited (Smith 1956:150). When in water, usually remains in submerged vegetation. Basks on small hummocks, logs, rocks, sometimes congregating in large groups. Hibernates by burrowing into mud or decayed vegetation of pond bottoms. Occurs from sea level to 1,100 ft (335 m) in Connecticut (Klemens 1993:158).

Special Habitat Requirements: Aquatic habitat with basking structures and areas of open water (15 m^2 minimum) (Whitlock et al. 1994:426).

Age/Size at Sexual Maturity: In Michigan, males exceeded 81 mm plastron length; females ranged from 110 to 120 mm (Gibbons 1968a). Turtles with straight-line carapace length (slc) greater than 90 mm can be sexed easily; most New England painted turtles larger than 130 mm slc are females.

Breeding Period: March to mid June and fall (Gibbons 1968a). Peak in April in Connecticut (Carr 1952:218).

Egg Deposition: May to July. Nest sites within a few yards of water (Cahn 1937, cited in Smith 1961:140), or up to one-half mile away (T. Tyning pers. commun.).

Clutch Size: 2 to 11 eggs, females may lay 2 clutches (Gibbons 1968a), typically 5 to 6 eggs.

Incubation Period: 72 to 80 days (Ernst and Barbour 1972:143), 63 days (Lynn and vonBrand 1945). Hatchlings from late clutches may overwinter in the nest. Nests are often destroyed by raccoons and skunks.

Eggs Hatch: Late August to early September, in Connecticut (Finneran 1948).

Home Range/Movement: Displays short-distance homing ability; fewer than 15 percent moved more than 100 m in a marsh in Michigan (Gibbons 1968a). Average distance traveled was 112 m in a shallow bay of a Wisconsin lake; 70 percent of the turtles did not travel. Individuals may remain in the same locality for years if conditions are favorable (Pearse 1923). For movements and hibernation sites in s. Ontario, see Taylor and Nol (1989).

Food Habits: Aquatic insects, snails, small fish, tadpoles, mussels, carrion, and aquatic plants taken by foraging along the bottom. Diet usually about 50 percent vegetation.

Comments: In New England there are no midland turtle populations per se. Individuals are part of an intergrade swarm. Information provided in this account is based on references for *Chrysemys picta marginata* where intergrades do not occur, but in New England, *Chrysemys picta marginata* and *C. p. picta* life history and habitat information are the same (Klemens 1993:158). Diurnal.

Blanding's Turtle

(*Emydoidea blandingii*)

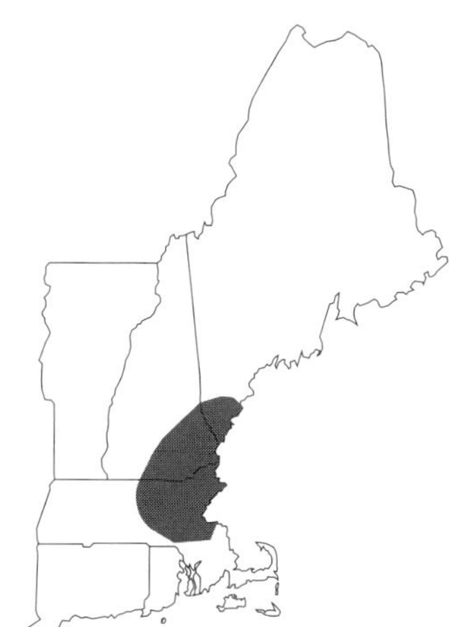

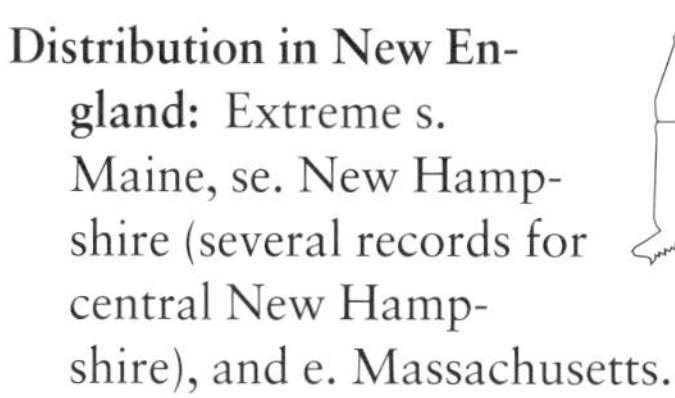

Range: Disjunct populations in New York, New Hampshire, and e. Massachusetts. Southern Quebec w. through the Lake States to central Minnesota, s. to Iowa and central Illinois. Scattered occurrences from Nova Scotia to Ohio.

Distribution in New England: Extreme s. Maine, se. New Hampshire (several records for central New Hampshire), and e. Massachusetts.

Status in New England: Populations are localized and distribution is spotty throughout its range (McCoy 1973:136.1). Generally scarce to rare, locally abundant in Massachusetts (Lazell 1972). An endangered species in Canada.

Habitat: Permanent, shallow, dark, waters with abundant vegetation; marshes, bogs, ditches, ponds, swamps, also in slow-moving rivers and protected coves and inlets of large lakes with abundant emergent and submerged vegetation (Whitlock et al. 1994:407). May wander overland. Basks on logs, stumps, banks. Active in winter or hibernates in mud or debris.

Special Habitat Requirements: Shallow waters with soft muddy bottoms and aquatic vegetation.

Age/Size at Sexual Maturity: At 12–15 yr for males with a plastron length of 181 to 190 mm, Massachusetts (Graham and Doyle 1977); males 131 to 190 mm in Michigan (Gibbons 1968b). Size differences between these two populations probably due to differences in food quality and availability (Graham and Doyle 1977).

Breeding Period: Early spring through October, most often from March to May, peak in late April (Ernst and Barbour 1972:181).

Egg Deposition: April or May. Nests made in sandy soils of upland areas, and plowed farmland in Massachusetts (T. Graham pers. commun.).

Clutch Size: 6 to 17 eggs (Carr 1952:136), typically 8 to 9 eggs, clutches of 9, 13, and 16 eggs for Massachusetts females (T. Graham pers. commun.). Clutch of 17 for a July nesting female (Graham and Doyle 1979). Two clutches may be laid each season.

Incubation Period: Unreported. Eggs maintained at temperatures below 78° F will produce almost all males; warmer temperatures may produce all females (Taylor 1993).

Eggs Hatch: Autumn (may emerge from nest next spring).

Home Range/Movement: Less than 100 m for 4 individuals in a marsh in sw. Michigan (Gibbons 1968b).

Food Habits: Crustaceans, insects, mollusks, fish, carrion, aquatic plants, succulent shoots, and berries. Crustaceans and crayfish account for about 50 percent of diet, insects more than 25 percent, and other invertebrates and vegetable matter 25 percent (Lagler 1943). Vernal pools are important feeding sites in spring and summer (Graham 1992).

Comments: Blanding's turtle is timid, shy, and reclusive, usually seen only when it leaves marshy wetlands in June to find a nest site on land. It commonly nests in plowed fields near wetlands (Graham 1992). Fragments of helmet-shaped Blanding's turtle carapaces have been reported from an Indian site dated 500 years BP on Hog Island, Muscongus Bay, Me. (French 1986).

Common Musk Turtle

(*Sternotherus odoratus*)

Range: Atlantic coast, s. Ontario, w. to the Mississippi River, s. to central Texas and s. Florida. Absent from n. New England.

Distribution in New England: Throughout Connecticut,

Rhode Island, and Massachusetts, s. New Hampshire and Maine, and in s. Vermont and the Lake Champlain drainage.

Status in New England: Common to local in Connecticut (Klemens 1993:201). Uncommon in Maine (Etchberger 1992).

Habitat: Permanent bodies of water: still, shallow, clear lake impoundments, ponds, and rivers, muddy bottoms preferred. Frequently found in weedy shallows of reservoirs (Klemens 1993:201). Refrains from using temporary water sources. Not found at higher elevations in the east, for example, below 274 m in Connecticut (Klemens 1993:201). Large populations found in areas with abundant aquatic vegetation (Pope 1939:39). Scattered records for occurrence in marshes, swamps, bogs, sloughs (Pope 1939:39). Usually gregarious when hibernating in bottom mud, debris, beneath rocks in river bottoms, or in river banks when the temperature falls below 10° C (50° F) (Cagle 1942).

Special Habitat Requirements: Riparian systems; slow-moving, muddy-bottom streams and rivers. Strictly aquatic except when laying eggs.

Age/Size at Sexual Maturity: Musk turtles in the northern portions of the range mature more slowly than individuals in the southern regions. Males at 3 or 4 yr, females at 2 to 7 yr (Tinkle 1961), or perhaps at 9 to 11 yr (Risely 1932).

Breeding Period: April to October, peak in April to May, September to October.

Egg Deposition: May to August, peak in June. Eggs laid in muck, rotted logs, stumps, sandy soil, grass, or on the ground at lake margins <15 m from the shore.

Clutch Size: 2 to 9 eggs (clutch size increases with latitude), typically 3 to 6 (Klemens 1993:203, Graham and Forsberg 1986).

Incubation Period: 60 to 90 days (Barbour 1971:162), 35 to 40 days (Edgren 1960).

Eggs Hatch: September to October (in north). Gregarious nesting habits, often malodorous.

Home Range/Movement: Overland movements probably seasonal or forced (Ernst and Barbour 1972:40). Average home range is 0.06 acre (0.02 ha) for males and 0.12 acre (0.05 ha) for females in Oklahoma. Overland movements ranged from 166 to 227 ft (35.4 to 69.2 m) for males, and 113 to 146 ft (34.4 to 44.5 m) for females (Mahmoud 1969). Exhibited homing behavior in Michigan: 13 out of 28 released individuals traveled up to 700 ft (213 m) to initial capture points (Williams 1952).

Food Habits: Primarily carnivorous, feeds along the bottom for snails, clams, aquatic insects and their larvae, particularly dragonfly nymphs and caddisfly larvae (Lagler 1943), minnows, worms, tadpoles, and fish eggs (Babcock 1919:36). While scavenging, plants and algae as well as carrion are eaten. Carrion accounted for 40 percent of the diet by volume for 73 individuals in Michigan (Lagler 1943).

Comments: Also called the stinkpot, the musk turtle often basks well out of water on horizontal limbs of slanting trees along the water's edge. Highly aquatic, nocturnal, and secretive (Klemens 1993:204). Individuals frequently covered with algae growth.

Eastern Spiny Softshell

(*Apalone s. spinifera*)

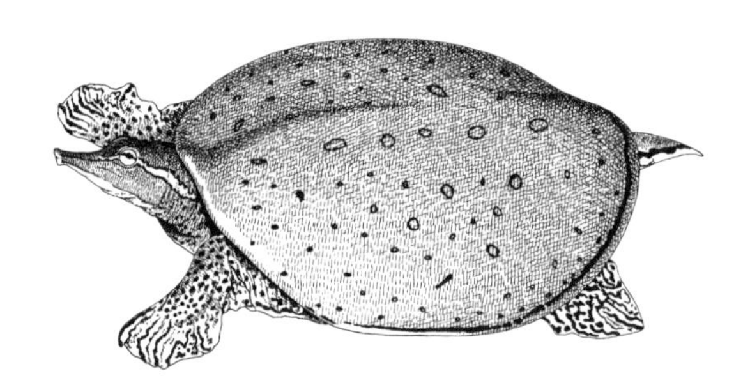

Range: Western New York w. to the Great Lakes and the Mississippi River, south to the Tennessee River and central Pennsylvania. A disjunct population occupies the Champlain Valley.

Distribution in New England: Western Vermont, along the shore of Lake Champlain and some associated river systems.

Status in New England: Uncommon.

Habitat: Aquatic, inhabiting large river systems. Also found in lakes and ponds that are connected to rivers during floods. Intolerant of pollution from sewage, industrial, or chemical wastes (Minton 1972:191). Basks on sand bars, mud flats, grassy beaches, but will use logs, rocks, and other objects when sandy or muddy banks are unavailable (Williams and Christiansen 1982). Hibernates beneath 2 to 3 in (5.1 to 7.6 cm) of river bottom mud from October to April in the north.

Special Habitat Requirements: Shallow muddy bottoms for burrowing. Some aquatic vegetation essential (N. Green pers. observ.).

Age/Size at Sexual Maturity: Females with plastron length of 180 to 200 mm, males at 90 to 100 mm.

Breeding Period: April or May.

Egg Deposition: May to August. Eggs laid in sandy soil or gravel beds near water's edge.

Clutch Size: Typically 12 to 18, with a range of 4 to 32 eggs (Ernst and Barbour 1972:264).

Eggs Hatch: August to October or hatchlings overwinter in nest.

Home Range Size: Turtles moved from their wintering grounds approximately 3 kilometers upriver in the Lamoille River, Vt., from Lake Champlain down into the lake in late April, and moved back upriver in September (Graham and Graham 1997).

Food Habits: Chiefly carnivorous. Crayfish and insects are the major food items, with tadpoles, frogs, mollusks, and fish eaten less frequently; vegetation and other plant materials also consumed. Primarily benthic feeders (Williams and Christiansen 1982).

Comments: Somewhat nocturnal.

Five-lined Skink

(*Eumeces fasciatus*)

Range: Southern end of Lake George, New York, se. New York and sw. Connecticut s. to n. Florida, and w. to central Texas.

Distribution in New England: Fairfield County, sw. Litchfield County, and nw. New Haven County, Conn. Also w. Rutland County, Vt.

Status in New England: Rare and localized in the Northeast through se. Connecticut. Records for Massachusetts are from Barre (Storer 1840:19) and New Bedford (Allen 1870:260).

Habitat: Steep, rocky areas with patchy tree and shrub cover in Connecticut (Klemens 1993:211). Mesic wooded areas, open or moderately dense with ground cover. Most abundant around old buildings and open woods. Frequently in damp spots, under logs, rock piles, leaf litter, sawdust piles. Suns for brief periods on warm days (Smith 1946:349). Found on open talus slopes in mixed deciduous woodlands, New York. Primarily terrestrial, but will climb snags to find insects. Hibernates from October until mid March in decaying logs or below the frost line, underground or under large rocks.

Special Habitat Requirements: Steep, rocky open woods with logs and slash piles.

Age/Size at Sexual Maturity: After second hibernation. Snout to vent length 53–74 mm for males, 61–67 mm for females.

Breeding Period: May.

Egg Deposition: Typically in June or July, 6 to 7 weeks after breeding (Smith 1956:193). Eggs laid under rocks, logs, in rotted stumps, in loose soil. Females usually guard eggs during incubation (Conant 1975:122). Addled eggs are ingested; it has been suggested that brooding females remove these eggs to reduce chances of predation (Groves 1982).

Clutch Size: 4 to 20 eggs (Barbour 1971:209), typically 9 to 12. Younger individuals lay fewer eggs (Fitch 1970).

Home Range/Movement: Male home-range diameter about 90 ft (27.4 m), female about 30 ft (9.1 m), in e. Kansas (Fitch 1954, cited in Minton 1972:210). Individuals may remain in same home range or move after emerging from hibernation.

Food Habits: Primarily insects and spiders; also snails, grubs, small vertebrates, beetles, wood roaches, flies, including young mice. Lizards occasionally eaten; will eat its own shed skin.

Northern Water Snake

(*Nerodia s. sipedon*)

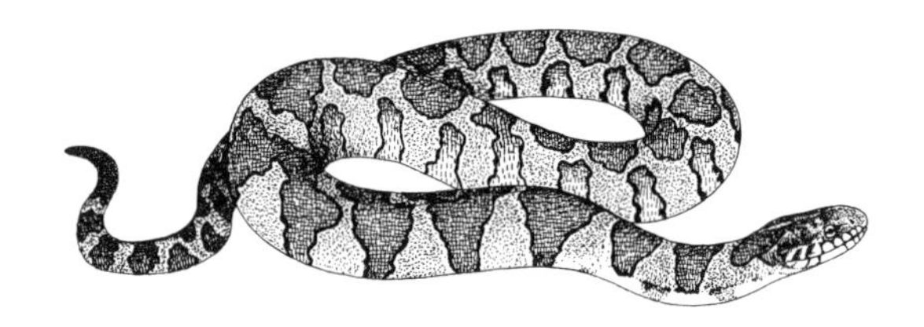

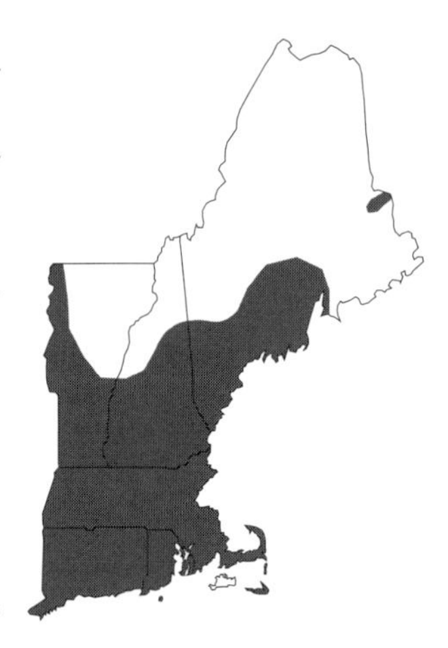

Range: Southern Maine, s. Ontario w. to n. Wisconsin, s. through Kansas to e. Colorado and n. Oklahoma, e. to central Indiana, Kentucky, Tennessee, and North Carolina.

Distribution in New England: Southwestern Maine, also e. Washington County around Moosehorn National Wildlife Refuge, s. New Hampshire and s. and w. Vermont, throughout Massachusetts, Connecticut, and Rhode Island.

Status in New England: Abundant in suitable habitat.

Habitat: Aquatic and semiaquatic habitats. Common around spillways and bridges where rocks provide cover, uncommon in deeply shaded woodland swamps and ponds, probably due to lack of basking sites (Klemens 1993:241). Found in the vicinity of rivers, brooks, wet meadows, ponds, swamps, bogs, old quarries to 1,400 ft (427 m) elevation in Connecticut. Inhabits brackish or fresh water (Wright and Wright 1957:513), absent from heavily polluted waters. Prefers still or slow-moving water, but also found in tidal creeks (Klemens 1993:241). Hibernates in crevices of rocky ledges, or in banks adjacent to aquatic habitat.

Special Habitat Requirements: Branches or logs overhanging the water, or boulders of dams and causeways in reservoirs.

Age/Size at Sexual Maturity: Males at 635 to 1,148 mm, females at 650 to 1,295 mm (Wright and Wright 1957:513).

Breeding Period: April to May and early fall.

Young Born: August to early October, usually during the last half of August. Viviparous.

No. of Young: 10 to 70 young, average 20 to 40. Larger females have larger litters.

Home Range/Movement: One individual moved 380 ft (115.8 m) along a river after 2 yr (Stickel and Cope 1947). In large ponds at an Indiana fish hatchery, 80 percent were recaptured in the same pond, 89 percent were in the same pond or an adjacent pond. Snakes along streams had larger home ranges (Fraker 1970).

Food Habits: Cold-blooded vertebrates: fish account for 61 percent of food items, frogs and toads 21 percent, salamanders 12 percent; also insects, crayfish, recently dead fish (Uhler et al. 1939). Fish account for more than 95 percent of diet (Raney and Roecker 1947). May occasionally take shrews and mice.

Comments: Frequently found basking. Active both day and night. Water snakes can be pugnacious. Their disposition and large size further the erroneous impression that they are dangerous; they are not.

Northern Brown Snake

(*Storeria d. dekayi*)

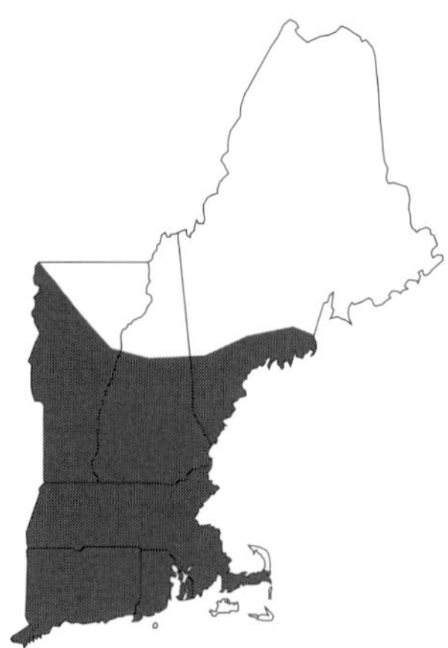

Range: Eastern United States from s. Maine and s. Canada w. to Michigan, s. to South Carolina. Range overlaps that of the midland brown snake.

Distribution in New England: Extreme s. Maine: Lincoln, Sagadahoc, Cumberland, and York Counties, reported from Somerset County, Me. (C. Baumgartner and R. Nemecek pers. commun.), s. New Hampshire and s. and w. Vermont, throughout Massachusetts except upper Cape Cod, throughout Connecticut and Rhode Island.

Status in New England: Common.

Habitat: Ubiquitous, found in urban and rural areas, dry or moist situations, from bogs to grasslands, vacant lots, parks, trash piles. May be abundant along railroad tracks. In the wild, found in damp woods, swamps, clearings, bogs, roadsides, open fields. Hides under stones, banks, logs, brush piles, leaves. Rare in old-growth forests (J. D. Lazell pers. commun.). Common in low-lying areas, but found from sea level to 1,300 ft (396 m) in Connecticut and over 1,600 ft (488 m) in Massachusetts. Hibernates in large groups from October to November until March or April; may use ant hills or abandoned mammal burrows.

Age/Size at Sexual Maturity: At 2 yr (Noble and Clausen 1936). At 251 mm total length for males, 242 mm for females (Wright and Wright 1957).

Breeding Period: Late March to April, possibly in the fall.

Young Born: Late July to August. Gestation period of 105 to 113 days (Clausen 1936). Viviparous.

No. of Young: 3 to 27 young (Fitch 1970), typically 14.

Home Range/Movement: Average daily movement of 10 to 15 ft (3.0 to 4.6 m) on Long Island. Thirteen of 32 individuals displayed homing behavior (Noble and Clausen 1936).

Food Habits: Slugs, snails, earthworms, insects, minnows; very small toads are occasionally eaten. Morphological skull adaptations allow extraction of land snails from thin shells (Rossman and Myer 1990).

Comments: Formerly called DeKay's snake. Commonly found in aggregations throughout the year (Noble and Clausen 1936). May seem to be scarce during July and August when it moves down into the soil to zones of lower temperature. Degree of fossorial tendency varies with microhabitat temperature (Elick et al. 1979). Active from evening to early morning; one of the few New England snakes that is active at night.

Northern Redbelly Snake

(Storeria o. occipitomaculata)

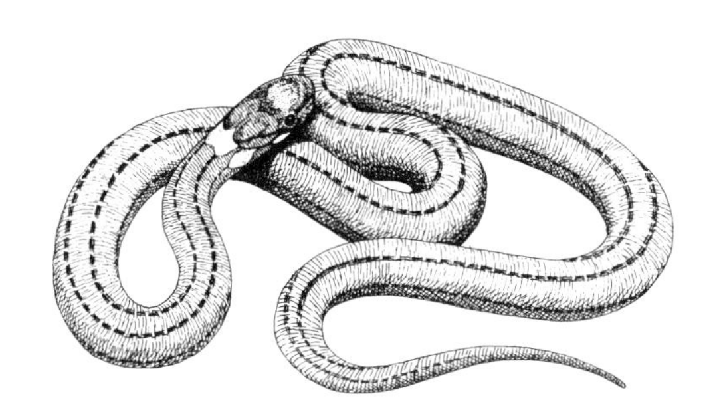

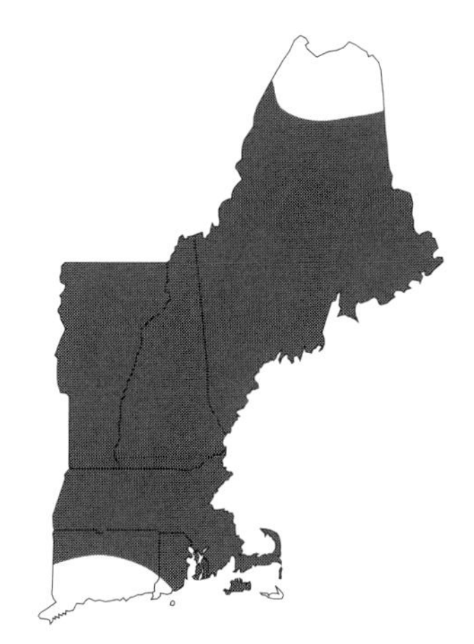

Range: Nova Scotia w. to s. Manitoba, s. to e. Texas, Georgia, and throughout the e. United States.

Distribution in New England: Coastal and central Maine (s. of Moosehead Lake), n.-central New Hampshire and nw. Vermont. South throughout Massachusetts except for outermost Cape Cod. Northwestern and ne. Connecticut and n. Rhode Island. Also reported from Martha's Vineyard (Lazell 1976).

Status in New England: Locally abundant. Rarely reported from the mountains of w. Maine (Burgason 1992a).

Habitat: Moist woods, hillsides, sphagnum bogs, fields, upland meadows and swamps, pond margins and valleys. Found under surface debris, also around abandoned buildings. Occurs at elevations from sea level (Massachusetts) to mountains, but above 500 ft (152 m) in Connecticut (Klemens 1993:254). Prefers woodlands: pine, oak-hickory, aspen, hemlock groves (Wright and Wright 1957:717). More frequently found in upland woody ridges. Occasionally found in damp meadows, marshy areas, swamp and bog edges. Hibernates from fall to March or April. Active through mid October in Connecticut (Klemens 1993:255).

Special Habitat Requirements: Woodland debris: bark, rotting wood.

Age/Size at Sexual Maturity: Males 182 to 359 mm, females 211 to 383 mm (Wright and Wright 1957:718), at 2 yr (Blanchard 1937a).

Breeding Period: Probably after emerging from

hibernation; a late summer or fall mating may also occur (Barbour 1971:287).

Young Born: August to September. Viviparous.

No. of Young: 1 to 14 young (Blanchard 1937a), typically 7 to 8.

Home Range/Movement: 1 adult found 100 ft (30.4 m) from release point in Michigan after 7 days (Blanchard 1937a).

Food Habits: Consumes slugs, earthworms, soft insects and larvae, sowbugs; occasionally small salamanders.

Comments: Has been found active at all times of day and evening. Degree of fossorial behavior varies (Elick et al. 1979). Young commonly mistaken for young ringneck or northern brown snakes. In southern New England, *S. occipitomaculata* tends to replace *S. dekayi,* except in disturbed habitats, where the latter is more common (Klemens 1993:253).

Common Garter Snake

(*Thamnophis sirtalis*) includes
Eastern Garter Snake (*Thamnophis s. sirtalis*) and
Maritime Garter Snake (*Thamnophis s. pallidulus*)

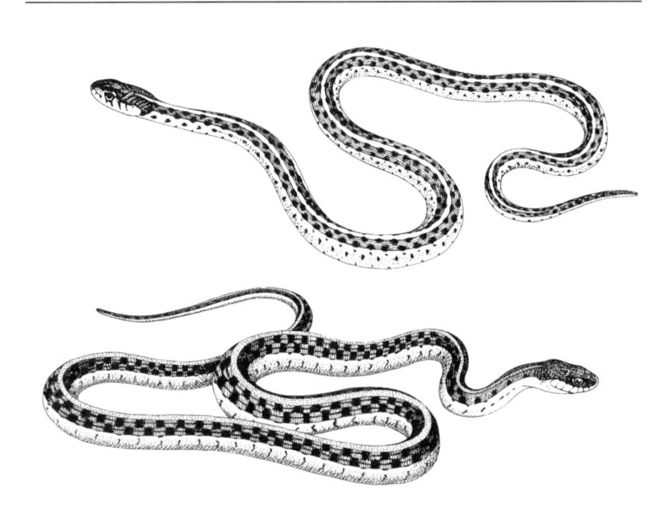

Range: Nova Scotia, w. to se. Manitoba, s. to e. Texas, and throughout the e. United States. Intergradation with *T. s. pallidula* occurs in n. New England (Fitch 1980:270.1).

Distribution in New England: Eastern garter snake: Coastal Maine w. to n.-central New Hampshire and nw. Vermont, s. throughout New England. Maritime garter snake: Generally north of this line, but intergrades where subspecies ranges overlap.

Status in New England: Very abundant; the most common and widespread New England snake. Occurs on many offshore islands.

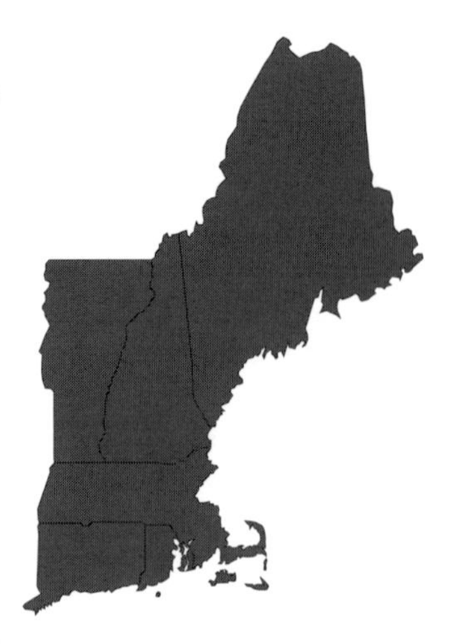

Habitat: Ubiquitous, terrestrial; found in moist areas, forest edges, stream edges, fence rows, vacant lots, bogs, swamps, overgrown yards. One specimen found under a rock in a stream through a dark hemlock grove (Klemens 1993:263). Found in almost all damp environments, from river bottoms to mountain elevations. Over 1,700 ft (518 m) in s. New England (Klemens 1993:264). Hibernates, often gregariously, in holes, rock crevices, mud, anthills, rotted wood, uprooted trees, house foundations, and sometimes partially or completely submerged under streambed rocks, from October to March or April. One of the earliest snakes to emerge from hibernation. Can survive the winter above frost line (Bailey 1949). Sustained submergence may promote survival (Costanzo 1986).

Age/Size at Sexual Maturity: Females in second year, some males the second spring after birth (Carpenter 1952a). At 400 mm snout to vent length for males and 500 mm for females in Kansas (Fitch 1965:531).

Breeding Period: Concentrated in the first few warm days after emergence from hibernation in early April (Connecticut) to May, also in fall before hibernation (Anderson 1965:169). Mates at or near hibernation site.

Young Born: July to early September. Gestation period of 3 to 4 months or longer in cooler climates (Blanchard and Blanchard 1942). Viviparous.

No. of Young: 3 to 85 young, typically 14 to 40 (Wright and Wright 1957). Zehr (1962) found 12 to 13 young was the average in New Hampshire. Number of young correlated with size and age of female (Fitch 1965:558). In Connecticut, broods of 8–15, number not correlated with total length of female (Klemens 1993:265).

Home Range/Movement: Approximately 5 acres (2.0 ha); most ranges were smaller in cutover agricultural fields in Indiana (Minton 1972). Activity range of about 2 acres (0.8 ha) in Michigan woodlands and open fields (Carpenter 1952a). Carpenter (1952b:250) defined activity range as an area covered by an animal in the course of its day-to-day existence, and that lacks definite home site or other center of activity. Home ranges of 35.0 acres (14 ha) for males and 22.2 acres (9.1 ha) for females were found in mixed habitat in Kansas (Fitch 1965:538). Many individuals migrate from hibernacula to summer ranges.

Food Habits: Earthworms account for 80 percent of food items; also amphibians, especially American toads, wood frogs, and redback and other salamanders. Also carrion, fish, leeches, caterpillars, other insects, small birds, rodents (Carpenter 1952b); also slugs, other snakes, mollusks, crayfish, sowbugs (Hamilton 1951).

Comments: Diurnal but sometimes active at night (Minton 1972:250). Seeks cover under objects on hot summer days. Pesticides have reduced local populations in New York (Gochfeld 1975). Garter snakes are apparently quite cold-tolerant, active from February through October in Connecticut (Klemens 1993:264). The subspecies *T. s. pallidulus* (Bleakney 1959) predominates east of the Penobscot River north of the mountains in western Maine; the eastern subspecies, *T. s. sirtalis* (Haskins 1992), predominates to the west, respectively.

Ribbon Snake

(*Thamnophis sauritus*) includes Eastern Ribbon Snake (*T. s. sauritus*) and Northern Ribbon Snake (*T. s. septentrionalis*)

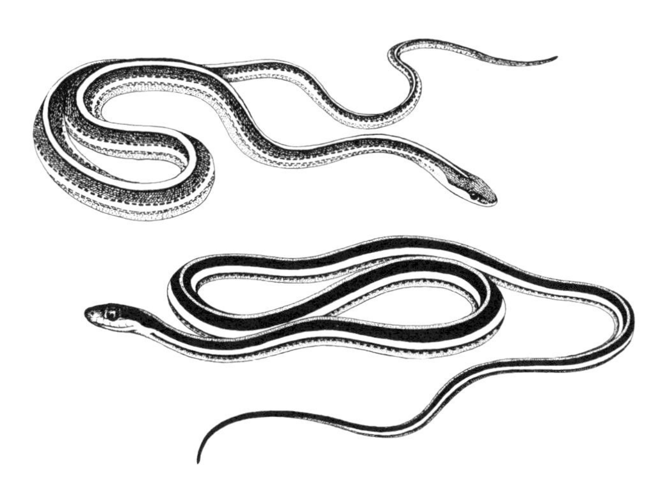

Range: Southern Maine to South Carolina and the Florida panhandle, w. to s. Indiana and e. Louisiana.

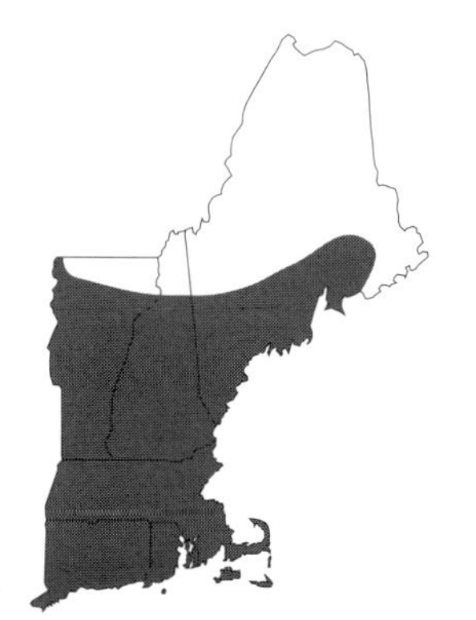

Distribution in New England: Southwestern Maine (York County with outlier records from Kennebec and Sagadahoc Counties [Lortie 1992]), e.-central and s. New Hampshire, s. and w. Vermont, throughout Massachusetts, all but sw. Connecticut; apparently rare in Rhode Island (C. Raithel pers. commun. in Klemens 1993:259).

Status in New England: Generally uncommon and localized, at least in southern New England (Klemens 1993:258).

Habitat: Semiaquatic, inhabiting grassy or shrubby stream edges, beaver ponds, wooded swamps, wet meadows, vernal pools, ponds, bogs, wet powerline cuts, and ditches. Prefers areas with brushy vegetation at water's edge for concealment. Found in highly disturbed as well as undisturbed areas (Klemens 1993:259). Also in damp or wet deciduous or northern pine forests. Seldom far from cover (Carpenter 1952b). May escape higher ground temperatures in summer by seeking shelter in shrubs or underground. Occurs from sea level to 900 ft (274 m) in Connecticut, but most observations made below 500 ft (152 m) (Klemens 1993:259). Two specimens collected at Mt. Washington, Mass., where most wetlands are 1,600 ft (488 m) or higher. Hibernates from October to March (Wright and Wright 1957:825).

Special Habitat Requirements: Aquatic habitats.

Age/Size at Sexual Maturity: Females during second year (Carpenter 1952a), males 400 to 819 mm, females 451 to 900 mm (Wright and Wright 1957:825).

Breeding Period: After emergence from hibernation.

Young Born: Late July to September, viviparous.

No. of Young: 3 to 20, typically 10 to 12.

Home Range/Movement: Average activity range of about 2 acres (0.8 ha); average distance traveled was approximately 280 ft (85.3 m) in open Michigan grassland and marsh (Carpenter 1952b).

Food Habits: Frogs, toads, and salamanders account for 90 percent of prey items; usually smaller or metamorphosing individuals were taken; also mice, spiders, minnows, and some insects (Carpenter 1952b).

Comments: Diurnal, commonly seen lying in shallow still waters.

Eastern Hognose Snake

(*Heterodon platyrhinos*)

Range: Southern Ontario w. to se. South Dakota, s. to central Texas and s. Florida.

Distribution in New England: Southwestern New Hampshire, principally along the Merrimack River (Michener and Lazell 1989). Southern Massachusetts, especially Cape Cod and other coastal areas, and inland habitats throughout Connecticut and Rhode Island. Unauthenticated records from Kittery, Me. (W. Chorman pers. commun.), and w. Maine (Albright 1992).

Status in New England: Locally common. Development at s. New England coasts has rendered the species uncommon there, despite abundance of sandy habitats (Klemens 1993:232).

Habitat: Where sandy soils predominate, such as beaches, open fields, dry, open pine or deciduous woods. Has been found on hillsides, farm fields, and around outbuildings. Low-lying areas of Connecticut (Klemens 1993:233), and in marshy woodlands in the Albany Pine Bush in New York, and wooded creek bottomlands (Stewart and Rossi 1981). Hibernates from late September to April or May under forest floor debris, stumps, trash piles (Wright and Wright 1957:308).

Special Habitat Requirements: Sandy soils, open woodlands. May be sensitive to disturbance.

Age/Size at Sexual Maturity: Male 400 to 1,050 mm, females 450 to 1,200 mm (Wright and Wright 1957:309).

Breeding Period: April to May and probably fall (Fitch 1970).

Egg Deposition: June to July. Eggs laid in earth, under or in pulpy wood of decaying logs.

Clutch Size: 4 to 61 eggs, typically 22 (Fitch 1970).

Incubation Period: 39 to 60 days (Anderson 1965:185).

Eggs Hatch: July to September, peak in August.

Home Range/Movement: After 5 months one individual in Maryland mixed habitat had moved 100 ft (30 m) (Stickel and Cope 1947).

Food Habits: Toads preferred, but frogs, fish, salamanders, insects, and worms are taken; rarely small birds and mammals and occasionally other snakes (Edgren 1955). Amphibians and reptiles accounted for 80 percent of the food items in 10 specimens in Virginia (Uhler et al. 1939).

Comments: Diurnal. Fossorial habits, probably seek cover by burrowing (Edgren 1955). Particularly vulnerable to heavy herbicide and pesticide use. Dramatic defense behavior includes head rearing, "hood" display, mock striking, and feigning death. A completely harmless snake.

Northern Ringneck Snake

(*Diadophis punctatus edwardsii*)

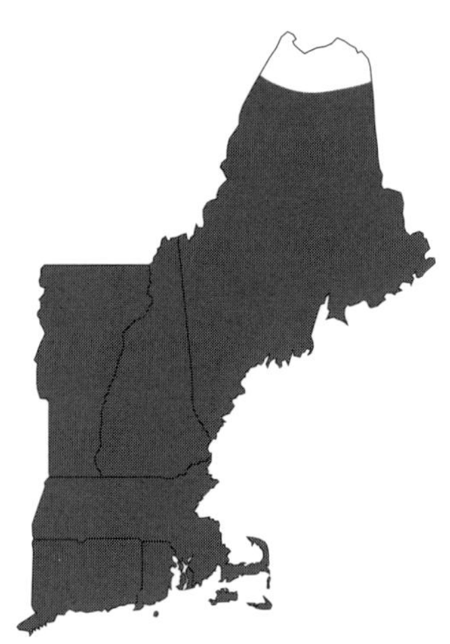

Range: Nova Scotia, w. New Brunswick, s. Quebec, s. Ontario w. to Wisconsin, s. to e. and s. Ohio, se. Illinois, n. Alabama, and central Virginia.

Distribution in New England: Throughout the region.

Status in New England: Common, but uncommonly encountered due to nocturnal activity and fossorial habit.

Habitat: Varied habitats: from pristine to disturbed, mesic to xeric, open to forested (Klemens 1993:223). Secretive, found under cover especially in moist shady woodlands with abundant hiding cover: stony woodland pastures, rocks, stone walls, old woodland log piles, debris, loose bark of logs and stumps (Burgason 1992b); shale banks in Maine (Fowler and Sutcliffe 1952) and boards are all used as cover. Hibernates from September to April or May. One individual found in a woodchuck den (Grizzel 1949). Occurs from sea level to mountaintops in southern New England (Klemens 1993:223).

Special Habitat Requirements: Mesic to xeric areas with abundant cover.

Age/Size at Sexual Maturity: Males at 13 to 14 months (Fitch 1960b), males 220 to 500 mm, females 220 to 550 mm (Wright and Wright 1957:187).

Breeding Period: Soon after emerging from hibernation in May. A female collected on May 3 in South Hadley, Mass., contained 5 eggs (Klemens 1993:224).

Egg Deposition: Late June to early July. Eggs laid in rotted logs, under logs or stones. Several females may use the same nest. A communal nest of 10 eggs found in a rotted railroad tie in Rhode Island (Klemens 1993:224).

Clutch Size: 1 to 10 eggs, typically 3 or 4 (Blanchard 1937b). 1 to 6 (average 3.5) for a sample of 130 ringneck snakes (Blanchard 1942). Smaller females lay fewer eggs (Fitch 1970).

Incubation Period: 4 to 6 weeks (Minton 1944). Average of 56 days in laboratory conditions (Blanchard 1930, cited in Wright and Wright 1957:188).

Eggs Hatch: Late August through September.

Home Range/Movement: Undocumented.

Food Habits: Small toads, frogs, salamanders, earthworms, lizards, small snakes, insects, and grubs.

Comments: Nocturnal. Degree of fossorial tendency varies with temperature preference (Elick et al. 1979). A rarely seen, beautiful snake.

Eastern Worm Snake

(Carphophis a. amoenus)

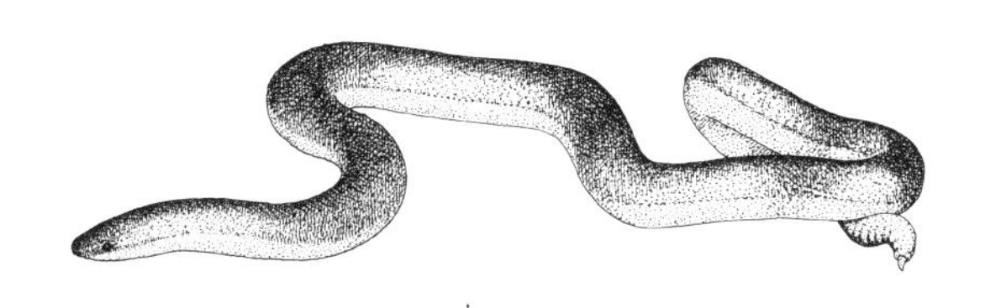

Range: Southcentral Massachusetts, se. New York through central Pennsylvania to s. Ohio, s. to central Alabama, n. Georgia and South Carolina.

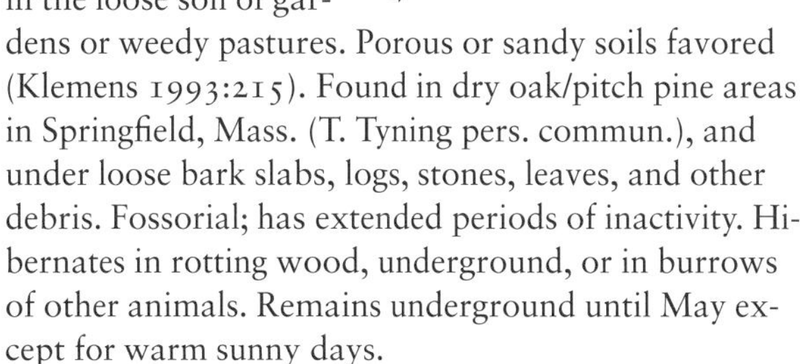

Distribution in New England: Connecticut Valley and s.-central Massachusetts, Rhode Island, and throughout Connecticut except the nw. uplands.

Status in New England: Locally abundant, but extremely secretive and rarely found.

Habitat: Dry to moist forests, often near streams; in the loose soil of gardens or weedy pastures. Porous or sandy soils favored (Klemens 1993:215). Found in dry oak/pitch pine areas in Springfield, Mass. (T. Tyning pers. commun.), and under loose bark slabs, logs, stones, leaves, and other debris. Fossorial; has extended periods of inactivity. Hibernates in rotting wood, underground, or in burrows of other animals. Remains underground until May except for warm sunny days.

Special Habitat Requirements: Well-drained, loose soil for burrowing, cover objects.

Age/Size at Sexual Maturity: 3 yr (Fitch 1970).

Breeding Period: Probably spring to early summer (McCauley 1945:97) and fall (Fitch 1970).

Egg Deposition: Mid through late June (Klemens 1993:216). Eggs probably laid in depressions under boulders or in hollow logs. Incubation period of 48 to 49 days in Kansas (Fitch 1970).

Clutch Size: 2 to 8 eggs (Wright and Wright 1957:106), typically 5 (McCauley 1945:55); 4 to 5 (Clark 1970).

Eggs Hatch: August to possibly September, but September hatching not confirmed (Klemens 1993:216).

Home Range/Movement: About 0.25 acre (0.1 ha) in Kentucky, varied from 23 m^2 for small juvenile female to 726 m^2 for adult male (Barbour 1971:240). Average for 10 individuals in a forested mountainous area of Kentucky was 253 m^2 or 0.025 ha (Barbour et al. 1969a).

Food Habits: Earthworms, soft-bodied insects and their larvae, grubs or slugs.

Comments: Nocturnal and secretive, worm snakes superficially resemble earthworms.

Northern Black Racer

(*Coluber c. constrictor*)

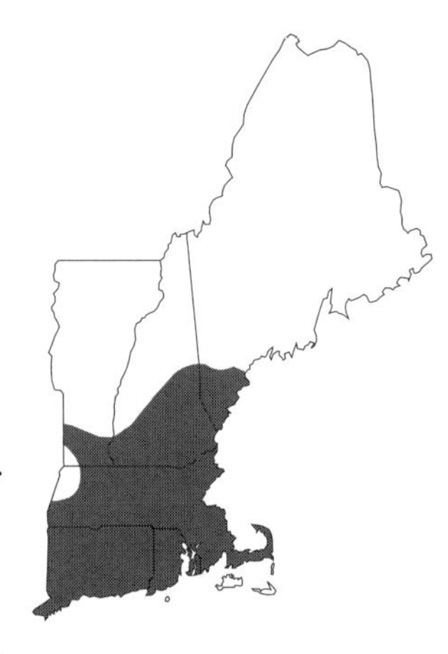

Range: Southern Maine w. to sw. Ohio, s. to central Alabama and South Carolina and throughout the e. United States.

Distribution in New England: From York and Oxford Counties, Me., w. through s. New Hampshire and s. and w. Vermont, s. throughout Massachusetts, Connecticut, and mainland Rhode Island. Rare in w. uplands of Massachusetts and nw. Connecticut.

Status in New England: Locally common.

Habitat: Moist or dry areas, lightly wooded areas, fields, roadsides, meadows, swamps, marshes, clearings, old fields, woodland edges, near old buildings, rock slopes from sea level to 1,300 feet (396 m) but rare and local above 500 ft (152 m) in Connecticut (Klemens 1993:218), stone walls, and farms. Has been found in deciduous and pine forests. Partially arboreal. Will use ledges for sunning. Hibernates in large congregations, sometimes with copperheads and rattlesnakes, often using deep rock crevices or abandoned woodchuck holes. Among the earliest snakes to emerge from hibernation.

Special Habitat Requirements: Racers thrive in areas that are periodically cleared or mowed.

Age/Size at Sexual Maturity: 2 to 3 yr (Wright and Wright 1957). Males at 13 to 14 months (Fitch 1960b), males 680 to 1,595 mm, females 710 to 1,683 mm (Wright and Wright 1957:135).

Breeding Period: Late April to early May in southern New England (Klemens 1993:220). To early June farther north (Vickery 1992).

Egg Deposition: June to early July. Laid in rotting wood, stumps, decaying vegetable matter, loose soil under stones, boards.

Clutch Size: 7 to 31 eggs, typically 16 to 17, clutch size proportional to size of female (Fitch 1963:420).

Incubation Period: Average of 51 days (Fitch 1970).

Eggs Hatch: Late August to early September.

Home Range/Movement: Very territorial; seems to have definite home range (Smith 1956:239). Average distance of 903 ft (275.2 m) in mixed Maryland habitat for 3 individuals after 2 yr (Stickel and Cope 1947). Requires large tracts of mixed old fields and woodlands (Klemens 1993:220).

Food Habits: Varied diet includes small mammals, insects, frogs, toads, small birds, birds' eggs, snakes, and lizards (Uhler et al. 1939). Small mammals and insects comprise 50 percent of diet (Surface 1906).

Comments: Diurnal. During the breeding season, racers are aggressive and will advance toward humans, but are very beneficial, preying on mice and voles.

Eastern Smooth Green Snake

(*Liochlorophis vernalis*)

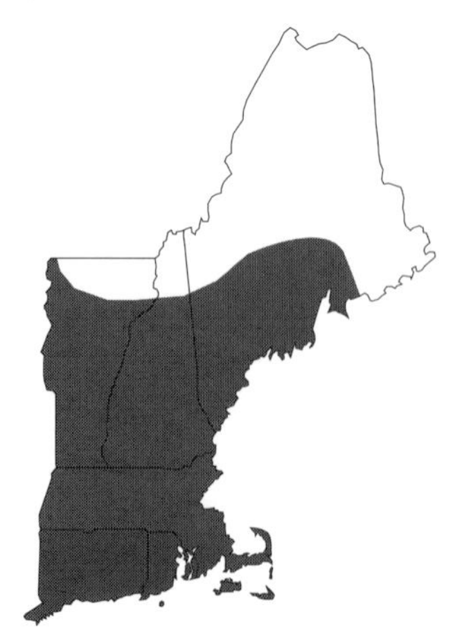

Range: Nova Scotia, s. Ontario, w. to s. Wisconsin, Michigan, and central Minnesota, s. to Ohio and New England, and in the Appalachians to Virginia and West Virginia.

Distribution in New England: Southern Maine (south of 45°30'), throughout New Hampshire (except northernmost areas) and Vermont (except ne. highlands), throughout s. New England, but localized in w. Massachusetts and Connecticut.

Status in New England: Common, but currently declining in s. New England. Rare in sw. Connecticut, common in coastal Rhode Island.

Habitat: Grassy fields, upland meadows, to 1,100 feet (335 m) in Connecticut; high-altitude areas with grassy, open spots. Also found in open aspen stands, sphagnum bogs, marshes, in vines and brambles, and hardwood stands.

Special Habitat Requirements: Upland grassy areas: fields, meadows, pastures, coastal grasslands, but also old fields, lightly wooded areas.

Age/Size at Sexual Maturity: Probably second year (Seibert and Hagen 1947). Mature at 300 mm for males, 280 mm for females total length (Wright and Wright 1957).

Breeding Period: Late August in Ontario (Smith 1956:236). Spring and late summer (Behler and King 1979:640).

Egg Deposition: Late July to August.

Clutch Size: 3 to 12 eggs (Wright and Wright 1957:558), typically 7 (Blanchard 1933a). Nest sites may be used by several females.

Incubation Period: Varies from 4 to 23 days (Blanchard 1933a).

Eggs Hatch: August to early September.

Home Range/Movement: Less than 30 yd (27.4 m) for 10 of 12 individuals studied in an uncultivated field in Illinois (Seibert and Hagen 1947).

Food Habits: Insects account for 73 percent of prey items; also spiders, snails (Surface 1906). Salamanders, millipedes, centipedes, particularly caterpillars, orthopterans, ants, flies (Uhler et al. 1939).

Comments: Hibernates early fall to April or May. Population decline may be related to insecticide spraying and loss of open fields and pasture. Inhabits abandoned farmland dominated by successional vegetation (Schlauch 1975).

Black Rat Snake

(*Elaphe o. obsoleta*)

Range: Southwestern New England w. through s. New York to s.-central Illinois and the Mississippi River area in Wisconsin, s. to Oklahoma, central Louisiana and Georgia. Range may be extending n. in the Connecticut River Valley (T. Tyning pers. commun.).

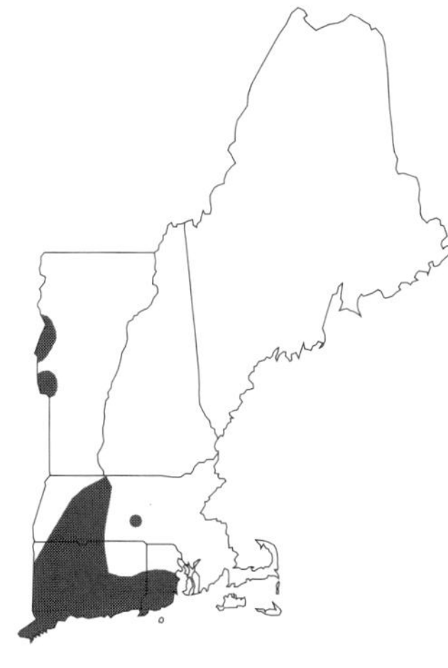

Distribution in New England: From w. Vermont, central Massachusetts (Connecticut River Valley and basalt ridges and rock outcroppings of the central Connecticut Lowland; Klemens 1993:227), and extreme e. Rhode Island w. through (primarily s.) Connecticut. Recently confirmed population in Sturbridge, Mass. (A. Richmond pers. commun.).

Status in New England: Locally common (Connecticut).

Habitat: Steep forested areas with ledges and rock outcroppings, but also a variety of habitats including woodlands, thickets, field edges, farmlands, rocky hillsides and mountaintops, river bottoms, old barns. Readily climbs trees. Found in dry oak and oak-hickory woods, and mesic bottomland forests; may occur in very dense woods (Wright and Wright 1957:232). In Connecticut, found in gorges and some coastal areas (Klemens 1993:228). Hibernates late November to April; may use talus slopes, cisterns or unused wells. Often found in groups with copperheads and rattlesnakes where these snakes occur.

Age/Size at Sexual Maturity: At 4 yr in Kansas (Fitch 1970). Males 1,095 to 1,835 mm, females 715 to 1,800 mm total length (Wright and Wright 1957:233).

Breeding Period: May to June.

Gestation Period: 8 to 12 weeks (Oliver 1955:243).

Egg Deposition: July to August. Laid in loose soil, decaying wood, manure piles, sawdust piles.

Clutch Size: 5 to 44 eggs, typically 8 to 12 in Kansas (Fitch 1970).

Home Range/Movement: Average at least 600 m in diameter for males, and at least 500 m for females in woods and fields in Maryland (Stickel et al. 1980).

Food Habits: Warm-blooded vertebrates account for 60 percent of prey items, particularly rodents, small birds and their eggs (30 percent); also amphibians, insects, spiders (Uhler et al. 1939). Young opossums, weasels, owls, and sparrow hawks have been captured as food (Minton 1972:272). Prey is killed by constriction. A beneficial snake, unfortunately often killed simply because of its large size.

Comments: Formerly called pilot or pilot black snake. Diurnal and arboreal. May reside in hollow trees

(Conant 1975:194). Frequently killed on roads, and sometimes collected for the pet trade or as pets.

Eastern Milk Snake

(*Lampropeltis t. triangulum*)

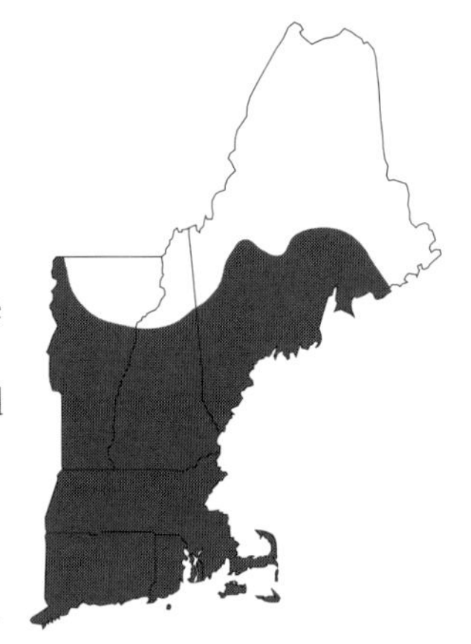

Range: Southwestern Maine and s. Ontario w. to central Minnesota, s. to Tennessee and w. North Carolina and throughout the Northeast. Intergrades with the scarlet king snake (*L. t. elapsoides*) in the sw. and se. portion of its range.

Distribution in New England: From Hancock County, Me., w. through New Hampshire to nw. Vermont, s. throughout the region. One unconfirmed report from Houlton, Aroostook County, Me.

Status in New England: Common.

Habitat: Various habitats, usually with brushy or woody cover, and found from sea level to mountain elevations. In w. Connecticut and Massachusetts, common between 700 and 1,700 ft (213–518 m) elevation on the Taconic Uplift (Klemens 1993:237). Usually found under cover. Farmlands, woods, barns, outbuildings, meadows, river bottoms, bogs, rocky hillsides, rodent runways (Klemens 1993:237). Found under logs, stones, boards, well covers, stones in creek bottoms, or other cover during the day. In pine forests, second-growth pine, bog woods, hardwoods, aspen stands. Hibernates from October or November to April, often in cellars of old houses with stone foundations.

Special Habitat Requirements: Suitable cover or loose soil for egg laying.

Age/Size at Sexual Maturity: Third or fourth year (Fitch and Fleet 1970), males to 1,115 mm, females 404 to 966 mm (Wright and Wright 1957:371).

Breeding Period: June (Wright and Wright 1957:371).

Egg Deposition: Mid June to July, in piles of soil, sawdust or manure, or under other cover, often in a communal nest site.

Clutch Size: 6 to 24 eggs, typically 13 (Wright and Wright 1957).

Incubation Period: 6 to 8 weeks (Wright and Wright 1957:371).

Eggs Hatch: August and September (Wright and Wright 1957).

Home Range/Movement: About 50 acres (20.25 ha) for *L. t. syspila*, movements of 250 to 1,300 ft (76.2 to 396.2 m) in open woodland in ne. Kansas (Fitch and Fleet 1970). Seasonal movement probable from drier hibernation sites to moist bottomlands for the summer (Breckenridge 1958, cited in Williams 1978:79).

Food Habits: Mice, other small mammals, other snakes, lizards, birds and their eggs, slugs. Mice accounted for 74 percent of the volume of stomach contents of 42 milk snakes in Pennsylvania (Surface 1908). Forages for food at night. Very beneficial around farms and other residential areas where outbuildings harbor mice.

Comments: Typically nocturnal. Numbers may be declining as abandoned fields revert to forests. Milk snakes thrive in disturbed, human-altered habitats, but vigorous defense when cornered and superficial resemblance to the venomous northern copperhead frequently result in this beneficial species being killed around human habitations (Klemens 1993:239).

Northern Copperhead

(*Agkistrodon contortrix mokasen*)

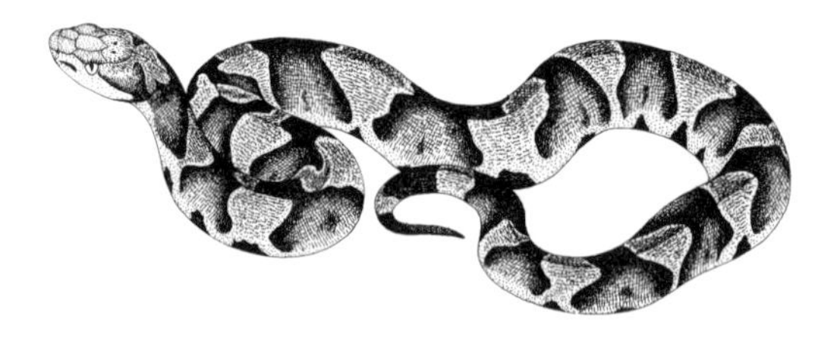

Range: Southwestern New England w. to sw. Illinois, s. to Georgia.

Distribution in New England: In Massachusetts, localized and rare, on three sites on basalt ridges in the central Connecticut Lowland as far north as s. Hampshire County (Klemens 1993:269). In Connecticut, widespread in sw. Fairfield and New Haven counties and se. Middlesex and New London counties, and along ridges

w. of Connecticut River (Klemens 1993:268).

Status in New England: Localized and rare (Massachusetts). Widespread in sw. and se. Connecticut; most abundant on basalt ridges of the central Connecticut Lowland.

Habitat: Usually associated with deciduous forests. Occupies varied habitats from swamps to low elevations in Connecticut: 100–700 ft (30–213 m) (Klemens 1993:269). Prefers areas with damp leaf litter (Fitch 1960a:116). Exposed mountainous, rocky hillsides, talus slopes, basalt ridges, ledges, open woods. Petersen and Fritsch (1986) reported a preference for traprock (basalt) ledges with extensive rock slides below, with a southerly exposure. Found in habitats with large rocks, rotting wood, and sawdust piles. During summer months may be found near swamps, ponds, or streams. Largely absent from white pine-northern hardwood, and beech-maple associations (Fitch 1960a:123). Reinert (1984) found this species used relatively open areas with high rock density and low-density surface vegetation.

Special Habitat Requirements: Rocky hillsides, talus slopes.

Age/Size at Sexual Maturity: Males during their second summer, females at 3 yr (Fitch 1960a:272).

Breeding Period: After emergence from hibernation in April (Connecticut, Finneran 1948) to May, peak in late May. Sperm may remain viable in the female for more than a year after copulation (Allen 1955). Gestation period of 105 to 110 days (Fitch 1960a:116).

Young Born: August to September, typically September in the northeast (Finneran 1953). Viviparous.

No. of Young: 1 to 17 young, typically 5 to 6 (Wright and Wright 1957:913). Litters produced in alternate years.

Home Range/Movement: In mixed habitat of woodlands, ledges and grassland in Kansas. Fitch (1960a:147–149) recorded 24.4 acres (9.7 ha) for males and 8.5 acres (3.4 ha) for females. Seasonal movements occur between hibernaculum and lowland areas.

Food Habits: Mice, other small rodents, insects, small birds, salamanders, lizards, small snakes, frogs, toads. Food obtained by ambush.

Comments: Nocturnal during summer months, diurnal in spring and fall. Has survived eradication in some areas due to cryptic coloration and retiring habits. Usually gregarious. During hibernation (from October to April), sometimes found with other species of snakes including rattlesnakes, but mutually exclusive in Connecticut (Petersen 1970). Den/sites are reused each year, a major limiting factor. Subspecific name sometimes spelled *mokasen* (e.g., Klemens 1993:272). This species is frequently confused with harmless species such as the milk snake, hognose snake, and water snake (Klemens 1993:268).

Timber Rattlesnake

(*Crotalus horridus*)

Range: Southern New Hampshire, the Champlain Valley to sw. New York, w. along the Ohio River Valley and n. to the Mississippi River in Wisconsin, s. to n. Texas and through the Appalachians to n. Georgia.

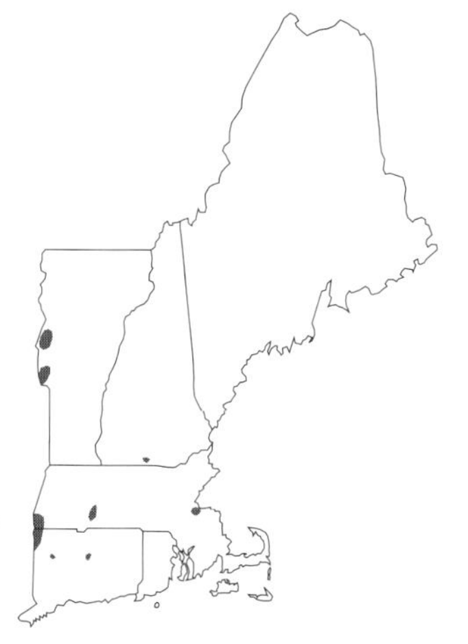

Distribution in New England: Southern New Hampshire, several areas in Berkshire County, Mass., the Blue Hills reservation in e. Massachusetts, two areas in Connecticut, nw. Litchfield County and in central Connecticut.

Status in New England: Localized and rare. Extirpated from Maine (Hunter et al. 1992).

Habitat: Remote timbered areas with rocky outcroppings, rock slides, dry ridges, and second-growth deciduous or coniferous forests with high rodent populations. Usually southern exposures. Sometimes in swamps, quarries, old stone walls, abandoned buildings. Often found near streams in late summer. Most common in areas not frequented by humans, few such sites remain. Reaches elevations of 6,000 ft (1,800 m) in the Southeast, but probably not found at highest elevations in the northeast due to harsh climatic conditions (Klauber 1972:511). Reinert (1984) found that in Pennsylvania this species frequented forested habitats rather than dry, rocky outcroppings. Hibernates from September to April in large numbers in rocky crevices usually overgrown with brush. Found with copperheads and other snakes, due to paucity of hibernacula (T. Tyning pers. commun.).

Special Habitat Requirements: Rock outcroppings on forested hillsides.

Age/Size at Sexual Maturity: Probably 3½ to 4 yr (Klauber 1972:335).

Breeding Period: Fall in Connecticut and Wisconsin (Messeling 1953). After emerging from hibernation (Fitch 1970). Gestation period probably about 5½ to 6 months.

Young Born: Late August to September, probably biennial cycles (Klauber 1972:691). Viviparous.

No. of Young: 5 to 17 young, typically 7 to 10 (Klauber 1972:733).

Home Range/Movement: Females return to hibernation dens to give birth to young. Hibernation dens may be used year after year. Home ranges and favored refuges probably exist, but few investigations have been conducted (Klauber 1972:606–607).

Food Habits: Prefers warm-blooded prey. Small mammals account for 87 percent of prey taken (Uhler et al. 1939), particularly mice, but includes rabbits, shrews, chipmunks, squirrels, bats, songbirds, and other snakes. Forages at night (Kimball 1978).

Comments: Formerly reported from widely separated areas in Connecticut, now extirpated from most of them (Klemens 1993:277).

Literature Cited

Adler, K. 1961. Egg-laying in the spotted turtle *Clemmys guttata* (Schneider). *Ohio J. Sci.* 61:180–182.

Albright, J. 1992. Hypotheticals, accidentals, and other oddities. Pages 144–146. In: Hunter, M. L., Jr.; Albright, J.; Arbuckle, J. (editors). *The amphibians and reptiles of Maine.* Maine Agric. Exp. Sta. Bull. 838. Orono: University of Maine. 188 pp.

Alexander, M. M. 1943. Food habits of the snapping turtle in Connecticut. *J. Wildlife Manage.* 7:278–282.

Allen, J. A. 1870. Notes on Massachusetts reptiles and batrachians. *Proc. Boston Soc. Nat. Hist.*13:260–263.

Allen, W. D. 1955. Some notes on reptiles. *Herpetologica* 11:228.

Anderson, J. D. 1967a. *Ambystoma maculatum. Catalogue of American Amphibians and Reptiles* 51.1–51.4.

Anderson, J. D. 1967b. *Ambystoma opacum. Catalogue of American Amphibians and Reptiles.* 46.1–46.2.

Anderson, J. D. 1972. Embryonic temperature tolerance and rate of development in some salamanders of the genus *Ambystoma. Herpetologica* 28:126–130.

Anderson, J. D.; Giacosie, R. V. 1967. *Ambystoma laterale* in New Jersey. *Herpetologica* 23:108–111.

Anderson, J. D.; Williamson, G. K. 1973. The breeding season of *Ambystoma opacum* in the northern and southern parts of its range. *J. Herpetol.* 7:320–321.

Anderson, P. 1965. *The reptiles of Missouri.* Columbia, Mo.: University of Missouri Press. 330 pp.

Ashton, R. E., Jr. 1975. A study of movement, home range, and winter behavior of *Desmognathus fuscus* (Rafinesque). *J. Herpetol.* 9:85–91.

Ashton, R. E.; Ashton, P. S. 1978. Movements and winter behavior of *Eurycea bislineata* (Amphibia, Urodela, Plethodontidae). *J. Herpetol.* 12:295–298.

Babbitt, L. H. 1937. The *Amphibia of Connecticut.* Bulletin 57. Hartford: Connecticut Geological and Natural History Survey. 50 pp.

Babcock, H. L. 1919. The turtles of New England. *Mem. Boston Soc. Nat. Hist.* 8:325–431.

Bailey, R. M. 1949. Temperature tolerance of garter snakes in hibernation. *Ecology* 30:238–242.

Ball, S. C. 1936. The distribution and behavior of the spadefoot toad in Connecticut. *Trans. Conn. Acad. Arts Sci.* 32:351–379.

Banasiak, C. F. 1974. *Population structure and reproductive ecology of the red-backed salamander in DDT-treated forests of northern Maine.* Ph.D. dissertation, University of Maine, Orono.

Barbour, R. W. 1971. *Amphibians and reptiles of Kentucky.* Lexington: University Press of Kentucky.

Barbour, R. W.; Harvey, M. J.; Hardin, J. W. 1969a. Home range, movement, and activity of the eastern worm snake, *Carphophis a. amoenus. Ecology* 50:470–476.

Barbour, R. W.; Hardin, J. W.; Schafer, J. P.; Harvey, M. J. 1969b. Home range, movement, and activity of the dusky salamander, *Desmognathus fuscus. Copeia.* 1969:293–297.

Barthalmus, G. T.; Bellis, E. D. 1969. Homing in the northern dusky salamander, *Desmognathus f. fuscus* (Rafinesque). *Copeia* 1969:148–153.

Barton, A. J.: Price, J. W. 1955. Our knowledge of the bog turtle, *Clemmys muhlenbergi,* surveyed and augmented. *Copeia* 1955:159–165.

Baumann, W. L.; Huels, M. 1982. Nests of the two-lined salamander, *Eurycea bislineata. J. Herpetol.* 16:81–83.

Behler, J. L.; King, F. W. 1979. *The Audubon Society field guide to North American reptiles and amphibians.* New York: Alfred A. Knopf. 719 pp.

Bellis, E. D. 1961. Growth of the wood frog, *Rana sylvatica. Copeia* 1961:74–77.

Bellis, E. D. 1965. Home range and movements of the wood frog in a northern bog. *Ecology* 46:90–98.

Bellis, E. D. 1968. Summer movement of red-spotted newts in a small pond. *J. Herpetol.* 1:86–91.

Bishop, S. C. 1926. Notes on the habits and development of the mudpuppy, *Necturus maculosus* (Rafinesque). *N.Y. State Mus. Bull.* 268:1–61.

Bishop, S. C. 1941. The salamanders of New York. *N.Y. State Mus. Bull.* 324:1–365.

Bishop, S. C. 1947. Curious behavior of a hoary bat. *J. Mammal.* 28:293–294.

Blanchard, F. N. 1923. The life history of the four-toed salamander. *Am. Naturalist* 57:262–268.

Blanchard, F. N. 1930. Further studies on the eggs and young of the eastern ringneck snake, *Diadophis punctatus edwardsii. Bull. Antivenin Inst. Am.* 4:4–10.

Blanchard, F. N. 1933a. Eggs and young of the smooth green snake, *Liopeltis vernalis* (Harlan). *Papers Mich. Acad. Sci. Arts Lett.* 17:493–508.

Blanchard, F. N. 1933b. Late autumn collections and hibernating situations of the salamander, *Hemicactylium scutatum* (Schlegel), in southern Michigan. *Copeia* 1933:216–217.

Blanchard, F. N. 1934. The relation of the four-toed salamander to her nest. *Copeia* 1934:136–137.

Blanchard, F. N. 1937a. Data on the natural history of the red-bellied snake, *Storeria occipitomaculata* (Storer), in northern Michigan. *Copeia* 1937:151–162.

Blanchard, F. N. 1937b. Eggs and natural nests of the eastern ring-neck snake, *Diadophis punctatus edwardsii. Papers Mich. Acad. Sci. Arts Lett.* 22:521–532.

Blanchard, F. N. 1942. The ring-neck snake, genus *Diadophis. Bull. Chicago Acad. Sci.* 7:1–144.

Blanchard, F. N.; Blanchard, F. C. 1942. Mating of the garter snake, *Thamnophis s. sirtalis* (Linnaeus). *Papers Mich. Acad. Sci. Arts Lett.* 27:215–234.

Bleakney, J. S. 1952. The amphibians and reptiles of Nova Scotia. *Can. Field-Naturalist* 66:125–129.

Bleakney, J. S. 1959. *Thamnophis s. sirtalis* (Linnaeus) in eastern Canada, redescription of *T. s. pallidula* (Allen). *Copeia* 1959:52–56.

Bragg, A. N. 1956. *Gnomes of the night, the spadefoot toads.* Philadelphia: University Pennsylvania Press.

Brandon, R. A. 1961. A comparison of the larvae of five northeastern species of *Ambystoma* (Amphibia, Caudata). *Copeia* 1961:377–383.

Breckenridge, W. J. 1958. *Reptiles and amphibians of Minnesota.* 2d ed. Minneapolis: University of Minnesota Press.

Breder, R. B. 1927. Turtle trailing: A new technique for studying the life habits of certain testudinata. *Zoologica* 9:231–243.

Breitenbach, G. L. 1982. The frequency of communal nesting and solitary brooding in the salamander, *Hemidactylium scutatum. J. Herpetol.* 16(4):341–346.

Brodie, E. D., Jr. 1968. Investigations on the skin toxin of the red-spotted newt, *Notophthalmus v. viridescens. Am. Midl. Naturalist* 80:276–280.

Brodie, E. D., Jr. 1977. Salamander antipredator postures. *Copeia* 1977:523–535.

Brodie, E. D., Jr.; Nowak, R. T.; Harvey, W. R. 1979. The effectiveness of antipredator secretions and behavior of selected salamanders against shrews. *Copei* 1979:270–274.

Bruce, R. C. 1969. Fecundity in primitive plethodontid salamanders. *Evolution* 23:50–54.

Bruce, R. C. 1972. Variation in the life cycle of the salamander, *Gyrinophilus porphyriticus. Herpetologica* 28:230–245.

Bruce, R. C. 1980. A model of the larval period of the spring salamander, *Gyrinophilus porphyriticus,* based on size-frequency distributions. *Herpetologica* 36:78–86.

Burgason, B. 1992a. Redbelly snake. Pages 144–146. In: Hunter, M. L., Jr.; Albright, J.; Arbuckle, J. (editors). *The amphibians and reptiles of Maine.* Maine Agric. Exp. Sta. Bull. 838. Orono: University of Maine. 188 pp.

Burgason, B. 1992b. Ringneck snake. Pages 128–130. In: Hunter, M. L., Jr.; Albright, J.; Arbuckle, J. (editors). *The amphibians and reptiles of Maine.* Maine Agric. Exp. Sta. Bull. 838. Orono: University of Maine. 188 pp.

Burger, J. W. 1935. *Plethodon cinereus* (Green) in eastern Pennsylvania and New Jersey. *Am. Naturalist* 69:578–586.

Burton, T. M. 1976. An analysis of the feeding ecology of the salamanders (*Amphibia, Urodela*) of the Hubbard Brook Experimental Forest, New Hampshire. *J. Herpetol.* 10:187–204.

Burton, T. M. 1977. Population estimates, feeding habits and nutrient and energy relationships of *Notophthalmus v. viridescens,* Mirror Lake, New Hampshire. *Copeia* 1977:139–143.

Burton, T. M.; Likens, G. E. 1975. Salamander populations and biomass in Hubbard Brook Experimental Forest, New Hampshire. *Copeia* 1975:541–546.

Cagle, F. R. 1942. Herpetological fauna of Jackson and Union Counties, Illinois. *Am. Midl. Naturalist* 28:164–200.

Cahn, A. R. 1937. The turtles of Illinois. *Ill. Biol. Monogr.* 16:1–218.

Caldwell, R. S.; Houtcooper, W. C. 1973. Food habits of larval *Eurycea bislineata. J. Herpetol.* 7:386–388.

Caldwell, R. S. 1975. Observations on the winter activity of the red-backed salamander, *Plethodon cinereus,* in Indiana. *Herpetologica* 31:21–22.

Carpenter, C. C. 1952a. Growth and maturity of three species of *Thamnophis* in Michigan. *Copeia* 1952:237–243.

Carpenter, C. C. 1952b. Comparative ecology of the common garter snake (*Thamnophis s. sirtalis*), the ribbon snake (*Thamnophis s. sauritus*), and Butler's garter snake (*Thamnophis butleri*) in mixed populations. *Ecol. Monogr.* 22:236–258.

Carr, A. F. 1952. *Handbook of turtles of the United States, Canada, and Baja California.* Ithaca, N.Y.: Comstock.

Carroll, D. M. 1991. *The year of the turtle: A natural history.* Charlotte, Vt.: Camden House.

Chadwick, C. S. 1944. Observations on the life cycle of the common newt in western North Carolina. *Am. Midl. Nat.* 32:491–494.

Clark, D. R. 1970. Ecological study of the worm snake *Carphophic vermis* (Kennicott). *Univ. Kansas Publ. Mus. Nat. Hist.* 19:85–194 + 48 fig., 4 pl.

Clarke, R. D. 1974a. Activity and movement patterns in a population of Fowler's toad, *Bufo woodhousei fowleri. Can. J. Zool.* 52:1489–1498.

Clarke, R. D. 1974b. Food habits of toads, genus *Bufo* (Amphibia: Bufonidae). *Am. Midl. Naturalist* 91:140–147.

Clausen, H. J. 1936. Observations on the brown snake *Storeria dekayi* (Holbrook), with special reference to the habits and birth of young. *Copeia* 1936:98–102.

Conant, R. 1938. The reptiles of Ohio. *Am. Midl. Naturalist* 20(1):1–200.

Conant, R. 1975. *A field guide to reptiles and amphibians of eastern and central North America.* 2d ed. Boston: Houghton Mifflin.

Conant, R.; Collins, J. T. 1991. *A field guide to reptiles and amphibians of eastern and central North America.* Boston: Houghton Mifflin. 450 pp.

Cook, R. P. 1983. Effects of acid precipitation on embryonic mortality of spotted salamanders (*Ambystoma maculatum*) and Jefferson salamanders (*A. jeffersonianum*) in the Connecticut Valley of Massachusetts. *Biol. Conserv.* 27:77–78.

Cooper, J. E. 1956. Aquatic hibernation of the red-backed salamander. *Herpetologica* 12:165–166.

Cortwright, S. 1986. The roles of dispersal and history in amphibian communities. *Proc. Indiana Acad. Sci.* 95:187.

Cortwright, S. A. 1988. Intraguild predation and competition. An analysis of net growth shifts in larval and amphibian prey. *Can. J. Zool.* 66:1813–1821.
Costanzo, J. P. 1986. Influences of hibernaculum microenvironment on the winter life history of the garter snake (*Thomnophis sirtalis*). *Ohio J. Sci.* 86:199–204.
Cowardin, L. M.; Carter, V.; Golet, F. C.; LaRoe, E. T. 1979. *Classification of wetlands and deepwater habitats of the United States.* FWS/OBS-79/31. Washington, D.C.: U.S. Department of Interior, Fish and Wildlife Service, Biological Services Program. 103 pp.
Currie, W.; Bellis, E. D. 1969. Home range and movements of the bullfrog, *Rana catesbeiana* (Shaw), in an Ontario Pond. *Copeia* 1969:688–692.
Dale, J. M.; Freedman, B.; Kerekes, J. 1985. Acidity and associated water chemistry of amphibian habitats in Nova Scotia. *Can. J. Zool.* 63:97–105.
Danstedt, R. T., Jr. 1975. Local geographic variation in demographic parameters and body size of *Desmognathus fuscus* (Amphibia, Plethodontidae). *Ecology* 56:1054–1067.
Davidson, J. A. 1956. Notes on the food habits of the slimy salamander, *Plethodon g. glutinosus. Herpetologica* 12:129–131.
Davis, S. L. 1992. Gray tree frog. Pages 56–60. In: Hunter, M. L., Jr.; Albright, J.; Arbuckle, J. (editors). *The amphibians and reptiles of Maine.* Maine Agric. Exp. Sta. Bull. 838. Orono: University of Maine. 188 pp.
DeGraaf, R. M.; Yamasaki, M. 1992. A nondestructive technique to monitor the relative abundance of terrestrial salamanders. *Wildl. Soc. Bull.* 20:260–264.
Delzell, D. E. 1958. Spatial movement and growth of *Hyla crucifer.* Ph.D. dissertation, University of Michigan, Ann Arbor.
Dickerson, M. C. 1969. *The frog book.* New York: Dover.
Dole, J. W. 1965. Summer movements of adult leopard frogs, *Rana pipiens* (Schreber), in northern Michigan. *Ecology* 46:236–255.
Dole, J. W. 1968. Homing in leopard frogs, *Rana pipiens. Ecology* 49:386–399.
Dole, J. W. 1973. Celestial orientation in recently metamorphosed *Bufo americanus. Herpetologica* 29:59–62.
Douglas, M. E.; Monroe, B. L. 1981. A comparative study of topographical orientation in *Ambystoma* (Amphibia: Caudata). *Copeia* 1981:460–463.
Downs, F. L. 1989. *Ambystoma laterale* Hallowell, blue-spotted salamander. Pages 102–107. In: Pfingsten, R. A.; Downs, F. L. (editors). *Salamanders of Ohio.* Columbus: Ohio State University Press. 315 pp.
Drake, C. J. 1914. The food of *Rana pipiens* (Schreber). *Ohio Naturalist* 14:257–269.
Driver, E. C. 1936. Observations on *Scaphiopus holbrooki* (Harlan). *Copeia* 1936:67–69.
Dunn, E. R. 1926. *The salamanders of the family Plethodontidae.* Northampton, Mass.: Smith College.
Edgren, R. A. 1955. The natural history of the hog-nosed snakes, genus *Heterodon:* A review. *Herpetologica* 11:105–117.
Edgren, R. A. 1960. Ovulation time in the musk turtle, *Sternotherus odeoatus* (Latreille). *Nat. Hist. Miscellanea* 152:1–3.
Elick, G. E.; Sealander, J. A.; Beumer, R. J. 1979. Temperature preferenda, body temperature tolerances, and habitat selection of small Colubrid snakes. *Trans. Missouri Acad. Sci.* 13:21–31.
Emlen, S. T. 1968. Territoriality in the bullfrog, *Rana catesbeiana. Copeia* 1968:240–243.
Ernst, C. H. 1968. A turtle's territory. *Int. Turtle Tortoise Soc. J.* 2(6):9, 34.
Ernst, C. H. 1970. Home range of the spotted turtle, *Clemmys guttata. Copeia* 1970:391–393.
Ernst, C. H. 1976. Ecology of the spotted turtle, *Clemmys guttata* in southeastern Pennsylvania. *J. Herpetol.* 10:25–33.
Ernst, C. H. 1977. Biological notes on the bog turtle, *Clemmys muhlenbergii. Herpetologica* 33:241–246.
Ernst, C. H.; Barbour, R. W. 1972. *Turtles of the United States.* Lexington: University Press of Kentucky.
Ernst, C. H.; Bury, R. B. 1977. *Clemmys muhlenbergii* (Schoepff). *Catalogue of American Amphibians and Reptiles* 204.1–204.2.
Ernst, C. H.; Lovich, J. E.; Barbour, R. W. 1994. *Turtles of the United States and Canada.* Washington, D.C.: Smithsonian Institution Press.
Etchberger, C. R. 1992. Common musk turtle, *Sternotherus odoratus.* Pages 97–100. In: Hunter, M. L., Jr.; Albright, J.; Arbuckle, J. (editors). *The amphibians and reptiles of Maine.* Maine Agric. Exp. Sta. Bull. 838. Orono: University of Maine. 188 pp.
Evans, H. E. 1947. Herpetology of Crystal Lake, Sullivan County, New York. *Herpetologica* 4:19–21.
Ewing, H. E. 1943. Continued fertility in the female box turtles following mating. *Copeia* 1943:112–114.
Farrell, R. F.; Graham, T. E. 1991. Ecological notes on the turtle *Clemmys insculpta* in northwestern New Jersey. *J. Herpetol.* 25:1–9.
Farrell, R. F.; Zappalorti, R. T. 1979. The ecology and distribution of the wood turtle, *Clemmys insculpta* (LeConte), New Jersey, Pt. 1. (Preliminary report on a research contract between the New Jersey Department of Environmental Protection, Endangered and Nongame Species Project, National Audubon Society, and Herpetological Association, No. 79.03) [Unpublished].
Farrell, R. F.; Zappalorti, R. T. 1980. An ecological study of the wood turtle, *Clemmys insculpta* (LeConte) (Reptilia, Testudines, Emydidae), in northern New Jersey, Pt. 2. (Report to the New Jersey Department of Environmental Protection, Endangered and Nongame Species Project, Herpetological Association, Rep. No. 80.02) [Unpublished].
Finneran, L. C. 1948. Reptiles at Branford, Connecticut. *Herpetologica* 4:123–126.
Finneran, L. C. 1953. Aggregation behavior of the female copperhead, *Agkistrodon contortrix mokeson,* during gestation. *Copeia* 1953:61–62.
Fitch, H. S. 1954. Life history and ecology of the five-lined skink, *Eumeces fasciatus. Univ. Kansas Publ. Mus. Nat. Hist.* 8:1–156.
Fitch, H. S. 1960a. Autecology of the copperhead. *Univ. Kansas Publ. Mus. Nat. Hist.* 13:185–288.
Fitch, H. S. 1960b. Criteria for determining sex and breeding maturity in snakes. *Herpetologica* 16:49–51.
Fitch, H. S. 1963. Natural history of the racer *Coluber constrictor. Univ. Kansas Publ. Mus. Nat. Hist.* 15:351–468.
Fitch, H. S. 1965. An ecological study of the garter snake, *Thamnophis sirtalis. Univ. Kansas Publ. Mus. Nat. Hist.* 15:493–564.
Fitch, H. S. 1970. *Reproductive cycles of lizards and snakes.* Lawrence: University of Kansas, Museum of Natural History. 52 pp.

Fitch, H. S. 1980. *Thamnophis sirtalis* (Linnaeus). *Catalogue of American Amphibians and Reptiles* 270.1–270.4.
Fitch, H. S.; Fleet, R. R. 1970. Natural history of the milk snake (*Lampropeltis triangulum*) in northeastern Kansas. *Herpetologica* 26:387–396.
Force, E. R. 1933. The age of attainment of sexual maturity of the leopard frog, *Rana pipiens* (Schreber), in northern Michigan. *Copeia* 1933:128–131.
Fowler, J. A.; Sutcliffee, R. 1952. An additional record for the purple salamander, *Gyrinophilus p. porphyriticus,* from Maine. *Copeia* 1952:48–49.
Fraker, M. A. 1970. Home range and homing in the watersnake, *Natrix s. sipedon. Copeia* 1970:665–673.
Freda, J.; Dunson, W. A. 1986. Effects of low pH and other chemical variables on the local distribution of amphibians. *Copeia* 1986:454–466.
French, T. W. 1986. Archaeological evidence of *Emydoidea blandingii* in Maine. *Herp. Rev.* 17:40.
French, T. W.; Master, L. L. 1986. Geographic distribution: *Ambystoma jeffersonianum* (Jefferson salamander). *SSAR Herpetol. Rev.* 17:26.
Friet, S. C. 1995. *Plethodon cinereus* (eastern red-backed salamander). Nest behavior. *Herp. Rev.* 26:198–199.
Gatz, A. J., Jr. 1971. Critical thermal maxima of *Ambystoma maculatum* and *A. jeffersonianum* in relation to time of breeding. *Herpetologica* 27:157–160.
Gibbons, J. W. 1968a. Reproductive potential, activity and cycles in the painted turtle, *Chrysemys picta. Ecology* 49:399–409.
Gibbons, J. W. 1968b. Observations on the ecology and population dynamics of the Blanding's turtle, *Emydoidea blandingii. Can. J. Zool.* 4:288–290.
Gilhen, J. 1974. Distribution, natural history, and morphology of the blue-spotted salamanders, *Ambystoma laterale* and *A. tremblayi* in Nova Scotia. *Nova Scotia Mus. Curatorial Rep.* No. 22. 38 pp.
Glowa, J. 1992. Spotted turtle. Pages 105–107. In: Hunter, M. L., Jr.; Albright, J.; Arbuckle, J. (editors). *The amphibians and reptiles of Maine.* Maine Agric. Exp. Sta. Bull. 838. Orono: University of Maine. 188 pp.
Gochfeld, M. 1975. The decline of the eastern garter snake, *Thamnophis s. sirtalis,* in a rural residential section of Westchester County, New York. *Engelhardtia* 6:23–24.
Gordon, D. M. 1980. An investigation of the ecology of the map turtle, *Grapfemys geographica* (LeSueur), in the northern part of its range. *Can. J. Zool.* 58:2210–2230.
Gordon, D. M.; MacCulloch, R. D. 1980. An investigation of the ecology of the map turtle, *Graptemys geographica* (Le Sueur), in the northern part of its range. *Can. J. Zool.* 58:2210–2219.
Gordon, R. E. 1968. Terrestrial activity of the spotted salamander, *Ambystoma maculatum. Copeia* 1968:879–880.
Gore, J. A. 1983. The distribution of Desmognathine larvae (Amphibia:Plethodontidae) in coal surface mine impacted streams of the Cumberland Pleateau, USA. *J. Freshwater Ecol.* 2(1):13–24.
Gosner, K.; Black, J. H. 1955. The effects of temperature and moisture on the reproductive cycle of *Scaphiopus h. holbrooki. Am. Midl. Naturalist* 54:192–203.
Graham, T. E. 1971a. Growth rate of the red-bellied turtle, *Chrysemys rubriventris,* at Plymouth, Massachusetts. *Copeia* 1971:353–356.
Graham, T. E. 1971b. Eggs and hatchlings of the red-bellied turtle, *Chrysemys rubriventris,* at Plymouth, Massachusetts. *J. Herpetol.* 5:59–60.
Graham, T. E. 1980. Red-belly blues. *Animals* 113:17–21.
Graham, T. E. 1982. Second find of *Pseudemys rubriventris* at Ipswich, Massachusetts, and refutation of the Naushon Island record. *Herpetol. Rev.* 12:82–83.
Graham, T. E. 1992. Blanding's turtle. Pages 112–115. In: Hunter, M. L., Jr.; Albright, J.; Arbuckle, J. (editors). *The amphibians and reptiles of Maine.* Maine Agric. Exp. Sta. Bull. 838. Orono: University of Maine. 188 pp.
Graham, T. E. 1995. Habitat use and population parameters of the spotted turtle *Clemmys guttata,* a species of special concern in Massachusetts. *Chelonian Conserv. Biol.* 1:207–214.
Graham, T. E.; Doyle, T. S. 1977. Growth and population characteristics of Blanding's turtle, *Emydoidea blandingii,* in Massachusetts. *Herpetologica* 33:410–414.
Graham, T. E.; Doyle, T. S. 1979. Dimorphism, courtship, eggs, and hatchlings of the Blanding's turtle, *Emydoidea blandingii* (Reptilia, Testudines, Emydidae), in Massachusetts. *J. Herpetol.* 13:125–127.
Graham, T. E.; Forsberg, J. E. 1986. Clutch size in some Maine turtles. *Bull. Maryland Herpetol. Soc.* 22:146–148.
Graham, T. E.; Graham, A. A. 1997. Ecology of the eastern spiny softshell, *Apalone spinifera spinifera,* in the Lamoille River, Vermont. *Chelonian Conserv. Biol.* 2:363–369.
Graham, T. E.; Hutchinson, V. H. 1969. Centenarian box turtles. *Int. Turtle Tortoise Soc. J.* 3:24–29.
Grant, W. C. 1955. Territorialism in two species of salamanders. *Science* 121:137–138.
Grizzell, R. A., Jr. 1949. The hibernation site of three snakes and a salamander. *Copeia* 1949:231–232.
Groves, J. D. 1982. Egg-eating behavior of brooding five-lined skinks, *Eumeces fasciatus. Copeia* 1982:969–971.
Hamilton, W. J., Jr. 1932. The food and feeding habits of some eastern salamanders. *Copeia* 1932:83–86.
Hamilton, W. J., Jr. 1934. The rate of growth of the toad *Bufo a. americanus* (Holbrook) under natural conditions. *Copeia* 1934:88–90.
Hamilton, W. J., Jr. 1948. The food and feeding behavior of the green frog, *Rana clamitans* (Latreille), in New York State. *Copeia* 1948:203–207.
Hamilton, W. J., Jr. 1951. The food and feeding behavior of the garter snake in New York. *Am. Midl. Naturalist* 46:385–390.
Hamilton, W. J., Jr. 1954. The economic status of the toad. *Herpetologica* 10:37–40.
Hammer, D. A. 1969. Parameters of a marsh snapping turtle population. Lacreek Refuge, South Dakota. *J. Wildl. Manage.* 33:995–1005.
Harding, J. H. 1977. Record egg clutches for *Clemmys insculpta. Herpetol. Rev.* 8:34.
Harding, J. H.; Bloomer, T. J. 1979. The wood turtle, *Clemmys insculpta* . . . a natural history. *Herp. Bull. N.Y. Herpetol. Soc.* 15(1):9–26.
Harris, J. P., Jr. 1959. The natural history of *Necturus:* 1. Habitats and habits. *Field Lab.* 27:11–20.
Harris, R. N. 1981. Intrapond homing behavior in *Notophthalmus viridescens. J. Herpetol.* 15:355–356.
Haskins, J. J. 1992. Common garter snake. Pages 150–153. In: Hunter, M. L., Jr.; Albright, J.; Arbuckle, J. (editors). *The amphibians and reptiles of Maine.* Maine Agric. Exp. Sta. Bull. 838. Orono: University of Maine, 188 pp.

Hassinger, D. D.; Anderson, J. D.; Dalrymple, G. H. 1970. The early life history and ecology of *Ambystoma tigrinum* and *Ambystoma opacum* in New Jersey. *Amer. Midl. Naturalist* 84:474–495.

Healy, W. R. 1974. Population consequences of alternative life histories in *Notophthalmus v. viridescens. Copeia* 1974:221–229.

Healy, W. R. 1975. Terrestrial activity and home range in efts of *Notophthalmus viridescens. Am. Midl. Naturalist* 93:131–138.

Heatwole, H. 1961. Habitat selection and activity of the wood frog, *Rana sylvatica* (LeConte). *Am. Midl. Naturalist* 66:301–313.

Hedeen, S. E. 1970. The ecology and life history of the mink frog, *Rana septentrionalis* (Braid). Ph.D. dissertation, University of Minnesota, St. Paul. 132 pp.

Hedeen, S. E. 1972. Postmetamorphic growth and reproduction of the mink frog. *Rana septentrionalis* (Braid). *Copeia* 1972:169–175.

Herreid, C. F.; Kinney, S. 1967. Temperature and development of the wood frog, *Rana sylvatica,* in Alaska. *Ecology* 48:579–589.

Highton, R. 1956. The life history of the slimy salamander, *Plethodon glutinosus,* in Florida. *Copeia* 1956:75–93.

Highton, R. 1962. Geographic variation in the life history of the slimy salamander. *Copeia* 1962:597–613.

Highton, R.; Savage, T. 1961. Functions of the brooding behavior in the female red-backed salamander, *Plethodon cinereus. Copeia* 1961:95–98.

Hinshaw, S. 1992. Northern leopard frog. Pages 77–81. In: Hunter, M. L., Jr.; Albright, J.; Arbuckle, J. (editors). *The amphibians and reptiles of Maine.* Maine Agric. Exp. Sta. Bull. 383. Orono: University of Maine. 188 pp.

Hoff, J. G. 1977. A Massachusetts hibernation site of the red-backed salamander, *Plethodon cinereus. Herpetol. Rev.* 8:33.

Hunter, M. L., Jr. 1992. Timber rattlesnake. Pages 154–157. In: Hunter, M. L., Jr.; Albright, J.; Arbuckle, J. (editors). *The amphibians and reptiles of Maine.* Maine Agric. Exp. Sta. Bull. 838. Orono: University of Maine. 188 pp.

Hurlbert, S. H. 1969. The breeding migrations and interhabitat wandering of the vermilion-spotted newt, *Notophthalmus viridescens* (Rafinesque). *Ecol. Monogr.* 39:465–488.

Hurlbert, S. H. 1970. The post-larval migration of the red-spotted newt, *Notophthalmus viridescens* (Rafinesque). *Copeia* 1970:515–528.

Husting, E. L. 1965. Survival and breeding structure in a population of *Ambystoma maculatum. Copeia* 1965:352–362.

Ingram, W. M.; Raney, E. C. 1943. Additional studies on the movement of tagged bullfrogs, *Rana catesbeiana* (Shaw). *Am. Midl. Naturalist* 29:239–241.

Jackson, S. D. 1990. Demography, migratory patterns, and effects of pond chemistry on two syntopic mole salamanders, *Ambystoma jeffersonianum* and *A. maculatum.* M.S. Thesis, University of Massachusetts, Amherst. 96 pp.

Jackson, S. 1994. The Jefferson salamander. *Mass. Wildl.* 44:2–6.

Jameson, E. W., Jr. 1944. Food of the red-backed salamander. *Copeia* 1944:145–147.

Johnson, J. E.; Goldberg, A. S. 1975. Movement of larval two-lined salamanders (*Eurycea bislineata*) in the Mill River, Massachusetts. *Copeia* 1975:588–589.

Kaplan, R. H.; Crump, M. L. 1978. The non-cost of brooding in *Ambystoma opacum. Copeia* 1978:99–103.

Kaplan, R. H. 1980. The implications of ovum size variability for offspring fitness and clutch size within several populations of salamanders (*Ambystoma*). *Evolution* 34:51–64.

Karns, D. R. 1980. Ecological risks for amphibians at toxic bog water breeding sites in northern Minnesota. In: *Proc. 1980 Joint Annual Herpetologists League/Society for the Study of Amphibians and Reptiles,* Milwaukee, Wis. Abstract.

Kimball, D. (editor). 1978. *The timber rattlesnake in New England.* A symposium: 17 September 1977, Springfield, Mass. Springfield, Mass.: Western Massachusetts Herpetological Society.

Kiviat, E. 1978. Bog turtle habitat ecology. *Bull. Chicago Herpetol. Soc.* 13:29–42.

Kiviat, E. 1980. A Hudson River tide-marsh snapping turtle population. *Trans. Northeast Fish Wildl. Conf.* 37: 158–168.

Klauber, L. M. 1972. *Rattlesnakes.* 2 vols. Berkeley and Los Angeles: University of California Press.

Kleeberger, S. R.; Werner, J. K. 1982. Home range and homing behavior of *Plethodon cinereus* in northern Michigan. *Copeia* 1982:409–415.

Klemens, M. W. 1993. *The amphibians and reptiles of Connecticut and adjacent regions.* Bull. 112. Hartford: Connecticut State Geological and Natural History Survey. 318 pp.

Kramek, W. C. 1972. Food of the frog *Rana septentrionalis* in New York. *Copeia* 1972:390–392.

Kramek, W. C. 1976. Feeding behavior of *Rana septentrionalis* (Amphibia, Anura, Ranidae). *J. Herpetol.* 10:251–252.

Krysik, A. J. 1980. Microhabitat selection and brooding phenology of *Desmognathus f. fuscus* in western Pennsylvania. *J. Herpetol.* 291–292.

Lagler, K. F. 1943. Food habits and economic relations of the turtles of Michigan with special reference to game management. *Am. Midl. Naturalist* 29:257–312.

Landre, E. 1980. The blue-spotted salamander. *Sanctuary: The Bulletin of the Massachusetts Audubon Society* 20:6–7.

Lazell, J. D., Jr. 1972. *Reptiles and amphibians in Massachusetts.* Lincoln: Massachusetts Audubon Society.

Lazell, J. D., Jr. 1976. *This broken archipelago.* New York: Demeter Press, Quadrangle, New York Times Book Co.

Lazell, C. D.; Raithel, C. 1986. *Gyrinophilus porphyriticus* (northern spring salamander). *SSAR Herpetol. Rev.* 17:26.

Linzey, D. W. 1967. Food of the leopard frog, *Rana p. pipiens,* in central New York. *Herpetologica* 23:11–17.

Logier, E. B. S. 1952. *The frogs, toads and salamanders of eastern Canada.* Toronto: University of Toronto Press.

Lortie, J. P. 1992. Ribbon snake. Pages 147–149. In: Hunter, M. L., Jr.; Albright, J.; Arbuckle, J. (editors). *The amphibians and reptiles of Maine.* Maine Agric. Exp. Sta. Bull. 838. Orono: University of Maine. 188 pp.

Lotter, F. 1978. Reproductive ecology of the salamander, *Plethodon cinereus* (Amphibia, Urodela, Plethodontidae) in Connecticut. *J. Herpetol.* 12:231–236.

Lucas, F. A. 1916. Occurrence of *Pseudemys* at Plymouth, Mass. *Copeia* 38:98–100.

Lynn, W. G.; Dent, J. N. 1941. Notes on *Plethodon cinereus* and *Hemidactylium scutatum* on Cape Cod. *Copeia* 1941:113–114.

Lynn, W. G.; vonBrand, T. 1945. Studies on the oxygen con-

sumption and water metabolism of turtle embryos. *Biol. Bull.* 88:112–125.
McCauley, R. H., Jr. 1945. *The reptiles of Maryland and District of Columbia.* Hagerstown, Md.: Published by author.
McCoy, C. J. 1973. *Emydoidea blandingii. Catalogue of American Amphibians and Reptiles* 136.1–136.4.
MacNamara, M. C. 1977. Food habits of terrestrial adult migrants and immature red efts of the red-spotted newt, *Notophtalmus viridescens. Herpetologica* 33:127–132.
Mahmoud, I. Y. 1969. Comparative ecology of the Kinsosternid turtles of Oklahoma. *Southwest Naturalist* 14:31–66.
Mairs, D. F. 1992. Mudpuppy. Pages 15–18. In: Hunter, M. L., Jr.; Albright, J.; Arbuckle, J. (editors). *The amphibians and reptiles of Maine.* Maine Agric. Exp. Sta. Bull. 838. Orono: University of Maine. 188 pp.
Markowsky, J. K. 1992. Dusky salamander, *Desmognathus fuscus.* Pages 33–35. In: Hunter, M. L., Jr.; Albright, J.; Arbuckle, J. (editors). *The amphibians and reptiles of Maine.* Maine Agric. Exp. Sta., Bull. 838. Orono: University of Maine. 188 pp.
Marshall, W. H.; Buell, M. F. 1955. A study of the occurrence of amphibians in relation to a bog succession. Ithasca State Park, Minnesota. *Ecology* 36:381–387.
Martof, B. S. 1953a. Territoriality in the green frog, *Rana clamitans. Ecology* 34:165–174.
Martof, B. S. 1953b. Home range and movements of the green frog, *Rana clamitans. Ecology* 34:529–543.
Martof, B. S. 1956. Factors influencing size and composition of populations of *Rana clamitans. Am. Midl. Naturalist* 56:224–245.
Martof, B. S. 1970. *Rana sylvatica* (LeConte). *Catalogue of American Amphibians and Reptiles* 86.1–86.4.
Martof, B. S.; Palmer, W. M.; Bailey, J. R.; Harrison, J. R. III. 1980. *Amphibians and reptiles of the Carolinas and Virginia.* Chapel Hill: University of North Carolina Press.
Matthews, R. C., Jr. 1982. Predator stoneflies: Role in freshwater stream communities. *J. Tennessee Acad. Sci.* 57:82–83.
Maynard, E. A. 1934. The aquatic migration of the toad, *Bufo americanus* (LeConte). *Copeia* 1934:174–177.
Messeling, E. 1953. Rattlesnakes in southwestern Wisconsin. *Conserv. Bull.* 18(10):21–23.
Michener, M. C.; Lazell, J. D. 1989. Distribution and relative abundance of the hognose snake, *Heterodon platyrhinos,* in eastern New England. *J. Herpetol.* 23:35–40.
Milam, J. C. 1997. Home range, habitat use, and conservation of spotted turtles, *Clemmys guttata* in central Massachusetts. M.S. thesis, University of Massachusetts, Amherst.
Minton, S. A., Jr. 1944. Introduction to the study of the reptiles of Indiana. *Am. Midl. Naturalist* 32:438–477.
Minton, S. A., Jr. 1954. Salamanders of the *Ambystoma jeffersonianum* complex in Indiana. *Herpetologica* 10:173–179.
Minton, S. A., Jr. 1972. *Amphibians and reptiles of Indiana.* Indianapolis: Indiana Academy of Science.
Mirick, P. G. 1985. The lower pH tolerance limit of *Ambystoma laterale* complex embryos in two populations from eastern Massachusetts. Westborough: Massachusetts Division of Fisheries and Wildlife. 33 pp. [Unpublished].
Moore, J. A. 1939. Temperature tolerance and rates of development in eggs of amphibia. *Ecology* 20:459–478.
Moore, J. E.; Strickland, E. H. 1955. Further notes on the food of Alberta amphibians. *Am. Midl. Naturalist* 54:253–256.
Morgan, A. H.; Grierson, M. C. 1932. Winter habits and yearly food consumption of adult spotted newts, *Triturus viridescens. Ecology* 13:54–62.
Neill, W. T. 1963. *Hemidactylium scutatum. Catalogue of American Amphibians and Reptiles* 2.1–2.2.
Nemuras, K. 1969. Survival of the Muhlenberg. *Int. Turtle Tortoise Soc. J.* 3:18–21.
Newman, H. H. 1906. The habits of certain tortoises. *J. Comp. Neurol. Psychol.* 16:126–152.
Nichols, J. T. 1939. Range and homing of individual box turtles. *Copeia* 1939:125–127.
Noble, G. K. 1926. The Long Island newt: A contribution to the life history of *Triturus viridescens. American Museum Novitates* 228:1–11.
Noble, G. K. 1929. Further observations on the life history of the newt, *Triturus viridescens. American Museum Novitates* 348:1–22.
Noble, G. K.; Brady, M. K. 1933. Observations on the life history of the marbled salamander, *Ambystoma opacum* (Gravenhorst). *Zoologica* 11:89–132.
Noble, G. K.; Clausen, H. J. 1936. The aggregation behavior of *Storeria dekayi* and other snakes, with special reference to the sense organs involved. *Ecol. Monogr.* 6:269–316.
Nyman, S.; Ryan, M. J.; Anderson, J. D. 1988. The distribution of the *Ambystoma jeffersonianum* complex in New Jersey. *J. Herpetol.* 22(2):224–228.
Nyman, S. 1991. Ecological aspects of syntopic larvae of *Ambystoma maculatum* and the *A. jeffersonianum* complex in two New Jersey ponds. *J. Herpetol.* 25:505–509.
Obbard, M. E.; Brooks, R. J. 1980. Nesting migrations of the snapping turtle (*Chelydra serpentia*). *Herpetologica* 36:158–162.
Oldham, R. S. 1966. Spring movements in the American toad, *Bufo americanus. Can. J. Zool.* 44:63–100.
Oliver, J. A. 1955. *The natural history of North American amphibians and reptiles.* Princeton, N.J.: C. Van Nostrand.
Oplinger, C. S. 1967. Food habits and feeding activity of recently transformed and adult *Hyla c. crucifer* (Wied). *Herpetologica* 23:209–217.
Organ, J. A. 1961. Studies of the local distribution, life history, and population dynamics of the salamander genus *Desmognathus* in Virginia. *Ecol. Monogr.* 31:189–220.
Palmer, E. L. 1949. *Fieldbook of natural history.* McGraw-Hill, New York.
Pearse, A. S. 1923. The abundance and migration of turtles. *Ecology* 4:24–28.
Pearson, P. G. 1955. Population ecology of the spadefoot toad, *Scaphiopus h. holbrooki. Ecol. Monogr.* 25:233–267.
Petersen, R. C. 1970. Connecticut's venomous snakes. Bull. 103. Hartford: Connecticut Geological and Natural History Survey. 40 pp.
Petersen, R. C.; Fritsch, R. W. 1986. Connecticut's venomous snakes: The timber rattlesnake and northern copperhead. Bull. 111. Hartford: Geological and Natural History Survey of Connecticut. 48 pp.
Petranka, J. W. 1989. Density-dependent growth and survival of larval *Ambystoma*: Evidence from whole-pond manipulations. *Ecology* 70:1752–1767.
Petranka, J. W. 1990. Observations on nest site selection, nest desertion, and embryonic survival in marbled salamander. *J. Herpetol.* 24:229–234.
Petranka, J. W. 1998. *Salamanders of the United States and*

Canada. Washington, D.C.: Smithsonian Institution Press. 587 pp.
Pluto, T. G.; Bellis, E. D. 1986. Habitat utilization by the turtle, *Graptemys geographica* along a river. *J. Herpetol.* 20:22–31.
Pluto, T. G.; Bellis, E. D. 1988. Seasonal and annual movements of riverine map turtles, *Graptemys geographica. J. Herpetol.* 22:152–158.
Pope, C. H. 1939. *Turtles of the United States and Canada.* New York: Alfred A. Knopf.
Pope, C. H. 1944. *Amphibians and reptiles of the Chicago area.* Chicago: Chicago Natural History Museum Press.
Possardt, E. E. 1974. The breeding biology and larval development of the wood frog (*Rana sylvatica*). Department of Forest and Wildlife Management, University of Massachusetts, Amherst [Unpublished].
Pough, F. H. 1976. Acid precipitation and embryonic mortality of spotted salamanders (*Ambystoma maculatum*). *Science* 192:68–70.
Pough, F. H.; Smith, E. M.; Rhodes, D. H.; Collazo, A. 1987. The abundance of salamanders in forest stands with different histories of disturbance. *Forest Ecol. Manage.* 20:1–9.
Pough, F. H.; Wilson, R. E. 1976. Acid precipitation and reproductive success of *Ambystoma* salamanders. Pages 531–544. In: *Proc. 1st International Symposium on Acid Precipitation and the Forest Ecosystem,* 1975, May 12–15, Columbus, Ohio. Gen. Tech. Rep. NE-23. Broomall, Pa.: U.S. Department of Agriculture, Forest Service, Northeastern Forest Experiment Station.
Powders, V. N.; Tietjen, W. L. 1974. The comparative food habits of sympatric and allopatric salamanders, *Plethodon glutinosus* and *Plethodon jordani* in eastern Tennessee and adjacent areas. *Herpetologica* 30:167–175.
Rand, A. S. 1950. Leopard frogs in caves in winter. *Copeia* 1950:324.
Raney, E. C. 1940. Summer movements of the bullfrog, *Rana catesbeiana* (Shaw), as determined by the jaw-tag method. *Am. Midl. Naturalist* 23:733–745.
Raney, E. C.; Roecker, R. M. 1947. Food and growth of two species of watersnakes from western New York. *Copeia* 1947:171–174.
Reinert, H. K. 1984. Habitat separation between sympatric snake populations. *Ecology* 65:478–486.
Richmond, N. D. 1947. Life history of *Scaphiopus h. holbrookii* (Harlan). Pt. I: Larval development and behavior. *Ecology* 28:53–67.
Ries, K. M.; Bellis, E. D. 1966. Spring food habitats of the red-spotted newt in Pennsylvania. *Herpetologica* 22:152–155.
Risely, P. L. 1932. Observations on the natural history of the common musk turtle, *Sternotherus odoratus* (Latreille). *Papers Mich. Acad. Sci. Arts Lett.* 17:685–711.
Rossman, D. A.; Myer, P. A. 1990. Behavioral and morphological adaptations for snail extraction in the North American brown snakes (Genus *Storeria*). *J. Herpetol.* 24:434–438.
Sayler, A. 1966. The reproductive ecology of the red-backed salamander, *Plethodon cinereus,* in Maryland. *Copeia* 1966:183–193.
Schlauch, F. C. 1975. Agonistic behavior in a suburban Long Island population of the smooth green snake, *Opheodrys vernalis. Engelhardtia* 6:25–26.
Seibert, H. C. 1989. *Ambystoma opacum* (Gravenhorst), marbled salamander. Pages 125–131. In: Pfingsten, R. A.; Downs, F. L. (editors). *Salamanders of Ohio.* Columbus: Ohio State University Press. 315 pp.
Seibert, H. C.; Hagen, C. W., Jr. 1947. Studies on a population of snakes in Illinois. *Copeia* 1947:1:6–22.
Semlitsch, R. D. 1980a. Geographic and local variation in population parameters of the slimy salamander (*Plethodon glutinosus*). *Herpetologica* 36:6–16.
Semlitsch, R. D. 1980b. Terrestrial activity and summer home range of the mole salamander (*Ambystoma talpoideum*). *Can. J. Zool.* 59:315–322.
Sever, D. M. 1979. Male secondary sexual characteristics of the *Eurycea bislineata* (Amphibia, Urodela, Plethodontidae) complex in the southern Appalachian Mountains. *J. Herpetol.* 13:245–253.
Shoop, C. R. 1965. Orientation of *Ambystoma maculatum:* Movements to and from breeding ponds. *Science* 149:558–559.
Shoop, C. R. 1968. Migratory orientation of *Ambystoma maculatum:* Movements near breeding ponds and displacements of migrating individuals. *Biol. Bull.* 135:230–238.
Shoop, C. R. 1974. Yearly variation in larval survival of *Ambystoma maculatum. Ecology* 55:440–444.
Shoop, C. R.; Gunning, G. E. 1967. Seasonal activity and movements of *Necturus* in Louisiana. *Copeia* 1967:732–737.
Smith, H. M. 1946. *Handbook of lizards.* Ithaca, N.Y.: Comstock.
Smith, H. M. 1956. *Handbook of amphibians and reptiles of Kansas.* 2nd ed. Misc. Publ. No. 9. Topeka: University of Kansas, Museum of Natural History.
Smith, P. W. 1961. The amphibians and reptiles of Illinois. *Ill. Nat. Hist. Surv. Bull.* 28:1–298.
Stenhouse, S. L. 1985. Migratory orientation and homing in *Ambystoma maculatum* and *Ambystoma opacum. Copeia* 1985:631–637.
Stewart, M. M. 1956a. Certain aspects of the natural history and development of the northern two-lined salamander, *Eurycea b. bislineata* (Green), in the Ithaca, New York region. Ph.D. dissertation, Cornell University, Ithaca, N.Y.
Stewart, M. M. 1956b. The separate effects of food and temperature differences on development of marbled salamander larvae. *J. Elisha Mitchell Sci. Soc.* 72:47–56.
Stewart, M. M. 1961. Biology of the Allegany Indian reservation and vicinity. Part 3. The amphibians, reptiles, and mammals. *N.Y. State Mus. Sci. Bull.* 383:63–88.
Stewart, M. M. 1975. Habitat management in the Adirondack Park. *N.Y. Environ. News* 2(17):1–2.
Stewart, M. M.; Rossi, J. 1981. The Albany Pine Bush: A northern outpost for southern species of amphibians and reptiles in New York. *Am. Midl. Naturalist* 106:282–292.
Stewart, M. M.; Sandison, P. 1972. Comparative food habits of sympatric mink frogs, bullfrogs, and green frogs. *J. Herpetol.* 6:241–244.
Stickel, L. F. 1950. Population and home range relationships of the box turtle, *Terrapene c. carolina* (Linnaeus). *Ecol. Monogr.* 20:351–378.
Stickel, L. F.; Stickel, W. H.; Schmid, F. C. 1980. Ecology of a Maryland population of black rat snakes (*Elaphe o. obsoleta*). *Am. Midl. Naturalist* 103:1–14.
Stickel, W. H.; Cope, J. B. 1947. The home ranges and wanderings of snakes. *Copeia* 1947:127–136.

Stille, W. T. 1952. The nocturnal amphibian fauna of the southern Lake Michigan beach. *Ecology* 33:149–162.
Stille, W. T. 1954. Eggs of the salamander *Ambystoma jeffersonianum* in the Chicago area. *Copeia* 1954:300.
Stockwell, S. S.; Hunter, M. L., Jr. 1989. Relative abundance of herpetofauna among eight types of Maine peatland vegetation. *J. Herpetol.* 23:409–414.
Stone, W. B.; Kiviat, E.; Butkas, S. A. 1980. Toxicants in snapping turtles. *N.Y. Fish Game J.* 27:39–50.
Storer, D. H. 1840. A report on the reptiles of Massachusetts. *Boston J. Nat. Hist.* 3:1–64.
Strang, C. A. 1983. Spatial and temporal activity patterns in two territorial turtles. *J. Herpetol.* 17:43–47.
Surface, H. A. 1906. The serpents of Pennsylvania. *Penn. Dept. Agric. Div. Zool. Bull.* 4:113–303.
Surface, H. A. 1908. First report on the economic features of the turtles of Pennsylvania. *Penn. Dept. Agric. Div. Zool. Bull.* 6:105–196.
Surface, H. A. 1913. The amphibians of Pennsylvania. *Penn. Dept. Agric. Div. Zool. Bull.* 3:65–152, 1–11.
Taub, F. B. 1961. The distribution of the red-backed salamander, *Plethodon c. cinereus,* within the soil. *Ecology* 42:681–698.
Taylor, G. M.; Nol, E. 1989. Movements and hibernation sites of overwintering painted turtles in southern Ontario.*Can. J. Zool.* 67:1877–1881.
Taylor, J. 1993. *The amphibians and reptiles of New Hampshire.* Concord: New Hampshire Fish and Game Department. 71 pp.
Thompson, E. L.; Gates, J. E.; Taylor, G. J. 1980. Distribution and breeding habitat selection of the Jefferson salamander, *Ambystoma jeffersonianum,* in Maryland. *J. Herpetol.* 14:13–20.
Thompson, E. L.; Gates, J. E. 1982. Breeding pool segregation by the mole salamanders, *Ambystoma jeffersonianum* and *A. maculatum,* in a region of sympatry. *Oikos* 38:273–279.
Tilley, S. G. 1968. Size-fecundity relationships and their evolutionary implications in five desmognathine salamanders. *Evolution* 22:806–816.
Tilley, S. G.; Lundrigan, B. L.; Brower, L. P. 1982. Erythrism and mimicry in the salamander *Plethodon cinereus. Herpetologica* 38(3):409–417.
Tinkle, D. W. 1961. Geographic variation in reproduction, size, sex ratio, and maturity of *Sternotherus odoratus* (Testudinata: Chelydridae). *Ecology* 42:68–76.
Trapido, H.; Clausen, R. T. 1938. Amphibians and reptiles of eastern Quebec. *Copeia* 1938:117–125.
Uhler, F. M.; Cottom, C.; Clarke, T. E. 1939. Food of snakes of the George Washington National Forest, Virginia. *Trans. Am. Nat. Resour. Wildl. Conf.* 4:605–622.
Uzzell, T. M., Jr. 1964. Relations of the diploid and triploid species of the *Ambystoma jeffersonianum* complex (Amphibia, Caudata). *Copeia* 1964:257–300.
Uzzell, T. M., Jr. 1967a. *Ambystoma jeffersonianum. Catalogue of American Amphibians and Reptiles* 47.1–47.2.
Uzzell, T. M., Jr. 1967b. *Ambystoma laterale. Catalogue of American Amphibians and Reptiles* 48.1–48.2.
Uzzell, T. M., Jr. 1967c. *Ambystoma platineum. Catalogue of American Amphibians and Reptiles* 49.1–49.2.
Vickery, P. 1992. Racer. Pages 125–127. In: Hunter, M. L., Jr.; Albright, J.; Arbuckle, J. (editors). *The amphibians and reptiles of Maine.* Maine Agric. Exp. Sta. Bull. 838. Orono: University of Maine. 188 pp.
Vinegar, A.; Friedman, M. 1967. *Necturus* in Rhode Island. *Herpetologica* 23:51.
Wacasey, J. W. 1961. An ecological study of two sympatric species of salamanders, *Ambystoma maculatum* and *Ambystoma jeffersonianum,* in southern Michigan. Ph.D. dissertation, Michigan State University, East Lansing.
Warfel, H. E. 1936. Notes on the occurrence of *Necturus maculosus* (Rafinesque) in Massachusetts. *Copeia* 1936:237.
Wells, K. D. 1976. Multiple egg clutches in the green frog (*Rana clamitans*). *Herpetologica* 32:85–87.
Wells, K. D. 1977. Territoriality and male mating success in the green frog (*Rana clamitans*). *Ecology* 58:750–762.
Wells, K. D.; Wells, R. A. 1976. Patterns of movement in a population of the slimy salamander, *Plethodon glutinosus,* with observations on aggregations. *Herpetologica* 32:156–162.
Werner, J. K. 1971. Notes on the reproductive cycle of *Plethodon cinereus* in Michigan. *Copeia* 1971:161–162.
Whitford, A. G.; Vinegar, A. 1966. Homing, survivorship, and overwintering of larvae in *Ambystoma maculatum. Copeia* 1966:515–519.
Whitlock, A. L.; Jarmon, N. M.; Medina, J. A.; Larson, J. S. 1994. *Wethings: Wetland habitat indicators for nongame species.* Amherst: The Environmental Institute, University of Massachusetts. 627 pp.
Wilbur, H. M. 1972. Competition, predation, and the structure of the *Ambystoma-Rana sylvatica* community. *Ecology* 53:3–21.
Wilbur, H. M. 1977. Propagule size, number, and dispersion pattern in *Ambystoma* and *Asclepias. The American Naturalist* 111:43–68.
Wilder, I. W. 1913. The life history of *Desmognathus fusca. Biol. Bull.* 24:251–342.
Wilder, I. W. 1917. On the breeding habits of *Desmognathus fusca. Biol. Bull.* 32:13–20.
Wilder, I. W. 1924. The relation of growth to metamorphosis in *Eurycea bislineata* (Green). *J. Exp. Zool.* 40:1–112.
Williams, J. E. 1952. Homing behavior of the painted turtle and musk turtle in a lake. *Copeia* 1952:76–82.
Williams, K. L. 1978. Systematics and natural history of the American milk snake, *Lampropeltis triangulum.* Milwaukee, Wis.: Milwaukee Public Museum Press.
Williams, P. K. 1973. Seasonal movements and population dynamics of four sympatric mole salamanders, Genus *Ambystoma.* Ph.D. dissertation, University of Indiana, Bloomington. 46 pp.
Williams, T. K.; Christiansen, J. L. 1982. The niches of two sympatric softshell turtles, *Trionyx muticus* and *Trionyx spiniferus,* in Iowa. *J. Herpetol.* 15:303–308.
Wilson, R. E. 1976. An ecological study of *Ambystoma maculatum* and *Ambystoma jeffersonianum.* Ph.D. dissertation, Cornell University, Ithaca, N.Y.
Witham, J. W. 1992. Redback salamander. Pages 46–50. In: Hunter, M. L., Jr.; Albright, J.; Arbuckle, J. (editors). *The amphibians and reptiles of Maine.* Maine Agric. Exp. Sta. Bull. 838. Orono: University of Maine. 188 pp.
Wood, T. J. 1953. Observations on the complements of ova and nesting of the four-toed salamander in Virginia. *Am. Naturalist* 87:77–86.
Woodward, B. S. 1982. Local intraspecific variation in clutch

parameters in the spotted salamander (*Ambystoma maculatum*). *Copeia* 1982:157–160.

Wright, A. H. 1914. *Life-histories of the anura of Ithaca, New York*. Washington, D.C.: Carnegie Institution of Washington. 98 pp.

Wright, A. H.; Allen, A. A. 1909. The early breeding habits of *Ambystoma punctatum*. *Am. Midl. Naturalist* 43:687–692.

Wright, A. H.; Wright, A. A. 1949. *Handbook of frogs and toads*. Ithaca, N.Y.: Comstock.

Wright, A. H.; Wright, A. A. 1957. *Handbook of snakes*. 2 vols. Ithaca, N.Y.: Comstock.

Wyman, R. L. 1988. Soil acidity and moisture and the distribution of amphibians in five forests of southcentral New York. *Copeia* 1988:394–399.

Zappalorti, R. T.; Farrell, R. F. 1980. An ecological study of the bog turtle, *Clemmys muhlenbergii* Schoepff (Reptilia, Testudines, Emydidae), in New Jersey. Pt. 3. Report to the New Jersey Department of Environmental Protection, Endangered and Nongame Species Project, Federal Aid Program and Herpetological Association, HA Rept. No. 80.01 [Unpublished].

Zappalorti, R. T.; Farrell, R. F.; Zanelli, E. M. 1979. The ecology and distribution of the bog turtle, *Clemmys muhlenbergii* (Schoepff), in New Jersey. Pt. 2. Report to the New Jersey Department of Environmental Protection, Endangered and Nongame Species Project, Federal Aid Program and Herpetological Association, HA Rept. No. 79.02, Vol. 1. 38 pp. [Unpublished].

Zehr, D. R. 1962. Stages in the development of the common garter snake, *Thamnophis sirtalis*. *Copeia* 1962:322–329.

Zenisek, C. J. 1964. A study of the natural history and ecology of the leopard frog, *Rana pipiens* Schreber. Ph.D. dissertation, Ohio State University, Columbus.

Birds

This section provides information on the life history, distribution, and habitat associations of birds that are resident, breed, or winter in New England. *Resident* species are present year round in the region, although different populations may occur from season to season. *Breeding* species leave the region during the nonbreeding season; species present during the (summer) breeding season may be birds in passage or summer visitors. *Wintering* species reside in the region during the nonbreeding season. Nomenclature follows the *Check-list of North American Birds*, 7th edition (American Ornithologists' Union 1998); species are arranged in phylogenetic order. All native and exotic inland species are included, as are marshland birds that occur along the coast and other coastal species that winter or occur as inland nonbreeders. We have included several species that regularly occur very locally or in low numbers, such as the gray partridge (*Perdix perdix*), which occurs only in extreme northwestern Vermont, and the yellow-breasted chat (*Icteria virens*), now much reduced and found predictably only in southwestern Connecticut and possibly in southern Rhode Island. We have omitted the prothonotary warbler (*Protonotaria citrea*), a straggler to New England and an exceedingly rare breeder. A single nesting record (in 1946) exists for Connecticut (Bull 1964). Occasional singing males are reported in eastern Massachusetts (see Veit and Petersen 1993:407), and a probable nesting occurred in Concord in 1886 (Griscom 1949). We have also omitted Henslow's sparrow (*Ammodramus henslowii*), which nested in the region in 1983 (Veit and Petersen 1993:443) and 1994 (Ells 1995).

For information on bird taxonomy and nomenclature, contact the American Ornithologists' Union at: <http://pica.wru.umt.edu/aou/birdlist.html>.

For information on breeding bird distribution and population trends, contact the USGS Breeding Bird Survey at: <www.mbr-pwrc.usgs.gov/bbs/bbs.html>.

Information in this section was obtained from extensive literature reviews and continuing research on forest bird habitat relationships, such as DeGraaf et al. (1998). Birds are the best-known class of vertebrates included in this volume. New England has a rich avifauna that has been studied more fully than that of any other region of the continental United States.

Literature Cited

American Ornithologists' Union. 1998. *Check-list of North American Birds*. 7th ed. Washington, D.C.: American Ornithologists' Union. 829 pp.

Bull, J. 1964. *Birds of the New York area*. New York: Harper and Row. 540 pp.

DeGraaf, R. M.; Hestbeck, J. B.; Yamasaki, M. 1998. Associations between breeding bird abundance and stand structure in the White Mountains, New Hampshire and Maine, USA. *Forest Ecol. Manage.* 103:217–233.

Ells, S. F. 1995. Breeding Henslow's sparrow in Massachusetts, 1994. *Bird Observer* 23:113–115.

Griscom, L. 1949. *The Birds of Concord*. Cambridge, Mass.: Harvard University Press. 340 pp.

Veit, R. R.; Petersen, W. R. 1993. *Birds of Massachusetts*. Lincoln: Massachusetts Audubon Society. 514 pp.

Sources used to prepare maps of species distributions and status in addition to expert reviews include:

Atlas of Breeding Birds in Connecticut. 1994. L. R. Bevier, editor. Bull. 113. Hartford: Geological and Natural History Survey of Connecticut. 461 pp.

Birds of Massachusetts. 1993. R. R. Veit and W. R. Petersen. Lincoln: Massachusetts Audubon Society. 514 pp.

Atlas of Breeding Birds in Maine. 1978–1983. P. Adamus. Augusta: Maine Department of Inland Fisheries and Wildlife. 366 pp.

The Atlas of Breeding Birds of Vermont. 1985. S. B. Laughlin and D. B. Kibbe (editors). Hanover, N.H.: University Press of New England. 456 pp.

The Atlas of Breeding Birds in Rhode Island. 1992. R. W. Enser. Providence: Rhode Island Department of Environmental Management. 206 pp.

Atlas of Breeding Birds in New Hampshire. 1994. C. R. Foss (editor). Dover, N.H: Arcadia. 414 pp.

Bird Species

The following list is made up of references on birds for further reading on life histories and distribution.

Citations by Region

North America

Askins, R. A. 2000. *Restoring North America's birds: Lessons from landscape ecology*. New Haven, Conn.: Yale University Press. 288 pp.

Bellrose, F. C. 1976. *Ducks, geese, and swans of North America*. Harrisburg, Pa.: Stackpole Books. 544 pp.

DeGraaf, R. M.; Rappole, J. H. 1995. *Neotropical migratory birds: Natural history, distribution, and population change*. Ithaca, N.Y.: Cornell University Press. 676 pp.

Godfrey, W. E. 1979. *The birds of Canada*. Ottawa, Ont.: National Museum of Natural Sciences. 428 pp.

Morse, D. H. 1989. *American warblers: An ecological and behavioral perspective*. Cambridge, Mass.: Harvard University Press. 406 pp.

Poole, A., et al. (editors). Various dates. *The birds of North America*. Philadelphia: Academy of Natural Sciences; Washington, D.C.: American Ornithologists' Union.

Northeastern United States

Andrle, R. F.; Carroll, J. R. (editors). 1987. *The atlas of breeding birds in New York State*. Ithaca, N.Y.: Cornell University Press. 551 pp.

Foss, C. R. (editor). 1994. *Atlas of breeding birds in New Hampshire*. Dover, N.H.: Arcadia. 414 pp.

Laughlin, S. B.; Kibbe, D. P. (editors). 1985. *The atlas of breeding birds of Vermont*. Hanover, N.H.: University Press of New England. 456 pp.

Veit, R. R.; Petersen, W. R. 1993. *Birds of Massachusetts*. Lincoln: Massachusetts Audubon Society. 514 pp.

Zeranski, J. D.; Baptist, T. R. 1990. *Connecticut birds*. Hanover, N.H.: University Press of New England. 328 pp.

Common Loon

(*Gavia immer*)

Range: Breeding: Alaska and n. Canada to Iceland, s. to central Massachusetts, Montana, and California. Winter: Atlantic Coast from Newfoundland to the Gulf of Mexico.

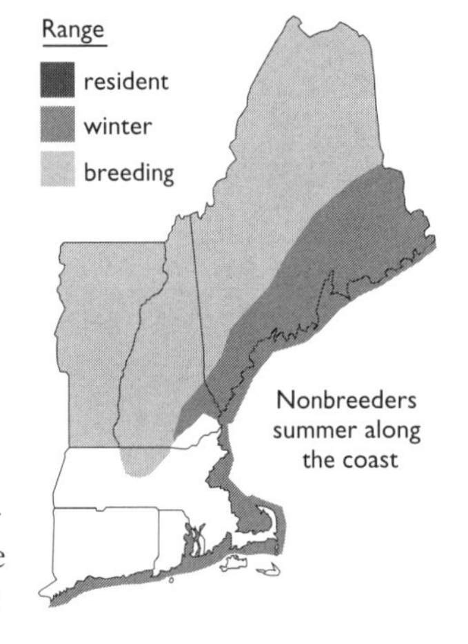

Distribution in New England: Throughout Maine, Vermont, and New Hampshire. In central Massachusetts in the Quabbin and Wachusett reservoirs (Rimmer 1992). Reports of nesting in this century in Connecticut (Litchfield County), but these reports lack supporting details (Zeranski and Baptist 1990:34–35).

Status in New England: Uncommon to locally common breeder; common migrant and coastal resident in winter. Subadults common along northeast coast in summer.

Habitat: Breeding: Large and small oligotrophic lakes with fish in both open and densely forested areas (McIntyre and Barr 1997). Nests on lakes as small as 2 acres, prefers lakes with islands. In Vermont, nesting habitat associated with lakes in spruce-fir or spruce-fir northern hardwood transition zones (Fichtel 1985:30 in Laughlin and Kibbe 1985). Wintering: Coastal bays and inlets from Maritime Provinces south. Occasional on fresh water inland in s. New England until freeze-up. After wintering on the coast, loons arrive on freshwater lakes soon after ice-out.

Special Habitat Requirements: Bodies of water with stable water levels and little or no human disturbance. Long stretch of water for flight take-off. Islets for nesting; shallow coves for rearing of young (Hammond and Wood 1977).

Nesting: Egg dates: June 2 to August 10 in Maine and New Hampshire (Veit and Petersen 1993:50). May 15 to July 16; peak: June, New York (Bull 1974:51). Clutch size: 1 to 3, typically 2. Incubation period: 28 to 29 days; 25 to 33 (average 28) days in Vermont (Fichtel 1985:30 in Laughlin and Kibbe 1985). Nestling period: 1 day (precocial). Broods per year: 1. Will renest after nest failure. Nest site: Nest is placed on ground at water's edge, usually on sand, rocks, or other firm substrate. Prefers small islands to shore (Sutcliffe 1980) but nests along protected bays, on promontories and small peninsulas. Islands provide better protection from mammalian predators than do shore sites.

Territory Size: 15 to 100 or more acres (6.1 to 40.5 ha) per pair in Minnesota (Olson and Marshall 1952), to 25 ha (62 acres) (Sjolander and Agren 1972). Although loons may attempt to nest on ponds as small as 5 to 6 ha (12 to 15 acres), lakes smaller than 80 ha (198 acres) generally support a single pair (Richards and Elkins 1994:2 in Foss 1994). On larger lakes, territories range from 22 to 415 acres (9 to 168 ha) (McIntyre 1988). Territories are larger on large lakes (Titus and VanDruff 1981). Nesting loons are strongly territorial.

Sample Densities: Average 2 birds per square mile (0.8 birds/km^2) over a 60 square mile (155 km^2) area in Minnesota (Olson and Marshall 1952), but highly variable depending on nest site availability.

Food Habits: Live fish (staple), also amphibians, insects, aquatic plants, crustaceans, mollusks, leeches, taken by diving in deep water.

Comments: Common loons nested in Connecticut as late as 1890 (Sage et al. 1913) but declined in the late 1800s due to hunting and recreation on lakes (Zeranski and Baptist 1990:35). Most feeding is done on the territory, which probably accounts for the large territory. Although fish is eaten in quantity when available, it does not seem to be a required food because loons are known to breed at fishless ponds (Pough 1951:3). The same nest sites are often occupied year after year, presumably by the same pair. Disturbance by canoeists and fishermen has been a serious problem during nesting. Nesting is often unsuccessful if water level fluctuates. Artificial nest rafts were placed on Squam Lake in New Hampshire by the Loon Preservation Committee from 1980 to 1989: 60 percent of the rafts were used (46/77) and hatched 55 percent of all chicks hatched during that time (Noon 1990). Conservation concerns include shoreline development, recreation, water-level fluctuation, and water quality (Fichtel 1985:30 in Laughlin and Kibbe 1985). The loon's yodel is the sound of quiet, remote lakes to many New Englanders; it is the territorial call, given only by the male (McIntyre and Barr 1997).

Pied-billed Grebe

(*Podilymbus podiceps*)

Range: Breeding: Southeastern Alaska and from central Canada s. locally through temperate North America. Winter: Through most of breeding range from s. British Columbia and the central United States s., casually farther n. Northern birds winter s. to Panama.

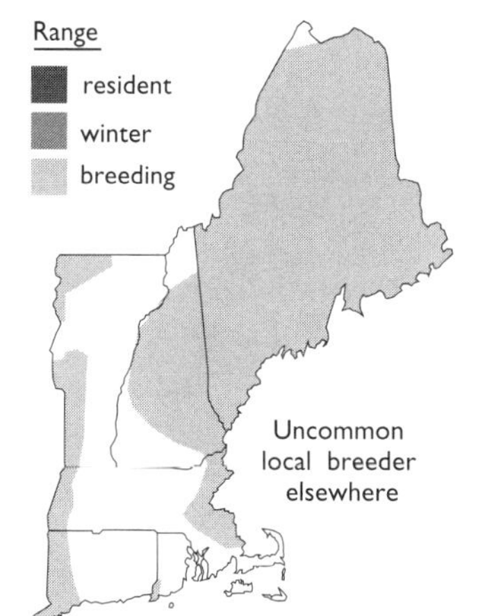

Distribution in New England: Nesting sites are locally distributed in Vermont and New Hampshire. Located throughout Maine except the most northern tip. Nests on the coast of ne. Massachusetts. In Connecticut, a rare nester w. of the Connecticut River (Zeranski and Baptist 1990). Extirpated as a breeding species in Rhode Island (Gibbs and Melvin 1992b). In mild winters, may occur on Cape Cod and the Islands.

Status in New England: Locally common; studies suggest that the population is declining throughout the Northeast (see Gibbs and Melvin 1992b). Rare in winter.

Habitat: Breeding: Freshwater ponds with large areas of emergent vegetation, marshes, and marshy inlets with areas of open water (Faaborg 1976), marshy edges of rivers and lakes, reed-bordered swamps with open water. Inhabits beaver ponds and human impoundments (Gibbs and Melvin 1992b). In Maine, found only on wetlands at least 12.5 acres (5 ha) (Gibbs and Melvin 1992b). Wintering: Interior rivers and open lakes, tidal creeks and estuaries.

Special Habitat Requirements: Fresh marshes, ponds, greater than 12.5 acres (5 ha).

Nesting: Egg dates: April 23 to June 28 in Massachusetts (Veit and Petersen 1993:51). April 21 to July 2, New York (Bull 1974:56). Clutch size: 2 to 10, typically 4 to 8. Incubation period: 23 to 24 days. Nestling period: 1 or 2 days (precocial). Broods per year: 1. Nest site: Usually built over shallow water anchored to the stems of emergent vegetation, less often located in shrubs such as sweet gale and buttonbush. A solitary nester, but occasionally several nests are widely spaced at large ponds (Palmer 1962:108).

Territory Size: Defends area within a 150-ft (45.7 m) radius (or more) of nest in Iowa potholes (Glover 1953).

Home Range: About twice the size of nesting territory (Glover 1953).

Sample Densities: Generally 1 pair on ponds up to 10 acres (4 ha) (Palmer 1962:108). Nests may be spaced 15 to 30 ft (4.6 to 9.1 m) apart on greater than 10 acres (4 ha) in size (Faaborg 1976).

Food Habits: Aquatic insects, small fish, snails, aquatic worms, crayfish, shrimp, amphibians, leeches, minor amounts of aquatic vegetation.

Comments: Pied-billed grebes have the widest distribution in the Americas of any grebe (Muller and Storer 1999). They are among the earliest arriving waterbirds in the spring; in mild winters they may winter on the Cape and Islands (Veit and Petersen 1993:51). This species was apparently never a common nester in Connecticut, and its abundance in s. New England probably declined after 1850 with the draining and filling of marshy ponds (see Zeranski and Baptist 1990:35 for review). Loss and alteration of wetland habitats through draining, filling, pollution, and siltation cited as the primary factors in population declines in the Northeast (Gibbs and Melvin 1992b). This is the only grebe that nests in New Hampshire (Vernon 1994:4 in Foss 1994). Birds are solitary or seen in pairs or small groups (to 10 individuals) outside of breeding season. Seldom seen in flight; it migrates at night, lands on the water at or before dawn, and prefers to escape danger by crash-diving or by stealthily sinking from view (Muller and Storer 1999).

Double-crested Cormorant

(*Phalacrocorax auritus*)

Range: Breeds in the se. Bering Sea, s. Alaska, and from sw. British Columbia and n. Alberta to Newfoundland, s. along the Atlantic and Pacific Coasts; very locally throughout the interior of North America. Winter: Along the Pacific Coast from the Aleutians and s. Alaska, s. to Baja California and Guerrero, Mexico; common throughout most of Mexico. On the Atlantic Coast from New England s.; in the Mississippi and Rio Grande Valleys; and along the Gulf Coast s. to Central America. Sightings are increasing in Puerto Rico. Northernmost Bahamas, Cuba, and Isle of Pines; vagrant elsewhere in the West Indies east to Guadeloupe.

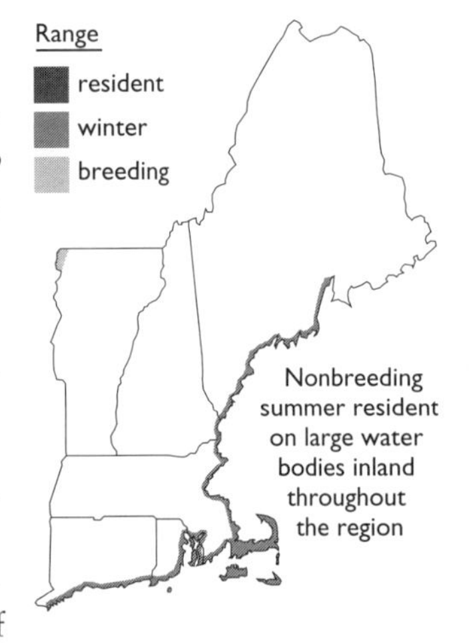

Distribution in New England: Coastal Maine (Krohn et al. 1995), Massachusetts, New Hampshire (Borror 1994a), Connecticut, and Rhode Island, except absent from Block Island. Also in Lake Champlain on Young Island and other islands (Kibbe and Laughlin 1985:391 in Laughlin and Kibbe 1985).

Status in New England: Abundant and increasing breeder and migrant, uncommon but increasing in winter.

Habitat: Coastal bays, estuaries, marine islands, freshwater lakes, ponds, rivers, sloughs, and swamps. Prefers perching on rocks, buoys, in trees, or on other objects that overhang or project from the water.

Special Habitat Requirements: Undisturbed nesting sites and a convenient, dependable food source within a foraging radius of 10 to 15 km.

Nesting: Egg dates: April 5 to late June in Massachusetts (Veit and Petersen 1993:70). Clutch size: 2 to 7, typically 3 to 4. Incubation period: 25 to 29 days. Nestling period: 5 to 6 weeks, independent at 10 weeks. Broods per year: 1. Age at sexual maturity: 2 or 3 yr. Nest height: On the ground or in an available tree. Nest site: In colonies on rocky islands, cliffs facing water, or in stands of live or dead trees in or near water. Nest consists of a foundation of sticks, seaweed, and debris, with a lining of finer material.

Food Habits: Primarily saltwater fishes including secondary consumption (Blackwell and Sinclair 1995), also some freshwater fishes, mollusks, crustaceans, salamanders, and seaworms taken by diving and swimming in deep water.

Comments: Double-crested cormorants nested on the Boston Harbor islands in the 16th century, as evidenced by bones of adults and young in Indian middens (Hatch 1982). The birds were persecuted from early colonial times because they were believed to compete for coastal fish resources; East Coast populations reached the low point in the 1920s (Zeranski and Baptist 1990:44) and were extirpated as breeders in s. New England by the early nineteenth century. A pronounced population increase began in New England in the mid twentieth century (Veit and Petersen 1993:69). The same nests are often occupied year after year. Destructive of trees and vegetation in nesting colonies (Nettleship and Duffy 1995). Early nesters comprise the center of a nesting colony; late nesters the periphery. Early nesters lay larger clutches and fledge more young per successful brood than do late nesters (McNeil and Leger 1987).

American Bittern

(Botaurus lentiginosus)

Range: Breeding: Extreme se. Alaska, central British Columbia, and sw. Northwest Territories e. to central Quebec and Newfoundland, s. to s. California, New Mexico, Texas, and Florida; breeds rarely s. of n. California, Utah, Great Plains states, Ohio River Valley, and Virginia. Winter: Southern British Columbia, Utah, New Mexico, central parts of Gulf States, and s. New England s. to s. California, Gulf of Mexico, and along Atlantic Coast. In Mexico winters locally mainly in the lowlands. A rare migrant and winter visitor in Guatemala. Winters in small numbers in the Bahamas and Greater Antilles, in particular Cuba; rare winter visitor to Puerto Rico; casual in Lesser Antilles.

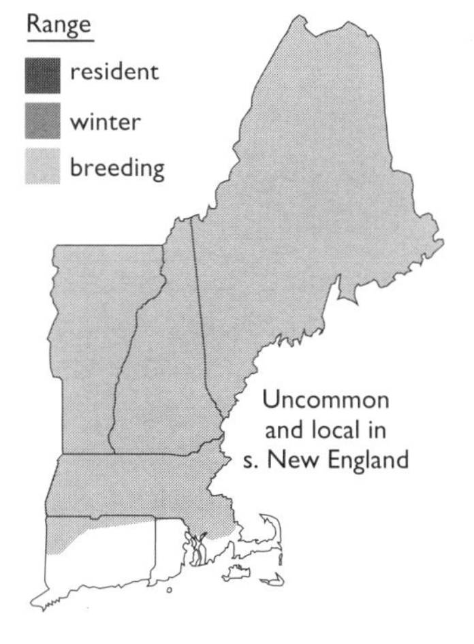

Distribution in New England: Northwestern Connecticut, Rhode Island on Block Island (Enser 1992), Massachusetts except Cape Cod and Franklin County, throughout Vermont (although most abundant in the Champlain Lowlands), New Hampshire, and Maine.

Status in New England: Locally common breeder; declining due to loss of freshwater and brackish wetlands. Uncommon migrant; rare in winter.

Habitat: Large freshwater or saltwater marshes, scrub-shrub swamps, emergent wetlands, and where tall, emergent vegetation is present such as cattails, bulrushes, sedges, and reeds. Nests almost exclusively in large cattail marshes in New England. Occasionally nests in wet fields or upland fields adjacent to water (Duebbert and Lokemoen 1977). Prefers impoundments and beaver-created wetlands to those created by glacial activity (Gibbs et al. 1992). Inhabited wetlands <2.5 to 62.5 acres (<1 to 25 ha), but were more abundant in the larger wetlands (Gibbs and Melvin 1990, 1992a).

Special Habitat Requirements: Large cattail or other freshwater wetlands with tall, dense emergent vegetation.

Nesting: Egg dates: May 1 to June 13, in Massachusetts (Veit and Petersen 1993:73). May 10 to June 29, New York (Bull 1974:90). Clutch size: 3 to 7, typically 4 to 5. Incubation period: About 24 days for each egg. Incubation commences with the first egg laid; consequently, the length of the incubation period depends on the number of eggs in the clutch (Kibbe 1985a:34 in Laughlin and Kibbe 1985). Nestling period: About 14 days. Broods per year: Probably 1. Age at sexual maturity: Unknown (Palmer 1962:502). Nest site: On a flimsy platform of cattails, reeds, or sedges placed in dense emergent vegetation just above water. Occasionally on the ground in wet meadows or in dry fields adjacent to water (Duebbert and Lokemoen 1977).

Sample Densities: 5 nests on one slough less than 160 acres (64.8 ha) in Saskatchewan (Bent 1926:75). 2 nests on 5 acres (2.0 ha) of cordgrass meadow in Minnesota (Vesall 1940).

Food Habits: Amphibians, small snakes, crayfish, insects, small fish (Gibbs et al. 1992b) taken by standing and waiting, walking slowly (Kushlan 1976).

Comments: The American bittern is a very shy, solitary, and elusive heron. Declining since the 1960s due to loss of extensive cattail marshes.

Least Bittern

(*Ixobrychus exilis*)

Range: Breeding: Locally in w. North America from s. Oregon s. through California to Mexico; in e. North America from s. Manitoba, Ontario, Quebec, and New Brunswick s. to Texas and Florida. Resident and winter populations extend through Middle America and the Greater Antilles; also resident in South America e. of the Andes from Colombia and Venezuela s. to central Argentina; w. of the Andes in coastal Peru.

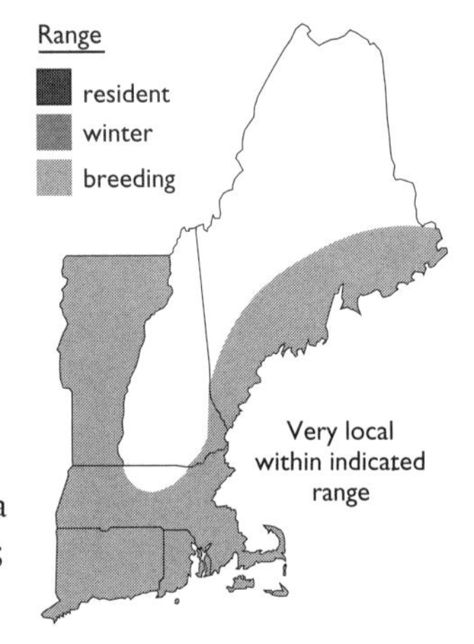

Distribution in New England: Primarily eastern Massachusetts and Rhode Island; also Connecticut (since 1982 has nested in Stratford, Waterford, Stonington, Durham, and South Windsor; Zeranski and Baptist 1990:45), Vermont (the Champlain lowlands, the n. Taconic Mountains, and the Eastern Foothills), and coastal Maine. No documented breeding records in New Hampshire since 1980, although possible nesting along the coast at Hampton and in Cascade Marsh, Sutton (Vernon 1994:370 in Foss 1994). Accidental in winter along the coasts of Connecticut and Rhode Island.

Status in New England: A rare and local breeder.

Habitat: Freshwater and brackish marshes with tall, dense vegetation such as cattails, sedges, reeds, bullrushes, sawgrass, smartweed, arrowhead, buttonbush, and other semiaquatic vegetation, and bogs. Also found at the edges of lakes and rivers with emergent and tall vegetation. Prefers marshes with scattered bushes or other woody growth. Will use managed impoundments and coves on lakes in Maine (Gibbs and Melvin 1990; Gibbs et al. 1991). Occasionally in coastal salt marshes and mangrove swamps.

Special Habitat Requirements: Freshwater wetlands with tall, dense vegetation.

Nesting: Egg dates: June 1–29, Massachusetts (Veit and Petersen 1993:73). May 15 to July 10, New York (Bull 1974:91). Clutch size: 3 to 6, typically 4 to 5. Incubation period: 17 to 18 days. Nestling period: 13 to 15 days. Broods per year: 1 or 2. One in New England (Forbush 1929 V.1:322). Age at sexual maturity: Unknown (Palmer 1962:496). Nest height: To 20 ft (6.1 m), usually 1–8 ft, typically 8 to 14 in (20.3 to 35.6 cm) above water level. Nest site: Single nests in dense stands of emergent vegetation approximately 6 in to 2.5 ft (15 to 76 cm) above water that is up to 3 ft (1 m) deep, and close to open water.

Territory Size: There is limited evidence of territoriality in least bitterns (Weller 1961).

Sample Densities: 15 nests were found in a 2-acre (0.8 ha) patch of rushes in Michigan (Wood 1951), but this was exceptional. 19 nests were found in a 44-acre (17.8 ha) marsh (Kent 1951 in Palmer 1962:496). 1 nest per 4 acres (1.6 ha) of useable vegetation (Beecher 1942). 0.4 calling males per 2.5 acres (1 ha) in a large marsh in Wisconsin (Manci and Rusch 1988).

Food Habits: Small fishes, crustations (mainly crayfish), frogs, snakes, salamander, leeches, slugs, occasionally small mammals (shrews and mice), and vegetable matter (Bent 1926, Weller 1961, Palmer 1962) taken by standing and waiting, walking slowly (Kushlan 1976), or by clinging to vegetation above water level and extending neck (Weller 1961).

Comments: The least bittern, the smallest member of the heron family, is considered to be among the most inconspicuous North American marsh birds. Vocal and colorful, it is seldom seen, even where common (Gibbs et al. 1992a). Preservation of wetlands larger than 12.5 acres (5 ha) with emergent vegetation at least 3 ft (>1 m) tall interspersed with open water 4 to 20 in (10 to 50 cm) deep is essential for this species to exist (Gibbs et al. 1992). Usually nests singly but may be colonial with other least bitterns in favorable habitat. Marsh filling and drainage, pollution, and insecticides have adversely affected parts of its range.

Great Blue Heron

(Ardea herodias)

Range: Breeding: From s. Alaska, coastal and s. British Columbia, se. Northwest Territories, and central Manitoba e. to Nova Scotia and s., locally, throughout the United States and much of Mexico to Guerrero, Veracruz, the Gulf Coast, and s. Florida, including the Keys, Cuba, the coast of the Yucatan Peninsula, and Los Roques of the n. Venezuelan coast. Winter: From s.-coastal Alaska, coastal British Columbia, central United States, and s. New England, s. throughout Mexico with stragglers to Colombia and Venezuela. Resident on islands from Aruba to Trinidad and Tobago and common during winter in Puerto Rico. A rare migrant or winter visitor in Trinidad, Tobago, the Caribbean lowlands, Honduras, Panama, Costa Rica, and Guatemala. Winters throughout the West Indies.

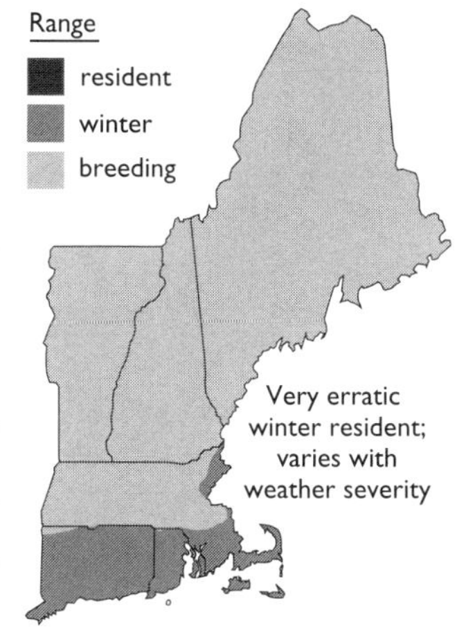

Distribution in New England: Breeding: Throughout Maine, New Hampshire, Vermont, and Connecticut, w. and central Massachusetts, and w. Rhode Island. Winter: In Massachusetts on the Outer Cape and Nantucket (Veit and Petersen 1993:74), coastal Rhode Island, and rare to uncommon along the coast of Connecticut, especially during severe winters (Zeranski and Baptist 1990:46).

Status in New England: Fairly common and increasing breeder; common migrant along the coast, uncommon in winter.

Habitat: Breeding: A variety of fresh and saltwater habitats including marshes, beaver impoundments, wet meadows, estuaries, tidal flats, sandbars, shallow bays, and the margins of lakes, ponds, streams, and rivers (Butler 1992). Winter: Coastal areas with bare (snow-free) ground and open water.

Special Habitat Requirements: Open water or wetland habitats, forested wetlands or tall trees near water in areas free from human disturbance.

Nesting: Egg dates: early April to late May (Veit and Petersen 1993:74). April 15 to June 9, New York (Bull 1974:73). Clutch size: 2 to 6, typically 4. Incubation period: About 28 days (Bent 1926:106). Nestling period: About 60 days (Pratt 1970). Broods per year: 1. Age at sexual maturity: 2 yr (Bent 1926:108). Nest height: To 130 ft (39.6 m), typically high in large trees. Nest site: Generally nests in colonies near water, preferably in isolated forested wetlands or on an island. Does not nest on offshore islands in Massachusetts. Builds nests at the tops of tall dead trees in New England, usually in isolated stands in beaver ponds. Birds are typically colonial nesters but may be solitary. Same nest often repaired for use each season. Nest may be up to 1.6 km from food sources.

Territory Size: Sizes vary with habitat and stage of reproductive cycle. Colonial nesters defend small areas often restricted to the distance the bird can extend its neck and bill from the nest (Cottrille and Cottrille 1958).

Sample Densities: Dozens of nests may be built in crown of a single tree. 131 active nests per 0.36 ha (0.9 acre) in Oregon (Werschkul et al. 1976).

Food Habits: Aquatic and terrestrial insects, fishes (primarily), amphibians, reptiles, crustaceans, occasionally small birds and mammals in shallow marshes, wet meadows, even upland fields in grasses and weeds. Stands and waits (Kushlan 1976, Willard 1977), walks slowly; active pursuit is infrequent.

Comments: The great blue heron, the largest heron in North America, was reported to be a "rare and extirpated summer resident" first found breeding at the Harvard Forest, Petersham, Mass. (Griscom and Snyder 1955:30). Individual feeding territories may be strongly defended during the nonbreeding season. Birds occasionally nest miles from feeding areas. Commonly seen feeding on tidal flats during fall migration.

Great Egret

(*Ardea alba*)

Range: Breeds from s. Oregon and Idaho s. to Arizona, and from Minnesota, s. Ontario, and Maine s. through Middle and South America to s. Chile and central Argentina. Resident in the West Indies. Winter: From n. California across the s. United States and s. along the Atlantic Coast from North Carolina s. through the breeding range.

Distribution in New England: Coastal Connecticut, Rhode Island in Narragansett Bay (Enser 1992:30), se. Massachusetts, Cape Cod and the Islands, and the Boston Harbor Islands (Veit and Petersen 1993:75), in breeding season n. to Plum Island.

Status in New England: Uncommon and local breeder; common migrant and summer visitor; range increasing in the East.

Habitat: Inhabits margins of lakes, ponds, streams, shrub swamps, freshwater and saltwater marshes and lagoons, tidal estuaries, and mudflats.

Special Habitat Requirements: Open water or wetland habitats near woodlands. Protection from human disturbance.

Nesting: Egg dates: May 3 to mid June. Clutch size: 1 to 6, typically 3 to 4. Incubation: 23 to 24 days. Age to first flight: 23 to 24 days. Broods per year: 1. Nest height: Up to 80 ft (24.4 m), typically 10 to 30 ft (3.0 to 9.1 m) above ground. Nest site: Nests singly or in colonies, often with other herons. Stick nests usually near water in tall trees or in thickets with adequate support for the nest.

Food Habits: Fishes, frogs, salamanders, snakes, snails, crustaceans, insects, small mammals, and small birds, taken in freshwater, brackish, or saltwater swamps, mudflats, and along streams and ponds.

Comments: The great egret was reported n. to Canada in the 1600s (Forbush 1925). Nearly rendered extinct by plume hunters, it was a very rare visitor to Connecticut by the mid 1800s and first nested there in 1901 (Bull 1964).

Snowy Egret

(*Egretta thula*)

Range: Breeding: From n. California and Montana s. to central and e. Texas, along the lower Mississippi Valley, and from Maine s. along the Atlantic and Gulf coasts to South America. After breeding, disperses n. to Oregon, Nebraska, the Great Lakes, and Atlantic Canada. Winter: From n. California, sw. Arizona, the Gulf Coast, and coastal South Carolina s. throughout the breeding range. Found virtually throughout the West Indies, although rare in the Lesser Antilles.

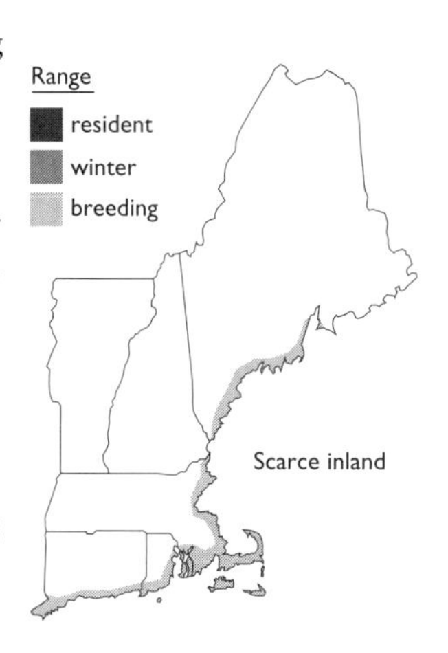

Distribution in New England: Southern-coastal Maine, coastal Massachusetts and Rhode Island including the islands, and coastal Connecticut. No known nesting in New Hampshire (Borror 1994:371 in Foss 1994) or Vermont. Scarce inland.

Status in New England: Locally common breeder; breeding range expanding northward. Common migrant and summer visitor.

Habitat: Borders of lakes, ponds, and rivers; fresh and brackish marshes, shallow bays and coves, cedar swamps, tidal creeks, estuaries, mudflats, mangroves, and rice fields. Sometimes in dry fields.

Special Habitat Requirements: Wetlands.

Nesting: Egg dates: Early May to July 18 (Veit and Petersen 1993:77). Clutch size: 3 to 6, typically 4 to 5. Incubation period: 18 to 24 days. Nestling period: 20 to 25 days. Broods per year: 1. Nest height: Averages 5 to 10 ft (1.5 to 3.0 m) above ground, up to 30 ft (9.1 m). Nest site: Nests in a tree or shrub in colonies (many are coastal), sometimes with other herons and egrets. May nest singly at the northern edge of the range. Nest is flat, elliptical, and loosely woven of slender twigs on a foundation of heavy sticks; additional material such as reeds, rushes, dead cane, etc. often used. May reuse nest in subsequent years (Davis 1986).

Food Habits: Small fish, frogs, lizards, snakes, shrimp, fiddler crabs, crayfish, grasshoppers, cutworms, and aquatic insects, taken in shallow water, mudflats. Shuffles its feet in shallow water to flush prey out of hiding. Occasionally follows other wading birds and captures food stirred up by their foraging, or hovers like a petrel, dropping into the water to catch prey in its bill.

Comments: Slaughter, by shooting and clubbing in great numbers, by plume hunters resulted in virtual extirpation in s. New England by the late 1800s. After an absence of 50 years, it was first reported again in Connecticut in 1931. Nesting resumed in Massachusetts in 1955 and in Connecticut by 1961 (see Zeranski and Baptist 1990:48).

Little Blue Heron

(*Egretta caerulea*)

Range: In w. North America, breeds in s. California and the Pacific slope of Middle America; in e. North America from Maine s. along the Atlantic coastal plain through Middle America; in the West Indies; in n. South America; and locally and sporadically in parts of the s. and central United States. Stragglers appear during the postbreeding period in North and South Dakota, Michigan, s. Ontario, s. Quebec, Nova Scotia, and other localities where the species is not known to breed. Winters from coastal Virginia s. throughout most of the breeding range.

Distribution in New England: Rare and restricted along coastal areas from Connecticut (Greenwich, Norwalk, and Stonington (Zeranski and Baptist 1990:49) n. to s. Cumberland County, Me. Breeding limited to Kettle Island (1994–1995) off Gloucester, Mass. (B. Blodgett pers. commun.). Not known to nest in New Hampshire.

Status in New England: Rare and local coastal breeder; late summer/fall visitor. Scarce inland.

Habitat: Tidal or freshwater marshes, bottomland hardwood swamps, riparian habitats, swamps, ponds, lakes, and impoundments.

Special Habitat Requirements: Shrubby or wooded sites isolated from terrestrial predators.

Nesting: Egg dates: Uncertain in New England, probably most in May (Veit and Petersen 1993:78); early May in New Jersey (Burger 1978). Clutch size: 1 to 6, typically 3 or 4 (Rodgers and Smith 1995). Incubation period: 22 to 23 days (Rodgers and Smith 1995). Nestling period: 42 to 49 days (Rodgers and Smith 1995). Broods per year: 1. Nest height: Sixteen nests in New Jersey ranged from 1.9 to 3 ft (0.62 to 1.12 m) in height (Burger 1978). Nest site: Low shrubs, bushes and small trees.

Territory Size: Colonial. Defends area 9 to 15 ft (3 to 5 m) around nest. May defend foraging territories (Palmer 1962).

Food Habits: Fish, crustaceans, frogs and grasshoppers, taken in shallow water 2 to 6 in (5 to 15 cm) deep (Willard 1977) by slow stalking (Jenni 1969).

Comments: This species was an "accidental" visitor to s. New England until about 1880 (Hill 1965), and first nested in Connecticut in 1971 (Hills 1978 cited in Zeranski and Baptist 1990) and in Massachusetts in 1940 (Hagar 1941). Commonly nests among snowy egrets.

Cattle Egret

(Bubulcus ibis)

Range: Breeding: Across the e. half of the United States and in the West from s. California, Arizona, and New Mexico s. through Middle America, the West Indies, and much of South America. Many individuals disperse n. after breeding, as far as s. British Columbia, s.-central Canada, and the Maritime Provinces. Winter: From the s. United States s. through the breeding range.

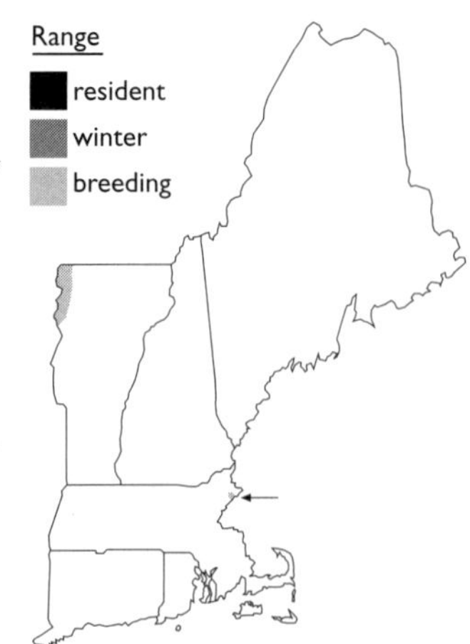

Distribution in New England: Vermont in the n. Lake Champlain region (Ellison 1985:40 in Laughlin and Kibbe 1985), Massachusetts, a small breeding population on Eagle Island off the n. coast (Veit and Petersen 1993:81), in Connecticut an occasional breeder on the Norwalk Islands (Zeranski and Baptist 1990:49), and in Rhode Island on the islands in Narragansett Bay (Enser 1992:33).

Status in New England: Rare and very local breeder; common but irregular visitor.

Habitat: A great variety of habitats including pastures, freshwater and salt marshes, fallow and plowed fields, orchards, road shoulders and other open grassy areas. Least aquatic of North American herons; usually found in close association with large hoofed mammals, particularly cattle. Nests colonially in heronries.

Nesting: Egg dates: May 13 to June 26 (Lake Champlain, Vt.); late May to early June (Massachusetts). Clutch size: 1 to 9, average 3 to 4. Incubation period: 21 to 26 days, average 24 days. Nesting period: 14 to 21 days (Telfair 1994), fly at 5 to 6 weeks. Broods per year: 1. Age at sexual maturity: Spring or summer of third year, few breed at 2 yr, but known to breed at 10 months (Telfair 1994, Maddock 1989 in Telfair 1994).

Food Habits: Grasshoppers, crickets, spiders, flies, frogs, voles (see comments below). Chick diet similar to that of adults, but greater proportion of grasshoppers (Telfair 1994). Older nestlings fed a high proportion of frogs (Siegfried 1978).

Comments: Cattle egrets apparently invaded South America from Africa in the late 1800s and reached Florida by at least 1930. The first specimen collected in North America was taken in Sudbury, Mass., April 23, 1952 (Veit and Petersen 1993:81). In Vermont, first record 1961; first nesting 1973 (Ellison 1985:40 in Laughlin and Kibbe 1985). These herons have been known to follow haying equipment to feed on voles (A. Pistorius in Ellison 1985c:40–41).

Green Heron

(*Butorides virescens*)

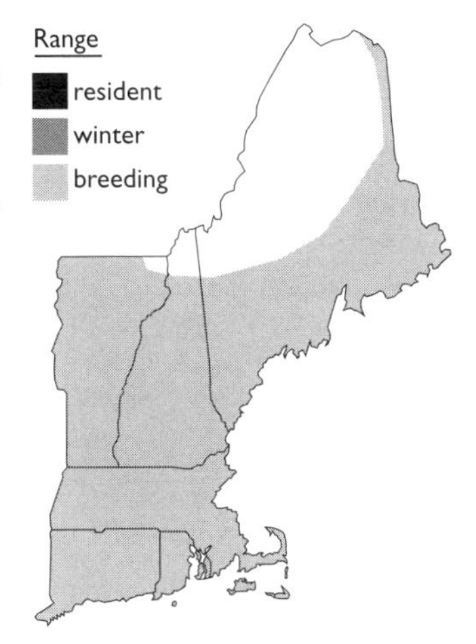

Range: Breeding: From Washington and Oregon s. through much of the sw. United States; in e. North America from the e. edge of the Great Plains, s. Ontario, and New Brunswick, s. through Middle America, the West Indies, and a large portion of South America. Stragglers appear during the post-breeding period in e. Washington, Idaho, s. Canada, and other localities where the species is not found as a breeder. Winter: From the s. United States, s. through the breeding range.

Distribution in New England: Eastern and s. Maine, central and s. New Hampshire, throughout Vermont except the most ne. region, and throughout Massachusetts, Connecticut, and Rhode Island.

Status in New England: Uncommon breeder, locally common migrant.

Habitat: Found in a variety of freshwater and saltwater habitats, primarily in shrub or forested wetlands, or in brushy areas along the margins of slow-moving rivers, lakes, and ponds. Frequently inhabits marshes, beaver ponds, salt marshes, mudflats, harbors, and human-created canals and ditches (Meyerriecks 1962). Perches on trees, stumps, or submerged debris; roosts on or close to the ground.

Special Habitat Requirements: Wooded wetlands, shallow water bodies for feeding.

Nesting: Egg dates: May 5 to June 17 in Massachusetts (Veit and Petersen 1993:84). April 29 to August 4, New York (Bull 1974:85). Clutch size: 3 to 6. Incubation period: 19 to 21 days. Nestling period: 16 to 17 days. Fledge when about 21 days old. Broods per year: 1 (rarely 2). Age at sexual maturity: Probably 2 yr (Davis and Kushlan 1994). Nest height: In large trees or shrubs 10 to 30 ft (3 to 9.1 m), typically 10 to 15 ft (3 to 4.5 m) above the ground, occasionally nests on the ground on a low tussock or muskrat house. Typically nests near water, but may nest in trees or shrubs away from water in dry woodlands or orchards. Nest site: Varies, often nests near water—on a hummock in a marsh or in a tree—but also may nest in trees or shrubs away from water. Sometimes uses old nest.

Territory Size: Male defends a large area upon arrival, but territory decreases to within a few feet of the nest as season progresses. Female helps defend the nest.

Sample Densities: Green herons are normally solitary nesters, but have been known to nest in small groups of up to 30 pairs (Pough 1951:47). The largest green heron colony in New England was 10 pairs at Fresh Pond, Cambridge, Mass., in 1896 (Brewster 1906). Larger groups are unusual, although 70 breeding pairs were found on an area 240 ft (73.2 m) by 1,500 ft (457.3 m) on Long Island, N.Y. Meyerriecks 1962:419).

Food Habits: Small fishes, crustaceans, mollusks, terrestrial and aquatic insects, reptiles, amphibians, spiders, leeches in shallow water, shallow bottoms, wetland vegetation. Stands and waits, walks slowly while foraging (Kushlan 1976).

Comments: Probably numerous in the 1800s; the gradual decline in population after 1910 attributed to the widespread filling of marshes (Hill 1965). Separate feeding territories are vigorously defended by some individuals. Occasionally many birds use a common feeding ground.

Black-crowned Night-Heron

(*Nycticorax nycticorax*)

Range: Breeds in the Western Hemisphere from central Washington and e.-central Alberta to s. Quebec, ne. New Brunswick, and Nova Scotia, s. locally through the United States, Middle America, the Bahamas, Greater Antilles, and South America to Tierra del Fuego. Wanders widely. After breeding, disperses throughout much of the United States, not restricted to its breeding range. Winter: In the sw. United States, lower Ohio Valley, Gulf Coast, and s. New England, s. throughout the breeding range. An uncommon and local resident in Puerto Rico. Winters virtually throughout the West Indies, most individuals being visitors from North America.

Distribution in New England: Along the coast of ne. and s. Maine, the Champlain Lowlands in Vermont, along the coast of se. New Hampshire (possibly breeding), in Massachusetts restricted to coastal areas of Plum Island, the North Shore, Cape Cod, Martha's Vineyard and Nantucket; Connecticut, and Rhode Island, primarily along the coastal regions, occasionally in secluded inland swamps.

Status in New England: Locally common breeder and migrant (Veit and Petersen 1993:34). Has declined since 1900 and particularly after World War II in the e. United States and Canada (Zeranski and Baptist 1990:52). Rare in winter.

Habitat: Inhabits a wide variety of freshwater, brackish, and saltwater wetlands, including margins along lakes, ponds, marshes, wooded swamps, slow-moving streams with pools, rivers, or human-constructed ditches, canals, or reservoirs (Davis 1993). Also found in wooded areas near coastal marshes, spruce groves on marine islands, hardwood forests on offshore islands, apple orchards, and sometimes city parks. Roosts by day in trees.

Special Habitat Requirements: Fresh and saltwater marshes, and wooded wetlands.

Nesting: Egg dates: April 10 to June 30 in Massachusetts (Veit and Petersen 1993:85). April 1 to July 12, New York (Bull 1974:86). Clutch size: 2 to 6, typically 3 to 5. Incubation period: 23 to 26 days. Age at first flight: About 42 days. Broods per year: 1 (2 if first nest fails). Age at sexual maturity: 2 to 3 yr (occasionally 1 yr, Gross 1923). Nest height: 7 to 15 ft (Mass. coast), typically 20 to 30 ft (6.1 to 9.1 m). Nest site: Varies from wooded swamps to shrub swamps, also known to breed in city parks. Occasionally constructs floating nest but most often builds in shrubs or trees.

Territory Size: Unknown. Birds defend area immediately surrounding nest. Some birds defend feeding territories and roosting sites (Palmer 1962:478, 479).

Sample Densities: 400 pairs on Pea Patch Island in the Delaware River in 1976 (Buckley et al. 1976).

Food Habits: Salt and freshwater fishes, crustaceans, mollusks, worms, leeches, aquatic and terrestrial insects, reptiles and amphibians, occasionally young birds and mammals. Forages in shallow water, muddy and sandy bottoms, or on dry ground by standing and waiting, walking slowly (Kushlan 1976).

Comments: Population declines probably due to DDT, shooting at fish hatcheries, and disturbance of rookeries (Veit and Petersen 1993:84, Tremblay and Ellison 1979). Feeding activity is greatest at dawn and dusk, with activity continuing into the early hours of darkness. Gregarious in all seasons, the black-crowned night-heron breeds on all continents except Australia and Antarctica (Davis 1993).

Yellow-crowned Night-Heron

(*Nyctanassa violacea*)

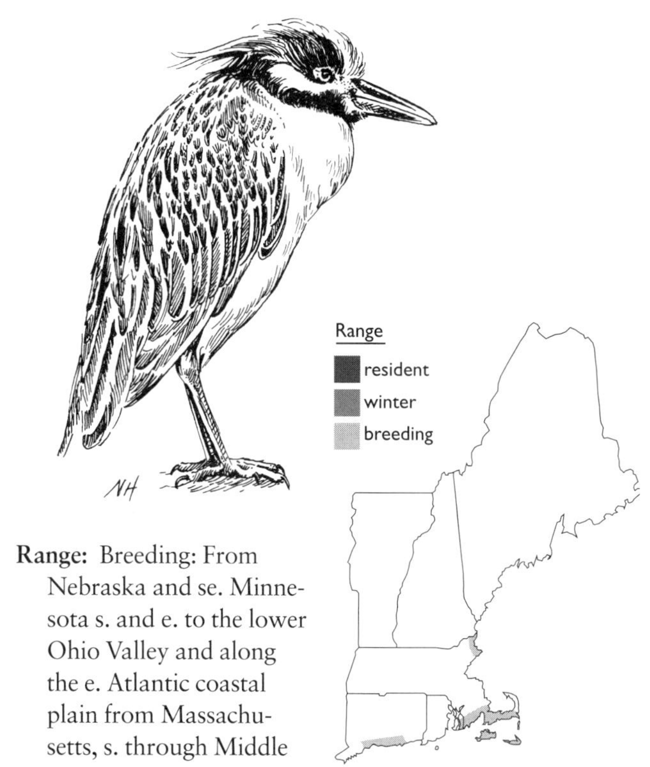

Range: Breeding: From Nebraska and se. Minnesota s. and e. to the lower Ohio Valley and along the e. Atlantic coastal plain from Massachusetts, s. through Middle America and the West Indies; in South America, s. and w. to Ecuador and along the n. and e. coast of the continent from Colombia and Venezuela to s. Brazil. Winter: From coastal South Carolina, s. throughout the breeding range.

Distribution in New England: Massachusetts, repeated summer occurrences of adult birds with young in late summer suggest breeding at Plum Island, Westport, Martha's Vineyard, Nantucket, and Centerville (Veit and Petersen 1993:86); in Connecticut along the coast in Norwalk, Fairfield, Bridgeport, and Milford (Zeranski and Baptist 1990:52).

Status in New England: Rare but probably regular coastal breeder; very rare inland (Veit and Petersen 1993:86).

Habitat: Inhabits shallow freshwater wetlands, such as forested and scrub-shrub wetlands, forested impoundments, and the margins of shallow creeks, rivers, lakes, and ponds, and saltwater wetlands including tidal marshes, bayous of large cypress swamps, mangrove swamps, or dry, rocky thickets on barrier, spoil and bay islands (Heightmeyer et al. 1989, Laubhan et al. 1991, Laubhan and Reid 1991, Watts 1989, 1995).

Special Habitat Requirements: Wooded swamps and riparian forests.

Nesting: Egg dates: Uncertain in Massachusetts, probably June (Veit and Petersen 1993:86). April 30 to June 10, New York (Bull 1974:90). Clutch size: 2 to 6, typically 3 to 4, sometimes 5 (Watts 1995). Incubation period: 24 days (Forbush 1929:342). Broods per year: Probably 1. Age at sexual maturity: Unknown (Palmer 1962:488). Nest height: To 50 ft (15.2 m). Nest site: Restricted to areas near water (Watts 1989, Laubhan and Reid 1991). Normally a solitary nester, not usually associating with black-crowned or other herons, which makes breeding confirmation difficult (Veit and Petersen 1993:86). Nests occurred in separate trees in s. Missouri (Laubhan and Reid 1991). Nests are placed in upper tree canopy on a limb away from the trunk (Watts 1989).

Territory Size: Sizes not known; birds defend an area for display, copulation, nesting, and feeding.

Sample Densities: 40 pairs on Pea Patch Island in Delaware River in 1976 (Buckley et al. 1976).

Food Habits: Mainly crustaceans, especially crabs and crayfish, mollusks (Reigner 1982). Sometimes takes leeches, reptiles, amphibians, small birds and mammals, insects, fishes in shallow water. Stands and waits; walks slowly (Kushlan 1976).

Comments: A southern species, yellow-crowned night-herons first nested in Connecticut in 1953 (Zeranski and Baptist 1990:52); in Massachusetts in 1928 (Townsend 1929). Birds feed during the day and at night, singly or in twos or threes.

Glossy Ibis

(*Plegadis falcinellus*)

Range: Cosmopolitan; breeds in North America locally from Maine and Rhode Island s. to Florida, and w. on the Gulf Coast to Louisiana. Also inland in Arkansas, the upper Midwest, and s. Canada. Winter: Northern Florida and the Gulf Coast of Louisiana, s. through the West Indies, nw. Costa Rica, and n. Venezuela.

Distribution in New England: Southern Maine, the islands off the coast of Massachusetts (Clark's, House, Monomoy, and the Boston Harbor islands), and isolated locations along the coast of Connecticut (Norwalk and Stonington) (Zeranski and Baptist 1990) and Rhode Island. Not known to nest in New Hampshire (Borror 1994b:372 in Foss 1994) or Vermont.

Status in New England: Locally common breeder; locally common migrant.

Habitat: Fresh, brackish, and salt water, primarily in marshes and estuaries. Favors shallow pools bordered by shrubs and emergent vegetation. On the New Jersey coast, birds nest in mixed stands of holly, red cedar, sumac, salt myrtle, bayberry, wild cherry, grape, and catbrier that grow on barrier beaches.

Special Habitat Requirements: Emergent and scrub-shrub wetlands.

Nesting: Egg dates: Probably June in Massachusetts (Veit and Petersen 1993:88). March to late May (Palmer 1962:521). Clutch size: 3 to 5, typically 3 to 4. Incubation period: 21 days. Nestling period: About 14 days. Broods per year: 1. Age at sexual maturity: Unknown (Palmer 1962:520). Nest height: To 10 ft (3.0 m). Nest site: On ground in tall, dense vegetation such as cattails and reeds, in low bushes or in tops of low trees growing in water. Birds are colonial nesters, commonly associating with herons.

Sample Densities: 600 pairs on Pea Patch Island in Delaware River in 1976 (Buckley et al. 1976).

Food Habits: Crustaceans (especially crayfish), small snakes, cutworms and other grubs, grasshoppers, leeches. Probes soft earth, shallow water, mud flats, flooded fields.

Comments: The glossy ibis is extending its breeding range in the Northeast. It first nested in Connecticut in 1971 (Zeranski and Baptist 1990:53), and in Massachusetts in 1973. Over 100 pairs nested on coastal islands of the state in 1977 (Massachusetts Audubon Society 1977), but colonies on House, Clark's, and Boston Harbor islands declined inexplicably in 1979–1981 (see Veit and Petersen 1993:88). Little is known about the bird's feeding habits in the Northeast.

Turkey Vulture

(Cathartes aura)

Range: Breeding: Southern British Columbia, w. Ontario, extreme s. Ontario, and Massachusetts, s. throughout the remaining continental United States to South America. North American individuals winter from n. California, Arizona, Texas, Nebraska, the Ohio Valley, and Pennsylvania, s. to the Gulf Coast, Florida, and nw. South America. In Guatemala, more birds are probably present during migration as the northern populations move through. In Colombia, numbers are augmented by migrants from the United States. Found in Cuba (including coastal cays), Isle of Pines, Cayman Islands, Jamaica, Hispaniola, Puerto Rico, and nw. Bahamas; casual on Bimini, New Providence, and St. Croix.

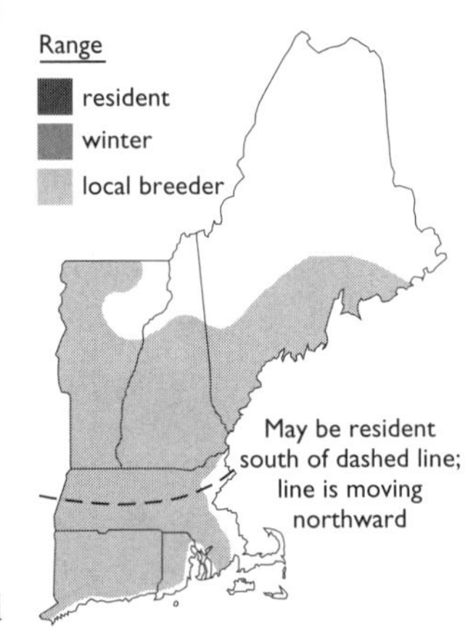

Distribution in New England: Throughout Connecticut, Rhode Island, and Massachusetts, except does not breed along the coastal regions; throughout Vermont except absent from the n.-central and Northeast Highland regions, common in New Hampshire s. of the White Mountains, and e. to s. Maine.

Status in New England: Common migrant and summer

visitor, but uncommon as a breeder. Some birds overwinter from Massachusetts s.

Habitat: Uses a wide variety of habitats, from tropical and temperate deciduous and mixed forests to open plains and deserts, to lowlands, and mountains. Also frequents highways and roads where carrion is abundant. Communal winter roosts in e. United States are characterized by large or dense conifers or large deciduous trees (Thompson et al. 1990).

Special Habitat Requirements: Mixed farmland and forest, which provides the best opportunity to forage on both domestic and wild carrion (Kirk and Mossman 1998).

Nesting: Egg dates: May 4 to June 30, New York (Bull 1974:166). Clutch size: 1 to 3, typically 2. Incubation period: 38 to 41 days. Nestling period: 70 to 80 days (Terres 1980:960). Broods per year: 1. Nest site: Does not build a nest. Lays eggs on the floor of caves (preferably with two entrances), on the ground on rocky outcrops or ledges, or in rocky caverns, hollow snags, old hawk nests, on the floor of abandoned buildings, or in clumps of dense shrubs. Eggs are usually well hidden from view and are fairly inaccessible to predators. Females show strong attachment to old nest sites.

Home Range Area: May exceed 400 square miles (1,040 km^2) in size (Coleman and Fraser 1989).

Sample Densities: Probably fewer than 2 or 3 birds per square mile (1 bird/km^2) at s. part of range (Forbush and May 1939:95). 0.3 pairs per 100 acres (40 ha) in mixed habitat (forest-brush-farmland) in Maryland (Stewart and Robbins 1958:105).

Food Habits: Carrion of amphibians, reptiles, birds, mammals, and fish; also eats small quantities of plant material (Prior 1990). Preferred feeding habitat: Open fields, ridges. The highly developed olfactory sense enables this bird to locate concealed carcasses under the forest canopy (Kirk and Mossman 1998).

Comments: The turkey vulture's dramatic recent range extension and population increase in the Northeast may be related to warmer climate, more deer, and more highways to cause road kills (Bull 1964:148, Bagg and Parker 1951). The species was recorded in the 1600s in Massachusetts (Josselyn 1672), and was more frequent in the 1820s than in the following 100 years. The first nesting in the Northeast in modern times occurred in New York in 1925, in Connecticut in 1930 (see Zeranski and Baptist 1990:80 for a review), in Massachusetts in 1954 (Veit and Petersen 1993:124), and in Vermont, New Hampshire, and Maine in 1979 (Foss 1994:40 in Foss 1994). Preens and roosts in tall snags or trees with open branches. May gather in groups of up to 70 birds to roost at night.

Canada Goose

(*Branta canadensis*)

Range: Breeding: The Arctic Coast of Alaska and n. Canada, e. to Baffin Island, s. to central California, e. to w. Tennessee, s. Ontario and Quebec, and Newfoundland. Winter: South-coastal and se. Alaska, British Columbia and s. Alberta, e. to the Atlantic Coast of Newfoundland, and s. to Mexico, the Gulf Coast, and n. Florida (Bellrose 1976).

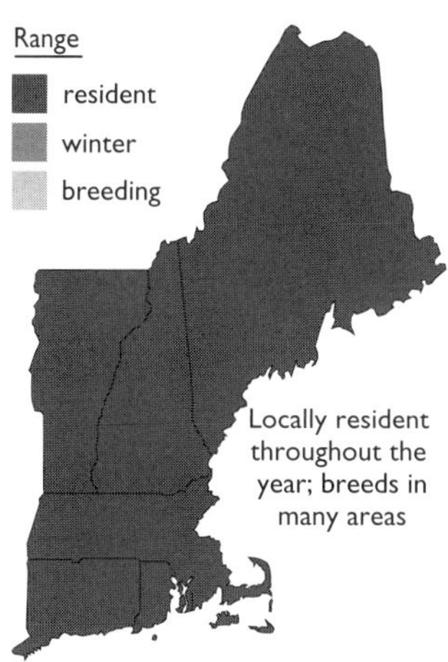

Distribution in New England: Breeding: Throughout s. New England, s. and w. Maine, s. and ne. New Hampshire, and w. and s. Vermont. Winter: s. New England, distance from coast depends on winter severity.

Status in New England: Common breeder, frequently a "nuisance" species; abundant migrant and common winter resident along s. New England coasts.

Habitat: Breeding: Freshwater and brackish marshes, shores of lakes, ponds, and large rivers; meadows, sloughs, emergent swamps, beaver or human impoundments, golf courses, agricultural fields, prairies, tundra, and on islands. Winter: Coastal marshes, inland ice-free lakes and rivers, agricultural land, and flooded fields.

Special Habitat Requirements: Elevated sites in marshes for nesting. Also grassy areas adjacent to lakes, ponds.

Nesting: Egg dates: April to June in Massachusetts (Veit and Petersen 1993:95). March 28 to May 14, New York (Bull 1974:99). Clutch size: 4 to 10, typically 5 or 6. Incubation period: 25 to 30 days. Nestling period: 1 day (precocial). Age at first flight: 40 to 73 days. Broods per year: 1. Age at sexual maturity: 2 or 3 yr. Nest site: Typically near water (up to 150 ft), preferably on a slightly elevated site, including beaver lodges, muskrat houses, old stumps, small islands, artificial platforms.

Territory Size: Varies considerably. Size seems to be influenced by aggressiveness of defending male, type of habitat, amount of protective cover, and density of breeding population (Bellrose 1976:159).

Sample Densities: Breeding densities vary greatly and seem to be influenced by availability of suitable nest sites rather than by territorial behavior (Johnsgard 1975:142).

Food Habits: Grazes tender shoots of grasses, sedges and other marsh plants, submerged vegetation, and agricultural crops. Also consumes small quantities of insects, insect larvae, seeds, mollusks, and small crustaceans in mud flats, agricultural fields, flooded meadows, salt marshes, shallow water with submergent vegetation.

Comments: In Connecticut Canada geese were "enormously abundant" in the Colonial period, principally as migrants (Hill 1965), decreasing to a "common" migrant by the 1870s (Merriam 1877); only four winter records exist between 1876 and 1909 (Sage et al. 1913). By the 1920s, Canada geese were again common in s. New England (Forbush 1925), with their decline and subsequent increase largely due to overhunting in the 1700s and 1800s followed by protection in 1908 (Zeranski and Baptist 1990:58–59). Nonmigratory "feral" goose populations appeared in the 1930s in the New York City area (Cruickshank 1942); resident flocks were abundant by the early 1960s in s. Connecticut. Of the 11 recognized subspecies of Canada goose (Bellrose 1976), resident flocks are of the introduced "giant" race, descended from released live decoys after their use was banned.

Mute Swan

(*Cygnus olor*)

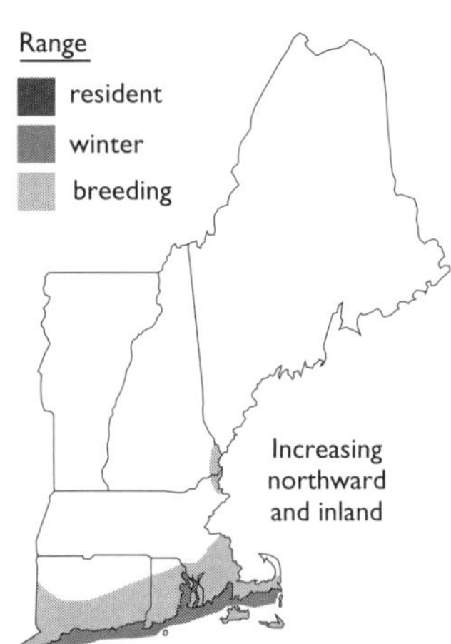

Range: Small breeding colonies exist in the e. United States, mainly along the ne. coastline and the Great Lakes, with largest numbers occurring in the Long Island area. Smaller groups range locally from New Hampshire and Massachusetts s. to Virginia. Winter: Primarily coastal.

Distribution in New England: Primarily Connecticut, Rhode Island, Massachusetts, and New Hampshire coasts, but scattered locations inland.

Status in New England: Locally common (coast) to uncommon but increasing resident inland. The east coast population shows few signs of slowing its rate of increase (Ciaranca et al. 1996).

Habitat: Nests on coastal bays, marshes and ponds having dense aquatic vegetation. Wintering inland birds may be forced to move to brackish or saltwater areas if fresh waters freeze. Otherwise, birds remain in breeding territories.

Special Habitat Requirements: Shallow waters with abundant aquatic vegetation.

Nesting: Egg dates: April 15 to early June in Massachusetts (Veit and Petersen 1993:91). Peak: May, Atlantic coast (Palmer 1976 V.2:45). Clutch size: 2 to 11, typically 5 or 7. Incubation period: 35 to 36 days. Nestling period: 1 day (precocial). Broods per year: 1. Age at sexual maturity: 2 to 6. Nest site: On the ground, preferably on small islands along secluded shores, in marshes, or at ponds. Favors shallow, clean, weed-filled waters. Sometimes colonial. Nest is typically placed in a clump of cattails surrounded by water.

Territory Size: 4 to 10 acres (1.6 to 4.0 ha) in Michigan (Wood and Gelston 1972). 12 territories ranged from 0.5 to 11.8 acres (0.2 to 4.8 ha) in Rhode Island (Willey 1968 in Palmer 1976). Highly territorial; when a pond is chosen for nesting, the male defends the entire pond against intruders (Palmer 1976 V.2:44).

Sample Densities: Many scores of semidomesticated pairs nested on 25 acres (10.1 ha) in England, indicating that these birds will usually tolerate close nesting (Bannerman 1958 in Palmer 1976 V.2:44).

Food Habits: Submerged vascular plants, crustaceans, and aquatic insects (adults and larvae) (Ciaranca et al. 1996).

Comments: Mute swans are native to Eurasia, and were introduced to the lower Hudson River Valley about 1910 as an adornment to parks and estates (Crosby 1922). By 1936 they were occurring in the wild, in the 1940s nested in Connecticut, and in the 1960s spread upland and up the coast to Salisbury, Mass. (Zeranski and Baptist 1990:54). Wherever they occur, mute swans are aggressive toward other waterfowl and destructive of waterfowl habitat because they feed at greater depth than other waterfowl and uproot water plants of value to native ducks and geese (HW Heusmann, pers. commun.). Several New England states either remove all mute swans as they occur, addle the eggs of nesting pairs, or limit permits to possess the birds.

Wood Duck

(Aix sponsa)

Range: Breeding: Western North America from s. British Columbia and sw. Alberta, s. to central California and w. Montana; in e. North America from e.-central Saskatchewan, e. to Prince Edward Island and Nova Scotia, s. to central and se. Texas and the Gulf Coast. In the West, winters irregularly throughout the breeding range; in the East, winters primarily in the southern parts of the breeding range. In Mexico, recorded in Sinaloa, Valley of Mexico, San Luis Potosi, and Tamaulipas. Winter: The West Indies, s. to the n. Bahamas and probably w. Cuba; casual on Jamaica, Puerto Rico, and Saba.

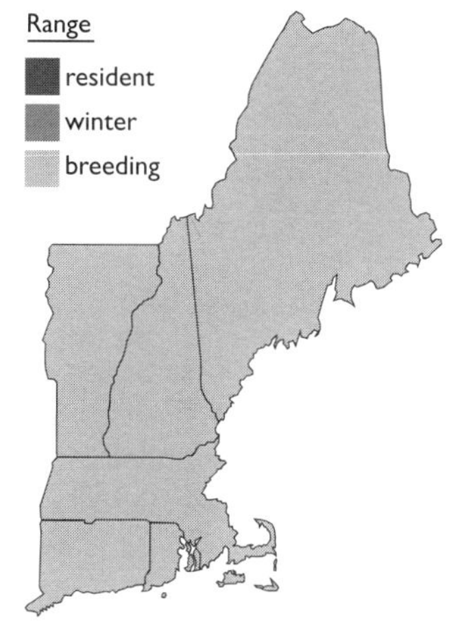

Distribution in New England: Throughout the region.

Status in New England: Fairly common and widespread breeder and migrant. Population has increased in recent years due to maturing trees with cavities and the increased availability of nest boxes.

Habitat: Inhabits woodlands near shallow, quiet inland lakes, ponds, streams, marshes, beaver impoundments, forested wetlands, and river bottomlands where nest sites are available and abundant plant and invertebrate food sources. Also uses scrub-shrub wetlands with woody shrubs (buttonbush, willow, alder) and an overhead cover of drowned trees (Hepp and Hair 1977). Stable water levels are important for successful brood rearing; stream channels may provide travel lanes for ducklings moving among wetlands (Hepp and Bellrose 1995).

Special Habitat Requirements: Natural or artificial nest cavities, persistent shallow water, and some brushy cover for retreat (McGilvrey 1968). Will use nest boxes if wood shavings and sawdust are provided. Nest boxes are usually placed in water 10 to 50 cm deep.

Nesting: Egg dates: March 23 to July 25, Massachusetts (Hepp and Bellrose 1995). Clutch size: 6 to 17, typically 6 to 10. Incubation period: 28 to 35 days. Nestling period: 1 day (precocial). Broods per year: 1, occasionally 2 at southern latitudes (Kennamer and Hepp 1987). Age at sexual maturity: 1 yr. Nest height: 16.5 to 49.5 ft (5 to 15 m) above ground with entrance holes 4 in (10 cm) in diameter, cavity depths of 23.6 in (60 cm), and cavity bottoms measuring 9.8 × 9.8 in (25 × 25 cm). Will use nest boxes 6 ft (2 m) above water. Nest site: In cavity of a mature, living (occasionally dead) deciduous or coniferous tree, or nest box over or close to still or slow-moving water. Will use natural cavities within 1.2 miles (2 km) of water (Bellrose 1976). Prefers water with a brushy overstory for concealment, and stumps and fallen logs for perching. Accepts nest boxes, but usually occupies natural cavities. Less frequently occupies cavities excavated by woodpeckers.

Territory Size: No feeding or nest territories, but male will defend mate if approached too closely (Hepp and Bellrose 1995).

Sample Densities: 63 nests (next boxes) per 150 acres (60.7 ha) in open marsh in Massachusetts (Grice and Rogers 1965:20). 1 nest per 12 acres (4.9 ha) (Bellrose et al. 1964).

Food Habits: Aquatic and terrestrial insects, fleshy fruits, acorns, hickory nuts, and waste grains. Lack of wetland foods drives wood ducks into uplands, groves, orchards, or fields, where they seek acorns, nuts, and grains (Bellrose and Holm 1994).

Comments: The wood duck is strictly a North American species. Once the most abundant duck in New England, the wood duck was rare by about 1910, and did not increase in population until the 1930s, possibly as a result of the return of beaver. See Zeranski and Baptist (1990:62) for review of historical trends. Wood ducks in Ontario find good breeding habitat in areas with beaver ponds and cavities made by pileated woodpeckers. Diets of Maine birds consisted of 95 percent vegetable and 5 percent animal material (Coulter 1957).

Gadwall

(Anas strepera)

Range: Breeding: Southern Alaska and s. Yukon to the New Brunswick-Nova Scotia border, s. locally to s. California, n. Texas, central Minnesota, and n. Pennsylvania and on the Atlantic Coast to North Carolina. Winter: Southern Alaska, s. British Columbia, and Colorado to s. South Dakota, Iowa, the s. Great Lakes, and Chesapeake Bay on the Atlantic Coast, s. to Mexico and the Gulf Coast. In Mexico, winters to Guerrero and Tobasco. Winters in Cuba and Jamaica; rare elsewhere in the West Indies.

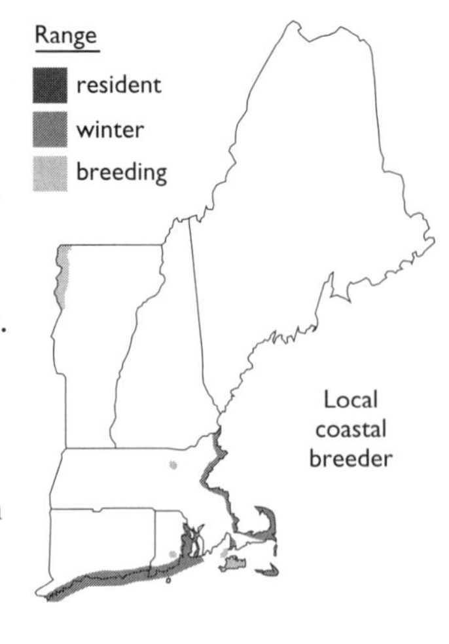

Distribution in New England: Breeding: Vermont, the n. Champlain Lowlands, and locally in e. Massachusetts, Ipswich, Monomoy, and Martha's Vineyard (Veit and Petersen 1993:105), s. to Rhode Island along the coast and Block Island (Enser 1992:44). Winter: Occasionally along the coast in e. Massachusetts (Plum Island, Salem, Plymouth), and Berkshire County, Mass. (Veit and Petersen 1993:105), s. to Rhode Island and Connecticut along the coast.

Status in New England: Uncommon breeder and common migrant, rare in winter.

Habitat: Breeding: Inhabits fresh and brackish marshes, sloughs, ponds, or small lakes. Generally avoids wetlands bordered by woodlands or thick brush; prefers those bordered by dense, low herbaceous vegetation or shrubby willows with grassy islands. Will use manmade ponds (Belanger and Couture 1988). Winter: Coastal bays and freshwater marshes that remain free of ice, but can be found on open water of any kind.

Special Habitat Requirements: Shallow water for feeding; marshes or grassy areas near water for nesting (DeGraaf and Rappole 1995:87).

Nesting: Egg dates: Late May to July in Massachusetts (Veit and Petersen 1993:105). May 30 to July 25, New York (Bull 1974:114). Clutch size: 7 to 13, typically 10 to 12. Incubation period: 24 to 27 days. Nestling period: 1 day (precocial). Age at first flight: 49 to 63 days (Johnsgard 1975). Broods per year: 1. Age at sexual maturity: 1 yr (late-hatched young may breed the second year). Nest site: Nests on dry ground on islands in lakes, upland meadows or pastures, and on prairies, usually within 165 ft (50 m) from water. Prefers to nest in uplands rather than over water, especially in dense herbaceous vegetation and under shrubby willows.

Sample Densities: This is primarily a prairie pothole resident. Sample densities are: average 6.3 birds per square mile (2 pairs/km^2) in mixed prairies of North Dakota (Bellrose 1976:209); 200 nests per acre (0.4 ha) (island), Lower Souris National Wildlife Refuge, North Dakota (Henry 1948 in Palmer 1976 V. 2:394); 1 pair per 60 to 80 acres (24.3 to 32.4 ha) in pothole areas of Dakotas

(Palmer 1976 V. 2:394). New England densities are unknown, but would not approach these concentrations.

Food Habits: Submerged aquatic plants (seeds or soft parts)—pondweeds, naiads, widgeon grass, eelgrass, filamentous algae, etc. in shallow water.

Comments: Accounts from the mid 1800s indicate that migrant gadwalls were common, but rare by the 1870s (Zeranski and Baptist 1990:65). Until about 1930, gadwalls were considered rare vagrants from the West. By the 1970s, they were nesting in Massachusetts at Plum Island and Penikese Island, Ipswich, and Martha's Vineyard (Veit and Petersen 1993:105). Mabbott (1920 in Palmer 1976 V. 2:400) found that 98 percent of the contents of 362 stomachs was vegetable and 2 percent animal matter. Gadwalls prefer succulent stems to seeds. Gadwalls rarely graze in pastures or grain fields.

American Wigeon

(Anas americana)

Range: Breeding: Central Alaska and central Yukon to New Brunswick and s. Nova Scotia, s. to ne. California, central Colorado, South Dakota, s. Ontario, and n. New York, sporadically to the Atlantic Coast. Winter: Throughout the Hawaiian Islands, and from s. Alaska to s. Nevada, sporadically across the central United States to the s. Great Lakes and Ohio Valley, and on the Atlantic Coast from Nova Scotia s. throughout the s. United States to Central America. Winters throughout Mexico. In Guatemala, a fairly common migrant and winter resident found along the Pacific Coast and on the interior lakes up to 1,500 m. In Honduras, a fairly common migrant and winter visitor on the Caribbean slope. In Costa Rica, Panama, and Colombia, a locally fairly common but somewhat irregular winter migrant. Occurs virtually throughout the West Indies.

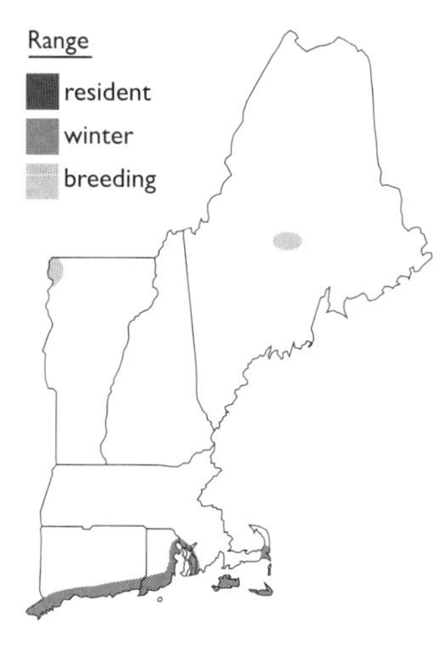

Distribution in New England: Breeding: Vermont (one confirmed nest in 1962, one probable nest in 1981), the n. Champlain Lowlands (Fichtel 1985:394), s. to Massachusetts on Penikese Island, 1972, and Monomoy 1981 and 1983 (Veit and Petersen 1993:106). Winter: Massachusetts on Nantucket and Martha's Vineyard (Veit and Petersen 1993:106), w. along the coastal region of Rhode Island and Connecticut. Fairly common winter resident in s. New England.

Status in New England: Rare and local breeder in Massachusetts. Common along the coast in winter.

Habitat: Breeding: Inhabits freshwater marshes, lakes, ponds, and prairie potholes from tundra to shortgrass and mixed prairies. Prefers bodies of water with exposed shorelines, and permanent to temporary waters. Winter: Shallow fresh and brackish ponds, wet meadows, coastal marshes, and bays.

Special Habitat Requirements: Large lakes, ponds, marshes, sluggish streams and rivers with open water, exposed shorelines, and emergent vegetation.

Nesting: Egg dates: June in Massachusetts (Veit and Petersen 1993:106). Clutch size: 6 to 12, typically 9 to 11 (Terres 1980). Incubation period: 22 to 24 days. Nestling period: Less than 1 day (precocial). Age at first flight: 45 to 58 days. Broods per yr: 1. Age at sexual maturity: 1 yr. Nest site: In a hollow on dry ground (Bellrose 1976:203), on an island or on shore, in tall vegetation. Nest may be as far as 400 yd (366 m) from water.

Sample Densities: 16.6 birds per square mile (6 birds/km^2) at Mackenzie River Delta (Bellrose 1976:199). 9.4 birds per square mile (4 birds/km^2) at Yukon Flats in Alaska (Bellrose 1976:199). 7.4 birds per square mile (3 birds/km^2) in parklands (Bellrose 1976:199). 1.66 birds per square mile (0.6 birds/km^2) in closed boreal zone of Ontario (Bellrose 1976).

Food Habits: Almost wholly vegetarian, eating quantities of leaves, stems, and buds on pondweed, widgeon grass (staples), and wild celery; also eats tender shoots of grasses and occasionally snails, beetles, and crickets.

Comments: American wigeon were apparently numerous before 1800, decreasing rapidly after 1870 (Griscom and Snyder 1955); one of the rarest s. New England dabbling ducks by 1900 (Forbush 1916).

American Black Duck

(*Anas rubripes*)

Range: Breeding: Northeastern Saskatchewan to n. Labrador and Newfoundland, s. to n. South Dakota, n. Illinois, central West Virginia, and on the Atlantic Coast to North Carolina. Winter: Southeastern Minnesota and central Wisconsin to New Brunswick and Nova Scotia, s. to s. Texas, the Gulf Coast, and south-central Florida. Winters as far n. as open water and food are available.

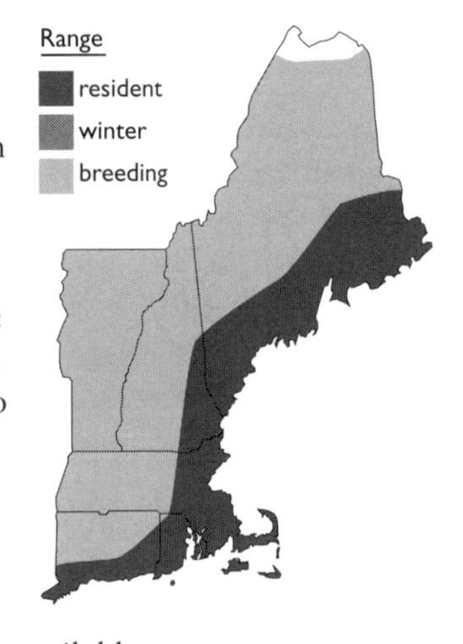

Distribution in New England: Breeding: Throughout the region except in northernmost Maine.

Status in New England: Common and widespread breeder, common migrant but declining winter resident.

Habitat: Breeding: Inhabits a wide variety of coastal and freshwater habitats, including brackish marshes, estuaries, edges of rivers, lakes, and ponds, forested swamps, beaver ponds, emergent wetlands, and open boreal and mixed hardwoods forests. Occurs in glacial kettle ponds surrounded by bog mats in Vermont (Ellison and Ellis 1985:52 in Laughlin and Kibbe 1985). Winter: Brackish marshes bordering bays, estuaries, and agricultural marshes.

Nesting: Egg dates: April 2 to May 24 in Massachusetts (Veit and Petersen 1993:99), April 2 to June 22, New York (Bull 1974:112). Clutch size: 6 to 11, typically 8 to 10. Incubation period: 26 to 28 days. Nestling period: Less than 1 day (precocial). Broods per year: 1. Age at sexual maturity: 1 yr. Nest height: Usually on the ground. Nest site: Generally in a hollow on dry ground in a site that is slightly elevated and well hidden in grasses, shrubs, or briers. Occasionally nests under live conifers, under logs, under tangles of dead vegetation, and in hollow tree boles (Coulter and Mendall 1968). Upland nests may be a mile or more from water (Palmer 1976 V.2:329). Occasionally uses old low crow or hawk nests and natural or excavated cavities in trees or tops of rotted stumps.

Territory Size: 6 acres (2.4 ha) in Lake Erie marsh (Trautman 1947 in Palmer 1976 V.2:338).

Sample Densities: 1 nest per 20 to 40 acres (8.1 to 16.2 ha) in bogs in Maine (Coulter and Miller 1968). 21.4 nests per acre (0.4 ha) on islands in Chesapeake Bay (Stotts 1957). 5 nests per acre (0.4 ha) on islands in Lake Champlain (Coulter and Miller 1968). 5.3 pairs per hundred acres (40 ha) of brackish marsh in Maryland (Stewart 1962).

Food Habits: Diet varies greatly with habitat. In marine environments eats mostly mollusks. In fresh and brackish environments, plants comprise most of the diet. Also consumes seeds, acorns, berries, crustaceans, amphibians, earthworms, and small fishes. In late fall and winter, birds may leave their rest areas early in the day and just before sunset to waste corn fields up to 25 miles (40 km) away (Bellrose 1976:261).

Comments: Once the most abundant duck in North America and prized as a table bird, the black duck declined with most waterfowl in the mid-1800s and was rare in the early twentieth century (Sage et al. 1913) owing to habitat loss and overhunting. It was again common by the late 1930s, but has declined again because of acidification (Luoma 1987) and hybridization with mallards (Bellrose 1976:253).

Mallard

(Anas platyrhynchos)

Range: Breeding: From n. Alaska, e. to s. Northwest Territories and across to s. Maine, s. to California, the s. Great Basin, and New Mexico, and from Oklahoma, e. through the Ohio Valley to Virginia. Winter: Generally from s. Alaska and s. Canada, s. to central Mexico. Introduced and established in the Hawaiian Islands. In Costa Rica, once a rare migrant but no recent reports. In Puerto Rico, an extremely rare winter migrant and resident. In Trinidad and Tobago, an occasional winter resident and migrant. In Honduras, a rare migrant sometimes seen in the Bay of Fonesca area and the Sula Valley. In Guatemala, a migrant and winter resident only rarely found in the Pacific lowland. A rare winter resident in Cuba and the Bahamas; casual to Jamaica, Puerto Rico, and St. Croix.

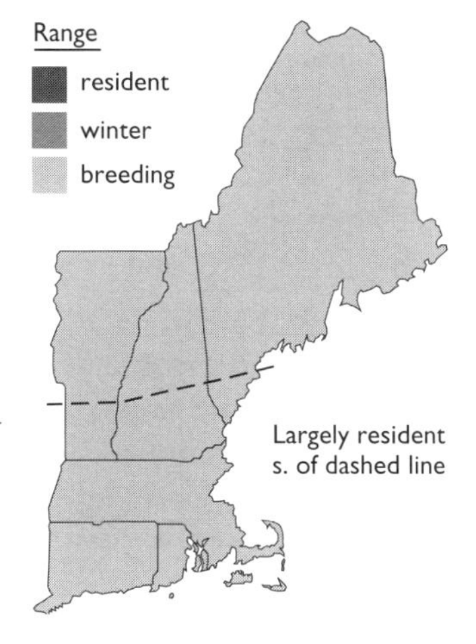

Distribution in New England: Breeding: Throughout the region. Winter: Throughout Massachusetts, Rhode Island, and Connecticut, s. Vermont, s. New Hampshire, and s. Maine.

Status in New England: Abundant resident and migrant.

Habitat: Breeding: Inhabits ponds, lakes, rivers, streams, marshes, wet meadows, wooded swamps, beaver impoundments, sloughs, reservoirs, cultivated fields, and barrier beaches; also boreal forest regions and subarctic deltas. Prefers water less than 16 in (41 cm) deep (Pough 1951:77). Avoids salt water. Winter: Inland ponds and rivers with open water, less often in coastal marshes.

Special Habitat Requirements: Shallow water less than 16 in (41 cm) deep that enables duck to feed on bottom by tipping up (Pough 1951:77).

Nesting: Egg dates: March 25 to July 1, New York (Bull 1974:110). Clutch size: 6 to 15, typically 8 to 12 (Harrison 1975). Incubation period: 23 to 29 days, usually 26. Nestling period: Less than 1 day (precocial). Broods per year: 1. Age at sexual maturity: 1 yr (some individuals breed later). Nest site: Typically nests on ground in dry or slightly marshy areas within 300 ft (100 m) of water, occasionally up to 1.2 miles (2 km) away in grasslands. Conceals nest well among weeds and grasses, in pastures, stubble in cultivated fields, or marsh vegetation; rarely nests in cavities, on hollowed tops of stubs, or in tree crotches.

Territory Size: About 2 square miles (5.2 km^2) at onset of breeding season (Palmer 1976 V.2:298). Territory coincides with home range early in breeding season and becomes progressively smaller as season advances. Drake probably defends mate rather than land area (Dzubin 1969 in Palmer 1976, V.2:298).

Sample Densities: 93 pairs per square mile (36 pairs/km^2) in Canadian Parklands (Dzubin 1969 in Palmer 1976, V.2:300). 6.1 pairs per square mile (2 pairs/km^2) in prairie pothole habitat (Drewien and Springer 1969). 29 nests per 5.7-acre (2.3-ha) island in Lake Champlain (Coulter and Miller 1968).

Food Habits: Seeds of sedges, grasses and smartweed are staples; also eats leaves, stems, and seeds of other marsh plants, waste grain, snails, insects, tadpoles, fishes, and fish eggs.

Comments: The most common and widely distributed duck in North America. Uncommon before 1920, mallards have been increasing since that time (Griscom and Snyder 1955). About 40,000 mallards winter between Massachusetts and Chesapeake Bay. Another 40,000 winter in Chesapeake Bay itself (Bellrose 1976:235). It is presently the most abundant duck in North America and ubiquitous throughout s. New England. The introduction of feral mallards has resulted in much hybridization with American black ducks and with various barnyard waterfowl, particularly Muscovy and Peking ducks (Veit and Petersen 1993:101).

Blue-winged Teal

(Anas discors)

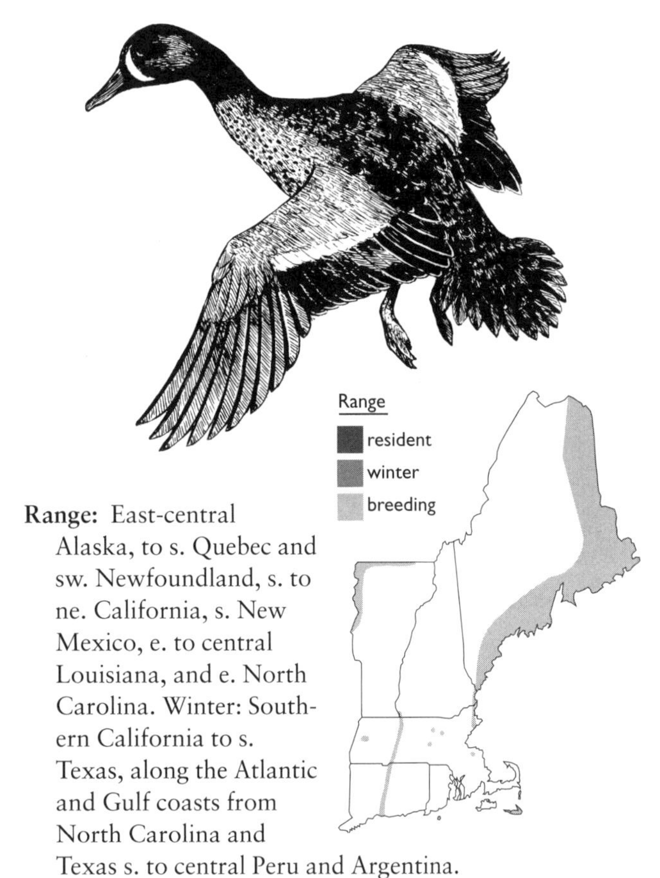

Range: East-central Alaska, to s. Quebec and sw. Newfoundland, s. to ne. California, s. New Mexico, e. to central Louisiana, and e. North Carolina. Winter: Southern California to s. Texas, along the Atlantic and Gulf coasts from North Carolina and Texas s. to central Peru and Argentina.

Distribution in New England: Central Connecticut along the Connecticut River, Rhode Island (one confirmed site on the coast in Little Compton; Enser 1992), widespread but uncommon in Massachusetts, most in e. Massachusetts, Nantucket and Monomoy, and on outer Cape Cod, n. Vermont in the Champlain Lowlands and n. to the Northeast Highlands, se. New Hampshire including one confirmed site in the Northeast in the Androscoggin drainage basin (Robinson 1994:28 in Foss 1994), and throughout s. and e. Maine.

Status in New England: Uncommon breeder; common migrant.

Habitat: Breeding: Freshwater marshes, marshy edges of lakes, streams, and ponds, sloughs, sedge meadows, prairie potholes, and occasionally in boreal and deciduous forests. Favors large freshwater marshes and ponds with emergent vegetation (Bull 1974:126). Prefers shoreline habitat to open water, and prefers calm water or sluggish water to fast water. Rarely uses salt or brackish areas (Pough 1951:86). Will use man-made ponds (Belanger and Couture 1988). Winter: Shallow inland freshwater marshes, coastal brackish and saltwater marshes.

Special Habitat Requirements: Freshwater marshes, sloughs, ponds, lakes, and sluggish streams.

Nesting: Egg dates: May 3 to July 4, New York (Bull 1974:126). Clutch size: 6 to 15, typically 8 to 11. Incubation period: 23 to 24 days. Nestling period: Less than 1 day (precocial). Broods per year: 1. Age at sexual maturity: 1 yr. Nest site: Prefers dense grassy sites, such as bluegrass, hayfields, and sedge meadows, or ground under bushes within a mile of water's edge. Occasionally builds on a sedge tussock or muskrat house surrounded by water.

Territory Size: Male appears to defend the female rather than an area of land (Johnsgard 1975:278, Bellrose 1976:281).

Sample Densities: 18 pairs per 100 acres (40 ha) on impoundments in Alberta (Keith 1961). 17.4 to 63.6 pairs per 100 acres (40 ha) on a variety of ponds in South Dakota (Drewien and Springer 1969). 4 to 22 pairs per 100 acres (40 ha) of wetland at 4 wetlands in Wisconsin (Jahn and Hunt 1964).

Food Habits: Seeds of sedges, grasses, pondweeds, and smartweeds (staple foods); also eats leaves of aquatic plants, snails, crustaceans, and insects in water, mud, short grasses. Preferred feeding habitat: Mud flats, shallow water, wet fields.

Comments: Blue-wings are the last waterfowl to arrive in spring and the first to leave in the fall. Populations in decline in many areas due to habitat destruction, and perhaps overhunting. Hens may lead broods over land distances up to 4,800 ft (1,463 m) to suitable marsh brooding areas (Robinson 1994:28 in Foss 1994).

Northern Shoveler

(Anas clypeata)

Range: Breeding: From n. Alaska to n. Manitoba, s. to nw. and e. Oregon, n. Utah, Colorado, Nebraska, Missouri, and central Wisconsin. Winter: From the coast of s. British Columbia to central Arizona; e. to the Gulf Coast and to South Carolina, on the Atlantic Coast s. to South America. In Mexico, a widespread winter visitor. In Honduras, a fairly common migrant and winter resident on the Caribbean slope; in Panama, a rare winter visitor. In Costa Rica, a fairly common winter resident in Guanacaste; elsewhere local and in small numbers. In Colombia, usually a regular migrant and winter resident found in the Magdalena and Cauca Valleys. Found virtually throughout the West Indies but uncommon in Puerto Rico and extremely rare in the U.S. Virgin Islands. In Trinidad, a rare winter migrant.

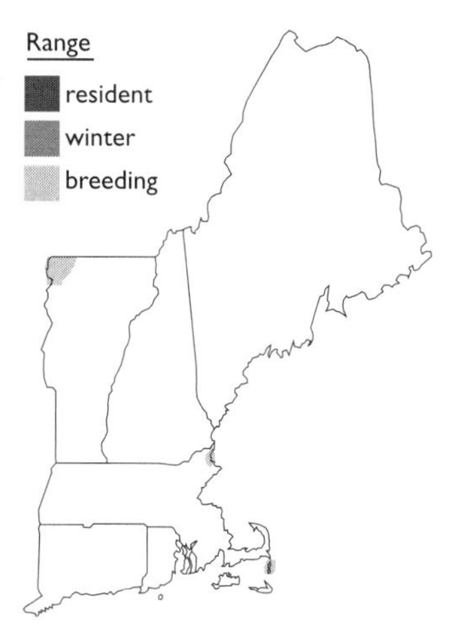

Distribution in New England: Breeding: A rare and local breeder in the Lake Champlain Lowlands in Vermont (Fichtel 1985:393 in Laughlin and Kibbe 1985), in Massachusetts on Plum Island (1975, 1977) and Monomoy (1974) (Veit and Petersen 1993:104). Winter: Rare and local on Cape Cod and Nantucket, very rare inland (Veit and Petersen 1993:104). Common migrant at Plum Island and Monomoy, Mass., fairly uncommon or rare elsewhere (Veit and Petersen 1993:104).

Status in New England: Rare breeder.

Habitat: Prefers shallow prairie marshes, margins of lakes and ponds, and saline wetlands with abundant plant and animal life floating on the surface, but also inhabits potholes, sloughs, marshes in taiga forests, and sewage-treatment ponds. Tolerates a wide range of water conditions from clear to stagnant, and highly alkaline.

Special Habitat Requirements: Shallow water with muddy bottoms, surrounded by dry grassy areas for nesting.

Nesting: Egg dates: Early April to early May, Delaware (Palmer 1976 V. 2:512). Clutch size: 6 to 14, typically 10 to 12 (DuBowy 1996). Incubation period: 22 to 24 days. Nestling period: Less than 1 day (precocial). Broods per year: 1. Age at sexual maturity: 1 yr. Nest site: A slight hollow on dry ground, preferably in short grasses within 100 m of water, but will nest in hay fields and meadows if grasses are not available. Seldom nests in weedy patches and avoids woody vegetation such as willows.

Territory Size: Males are not truly territorial, but will defend their mates and a core area of their home range during breeding season (DuBowy 1996).

Home Range: 8 home ranges ranged from 20 to 128 acres (8.1 to 51.8 ha), average 76 acres (30.8 ha) (Poston 1969). 6 home ranges ranged from 15 to 90 acres (6.1 to 36.4 ha), average 49.7 acres (20.1 ha) (Poston 1969).

Sample Densities: 4.5 birds per square mile (2 pairs/km^2) in mixed prairie habitat (Bellrose 1976:293). 12.7 pairs per square mile (5 pairs/km^2) on a 3-square-mile (7.8 km^2) study area (Poston 1969). 44 pairs per square mile (17 pairs/km^2) in favorable habitat in North Dakota (Stewart and Kantrud 1972).

Food Habits: Aquatic insects, ostracods, copepods, snails, fish, seeds of aquatic plants, plankton, also some sedges, grasses, water lilies, pondweeds, bulrush seeds, algae, and smartweeds in shallow water (surface and bottom). Strains water through the lamellae of mandibles in order to separate and consume small aquatic invertebrates and seeds.

Comments: Probably much more numerous in New England in the early days of settlement (Forbush 1916 in Zeranski and Baptist 1990:64).

Northern Pintail

(Anas acuta)

Range: Breeding: Northern Alaska across Canada to n. and e. Quebec, New Brunswick, and Nova Scotia to California, across to the Great Lakes, St. Lawrence River, and Maine. Winter: Southern Alaska, s. to n. New Mexico, and e. to central Missouri and the Ohio Valley (uncommonly); along the Atlantic Coast from Massachusetts, s. throughout the s. United States, throughout Mexico, and to South America. In Honduras, a regular migrant and winter resident. In Panama, a rare migrant and winter resident; no longer occurs in the Canal Zone with any regularity. A fairly common winter resident in the West Indies e. to Hispaniola, but rare in the Virgin Islands.

Distribution in New England: Breeding: Vermont, restricted to the Champlain Lowlands (Ellis and Ellison 1985:56 in Laughlin and Kibbe 1985); occasionally at Plum Island and Ipswich, Mass., since introduction in Topsfield in the late 1950s (Veit and Petersen 1993:102). In mid 1980s, several pairs at Monomoy I. Winter: Massachusetts, Connecticut, and Rhode Island along the coast.

Status in New England: Uncommon local breeder; has been declining for many years, especially in the w. and central United States (DeGraaf and Rappole 1995:80).

Habitat: Breeding: Typically inhabits open country with low vegetation interspersed with shallow, seasonal, or intermittent wetlands. Frequents lakes, rivers, marshes, and ponds in grasslands, barrens, dry tundra, open boreal forest, and cultivated fields. Nests up to 1 mile (1.6 km) from water (Bellrose 1980). Winter: Inhabits shallow inland fresh and brackish wetlands, flooded agricultural fields, tidal wetlands, mudflats along rivers, and sounds and bays.

Special Habitat Requirements: Mudbanks or exposed water margins (Palmer 1976 V. 2:446) and shallow wetlands for feeding.

Nesting: Egg dates: May 17 to July in Massachusetts (Veit and Petersen 1993:102). April to July (Bellrose 1976). Clutch size: 3 to 12, typically 7 to 9 (Austin and Miller 1995). Incubation period: 22 to 24 days. Nestling period: Less than 1 day (precocial). Broods per year: 1. Age at sexual maturity: 1 yr. Nest site: Often builds a nest in a dry hollow on dry ground, may be concealed by grasses or shrubs. Nest usually in stubble fields, in a dry portion within a large marsh, or in lightly grazed pastures; rarely at the edge or over water (Austin and Miller 1995). Generally avoids extensively wooded or brushy areas.

Territory Size: No evidence of territoriality (Austin and Miller 1995).

Sample Densities: 5.6 pairs per square mile (2 pairs/km^2) in South Dakota (Drewien and Springer 1969). About 12 pairs per 100 acres (40 ha) of water in Alberta (Keith 1961). 32 pairs per square mile (12 pairs/km^2) in favorable habitat in North Dakota (Stewart and Kantrud 1972).

Food Habits: The pintail is chiefly a seed eater, preferring seeds of pondweeds, sedges, grasses, smartweeds, and cultivated grains. Also consumes aquatic insects, crustations, and snails (Austin and Miller 1995). Preferred feeding habitat: Shallow waters of marshes, ponds, meadows, and grain fields.

Comments: In s. New England, northern pintails were common, even abundant migrants until the 1850s; overhunting rendered them rare by 1900 (Zeranski and Baptist 1990:63). On wintering grounds, pintails feed in uplands on mast or grain or on tidal flats where they pick up marine animals (Pough 1951:82). Anderson (1959 in Palmer 1976 V. 2:458) found that the contents of 881 stomachs taken in Illinois consisted of 97 percent vegetable and 3 percent animal matter.

Green-winged Teal

(*Anas crecca*)

Range: Breeding: Alaska and nw. and s. Northwest Territories to n.-central Labrador and Newfoundland, s. to central Oregon, Colorado, s. Ontario and Quebec, and Nova Scotia; breeds locally from s. California, e. to s. New Mexico, Iowa, and Pennsylvania, and on the Atlantic Coast to Delaware. Winter: Southern Alaska and s. British Co-

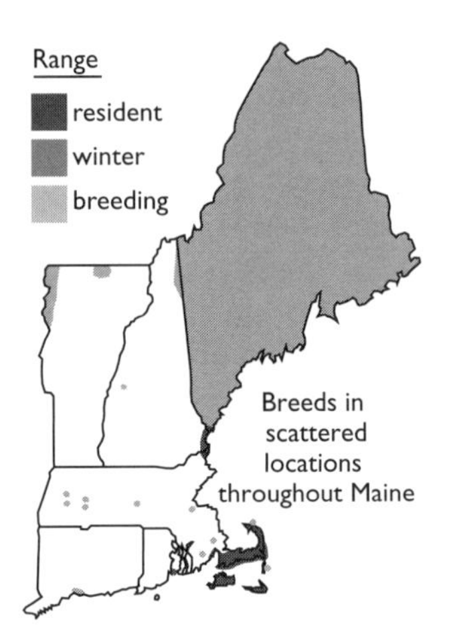

lumbia to New Brunswick and Nova Scotia, s. to Central America; also in the Hawaiian Islands. In Honduras, a rare winter migrant from the United States. In Tobago, a common migrant and winter resident. In Guatemala, a rare migrant and winter resident found in the Pacific lowland and on the volcanic lakes; could occur in the Caribbean lowland. Migrates and winters throughout Mexico. A rare migrant and winter resident in the West Indies, apparently most numerous in the Caribbean in Cuba (October–April).

Distribution in New England: Breeding: Throughout Maine, w. to Vermont in the n.-central region and Northeast Highlands and the Champlain Lowlands (Ellis 1985:50 in Laughlin and Kibbe 1985), s. to Massachusetts locally in swamps and ponds throughout the state (Veit and Petersen 1993:98), to extreme ne. Rhode Island (Enser 1992:40). Winter: Coastal New Hampshire (Staats and Richards 1994:22 in Foss 1994), s. to Massachusetts, especially on Cape Cod and Nantucket (Veit and Petersen 1993:98).

Status in New England: Relatively uncommon and local breeder (Vermont, New Hampshire, Massachusetts) to common (Maine). Irregular winter resident.

Habitat: Breeding: Inland lakes, ponds, sloughs, mudflats, and prairie potholes with dense rushes or other emergent vegetation, adjacent to grasslands with thickets or woods, dry hillsides, or sedge meadows for nesting (Bellrose 1980, DeGraaf and Rappole 1995:77, Johnson 1995). Will use beaver and human impoundments in wooded areas (Baldassarre and Bolen 1994); also uses northern boreal forests, forested wetlands, and ponds on mixed and shortgrass prairies (Bellrose 1980). Winter: Fresh and brackish marshes, lakes, ponds, shallow streams, sloughs, forested wetlands, brackish estuaries, and flooded agricultural fields. Uses mudbanks, stumps, logs, and low limbs of dead trees for resting and perching.

Special Habitat Requirements: Shallow streams, marshes, ponds, pools, and sloughs.

Nesting: Egg dates: Mid May to June 30 in Massachusetts (Veit and Petersen 1993:98). May 25 to July 15, New York (Bull 1974:123). Clutch size: 6 to 18, typically 10 to 12 (Harrison 1975). Incubation period: 21 to 23 days. Nestling period: Less than 1 day (precocial). Broods per year: 1. Age at sexual maturity: 1 yr. Nest site: Nests in a depression on dry ground in dense grass, at the base of shrubs, or under a log, may be 0.25 miles (400 m) or more from water, usually 2 to 300 ft (1 to 91 m).

Sample Densities: 1 pair per 60 acres (24.3 ha) in grasslands in Alberta (Keith 1961). 20 pairs per square mile (8 pairs/km^2) in favorable habitat in North Dakota (Stewart and Kantrud 1972).

Food Habits: Seeds (staple) of wetland plants, especially millets, smartweed, and nutgrasses; insects, crustaceans, mollusks. Preferred feeding habitat: Mudflats. Generally forages in flooded or dry grain fields and woodlands in the spring and fall (Nummi 1993, Johnson 1995).

Comments: Green-winged teal were common migrants until about 1850 in s. New England (Cruickshank 1942), even abundant at times (Forbush 1916). Greatly reduced by hunting in the late 1800s, they were not again common migrants until the 1940s. Green-winged teal were first confirmed nesting in Massachusetts in 1954 (see Veit and Petersen 1993:93); no nesting yet confirmed in Connecticut (Zeranski and Baptist 1990:61). Birds do not commonly nest in much of the Northeast. They are uncommon in winter but may appear in large groups locally. Occasionally they are seen in company of Eurasian green-winged teal (*A. c. crecca*).

Canvasback

(*Aythya valisineria*)

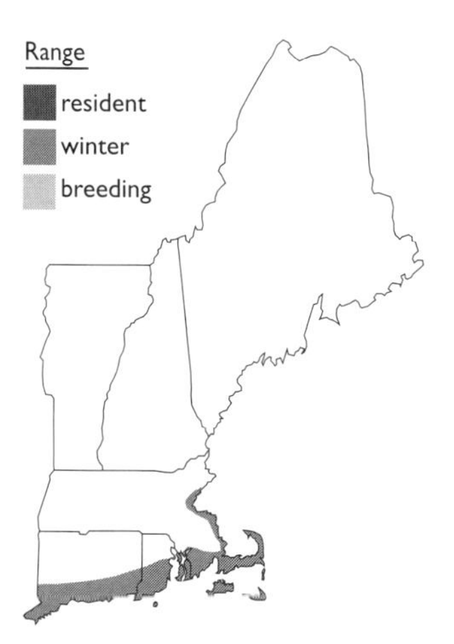

Range: Breeding: Central Alaska and n. Yukon to e. Ontario and s. to s.-coastal Alaska; locally in inland areas to ne. California across to n. Utah, central New Mexico, nw. Iowa, and s. Ontario. Winter: Along the Pacific Coast from the central Aleutians and s.-coastal Alaska, s. to Baja California, from Arizona and New Mexico to the Great Lakes, and on the Atlantic Coast from New England s. to the Gulf Coast and Mexico. In Mexico in winter, s. to central Mexico (Michoacan, Federal District, and Veracruz). In Guatemala and Honduras, a rare winter migrant from the United States. A rare fall and winter visitor to Cuba and Puerto Rico.

Distribution in New England: Winters in Massachusetts along the south shore, Cape Cod and the Islands, and Cambridge (Veit and Petersen 1993:107), w. to s. Rhode Island and s. Connecticut along the coast.

Status in New England: Uncommon to common and local winter resident.

Habitat: Breeding: Prefers shallow, fresh and brackish marshes, ponds, lakes and rivers bordered by emergent vegetation (especially cattails and bulrushes) and little if any woody vegetation around the shoreline. Winter: primarily estuaries, sheltered bays, and inland lakes free from ice and that contain wild celery and pondweeds.

Special Habitat Requirements: Marshes, ponds, lakes, and rivers bordered by emergent vegetation and with stretches of open water for taking off and landing.

Nesting: Egg dates: Usually May to June (Terres 1980). Clutch size: 7 to 12, typically 9 to 10 eggs (Bent 1923). Incubation period: 23 to 29 days. Nestling period: 2 days. Age at first flight: 63 to 77 days (Hochbaum 1944). Broods per year: 1. Age at sexual maturity: 1 yr. Nest site: Typically on floating mats of emergent vegetation, over water 6 to 12 in deep, and in vegetation (bullrushes, reeds, or cattails), occasionally uses muskrat houses. Nests are generally 3 to 60 ft (1 to 18 m) from the edge of open water. Rarely nests on dry ground.

Food Habits: Seeds and vegetative parts of aquatic plants, especially wild celery (*Vallisneria* spp.) and pondweeds (*Potomogeton* spp.), aquatic invertebrates, small fishes, and some small mollusks (Noyes and Jarvis 1985).

Comments: Winter surveys in the United States have shown a decline in canvasbacks over the past 20 years. Declines may be due to loss of breeding habitat through drainage and drought (Hohman et al. 1990). The abundance of canvasbacks has varied greatly in Massachusetts during the past century. Scarce before 1900, between 1910 and 1950 flocks of 100 to 150 were found on Cape Cod and the Islands in fall and winter (see Veit and Petersen 1993 for review).

Ring-necked Duck

(*Aythya collaris*)

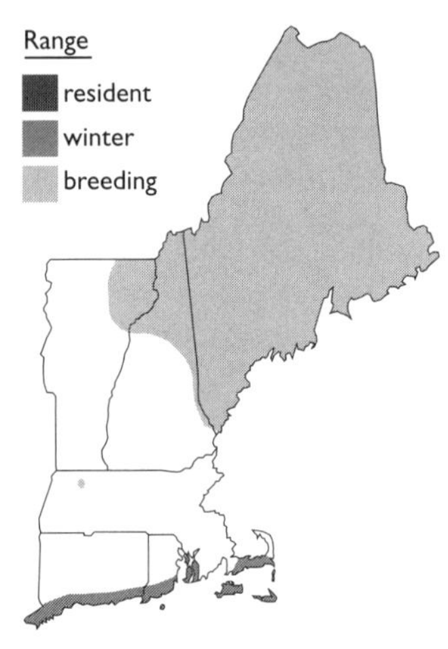

Range: East-central and se. Alaska, and from central British Columbia and nw. and s. Northwest Territories to Newfoundland and Nova Scotia, s. to ne. California, se. Arizona, n. Illinois, and Massachusetts. Winter: On the Pacific Coast from se. Alaska, in the interior from s. Nevada to the lower Mississippi and Ohio Valleys, and on the Atlantic Coast from New England s. through the s. United States, throughout Mexico, and to Panama. A rare migrant and winter visitor throughout Guatemala and in Honduras in the Caribbean lowlands. In Costa Rica and Panama an uncommon to locally common winter resident. An uncommon and local visitor to Puerto Rico; rare on St. Croix, and extremely rare on St. John and St. Thomas. Winters as far s. as Barbados, St. Lucia, and St. Andres; also Trinidad and Margarita Island.

Distribution in New England: Breeds throughout Maine, w. to n. and central New Hampshire and the Northeast Highlands in Vermont. Last reports of breeding in Massachusetts in Ashfield in 1978 and 1979. Winter: Along the coast of Connecticut, Rhode Island, and mid Cape Cod and the Islands.

Status in New England: Common (Maine) to rare and local breeder (Massachusetts). Common spring and abundant fall migrant (Veit and Petersen 1993:109).

Habitat: Commonly breeds in shallow (<1.5 m), dense

swamps, marshes, and bogs with low vegetation cover (especially sweetgale or leatherleaf), typically with a pH range of 5.5 to 6.8, and preferably near or in woodlands. Also uses sloughs, beaver flowages, and small potholes near larger wooded lakes, rivers, or reservoirs with submerged and emergent vegetation. In winter, fresh or brackish marshes, lakes, rivers, and estuaries. Seldom uses strictly saline water.

Special Habitat Requirements: In both summer and winter, prefers shallow freshwater wetlands with stable water levels and abundant emergent and submerged or floating plants (Hohman and Eberhardt 1998).

Nesting: Egg dates: Mid May to early July. Peak: Mid May (Maine) (Mendall 1958). Clutch size: 6 to 14, typically 8 to 10. Incubation period: 25 to 29 days. Nestling period: 1 day (precocial). Age at first flight: 49 to 56 days. Broods per year: 1. Age at sexual maturity: 1 yr. Nest site: In Maine, nests are typically on a floating mat of vegetation, but often in clumps of herbaceous or shrubby growth or on islands. Common cover plants are sedges, sweet gale, and leather leaf. Most nests are within a few feet of open areas with water (Bellrose 1976:332). Occasionally nests in upland habitats (Evrard et al. 1987).

Territory Size: Pairs space themselves but show little aggression. In Maine, Mendall (1958) observed ducks nesting as closely as 5 or 6 ft (1.5 to 1.8 m).

Sample Densities: 1 pair per 6 acres (2.4 ha) to 1 pair per 23 acres (9.3 ha) in various habitats (Mendall 1958:65). 6 nests were found on a ¼-acre (0.1 ha) island in Maine (Mendall 1958). Average (6 yr) 9 pairs per 100 acres (40 ha) in n. Wisconsin wetlands (Jahn and Hunt 1964).

Food Habits: Seeds and vegetative parts (tubers) of submergent and emergent plants, 75 percent of diet; aquatic invertebrates 25 percent of diet (Bellrose 1976:334). Preferred feeding habitat: Shallow water, usually less than 6 ft (2 m) deep (Bellrose 1976:334).

Comments: May have been numerous in Colonial times (Forbush 1916) but a very rare migrant in the 1800s (Griscom and Snyder 1955). First wintered in Connecticut in the early 1960s (Zeranski and Baptist 1990:68).

Bufflehead

(*Bucephala albeola*)

Range: Breeding: From central Alaska to ne. Manitoba and n. Ontario, s. to n. Washington, s. Manitoba, and locally in s. Ontario; also locally s. to the mountains of n. California, and to Wyoming, Iowa, and Wisconsin. Winter: From the Aleutian Islands on the Pacific Coast, the Great Lakes, and Newfoundland on the Atlantic, s. in coastal states and the Ohio and Mississippi Valleys, to the s. United States and Mexico.

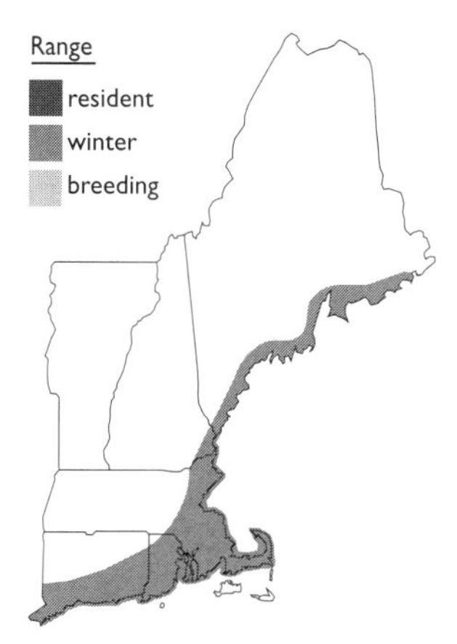

Distribution in New England: The coast of Maine, coastal New Hampshire, s. to e. Massachusetts, central and s. Rhode Island, and s. Connecticut.

Status in New England: Abundant migrant and common resident on the coast in winter.

Habitat: Saltwater in sheltered coves, harbors, and bays, and in estuaries, and inland on rivers, ponds, lakes, and reservoirs with open water (Gauthier 1993).

Food Habits: Aquatic invertebrates—crustaceans and mollusks (90 percent of diet), also fish, seeds of pondweeds (*Potomogeton*), widgeon grass (*Ruppia*), and bulrush (*Scirpus*) (Gauthier 1993, Erskine 1971). Preferred feeding habitat: Shallow water 4 to 15 ft (1.2 to 4.6 m) deep over tidal flats and in large open bodies of water (Gauthier 1993).

Comments: The smallest diving duck in North America, bufflehead are fairly common inland migrants on rivers, lakes, and ponds. Winter populations have increased greatly over the past 50 years (Veit and Petersen

1993:120); one of the few species of ducks whose numbers have increased markedly since the 1950s (Gauthier 1993). In its boreal forest breeding grounds, nests in old flicker holes.

Common Goldeneye

(*Bucephala clangula*)

Range: Breeding: Western Alaska and n. Yukon to central Labrador and Newfoundland, s. to central Alaska, n. Washington across to n. Michigan, Maine, New Brunswick, and s. to s. California, on the Great Lakes, in the Mississippi and Ohio Valleys, and s. to the Gulf Coast, and on the Atlantic Coast from Newfoundland s. to Florida; irregularly elsewhere in the interior of the United States (DeGraaf et al. 1991:60). Winter: Coastal Newfoundland, Gulf of St. Lawrence, to w. British Columbia, and coastal Alaska, including the Aleutian Islands, s. throughout North America to Central Baja California, central Arizona, New Mexico, s. to the Gulf Coast, central Florida, and Tamaulipas, Mexico. Rare in inland waters of Georgia, South Carolina, and North Carolina. Highest concentrations in coastal bays from n. New England s. to Chesapeake Bay and on the coasts of se. Alaska and British Columbia. Remains on inland water as long as it stays open (Eadie et al. 1995:2).

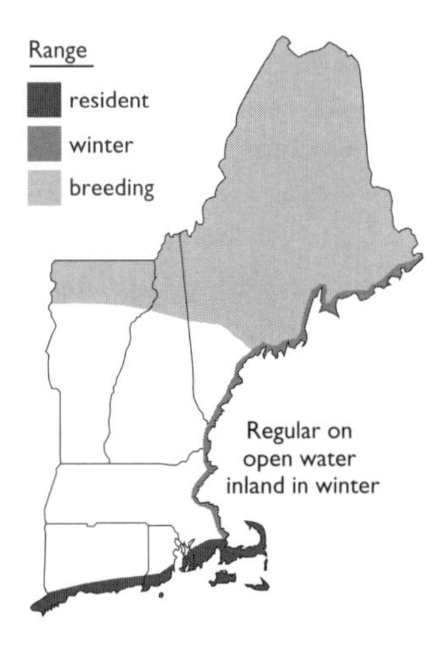

Distribution in New England: Breeding: Throughout most of Maine, n. New Hampshire, and n. Vermont e. to the Champlain Lowlands. Does not breed in Massachusetts, Connecticut, or Rhode Island. Winter: Coastal regions (bays, estuaries, tidal rivers) from Maine s. to Rhode Island and Connecticut, and inland as far n. as open water and food are available. Very common to abundant migrant and winter resident in large coastal bays.

Status in New England: Locally common winter resident on the coast, common migrant inland.

Habitat: Breeding: Lakes, ponds, shallow rivers, forested wetlands, and bogs bordered by open mature forests with large cavity trees. Prefers lakes with clean water, abundant aquatic invertebrates (insects, mollusks, crustaceans), and moderate densities of emergent or submergent vegetation (Eadie et al. 1995). Will nest in both coniferous and deciduous trees (Eadie et al. 1995). May avoid lakes where fish (e.g., yellow perch) compete for invertebrate prey (Eriksson 1978, 1983; Eadie and Keast 1982), although does nest successfully on lakes containing fish (Zicus et al. 1995). Abundant on acidic, fishless lakes (Mallory et al. 1994). Appears tolerant of moderate to heavy year-round human disturbance on large (ca. 1,250 ha) lakes in Minnesota (Zicus et al. 1995). Winter: Bays, estuaries, rivers; and inland as far n. as open water and food are available.

Special Habitat Requirements: Large trees with cavities for nesting, near clear, clean water with abundant aquatic invertebrates for feeding.

Nesting: Egg dates: April 7 to May 25, New Brunswick (Carter 1958). Clutch size: 5 to 15, typically 8 to 12. Incubation period: 28 to 32 days. Nestling period: 1 to 2 days (precocial). Age at first flight: 56 (Carter 1958) to 62 days (Johnsgard 1975, Bellrose 1980). Broods per year: 1. Age at sexual maturity: 2 yr. Nest height: 6 to 60 ft (1.8 to 18.3 m), typically 20 ft (6.1 m). Nest site: Cavities in trees (live or dead) in or near water, but up to 1.3 km from the shoreline. Prefers hollowed tops of standing stubs, but also uses cavities formed by broken limbs or holes made by pileated woodpeckers. When cavities are not available, may use abandoned buildings or cavities in rocks. Readily uses nest boxes; prefers those with dark interior and lined with wood chips, and placed high above the ground. Females often use previous year's nest (Lumsden et al. 1986).

Territory Size: No size information. The male defends a small area surrounding the nest site and females defend a brood territory.

Sample Densities: Approximately 1 pair per 100 acres (40 ha) of hardwood swamp in New Brunswick (Carter 1958), approximately 4 birds per 100 acres (40 ha) in Maine (see Eadie et al. 1995:18).

Food Habits: Primarily aquatic invertebrates (insects, mollusks, crustaceans), occasionally small fish or spawn (Eadie et al. 1995). Seeds, tubers, and leafy parts of pondweeds make up a small proportion of the diet (Bellrose 1980).

Comments: 16 cavities in New Brunswick were located at an average height of 23 ft (7.0 m) above the ground in trees that averaged 26 in (67 cm) in diameter. The trees were about 250 yr old (Prince 1968). Cottam (1939 in Palmer 1976 V. 3:397) found that the contents of 395 stomachs contained 74 percent animal and 26 percent vegetable matter.

Hooded Merganser

(Lophodytes cucullatus)

Range: Breeds from s. Alaska to Nova Scotia, s. Oregon and Idaho, e. to Maine and Massachusetts, and locally in the Mississippi Valley and se. United States. Winters from British Columbia and New England s. to California, Texas, and Florida. Rare migrant and winter visitor to n. Mexico, recorded to Baja California, Tamaulipas, Veracruz, Michoacan, and the Valley of Mexico. A rare winter visitor in the West Indies, where it has been recorded from the Bahamas, Cuba, Puerto Rico, St. Croix, and Martinique.

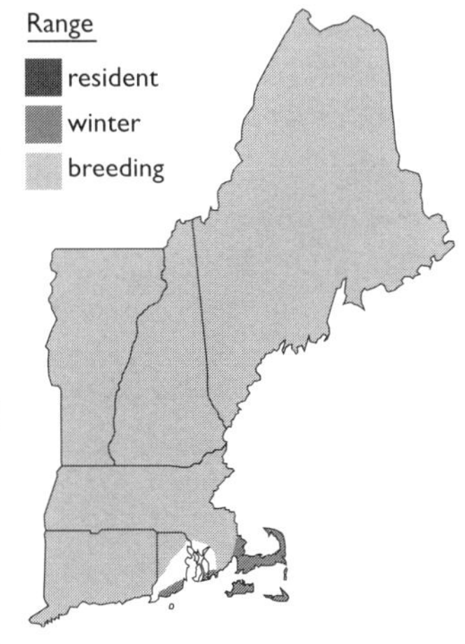

Distribution in New England: Breeding: Throughout Maine, New Hampshire, Vermont, and Massachusetts, except absent from Cape Cod and the Islands (Veit and Petersen 1993:121); Connecticut in the n. Farmington River watershed (Zeranski and Baptist 1990:77), and nw. Rhode Island (Enser 1992:45). Winter: Massachusetts, uncommon but regular on Cape Cod and the Islands (Veit and Petersen 1993:121), occasional in winter in New Hampshire and Vermont on open water (Richards 1994:36 in Foss 1994, Norse and Fichtel 1985:64 in Laughlin and Kibbe 1985). Rare in Maine.

Status in New England: Uncommon and local breeder; common migrant, especially in fall. Uncommon but regular in winter on Cape Cod and the Islands (Veit and Petersen 1993:121). Increasing in the e. part of its range.

Habitat: Breeding: Inhabits freshwater forested wetlands, forested margins of rivers, streams, small lakes, ponds, reservoirs, and emergent marshes. Prefers clear freshwater habitats, especially those with sandy, gravelly, or cobbled bottoms, and abundant small fish. Most abundant on wide 50–66 ft (15–20 m), swift 0.7–1 ft/sec (0.2–0.3 m/sec), shallow 1.7 ft (0.5 m), forested rivers with cobbled bottoms in Wisconsin (Kitchen and Hunt 1969). Roosts on exposed rocks, logs, and unvegetated sandbars (Beard 1964). Tends to avoid areas of human activity.

Special Habitat Requirements: Clear forest streams, rivers, ponds, lakes, and swamps with nearby cavity trees (Dugger et al. 1994).

Nesting: Egg dates: March 30 to early June (Veit and Petersen 1993:121). April 25 to July 2, New York (Bull 1974:152). Clutch size: 8 to 12, typically 10. Incubation period: 28 to 41 days. Nestling period: 1 day (precocial). Broods per year: 1. Age at sexual maturity: Probably 2 yr. Nest site: Natural tree cavities, old pileated woodpecker holes, or nest boxes. Prefers tree cavities or nest boxes in or near water (Morse et al. 1969).

Sample Densities: Average 2.14 broods per mile (0.6 km) of river. Highest densities occurred on heavily wooded rivers—lowest densities were found on marshy rivers (Kitchen and Hunt 1969). Density may be related to availability of food rather than nest sites (Kitchen and Hunt 1969).

Food Habits: Cottam and Uhler (1937 in Palmer 1976:459) found that the contents of 138 stomachs contained fishes (44 percent), crayfish (22 percent), other crustaceans (10 percent), aquatic insects (13 percent), and vegetable matter (4 percent). Preferred feeding habitat: Clear shallow water, usually less than 24 in (60 cm) deep.

Comments: Probably once common in presettlement times, hooded mergansers declined greatly in the 1800s as forests were cleared and beaver extirpated. Return of forests and beaver and provision of nest boxes have enabled the bird to recover. The most common nesting duck at Umbagog Lake in n. New Hampshire in 1870 (Brewster 1924), it was the rarest after a few years of heavy shooting (Richards 1994:36 in Foss 1994).

Common Merganser

(*Mergus merganser*)

Range: Breeding: Central and s.-coastal Alaska, n. Saskatchewan, and Newfoundland, s. to the mountains of central California and n. New Mexico. East of the Rockies, breeds s. to s. Saskatchewan, central Michigan, s. Maine, and to w.-central Nova Scotia. Winter: The Aleutian Islands and south-coastal Alaska, e. across to s. Canada to Newfoundland and s. to s. California and the Gulf Coast to Central Florida (DeGraaf et al. 1991:64).

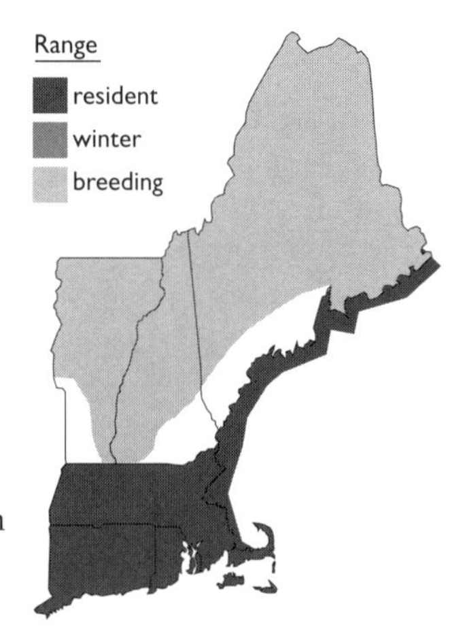

Distribution in New England: Breeding: Rare in Connecticut (n. Farmington River watershed), Rhode Island, Massachusetts (Berkshire and Worcester counties, and in the Connecticut River Valley; Veit and Petersen 1993:122), common in n. Vermont, n. New Hampshire (Richards and Staats 1994), and throughout Maine except the s. coast. Winter: Southern New England, coastal Maine.

Status in New England: Locally common to uncommon breeder in n. New England, rare in Massachusetts. Common winter resident in s. New England.

Habitat: Breeding: Clear ponds, lakes, rivers, and reservoirs with forested shorelines. Rare on saltwater. Winter: Prefers fresh and brackish waters of lakes, rivers, ponds, and estuaries.

Special Habitat Requirements: Clear forested lakes, rivers, and remote ponds and streams, large trees with cavities for nesting.

Nesting: Egg dates: June to early July in Massachusetts (Veit and Petersen 1993: 122), May 20 to June 25 in Vermont (Norse and Fichlel 1985:66 in Laughlin and Kibbe 1985). May 5 to July 10, New York (Bull 1974:155). Clutch size: 6 to 17, typically 6 to 10. Incubation period: 28 to 35 days. Nestling period: 1 to 2 days (precocial). Broods per year: 1. Age at sexual maturity: Probably 2 yr. Nest height: To 50 ft. Nest site: Usually in a natural tree cavity or abandoned pileated woodpecker hole at any height. Sometimes on ground in holes in banks, on cliffs or piles or rocks. Ground nests are well hidden under low limbs, overhanging rocks or dense shrubs. Nests are usually placed close to water. Accepts nest boxes.

Territory Size: No size information. Pairs are generally widely spaced, probably because of feeding requirements.

Sample Densities: 1 or 2 pairs per 16-mile (25.6-km) stretch of river in Michigan (Parmalee 1954). 34 pairs per 2.3 square miles (34 pairs/6 km^2) on islands off Finnish coast (Hilden 1964).

Food Habits: Fishes (staple), mollusks (winter). Roots and stems of aquatic plants, leeches, frogs, aquatic salamanders, worms, aquatic insects (Timken and Anderson 1969). Preferred feeding habitat: Calm to rapid flowing shallow water, 1.5 to 6 ft (0.5 to 1.8 m) deep (Johnsgard 1975:514).

Comments: Small groups of common mergansers have been known to cooperatively drive small fish into shallow water for easy capture (Johnsgard 1975:514). Common mergansers and loons often nest on the same lakes, and confrontations are frequent (Richards and Staats 1994:38 in Foss 1994). Apparently first nested in Connecticut in 1962 (Zeranski and Baptist 1990:78).

Red-breasted Merganser

(*Mergus serrator*)

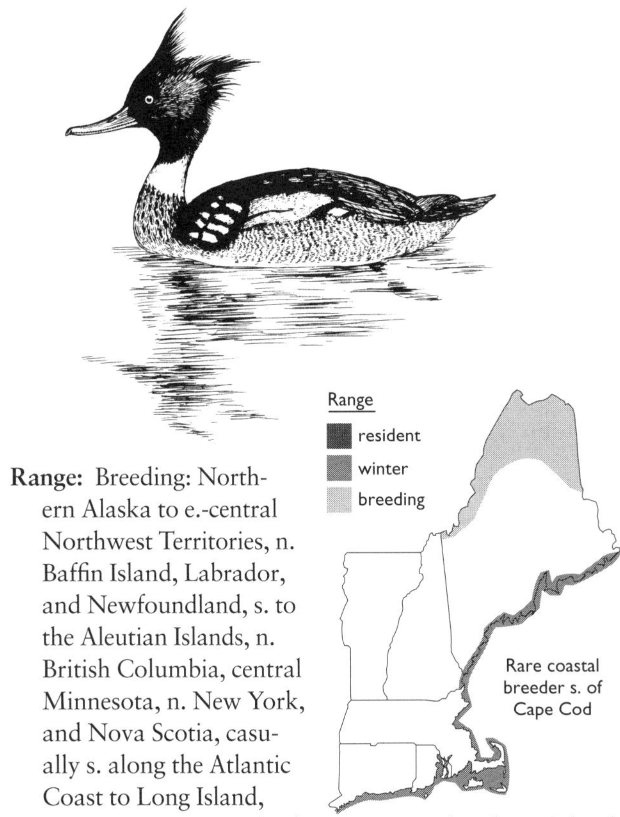

Range: Breeding: Northern Alaska to e.-central Northwest Territories, n. Baffin Island, Labrador, and Newfoundland, s. to the Aleutian Islands, n. British Columbia, central Minnesota, n. New York, and Nova Scotia, casually s. along the Atlantic Coast to Long Island, N.Y. Winter: Primarily along coasts and on large inland bodies of water from s. Alaska, the Great Lakes, and Nova Scotia, s. to Baja California, s. Texas, and the Gulf Coast. In Mexico, winters mainly on both coasts of Baja California and in Sonora, Sinaloa, Chihuahua, and Tamaulipas. Casual in the West Indies, where it has been recorded from the Bahamas, Cuba, and Puerto Rico.

Distribution in New England: Breeding: Northern and coastal Maine. One nest record from Vermont (1980 Calais, Washington County; Fichtel 1985:68 in Laughlin and Kibbe 1985); an occasional breeder in Massachusetts, most recent nest record in 1989, Duxbury Beach, may breed elsewhere in Essex County and around the Elizabeth Islands (Veit and Petersen 1993:122). Winter: Along the e. and s. coasts.

Status in New England: Rare breeder in Massachusetts, uncommon in n. New England. Common to abundant in winter along the coast.

Habitat: Breeding: Borders of rivers, ponds, lakes, and manmade impoundments with open shores, gravel bars, or rocks for roosting and preening areas. Also on coastal shores, bays, and tidal channels. Prefers to breed on small vegetated islands or islets (Bellrose 1980). Winter: Mainly in estuaries, sheltered bays, inlets, coves, and the mouths of rivers, less frequently on inland bodies of water.

Special Habitat Requirements: Clear water.

Nesting: Egg dates: June to July (Terres 1980:196). Clutch size: 5 to 16, typically 8 to 10. Incubation period: 26 to 28 days. Nestling period: 1 day (precocial). Age at first flight: about 59 days. Broods per year: 1. Age at sexual maturity: 2 yr. Nest site: Nests on the ground in a scooped-out hollow, may be under conifer boughs or shrubs, or under or between boulders or driftwood in shallow cavities. Usually within 30 ft (10 m) of water. May nest within tern and gull colonies (Young and Titman 1986). Favors areas with scattered boulders. Accepts nest boxes.

Sample Densities: Birds are usually solitary, but at an ideal site 6 to 10 birds will nest closely. Greater densities have been found on islands than on mainland of Iceland (Bengtson 1970).

Food Habits: Fishes (staple), crustaceans, aquatic insects, worms, fish eggs. Preferred feeding habitat: Shallow, sandy coastal shores just beyond the breakers, inlets and river mouths.

Comments: Red-breasted mergansers apparently nested regularly on Monomoy Island, Mass., between 1877 and 1955 (Griscom and Snyder 1955) and again in 1978 (Veit and Petersen 1993:122). Groups engage in cooperative feeding by driving schools of fish into shallow water where they can be caught more easily (Bellrose 1976:453).

Osprey

(*Pandion haliaetus*)

Range: Breeding: Northwest Alaska and n. Yukon to central Labrador and Newfoundland, s. locally to Baja California, central Arizona, s. Texas, the Gulf Coast, and s. Florida. Winter: Central California, s. Texas, the Gulf Coast, and Florida, s. to Argentina. In Mexico, a widespread winter visitor. In Guatemala, Costa Rica, Honduras, and Panama, a common migrant and winter resident with the local population being augmented by n. birds. Mainly a winter resident s. to n. Chile, n. Argentina, and Uruguay. A fairly common winter resident in Puerto Rico and the Virgin Islands. Breeds in the Bahamas and on cays of Cuba. North American individuals winter throughout the West Indies.

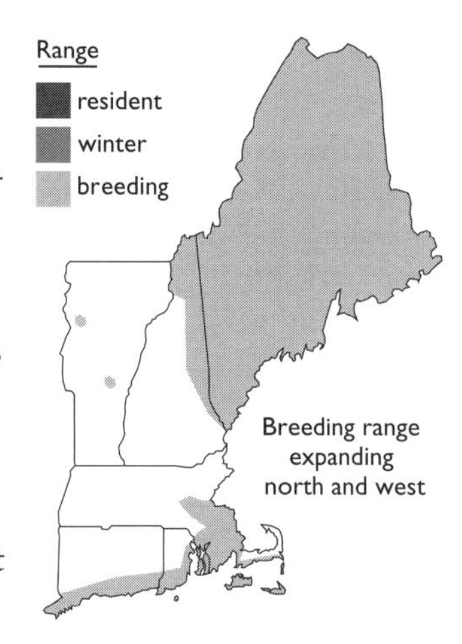

Distribution in New England: Breeds throughout Maine, w. to n. New Hampshire and Great Bay, s. to se. Massachusetts and the islands s. of Cape Cod (Veit and Petersen 1993:126), w. along the coastal regions of Rhode Island and Connecticut from the Pawtucket River w. to Guildford (Zeranski and Baptist 1990:81).

Status in New England: Locally common to uncommon breeder, numbers increasing in many areas where nesting platforms have been installed; fairly common migrant.

Habitat: Breeding: Occupies a wide range of habitats in close proximity to large bodies of water, primarily lakes and rivers, and along coastal waters.

Special Habitat Requirements: Elevated nest sites near water with abundant fish resources; preferably with little human disturbance.

Nesting: Egg dates: May 2 to early July in Massachusetts (Veit and Petersen 1993:127). April 27 to June 21, New York (Bull 1974:169). Clutch size: 3 to 4, typically 3. Incubation period: 35 to 40 days (Poole 1989). Age at first flight: 8 to 10 weeks. Broods per year: 1. Age at sexual maturity: 3 yr. Nest height: Elevated nest sites to 60 ft (18.3 m). Nest site: A variety of structures ranging from natural snags in trees to rocky ledges, sand dunes, telephone pole cross-arms, artificial platforms, billboards, channel buoys, and sometimes on the ground. Prefers nest sites in or near water, that provide good visibility and security. Occasionally nest in loose colonies.

Territory Size: Undetermined. A pair defends the immediate nest site from other ospreys (Ogden 1975).

Sample Densities: Nests may be grouped as close as 65 ft (20 m) (Ogden 1975). Active nests in the Adirondacks (New York) ranged from 4.5 to 20 miles (7.2 to 32 km) apart (Singer 1974).

Food Habits: Fish, also snakes, frogs, storm-petrels, sandpipers, ducks, when fish not available. Preferred feeding habitat: Shallow-water areas of rivers, shoals of lakes where fish are close to the surface.

Comments: In the early 1800s and until about 1870, ospreys nested abundantly along the s. New England coast, especially east of the Connecticut River; they began to decline by the turn of century due to development, but use of chlorinated hydrocarbons, especially DDT after World War II, caused precipitous declines by the 1960s (see Zeranski and Baptist 1990:82 for a historical review). Ospreys show strong attachment to breeding grounds, returning to the same nest or area year after year. The osprey population in the lower Connecticut River–Long Island Sound area is increasing in response to the ban on persistent pesticides and egg transplants from Maryland birds.

Bald Eagle

(*Haliaeetus leucocephalus*)

Range: Breeding: Central Alaska and n. Yukon across Canada to Labrador and Newfoundland, s. locally to the Aleutian Islands, s. Alaska, central Arizona, sw. and central New Mexico, Baja California, and the Gulf Coast; locally distributed in the interior of North America. Winter: Generally throughout the breeding range, but most frequently from s. Alaska and s. Canada southward (DeGraaf et al. 1991:75).

Distribution in New England: Breeding: Throughout Maine, w. to New Hampshire, with a nesting pair in 1989, Umbagog Lake (Evans 1994:373 in Foss 1994) and since 1998 in Hancock (J. Kanter, pers. commun.), s. to w. Massachusetts in the Quabbin Reservoir, to n. Connecticut in the Northwest Hills in habitat associated with large reservoirs. Not nesting in Vermont or

Rhode Island. Winter: Coastal regions throughout New England, also inland if open water; Connecticut along the Connecticut and Housatonic rivers, and large reservoirs in the Northwest Hills (Zeranski and Baptist 1990:83), n. to Massachusetts (Quabbin Reservoir and along the Connecticut River) to n. New Hampshire and along the Merrimack River and on Great Bay (Evans 1994:373 in Foss 1994).

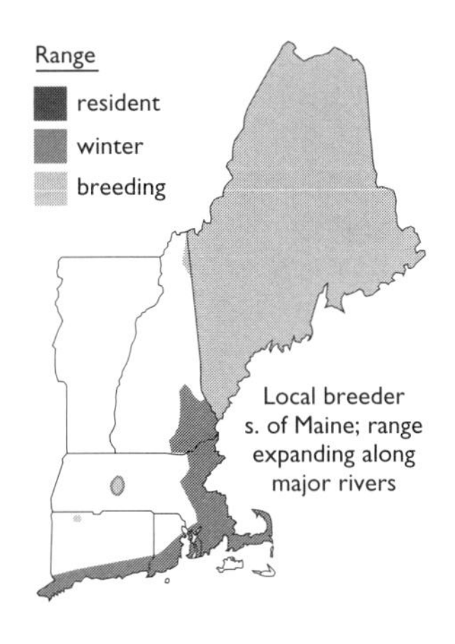

Status in New England: Uncommon migrant and breeder; locally common winter resident.

Habitat: Breeding: Large lakes, river, and estuaries in open areas, forests, and mountains. Commonly uses large trees adjacent to water for nesting, perching, and roosting. Prefers areas with minimal human disturbance; may abandon nest if human activity around the nest site. Winter: Coastal regions throughout New England, also on large inland bodies of water if open water persists, or if food resources such as deer carcasses are present.

Special Habitat Requirements: Large bodies of water containing abundant fish resources, large trees for nesting, perching, and roosting, and minimal human disturbance (DeGraaf et al. 1991:75).

Nesting: Egg dates: March 6 to May 14, New York (Bull 1974:174). Clutch size: 1 to 3, typically 2. Incubation period: About 35 days. Nestling period: 72 to 74 days (Bent 1937). Broods per year: 1. Age at sexual maturity: 3 yr. Nest height: 35 to 100 ft (10.7 to 30.5 m). Nest site: Nest placed 5 to 30 ft (1.5 to 9.1 m) below the top of a tall living tree near the water's edge. Frequently uses a white pine (Bent 1937), but tree species is not as important as size, shape, and proximity to other nesting eagles and water. Same nest often used year after year (up to 35 yr documented). Occasionally uses cliffs for nesting. May use artificial nests (Grubb 1995).

Territory Size: A pair of southern bald eagles defended an area that extended 0.5 mile (0.8 km) in all directions from the nest (Broley 1947). Usually a minimum of several square miles in area (Pough 1951:158).

Food Habits: Fish (staple), small to medium mammals, large birds, turtles, carrion. Preferred feeding habitat: Lakes, rivers, coastal bays, and inlets.

Comments: Bald eagles nested in Connecticut and Massachusetts in the 1800s and early in this century; they fell victim to pesticide poisoning and were extirpated since the 1950s in New England. Introduced through a hacking program at Quabbin Reservoir from 1982 to 1986, where birds had been seen in winter since 1951; two pairs first fledged young there in 1989 (Veit and Petersen 1993:130). More recently, birds have attempted nesting at Barton's Cove on the Connecticut River. Bald eagles show strong attachment to nesting territories and nest sites (Broley 1947).

Northern Harrier

(*Circus cyaneus*)

Range: Breeds in n. Alaska to s. Quebec and Newfoundland, s. to n. Mexico, Illinois, and Virginia. Winter: Southern British Columbia, s. Ontario, and Massachusetts, s. through the United States, Middle America, and the West Indies to n. South America. In Guatemala, Costa Rica, and Panama, a rare migrant and winter resident most numerous on the Pacific slope. In Colombia, a rare migrant and winter resident. A casual migrant to nw. Venezuela. In Puerto Rico and the Virgin Islands, an uncommon to rare winter resident.

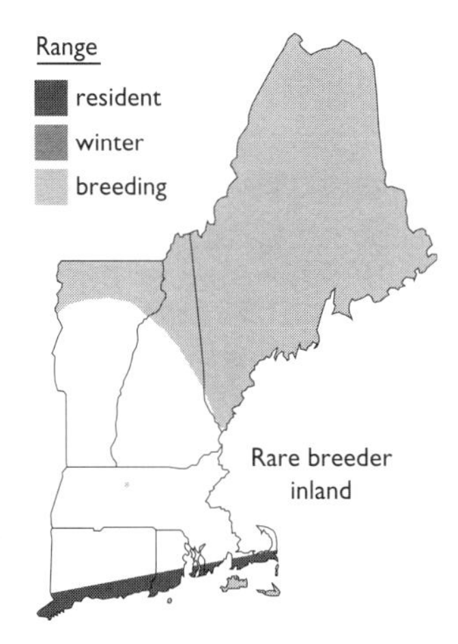

Distribution in New England: Breeds throughout Maine, w. to n. and w. New Hampshire, to Vermont in the Champlain Lowlands and the North Central and Northeast Highlands (Ellison 1985:72 in Laughlin and Kibbe 1985), s. to Massachusetts where nests on Martha's Vineyard, Nantucket, and several of the smaller islands off Cape Cod; one nest in Worcester County, Mass., in 1990 and 1991 (Veit and Petersen 1993:131); and s. to Rhode Island, presently known to nest only on Block Island (Enser 1992:48). Extirpated as a breeding species from Connecticut since the 1960s (Zeranski and Baptist 1990:84).

Status in New England: Rare and local breeder primarily at coastal marshes, occasionally inland; fairly common migrant and winter resident.

Habitat: Breeding: Typically inhabits wet meadows and sloughs in open country, fresh or salt marshes, and emergent swamps (Serrentino 1992). Winter: Primarily coastal dunes and salt marshes, pastures; also upland grasslands, sloughs (MacWhirter and Bildstein 1996).

Special Habitat Requirements: Open country.

Nesting: Egg dates: Mid May to mid June in Massachusetts (Veit and Petersen 1993:131). April 20 to June 25, New York (Bull 1974:177). Clutch size: 5, frequently 4 to 6, occasionally 7 to 9 (Hecht 1951). Incubation period: 28 to 36 days. Age at first flight: 30 to 35 days. Age at maturity: Usually 2 yr of age. Broods per year: 1. Nest site: On the ground, usually in tall, dense clumps of vegetation in dry fields, cutover areas, swamps with low shrubs and clearings; sometimes builds over water on a stick foundation, sedge tussock, or willow clump, or on a knoll of dry ground. Foundation of nest may be 18 in (45.7 cm) high in wet areas (Harrison 1975:39). Rarely nests in brackish or saltwater marshes. Uses large undisturbed habitat tracts (Apfelbaum and Seelbach 1983).

Territory Size: Male territory 2 acres (0.8 ha) in Idaho (Martin 1987).

Home Range: 0.38 square mile (0.98 km^2) to 3.89 square miles (10.1 km^2) in Michigan (Craighead and Craighead 1969:259, 260). Winter range: 0.55 square mile (1.4 km^2) and 0.63 square mile (1.6 km^2) for a pair and a single harrier, respectively (Craighead and Craighead 1969:26).

Sample Densities: 3 pairs nested within 400 yd (366 m) of each other (Bent 1937). 4 pairs per square mile (1.5 pairs/km^2) in favorable habitat in North Dakota (Stewart and Kantrud 1972). Densities vary with food availability.

Food Habits: Small mammals (staple), especially rodents, shrews and lagomorphs, small birds, amphibians, reptiles, insects, and occasionally carrion. Hunts meadows and fields, quartering low over the ground.

Comments: In n. New England, northern harriers are most common in agricultural areas of Coos County, N.H. (Serrentino 1994:44 in Foss 1994) and the Champlain Lowlands of Vermont. In Massachusetts, most nesting is on the islands off Cape Cod (Veit and Petersen 1993:131), and is extirpated as a nester in Connecticut (Zeranski and Baptist 1990:84). Generally roosts on the ground or perches on very low objects such as fence posts or tree stumps. During the nonbreeding season, inhabits areas far removed from nesting habitat. Roosts in undisturbed fields or marshes in winter. Population decline due to habitat loss from development and old fields reverting to forests. This bird likely increased in interior New England as land was cleared for agriculture, and then declined with habitat loss and shooting.

Sharp-shinned Hawk

(Accipiter striatus)

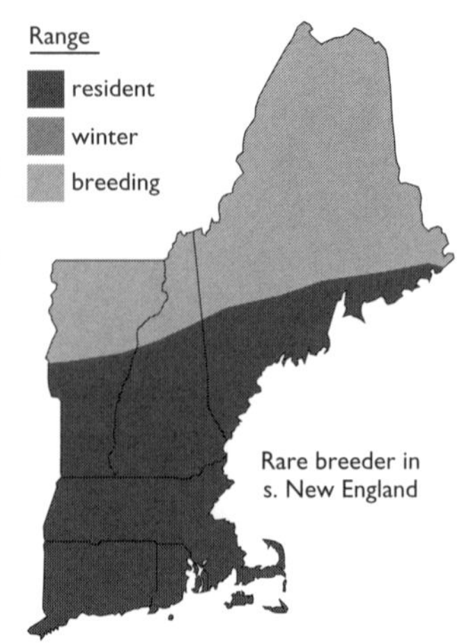

Range: Breeding: From w. and central Alaska and n. Yukon to s. Labrador and Newfoundland, s. to central California, s. Texas, the n. parts of the Gulf States, and South Carolina. Winter: From s. Alaska and the southernmost portions of the Canadian provinces, s. through the United States to Panama. In Mexico, U.S. birds migrate and winter through most of the country. In Guatemala and Costa

Rica, an uncommon migrant and winter resident on both the Caribbean and Pacific slopes. In Panama, an uncommon winter visitor. Birds south of Panama are residents there. A few North American individuals winter in the Bahamas and in the Greater Antilles.

Distribution in New England: Breeding: Throughout the region, except in the Northwest Hills of Connecticut and in Rhode Island, where there are no current nesting records (Enser 1992:49). Winter: Southern Maine, all but the most n. part of New Hampshire, throughout Massachusetts, Rhode Island, and Connecticut.

Status in New England: Rare and local breeder, common migrant, uncommon winter resident.

Habitat: Coniferous or mixed forests with clearings or edges bordering on brushy meadows. Uses red spruce and hemlock forests in Vermont (Nichols 1985:74 in Laughlin and Kibbe 1985), white pine in Massachusetts for nesting (Veit and Petersen 1993:132).

Special Habitat Requirements: Mature coniferous or mixed conifer-deciduous forests with clearings; woodlands bordering brushy openings.

Nesting: Egg dates: May 3 to June 24 in Massachusetts (Veit and Petersen 1993:132). April 16 to June 21, New York (Bull 1974:179). Clutch size: 3 to 8, typically 4 or 5. Incubation period: 34 to 35 days (Nice 1954). Nestling period: 24 to 27 days (males 24 days, females 27 days) (Platt 1976). Broods per year: 1. Nest height: 10 to 60 ft (3 to 18 m), typically 30 to 35 ft (9.1 to 10.7 m). Nest site: Most often in a conifer (white pine in Massachusetts, hemlock in New York) (Bent 1937:96, Bull 1974:179). Seldom in a deciduous tree (oak, beech). Nest is typically placed on a limb against the trunk of a medium-sized tree and is well concealed. Nest tree is often at the edge of a clearing. Sometimes refurbishes and uses an old nest.

Home Range: 0.26 square mile to 0.51 square mile in Moose, Wyo. (Craighead and Craighead 1969:263).

Food Habits: Small to medium birds (staple); small mammals, mainly rodents, shrews, moles and young lagomorphs, frogs, lizards, insects on the forest floor, in meadow grasses, bushy pastures. Preferred feeding habitat: Generally forages over open areas—avoids hunting in extensive forests.

Comments: Sharp-shinned hawks have markedly declined as breeding birds in s. New England since 1900–1920, with only a few nests reported in Connecticut (Zeranski and Baptist 1990:85) and Massachusetts (Veit and Petersen 1993:32) since the 1950s. Once believed to have declined from forest clearing (Griscom and Snyder 1955), the species has not returned with the forest. May be more of a forest edge species than previously thought. There is some evidence of an increase since the 1970s. Like most hawks, sharp-shins were persecuted to preserve song and game birds, a mistaken belief even of early ornithologists (Eaton 1914). Usually does not live in the same forests with its competitor, the Cooper's hawk. DDT may have caused eggshell thinning (Henny et al. 1973).

Cooper's Hawk

(Accipiter cooperii)

Range: Breeding: From s. British Columbia and central Alberta, e. to s. Quebec and Maine (rare in New Brunswick, Prince Edward Island, and Nova Scotia), s. to Baja California, Sinaloa, Chihuahua, Nuevo Leon, s. Texas, Louisiana, central Mississippi, central Alabama, and Florida. Winter: From Washington, Colorado, and s. Minnesota, to New England, s. through the s. United States to Costa Rica. Winters through much of Mexico. In Guatemala, a rare migrant and winter resident, recorded only in the highlands w. of Guatemala City and at Coban. In Costa Rica, a very rare migrant and winter resident in the highlands; winters from Cordillera de Tilaran to

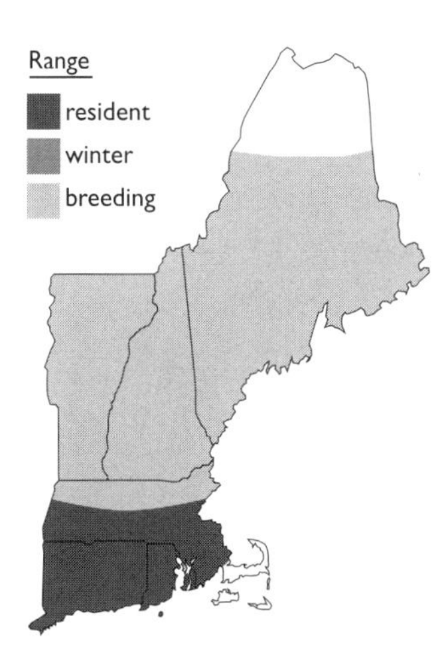

Cordillera de Talmanca (Cerro de la Muerte) from 4,950 to 9,900 ft (1,500 to 3,000 m). Accidental in Colombia.

Distribution in New England: Breeding: A rare and local breeder throughout New Hampshire, Vermont, Massachusetts, Connecticut, and Rhode Island, and in s. Maine. Winter: An uncommon winter resident in Massachusetts (Veit and Petersen 1993:133), Connecticut (Zeranski and Baptist 1990:86), and Rhode Island.

Status in New England: Rare and local breeder; uncommon migrant and winter resident (Veit and Petersen 1993:133).

Habitat: Breeding: Inhabits mature stands of coniferous, deciduous, and mixed forests in semi-open country, open woodlands, including small woodlots, pine plantations, and suburban forests (Titus and Mosher 1981). Tolerant of forest fragmentation and human disturbance; will nest in urban areas (Rosenfield et al. 1992, Rosenfield and Bielefeldt 1993, Titus and Mosher 1988, Murphy et al. 1988). Winter: Similar to breeding habitat.

Special Habitat Requirements: Mature coniferous or deciduous woodlands in otherwise open or semi-open country.

Nesting: Egg dates: April 20 to June 16, New York (Bull 1974:179). Clutch size: 3 to 6, typically 4. Incubation period: Begins after 3rd egg laid, 30 to 36 days typical (Meng and Rosenfield 1988). Nestling period: 30 to 36 days, 34 to 36 typical (Meng and Rosenfield 1988). Broods per year: 1. Age at sexual maturity: $\geq$2 yr. Nest height: 20 to 60 ft (6.1 to 18.3 m), typically 35 to 45 ft (10.7 to 13.7 m). Nest site: The nest is commonly placed in the canopy of a deciduous or coniferous tree on a horizontal branch or in a crotch near the trunk. The selected tree is often near the edge of a wooded area, with large open fields and water nearby. Commonly uses old crow nests. Cooper's hawks frequently return to the same nest site year after year (Bull 1974:179).

Home Range: 0.07 square mile (0.2 km^2) to 2.05 square miles (5.3 km^2) in Michigan (Craighead and Craighead 1969:258, 260). Average winter range: 1.5 to 2 miles (2.4 to 3.2 km) in diameter (Craighead and Craighead 1969).

Sample Densities: 0.2 pairs per 100 acres (40 ha) in mixed forest-farmland habitat in Maryland (Stewart and Robbins 1958:110).

Food Habits: Small to medium-sized birds and small mammals, especially jays and robins, rodents and young lagomorphs; occasionally eats insects and amphibians. Cooper's hawks hunt primarily in woodlots away from the nest area and in open areas near woodlands.

Comments: Only two pairs of Cooper's hawks were found in Massachusetts during 1974–1979 (Veit and Petersen 1993:133). Once one of the most abundant birds of prey in the mid 1800s, Cooper's hawks declined in New England since the late 1800s, were long persecuted because they preyed on chickens at farms (Zeranski and Baptist 1990:86). Cooper's hawks usually hunt on the wing low to the ground or just above treetop level.

Northern Goshawk

(Accipiter gentilis)

Range: Breeding: From w. and central Alaska and n. Yukon, to Labrador and Newfoundland, s. to s. Alaska, central California, s. New Mexico, w. South Dakota, n. Minnesota, and nw. Connecticut, and in the n. Appalachian Mountains. Winter: Throughout the breeding range, may extend as far s. as the Gulf States during periodic invasions related to food shortage (DeGraaf et al. 1991:79).

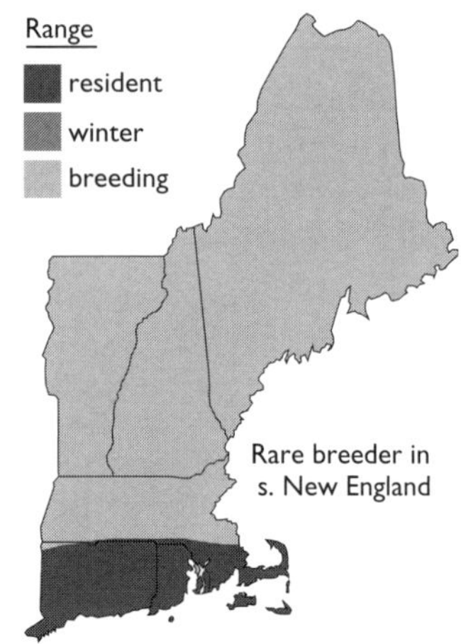

Distribution in New England: Breeding: Throughout the region. Winter: Throughout the region except absent in northernmost Maine.

Status in New England: Uncommon to rare resident but increasing.

Habitat: Breeding: Inhabits the interior of mature, coniferous (hemlock and white pine) and mixed forests in temperate and boreal regions, from sea level to treeline. Seems to prefer mature forests with large trees with open understories (Squires and Reynolds 1997).

Special Habitat Requirements: Lack of human disturbance at nest site.

Nesting: Egg dates: April 1 to May 10, Massachusetts (Veit and Petersen 1993:134). Clutch size: 2 to 5, typically 3 or 4. Incubation period: 36 to 38 days (Brown and Amadon 1968). Nestling period: 34 to 37 days (Reynolds et al. 1982). Broods per year: 1. Nest height: 20 to 75 ft (6.1 to 22.9 m), typically 30 to 40 ft (9.1 to 12.2 m). Nest site: Nest usually placed in a crotch or on a limb close to the trunk at the bottom of the canopy; nest usually in a large tree at the bottom of a wooded slope (Yamasaki et al. unpubl.). Often nests in beech, birch, pine, or hemlock. Builds on top of old nest (own or other hawk's) or makes a new nest.

Home Range: 0.82 square mile (2.1 km^2) (observed area) in Moose, Wyo., in 1947 (Craighead and Craighead 1969:263). May be area-sensitive; small forest tracts (N.J.) bounded by roads were not used for nesting (Bosakowski and Speiser 1994).

Food Habits: Small to medium birds (staple); mammals, especially rodents and lagomorphs. Preferred feeding habitat: Clearings and brushy openings in forests.

Comments: The nesting range of northern goshawks decreased as forests were cleared for settlement and the passenger pigeon declined toward extinction (Bent 1937). Since about 1955 they have steadily increased e. and s. from w. Massachusetts, to which they were restricted in s. New England. In Vermont only three nesting records existed before 1933 (Ellison 1985:78 in Laughlin and Kibbe 1985), but now goshawks nest throughout n. New England. Range expansion and an increase in population are generally attributed to the regrowth of New England forests. Numbers of breeding birds have recently increased greatly in the mountains of New York (Bull 1974:180).

Red-Shouldered Hawk

(*Buteo lineatus*)

Range: Breeding: From n. California s., w. of the Sierra Nevada divide, to Baja California, and from e. Nebraska, central Minnesota, s. Ontario, and s. New Brunswick, s. to Mexico to Veracruz in the e. and Sinoloa in the w. Winter: Primarily from e. Kansas and central Missouri to s. New England, s. to n. Mexico, but also sporadically throughout the breeding range.

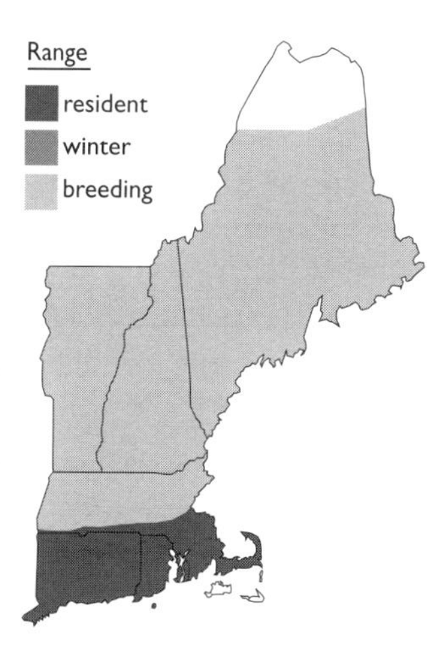

Distribution in New England: Breeds throughout central and s. Maine, New Hampshire (although more common s. of the White Mountains), Vermont, s. to Massachusetts, except absent from Cape Cod and the Islands (Veit and Petersen 1993:135). Connecticut locally inland and more frequent e. of the Connecticut River Valley (Zeranski and Baptist 1990:87), and w. Rhode Island (Enser 1992:52).

Status in New England: Uncommon breeder and migrant.

Habitat: Inhabits mature forested wetlands (especially deciduous), wooded river swamps, bottomlands, and wooded margins of marshes and beaver ponds, often near natural openings or fields for foraging; also inhabits upland forests adjacent to wetlands. Nests in mixed deciduous-coniferous forests on a slope above forested swamps in New Hampshire (Gavutis 1994:52 in Foss 1994) and Massachusetts (Portnoy and Dodge 1979).

Special Habitat Requirements: Riparian deciduous woodlands with tall trees for nesting.

Nesting: Egg dates: April 3 to June 5 in Massachusetts (Veit and Petersen 1993:135). March 25 to May 26, New York (Bull 1974:186). Clutch size: 2 to 6, typically 3 to 4. Incubation period: 28 days. Nestling period: 35 to 42 days. Broods per year: 1. Age at sexual maturity: Usually ≥1 yr (Henny et al. 1973). Nest height: 20 to 60 ft (6.1 to 18.3 m), typically 35 to 45 ft (10.4 to 13.7 m). Nest site: Located below the canopy, in the top half of the tree. Nest is usually in a main fork or close to the tree trunk, and always near water such as a river, pond, or swamp (Titus and Mosher 1981, Crocoll and Parker 1989). Has built nests in oak, pine, bald cypress, mangrove, cottonwood, birch, beech, sycamore, yellow poplar, ash, sweetgum, and maple (DeGraaf and Rappole 1995:119). May use an abandoned hawk, crow, or squirrel nest as a foundation. Often uses the same nest in successive years. Nests in large trees with large support branches are most likely to fledge at least one bird; hawks using old nests may have a higher nesting success rate than those building new nests (Dijak et al. 1990).

Territory Size: 180 acres (72.9 ha) in Kansas (Fitch 1958). 0.03 square mile to 0.60 square mile (0.07 to 1.6 km^2) in Michigan. Winter ranges in Michigan were usually between 1.5 and 2 square miles (3.8 to 5.2 km^2) (Craighead and Craighead 1969:24). Average home range area 480 acres (192 ha) in Maryland (Stewart 1949).

Sample Densities: 1 pair per 0.8 square mile (1 pair/2.1 km^2) in floodplain forest in Maryland (Stewart 1949). About 1 pair per 120 acres (48.6 ha) of floodplain in Maryland (Henny et al. 1973). 1 pair per 428 acres (171 ha) in w. New York (Crocoll and Parker 1989).

Food Habits: Amphibians, reptiles, crustaceans (crayfish), insects, mammals such as small rodents, shrews, and moles (Ernst 1945). Also takes young birds of many species. In addition to foraging in wooded swamp nesting habitats, also hunts in drier woodland clearings and fields.

Comments: Red-shouldered hawks are among the first hawks to return to their nesting grounds in the spring. They have declined greatly in abundance since 1900–1920 in s. New England as the wet woods they prefer have been drained or filled for development (Peterson and Crocoll 1992). In the early 1890s red-shouldered hawks were more abundant nesters than red-tailed hawks in Connecticut (Averill 1892). Pairs may remain mated for life. The call is closely mimicked by the blue jay (Crocoll 1994).

Broad-winged Hawk

(Buteo platypterus)

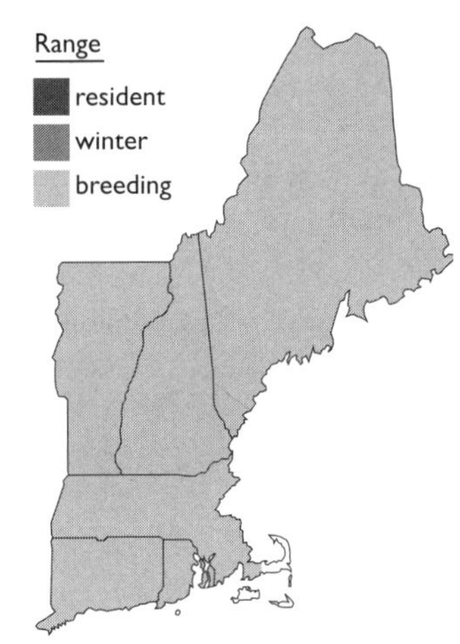

Range: Breeds in central Alberta and central Saskatchewan, and from central Manitoba to New Brunswick and Nova Scotia s. to e. Texas, the Gulf Coast, and Florida. Winter: From s. Florida and from Mexico to South America. In Mexico, a common migrant throughout the ne. and s.; a few birds may stay for the winter. In Guatemala, a common migrant and winter resident. It is perhaps the most abundant hawk in Panamanian and Costa Rican woodlands during the winter months. In Colombia, a common migrant and winter resident w. of the Andes and on the e. slope of the e. Andes. A winter resident in French Guiana, Venezuela, Colombia, and w. South America, s. to w. Peru, e. of the Andes to Bolivia, nw. Brazil, and sw. Mato Grosso. Resident in Tobago, migrant to Trinidad. Continental North American individuals winter for the most part in central and n. South America and Trinidad.

Distribution in New England: Breeds throughout the region, except absent on outer Cape Cod, Martha's Vineyard, and Nantucket; a rare breeder on coastal lowlands in Rhode Island and absent from Block Island (Enser 1992:53).

Status in New England: Fairly common breeder, abundant migrant inland, uncommon along the coast.

Habitat: Deciduous and mixed deciduous-coniferous forests with openings, also near lakes, ponds, and marshes. Black and yellow birches are commonly selected for nesting trees in New England. Nests at elevations up to 2,500 ft (760 m) in the White Mountains in New Hampshire (Elkins 1994:54 in Foss 1994). Often nests along country roads with little traffic, generally away from human dwellings (Armstrong and Euler 1983). Inhabits young deciduous forests with a mean canopy height of 63.3 ft (21 m) up to elevations of 1317.4 ft (399.2 m) in Ontario (Armstrong and Euler 1983).

Special Habitat Requirements: Forests with openings.

Nesting: Egg dates: April 27 to June 26, New York (Bull 1974:188). Clutch size: 1 to 4, typically 2 or 3. Incubation period: 28 to 35 days. Nestling period: 29 to 30 days. Broods per year: 1. Age at sexual maturity: ≥1 yr (Burns 1911, Crocoll and Parker 1989). Nest height: 3 to 80 ft (1.0 to 24.4 m). Typically 25 to 40 ft (7.6 to 12.2 m). Nest site: Uses coniferous or deciduous trees, generally chooses larger trees of the most abundant species (Burns 1911, Crocoll and Parker 1989, Titus and Mosher 1981). Nests on wooded slopes, often near ponds and streams (Matray 1974). Typically locates nest in first main crotch in tree next to the trunk (Burns 1911, Crocoll and Parker 1989). Sometimes uses old crow or hawk nests.

Territory Size: Few data available, but does defend a home range (Goodrich et al. 1996).

Food Habits: Amphibians, reptiles, insects, small mammals such as shrews (staple) and mice, occasionally takes young birds, usually from perches below the forest canopy at the edge of forest openings, frequently near open water and marshes (Toland 1986, Goodrich et al. 1996). Will hunt over meadows and wetlands.

Comments: Now one of the most common New England hawks, broad-wings were very rare at the peak of land clearing in the mid 1800s. Also, the shooting of hawks was a popular activity in the late 1800s. The reversion of former farmland to forest has provided abundant habitat for the bird. The fall migration of broad-winged hawks (September 12–20 in Massachusetts) can be spectacular when ne. winds and low cloud cover prevail over inland mountain ranges (Veit and Petersen 1993:137).

Red-tailed Hawk

(*Buteo jamaicensis*)

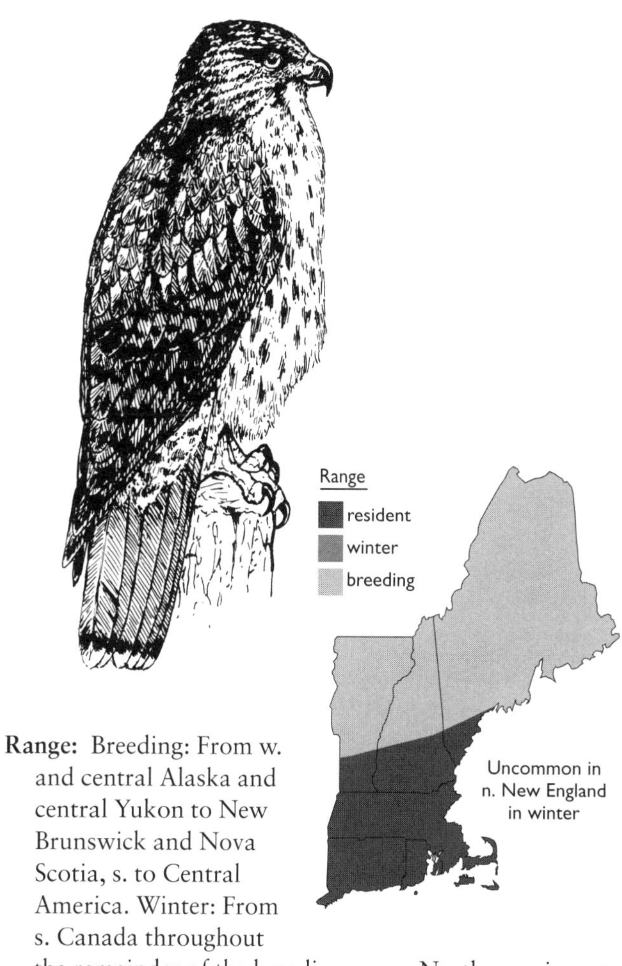

Range: Breeding: From w. and central Alaska and central Yukon to New Brunswick and Nova Scotia, s. to Central America. Winter: From s. Canada throughout the remainder of the breeding range. Northern migrants occur through much of Mexico except the Yucatan Peninsula. In Guatemala and Honduras, the resident population is augmented by U.S. migrants and winter residents. In Costa Rica and Panama, a few individuals are sometimes seen migrating in spring and fall. A casual and possible winter visitor from North America on New Providence, Eleuthera, and Great Inagua. Also occurs on Grand Bahama, Abaco, Andros, the Greater Antilles, and some of the n. Lesser Antilles.

Distribution in New England: Breeding: Throughout the region. Winter: Throughout Connecticut, Rhode Island, Massachusetts, and in s. Maine, s. New Hampshire, s. Vermont.

Status in New England: Regular but uncommon breeder, common in winter.

Habitat: Breeding: Inhabits a wide variety of primarily open habitats throughout its range, from scrub deserts and plains with scattered trees, pastures, urban parks, and mixed deciduous and coniferous forests to tropical rain forest (Preston and Beane 1993). Prefers a matrix of open pasture, fields, meadows, or swampy areas interspersed with coniferous or deciduous woods (Freemark and Collins 1989). In much of n. New England, favors fallow pastures interspersed with woodlots (Ellison 1985e): 84–85 in Laughlin and Kibbe 1985). Among buteos, the most tolerant of human disturbance (Dobkin 1992). Winter: In the Northeast, similar to breeding habitat.

Special Habitat Requirements: Large trees for nesting and perching.

Nesting: Egg dates: March 30 to April 30 in Massachusetts (Veit and Petersen 1993:139). March 8 to May 16, New York (Bull 1974:184). Clutch size: 1 to 5, typically 2 or 3. Incubation period: 28 to 35 days (Bent 1937, Hardy 1939). Age at first flight: 42 to 46 days (Preston and Beane 1993). Broods per year: 1. Age at sexual maturity: Generally at least 2 yr. Nest height: 35 to 90 ft (10.7 to 27.4 m). Nest site: Usually nests in a tall tree at the edge of a woodland near open areas for hunting, or in an isolated tree in an open area (Minor et al. 1993, Speiser and Bosakowski 1988, DeGraaf and Rappole 1995).

Territory Size: Defended territories are three-dimensional; 80 to 200 acres (32.4 to 81.0 ha) excluding peripheral areas in California (Fitch et al. 1946).

Sample Densities: 1 pair per 2.2 square miles (1 pair/5.7 km^2) in deciduous woodland in New York (Hagar 1957). 1 pair per 4.1 square miles (1 pair/10.6 km^2) in fields and woodlands (Gates 1972). 1 pair per 0.5 square mile (1 pair/1.3 km^2) in pine-oak habitat in California (Fitch et al. 1946).

Food Habits: Small to medium-sized mammals, especially voles, mice, rats, cottontails, chipmunks, and squirrels; also takes amphibians, reptiles, small birds, insects, and carrion in short meadow grasses; 60 to 80 percent of hunting done from perches (Fitch et al. 1946, Ballam 1984).

Comments: Inconspicuous as breeding birds, red-tailed hawks suffered a long-term decline in s. New England until recently. Scarce when the landscape was largely cleared for agriculture, they have increased as a winter resident since 1965 (Zeranski and Baptist 1996:90, possibly now the most common buteo in Vermont (Ellison 1985e:84 in Laughlin and Kibbe 1985). Birds possibly mate for life. Red-tails generally select the largest and tallest trees for nesting, over 35 ft (10.7 m) tall. They can utilize small woodlots for nesting, the smallest being about 15 acres (6.1 ha). Occasionally they use isolated trees up to 50 yd (45.7 m) from woods (Hagar 1957). Conspicuous in winter, especially in the Connecticut River valley and along major highways.

Rough-legged Hawk

(Buteo lagopus)

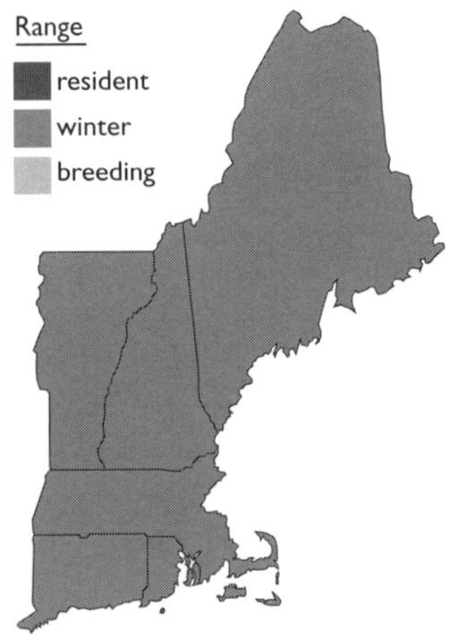

Range: Breeding: From w. and n. Alaska, n. Yukon, and n. Labrador, s. to n. and se. Mackenzie, e. to n. Quebec and Newfoundland; also from Kodiak Island and Umnak in the e. Aleutian Islands and the Arctic Islands, n. to Prince Patrick, Victoria, Bylot, and sw. Baffin Islands. Winter: From s.-central Alaska (casually) and s. Canada, s. to s. California, s. Arizona, e. to s. Texas, Missouri, Tennessee, and Virginia, casually to e. Texas and the Gulf Coast. Concentrates in areas of high prey density during winter (DeGraaf et al. 1991:91).

Distribution in New England: Winter: throughout the region.

Status in New England: Irruptive, rare to occasionally fairly common winter resident (Veit and Petersen 1993:139) in coastal areas and in the Connecticut River Valley farmlands.

Habitat: Breeding: Inhabits open tundra and mountainsides; does not inhabit forest unless much open ground is present. Hunts over fields, pasture, riparian areas, and marshes. Winter: Prefers conifer groves for roosting; will roost communally (Schnell 1969). Hunts over open, treeless areas: extensive coastal marshes, dunes, extensive croplands and pasture.

Special Habitat Requirements: Open areas for hunting.

Food Habits: Summer and winter—mainly mammals, primarily mice, lemmings, and pocket gophers. Also consumes young ptarmigan, carrion, and occasionally small hares and rabbits.

Comments: Numbers declined in the early part of the twentieth century because of wholesale slaughter in the belief that all raptors were chicken and gamebird killers. Rough-legged hawk populations are highly irruptive, with the largest numbers generally occurring during years of meadow vole (*Microtus*) abundance (Bull 1974:189). This species is more numerous from central New York w. to the Lakes Region.

Golden Eagle

(*Aquila chrysaetos*)

Range: Breeding: From n. and w. Alaska, e. to Labrador, s. to s. Alaska, Baja California, w. and central Texas, New York and New England. Winter: From s.-central Alaska and the s. portions of the Canadian provinces, s. throughout the breeding range, rarely to coastal South Carolina (DeGraaf et al. 1991:92).

Distribution in New England: Breeding: Central Maine. Wintering: Quabbin Reservoir, Mass., and occasionally Martha's Vineyard and Nantucket.

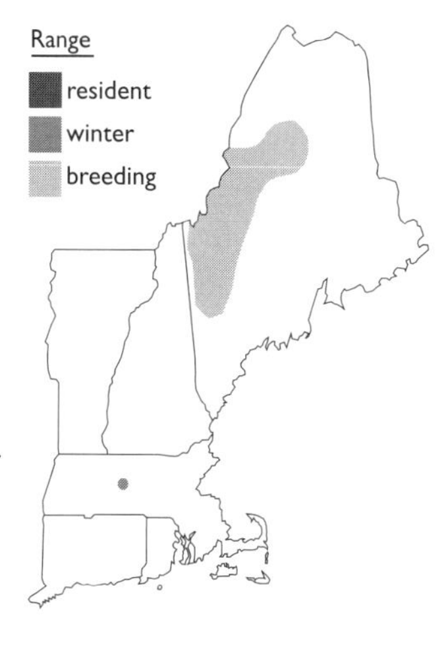

Status in New England: Rare; one pair occ. nests in Maine. A regular migrant and winter resident.

Habitat: Breeding and wintering: Uses a variety of open habitats from barren areas to open coniferous forests, especially in mountainous terrain. Typically nests on mountain cliffs associated with coniferous forests in Maine (Boone and Krohn 1996:116). Hunts over open areas such as bogs, marshes, meadows, pastures, clearcuts, and burned areas. Prefers remote areas with little human disturbance.

Special Habitat Requirements: Cliffs for nesting, large expanses of remote open area for hunting.

Nesting: Egg dates: May to June. Clutch size: 1 to 4, typically 2, occasionally 3. Incubation period: 43 to 45 days. Age at first flight: 65 to 70 days. Broods per year: 1. Nest site: Usually on cliff ledges, especially those overlooking grasslands; less commonly in a large tree (Menkens and Anderson 1987).

Territory Size: 20 to 60 square miles (51.8 to 155.4 km^2) with an average of about 36 square miles (93.2 km^2) (Pough 1951:155). 50 to 100 square miles (130 to 260 km^2) in the Appalachians (Spofford 1971).

Sample Densities: 56 breeding pairs per 240-km (149-mile) stretch of Snake River in Idaho. 1 pair per 8 km (5.0 miles) of river (Craighead and Craighead 1969). 1 pair per 5 km (3.1 miles) of river (Spofford 1971). Density is probably a function of availability of suitable nest sites, adequate prey, and minimum nesting territory size (Beecham and Kochert 1975).

Food Habits: Primarily mammals (primarily lagomorphs), but also marmots, prairie dogs, ground squirrels, weasels, woodrats, skunks, and mice, rarely larger mammals (Collopy 1984). Also eats insects, medium to large birds, reptiles, and some carrion.

Comments: Pairs probably mate for life. A few golden eagles breed in the Adirondacks of New York, but very few fledge young (Spofford 1971). May use the same nest year after year, or pairs may use alternate nests in successive years. Populations are limited by availability of suitable nesting cliffs and adequate food resources (Foss 1994:374). Fairly regularly seen in small numbers in winter at Quabbin Reservoir, Mass.

American Kestrel

(*Falco sparverius*)

Range: Breeding: From w. and central Alaska and s. Yukon to n. Ontario, s. Quebec, and s. Newfoundland, s. through Middle America, the West Indies, and South America, excluding the Amazon basin. Winter: From se. Alaska, s. British Columbia, and the central United States, s. throughout the breeding range.

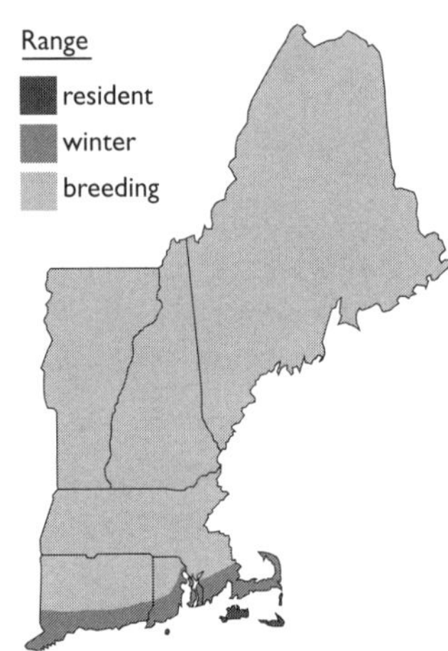

Distribution in New England: Breeding: Throughout the region. Winter: s. coastal areas.

Status in New England: Fairly common breeder from Massachusetts n., uncommon and local in Connecticut (Zeranski and Baptist 1990:91).

Habitat: Breeding: Occupies a wide range of habitats including deserts, forested edges, grasslands, pastures, utility rights-of-way, marshes, beaver flowages, and suburban areas. Winter: Occupies the same habitat types as those used during the breeding season.

Special Habitat Requirements: Nest cavities in trees with diameter breast height greater than 12 in (30.5 cm), and elevated perches from which to sight prey.

Nesting: Egg dates: April 27 to 26 May in Massachusetts (Veit and Petersen 1993:142). April 5 to June 29, New York (Bull 1974:202). Clutch size: 3 to 7, typically 4 or 5. Incubation period: 29 to 31 days. Nestling period: 30 to 31 days. Broods per year: 1, 2 in the s. when prey very abundant (Toland 1985). Nest height: 4 to 50 ft (1.2 to 15.2 m), typically 10 to 35 ft (3.0 to 10 m). Nest site: Prefers to nest in tree cavities with small entrances, especially dead elms with flicker holes in fencerows and pastures (Wheeler 1992). If natural cavities are unavailable will nest under the eaves of buildings, in niches in rocky cliffs, old nests of other birds, or unused chimneys. Will use nest boxes (Bortolotti 1994, Wheeler 1992); in fact, nest boxes are an effective management tool (Hammerstrom et al. 1973). Nest sites usually along roadways, streams, ponds, or forest edges, from 3.3 to 66 ft (1 to 20 m) above the ground, though typically from 10 to 33 ft (3 to 10 m).

Territory Size: 351 acres (142 ha) (Hardin and Evans 1977).

Sample Densities: 6 pairs per 0.5 square mile (6 pairs/1.3 km^2) in nest boxes in Pennsylvania (Nagy 1963). Maximum pairs (0.28 per 100 acres (40 ha) (Hardin and Evans 1977). 45 pairs per square mile (45 pairs/2.6 km^2) in central Utah (Smith et al. 1972). 0.44 pairs per square mile (0.16 pairs/km^2) in Michigan (Craighead and Craighead 1969).

Food Habits: Insects (staple), especially grasshoppers, crickets, and beetles; mammals such as small mice, shrews; small birds; reptiles and amphibians.

Comments: American kestrels are one of the few raptors that apparently increased in population since 1900, although wintering numbers have declined in Connecticut since the 1970s (Zeranski and Baptist 1990:92).

Merlin

(*Falco columbarius*)

Range: Breeds from nw. Alaska and n. Yukon to Labrador and Newfoundland, s. to s. Alaska, e. Oregon, n. Minnesota, s. Quebec, New Brunswick, and Nova Scotia. Winters w. of the Rockies from south-central Alaska, s. British Columbia, Wyoming, and Colorado s., locally across s. Canada, and in the e. United States from Maryland, the Gulf Coast, and s. Texas, s. through Middle America and the West Indies to n. South America, from nw. Peru to n. Venezuela and Trinidad.

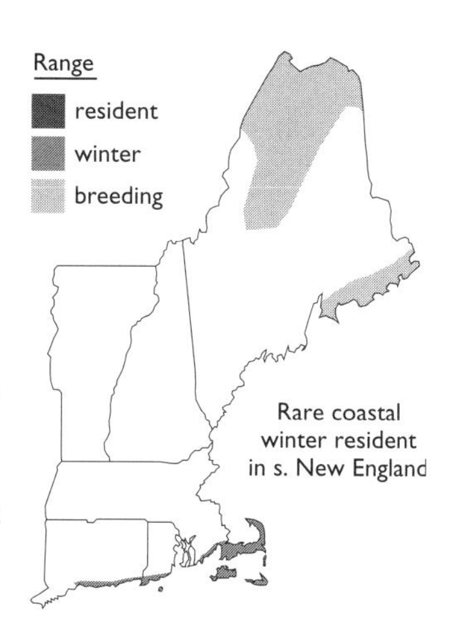

Distribution in New England: Breeding: In n. Maine and the Downeast coast. Winter: Rare along the se. coast of Massachusetts and coastal Connecticut and Rhode Island.

Status in New England: Uncommon migrant primarily along the coast; rare coastal winter resident.

Habitat: Breeding: Inhabits open to semi-open mixed or coniferous forests interspersed with lakes, marshes, bogs, pastures, and fields. Frequently nests on islands in large lakes (Craighead and Craighead 1940). In urban areas (Canada), nests in conifers in residential areas, school yards, parks, and cemeteries (Sodhi et al. 1993, Warkentin and James 1988, Sodhi et al. 1992).

Special Habitat Requirements: Open forests adjacent to open areas for foraging.

Nesting: Egg dates: May through June. Clutch size: 2 to 7, typically 4 to 5. Incubation period: average 30 days, range 28 to 32 days. Nestling period: About 29 days, range 26 to 32 days. Broods per year: 1 (Sodhi et al. 1993). Age at sexual maturity: At least 2 yr. Nest height: 35 to 60 ft (10.7 to 18.3 m) above ground. Nest site: Often nests in old stick nests of corvids or other raptors, occasionally nests on the bare ledge of a cliff or in a tree cavity (Harrison 1975). Frequently nests near water.

Home Range: Average 9.3 per square mile (23.3/km^2) in Montana (Becker and Sieg 1987). Average 3.3 per square mile (8.2/km^2) in Saskatchewan (Sodhi and Oliphant 1992).

Food Habits: Primarily small birds, also small mammals, and insects. Hunts in a variety of open habitats—marshes, beaches, mudflats, and fields (Godfrey 1979:103).

Comments: Habitat loss may be a major factor in declining merlin populations (Oliphant 1985). Formerly called the "pigeon hawk" because in flight it can be mistaken for a member of the pigeon family (Sodhi et al. 1993).

Peregrine Falcon

(*Falco peregrinus*)

Range: Breeding: From n. Alaska, Banks, Victoria, s. Melville, Somerset, and n. Baffin Islands and Labrador, s. to Baja California, s. Arizona, New Mexico, w. and central Texas, and Colorado; recently reintroduced and reestablished as a breeding bird in parts of the ne. United States. Winter: From s. Alaska, the Queen Charlotte Islands, coastal British Columbia, the central and s. United States, and New Brunswick, s. to South America. In Guatemala, Costa Rica, Honduras, and Panama, a rare migrant and winter resident. In Colombia, an uncommon migrant and winter resident w. of the Andes to at least 2,800 m. A winter visitor throughout the West Indies. In Trinidad and Tobago, a fairly common winter visitor seen in Caroni Swamp and St. Giles Islet.

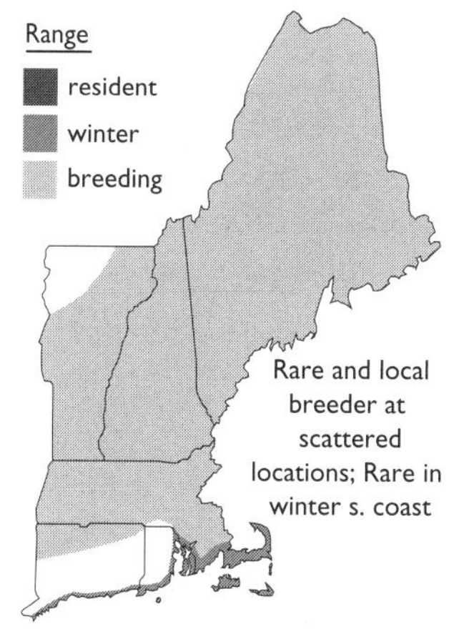

Distribution in New England: Breeding: The White Mountains, and locally throughout the region.

Status in New England: Rare and local breeder, migrant, and winter resident, primarily along the coast.

Habitat: Breeding: Open habitats, from tundra and seacoasts to high mountains and open forested regions, typically where there are rocky cliffs with ledges overlooking rivers, streams, lakes, or coastal bays, and an abundance of birds. Sometimes breeds in cities on tall buildings. Winter: Along the coast.

Special Habitat Requirements: Cliffs or other suitable nesting sites near water, and an abundance of prey.

Nesting: Egg dates: March 26 to May 31, New York (Bull 1974). Clutch size: 2 to 5, typically 4. Incubation period: 28 to 35 days (Harrison 1975, Terres 1980). Age at first flight: 35 to 42 days (Brown and Amadon 1968). Broods per year: 1. Age at maturity: Usually in the third year after hatching (Nelson 1972). Nest height: Cliffs, 800–3,000 ft (240 to 910 m) in New Hampshire (Lanier and Bollengier 1994:60 in Foss 1994). Nest site: A shallow depression scraped in gravel and debris on a cliff ledge, pothole, or small cave (Ponton 1983). Also nests on bluffs, slopes, pinnacles, cutbanks, seastacks, and ledges of tall buildings (Bird and Bird 1990). Historically nested in holes and stubs of large trees. Will use the same nest in successive years (Ratcliffe 1980).

Sample Densities: Hickey (1942) listed 19 pairs per 10,000 square miles (19 pairs/25,900 km^2) around New York City. Average distance between pairs was about 1 mile (1.6 km) in an unusually dense island population along coast of British Columbia (Beebe 1960 in Hickey and Anderson 1969). About 1 pair per 2,000 square miles (1 pair/5,180 km^2) was estimated for parts of w. North America where peregrine was considered common, and less than 1 pair per 20,000 square miles (1 pair/51,800 km^2) where rare (Bond 1946 in Hickey and Anderson 1969).

Food Habits: Small to large birds (staple), occasionally takes mammals or dead fish. Strikes birds in the air and recovers prey on the ground.

Comments: Peregrine falcons nested in Berkshire County, Mass., until 1951 (Hagar 1969) and on Mt. Tom (Mass.) until 1955; they disappeared in the e. United States by the early 1960s as a result of pesticide contamination (Hickey 1969). Peregrines did not breed again in Massachusetts or New England until hacking programs in the late 1970s to mid 1980s resulted in natural nestings in the late 1980s in Massachusetts and New Hampshire.

Gray Partridge

(Perdix perdix)

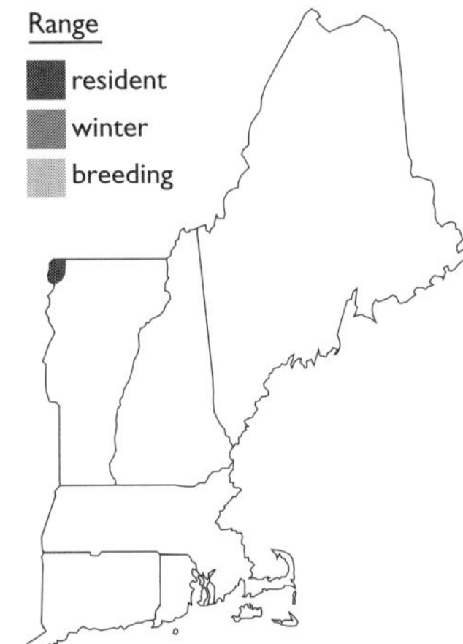

Range: Breeding: Widely introduced in North America and established locally from s. British Columbia to sw. Quebec, New Brunswick, and Nova Scotia, s. to n. Nevada, n. South Dakota, central Indiana, and n. Vermont (DeGraaf et al. 1991:98).

Distribution in New England: Breeding: The n. Champlain Valley in Vermont (Kibbe 1985c:88 in Laughlin and Kibbe 1985).

Status in New England: Uncommon local resident.

Habitat: Breeding: Near edges of large agricultural fields, pastures, and grasslands. Prefers cropland interspersed with native grasslands. Winter: Same as breeding habitat. Typically remain in open areas, but may seek shelter during periods of strong winds.

Special Habitat Requirements: Herbaceous cover for nesting and brood cover.

Nesting: Egg dates: Mid to late June (Edminster 1954:372). Clutch size: 5 to 20, typically 15 to 17, among the largest single hen clutches of any bird (Carroll 1993). Incubation period: 24 to 25 days. Nestling period: 1 day, young are fully plumaged and grown in about 10 weeks (Edminster 1954:373). Nest site: A shallow depression on the ground usually concealed by grasses; prefers hayfields, also the edges of grain fields, pastures, fence rows, and roadsides (Carroll 1993).

Territory Size: Varies with the season: Spring 24 to 42 acres (9.7 to 17 ha), the remainder of the year 237 acres (96 ha) in e. South Dakota (Smith et al. 1982).

Sample Densities: On three 160-acre study areas in Michigan, Yeatter (1934) reported one bird for 4.4, 11, and 13.3 acres. Gray partridge populations may follow 10-yr cycles of abundance, similar to those of native grouse species (Aldrich 1947).

Food Habits: Green leafy materials, seeds of domestic crops and weedy herbs and cultivated grains. Insects occasionally taken by adults. Chicks are insectivorous during the first 2 weeks of life (Edminster 1954:378). Grit required throughout the year.

Comments: A native species of Europe and e. Russia, gray partridge were originally introduced to Vermont before 1893 without success; they disappeared after the severe winter of 1904–1905 (Foote 1946). They were introduced in New York in large numbers around 1930 (Wilson 1959) and gradually spread through the St. Lawrence River Valley to the n. Champlain Valley's flat open grasslands. Gray (formerly "Hungarian") partridge are able to survive extreme winter cold, and they will "snow roost" as do ruffed grouse; ice and sleet storms deprive the birds of food and cover and cause periodic declines (Kibbe 1985:88 in Laughlin and Kibbe 1985).

Ring-necked Pheasant

(*Phasianus colchicus*)

Range: Breeding: Southeastern New Hampshire, w. to British Columbia, s. to s. New Jersey and se. Pennsylvania. Winter: Same as breeding range.

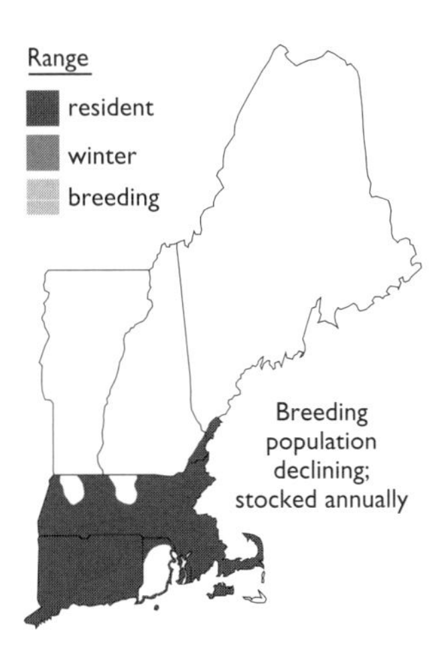

Distribution in New England: A stocked gamebird in New Hampshire, Massachusetts, Connecticut, and Rhode Island, ring-necked pheasants overwinter and breed in low to moderate numbers in the New Hampshire Seacoast area, e. Massachusetts, s. and w. Connecticut, and the coastal lowlands on Narragansett Bay, R.I.

Status in New England: Common (Massachusetts, Connecticut and Rhode Island) to uncommon (New Hampshire and s. Maine). Block Island supports one of the densest breeding populations in New England (Raithel 1992). Populations are regularly restocked for hunting in the fall.

Habitat: Breeding: Open cultivated fields of grass or grain, fallow fields, brushy pastures, hedgerows by roadsides, cut-over land, open ungrazed woodlots. Agricultural lands provide the best habitat. Absent in mountains, avoids forested areas. Wintering: Birds seek areas with dense protective cover, often cattail swamps interspersed with thickets.

Nesting: Egg dates: April 14 to July 28, New York (Bull 1974:205), late April to late June, Massachusetts (Veit and Petersen 1993:144). Clutch size: 6 to 15, typically 10 to 12. Incubation period: 23 to 25 days. Nestling period: Less than 1 day (precocial). Broods per year: 1 or 2. Age at sexual maturity: 1 yr. Nest site: On the ground among weeds or cultivated hay or edges of pastures.

Territory Size: The male defends an area for courting, mating, and feeding (females may nest nearby or elsewhere). Eleven territories ranged from 3 to 13 acres (1.2 to 4.3 ha) (Twining 1946).

Sample Densities: In the Northeast, densities ranged from fewer than 1 bird per 100 acres (40 ha) in poor habitat to more than 50 birds per 100 acres (40 ha) in optimum habitat (Allen 1956). At Pelee Island, Ontario, Stokes (1956:367) noted the following densities: 6.2 birds per acre (0.4 ha) in hayfield, 5.7 birds per acre (0.4 ha) in abandoned pasture, and 12.4 birds per acre (0.4 ha) in weeds. In autumn, good habitat probably supports 10 to 15 birds per 100 acres (40 ha) (Studholme and Benson 1956).

Food Habits: Cultivated grains and weed seeds (staple), buds and soft parts of herbaceous vegetation, fleshy fruits, insects. Feeds in grain fields and weed fields bordered by hedgerows and swamps that afford protection (Allen 1956).

Comments: Pheasants were first successfully introduced from China to Oregon in 1881, but were first introduced from England to the Northeast (Governor's Island, N.Y.) as early as 1733. Gov. Wentworth brought the first pheasants from England for release in Wolfeboro, N.H., around 1790 (Nevus 1944). Stocked birds in the Champlain lowlands were first given protection in 1892 (Foote 1946). Vermont ceased stocking birds after a trial in 1966–1971, but private individuals there and elsewhere continue to release birds for hunting in the fall. Pheasants were released in Connecticut in the late 1800s and were first reported in the wild in 1897 (Sage et al. 1913). After peak populations in the mid-1930s, the population declined markedly in 1943, and reproduction in the wild has remained low since (Loranger 1980). Ring-necked pheasants were abundant breeding birds in the Connecticut Valley in Massachusetts until the early 1970s; the switch from picked to chopped corn was a major factor in their decline to a very sporadic breeder there. Pheasants have declined along with the decline of agriculture throughout their range in New England.

Ruffed Grouse

(*Bonasa umbellus*)

Range: Resident from central Alaska and n. Yukon to s. Labrador, s. to nw. California, central and e. Idaho, central Utah, Wyoming and Montana, central and se. Minnesota, Ohio, in the Appalachian Mountains to n. Georgia and ne. Virginia; locally to w. South Dakota; introduced and established in Iowa and Newfoundland (DeGraaf et al. 1991:104).

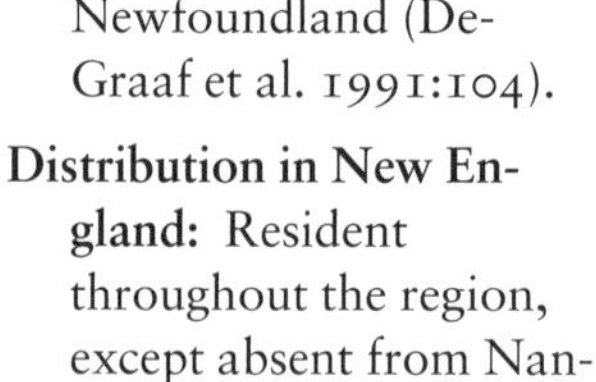

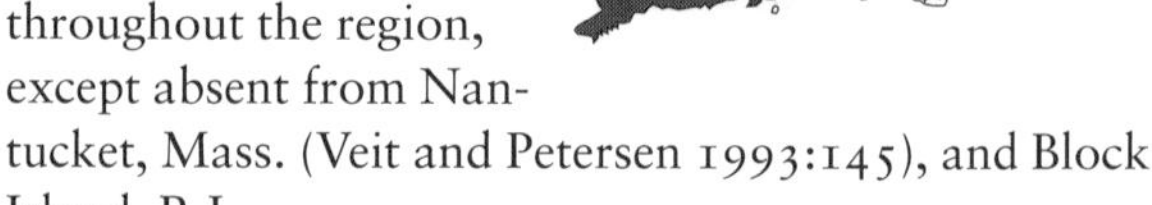

Distribution in New England: Resident throughout the region, except absent from Nantucket, Mass. (Veit and Petersen 1993:145), and Block Island, R.I.

Status in New England: Fairly common inland, less common along the s. coasts.

Habitat: Inhabits brushy, mixed-age woodlands, successional to mature hardwood and mixed forests, commonly with birch or aspen present. Hens and broods prefer areas with a dense understory and fairly open herbaceous ground cover. Early successional habitat—regenerating clear-cuts or old burns—are ideal. More mature woodlands, especially coniferous forests, used in winter for cover and roosting. Birds roost in snow when it is soft and deep, or they roost on the ground or in trees.

Special Habitat Requirements: Drumming sites (logs or stone walls), in relatively dense hardwood saplings, small poles or brushy escape cover (Boag and Sumanik 1969), hardwood stands for nesting and feeding, sunny openings for dusting (Pough 1951:176). Strongly associated with the aspen type in the Lake States (Gullion 1972). Grouse also occur in New England woodlands in which aspens exist only as scattered trees or are absent. Old orchards are ideal fall habitat in New England. Catkin-bearing trees and shrubs are characteristic of grouse habitat. Ruffed grouse need four types of cover—drumming site, nesting, brood, and winter cover—to survive [see Atwater and Schnell (1989) for a review].

Nesting: Egg dates: April 20 to June 9 in Vermont (Ellison 1985:92 in Laughlin and Kibbe 1985). April 1 to June 22, New York (Bull 1974:206). Clutch size: 8 to 15, typically 9 to 12. Incubation period: About 24 days. Nestling period: Less than 1 day (precocial). Broods per year: 1. Age at sexual maturity: 1 yr. Nest site: An excavated bowl, lined with bits of vegetation within reach of the hen as she sits on the nest (Storm and Scott 1989); usually on dry ground in dense cover at the base of a solid object such as a tree or stump, a fallen log, or rock. Very often located at edge of a clearing and close to a source of water.

Home Range: For males, the home range may be as small as 6 to 10 acres (2.4 to 4.0 ha) (Gullion 1972).

Sample Densities: Maximum density under optimum conditions seems to be about 1 pair per 6 to 8 acres (2.4 to 3.2 ha) (Gullion 1972).

Food Habits: Seeds, insects, fruit, leaves; buds of birch, aspen, hazel, hophornbeam, and cherry are staples in fall and winter (Bump et al. 1947). Aspen stands are favorite feeding spots in winter, especially in the boreal forest zone (Svoboda and Gullion 1972).

Comments: The familiar drumming of male grouse (called pa'tridge in n. New England) is caused by a series of compression waves created by the beating of the bird's wings while he remains stationary on an elevated display site—a log, stump, or stone wall. Drumming begins right after snow-melt, and is most frequent from March through May. Males are polygamous. In winter, birds may roost in small groups.

Spruce Grouse

(*Falcipennis canadensis*)

Range: Resident: From n. Alaska to n. Quebec, Labrador, New Brunswick, and Nova Scotia, s. to s.-coastal and se. Alaska, n. Oregon, se. Idaho, nw. Wyoming, w. Montana, and se. Alberta, to n. Minnesota, e. to ne. Vermont, n. New Hampshire, and n. Maine. Range is generally congruent with that of the n. coniferous forest (DeGraaf et al. 1991:101).

Distribution in New England: The Northeast Highlands of Vermont (Oatman 1985:90 in Laughlin and Kibbe 1985), n. New Hampshire (Smith 1994:64 in Foss 1994), and n. Maine (Boone and Krohn 1996).

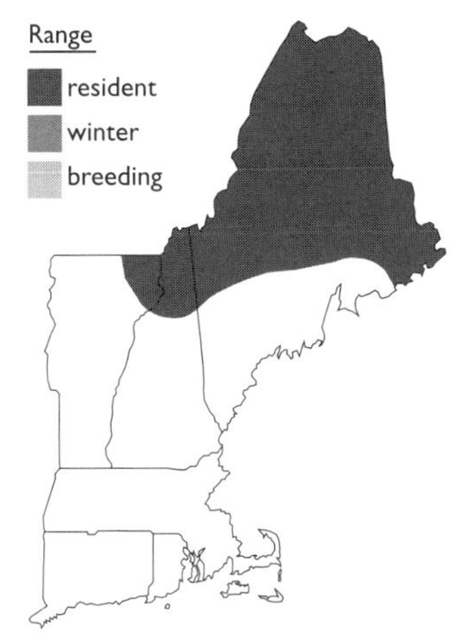

Status in New England: Rare resident in ne. Vermont, uncommon in the White Mountains, n. New Hampshire, and n. Maine.

Habitat: Inhabits extensive, dense stands of coniferous forests, especially where living branches reach the ground and where there are numerous scattered forest openings of a few hundred square feet such as those created by bogs or swamps. In the Northeast, utilized conifer stands generally range from 23 to 46 ft (7 to 14 m) in height (Boag and Schroeder 1992). Associated with spruce-fir forests on mountain ridges and peaks and in low-elevation bogs in New Hampshire (Smith 1994:64 in Foss 1994), and balsam fir forests in Maine (Allan 1985). Birds use more open areas such as bogs and heaths in the summer (Boag and Schroeder 1992). In winter, birds seek denser areas of forest than those used during warmer months.

Special Habitat Requirements: Large stands of dense coniferous forest for food and shelter.

Nesting: Egg dates: May 5 to June 24, se. Canada and Maine (Bent 1932:129). Clutch size: 4 to 10, typically 6 to 8. Incubation period: 22 to 25 days. Nestling period: Less than 1 day (precocial). Broods per year: 1. Age at sexual maturity: 1 yr. Nest site: On ground, often in moss in sites well concealed by overhead conifer branches, brush, or adjacent to a tree or stump.

Territory Size: 3 to 21 acres (1.2 to 8.5 ha) per territorial male in white spruce and paper birch habitat in Alaska (Ellison 1971); these areas were used exclusively by males but were not necessarily defended.

Home Range: May be as large as 100 ha (247 acres) for territorial males and 100 to 150 ha (247 to 370 acres) for hens that have nested (Ellison 1973). Individual birds use home ranges averaging less than 24 acres (10 ha).

Sample Densities: Possibly 10 males per square mile (4 males/km^2) in spruce-birch habitat in Alaska (Ellison 1971). 7 to 11 birds per 0.4 square mile (7 to 11 birds/km^2) (spring density); 20 to 36 birds per 0.4 square mile (20 to 36 birds/km^2) (autumn density) (Ellison 1973).

Food Habits: Buds and needles of conifers (winter staple), insects, seeds, fruits and tender leaves of herbaceous plants, mushrooms (Pendergast and Boag 1970).

Comments: The spruce grouse is one of Vermont's rarest resident birds, restricted to the small boreal zone in the

Northeast Highlands since 1932 except for one site outside its current Essex County range, on Wheeler Mountain in Sutton. It is more common in New Hampshire from the White Mountains n. at elevations up to 4,500 ft (Smith 1994:64 in Foss 1994). Diet shifts in late summer and early autumn from nonconiferous materials to primarily coniferous ones. Spruce grouse do not erect the crown feathers as do ruffed grouse when alarmed (Boag and Schroeder 1992).

Wild Turkey

(Meleagris gallopavo)

Range: Resident locally in every state except Alaska, also present in s. Canada and Mexico. Has been reintroduced into much of its former range, and successfully introduced in 10 states outside the historic range (Kennamer et al. 1992). Birds in the e. United States n. of Florida are the *silvestris* subspecies.

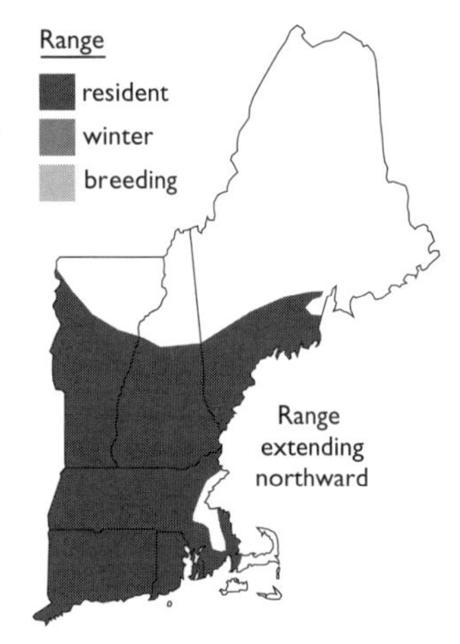

Distribution in New England: Southern Maine, central New Hampshire, w. and s. Vermont, central and w. Massachusetts and expanding into Plymouth County, w. Rhode Island, and throughout Connecticut. The n. limit of eastern wild turkeys seems to be determined by the condition, depth, and duration of snow cover (Markley 1967). Deep powdery or fluffy snow is most detrimental; turkeys cannot scratch through more than 6 in of snow or walk far if it exceeds 1 ft (30 cm) in depth. The limiting snow line approximately corresponds with the transition zone between oak-hickory forests to the s. and northern hardwoods and aspen-birch to the n. (Healy 1992a).

Status in New England: Common resident.

Habitat: In summer in the East prefers open, mature hardwood forests containing mast-bearing trees and scattered openings such as forest clearings and agricultural land. In winter, mature forests, especially south-facing slopes with mast-bearing trees and abundant springs and seeps. In n. areas where snows can be deep and fluffy, turkeys use seeps and dairy farm manure spreads, silage, and waste grain as primary feeding sites (Healy 1992a).

Special Habitat Requirements: Forests with mast-producing trees, forest openings, and dense coniferous or mixed forests for roosting.

Nesting: Egg dates: April in Massachusetts (Veit and Petersen 1993:146). April 26 to July 9, New York (Bull 1974:212). Clutch size: 8 to 15, typically 10 to 11. Incubation period: 26 to 28 days (Eaton 1992). Nestling period: 1 day (precocial). Broods per year: 1, but hens renest if the first clutch is destroyed (Eaton 1992). Age at sexual maturity: Usually 1 yr for females and 2 yr for males. Nest site: Turkeys lay their eggs in a simple depression on dry ground, usually in dead leaves. Nests are usually under a fallen tree crown or in low, shrubby cover, near water, and next to a tree or stump. Hens exhibit no forest-type preference, and often nest in cutover areas.

Home Range: Highly variable owing to flock segregation by sex and annual food supply. Hens with broods range over an area of 100 to 200 ha (250 to 500 acres) during summer, but use smaller centers of activity within the larger range depending on habitat quality (Healy 1992a). Large movements often occur in early spring, late summer, and fall (Healy 1992b).

Sample Densities: 7.7 birds per square mile (3 birds/km^2) on forested refuge land in West Virginia (Uhlig 1950). 15 to 20 birds per square mile (6 to 8 birds/km^2) under ideal conditions (Pough 1951:189). 16 to 25 birds per square mile (6 to 9 birds/km^2) for good range in the East (National Wild Turkey Federation 1986).

Food Habits: Mast (acorns and beechnuts staples); also fruits and seeds of trees, shrubs, and herbaceous vegetation, tubers, roots, and insects (Korschgen 1967). Generally feeds in mast-producing woodlands during fall and winter; fields, pastures, and woodlands with rich, herbaceous ground cover during summer.

Comments: Wild turkeys were extirpated in Connecticut by about 1811 (Zeranski and Baptist 1990:96) and in Massachusetts by about 1850 (Veit and Petersen 1993:146). Prior to that time, turkeys were both abundant and quite tame birds, so easily taken that their capture was work for Indian children (Kennamer et al. 1992). Turkeys were greatly reduced in the mid 1600s by the colonists in s. New England (Josselyn 1672). Wild turkeys from game farm stock were repeatedly introduced beginning in 1929 in e. Massachusetts and in the Quabbin area in the 1950s but none were successful. The introduction of wild-trapped birds from New York in w. Massachusetts, Vermont, and New Hampshire in the 1970s resulted in the robust populations present today. Turkeys have recently expanded their range northward into central New Hampshire; hens with broods were regularly seen along the Kancamagus Highway in Lincoln, N.H. in summer 1999. The wild turkey is the only bird in the western hemisphere to receive worldwide importance through domestication (Eaton 1992).

Northern Bobwhite

(*Colinus virginianus*)

Range: Breeding: Resident from se. Wyoming and central South Dakota to s. through the central and e. United States to Florida and s. Arizona into Mexico; also se. Massachusetts and adjacent Rhode Island. Introduced and established in w. North America (DeGraaf et al. 1991:111).

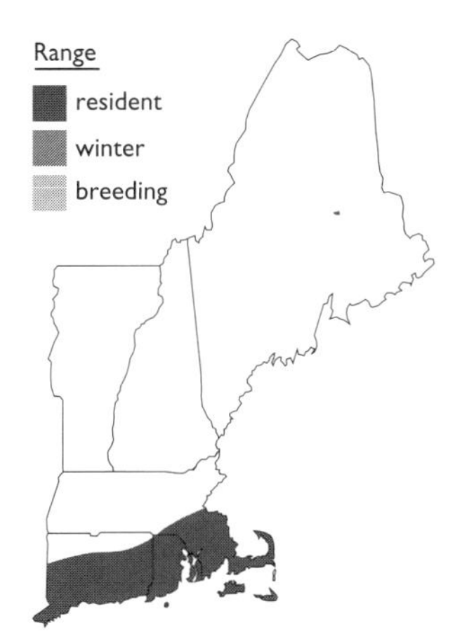

Distribution in New England: Southeastern Massachusetts, including Cape Cod, Martha's Vineyard and Nantucket, e. and s. Connecticut, and Rhode Island.

Status in New England: Locally common to uncommon resident. Commonly replenished with released stock. Until the early 1970s, bobwhites locally inhabited the Connecticut River Valley n. to Sunderland, Mass. Development and decline of agriculture has eliminated the bird's habitat in much of its former range.

Habitat: Fields, early successional habitats and open woodlands adjacent to pastures, meadows, and agricultural fields with abundant weedy growth (Brennan 1999). Prefers cultivated and fallow agricultural lands with hedgerows and dense brush for cover. Open pitch-pines and barrens on Cape Cod and the islands. Avoids deep woods.

Special Habitat Requirements: Open woodlands adjacent to fields and brushy cover. In winter, dense cover within 150 ft of feeding areas (Pough 1951:184).

Nesting: Egg dates: May 20 to September 2 in Massachusetts (Veit and Petersen 1993:147). May 25 to September 24, New York (Bull 1974:203). Clutch size: 12 to 20, typically 14 to 16. Incubation period: 23 to 24 days. Nestling period: 1 or less days (precocial). Broods per year: 1 or 2. Age at sexual maturity: 1 yr. Nest site: Usually among dead or growing grasses surrounded by patches of bare ground, often along fence rows or in neglected corners of pastures within 50 ft (15 m) of a clearing. Standing vegetation is usually less than 20 in (0.5 m) high with upright stems sparse enough for birds to pass between (Rosene 1969:63).

Territory Size: Approximately 1.1 acres (0.4 ha) in fallow field in Kansas (Fitch 1958). During several hundred observations, whistling males were seen no closer than 50 ft (15 m) (Rosene 1969:61).

Home Range: Coveys generally remain within an area about 0.5 mile (0.8 km) in diameter (Stoddard 1931). Winter ranges are from 4 to 77 acres (1.6 to 31.2 ha), averages 8.2 to 17.9 acres (3.3 to 7.2 ha), on two plantations in South Carolina (Rosene 1969:88).

Sample Densities: Stockard (1905:149) found 16 nests in a 30-acre (12.1-ha) field of sedge in Mississippi. 5 pairs per 100 acres (40 ha) in field and edge habitat. 1.5 pairs per 100 acres (40 ha) in pine-deciduous forest and farmland in Maryland (Stewart and Robbins 1958:125).

Food Habits: Vegetation: soft herbaceous parts of plants, buds, seeds, and fruits. Young are almost totally insectivorous (Brennan 1999). Commonly feeds in cultivated fields and open areas near protective brushy cover. Judd (1905 in Bent 1932:18) found that 917 stomachs contained 84 percent vegetable and 16 percent animal matter.

Comments: Bobwhite quail were abundant in the early 1800s in s. New England, especially along the coast, after clearing for agriculture began. The birds were greatly reduced in abundance by severe winters; the winter of 1812–1813 was especially severe and the bird was called "no longer common" in New England by Timothy Dwight in 1821 (Forbush 1916). Also, Forbush detailed one snowstorm in 1857 that essentially wiped out the bobwhite on Cape Cod; they were later replaced by southern birds. The large, pale native New England bobwhite was extinct by 1872 (Griscom and Snyder 1955:80). All bobwhites now in New England are of southern stock, which have been repeatedly introduced since about the time of the Civil War. Formerly found as far north as extreme s. Maine (Johnsgard 1973:412), bobwhites now do not occur north of se. Massachusetts in New England.

King Rail

(Rallus elegans)

Range: Breeding: Locally from e. Nebraska and central Minnesota to Connecticut, s. through nw. and central Kansas, central Oklahoma, and most of the e. United States, to w. and s. Texas, central Mississippi and Alabama, and s. Florida. Winter: Primarily from s. Georgia, Florida, the s. portions of the Gulf States, and s. Texas, s. to Mexico. In Mexico, a casual visitor reported in Tamaulipus, Guanajuato, and Veracruz. Found on Cuba and the Isle of Pines.

Distribution in New England: Breeding: Rare in Connecticut, occurs in the lower Connecticut River and tributaries at Durham, Lyme, Old Lyme, and estuaries in Stonington and Westbrook (Zeranski and Baptist 1990:100); Rhode Island, sw. Washington County and ne. Bristol County (Enser 1992:61). Massachusetts, critically rare at the n. periphery of its range (Crowley 1994), Vermont breeding not confirmed since 1981. Does not occur in New Hampshire or Maine.

Status in New England: Rare breeder in coastal marshes in s. New England.

Habitat: Coastal brackish and freshwater marshes and inland freshwater marshes with abundant vegetation (especially grasses, sedges, rushes, and cattails), roadside ditches, tidal rivers, shrub-swamps, rice fields. Forages and nests along waterways made by muskrats, and consequently they are often killed in traps set for muskrats (Meanley 1992). Nest sites reported from shallow tidal and nontidal marshes (Meanley 1992). Wintering birds use coastal brackish, saltwater (rarely), and freshwater marshes.

Special Habitat Requirements: Wetlands with abundant vegetation and fairly stable water levels during the breeding season.

Nesting: Egg dates: Probably late May and June in Massachusetts (Veit and Petersen 1993:150). May 24 to July 3, New York (Bull 1974:218). Clutch size: 6 to 15, typically 10 or 11. Incubation period: 21 to 23 days. Nestling period: Less than 1 day (precocial). Broods per year: 1 (possibly 2 in s. states). Nest height: To 1.5 ft (0.5 m). Typically 0.5 to 1.5 ft (0.2 to 0.5 m). Nest site: Usually 6 to 18 in (0.1 to 0.5 m) above water, often on a hummock among cattails, marsh grasses, rushes, or other aquatic vegetation, the stalks and leaves of which form a natural canopy. Water depth surrounding nest is usually less than 2 ft (0.6 m).

Territory Size: Size not known. Male establishes and defends territory against king rail and other rail species.

Sample Densities: 3 nests per 464 ft (141.5 m) of roadside ditch 30 ft (9.1 m) wide in Arkansas (Meanley 1969:49). 30 birds per 100 acres (40 ha) of inland fresh marsh in Florida (Bateman 1977).

Food Habits: Largely crustations and aquatic insects, also amphibians, mollusks, seeds of marsh plants, and waste grain in marsh vegetation, very shallow water, 2 or 3 in deep (5.1 to 7.6 cm), mudflats exposed by low tide.

Comments: A rare and secretive breeder in s. Connecticut and Rhode Island marshes and estuaries, the king rail has not been confirmed as a breeder in Massachusetts since 1979. Wetland protection measures that prevent loss of habitat are critical for the maintenance of king rail populations in New England.

Virginia Rail

(Rallus limicola)

Range: Breeding: Locally from s. British Columbia and nw. Alberta to s. Quebec, New Brunswick, and sw. Newfoundland, s. to Baja California, n. Texas, Kansas, Ohio, Virginia, and along the Atlantic Coast to North Carolina. Winter: Southern British Columbia and Washington, n. Mexico, and the Gulf states, and Georgia s. to Guatemala. Resident in central Mexico, Ecuador, Peru, s. Chile, and Argentina.

Distribution in New England: Throughout the region in freshwater marshes.

Status in New England: Locally common breeder, uncommon migrant, rare in winter.

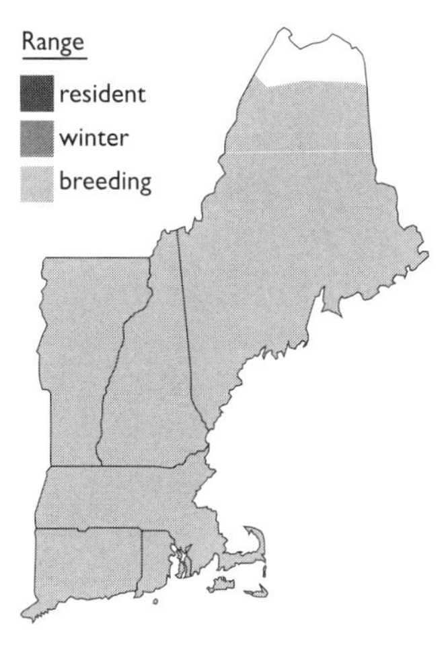

Habitat: Breeding: Primarily shallow freshwater marshes, marshy borders of lakes and streams, occasionally in brackish and salt marshes. Associated with wetlands with robust stands of emergent vegetation such as cattails (Johnson and Dinsmore 1986) and bulrush and a high abundance of invertebrates (Taylor 1971, Sayer and Rundle 1984, Gibbs et al. 1991). Requires standing water with mudflats or moist soil for foraging (Gibbs et al. 1991). In deepwater habitats will walk on floating vegetation to forage (Sayer and Rundle 1984, Johnson and Dinsmore 1985). Absent from wetlands lacking adequate areas of shallow water or mudflats (Conway 1995). Associated with large wetlands with abundant emergent vegetation, and relatively high pH and conductivity in Maine (Gibbs et al. 1991).

Special Habitat Requirements: Freshwater marshes with emergent vegetation, especially sedge and cattail.

Nesting: Egg dates: May 12 to July 18 in Massachusetts (Veit and Petersen 1993:151); May 5 to July 15, New York (Bull 1974:219). Clutch size: 4 to 13, typically 8 to 10 (Walkinshaw 1937). Incubation period: 13 to 22 days (Zimmerman 1977). Nestling period: 3 to 4 days (Kaufmann 1987, 1988). Broods per year: 1 or 2. Nest height: Usually constructs nests in cattails and sedges 2 to 12 in (5.1 to 30.5 cm) above shallow water. Often covered with a loose canopy of vegetation and attached to surrounding vegetation or on a clump of grass or a tussock. Nest site: Nest is usually 2 to 12 in (5 to 30 cm) above water level in marsh vegetation, usually well anchored to plant stalks and protected by leaves that form a canopy. Water depth near nest typically 3 to 10 in (7.6 to 25.4 cm). Adults build numerous dummy nests within their territory (Conway 1995).

Territory Size: Territorial during pair formation and nest establishment (Conway 1995).

Sample Densities: 1 bird per acre (0.4 ha) in marsh in Colorado (Zimmerman 1977:50). 1.2 pairs per 2.5 acres (1 ha) in marsh in New York (Post and Enders 1970). 25 pairs per 2.5 acres (1 ha) in Michigan (Berger 1951). One or more pairs per 0.3 acre (0.1 ha) in Vermont (Kibbe 1985:96 in Laughlin and Kibbe 1985). New Hampshire considered within the species area of highest breeding density (Zimmerman 1977).

Food Habits: Invertebrates, worms, slugs, snails, crustaceans, small fish, seeds of marsh plants, berries, and occasionally frogs, and small snakes, taken in soft mud or the tops and undersides of floating plants, pond debris. Preferred feeding habitat: Weed fields adjacent to breeding area.

Comments: Protection of existing wetlands is essential to the welfare of the Virginia rail, which declined in the late 1800s to early 1900s due to filling and draining of marshes (Eddleman et al. 1988).

Sora

(*Porzana carolina*)

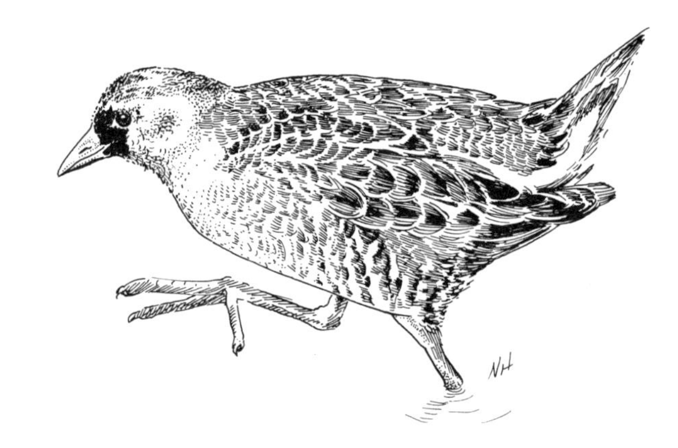

Range: Breeding: From s. Yukon and west-central and sw. Northwest Territories to w.-central and s. Quebec and sw. Newfoundland, s. locally to Baja California, central Arizona, New Mexico, central Illinois, and Maryland. Winter: From central California to the Gulf Coast and s. South Carolina, s. to South America; occasionally n. to s. Canada. In Mexico, winter residents can be found (probably widely) in marshes and mangroves. In Guatemala, an uncommon and local migrant and winter visitor on the Pacific slope and in the Peten; in Costa Rica, a widespread to local migrant and winter resident found in the lowlands and casually to 4,950 ft (1,500 m). Winter resident in the tropical to temperate zone in Guyana, Venezuela, and Columbia s. to Ecuador and central Peru. Generally an uncommon visitor to Puerto Rico and the Virgin Islands from October to April. Occurs as a migrant from North and Central America in both Trinidad and Tobago.

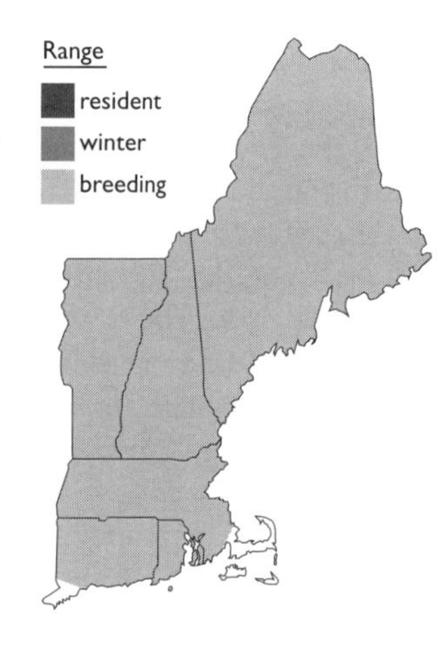

Distribution in New England: Breeding: Southern Maine, se. New Hampshire, and throughout Vermont, Massachusetts, Connecticut, and Rhode Island where suitable wetland habitat exists, except absent from Cape Cod, Martha's Vineyard, and Nantucket in Massachusetts and Block Island in Rhode Island. Winter: A few may overwinter in unfrozen marshes in s. New England.

Status in New England: Rare breeder and winter resident.

Habitat: Habitat use is highly correlated with amount of emergent vegetation (Johnson and Dinsmore 1986). Prefers freshwater marshes with shallow to intermediate water depths and dominated by emergent vegetation (cattails, bulrushes, sedges, burreeds) (Walkinshaw 1940), but also inhabits brackish and salt marshes, ponds, swamps, bogs, and sloughs with emergent vegetation. Occasionally uses wet meadows or upland fields adjacent to wetlands (Johnson and Dinsmore 1985, Griese et al. 1980, Gibbs and Melvin 1990, Melvin and Gibbs 1996, Crowley 1994, Low and Mansell 1983).

Special Habitat Requirements: Large marshes and wetlands with abundant emergent vegetation.

Nesting: Egg dates: May 20 to June 11 in Massachusetts (Veit and Petersen 1993:152). April 30 to July 17, New York (Bull 1974:220). Clutch size: 5 to 16, typically 8 to 11 (Melvin and Gibbs 1996). Incubation period: 16 to 19 days [about 14 days (Odum 1977)]. Nestling period: Less than 1 day (precocial), independent around 4 weeks. Broods per year: 1. Nest site: Constructs a well-concealed nest on a raised platform over shallow, 6 to 8 in (15 to 20 cm) water, among dense emergent vegetation. Occasionally nests on the ground.

Territory Size: Distances between sora nests ranged from 4 to 83 ft (1.2 to 25 m) in Minnesota, Michigan, and Colorado (Pospichal and Marshall 1954, Berger 1951, Glahn 1974).

Sample Densities: Thought to be about 12 birds per acre (0.4 ha) at a reservoir in Colorado (Odum 1977). 35 nests per 107 acres (43.3 ha) in Iowa (Tanner and Hendrickson 1956). 4 pairs per 0.5 acre (0.2 ha) in cattail and sedge marsh in Michigan (Berger 1951).

Food Habits: Seeds of wetland plants, aquatic insects, mollusks, and crustaceans. In fall, consumes grains and seeds, especially sedges, bulrush (Webster 1964), wild rice and cultivated grains.

Comments: The sora has been steadily declining in the Northeast since the turn of the century. Loss of emergent wetland habitat to development in the late 1800s was a major factor in sora population declines. Currently, changing conditions and plant composition within wetlands may be deleterious to the species.

Common Moorhen

(*Gallinula chloropus*)

Range: Breeding: Locally in California, central Arizona, and n. New Mexico, and from central Minnesota and s. Wisconsin to Maine and Massachusetts, s. to South America. Winter: Eastern North America from South Carolina and the Gulf Coast s., elsewhere throughout the breeding range, occasionally n. to Utah, Minnesota, s. Ontario, and New England. In Mexico, a local resident or migrant found in marshes nearly throughout the country. In Guatemala, an uncommon and local migrant and winter visitor on the Pacific slope. In Costa Rica, a resident, and possibly migrant, during the northern winter. In Panama, locally common in the lowlands on both slopes; widespread and increasing in the Canal area. A fairly common resident of Caroni Marsh in Trinidad. In Colombia, a local resident and possible northern migrant found at up to 3,100 m in the Caribbean region. Found virtually throughout the West Indies; a common permanent resident in Puerto Rico, uncommon in the Virgin Islands.

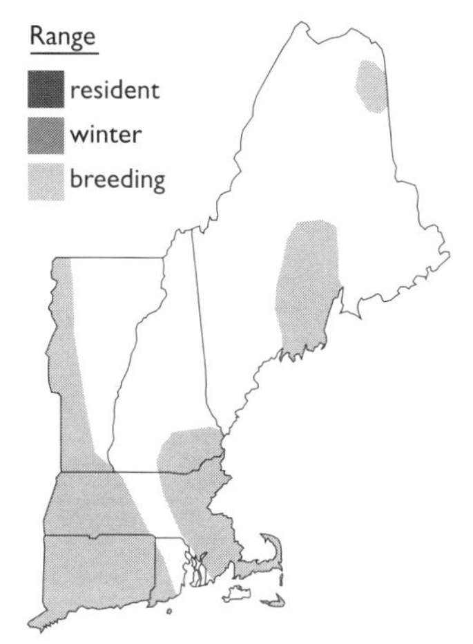

Distribution in New England: Breeding: Northeastern and s.-central Maine, se. New Hampshire, w. Vermont, locally throughout Massachusetts and Connecticut, except no records from Martha's Vineyard. No recent confirmed nesting in w. Rhode Island (Enser 1992:64).

Status in New England: Rare to uncommon local breeder and migrant; rare in winter.

Habitat: Inhabits fresh and brackish marshes, margins of lakes, ponds, slow-flowing rivers and streams, and sewage lagoons. Suitable bodies of water have emergent vegetation such as cattails, bulrushes, reeds, sedges, and burreeds growing in water more than 12 in (30 cm) deep, interspersed with open water (Strohmeyer 1977).

Special Habitat Requirements: Shallow bodies of water with emergent vegetation interspersed with areas of open water.

Nesting: Egg dates: May 22 to July 17 in Massachusetts (Veit and Petersen 1993:155). May 14 to July 25, New York (Bull 1974:224). Clutch size: 6 to 17, typically 10 to 12 (Fredrickson 1971). Incubation period: 21 days. Nestling period: Probably less than 1 day (precocial). Broods per year: 1 or 2. Nest height: To 2 ft (0.6 m), typically less than 1 ft (0.3 m). Nest site: Typically nests on a hummock or other clump of emergent vegetation. Occasionally nests in shrubs such as willow or alder. Nest is usually over water 1 to 3 ft deep (0.3 to 0.9 m) and is well concealed by a canopy formed from surrounding taller plants.

Food Habits: Leaves and stems of underwater plants, duckweed, leaves of grass and herbs, seeds and berries are staples. Animal foods include snails, insects, and worms.

Comments: Common moorhens are now less abundant and less widespread than in the early 1900s due to the filling of wetlands. These birds are habitat specialists, restricted to cattail marshes bordering large freshwater ponds (Allen 1939). Formerly called the common gallinule.

American Coot

(*Fulica americana*)

Range: Breeding: Breeds from the s. Yukon, sw. Quebec, s. New Brunswick, and Nova Scotia, s. locally through Middle America, the West Indies, and Andean South

America. Winters from se. Alaska and British Columbia, s. through the Pacific States, and from Colorado and n. Arizona to the lower Mississippi and Ohio Valleys and Maryland, s. through the se. United States through the breeding range. In Mexico, winters along both coasts and locally in the interior throughout the country. In Central America, a common migrant and winter visitor most abundant on the Caribbean slope. Locally common resident in Colombia. A fairly common visitor to Puerto Rico but uncommon in the Virgin Islands during most months except summer. Found in the Bahamas and the Greater Antilles, vagrant to the Lesser Antilles.

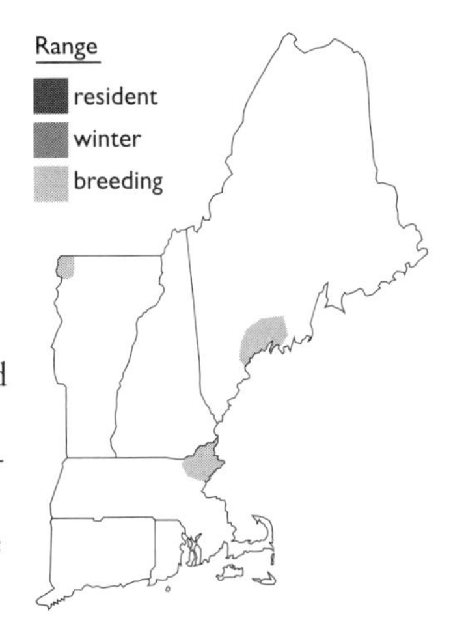

Distribution in New England: Along the s. Maine coast, the n. Champlain Lowlands in Vermont (Kibbe 1985b:102–103 in Laughlin and Kibbe 1985), and ne. Massachusetts; isolated breeding records from Plum Island and the Sudbury River Valley, Mass. (Veit and Petersen 1993:155).

Status in New England: Rare and local breeder. Occasionally abundant fall migrant and uncommon winter resident in s. New England. Long-term decline noted in the e. United States (DeGraaf and Rappole 1995:145).

Habitat: Inhabits marshes, ponds, lakes, reservoirs, sewage lagoons, sloughs, wet meadows, and marshy borders of creeks and rivers.

Special Habitat Requirements: Shallow, freshwater wetlands with water about 3 ft (1 m) deep and emergent vegetation interspersed with areas of open water (Fredrickson 1977).

Nesting: Egg dates: Mid May in Massachusetts (Veit and Petersen 1993:156). April 25 to July 14, New York (Bull 1974:224). Clutch size: 6 to 22, typically 8 to 12 (Fredrickson 1970). Incubation period: 23 to 24 days. Nestling period: 1 to several days (precocial). Age at first flight: Probably 7 to 8 weeks. Age at sexual maturity: Probably 1 yr. Nest site: Usually floating on surface of water 1 to 4 ft (0.3 to 1.2 m) deep and anchored to surrounding emergent vegetation (often cattails or bulrushes). Coots build platforms of vegetation for resting and brood rearing, and use muskrat houses for the same purposes. There are few records of coots nesting in New England; there are no confirmed nestings in Massachusetts before 1959 (see Veit and Petersen 1993:156 for review) and two records for Vermont (Kibbe 1985b:102 in Laughlin and Kibbe 1985).

Territory Size: Average about 1 acre (0.4 ha) for 5 pairs in California (Gullion 1953). Coots are strongly territorial during the breeding season, defending nest against both coots and other marsh birds.

Sample Densities: 432 pairs per square mile (166 pairs/km^2) under ideal conditions in the prairies of North Dakota (Stewart and Kantrud 1972). 1 nest per 0.54 acre (0.2 ha) in Iowa (Friley et al. 1938).

Food Habits: Underwater plants are staples as well as algae, grass shoots, grains, aquatic insects; bulk of diet is vegetable matter but also takes fish, tadpoles, worms, and crustaceans in shallow water.

Comments: American coots are primarily western and midwestern birds, never common in New England. Highest breeding densities of coots in Iowa were noted where 50 percent of marsh was open water and remaining 50 percent was emergent vegetation (Weller and Fredrickson 1973).

Killdeer

(*Charadrius vociferus*)

Range: Breeds from e.-central and se. Alaska and s. Yukon to central Quebec, w. Nova Scotia, Prince Edward Island, and w. Newfoundland, s. to Southern Baja California, central Mexico, Tamaulipas, the Gulf Coast, and s. Florida; also in the s. Bahamas and Greater Antilles, and in w. South America along the coast of Peru. Winters from s. British Columbia across the central United States to New England and s. throughout the remainder of North America to Colombia and Ecuador. Winters widely throughout Mexico. Found throughout the West Indies as a transient or winter visitor; resident in the Bahamas and Greater Antilles; uncommon in the Virgin Islands.

Distribution in New England: Breeds throughout the

region. Winters along s. coast.

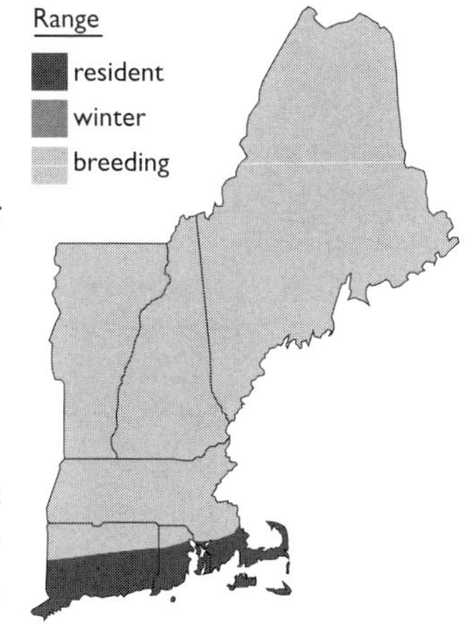

Status in New England: Fairly common and widespread local breeder throughout much of the region; uncommon on Cape Cod and Nantucket. Common spring and fall migrant.

Habitat: Breeding: Inhabits open dry uplands, meadows, pastures, cultivated fields, airports, golf courses, and disturbed or heavily grazed or managed areas where grasses are short, sparse, or absent. Winter: Plowed or sparsely vegetated moist fields, wetlands, coastal flats and beaches, and the margins of rivers and lakes that are free of ice.

Special Habitat Requirements: Open fields or waste areas with closely cropped or sparse vegetation.

Nesting: Egg dates: April 3 to July 4, New York (Bull 1974:239). Clutch size: 3 to 5, typically 4. Incubation period: 24 to 29 days. Nestling period: Less than 1 day (precocial) (Nickell 1943). Broods per year: 1 or 2. Age at sexual maturity: 1 yr. Nest site: Eggs are deposited on bare, often gravelly ground in a depression hollowed out by the female. Often a few small stones, wood chips, or other plant debris are placed inside. Pastures, meadows, golf courses, and cultivated fields are the most common sites. Even flat graveled roofs are sometimes used for nesting.

Sample Densities: 24 pairs per square mile (9 birds/km^2) maximum density in North Dakota (Stewart and Kantrud 1972). 3.9 pairs per 100 acres (40 ha) in plowed field and wheat field in Maryland (Stewart and Robbins 1958:136).

Food Habits: Insects—especially beetles and grasshoppers; centipedes, spiders, worms, snails, crayfish, weed seeds. Feeds on bare soil, short grasses, with robin-like running and pausing.

Comments: Killdeer were common nesters in New England in the early 1800s (Forbush 1916); they declined after 1850 with the decline of agriculture and the heavy hunting of shorebirds in the late 1800s. Killdeer were "practically extirpated" in Connecticut by about 1900 (Forbush 1916), not common again until after the 1940s. Killdeer are solitary feeders. They spread out over an area rather than feed in compact groups like sandpipers. Habitat trends: Increasing in Maine, Ohio; decreasing in Massachusetts, New Hampshire, Rhode Island; static in Connecticut, Delaware, and Vermont (Jurek and Leach 1977).

Willet

(*Catoptrophorus semipalmatus*)

Range: Breeds locally from e. Oregon, central Alberta, and sw. Manitoba, s. to ne. and e.-central California, w. and n. Nebraska, and e. South Dakota; locally along the Atlantic and Gulf Coasts from Prince Edward Island, s. to s. Florida, and w. to s. Texas. Occurs sporadically (nonbreeding birds) in summer as far s. as n. South America. Winters along the Pacific and Atlantic Coasts from n. California and Virginia, s. through Central America and n. South America to Brazil on the Atlantic Coast and to Peru (occasionally n. Chile) on the Pacific Coast. In Mexico, breeds in Tamaulipas and winters along the coasts, mainly on the Pacific side. A common migrant and fairly common winter resident throughout Central America. In Colombia, a common migrant and winter resident on both coasts and occasionally inland. Fairly common in Puerto Rico and the Virgin Islands. A winter visitor to both Trinidad and Tobago.

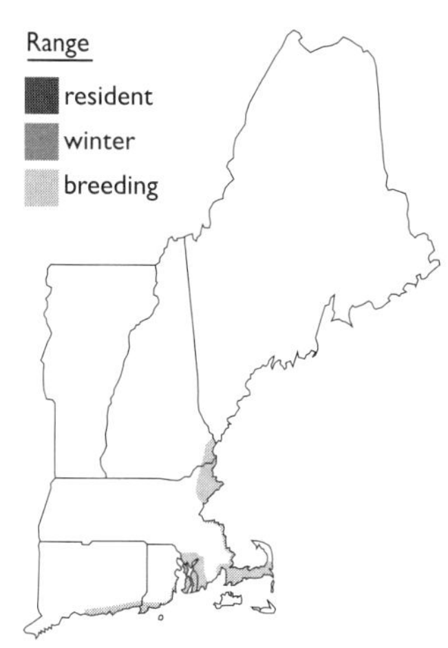

Distribution in New England: Southeastern (Hampton and Seabrook) New Hampshire (Gavutis 1994:82 in Foss 1994), s. Maine, coastal ne. and se. Massachusetts, Connecticut along the coast from Gilford e., and Hog Island in Rhode Island (Enser 1992:68).

Status in New England: Locally common breeder and migrant in s. coastal New England.

Habitat: Salt marshes, margins of marshy lakes and

ponds, exposed mudflats, sandbars, tidal creek banks, beaches, sandy islands, dunes, open pastures, and dry uplands near water. Often perches on bushes, trees, fences, posts, and buildings.

Special Habitat Requirements: Coastal marshes and nearby grassy areas in the e. United States, moist plains and prairies in w. North America.

Nesting: Egg dates: Mid April to mid May. Clutch size: 4 to 5, typically 4. Incubation period: 22 to 29 days. Age at first flight: 14 to 21 days. Broods per year: 1. Nest site: In a depression on the open ground or in a thick clump of vegetation, sometimes far from water. Nest sometimes located in open areas on a sandy beach or well hidden in low grasses (Wilcox 1980).

Food Habits: Fiddler crabs, mollusks, marine worms, small fishes, insects, and some seeds and vegetation (Stenzel et al. 1976). Forages on exposed tidal flats, salt or brackish marshes, sand bars, water-soaked pastures, muddy creek banks, and salt flats.

Comments: Before 1880, willets nested along the entire Atlantic Coast from Nova Scotia to Florida. Egging and shooting so reduced the population that by the early twentieth century there were none nesting between Nova Scotia and New Jersey. Willets now nest along the New England coast from Connecticut to Maine. The first Massachusetts nesting since 1877 occurred at Monomoy in 1976 (see Veit and Petersen 1993:170–171 for review). It may nest inland in New England one day as it does in the West.

Spotted Sandpiper

(Actitis macularia)

Range: Breeding: Breeds from central Alaska and central Yukon to Labrador and Newfoundland, s. to s. Alaska, s. California, and central Arizona, and e. to the n. portions of the Gulf States, North Carolina, Virginia, and e. Maryland. Winters from sw. British Columbia, s. Arizona, s. New Mexico, s. Texas, the s. portions of the Gulf States, and coastal South Carolina to Chile and Argentina. Occasionally nonbreeding birds remain on the wintering ground in summer. A migrant throughout Mexico, wintering on both coasts and at low altitudes inland. The most widespread shorebird in Costa Rica, and probably in the rest of Central America as well. In South America, winters chiefly inland, regularly s. to Bolivia and s. Brazil, occasionally to Chile, Uruguay, and Argentina. A common winter visitor to the West Indies.

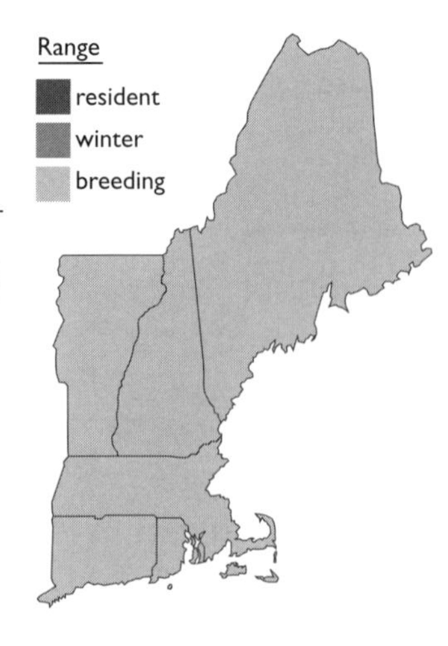

Distribution in New England: Generally throughout the region, except absent from Block Island in Rhode Island.

Status in New England: Fairly common but declining breeder in s. New England, common migrant.

Habitat: Inhabits edges of ponds, lakes, rivers, and streams. Occasionally occurs at edges of tidal creeks, on sandy beaches on offshore islands, or in grass at the edges of salt marshes (Veit and Petersen 1993:172, Zeranski and Baptist 1990:114). Sometimes found far from water in dry fields, pastures, weedy shoulders of roads, and at the edges of gravel parking lots. Prefers open habitat.

Special Habitat Requirements: Open margins of freshwater bodies.

Nesting: Egg dates: May 6 to July 26, New York (Bull 1974:251). Clutch size: 3 to 5, typically 4. Incubation period: 20 to 24 days. Nestling period: Less than 1 day (precocial). Age at first flight: 15 to 16 days. Age at sexual maturity: 1 yr. Nest site: Nests are solitary or in loose colonies. Eggs are laid in a depression in the ground that is lined with grass. Often under shrubs or weeds or in tall grass up to 30 in high (76.2 cm).

Territory Size: Little or no defended area.

Sample Densities: 43 pairs per 17.6 acres (7.1 ha) of dry meadow-rocky shore-sandy beach habitat in Michigan (Miller and Miller 1948).

Food Habits: With the exception of toad tadpoles, almost all animals small enough to be eaten (Oring et al. 1997); insects, especially grasshoppers and crickets, small fish (occasionally), and crustaceans. Walks slowly and gleans, occasionally catching insects on the wing.

Comments: The spotted sandiper is the most widespread breeding sandpiper in North America (Oring et al. 1997). The spotted sandpiper is perhaps declining as a breeding bird, although it is still common in New England. It was not shot in great numbers as were many other shorebirds; its decline may be due to the decline of agriculture and farm ponds (Zeranski and Baptist 1990:115). Adults may perform injury-feigning distraction displays if the nest is approached, as do killdeer.

Upland Sandpiper

(*Bartramia longicauda*)

Range: Breeds locally from n.-central Alaska, n. Yukon, and n. Alberta to s. Quebec, central Maine, and s. New Brunswick, s. to ne. Oregon, central Colorado, n.-central Texas, central Missouri, West Virginia, and Maryland. Migrates through Central America, the Caribbean, and n. South America. Winters in Brazil, Paraguay, Uruguay, and Argentina. In Mexico, a spring and fall migrant through the Valley of Mexico. An uncommon and local migrant in much of Central America. A transient in the West Indies.

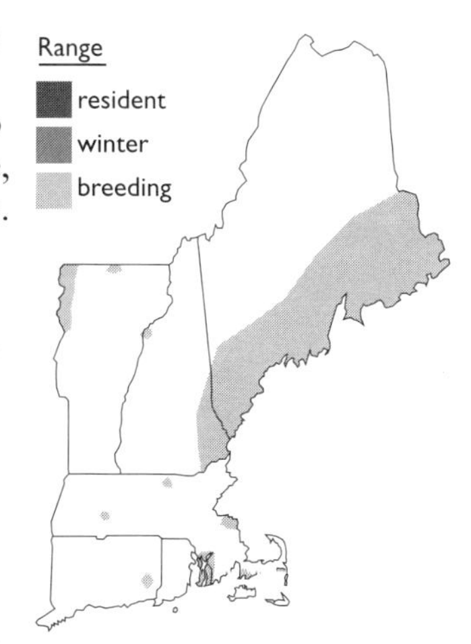

Distribution in New England: Breeds very locally throughout its range: Maine blueberry barrens, New Hampshire (Pease International Tradeport, Newington, Rochester, Dover, and Haverhill) (Foss 1994:86), w. Vermont in the Champlain Lowlands and n. on Lake Memphremagog, Massachusetts (Bridgewater, Middleboro, and Westover Air Force Base) (Veit and Petersen 1993:174, Blodgett pers. commun.), s. to Hartford County in Connecticut (Zeranski and Baptist 1990:115), and e. Rhode Island (Enser 1992:70).

Status in New England: Very uncommon local breeder, mostly at large airfields, rare to locally common migrant (Carter 1992).

Habitat: Breeding: Inhabits large grassy areas such as grazed pastures, meadows, hay fields, air fields, open grassy areas on military bases, and highway rights-of-way. Associated with low vegetation; 9.8 to 15.7 in (25 to 40 cm) in Wisconsin (Ailes 1980), and 3.1 to 9.8 in (8.0 to 25.0 cm) in Massachusetts (Carter 1992).

Special Habitat Requirements: Open habitats with low vegetation.

Nesting: Egg dates: April 23 to June 15, New York (Bull 1974:254). Clutch size: 4 to 5, typically 4. Incubation period: 21 to 24 days. Nestling period: Less than 1 day (precocial). Age at first flight: About 30 to 31 days. Broods per year: 1. Age at sexual maturity: 1 yr. Nest site: Well hidden in a depression in grass covered by nearby vegetation. Usually nests in loosely spaced colonies.

Territory Size: Two pairs had territories of 20 to 30 acres (8.1 to 12.1 ha) each in grassland in Wisconsin (Wiens 1969:41).

Sample Densities: 1 nest per 1.5 to 15 acres (0.6 to 6.1 ha) (Harrison 1975:70). 20 pairs per square mile (8 pairs/km^2—maximum density) in North Dakota (Stewart and Kantrud 1972). 7 nests were found in a 17-acre (6.9-ha) timothy field (Buss and Hawkins 1939).

Food Habits: Insects—especially grasshoppers and crickets, waste grains, and seeds of grasses and weeds. Animal 97 percent, vegetable 3 percent (McAtee 1912 in Bent 1929). Forages with a robin-like alternate running and pausing.

Comments: Clearing for agriculture created good habitat for the upland sandpiper in New England and enabled the bird to expand its range e. from the prairies (Bagg and Eliot 1937). Its abundance peaked in the mid 1800s, and the bird was shot in great numbers because of its table quality. The decline of agriculture and reversion of fields to forest have eliminated much of the former hayfield and old pasture habitat in New England. Extensive plowing and cultivating have destroyed much habitat in the prairies of the Midwest. Major airports now constitute the bird's habitats in the Northeast.

Common Snipe

(*Gallinago gallinago*)

Range: Breeding: Breeds from n. Alaska and n. Yukon to s. Northwest Territories, n. Quebec, and central Labrador, s. to central California, Arizona, and Colorado across to West Virginia, New England, and the Maritime Provinces. Winters from se. Alaska, s. British Columbia, the central United States, and Virginia, s. through Middle America and the West Indies to central South America. Populations in n. and central South America are resident; those in s. South America are austral migrants.

Distribution in New England: Breeding: Throughout Maine, New Hampshire, and Vermont, to w. and central Massachusetts. Not reported as a breeding bird in Rhode Island nor since the 1970s in Connecticut. Winter: Coastal Connecticut and Rhode Island, except absent from Block Island, and se. Massachusetts, except absent from Martha's Vineyard and Nantucket.

Status in New England: An uncommon to rare breeder, common migrant, rare winter resident in s. New England. It is not clear whether snipe were ever common in New England as breeding birds, although they have long been common migrants. Snipe were never common nesters in Connecticut (Zeranski and Baptist 1990:129). The southern edge of the breeding range lies essentially in Massachusetts; the species may have been more widespread as a breeder before 1900 (Forbush 1916). In Vermont, snipe have been historically considered uncommon (Perkins and Howe 1901); in New Hampshire, Brewster (1924) recorded the bird only as a migrant at Umbagog Lake.

Habitat: Breeding: Inhabits wet thickets, emergent and shrub wetlands, wet meadows, sedge bogs, fens, pond margins, and lowlands associated with rivers and brooks where soils are muddy and vegetation is sparse (Tuck 1972). Occupies areas with low woody growth such as willows and alders, and a ground cover of sphagnum, sedges, and grasses, preferably near open habitats such as pastures and fields. Winter: Wet marshes, meadows, fallow agricultural fields, and stream edges.

Special Habitat Requirements: Wet thickets and meadows, bogs, and shorelines with low vegetation, moist organic soils.

Nesting: Egg dates: April 20 to June 16, New York (Bull 1974:242). Clutch size: 3 to 5, typically 4. Incubation period: 18 to 20 days. Nestling period: Less than 1 day (precocial). Broods per year: 1. Age at sexual maturity: 1 yr. Nest site: Nest is concealed among grasses or other vegetation on dry ground, sometimes on a tussock of grass or sedge.

Territory Size: Unknown. Male selects and defends an area against other snipes. Size decreases as incubation advances. Defense ceases after chicks hatch.

Sample Densities: Breeding: Up to 17 pairs per 100 ha (247 acres) on peatland in Newfoundland (Tuck 1972). Spring: 11.6 birds per 100 acres (40 ha) in Oregon (Fogarty and Arnold 1977). Winter: 275 birds per 100 acres (40 ha) in Florida (Fogarty and Arnold 1977).

Food Habits: Larvae of aquatic insects (about 50 percent of diet), earthworms (staple), snails, small crustaceans, seeds of marsh plants. Animal matter constitutes 61.5% of the diet (Erickson 1941). In winter, forages in unfrozen marshes where taller vegetation dies back, exposing bare mud.

Comments: Males return in early spring—March or early April in Vermont and New Hampshire—about 2 weeks before the females. Males commence aerial displays upon arrival; peak activity occurs just after the females arrive—a wavering, winnowing sound best heard at dusk and on moonlit nights (Kibbe 1985:110 in Laughlin and Kibbe 1985).

During both spring and fall migration, snipe congregate when heavy rainfall floods cornfields and grassy meadows (Veit and Petersen 1993:200). Snipe are one of the few shorebirds that perch on fenceposts or even in the tops of trees.

American Woodcock

(*Scolopax minor*)

Range: Breeds from s. Manitoba, n. Minnesota, and south-central and s. Ontario to n. New Brunswick and Newfoundland, s. throughout e. North America to the Gulf States, and s. Florida, and w. Winters from e. Oklahoma, s. Missouri, Tennessee, and Virginia, s. to east-central Texas, the Gulf Coast, and s. Florida (DeGraaf et al. 1991:160).

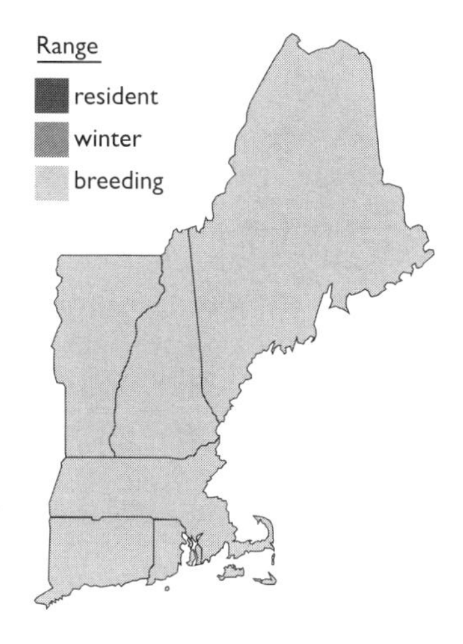

Distribution in New England: Throughout the region. Uncommon but regular in winter along the Connecticut and Rhode Island coasts, occasionally in Massachusetts.

Status in New England: Locally fairly common breeder, but declining due to habitat loss through forest succession. Fairly common migrant, uncommon but regular breeder in southern New England. Further declines may result if forest management practices increasingly favor uneven-aged stands, which do not provide large areas of young forest and shrubland that the bird needs for breeding (Keppie and Whiting 1994).

Habitat: Inhabits young forest and old fields, low wet thickets and moist woodlands in early stages of succession near fields and clearings; the best cover has scatted evergreens in the understory. Also found in agricultural areas, bottomland forests, forests along streams and ponds, brushy edges of woods, overgrown fields, and abandoned orchards, and pine stands, especially in hot dry summers and in spring and fall heavy rains or snow.

Habitat varies with activity, time of day, and season. Forest openings and old fields provide display areas (singing grounds) for males as grasses and herbs are replaced by (clumps of) young trees, shrubs. Adjacent young hardwoods and shrubs, especially alder (*Alnus* sp.) <20 yrs old provide moist ground for daytime feeding. Nests and young broods inhabit young to mid-age forest interspersed with openings; older broods in more mature forest. In summer, many birds roost in fields. Fall migration diurnal habitat is young hardwoods (Keppie and Whiting 1994).

Special Habitat Requirements: Fertile, moist soil that contains earthworms. Fields or forest openings at least 1.2 ha (Sepik and Derleth 1993) for courtship activities and nocturnal roosting. Dense brushy swales for diurnal cover (Sepik et al. 1981). Old farmland reverting to forest is ideal New England woodcock habitat.

Nesting: Egg dates: April 11 to May 6 in Massachusetts (Veit and Petersen 1993:201). March 24 to June 17, New York (Bull 1974:240). Perhaps the earliest-nesting ground bird in the region (Kippie and Whiting 1994). Clutch size: 3 to 5, typically 4. Incubation period: 20 to 21 days. Nestling period: All leave nest together a few hours after hatching (precocial) (Keppie and Whiting 1994). Broods per year: 1 unless first clutch or brood lost. Age at sexual maturity: 1 yr. Nest site: On forest floor or abandoned field in slight depression lined with a few dead leaves. Nest is usually located within 50 yd (45.7 m) of an edge, hidden in a variety of cover from grasses to young or middle-aged hardwoods of light to medium density (Owen 1977).

Territory Size: Females do not defend nests. The singing ground of the male may range in size from about 0.25 acre (0.1 ha) to more than 100 acres (40 ha) (Owen 1977).

Sample Densities: 4 to 7 males per mile (1.6 km) in New Hampshire and Maine singing ground surveys in 1971 and 1972 (Owen 1977). 5.6 territorial males per 100 acres (40 ha) in brushy abandoned farmland in Maryland. 1.5 territorial males per 100 acres (40 ha) in cut and burned woodland in Maryland (Stewart and Robbins 1958:139).

Food Habits: Earthworms account for 50 to 90 percent of diet (Sperry 1940). Larvae of beetles, flies, and other insects form the balance. Leaves, seeds, and fruits are occasionally taken. Feeds in open pastures, cultivated fields, stream banks.

Comments: Woodcock were abundant in the mid 1800s when much of the New England landscape was in fields and thickets (Bailey 1955); hunters took "hundreds" in a single day (Bull 1964); and market gunning caused a marked decline in the 1880s and 1890s (Bagg and Eliot 1937). Hunting regulations permitted the bird to again become common in the 1930s, but forest successional trends are diminishing available habitat.

Woodcock feed primarily on earthworms; their highly specialized bills are prehensile at the tip to seize worms found by probing soft wet soil. Individual woodcock commonly overwinter as far north as central Vermont in wet places that stay open throughout the winter.

Males typically have polygynous mating habits. Courtship takes place in fields or other openings where there are scattered woody plants 1 to 2 ft (0.3 to 0.6 m) high, in early succession (Sheldon 1967:64). Estimated carrying capacity under best breeding conditions is about 1 pair per 5.5 to 6.0 acres (2.2 to 2.4 ha) (Mendall and Aldous 1943).

Ring-billed Gull

(*Larus delawarensis*)

Range: Breeding: In w. North America from s. interior British Columbia and ne. Alberta to n.-central Manitoba, s. to ne. California, s.-central Colorado, and ne. South Dakota; in e. North America from n.-central Ontario to s. Labrador, s. to e. Wisconsin and n. Illinois, across to nw. Vermont. Winter: Along the Pacific Coast from s. British Columbia, s. to Mexico, Guatemala, El Salvador, and Panama; in the interior, from the Great Lakes to Mexico and the Gulf Coast; and along the Atlantic Coast, from the Gulf of St. Lawrence to Cuba, Jamaica, the Lesser Antilles, Trinidad, and Tobago.

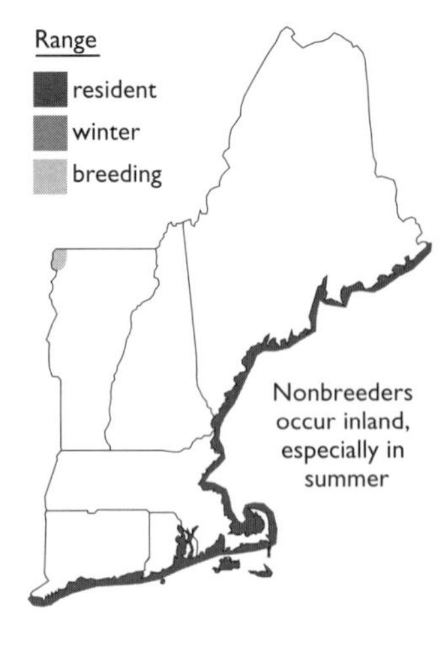

Distribution in New England: The n. Lake Champlain region in Vermont (Ellison 1985:114 in Laughlin and Kibbe 1985); Lake Umbagog, N.H.; and Massachusetts (Veit and Petersen 1993:217). Winters in coastal regions of s. Maine, New Hampshire, Massachusetts—including Nantucket but not Martha's Vineyard—Rhode Island, and Connecticut.

Status in New England: Locally common breeder. Abundant migrant and locally common in winter on the coast; uncommon migrant and in winter inland. Increasing in New England.

Habitat: Breeds on small to moderately sized rocky islands and occasionally peninsulas in large freshwater lakes or ponds. Rarely nests on islands in rivers or oceanic islands. Usually avoids densely settled areas. Winters along the coast in harbors, bays, estuaries, beaches, mudflats, rivers, lakes, ponds, reservoirs, refuse dumps, sewage outlets, and irrigated and plowed fields. Roosts on exposed sandbars and islands. Congregates on plowed fields, golf courses, and athletic fields after rains to feed on earthworms.

Special Habitat Requirements: Sparsely vegetated islands and peninsulas for nesting.

Nesting: Egg dates: May 1 to June 17, Vermont (Ellison 1985:114 in Laughlin and Kibbe 1985). Clutch size: 2 to 4, typically 3. Incubation period: 20 to 21 days. Broods per year: 1. Age at sexual maturity: 3–4 yr (Ryder 1993). Nest height: Typically on the ground in flat, elevated, sparsely vegetated areas, occasionally in low trees. Nest site: Primarily on islands, next to or under vegetation, or rocks on sandbars, beaches with rocks, driftwood, and bare rocky areas (Ryder 1993).

Territory Size: Colonial nester, 15,000 pairs on Four Brothers Islands in Lake Champlain (N.Y.) in 1982 (Ellison 1985:114 in Laughlin and Kibbe 1985).

Food Habits: Insects, earthworms, grubs, fish, rodents, grain, and sometimes birds eggs (Jarvis and Southern 1976); they also scavenge. Plunges on the surface of the water, picks up food while swimming, forages in flocks in pastures, golf courses, athletic fields (esp. after rain) and plowed fields (Ryder 1993).

Comments: An inland gull that breeds largely on fresh water, the ring-billed gull first nested in Vermont on

Young Island, Lake Champlain, in 1951, when 100 pairs nested. By 1983, there were 9,000 birds and 4,000 nests. Frequents dumps and parking lots in winter (Ryder 1993).

Herring Gull

(*Larus argentatus*)

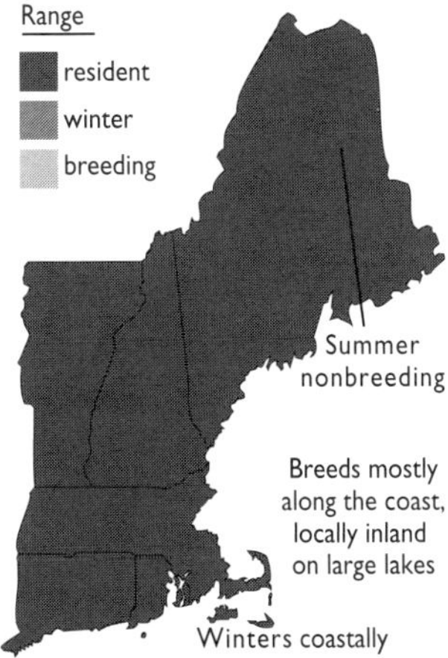

Range: Breeding: North America from n. Alaska and n. Yukon to east-central Northwest Territories, w. Baffin Island, and n. Labrador, s. to south-central British Columbia, central Alberta, n. Minnesota, ne. Illinois, n. Ohio, n. New York, and along the Atlantic Coast to ne. South Carolina. Winter: From s. Alaska, the Great Lakes region, and Newfoundland s., mostly at sea and along coasts, large rivers, and lakes s. to Panama and throughout the Caribbean.

Distribution in New England: Breeding: Throughout Maine, mainly on the coast and inland islands in large lakes and rivers (Boone and Krohn 1996), n. Lake Champlain in Vermont (Ellison 1985:116 in Laughlin and Kibbe 1985), e. New Hampshire on the coast and inland islands in large lakes and rivers, e. Massachusetts along the coast, including Martha's Vineyard and Nantucket Islands, and inland on Wachusett Reservoir (Veit and Petersen 1993:218); throughout Rhode Island, including Block Island (Enser 1992:72), and w. to e. and s. Connecticut (Zeranski and Baptist 1990:137). Winter: Throughout Maine and e. New Hampshire on the coast and large inland lakes and rivers with open water (Boone and Krohn 1996, Elkins 1985:116 in Laughlin and Kibbe 1985), e. Massachusetts including Martha's Vineyard and Nantucket (Veit and Petersen 1993:218), Rhode Island including Block Island (Enser 1992:72), and Connecticut along the coast and the Connecticut River Valley to Hartford, and inland on large lakes and rivers (Zeranski and Baptist 1990:137).

Status in New England: Locally very abundant breeder, migrant, and winter resident, less abundant inland than along the coast.

Habitat: Nests on sandy, rocky, and wooded islands, sand dunes, shorelines of oceans, lakes, and large rivers, barrier beaches, mudflats, salt marshes, cliffs, buildings, fields, grassy airport strips, and open dumps. Winters primarily along the coast or inland bodies of open water where food is likely to be abundant.

Special Habitat Requirements: Nest sites must be free of terrestrial predators.

Nesting: Egg dates: 24 April to 26 July (Veit and Petersen 1993:218). Clutch size: 2 to 4, typically 3. Incubation period: 25 to 28 days. Nestling period: Will brood chicks from the cold up to 10 days after hatching (Pierotti and Good 1994). Broods per year: 1. Nest height: Prefers to nest in low sites, occasionally in trees. Nest site: On grassy hummocks, drift adjacent to salt marshes, rock terraces, rocky cliffs, tall clumps of vegetation, marine terraces and beaches above high tide, and on rooftops; placed next to a large object (log, rock, bush) to create a barrier between the nest and the closest neighbor (DeGraaf and Rappole 1995, Pierotti and Good 1994).

Territory Size: Maintains a personal space of less than 3 ft (1 m) on roosting areas (Pierotti and Good 1994).

Food Habits: Fish, crustaceans, offal from fishing boats, fish processing plants, and open dumps (Belant et al. 1993). May forage on eggs and young of conspecifics and other seabirds within the colony (Pierotti and Annett 1987, 1991).

Comments: During the 1800s herring gulls did not breed s. of e. Maine (Drury 1973). The first New Hampshire nesting was at Umbagog Lake in the 1870s (see Ellison 1985:116 in Laughlin and Kibbe 1985). The first Massachusetts nesting was in 1912 on Martha's Vineyard; between 1930 and 1970, herring gulls underwent one of the most remarkable population expansions of any New England bird (Veit and Petersen 1993:219), due to open landfills and fish process waste dumped just offshore. Controls on landfills have slowed the pace of population growth. In recent years the herring gull has extended its range southward along the Atlantic Coast. Is often a detriment to colonial birds such as terns and laughing gulls (Bull and Farrand 1977:445, Blodget 1988, Drury 1973, Drury and Nisbet 1972).

Great Black-backed Gull

(*Larus marinus*)

Range: Holarctic, restricted to the North Atlantic Ocean. Breeding: In North America from Labrador to North Carolina. Winter: From Newfoundland and s. to North Carolina, rarely to Florida (Veit and Petersen 1993:223).

Distribution in New England: Breeding: Along the immediate coasts of Maine, New Hampshire, Rhode Island, Connecticut, and Massachusetts.

Status in New England: Locally abundant coastal breeder, abundant in winter on the coast, locally common and increasing inland (Slate 1994:94–95 in Foss 1994).

Habitat: Breeding: Coastal rocks, beaches. Wintering: Usually rivers and freshwater lakes near the coast and coastal islands.

Nesting: Egg dates: Early May through June. Clutch size: 3, sometimes 2. Incubation period: 27 days; incubation begins before clutch complete; eggs hatch over several days (Good 1998). Nestling period: 42 to 56 days. Broods per year: 1. Nest site: On the ground on small coastal islands or on cliff ledges. Great black-backed gulls nest about 1 week earlier than herring gulls. Male and female dig various scrapes and fill them with vegetation and feathers; the ultimate choice of nest site is where the female lays eggs. The chosen site is usually protected (but not on Appledore I.) from prevailing wind(s), placed next to a large object (rock, bush, clump of beach grass) that acts as a visual barrier between the nest and closest neighbors (Good 1998).

Territory Size: Nests may be solitary or in colonies of various sizes, usually among herring gull nests.

Food Habits: Fish, shellfish, eggs and young of seabirds, carrion, fish processing waste and garbage.

Comments: Before 1930, great black-backed gulls occurred primarily as winter residents in New England; the southernmost breeding area was the Bay of Fundy in 1900. Bent (1921) included Isle au Haut as the s. breeding limit, and by 1928 the species nested on Duck Island, Kittery, Me. (Slate 1994:94 in Foss 1994). Since the 1920s, great black-backed gulls have undergone a dramatic population growth and range expansion southward and inland, and have also increased in proportion to herring gulls, possibly reflecting earlier nesting or dietary preference (Veit and Petersen 1993:224). The great black-backed gull, the "Minister" gull to maritime Canadians, is the largest and heaviest gull in North America, and one of the largest in the world (Good 1998).

Common Tern

(*Sterna hirundo*)

Range: Breeding: Circumpolar; in North America, breeds from s.-central Northwest Territories to s. Quebec and

s. Labrador, Newfoundland, Nova Scotia, and Maine, s. to e. Washington and se. Alberta, across to central Minnesota, n. Ohio, and nw. Vermont, and locally along the Atlantic Coast to North Carolina; locally on the Gulf Coast in Texas, Mississippi, and w. Florida. Winters in South America from Colombia and Venezuela, s. to Brazil, Peru, and Argentina. Migrates through the Caribbean, rarely along the coasts of s. California, South Carolina, Florida, and the Gulf Coast.

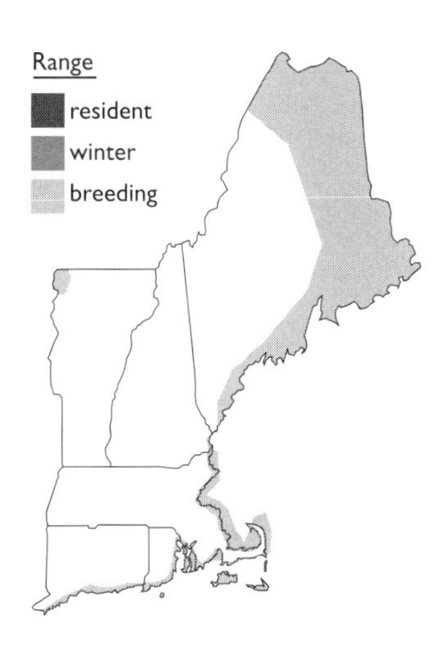

Distribution in New England: Breeding: Northern and coastal Maine, nw. Vermont on Lake Champlain, and the coastal regions of New Hampshire, Massachusetts, Rhode Island, and Connecticut.

Status in New England: Locally common breeder, and abundant coastal migrant, rare inland; declining in e. North America.

Habitat: Breeding: Inhabits a variety of open, sparsely vegetated habitats near water, including barrier beaches, marshy islands, low, small, rocky islands in lakes and rivers, and gravelly or sandy beaches.

Special Habitat Requirements: Nest areas with sparse vegetation that are isolated from predation and disturbance.

Nesting: Egg dates: May through July. Clutch size: 2 or 4, usually 3. Incubation period: 24 to 26 days. Nestling period: 3 to 4 days. Broods per year: 1, possibly 2 (Nisbet 1994). Nest site: A slight hollow in sand or among pebbles, or a hollowed mound of grasses and seaweeds; built either in the open or near weeds, grasses, or bushes.

Territory Size: The vicinity of the nest.

Food Habits: Small fish taken by plunging to the water surface.

Comments: Common terns were very abundant coastal breeders before 1870, with populations in the hundreds of thousands of pairs. Beginning about 1870, common, arctic, and roseate terns were taken in great numbers for the millinery trade; populations were reduced to tens of thousands of pairs by the 1890s, mostly on offshore islands. After protection from 1900 to 1910, the population of common terns increased until about 1970, when it again declined due to the great increase in the herring gull population, which competes for nest sites and disturbs nesting terns (Kress et al. 1983). Decline associated with loss of suitable nesting habitat, competition with gulls for nest sites, predation, human disturbance, and pollution.

Black Tern

(*Chlidonias niger*)

Range: Breeds from sw. and e.-central British Columbia and sw. Northwest Territories to s. Quebec and New Brunswick, s. locally to central California and Utah, across to Nebraska, Illinois, Pennsylvania, and Maine. Nonbreeding birds occur in summer on the Pacific Coast s. to Panama, and in e. North America to the Gulf Coast. Winters from Panama s. to Peru in the w. and Surinam in the e. Occurs as a migrant in the Caribbean region.

Distribution in New England: Lake Champlain and Lake Memphremagog in Vermont, Umbagog Lake in New Hampshire, and Maine.

Status in New England: locally common breeder, fairly common inland spring migrant, very common coastal fall migrant.

Habitat: Riparian marshlands of inland lakes with about 50% cover of emergent vegetation, such as cattails, rushes, and burreed (Novak 1992).

Nesting: Egg dates: 24 May to 12 July in New York

(Firstencel in Novak 1992). Clutch size: 2 to 4, typically 3 (Bent 1921). Incubation period: 21 to 24 days. Nestling period: 2 to 3 days (Dunn 1979). Broods per year: 1. Nest site: Often on stumps, logs, planks or floating islands of rotting vegetation, or on old muskrat houses. Eggs are often wet (Harrison 1975:78). Aggressively defend the nest.

Food Habits: Insects, largely dragonflies, damselflies, and mayflies (highly insectivorous on the breeding grounds), fish, and small crustaceans (Dunn and Agro 1995).

Comments: Black-terns are semicolonial nesters (Cuthbert 1954). Black terns did not nest in New England before the early 1920s (Forbush 1925); the first nesting colony in Vermont was at the n. end of Lake Champlain in 1950 (Smith 1950) and at Lake Memphremagog in 1963.

Rock Dove

(*Columba livia*)

Range: Introduced and resident throughout the urbanized world, in North America resident from s. Alaska and s. Canada, s. throughout the continent (DeGraaf et al. 1991:174).

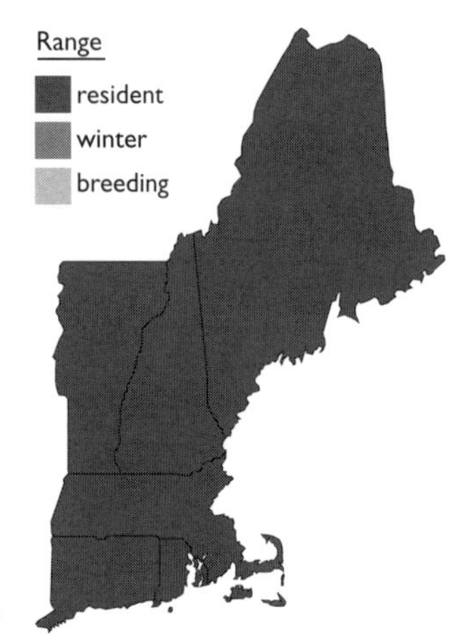

Distribution in New England: Resident throughout the region.

Status in New England: Abundant resident especially around cities and farms.

Habitat: Inhabits open country, sometimes near cliffs and ledges that have roosting sites, but is more common near human habitations, especially cities and farms.

Nesting: Egg dates: Throughout the year (Bull 1974:316), but in New England usually March to June and from August to November (Zeranski and Baptist 1990:151). Clutch size: 1 to 2, typically 2. Incubation period: 17 to 19 days. Nestling period: 25 to 32 days (Johnston 1992). Broods per year: 2 or 3. Age at sexual maturity: 1 yr. Nest site: Usually on or in buildings or bridges or other structures in semidark cavities. Rock doves nest singly or in colonies.

Food Habits: Seeds of grasses and grains, fruit, and human handouts.

Comments: Originally a Eurasian species, domestic rock doves were introduced to North America by the French in 1606 at Port Royale, Nova Scotia, at Virginia about 1621, and in Massachusetts about 1642 (Terres 1980). They prefer to roost in groups in areas that are sunny and sheltered from winds. May mate for life. Domestic and feral pigeons (rock doves) are among the most intensively studied of all birds. Knowledge of flight mechanics, sensory perception, orientation, navigation, and learning, among other aspects of avian biology, has depended heavily on research using the rock dove (Johnston 1992).

Mourning Dove

(*Zenaida macroura*)

Range: Breeds from s. and central Alberta to s. New Brunswick and Nova Scotia, s. to Mexico and the

Greater Antilles. Winters primarily from n. California, e. across the central United States to Iowa, s. Michigan, s. Ontario, New York, and New England, s. throughout the breeding range to central Panama. Portions of the ne. population migrate to the se. United States in winter (Veit and Petersen 1993:254).

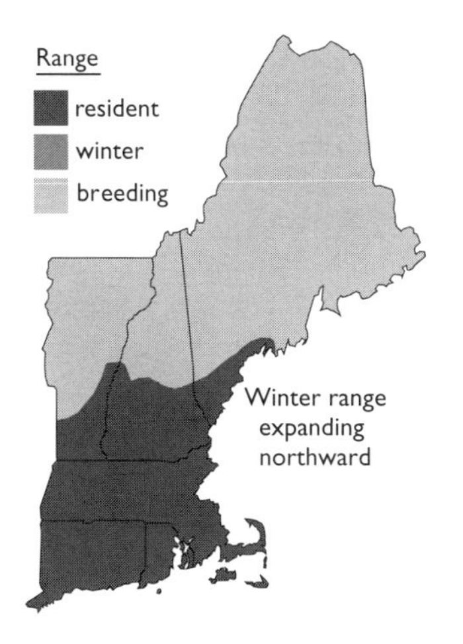

Distribution in New England: Breeds throughout the region. Winters in s. Maine, s. New Hampshire, s. Vermont, and throughout Massachusetts, Connecticut, and Rhode Island.

Status in New England: Common and widespread breeder, abundant migrant, increasingly common winter resident in s. New England.

Habitat: Inhabits a broad range of open habitats, including fields, orchards, open mixed woodlands and wood edges, shelterbelts, agricultural areas, cemeteries, evergreen plantations, suburbs, and cities. Avoids dense forests, readily uses burned woodlands (Mirarchi and Baskett 1994).

Special Habitat Requirements: Open country with some bare ground and seed-producing vegetation.

Nesting: Egg dates: About April 17 to July 27 in Massachusetts (Veit and Petersen 1993:155) but some nesting in March. March 9 to September 28, New York (Bull 1974:320). Clutch size: 1 to 3, typically 2, New York (Lehner 1965). Incubation period: 13 to 14 days. Nestling period: 12 to 14 days. Broods per year: 2 or more (as many as 6 clutches per season in temperate areas (Keeler 1977). Age at sexual maturity: 1 yr. Nest height: To 50 ft (15.2 m), typically 10 to 25 ft (3.0 to 7.6 m). Nest site: Extremely variable: often in a coniferous tree, also in tangles of shrubs or vines, occasionally uses old nest of other bird to support its twig platform. Nest is typically placed on a horizontal limb. Solitary or loosely colonial. DeGraaf (1975:29) found coniferous vegetation 0 to 35 ft (10.7 m) high important to mourning dove occurrence in suburban Massachusetts.

Sample Densities: Relative densities of breeding mourning doves based on the mean number heard per 20-mile (32-km) survey route: 0 to 9.9 birds—Maine, New Hampshire, Vermont, w. Pennsylvania, e. Ohio, West Virginia (Keeler 1977); 10.0 to 29.9 birds—Massachusetts, Rhode Island, Connecticut, New Jersey, e. Pennsylvania, Maryland (Keeler 1977); 30.0 to 59.9 birds—w. Ohio (Keeler 1977). 76 pairs per square mile (29 pairs/km^2) in favorable habitat in North Dakota (Stewart and Kantrud 1972).

Food Habits: Weed seeds and waste cereal grain of agriculture, occasionally takes small snails (Mirarchi and Baskett 1994). Commonly feeds in cultivated fields.

Comments: At the turn of the century, mourning doves were rare summer residents in s. New England, having been greatly reduced by extensive hunting during the late 1800s (Zeranski and Baptist 1990:152). Winter bird feeding may have contributed to the increase which occurred since 1920. Pair bond is usually lifelong monogamy. Birds generally increase their range and numbers in areas with secondary growth, cultivated fields and pastures (Goodwin 1977:206).

Black-billed Cuckoo

(*Coccyzus erythropthalmus*)

Range: Breeding: From e.-central and se. Alberta and s. Saskatchewan to New Brunswick and Nova Scotia s., at least locally, to e. Colorado, north-central Texas, n. Arkansas, n. Alabama, and the Carolinas. Migrates through Central America and the Caribbean. Winter: Colombia, Venezuela, Ecuador, and Peru.

Distribution in New England: Throughout the region, more widespread than the yellow-billed cuckoo (Veit and Petersen 1993:256).

Status in New England: Uncommon but widespread

breeder and uncommon migrant. More common away from the coast in Connecticut (Zeranski and Baptist 1990:154).

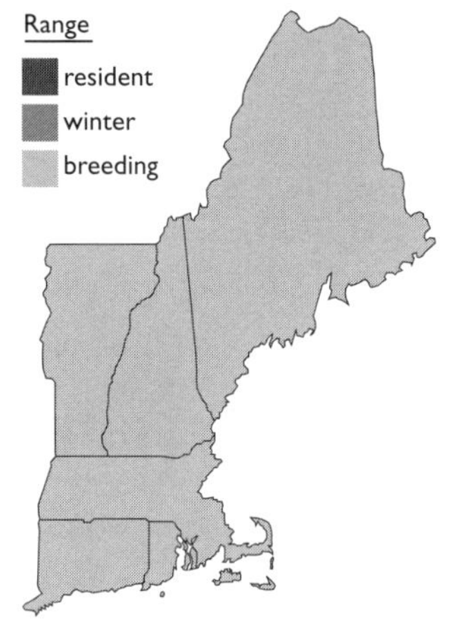

Habitat: Extensive areas of upland woods with an understory of shrubs, and vines. Also occurs in brushy pastures, shrubby hedgerows, shelterbelts, open woodlands, orchards, thickets, and along wooded roadsides. Common in scrub-oak forests on Cape Cod (Veit and Petersen 1993:256).

Special Habitat Requirements: Low, dense, shrubby vegetation.

Nesting: Egg dates: May 20 to August 29, Massachusetts (Veit and Petersen 1993:257) and New York (Bull 1974:325). Clutch size: 2 to 5 (Sealy 1978), typically 2 to 3. Incubation period: 11 to 12 days. Nestling period: 7 to 9 days (young perch on branches, unable to fly). Broods per year: Probably 1. Age at first flight: 21 to 24 days. Age at sexual maturity: 1 yr. Nest height: 2 to 20 ft (0.6 to 6.1 m), typically 4 to 6 ft (1.2 to 1.8 m). Nest site: Usually low in shrub or on branch of deciduous or coniferous tree, well concealed among the leaves (Spencer 1943).

Food Habits: Caterpillars (staple); also eats beetles, grasshoppers, crickets, and other insects, and fleshy fruits.

Comments: The feeding habits of the yellow-billed and black-billed cuckoos appear to be similar. Both species may respond to outbreaks of caterpillars, particular eastern tent caterpillars and gypsy moths (Miller 1934). The relationship is not straightforward, however. Cuckoo populations fluctuate erratically, and tent caterpillars and gypsy moths pupate by late June (Griscom 1949), before cuckoos normally fledge young. However, the great gypsy moth infestation of 1953 was followed by abundant migrant cuckoos in 1954 (Veit and Petersen 1993:259). Black-billed cuckoos seem to use extensive woodlands more than do yellow-billed cuckoos (Pough 1949:5).

Yellow-billed Cuckoo

(*Coccyzus americanus*)

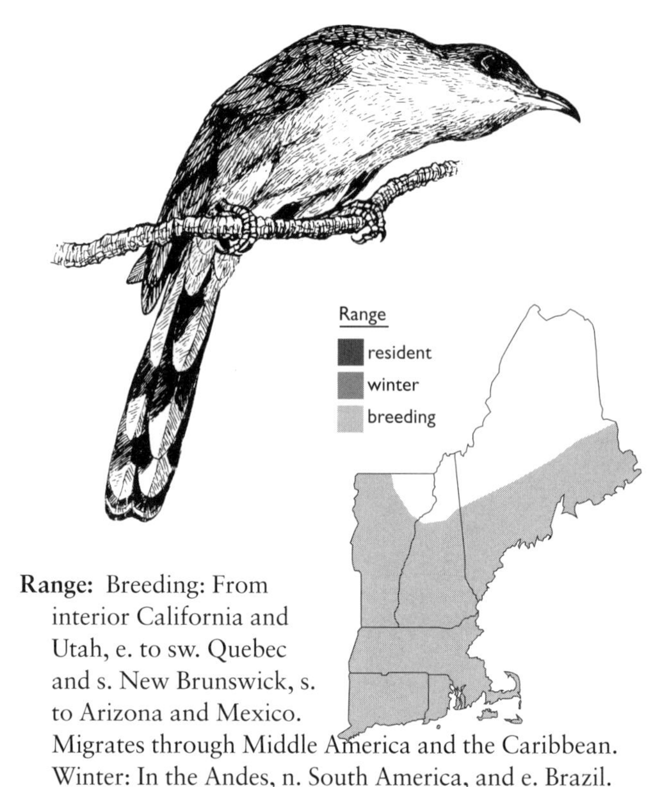

Range: Breeding: From interior California and Utah, e. to sw. Quebec and s. New Brunswick, s. to Arizona and Mexico. Migrates through Middle America and the Caribbean. Winter: In the Andes, n. South America, and e. Brazil.

Distribution in New England: Southern Maine, s. New Hampshire, w. and s. Vermont, and throughout Massachusetts, Connecticut, and Rhode Island. More of an inland than a coastal species.

Status in New England: Uncommon but fairly widespread breeder, very variable from year to year. Long-term and recent declines noted throughout its range. Fairly common inland migrant, especially in fall.

Habitat: Most abundant in second-growth deciduous forests with scattered openings; also in dense thickets along watercourses, woodland edges, country roadsides, abandoned orchards and fields overgrown with shrubs, logged areas, and thickets in open woods. Avoids dense woods and does not occur at high elevations. Scarce along the coast in Massachusetts (Veit and Petersen 1993:257), more abundant along the coast in Connecticut (Zeranski and Baptist 1990:154).

Special Habitat Requirements: Low, dense shrubby vegetation.

Nesting: Egg dates: May 20 to June 30 in Massachusetts (Veit and Petersen 1993:257). May 26 to August 19, New York (Bull 1974:324). Clutch size: 1 to 5, typically 3 or 4 (Preble 1957). Incubation period: About 14 days (Potter 1980). Nestling period: 7 to 9 days (young perch on branches, unable to fly). Broods per year: Probably 1. Age at sexual maturity: 1 yr. Nest height: 2 to 20 ft (0.6 to 6.1 m), typically 4 to 10 ft (1.2 to 3.0 m). Nest site: Prefers to nest in thick bushes overgrown with grape vines or in trees on horizontal limbs. Nest is usually well concealed by surrounding foliage.

Sample Densities: 8 territorial males per 100 acres (40 ha) in upland oak forest in Maryland. 6 territorial males per 100 acres (40 ha) in floodplain forest in Maryland. 4 territorial males per 100 acres (40 ha) in hedgerows, active and abandoned farmland in Maryland (Stewart and Robbins 1958:177). 2.8 territorial males per 100 acres (40 ha) in beech-oak forests in Ohio (Samson 1979).

Food Habits: Caterpillars (staple) and other insects; many kinds of fruits such as grapes, mulberries, and elderberries.

Comments: Secretive and shy, the yellow-billed cuckoo's presence is likely revealed by its "hollow, wooden call: *ka ka ka ka kow kow kow kow kowlp kowlp kowlp*" (Hughes 1999). The onset of breeding is apparently related to an abundant local food supply. Once begun, the breeding cycle is extremely rapid, 17 days from egg laying to fledging of young; bursting feather sheaths allow nestlings to become fully feathered within two hours of hatching (Hughes 1999). Frequently consumes larvae of gypsy moth tent caterpillars and fall webworms (Miller 1934). A decline noted after the Rhode Island gypsy moth infestation in 1987 (Enser 1992). Parasitism is not common as it is in the European (common) cuckoo (*Cuculus canorus*).

Barn Owl

(*Tyto alba*)

Range: Breeding: Nearly cosmopolitan; in North America, resident from sw. British Columbia, s. Idaho and Montana, e. to w. Vermont and Massachusetts, s. through the United States to South America (DeGraaf et al. 1991:188). Northernmost populations are partially migratory, wintering s. to s. Mexico and the West Indies (Stewart 1952).

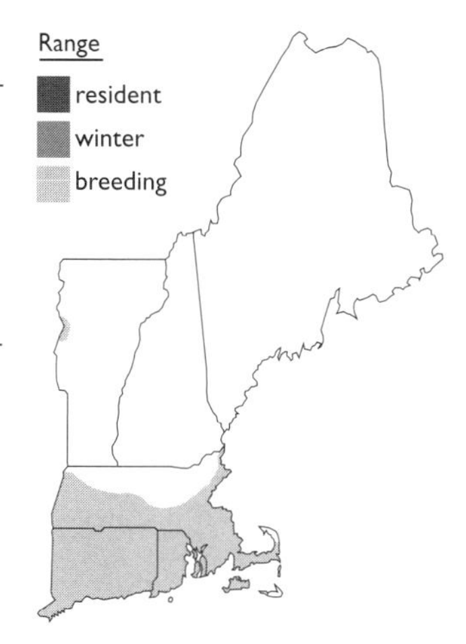

Distribution in New England: Vermont (s. Champlain Lowlands), s. New Hampshire (one documented nest record in Hollis, N.H., 1977) (Elkins 1978), Massachusetts, in river valleys and the se. coastal plain (Veit and Petersen 1993:259), w. Connecticut and coastal regions of the Connecticut River, and Rhode Island on Aquidneck and Block Islands (Enser 1992:85).

Status in New England: Rare to uncommon and local breeder and resident. Rare in winter.

Habitat: Semiopen, low-elevation habitats such as agricultural areas, pastures, salt marshes, grasslands, wet meadows, deserts, typically including old barns, silos, church steeples, or abandoned buildings. Natural nest sites include hollow trees, burrows in the ground, protected holes or crevices in banks or cliffs, and abandoned mammal burrows (e.g., foxes, woodchucks). Will use nest boxes (see Looman et al. 1996).

Special Habitat Requirements: Nest cavities, either natural (hollow trees, steep bluffs) or man-made (barns, silos, nest boxes) in extensive open, grassy areas (Rosenburg 1992). An abundant supply of small mammals for food.

Nesting: Egg dates: Late February to early July in Massachusetts (Veit and Petersen 1993:260). February to December. Peak: April to June, New York (Bull 1974:328). Clutch size: 2 to 11, typically 5 to 7 (Johnsgard 1988). Incubation period: 32 to 34 days (Harrison 1975). Nestling period: 50 to 60 days. Broods per year: 1, occasionally 2, although uncommon in temperate regions (Marti 1992). Age at sexual maturity: 1 yr. Nest site: Usually man-made structures—barns, abandoned buildings, silos, church steeples, and sometimes tree cavities. Also nests in holes in bluffs facing the sea on Martha's Vineyard and Block Island, R.I.

Food Habits: Rodents (especially voles [*Microtus*]) are staples; also takes other small mammals and occasionally takes small birds. Hunts meadow grasses, quartering low (a few meters above ground).

Comments: The barn owl is the most widely distributed owl, but was "accidental" in New England before

1900; although a nesting was confirmed in Connecticut (Winchester) in 1892, it was not until 1920 that the range extended through Connecticut (Zeranski and Baptist 1990:155). The first nesting in Massachusetts was on Martha's Vineyard in 1928 (Keith 1964). The first Vermont nests were found in the early 1980s. Birds are nocturnal, roosting by day and hunting by night; hence, they are seldom observed even in thickly settled towns and cities. Barn owls have been observed roosting in cedar groves (Georgia) and pine plantations (Michigan). Breeding is irregular, depending on availability of food. Wallace (1948) in Michigan found that 80 to 90 percent of diet consisted of meadow voles (*Microtus*). Populations are declining throughout its range due to loss of farmland and grasslands (Colvin 1985, Marti 1992). Barn owls are not tolerant of extremely cold weather, and suffer high mortality in extreme winters in New England. Like other owls, barn owls congregate in winter roosts, commonly in dense groves of conifers (Veit and Petersen 1993:260).

Eastern Screech-Owl

(*Otus asio*)

Range: Resident: From s. Manitoba and n. Minnesota to sw. Quebec and central Vermont, s. through the e. United States, e. Montana, e. Colorado, and w. Oklahoma to s. Texas, and across to s. Florida (DeGraaf et al. 1991:190).

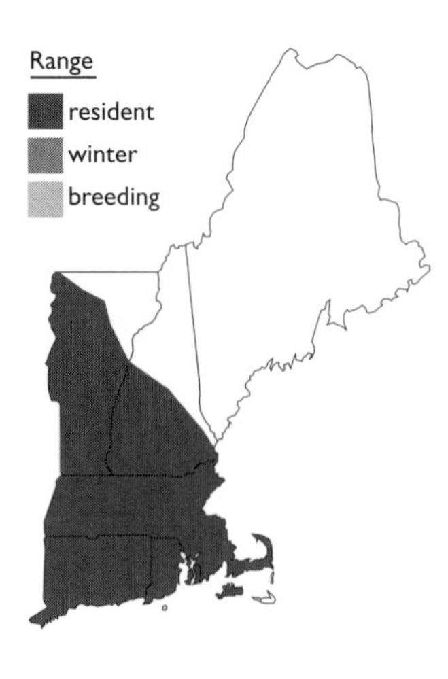

Distribution in New England: Resident in Vermont, except the Northeast Highlands (Laughlin 1985a:132 in Laughlin and Kibbe 1985); central New Hampshire; throughout Massachusetts, but uncommon on Cape Cod, Nantucket, and most of the higher elevations (Veit and Petersen 1993:260); Rhode Island, primarily the coastal lowlands (Enser 1992:83); and Connecticut, generally along the coast (Zeranski and Baptist 1990:155). Does not breed in Maine (Boone and Krohn 1996).

Status in New England: Fairly common at lower elevations in s. New England. Uncommon on Cape Cod.

Habitat: Open deciduous forests and woodlands with sparse understories (Robbins et al. 1989, Gehlbach 1994). From mountain slopes generally below 5,000 ft (1,500 m) to river valleys (Gehlbach 1995). Favors open woodlands adjacent to meadows, marshes, or fields in the East, oak and riparian woodlands in the West. Also inhabits trees in residential areas, orchards, small woodlots, and forest edges.

Special Habitat Requirements: Cavities for nesting and roosting.

Nesting: Egg dates: April 7 to May 5, Massachusetts (Veit and Petersen 1993:261). March 23 to May 11, New York (Bull 1974:329). Clutch size: 3 to 7, typically 4 or 5. Incubation period: 27 to 34 days, typically 30. Nestling period: About 30 days. Broods per year: 1. Age at sexual maturity: 1 yr (a small percentage breed at 2 yr of age). Nest height: 5 to 50 ft (1.5 to 15.2 m), typically 5 to 30 ft (1.5 to 9.1 m). Nest site: Natural cavities and abandoned woodpecker holes, especially those of flickers and pileated woodpeckers. Cavities are also used for roosting and caching food.

Territory Size: Averages 6 ha during breeding season, increases to 11 ha in winter; may overlap with neighboring males by 25% in Texas (Gehlbach 1994). For nesting males 8 ha, for females 6 to 9 ha in Connecticut (Smith and Gilbert 1984). Territory size increases as food resources decrease (Gehlbach 1994).

Sample Densities: 1 pair per 2.5 square miles (1 pair/6.5 km^2) in Michigan (Craighead and Craighead 1969:214). 1 pair per 4 square miles (1 pair/10.4 km^2) in Wyoming (Craighead and Craighead 1969:215).

Food Habits: Rodents (especially meadow voles) and insects are staples (Van Camp and Henny 1975); crayfish, snails, reptiles, amphibians, birds, and fish are also taken. Preferred feeding habitat: Grassy openings among widely spaced trees, open fields, meadows, and, in New England, along wooded field margins or marshy streams.

Comments: Cavities and nest boxes are used by the owls during winter months as feeding stations or food caches. The birds are opportunistic predators, generally consuming animal prey most readily available. Screech-owls may have historically occupied n. New Hampshire, but are now rarely found in s. New Hampshire. In Massachusetts, just to the south, they are the most common breeding bird of prey (Veit and Petersen 1993:261). The decline of agriculture may be related to declines of screech-owls, which seem to prefer small woodlots. Both red- and gray-phase birds occur in New England; gray-phase birds are more common in the north, red-phase more common in the south.

Great Horned Owl

(*Bubo virginianus*)

Range: Breeds from w. and central Alaska and central Yukon to Labrador and Newfoundland, s. throughout the Americas to Tierra del Fuego. Winters generally throughout the breeding range, with the northernmost populations being partially migratory (DeGraaf et al. 1991:192). One of the most widespread birds in North America.

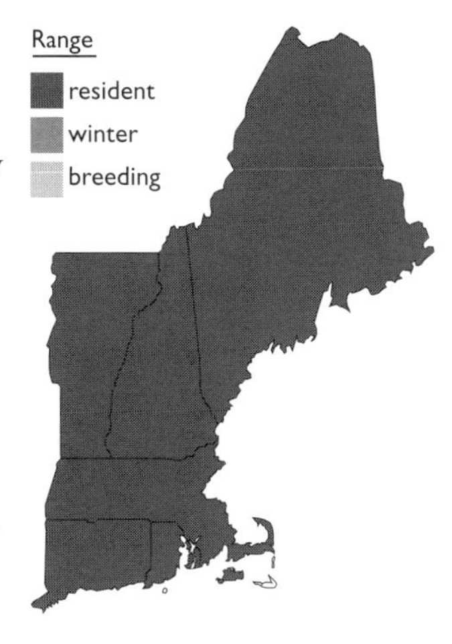

Distribution in New England: Resident throughout the region, except absent from Nantucket in Massachusetts and Block Island in Rhode Island.

Status in New England: Uncommon but widespread resident; absent on Nantucket.

Habitat: Inhabits mature forests, orchards, second-growth forests, forested wetlands, riverine forests, agricultural areas, and large suburban parks.

Special Habitat Requirements: Large abandoned birds' nests or large cavities for nesting.

Nesting: Egg dates: February 17 to April 20, Massachusetts (Veit and Petersen 1993:262). January 28 to April 18, New York (Bull 1974:331). Clutch size: 1 to 3, typically 2. Incubation period: 28 to 35 days (various reports of 28, 30, and 35 days). Nestling period: 40 to 45 days. Broods per year: 1. Age at sexual maturity: 2 yr (about 25 percent breed when 1 yr old). Nest height: 30 to 70 ft (10.1 to 21.3 m). Nest site: Commonly uses the old nest of a large bird such as heron, crow, or hawk. Also nests in large natural cavities in trees and on ledges.

Territory Size: 7.8 to 10.4 km^2 (3 to 4 square miles) in New York State (Baumgartner 1939). A pair used the same 15.5 km^2 (6 square miles) in Vermont for 10 yr (Laughlin 1985b:134 in Laughlin and Kibbe 1985).

Sample Densities: 1 pair per 5.3 square miles (1 pair/13.7 km^2) in Michigan (Craighead and Craighead 1969:214). 1 pair per 3 square miles (1 pair/7.8 km^2) in Wyoming (Craighead and Craighead 1969:215). 1 pair per 1.1 square miles (1 pair/2.8 km^2) in Kansas. Optimum habitat probably supports from 1 to 3 pairs per square mile (0.4 to 1 pair/km^2) (Baumgartner 1939). 1 pair per 4.4 square miles (1 pair/11.4 km^2) in deciduous woodland in New York (Hagar 1957).

Food Habits: Lagomorphs and rodents are staple foods; other prey includes birds, small carnivorous mammals, reptiles on the forest floor. A perch-and-pounce hunter, the great horned owl's approach to prey is silent, direct, and rapid. Sometimes leaves woodlands to hunt over meadows and salt marshes, occasionally preys on nesting terns on Cape Cod (Veit and Petersen 1993:262).

Comments: This owl is a crepuscular hunter in extensive forests with openings, forest edges, and small woodlots. Birds may have become more tolerant of human activity and occasionally are seen at parks in cities and towns. The great horned owl is one of the largest native owls and the earliest species to nest in New England, often brooding eggs in January or early February, even in Vermont. It was greatly decreased as forests gave way to settlement and thought to be "fast nearing extermination . . . as a permanent resident" (Brockway 1918 in *Auk* 35:351 as cited in Zeranski and Baptist 1990). Great horned owls have increased in this century with the return of forests. Large, powerful, and long-lived, the great horned owl is adapted to survive in any climate except arctic or alpine regions. Equally at home in deserts, grasslands, suburban, and forest habitats north to tree line, it has the most extensive range, widest prey base, and most variable nesting sites of any American owl (Houston et al. 1998).

Snowy Owl

(*Nyctea scandiaca*)

Range: Holarctic; in North America breeds from n. Alaska and n. Yukon to Prince Patrick and n. Ellesmere islands, s. to coastal w. Alaska, n. Mackenzie and s. Keewatin, across to n. Quebec and n. Labrador. Winter: Irregularly s. to s. Canada, Minnesota, and New York, and occasionally further s. (DeGraaf et al. 1991:193).

Distribution in New England: Winters in n. Maine, n. New Hampshire, and n. Vermont; an irregular winter visitor in Massachusetts on Cape Cod and the Islands, Connecticut River Valley farmlands, and in salt marshes around Plum Island, the Boston Harbor Islands (Veit and Petersen 1993:263); very rare inland in Connecticut (Zeranski and Baptist 1990:157) and Rhode Island (Parmelee 1992).

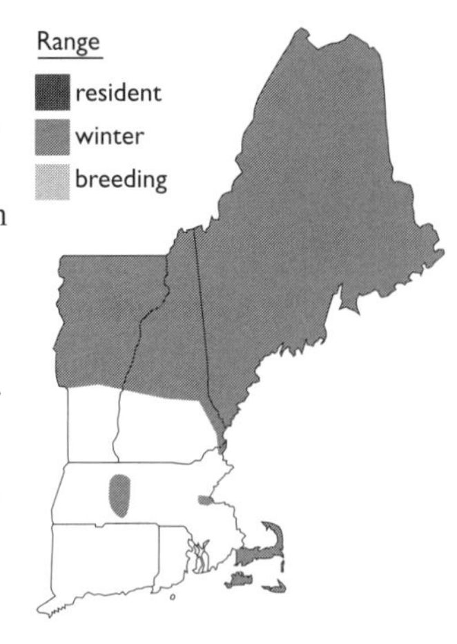

Status in New England: Irregular winter resident, in some years fairly common.

Habitat: Open areas along the coast, including sandy beaches, barrier islands, and marshes, and inland in open fields and airports.

Food Habits: In New England, feeds on rats, meadow voles, and other small mammals and a variety of birds at inland farmlands and coastal marshes.

Comments: Snowy owls are diurnal predators of open country. Flights invade New England about every 4 to 5 years. In the winters of 1876–1877, 1901–1902, and 1926–1927, great numbers occurred (Veit and Petersen 1993:263). The snowy owl is probably the earliest bird species recognizable in prehistoric cave art (Parmelee 1992).

Northern Hawk Owl

(*Surnia ulula*)

Range: Holarctic; breeding in North America from treeline in w. and central Alaska to s. Keewatin, central

Labrador, s. to s. Alaska, n. Minnesota, n. Michigan, and New Brunswick. Winter: From the breeding range s. irregularly to s. Canada and n. Minnesota, and casually to the n. United States including New England and the Adirondacks.

Distribution in New England: Fairly regular winter visitor in n. Maine. Accidental winter visitor to Massachusetts and Connecticut.

Status in New England: Rare to uncommon in n. Maine.

Habitat: Northern, coniferous or mixed forests, forest edges, old burns, clear-cuts, early successional forests, and muskeg.

Special Habitat Requirements: Forests with openings or adjacent to open habitat for foraging.

Food Habits: Small mammals including hares, and small to medium-sized birds including grouse and ptarmigan.

Comments: Northern hawk owls were more frequently seen in New England a century ago than they are today. In Massachusetts, 24 specimens were collected—all before 1890 (Griscom and Snyder 1955); there are five documented records for the state since 1900 (Veit and Petersen 1993:264). There are five records for Connecticut (see Zeranski and Baptist 1990:157). Most s. New England specimens and records are from the pronounced flight year of 1867–1868 and also the period from 1884 to 1888 (see Veit and Petersen 1993:264–265). The hawk owl resembles the *Accipiter* hawks, thus its name. Primarily diurnal, it usually perches atop prominent trees. When flying, either glides low over the ground at high speed or flaps its pointed wings in powerful, falcon-like strokes (Duncan and Duncan 1998).

Barred Owl

(Strix varia)

Range: Resident from s. and e. British Columbia, e. Washington, and extreme nw. Montana, e. to central Saskatchewan, and from s. Manitoba and central Ontario to New Brunswick and Nova Scotia, s. to central and s. Texas, the Gulf Coast, s. Florida, and n. Mexico. Northernmost populations are partially migratory.

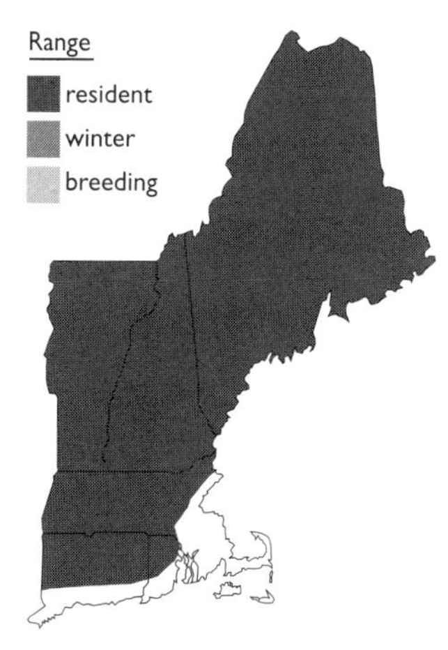

Distribution in New England: Resident throughout the region, except absent from se. Massachusetts, Cape Cod and the Islands, and e. Rhode Island including Block Island.

Status in New England: Common (n. New England and w. Massachusetts) to uncommon (Connecticut, Rhode Island) resident.

Habitat: Inhabits extensive, mature, moist mixed woodlands, especially those bordering lakes, streams, marshes, or low meadows. Prefers forests with an open understory for nesting and foraging, similar to that of the red-shouldered hawk.

Special Habitat Requirements: Moist mature forests or forested wetlands with large cavity trees or stubs.

Nesting: Egg dates: March 23 to May 3, New York (Bull 1974:334). Clutch size: 2 to 4, typically 2 or 3. Incubation period: 28 to 33 days. Nestling period: 28 to 35 days. Broods per year: 1. Nest height: To 80 ft (24.4 m). Nest site: Typically in a large natural cavity in a dead or living tree or in a hollow-topped stub. Where cavities are scarce, barred owls may use an old hawk nest. Pairs show strong attachment to same nest area from year to year. One Massachusetts pair nested in the same pine woods for 34 yr, another for 26 yr (Harrison 1975:100). In New York, owls often roost in dense stands of hemlock or pines (Bull 1974:333).

Home Range: Average size for 9 owls was 565 acres (228.7 ha) (range 213 to 912 acres, 86.2 to 369.2 ha) in deciduous woodland-open field-marsh habitat in Minnesota (Nicholls and Warner 1972).

Sample Densities: 3 pairs per 36 square miles (3 pairs/93.2 km^2) in extensive deciduous woodlots in Michigan (Craighead and Craighead 1969:92). 0.5 pair per 100 acres (40 ha) in lowland forest in Maryland (Stewart and Robbins 1958:180).

Food Habits: Mice (staple) and other small mammals, frogs, birds, insects, crayfish. Feeds in forests and small forest openings.

Comments: Barred owls are nocturnal hunters, but birds with broods may hunt during daylight hours. Commonly calls during daylight hours, especially in late summer and early fall. Will respond to imitations of its call. A bird of extensive forests, the barred owl is probably more abundant in much of New England now than it was a century ago. It has declined in Connecticut with the loss of continuous forest cover due to urbanization. It is the most common owl in Vermont and New Hampshire.

Great Gray Owl

(Strix nebulosa)

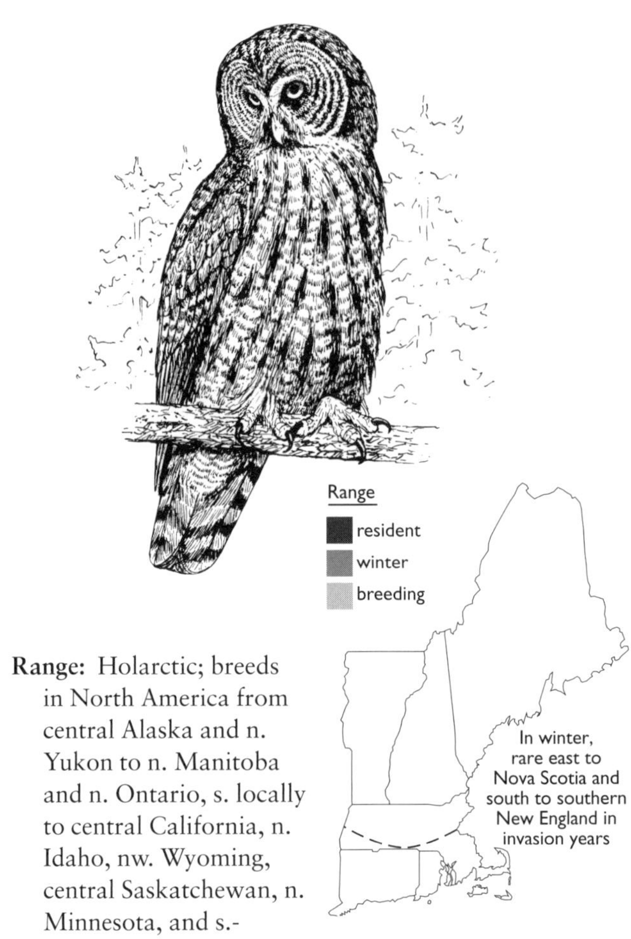

Range: Holarctic; breeds in North America from central Alaska and n. Yukon to n. Manitoba and n. Ontario, s. locally to central California, n. Idaho, nw. Wyoming, central Saskatchewan, n. Minnesota, and s.-central Ontario. Winters generally throughout the breeding range, wandering s. irregularly to the northern tier of the United States as far s. as s. New England, Long Island, New York, and New Jersey (Veit and Petersen 1993:268).

Distribution in New England: Open country; extensive fields with some forest patches.

Status in New England: Very rare and irregular winter visitor.

Habitat: Large, open, wet fields bordered by woodland patches (Veit and Petersen 1993:268). Also, northern coniferous forests with openings created by sphagnum and tamarack bogs, muskegs, and fields (Bull and Duncan 1993).

Food Habits: Small mammals (primarily rodents), small birds (Bull and Duncan 1993). Feeds in bogs, meadows, and frozen marshes.

Comments: A major flight of great gray owls occurred in 1842–1843, and single specimens were taken about 1 per decade in Massachusetts through the 1890s, but from 1900 through 1978, no flights reached Massachusetts (Veit and Petersen 1993:268). In the 1970s, individual birds were recorded in 1972 and 1977, but during the winter of 1978–1979, the first major flight of the century occurred, with birds found across the state (see Veit and Petersen 1993 for details). In all, 92 individual birds were reported from New Brunswick, New England, and Long Island (see Zeranski and Baptist 1990:159).

Long-eared Owl

(Asio otus)

Range: Holarctic; breeds in North America from n. Yukon, sw. Northwest Territories, n. Saskatchewan and Nova Scotia, s. to n. Baja California, s. Arizona, w. and central Texas, Arkansas, n. Ohio, w. Virginia, and New England. Winters from s. Canada, s. to Baja California, central Mexico, s. Texas, the Gulf Coast, and Georgia; casually to Florida.

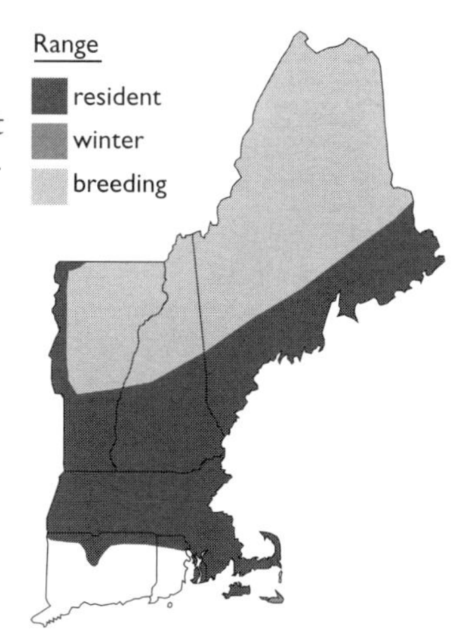

Distribution in New England: Nesting reports from Maine; the s. Champlain Lowlands and s. Green Mountain region in Vermont (Pistorius 1985:138 in Laughlin and Kibbe 1985); e. Massachusetts, including Martha's Vineyard (Veit and Petersen 1993:269); n.-central Connecticut (Zeranski and Baptist 1990:159). A calling individual was reported from w. Rhode Island (Enser 1992:86). One in Bartlett, N.H., Feb. 1995 (C. Costello, pers commun.).

Status in New England: Rare and local breeder and migrant. Uncommon to occasionally fairly common in winter.

Habitat: Breeds in dense coniferous or mixed forests or groves adjacent to open habitat for foraging; occasionally uses deciduous forests. Also inhabits open or dense thickets, forest edges, parks, orchards, isolated woodlots, forested wetlands, riparian woodlands (Marks et al. 1994), and shorelines of reservoirs and lakes. Winters in communal roosts in dense forests or coniferous thickets adjacent to open habitat (Bosakowski 1984, Smith and Devine 1993). Occurs at elevations up to 10,800 ft (3,300 m).

Special Habitat Requirements: Dense (usually coniferous) forests or groves for nesting and roosting cover.

Nesting: Egg dates: March 15 to May 14 in Massachusetts (Veit and Petersen 1993:269). March 31 to May 31 in New York (Bent 1938). Clutch size: 3 to 10, typically 4 or 5. Incubation period: About 26 to 28 days. Female incubates and broods young nestlings (Craig et al. 1988). Nestling period: 21 days (Marks 1986). Broods per year: 1. Age at sexual maturity: 1 yr (Marks 1985). Nest height: Usually 16.5 to 33 ft (5 to 10 m). Nest site: Most often uses old nests of crows or hawks. Rarely nests in natural tree cavities, squirrel nests, or on the ground or ledges.

Territory Size: Does not seem to defend an area other than the space around the nest. Home ranges of adjacent nesting pairs (Idaho) overlapped (Craig et al. 1988).

Sample Densities: 1 pair per 37 square miles (1 pair/95.8 km^2) in Michigan. 1 pair per 4 square miles (1 pair/10.4 km^2) in Wyoming (Craighead and Craighead 1969:212,215). 1 pair per 0.1 to 0.4 square mile (1 pair/0.3 to 1.0 km^2) in Wyoming (Craighead and Craighead 1969:264).

Food Habits: Mice and voles (staples), also small mammals, reptiles, amphibians, insects, and small birds (Armstrong 1958). Meadow voles and white-footed mice composed the preponderance of prey detected in pellets collected in fall and winter in Massachusetts (Nantucket and Belmont) (Holt and Childs 1991).

Comments: Birds are gregarious in winter, with flocks of 5 to 25 occupying communal roosts (Pough 1949:19). Long-eared owls are quiet and their nocturnal habits make them difficult to observe in all seasons; they can produce a great variety of vocalizations: doglike barks and yelps, catlike meows, and shrieks, cackles and whistles (Pistorius 1985:138 in Laughlin and Kibbe 1985), also vocalizations that are easily confused with those of other owls (Veit and Petersen 1993:269). The bird must be seen to be identified. The long-eared owl was formerly much more numerous in s. New England in the nineteenth century than it is at present, having declined as a nesting species around the turn of the century (Zeranski and Baptist 1990:160).

Short-eared Owl

(*Asio flammeus*)

Range: Cosmopolitan except absent from Australia. Breeds in North America from n. Alaska and n. Yukon to n. Quebec and Labrador, s. to central California, n. Nevada, Utah, Kansas, Missouri, n. Ohio, n. Virginia, New Jersey, nw. Vermont, and se. Massachusetts. Winters generally in the breeding range from s. Canada, s. to Mexico. Individuals in Puerto Rico, Hispaniola, and South America are resident.

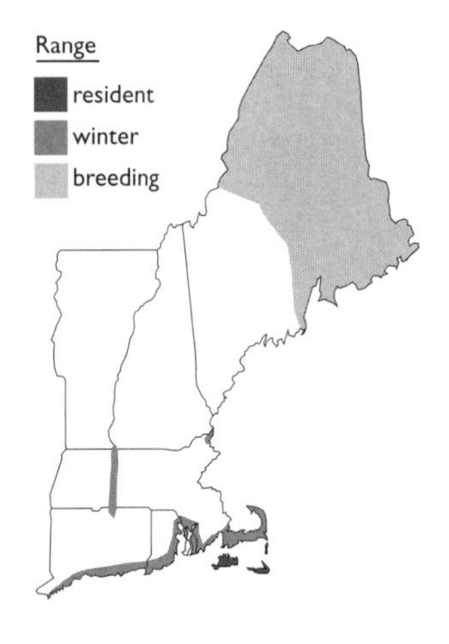

Distribution in New England: The Champlain Lowlands in Vermont and the islands off Cape Cod, Mass. Winter: Throughout the breeding range and coastal, or near the coast, in New Hampshire, Massachusetts, Connecticut, and Rhode Island.

Status in New England: Rare breeder, especially on the islands off Cape Cod. Fairly common late fall migrant and winter resident along s. New England coast.

Habitat: Breeding: Open country, primarily marshlands (fresh and saltwater) and open grasslands, but also heathlands, dunes, old fields, agricultural areas, tundra, extensive forest clearings, sagelands, deserts, sparse shrub-steppe, pastures, meadows, prairies, lower mountain slopes, and other open habitat. Winter: Same as breeding habitat, but with a preference for open areas with little or no snow.

Special Habitat Requirements: Extensive open marshes or grasslands.

Nesting: Egg dates: Late May to mid June, Massachusetts (Veit and Petersen 1993:270). April 12 to May 19, New York (Bull 1974:338). Clutch size: 4 to 9, linked to prey density typically 5 to 7 (Holt and Leasure 1993). Incubation period: About 21 days. Age at first flight: About 28 days. Nestling period: 12 to 16 days. Broods per year: Occasionally 2 (Holt and Leasure 1993). Nest site: On the ground in a shallow depression, either obscured by grasses or clumps of weeds or in exposed situations. Occasionally nests in small loose colonies, rarely nests in an excavated burrow.

Territory Size: 73.9 to 121.4 ha (182.5 to 299.8 acres) in sedges and rushes in New York (Clark 1975:43).

Food Habits: Primarily small mammals, also insects, occasionally birds (Short and Drew 1962; Stegeman 1957). Active day and night, hunting low over the ground (Holt and Leasure 1993).

Comments: Loss of marshes and grasslands has resulted in dramatic declines since the 1930s (Tate 1992). Extirpated from Connecticut as a nesting species (Zeranski and Baptist 1990:160). In winter, birds tend to roost in groups in open field or close to ground in conifers or brush if snow is deep. Birds often hunt by day (especially dawn and dusk) and depend almost totally on rodents.

Boreal Owl

(*Aegolius funereus*)

Range: Holarctic; breeds in North America from central Alaska and central Yukon to central Quebec and Labrador, s. to n. British Columbia and central Alberta, se. to ne. Minnesota, w. and central Ontario, s. Quebec and New Brunswick; also to central Colorado and ne. Wyoming in the Rocky Mountains. Winters generally throughout the breeding range.

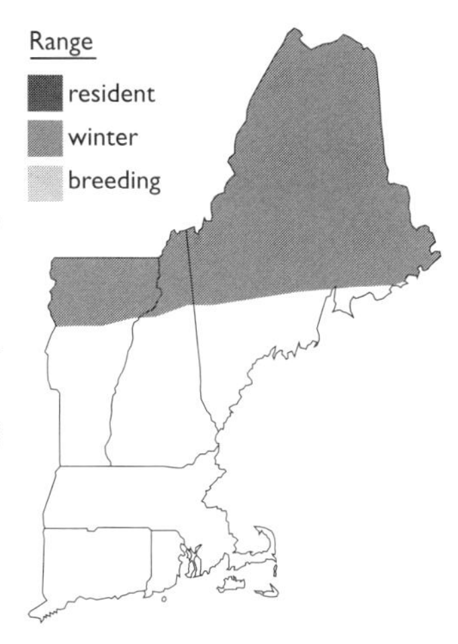

Distribution in New England: A very rare and irregular winter visitor: n. Maine, New Hampshire, Vermont, and e. Massachusetts; accidental in Connecticut.

Status in New England: Very rare and irregular winter visitor.

Habitat: Dense coniferous and mixed hardwood forests or woodlots.

Food Habits: Small mammals (particularly voles and mice up to 50 g) and small birds (Hayward and Hayward 1993).

Comments: Incursions of boreal owls well south of their normal winter range occurred in the winters of 1859–1860 and 1922–1923 (Zeranski and Baptist 1990:161). A major "flight" occurred in 1991–1992 when six birds were recorded in Massachusetts (Veit and Petersen 1993:271). Known as Tengmalm's owl in Europe, where it is the most abundant forest owl (Hayward and Hayward 1993).

Northern Saw-whet Owl

(*Aegolius acadicus*)

Range: Breeds from s. Alaska, central British Columbia, and central Alberta to s. Quebec and n. New Brunswick, s. to s. California, central Mexico, extreme w. Texas, central Missouri, s. Wisconsin, central Ohio, West Virginia, and New York; also in the mountains of e. Tennessee and w. North Carolina. Winters generally throughout the breeding range, s. irregularly to s. Arizona, the Gulf Coast, and central Florida.

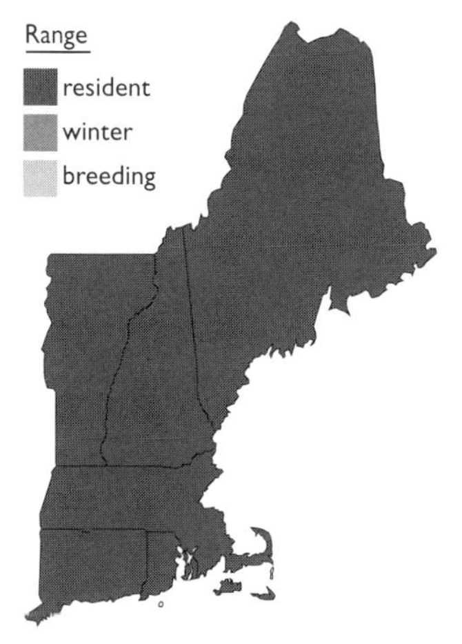

Distribution in New England: Resident throughout the region.

Status in New England: Fairly common resident in n. New England, uncommon in Massachusetts, rare in Connecticut.

Habitat: Prefers moist mature woods and dense forested wetlands (white cedar, hemlock, or red maple). Common at forest edges. Also inhabits tamarack bogs, alder thickets, pine forests, pitch-pine barrens, cedar groves, and roadside shade trees. Similar habitats in winter.

Special Habitat Requirements: Large trees with large woodpecker holes or natural cavities. Also dense vegetation for roosting, perches for foraging (Cannings 1993).

Nesting: Egg dates: April 4 to May 31, Massachusetts (Veit and Petersen 1993:272). March 31 to June 11, New York (Bull 1974:340). Clutch size: 4 to 7, typically 5 or 6. Incubation period: 21 to 28 days (females incubate eggs and brood young; males provide almost all the food for the female and young—Cannings 1993). Broods per year: 1 (Terrill 1931). Nest height: 14 to 60 ft (4.3 to 18.3 m). Typically 20 to 40 ft (6.1 to 12.2 m). Nest site: Usually in an abandoned woodpecker nest hole (especially those of the northern flicker), but will use natural cavities or nest boxes with a layer of straw or sawdust.

Territory Size: 378 acres (151 ha) in s. British Columbia (Cannings 1987). Average 281 acres (114 ha) in Minnesota (Forbes and Warner 1974).

Sample Densities: 1 bird per 1.86 square miles (1 bird/4.8 km^2) in spruce-fir in mountains of North Carolina (Simpson 1972). Maximum 1 pair per 40 ha (100 acres) (Hardin and Evans 1977).

Food Habits: Mainly small mammals—especially deer mice, young squirrels, shrews, chipmunks; also takes insects and occasionally small birds. During the nonbreeding season on Nantucket (Mass.), white-footed mice, meadow voles, short-tailed shrews, one eastern mole, and one masked shrew composed the identified diets of eight owls over four years; white-footed mice and meadow voles were the dominant prey (Holt et al. 1991).

Comments: Saw-whet owls are New England's smallest owls. Males weigh about as much as the American robin, females 25 percent more (Cannings 1993). They hunt and roost close to the ground. They are nocturnal and so are seldom seen but, when found on their daytime roost, they usually allow very close approach.

Common Nighthawk

(*Chordeiles minor*)

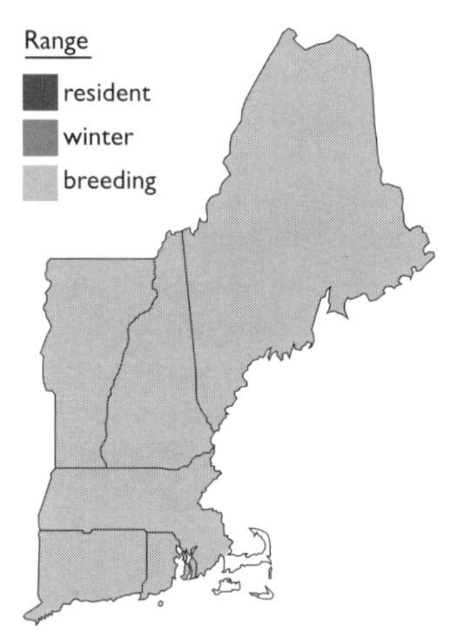

Range: Breeding: Southern Yukon and s. Northwest Territories to central Quebec and s. Labrador, s. to s. California, s. Nevada, s. Arizona, Texas, the Gulf Coast, Florida, Mexico, and Central America. Winter: South America, s. to Argentina. Probably a migrant throughout except Baja California; rare on Yucatan Peninsula. The small numbers of breeding individuals in Costa Rica move s. with the abundant northern migrants in the fall; a sporadically common spring migrant. An uncommon to fairly common migrant in Colombia in early fall. Migrant and in part resident over most of South America, s. to Buenos Aires, Argentina. North American individuals occur as transients throughout the West Indies.

Distribution in New England: Throughout Maine, New Hampshire, Vermont, and Massachusetts, except absent from Cape Cod and the Islands, Rhode Island, and locally in Connecticut in larger towns and cities along the coast and inland in rural areas (Zeranski and Baptist 1990:161).

Status in New England: Locally common to rare breeder; common to abundant fall migrant. Declining in the e. United States and Canada.

Habitat: Prefers open habitats such as grasslands,

cultivated fields, burned-over woodlands, large woodland clearings, prairies, dry barren plains, rocky outcrops, gravel beaches, coastal sand dunes, railroad rights-of-way, and flat gravel roofs (Poulin et al. 1996). Always in open areas.

Nesting: Egg dates: July 5 to 15, Massachusetts (Veit and Petersen 1993: 273). May 25 to July 25, New York (Bull 1974:345). Clutch size: 1 to 3, typically 2. Incubation period: 19 to 20 days (Gross 1940). Age at first flight: 21 days. Broods per year: 1 or 2. Age at sexual maturity: 1 yr. Nest site: Builds no nest. Lays eggs on bare ground, usually on gravel or partially vegetated soil, gravel roofs.

Territory Size: 10.1 to 56.3 acres (4.0 to 22.5 ha), average 25.7 acres (10.3 ha), in Detroit, MI (Armstrong 1965).

Sample Densities: 13 pairs per 321.1 acres (128.4 ha) in Michigan (Armstrong 1965).

Food Habits: Flying insects, especially flying ants, mosquitos, moths, grasshoppers.

Comments: Common nighthawks are mainly crepuscular and nocturnal but occasionally feed during the day. Declining in New Hampshire except in urban areas. After the mid 1800s they began to nest on flat gravel roofs in urban areas (Sage et al. 1913). The use of rubber instead of gravel on flat roofs may be responsible for declines in urban populations (Brigham 1989, Marzilli 1989).

Whip-poor-will

(Caprimulgus vociferus)

Range: Breeding: Breeds from n.-central Saskatchewan and s. Manitoba to s. Quebec and Nova Scotia, s. to e. Kansas, ne. Texas, and n. Louisiana, across to central Georgia; and from s. California, s. Nevada, central Arizona, and extreme w. Texas, s. to Mexico. Winters from s. Texas, the Gulf Coast, and e.-central South Carolina, s. to Central America. A migrant and permanent resident in much of Mexico. Wintering birds are found in the highlands of Guatemala. The whip-poor-will is an uncommon resident and a fairly common winter visitor in Honduras. The e. North American breeding population winters in much of Honduras. A casual to very rare winter resident along the Pacific Coast of Costa Rica. Winters casually to western Panama. Rare or casual as a winter resident in Cuba.

Distribution in New England: Breeds throughout the region except absent from n. Maine and absent from Block Island in Rhode Island.

Status in New England: Locally fairly common; long-term and recent precipitous declines in e. North America. Uncommon and inconspicuous migrant.

Habitat: Prefers fairly dry open deciduous or mixed woodlands of pine, oak, and beech, especially early successional forests. Also favors stands adjacent to large clearings.

Special Habitat Requirements: Dry, open woodland, early successional forests, adjacent to large clearings or brushy field edges.

Nesting: Egg dates: May 16 to June 30, New York (Bull 1974:344). Clutch size: 1 or 2, typically 2 (Raynor 1941). Incubation period: 20 days. Broods per year: 1 or 2 (Bull 1974:344). Age at sexual maturity: 1 yr. Nest site: Builds no nest; eggs are laid on well-drained ground in the open or under a bush (Raynor 1941).

Territory Size: 14.9 acres (6 ha), 25.5 acres (11.1 ha), and 6.9 acres (2.8 ha) in oak, hickory, elm woodlands in Kansas (Fitch 1958).

Sample Densities: 1.4 territorial males per 100 acres (40 ha) in upland forest and brush habitat in Maryland (Stewart and Robbins 1958:184).

Food Habits: Mainly flying insects but occasionally takes crickets, ants, and beetles from the ground (Bent 1940).

Comments: Still common in pine barrens and on Cape Cod and the Islands, whip-poor-wills have declined drastically over the last 30 to 40 years. Well known by voice but seldom seen, the whip-poor-will was known by its descriptive name in New Hampshire in 1792 when Jeremy Belknap included it in a list of the state's birds. The northern limit of the breeding range in Vermont and New Hampshire coincides with the 68°F (18°C) isotherm (Robbins 1994b:120 in Foss 1994).

Chimney Swift

(*Chaetura pelagica*)

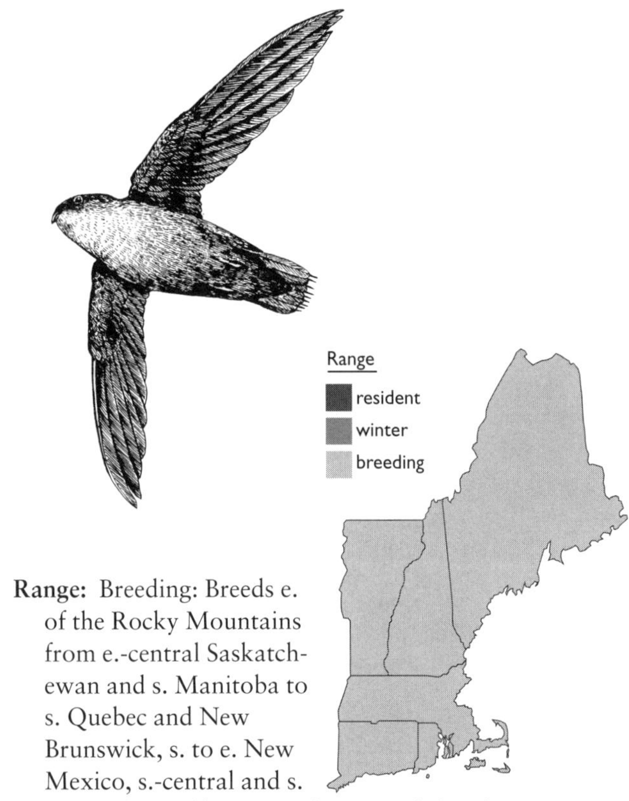

Range: Breeding: Breeds e. of the Rocky Mountains from e.-central Saskatchewan and s. Manitoba to s. Quebec and New Brunswick, s. to e. New Mexico, s.-central and s. Texas, the Gulf Coast and s.-central Florida. Winters in South America. A migrant through the east slope of Mexico. A migrant throughout much of Guatemala. The chimney swift is an uncommon to common fall migrant and a common spring migrant in the lowlands and islands off the n. coast of Honduras. A sporadically common to abundant fall migrant on Costa Rica's Caribbean slope. Spring migration is along the Caribbean coast. Winters mostly in w. Amazonia, also apparently in w. Peru and n. Chile. Extremely rare in the Virgin Islands. No records from Puerto Rico. A fall and spring migrant recorded from the Bahamas, Cuba, Hispaniola, Jamaica, and the Cayman Islands.

Distribution in New England: Breeds throughout the region, except absent from Block Island in Rhode Island.

Status in New England: Common, but declining throughout the breeding range. An uncommon nester in Connecticut (Zeranski and Baptist 1990:163). More common in inland habitats than along the coast in Massachusetts and Rhode Island (Veit and Petersen 1993:163, Enser 1992:90).

Habitat: Nests in unused chimneys and structures suitable for nesting in the vicinity of buildings in towns, cities, and farms.

Special Habitat Requirements: Chimney swifts have nested in chimneys almost exclusively throughout the historical period (Veit and Petersen 1993:277), but formerly used hollow trees for nest sites (DeGraaf and Rappole 1995:279).

Nesting: Egg dates: May 30 to July 27, New York (Bull 1974:347). Clutch size: 3 to 6, typically 4 or 5 (Fischer 1958). Incubation period: 18 to 21 days. Nestling period: 1 or 2 days (able to crawl out of nest but unable to fly). Broods per year: 1. Age at sexual maturity: 1 yr. Age at first flight: 28 days. Nest site: Formerly nested in hollow trees but has adapted to chimneys, silos, building walls, airshafts, even old wells. At Kent, Ohio, birds typically nested in ventilation shafts at an average depth of about 20 ft (6 m) (Dexter 1977). Solitary or colonial nesters. Nesting success is related to the number and age of adults at the nest (Dexter 1981).

Sample Densities: 0.6 pair per 100 acres (40 ha) in mixed forest, brush, and fields, and near buildings with chimneys (Stewart and Robbins 1958:187).

Food Habits: Flying insects.

Comments: Chimney swifts apparently became common during earliest settlement as the colonists built chimneys; John Josselyn (1672) clearly described nesting in chimneys. Referring to the bird's pre-Colonial nest sites, Cruickshank (1942) noted that it no longer nested in hollow trees.

Ruby-throated Hummingbird

(*Archilochus colubris*)

Range: Breeding: East of the Rocky Mountains, from central Alberta and central Saskatchewan to s. Quebec and New Brunswick, s. to s. Texas, the Gulf Coast, and

Florida. Winter: Southern Texas, s. to Central America, also in s. Florida, but mainly in Mexico and Central America. Rare winter resident in the Bahamas and Cuba. Causal in Hispaniola, Grand Cayman, and Jamaica. Accidental in Puerto Rico. A fairly common migrant and winter visitor in Guatemala. An uncommon to locally common winter resident in n. Pacific lowlands of Costa Rica. A rare winter visitor to Panama.

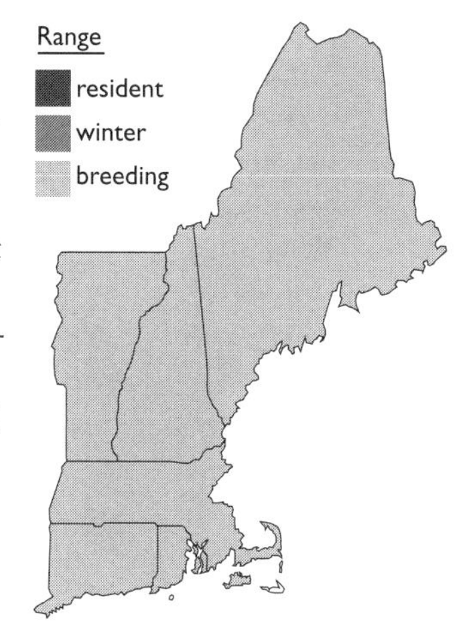

Distribution in New England: Breeds throughout the region, except absent from Block Island, R.I. (Enser 1992:91).

Status in New England: Fairly common, more abundant inland than near the coast. Declining in Connecticut due to increased urbanization (Zeranski and Baptist 1990:164), numbers decreasing in Massachusetts in past 30 yr (Veit and Petersen 1993:278), declines not noted in other New England states, but may be declining steadily across its range, perhaps because of declining area or quality of tropical wintering grounds (Dobkin 1992).

Habitat: Inhabits a variety of wooded habitats, from rather dense to open coniferous and deciduous woodlands, orchards, and shade trees in yards. Appears to prefer hardwoods to conifers, especially in regenerating clear-cuts and mature stands with rough lichen (DeGraaf and Rappole 1995:289). Also inhabits mixed woodlands, parks, gardens, woodland clearings, and forest edges. Often breeds in woodlands near streams and in wooded swamps, especially balsam fir flats in n. New England.

Special Habitat Requirements: Plants that provide tubular nectar-bearing (especially red) flowers such as honeysuckle, lantana, gilia, and trumpet vine (DeGraaf and Rappole 1995:290).

Nesting: Egg dates: May 24 to July 22, Massachusetts (Veit and Petersen 1993:278). May 29 to August 6, Vermont (Norse and Laughlin 1985:150 in Laughlin and Kibbe 1985). May 21 to August 16, New York (Bull 1974:348). Clutch size: typically 2, rarely 3 (Bent 1940). Incubation period: 14 to 17 days (Harrison 1975). Nestling period: 19 to 22 days. Broods per year: 1, 2 common in Alabama (Robinson et al. 1996). Nest height: 1.7 to 49.5 ft (0.5 to 15 m), typically 16.5 to 23.1 ft (5 to 7 m) above ground (Robbinson et al. 1996). Nest site: Often on a down-sloping branch of a tree such as maple, beech, birch, hornbeam, hemlock; frequently over a brook and sheltered overhead by leaves.

Territory Size: Female alone defends immediate area around the nest. A male in Ohio defended a feeding territory of 0.25 acre (0.1 ha) (Pitelka 1942).

Sample Densities: Maryland—15 pairs per 100 acres (40 ha) in well-drained floodplain forest. 8 pairs per 100 acres (40 ha) in upland oak forest. 6 pairs per 100 acres (40 ha) in mature northern hardwood forest. 4 pairs per 100 acres (40 ha) in hedgerows and active and abandoned farmland (Stewart and Robbins 1958:188).

Food Habits: Nectar from tubular flowers, small insects, sap at yellow-bellied sapsucker wells (Miller and Nero 1983). Occasionally willow catkins when food is scarce (Sealey 1989).

Comments: The only hummingbird to nest e. of the Mississippi River, the ruby-throated is widely distributed in forests as well as gardens. Males are polygynous. Sexes migrate separately, with males arriving in the Northeast several days before the females. In fall, males leave for the wintering grounds a month before females and young.

Belted Kingfisher

(*Ceryle alcyon*)

Range: Breeding: From w. and central Alaska, central Yukon, and w. and s.-central Northwest Territories, to central Quebec and e.-central Labrador, s. to s. California, s. Texas, the Gulf Coast, and central Florida. Winter: From s.-coastal and se. Alaska, central and s. British Columbia, and w. Montana, across to Nebraska, the

s. Great Lakes, and New England, s. throughout Central America. Occasional in winter as far s. as Guyana, coastal Venezuela, and Colombia. Found throughout the year in the West Indies, with some northern birds augmenting local populations.

Distribution in New England: Breeding: Widespread throughout the region. Winter: Near open water in Connecticut, Rhode Island, Massachusetts, coastal New Hampshire (occ.), and Vermont (Ellison 1985:152 in Laughlin and Kibbe 1985).

Status in New England: Uncommon but widespread breeder. Uncommon but variable winter resident depending on winter severity. Belted kingfishers are about as common in Vermont as they were in the mid 1800s (Thompson 1853, Spear 1976).

Habitat: Occurs near ponds, lakes, rivers, streams, reservoirs, large vernal pools, estuaries, and harbors near exposed vertical banks for nesting. Prefers small clear bodies of water. Stream riffles considered an important habitat feature (Davis 1982).

Special Habitat Requirements: Nest sites (more than 75 percent sand, less than 7 percent clay in high, steep banks—Brooks and Davis 1987) preferably within a mile (1.5 km) of clear water with abundant aquatic prey, and perches above the water to sight prey.

Nesting: Egg dates: May 14 to June 6, Massachusetts (Veit and Petersen 1993:280), May 11 to June 15 in Vermont (Ellison 1985:152 in Laughlin and Kibbe 1985). Clutch size: 5 to 8, typically 6 or 7. Incubation period: 23 to 24 days (Bent 1940). Nestling period: 27 to 29 days (Hamas 1994). Broods per year: 1. Age at sexual maturity: 1 yr. Nest site: Typically a burrow 3 to 6 ft (1 to 2 m) deep in a bank of sandy clay fairly near water, but up to a mile from water. Both sexes excavate the burrow, 35–64 cm below the top of the bank (Hamas 1994).

Territory Size: 2 pairs on 2 lakes used 0.5 mile (0.8 km) of shoreline (Sayler and Lagler 1946). 0.5 to 5 miles (0.8 to 8 km) from nest site (Cornwell 1963).

Sample Densities: 1 pair per 1.8 square miles (1 pair/4.7 km^2) in Minnesota (Cornwell 1963).

Food Habits: Major foods: Fish (staple) (Bent 1940), crayfish, insects, mollusks, amphibians, reptiles, small mammals, and occasionally fruits (Forbush 1927). Preferred feeding habitat: Shallow borders of clear bodies of water.

Comments: Large lakes that become turbid as the result of wave action have fewer kingfishers than small, clear bodies of water.

Red-headed Woodpecker

(*Melanerpes erythrocephalus*)

Range: Breeding: From s. Saskatchewan, s. Ontario, s. New Hampshire, and s. New Brunswick, s. to central Texas, the Gulf Coast, and Florida, extending w. to central Montana, e. Wyoming, e. Colorado, and central New Mexico, rarely to ne. Utah. Winter: Regularly through the s. two-thirds of the breeding range, rarely or casually n. to the limits of the breeding range.

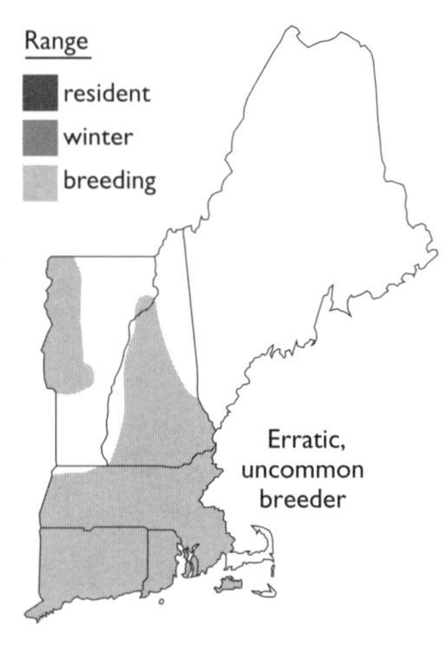

Distribution in New England: Breeds in w. New Hampshire (Rumney and Monroe), w. to the Champlain Lowlands in Vermont, Martha's Vineyard and central Massachusetts, and two sites in Connecticut (Suffield and nw. Fairfield County); not reported from Rhode Island. The distribution of red-headed woodpeckers varies widely from year to year. An occasional winter resident in Massachusetts (Veit and Petersen 1993:282).

Status in New England: Rare, local, and irregular breeder; declining in the East. Occasionally winters in s. New England.

Habitat: Breeding: Inhabits open lowland or upland forests or woodlots with low stem density. Prefers savannahlike grasslands with scattered trees and forest edges. Attracted to areas with dead trees that provide nesting and roosting sites, and rank, herbaceous ground cover that fosters abundant insect populations. Breeding birds in Massachusetts use dead trees in open deciduous forests bordering in fields (Veit and Petersen 1993:282). Avoids forests with closed canopies. Winter: Tends to move to forest interior; oaks and beech that provide mast may be important components of winter habitat (Willson 1970, Reller 1972).

Special Habitat Requirements: Relatively open areas with snags and lush herbaceous ground cover (Hardin and Evans 1977).

Nesting: Egg dates: May 28 to June 17, Massachusetts. May 16 to June 19, New York (Bull 1974:355). Clutch size: 3 to 8, typically 5. Incubation period: 12 to 14 days. Nestling period: 27 to 30 days. Broods per year: 1 or 2. Age at sexual maturity: 1 yr. Nest height: 8 to 80 ft (2.4 to 24.4 m), typically 23 to 40 ft (7 to 12.4 m). Nest site: Tree cavity usually excavated in dead tree or limb without bark that is surrounded by open space.

Territory Size: In winter the birds restrict their activities to small, well-defined territories (Kilham 1958b).

Sample Densities: 9 to 12 birds per 100 acres (40 ha) in bottomland woods with much edge and large internal openings (oak-hickory-hackberry-elm community) (Graber et al. 1977). 25 birds per 100 acres (40 ha) in suburban-residential habitat (Cooke 1916 in Graber et al. 1977). 28 birds per 100 acres (40 ha) in shrub area (Graber and Graber 1963).

Food Habits: Insect larvae and adults, wild fruits, acorns (especially those of pin oak), beechnuts, corn. Feeds in open areas adjacent to woodlots (Conner 1976); feeds in upper parts of trees in winter (Williams 1975b).

Comments: Red-headed woodpeckers were more abundant before 1900 than at any time since. Red-headed woodpeckers tend to excavate cavities in trunks rather than limbs (Reller 1972). Red-headed populations have increased following the death of trees over a large area by fire, flood, or disease (Graber et al. 1977). Woodlots used for nesting in sw. Virginia ranged from 0.5 to 20.0 ha (1.2 to 50 acres) (Conner 1976). Factors contributing to this species' decline include competition with starlings for nest sites, and dead tree removal for firewood (Arbib 1982).

Red-bellied Woodpecker

(*Melanerpes carolinus*)

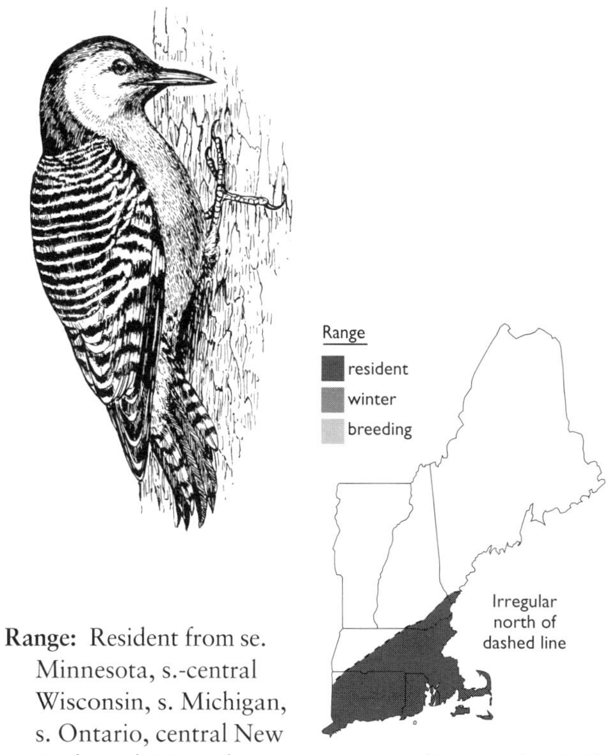

Range: Resident from se. Minnesota, s.-central Wisconsin, s. Michigan, s. Ontario, central New York, and Massachusetts, s. to central Texas, the Gulf Coast, and s. Florida, and w. to Iowa, e. Nebraska, w. Kansas, central Oklahoma, and n.-central Texas.

Distribution in New England: Connecticut, Rhode Island, central and e. Massachusetts. Infrequent at elevations above 800 ft (242 m) in Connecticut (Zeranski and Baptist 1990:166).

Status in New England: Locally rare to uncommon resident, but increasing.

Habitat: Inhabits mature deciduous and coniferous forests and edges. Frequents uplands but prefers bottomlands. Also uses woodlots near farms and orchards.

Special Habitat Requirements: Extensive open, mature woodlands with dead trees or trees with large dead limbs for nesting.

Nesting: Egg dates: April 26 to June 28, New York (Bull 1974:354). Clutch size: 3 to 8, typically 4 to 6. Incubation period: 14 days. Broods per year: 1 (north), 2 (south). Age at sexual maturity: 1 yr. Nest height: 5 to 70 ft (1.5 to 21.3 m). Nest site: Cavity in sound or soft wood, often in limb at edge of woodland, less often in trunk of dying or dead tree, building, utility pole, or stump. May excavate a cavity or occupy an existing one (Kilham 1958a). Also uses nesting boxes.

Territory Size: Average 6.1 acres (2.5 ha) (3 territories) in virgin floodplain forest in Illinois. Average 4.4 acres (1.8 ha) (2 territories) in mature upland forest in Illinois (Graber et al. 1977). Winter: 3 to 4 acres (1.2 to 1.6 ha) (Kilham 1963).

Sample Densities: 23 birds per 100 acres (40 ha) in virgin floodplain (elm-maple) forest in Illinois. 6 birds per 100 acres (40 ha) in bottomland forest in Illinois (Graber et al. 1977). 19 pairs per 100 acres (40 ha) in white oak, tulip-poplar forest in Maryland (Stewart and Robbins 1958:193). These densities are high for New England. Martha's Vineyard has the densest population, perhaps 40 pairs (Laux in Veit and Petersen 1993:284).

Food Habits: Insects, especially ants and beetles; beech and acorn mast, corn, wild fruits. Feeds in lowlands (Williams 1975b); also upland forest edges. Birds may seek food in areas outside of breeding habitat such as cornfields (Reller 1972).

Comments: In the mid nineteenth century, red-bellied woodpeckers were accidental in s. New England, and until 1955 were rare vagrants (Griscom and Snyder 1955). In the 1960s they began to extend their range northward, first nesting in Connecticut in 1962 (Zeranski and Baptist 1990:166) and Massachusetts in 1977 (Veit and Petersen 1993:284). They are most abundant on the s. and se. coasts and in the Connecticut River Valley. Birds often store food in crevices for later use (Kilham 1963). Yeager (1955) noted that populations increased where flooding had killed trees.

Yellow-bellied Sapsucker

(Sphyrapicus varius)

Range: Breeds from e. Alaska, e. to central Newfoundland, s. to ne. British Columbia, e. North Dakota, New Hampshire, and locally in the Appalachians s. to e. Tennessee and w. North Carolina; also in the Rocky Mountain region from s.-central British Columbia to w. Montana s., e. of the Cascades, to e.-central California and w. Texas. Winters from Missouri, the Ohio Valley, and New Jersey, s. through Texas and the se. United States to central Panama; also from s. California, central Arizona, and central New Mexico, s. to s. Baja California and Jalisco. A transient and winter visitor to the highlands of Central America. Not uncommon in most of the West Indies, but rare e. of Hispaniola.

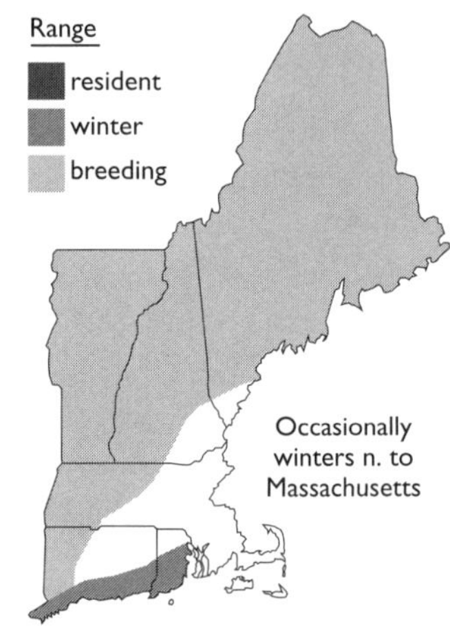

Distribution in New England: Breeding: Throughout Vermont, Maine, except absent from the most sw. portion of the state, New Hampshire, except absent from the se., s. to w. Massachusetts and nw. Connecticut. Winter: Coastal Connecticut and Rhode Island.

Status in New England: A rare breeder and winter resident in Connecticut (Zeranski and Baptist 1990:166), fairly common breeder w. of the Connecticut River in Massachusetts (Veit and Petersen 1993:284), common throughout most of Vermont, New Hampshire, and Maine.

Habitat: Nests in deciduous and mixed deciduous-coniferous forests, especially in second-growth woodlands with aspen, paper birch, American beech, white pine, and hemlock (Norse and Fichtel 1985:156 in Laughlin and Kibbe 1985). Occupies orchards, parks, woodlands, and floodplain forests in winter.

Special Habitat Requirements: Dead or live trees with a central decay column. Excavates nests in snags or live trees with a decayed center. May use the same tree for several years, but excavates a new nest cavity each year (Kilham 1971, Lawrence 1967). Prefers aspens, but will nest in pines, birch, elm, butternut, cottonwood, alder, willow, beech, maple, and fir.

Nesting: Egg dates: May 15 through June, Massachusetts (Veit and Petersen 1990:285). April 29 to June 19, New York (Bull 1974:358). Clutch size: 4 to 7, typically 5 or 6. Incubation period: 12 to 13 days. Nestling period: 24 to 26 days. Broods per year: 1. Age at sexual maturity: 1 yr. Nest height: 8 to 40 ft (2.4 to 12.2 m). Nest site: Excavates a cavity in a dead or living tree with rotten heartwood. Nests in a variety of trees but prefers aspen when available (Lawrence 1967, Howell 1952). Favors trees infected with *Fomes* (Kilham 1971).

Territory Size: Varies from immediate vicinity of nest to 150 yd (137.2 m) or more (Howell 1952).

Home Range: 5.1 acres (2.1 ha) and 5.4 acres (2.2 ha) for 2 pairs in mature second-growth forest in Ontario (Lawrence 1967).

Sample Densities: 5.6 territorial males per 100 acres (40 ha) in oak-hickory forest in Ohio (Samson 1979). Winter—12 birds per 100 acres (40 ha) in bottomland forest in Illinois (Graber et al. 1977).

Food Habits: Inner bark of trees, sap, insects (excluding wood-boring larvae), fruits and berries. Drills sap wells in the smooth bark on trunks and limbs of live trees.

Comments: Yellow-bellied sapsuckers are migratory woodpeckers, common in orchards in spring and fall. They leave small, evenly spaced holes around tree trunks, especially apple, birch, hemlock, and aspen, and eat the sap as well as insects attracted to it (Norse and Fichtel 1985:156 in Laughlin and Kibbe 1985). Beal (1911 in Graber et al. 1977) found that the contents of 313 stomachs contained 49 percent animal and 51 percent vegetable matter. Poison ivy berries are an important winter food during prolonged subfreezing weather (C. S. Robbins pers. commun.).

Downy Woodpecker

(Picoides pubescens)

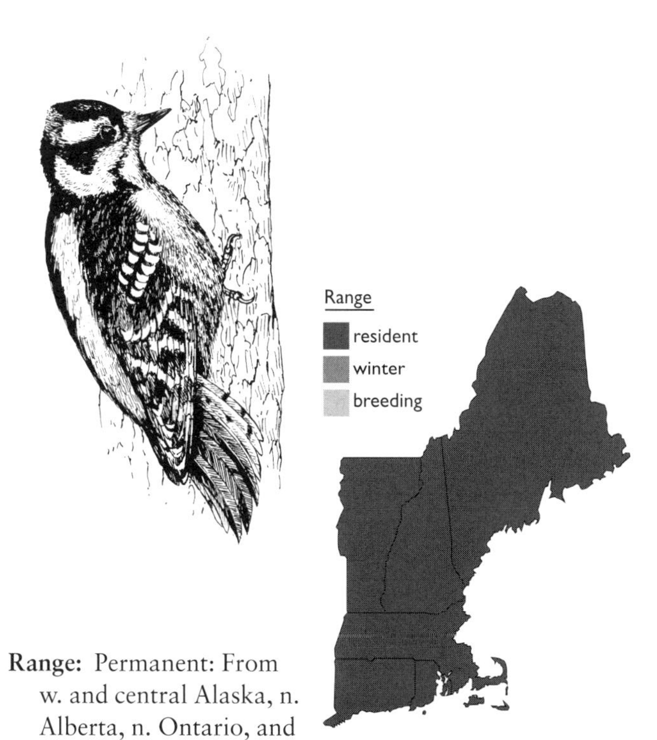

Range: Permanent: From w. and central Alaska, n. Alberta, n. Ontario, and Newfoundland, s. to s. California, central Arizona, the Gulf Coast, and s. Florida. In more n. populations, is mostly migratory, and occurs irregularly southward.

Distribution in New England: Resident throughout its range.

Status in New England: Common resident.

Habitat: Inhabits deciduous, mixed, and urban forests; prefers bottomland forests. Also in orchards, forest edges, and farmyards. Associated with woodlands with living and dead trees ranging from 10 to 22 in (25.4 to 55.9 cm) diameter breast height (dbh). Shugart et al. (1974) found a high correlation between downy woodpecker distribution and sapling density, indicating that sapling removal may decrease downy habitat quality.

Special Habitat Requirements: Dead or living trees greater than 6 in (15.2 cm) dbh for nesting.

Nesting: Egg dates: May 20 to June 21 in Massachusetts (Veit and Petersen 1993:285). Clutch size: 3 to 8 eggs, typically 4 or 5. Incubation period: 12 days. Nestling period: 20 to 22 days (postfledgling care continues for 3 weeks) (Lawrence 1967). Broods per year: 1, 2 in the south. Age at sexual maturity: 1 yr. Nest height: 3 to 50 ft (0.9 to 15.2 m), typically 20 ft (6.1 m). Nest site: A cavity in living or dead tree. Prefers to excavate cavity nests near the tops of dead trees or dead limbs of live trees in fairly open stands or at forest edges. Often nests in same tree in successive years, but excavates a new nest cavity each year (Hardin and Evans 1977). In the fall excavates a fresh hole for winter roosts (Harrison 1975).

Territory Size: 1.3 to 3.1 acres (0.5 to 1.3 ha) (average 2.0 acres, 0.8 ha) for 9 pairs in mature lowland forest in Illinois (Calef 1953 in Graber et al. 1977).

Home Range: 5 to 8 acres (2.0 to 3.2 ha) is estimated size for 2 pairs in second-growth forest in Ontario (Lawrence 1967).

Sample Densities: 36 birds per 100 acres (40 ha) in virgin floodplain forests in Illinois (Snyder et al. 1948). 3.6 territorial males per 100 acres (40 ha) in birch-basswood forests in Pennsylvania (Samson 1979).

Food Habits: Insects, especially wood-boring ants and beetle larvae in bark crevices of living and dead trees, under loose bark in woodlots, forest edges, forested parks, cemeteries.

Comments: The abundance and distribution of New England's smallest woodpecker have not changed in the historical period, unlike those of some other woodpeckers. Females tend to forage on small branches less than 5 cm (2 in) in diameter; males tend to forage on trunks (Jackson 1970). Beal (1911 in Bent 1939) found 76 percent animal and 24 percent vegetable material in 723 stomachs.

Hairy Woodpecker

(*Picoides villosus*)

Range: Resident from w. and central Alaska, n. Saskatchewan, and Newfoundland, s. throughout most of North America to Central America and the Bahamas. Northerly populations generally migrate southward.

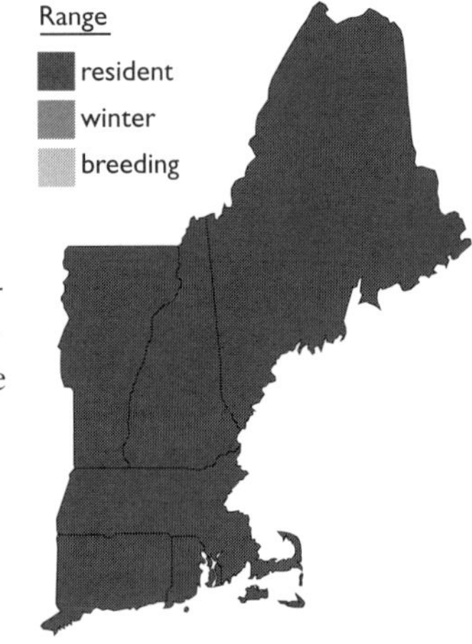

Distribution in New England: Resident throughout the region.

Status in New England: Common but variable year to year. Less abundant on outer Cape Cod in Massachusetts and rare on Nantucket (Veit and Petersen 1993:286) and the coastal region of Rhode Island (Enser 1992:95).

Habitat: Inhabits extensive forests of all types and conditions, preferring bottomlands with mature trees, especially along forest edges. Also occurs in burned areas, red maple swamps, and orchards.

Special Habitat Requirements: Nest trees greater than 10 in (25.4 cm) diameter breast height.

Nesting: Egg dates: April 23 to May 19, New York (Bull 1974:359). Clutch size: 3 to 6, typically 4. Incubation period: 11 to 12 days. Nestling period: 21 to 30 days. Broods per year: 1 (north), 2 (south). Age at sexual maturity: 1 yr. Nest height: 15 to 45 ft (4.6 to 13.7 m), typically 35 ft (10.7 m). Nest site: Excavates cavities in live or dead trees, in the trunk or on the underside of a large limb. Favors trees with decayed interiors.

Territory Size: 6.5 acres (2.6 ha) (one territory) in mature upland forest in Illinois (Allison 1947 in Graber et al. 1977). 1.6 to 3.7 acres (0.6 to 1.5 ha) (average 2.6 acres, 1.1 ha) in mature bottomland in Illinois (Calef 1953 in Graber et al. 1977).

Home Range: 6 to 8 acres (2.4 to 3.2 ha) (estimated minimum sizes of 2 ranges) in second-growth forest in Ontario (Lawrence 1967).

Sample Densities: 17 to 24 birds per 100 acres (40 ha) in mature bottomland forest in Illinois (Calef 1953 in Graber et al. 1977). 4 birds per 100 acres (40 ha) in upland oak-hickory forest in Illinois (Franks and Martin 1967). 3.6 territorial males per 100 acres (40 ha) in birch-basswood in Pennsylvania (Samson 1979).

Food Habits: Adults and larvae of wood-boring beetles, ants, and caterpillars are staples, but also eats fruits, nuts, corn.

Comments: The hairy woodpecker is one of the most widely distributed woodpeckers in North America. Birds are highly sedentary and may remain on home range for life. Females tend to feed lower and on different species of trees than males (Kilham 1965, 1968a), a behavior that apparently lessens competition.

Three-toed Woodpecker

(*Picoides tridactylus*)

Range: Holarctic; resident from nw. and central Alaska, n. Manitoba, n. Quebec, and Newfoundland, s. to w. and s. Alaska, central Washington, and s. Oregon, in the

Rocky Mountains to e. Nevada, central Arizona, and s.-central New Mexico, and to sw. and central Alberta, s. Manitoba, ne. Minnesota, central Ontario, n. New York, n. New England, and s. Quebec.

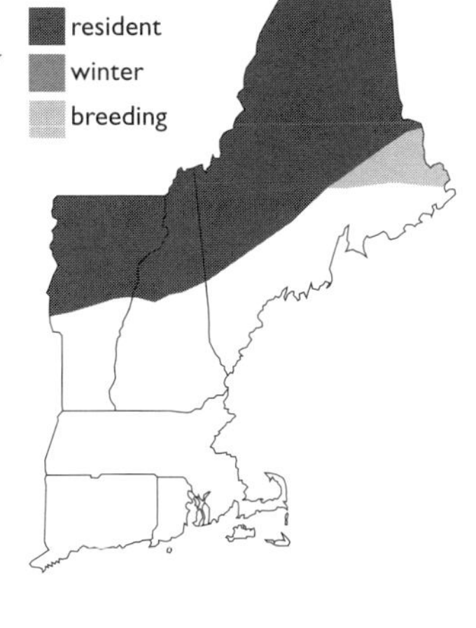

Distribution in New England: Breeds in n. Maine, n. New Hampshire, and n. Vermont. Very rare and irregular in Massachusetts in the winter (Veit and Petersen 1993:287).

Status in New England: Rare resident in n. New England; rare and irregular winter visitor in s. New England.

Habitat: Coniferous forests, especially burned areas with large stands of dead trees. Also favors logged areas and bogs where dead trees occur.

Special Habitat Requirements: Dead limbs for nesting.

Nesting: Egg dates: May 14 to June 14, New York (Bull 1974:363). Clutch size: Typically 4. Incubation period: About 14 days (England 1940). Nestling period: 22 to 26 days. Broods per year: 1. Age at sexual maturity: Probably 1 yr. Nest height: 5 to 12 ft (1.5 to 3.7 m), rarely to 40 ft (12.2 m). Nest site: A cavity in a living or dead tree, often in a burned stand. Loosely colonial in areas with abundant food.

Food Habits: Wood-boring larvae of moths and beetles. Short (1974) observed birds in New York feeding mainly in the upper parts of live conifers.

Comments: Three-toed woodpeckers may occur as far s. as Massachusetts following bark beetle outbreaks in e. Canada, as do black-backed woodpeckers (Veit and Petersen 1993:288). Birds are sedentary, rarely leave home ranges. Seldom venture far from deep woods. Diet is greater than 90 percent animal matter and less than 10 percent vegetable matter; 75 percent of the animal portion consists of wood-boring larvae (Bent 1939:118).

Black-backed Woodpecker

(Picoides arcticus)

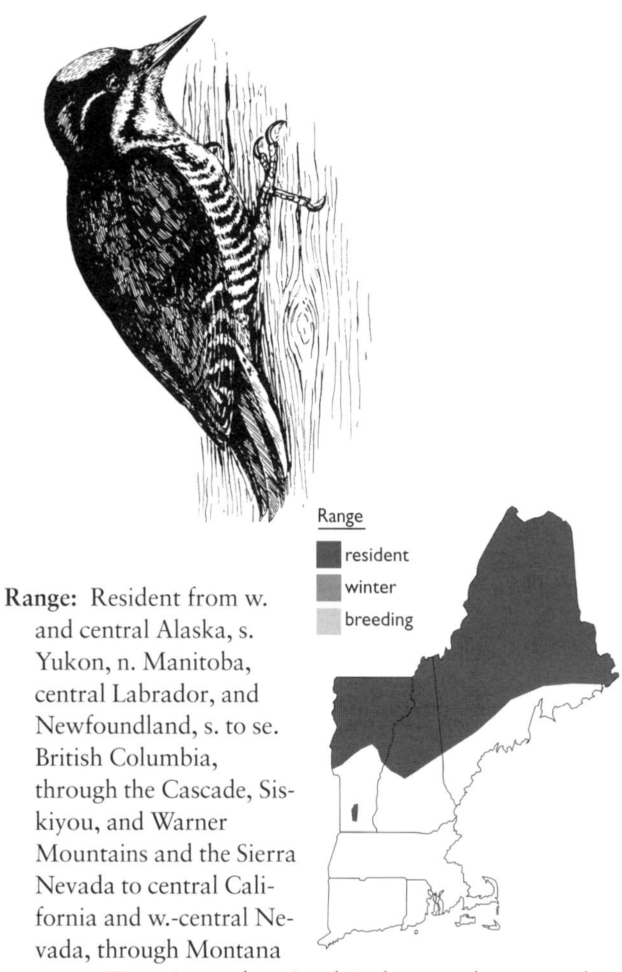

Range: Resident from w. and central Alaska, s. Yukon, n. Manitoba, central Labrador, and Newfoundland, s. to se. British Columbia, through the Cascade, Siskiyou, and Warner Mountains and the Sierra Nevada to central California and w.-central Nevada, through Montana to nw. Wyoming and se. South Dakota, and to sw. and central Alberta, se. Manitoba, n. Minnesota, n.-central Michigan, n. New York, and n. New England. Moves s. irregularly in winter to s. New England and New Jersey.

Distribution in New England: Resident in n. and central Maine, n. New Hampshire, and n. Vermont.

Status in New England: Uncommon breeder in n. New Hampshire and Maine, rare in Vermont. Rare irregular winter visitor in s. New England.

Habitat: Boreal coniferous forests, especially in burned, logged, boggy, or beetle-killed or budworm-killed forests where dead trees are numerous. Favors spruce, fir, and larch.

Special Habitat Requirements: Dead or live trees with decay columns for nesting, dead trees with loose bark for feeding.

Nesting: Egg dates: May 18 to June 12, New York (Bull 1974:360). Clutch size: 2 to 6, typically 4. Incubation period: About 14 days. Broods per year: 1. Age at sexual maturity: 1 yr. Nest height: 2 to 15 ft (0.6 to 4.6 m). Usually excavates a new cavity each year as part of courtship activity, as well as for roosting. Nest site: Excavates a cavity in a living tree with a decayed interior or a dead tree or stub (often balsam fir). Sometimes uses old utility poles. Birds in New York nested in small open areas with windfalls and dead trees (Bent 1939:106). Birds in New York almost invariably nest in dead trees (Bull 1974:361).

Food Habits: Bark-boring beetle larvae and other insects under loose bark, in decayed heartwood, bark crevices, lower parts of dead trees. Commonly scales off loose bark. Commonly feeds in beaver swamps and other places where there are recently killed trees with loose bark. Short (1974) observed black-backs feeding mainly in dead trees bordering a bog.

Comments: Following years of high food abundance, especially of spruce budworm in northern forests, black-backed woodpeckers commonly occur much farther s. of their normal winter range. Major flights have been recorded since 1860–1861; 1923–1926, during the 1950s and 1960s, but none since 1975 (Veit and Petersen 1993:288–289). Bark-boring beetle larvae account for 75 percent of the volume of animal food. The remaining 25 percent consists of other insects and spiders, and plant materials.

Northern Flicker

(Colaptes auratus)

Range: Breeding: From central Alaska, n. Manitoba, n.-central Quebec, and Newfoundland, s. throughout most of North America and nw. Mexico. Winter: From s. Canada, s. through the breeding range.

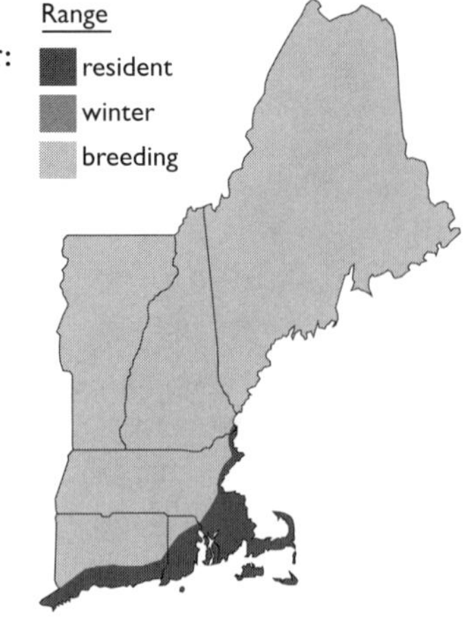

Distribution in New England: Breeding: Throughout the region. Winter: Primarily along the coasts.

Status in New England: Common breeder; common in spring and very common fall migrant. Uncommon winter resident along the coasts; rare inland.

Habitat: Breeding: Commonly inhabits open woodlands, forest edges, savannas, and riparian woodlands; also found in swamps and beaver ponds with standing trees, burned forests, and clear-cuts with tree stubs left standing, orchards, and suburban and urban areas with large trees (Moore 1985:166 in Laughlin and Kibbe 1985, Sedgwick and Knopf 1990, Conner and Adkisson 1977, Kilham 1983). Winter: Generally inhabits coastal coniferous forests or swamps.

Special Habitat Requirements: Cavity nest sites in large trees (preferably dead or dying) in open woodlands or along forest edges.

Nesting: Egg dates: April 20 to 14 June, Massachusetts (Veit and Petersen 1993:290). April 20 to June 19, New York (Bull 1974:351). Clutch size: 3 to 10, typically 6 to 8. Incubation period: 11 to 12 days. Nestling period: About 23 days. Broods per year: 1. Age at sexual maturity: 1 yr. Nest height: 2 to 60 ft (0.6 to 18.3 m). Typically 10 to 30 ft (3.0 to 9.1 m). Nest site: Cavity usually excavated in dead or diseased tree trunks or large limbs (Moore 1995). Also nests in buildings, telephone poles, and even in kingfisher holes. One record of a ground nest in Massachusetts (Veit and Petersen 1993:290).

Territory Size: Both male and female defend the nesting area. 1.55 acres (0.6 ha) in Illinois (Calef 1953 in Graber et al. 1977). Territorial defense is limited to nest site during the incubation period. During this period, other flickers may occupy original territory (Kilham 1973). Kilham (1973) observed 3 pairs on less than 1 acre (0.4 ha).

Sample Densities: Average 33.3 birds per 100 acres (40 ha) in second-growth hardwoods in Illinois (Fawks 1937, 1938 in Graber et al. 1977). 13 birds per 100 acres (40 ha) in oak-hickory type in Illinois (Franks and Martin 1967). 19 pairs per 100 acres (40 ha) in white pine woodland on Nantucket (Dennis 1969).

Food Habits: Ants (staple) and a variety of other insects, especially ground beetles, crickets, and grasshoppers.

Also commonly takes wild fruits. Commonly forages in grassy areas such as lawns, pastures, openings in woods, cornfields (especially in fall and winter).

Comments: The flicker is New England's only brown woodpecker, and is largely a ground feeder. The flicker is the most common breeding woodpecker in Connecticut (Zeranski and Baptist 1990:168). About 61 percent of the diet consisted of animal food and about 39 percent vegetable food. Ants represented about 75 percent of the volume of animal material (Beal 1911 in Bent 1939:277).

Pileated Woodpecker

(*Dryocopus pileatus*)

Range: Resident from s. and e. British Columbia, sw. Mackenzie, central Manitoba, New Brunswick, and Nova Scotia, s. through Alberta, Washington, s.-central Idaho, w. Montana, and Oregon, to n. California, w. to e. Dakotas, Missouri, and Oklahoma, and s. to e.-central Texas, the Gulf Coast, and s. Florida (DeGraaf et al. 1991:253).

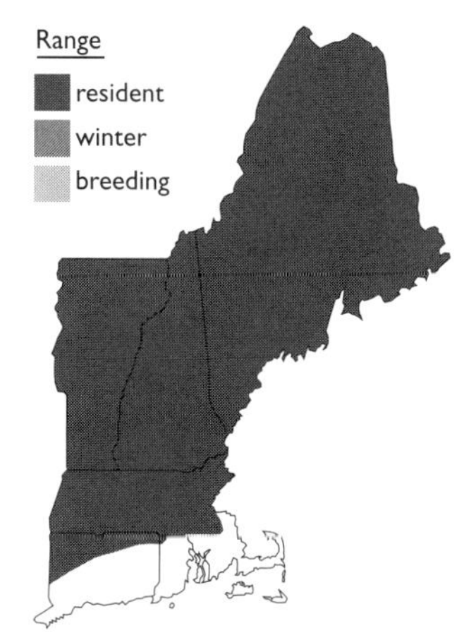

Distribution in New England: Throughout Maine, New Hampshire, Vermont, much of Connecticut (although rare along the coast), and Massachusetts except virtually absent from the se. part of the state and the islands (Veit and Petersen 1993:291), and in nw. Rhode Island (Enser 1992:97).

Status in New England: Common resident.

Habitat: Mature deciduous, mixed, and coniferous forests, preferably near water. Associated with large living trees [≥18 in (≥45 cm) diameter breast height (dbh)], rather than with extensive forests; often in quite open areas with large old trees, commonly along roads. Occur at elevations below 2,000 feet (610 m) in New Hampshire (Kilham and Foss 1994:142 in Foss 1994).

Special Habitat Requirements: Large trees (typically dead or with large dead limbs) for nesting and feeding (Bull and Jackson 1995).

Nesting: Egg dates: May 11–28, Massachusetts (Veit and Petersen 1993:291). April 22 to May 19, New York (Bull 1974:352). Clutch size: 3 to 4, typically 4 (Bull and Jackson 1995). Incubation period: 18 days. Nestling period: 26 to 31 days. Broods per year: 1. Age at sexual maturity: 1 yr. Nest height: 15 to 70 ft (4.6 to 21.3 m), typically 45 ft (13.7 m). Nest site: Cavity in trunk of dead or living tree; sometimes in large dead limbs, preferably near water. Conner et al. (1975) found pileated nests in trees with a diameter breast height range of 13 to 35.8 in (33 to 91 cm) (average 21.5 in, 54.6 cm). Hoyt (1957) observed that they rarely reuse old nest holes.

Territory Size: Feeding territory of 98.8 to 197.6 acres (40 to 60 ha) in New York (Hoyt 1957).

Sample Densities: 3 birds per 100 acres (40 ha) in virgin bottomland forest in Illinois (Snyder et al. 1948). 3 birds per 100 acres (40 ha) in mature bottomland forest in Illinois (Graber et al. 1977). 0 to 0.5 birds per 100 acres (40 ha) in mature upland forest in Illinois (Graber et al. 1977). 1 pair per 1,643 acres (665 ha) in ponderosa pine, Douglas fir, and larch habitat in Oregon (Bull 1975). Maximum 1 pair per 98.8 acres (40 ha) (Hardin and Evans 1977).

Food Habits: Larvae and adults of many kinds of insects, especially ants, which account for more than 50 percent of the diet. Also eats wild fruits, acorns, beechnuts (Bent 1939:183). Most feeding is done in decayed wood (Tanner 1942), especially carpenter ant colonies in large trees, both in forest interiors and at edges (Bent 1939:184).

Comments: After extensive clearing in the eighteenth and nineteenth centuries, the pileated woodpecker was rare in New England by 1900 (Howe and Allen 1901), was a regular but local nester in the Connecticut River Valley by the 1920s, and by the 1970s was occurring e. of the Connecticut River in s. New England. Pileated woodpeckers need large trees, not necessarily unbroken forests. They are fairly common in suburban areas where large trees line the roads.

Olive-sided Flycatcher

(*Contopus cooperi*)

Range: Breeds from w. and central Alaska and central Yukon to n. Ontario, s.-central Quebec, and s. Labrador, s. to s. California across to w. Texas, and e. of the Rocky Mountains, to central Saskatchewan, n. Wisconsin, ne. Ohio, and Massachusetts; also locally in the Appalachians to w. North Carolina. Winters from Costa Rica to South America and, casually, in s. California. Breeds in n. Baja California and migrates throughout Mexico and central America. Winters chiefly in mountains of n. and w. South America from Venezuela s. to se. Peru and n. Bolivia.

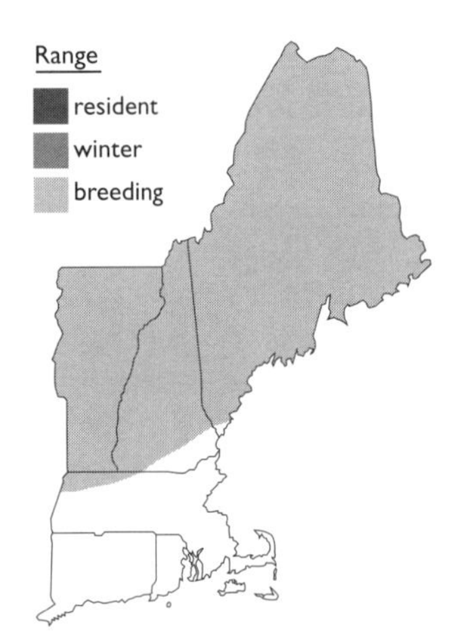

Distribution in New England: Throughout Maine, n. and central New Hampshire, and Vermont; s. to w. Massachusetts, and very rare in nw. Connecticut with one nesting record from Norfolk (Zeranski and Baptist 1990:170). Uncommon migrant.

Status in New England: Locally to fairly common breeder in Maine, New Hampshire, and Vermont, a very uncommon and local breeder in Berkshire County (Mass.) e. to Worcester County at high elevations in Massachusetts (Peterson and Fichtel 1992, Veit and Petersen 1993:292).

Habitat: Inhabits montane coniferous forests up to 9,900 ft (3,000 m) in elevation. Prefers forests of tall spruces, firs, balsams, and pines; burned-over areas with tall standing dead trees, wooded streams, and borders of northern bogs and muskegs. Prefers very open stands, and generally responds favorably to openings in coniferous forests and forest edges.

Special Habitat Requirements: Tall, exposed perches, typically near bogs, swamps, clearcuts, or beaver ponds.

Nesting: Egg dates: June 5 to July 1, Massachusetts (Veit and Petersen 1993:293). June 9 to June 27, New York (Bull 1974:378). Clutch size: 2 to 4, typically 3. Incubation period: 16 to 17 days. Nestling period: 21 to 23 days. Broods per year: 1. Age at sexual maturity: 1 yr. Nest height: 5 to 50 ft (2.1 to 15.2 m). Nest site: Usually well hidden on a horizontal branch in a conifer, usually far out from the trunk.

Territory Size: Breeding birds require an area of several acres (Harrison 1975:127). 4 to 8 acres (1.6 to 3.2 ha) in Maryland (Stewart and Robbins 1958).

Food Habits: Insects, especially hymenopterans, taken by sallying from a prominent perch.

Comments: The olive-sided flycatcher was widely distributed in Massachusetts in the mid 1800s, nesting throughout the state except on Nantucket. By 1927 it was restricted to high elevation coniferous forests in Worcester County and the pitch pine barrens of se. Massachusetts, including Cape Cod. Currently it is restricted to a few sites in the Berkshires and the Worcester Plateau and is no longer found in the se. pitch pine forests (Veit and Petersen 1993:292–293). Typically perches on high, exposed or dead-topped tree limbs, hawking insects and calling the distinctive "quick three beers."

Eastern Wood-Pewee

(*Contopus virens*)

Range: Breeding: From se. Saskatchewan to s. Quebec and New Brunswick, s. to Texas, the Gulf Coast, and central Florida, and w. to the e. Dakotas, central Oklahoma, and s.-central Texas. Migrates through Mexico and Central America on the Caribbean coast. Winters from Nicaragua to South America, mainly from Venezuela to Peru.

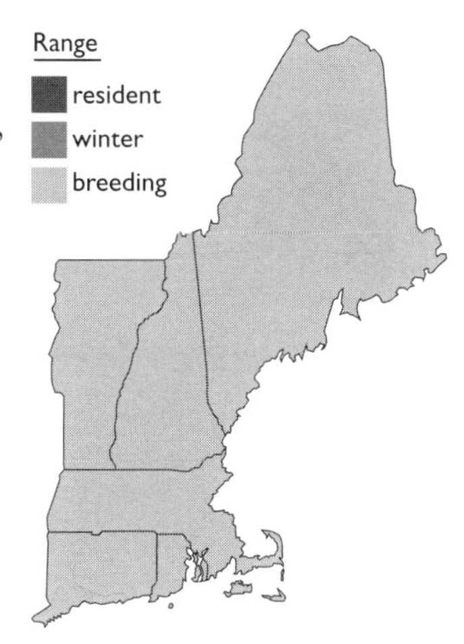

Distribution in New England: Breeds throughout the region.

Status in New England: Uncommon to common breeder; fairly common spring and uncommon fall migrant.

Habitat: Generally inhabits deciduous forests, primarily edges and clearings. Associated with deciduous forests with (87 percent) closed canopy and an open (22.2 percent) understory in Grafton, Vt. (Thompson and Capen 1988). Will use areas with a dense understory if the canopy above is incomplete or open. Also inhabits mixed forests, bottomlands, woodlots, orchards, parks, roadsides, uplands, and residential shade trees (McCarty 1996). In the Midwest strongly associated with oaks (Graber et al. 1974), and probably requires a predominance of hardwoods throughout its range. Rare or absent in pure spruce stands in Vermont (Ellison 1985:172 in Laughlin and Kibbe 1985), but inhabits coastal pitch pine forests in Massachusetts (Veit and Petersen 1993:293), and pole stands of eastern white pine and spruce-fir in the White Mountains in New Hampshire (DeGraaf and Chadwick 1987). Occurs in riparian habitat in the Midwest, but tends to avoid streams in eastern forests (Murray and Stauffer 1995).

Special Habitat Requirements: Open deciduous or mixed forests or forest edge.

Nesting: Egg dates: June 10 to July 25, Massachusetts (Veit and Petersen 1993:294). May 30 to July 20, New York (Bull 1974:378). Clutch size: 2 to 4, typically 3. Incubation period: 12 to 13 days. Nestling period: 15 to 18 days. Broods per year: 1, possibly 2. Nest height: 9 to 65 ft (2.7 to 19.7 m), typically 25 ft (7.6 m). Nest site: Nests primarily in deciduous trees or saplings, occasionally in coniferous trees or saplings. Nest located on a horizontal limb, usually well out from the trunk, often on a lower dead twig on a live tree. Nest camouflaged with spiderwebs and lichens.

Territory Size: 1.4 to 3.1 acres (0.6 to 1.3 ha) in lowland forest in Illinois (Fawver 1947, Calef 1953 in Graber et al. 1974). Average territory size of 19.3 acres (7.7 ha) in forest stands in Wisconsin (Bond 1957). Average territory size of 5.5 acres (2.2 ha) in Iowa (Best and Stauffer 1986).

Home Range: 10.8 acres (4.4 ha) (Odum and Kuenzler 1955).

Sample Densities: 54.2 territorial males per 100 acres (40 ha) in ash-basswood forests in New York (Samson 1979). Maryland—19 pairs per 100 acres (40 ha) in virgin hardwood forest. 7 pairs per 100 acres (40 ha) in unsprayed apple orchard. 6 pairs per 100 acres (40 ha) in upland oak forest. 5 pairs per 100 acres (40 ha) in pine-oak forest (Stewart and Robbins 1958:207).

Food Habits: Small flying insects taken from leaf surfaces as well as in the air. Commonly feeds in woodland clearings, edges of fields, marshes; generally feeds in mid to lower tree canopy.

Comments: Common through the historical period, wood-pewees occur in deciduous woodlands with relatively open understories. The nests are usually associated with openings. They will nest in a forest with a dense understory if canopy above is incomplete or sparse. One of the few open-nesting birds that will nest on a dead twig or tree.

Yellow-bellied Flycatcher

(*Empidonax flaviventris*)

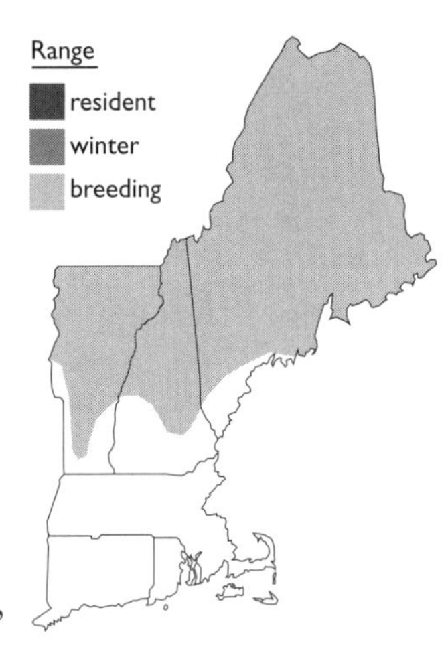

Range: Breeds from n. British Columbia and w.-central and s. Northwest Territories to s. Labrador and Newfoundland, s. to central Alberta, n. North Dakota, and n. Minnesota, across to s. Ontario, ne. Pennsylvania, and Nova Scotia. Winters from Mexico to Panama. Migrant and winter visitor to e. portions of Mexico, except the Yucatan Peninsula.

Distribution in New England: Northern and central Maine, the White Mountains and n. in New Hampshire, and central and n. Vermont.

Status in New England: Uncommon (n. Vermont and New Hampshire) to common (Maine) breeder, uncommon migrant in s. New England.

Habitat: Northern forests: dense, moist coniferous forests of spruce and fir, frequently occurs in low, cedar, tamarack, spruce, and sphagnum bogs; wet mossy glades, swampy thickets bordering ponds and streams, and cool, moist mountainsides. Up to 4,000 ft (1,220 m) in elevation in the White Mountains in New Hampshire (Janeway 1994:148 in Foss 1994).

Special Habitat Requirements: Low, wet areas within coniferous forests.

Nesting: Egg dates: June 10 to June 27, New York (Bull 1974:371). Clutch size: 3 to 5, typically 3 or 4 (Harrison 1975:121). Incubation period: 15 days. Nestling period: 13 days. Age at sexual maturity: 1 yr. Nest site: On or near the ground, sometimes at the base of a tree in a cavity formed by roots, but more often beside a hummock or mound and well hidden in sphagnum moss or other vegetation (Walkinshaw 1957).

Food Habits: Flying insects, fruits (occasionally).

Comments: Yellow-bellied flycatchers breed primarily in Canada, and are one of New England's latest spring migrants. Birds perch and feed close to ground. Food habit studies indicate a predominance of animal food (97 percent) over vegetable food (3 percent) (Beal 1912 in Bent 1942:178).

Acadian Flycatcher

(*Empidonax virescens*)

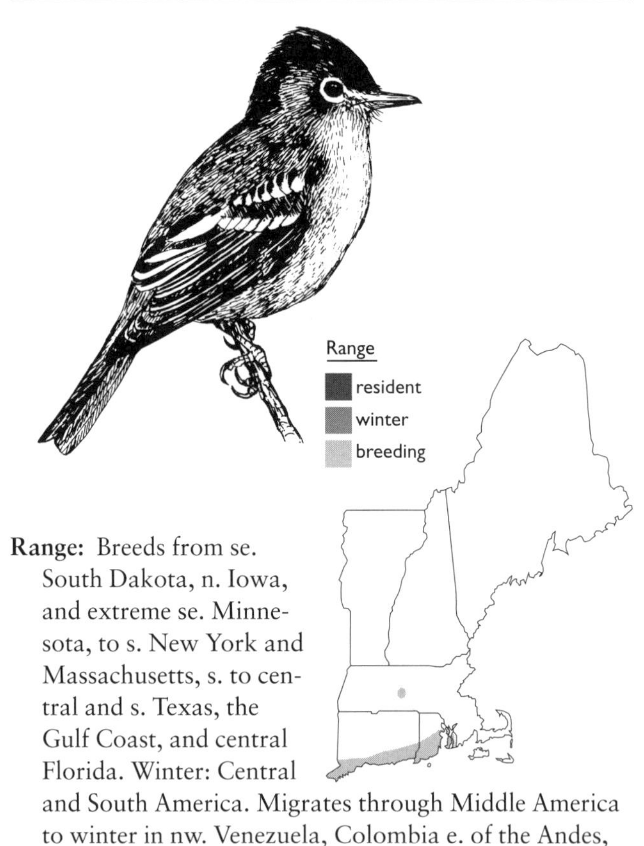

Range: Breeds from se. South Dakota, n. Iowa, and extreme se. Minnesota, to s. New York and Massachusetts, s. to central and s. Texas, the Gulf Coast, and central Florida. Winter: Central and South America. Migrates through Middle America to winter in nw. Venezuela, Colombia e. of the Andes, and Ecuador. A rare transient in the Bahamas, w. Cuba, and Isle of Pines.

Distribution in New England: Southeasten Massachusetts and the Quabbin Reservoir, central and s. Rhode Island, and locally throughout s. Connecticut (Zeranski and Baptist 1990:172).

Status in New England: Rare and local breeder in southern New England, range expanding northward.

Habitat: Mature deciduous and mixed forests, especially beech and hemlock forests in New England. Prefers dense lowland forests with an understory, wooded swamps, and forested ravines near streams.

Special Habitat Requirements: Mature, extensive deciduous and mixed forests with tall trees, a closed canopy, and open spaces in the understory for feeding.

Nesting: Egg dates: June, Massachusetts (Veit and Petersen 1993:296). Clutch size: 2 to 4, usually 3. Incubation period: 13 to 15 days (Mumford 1964). Nestling period: 13 to 15 days (Newman 1958). Broods per year: 2. Nest height: 8 to 20 ft (2.4 to 6 m). Nest site: Nests on a fork of a horizontal branch well away from the main trunk, often along a stream and sometimes over water. Prefers an open space below the nest to approach the nest easily. Favors the lower branches of beech, dogwood, and witch-hazel, but also nests in oak, hickory, maple, basswood, and cherry.

Territory Size: 21.8 territorial males per 100 acres (40 ha) in birch-basswood forests in Pennsylvania (Samson 1979).

Food Habits: Moths, caterpillars, beetles, wasps, bees, and some wild berries (Terres 1980:381).

Comments: The acadian flycatcher was first collected in Connecticut in 1874 and first nested in 1875; in the early 1900s to about 1950 it was restricted to the sw. part of the state (Zeranski and Baptist 1990:172). Absent from s. New England until the 1960s, acadian flycatchers are expanding their range northward, and a nesting attempt was recorded in central New Hampshire in 1986 (Andrews 1994:150 in Foss 1994).

Alder Flycatcher

(*Empidonax alnorum*)

Range: Breeding: From central Alaska and central Yukon to central and e. Quebec, s. Labrador and s. Newfoundland, s. to s.-central British Columbia and s. Alberta, across to s.-central Minnesota, e. Pennsylvania, Massachusetts, and nw. Connecticut; also in the Appalachians s. to w. North Carolina. Migrates through Central America. Winter: From Colombia and Venezuela, s. to n. Argentina.

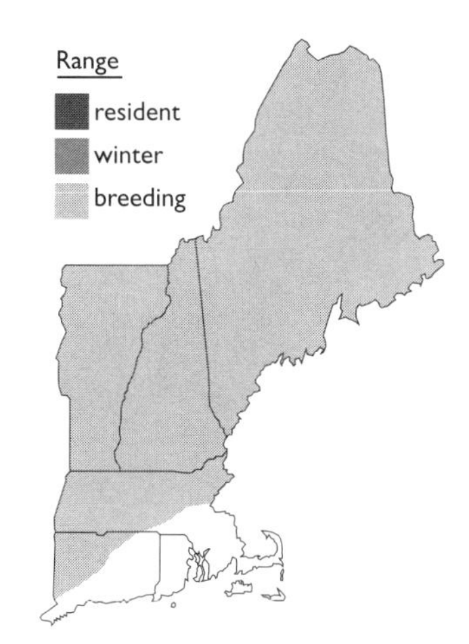

Distribution in New England: Breeds throughout Maine, New Hampshire, and Vermont, s. to w. and central Massachusetts and nw. Connecticut. Not reported from Rhode Island.

Status in New England: Common (Maine) to uncommon (s. New England) breeder. More abundant in w. than in e. Connecticut and Massachusetts (Zeranski and Baptist 1990:173, Veit and Petersen 1993:298).

Habitat: Typically inhabits low, damp alder swamps. Usually found near water in dense, low, damp thickets of alders, willows, sumacs, viburnum, elderberries, and red-osier dogwood bordering bogs, swamps, marshes, and along the edges of streams and ponds.

Special Habitat Requirements: Wet areas with dense, low shrubs and clearings with wet, shrubby edges.

Nesting: Egg dates: June 19 to July 3, Massachusetts (Veit and Petersen 1993:297). June 11 to July 29, New York (Bull 1974:376). Clutch size: 3 to 4. Incubation period: 12 to 14 days. Nestling period: 13 to 16 days. Broods per year: 1. Age at sexual maturity: 1 yr. Nest height: 1 to 6 ft (0.3 to 1.8 m), typically 3 to 4 ft (0.9 to 1.2 m). Nest site: In low tree or shrub saddled on a branch or in an upright fork.

Territory Size: Singing males of three separate populations had average territory sizes of 3.2, 3.8, 7.7 acres (1.2, 1.5, 3.1 ha) (Stein 1958).

Sample Densities: 4.6 territorial males per 100 acres (40 ha) in shrub swamps and thickets in New York (Samson 1979).

Food Habits: Flying insects.

Comments: Alder and willow flycatchers were once considered one species—Traill's flycatcher—so historical information on their respective populations is uncertain. Visually identical, the alder flycatcher's song is "fee-bee' o," and the willow's is "fitz'-bew." Although willow flycatchers have become increasingly common, the alder flycatcher is more common in w. Massachusetts and n. New Hampshire.

Willow Flycatcher

(*Empidonax traillii*)

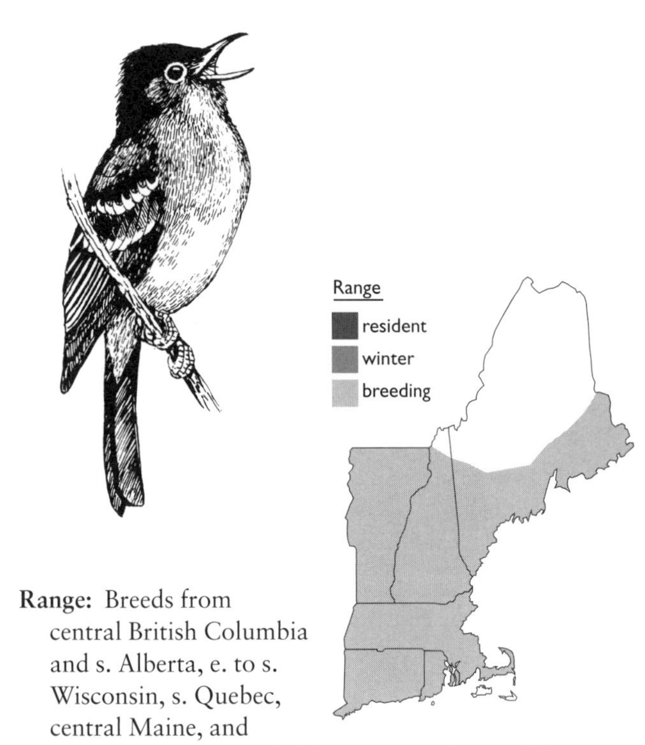

Range: Breeds from central British Columbia and s. Alberta, e. to s. Wisconsin, s. Quebec, central Maine, and Nova Scotia, s. to s. California, w. and central Texas, Arkansas, n. Georgia, and central and e. Virginia. Winters from Mexico to nw. Colombia.

Distribution in New England: Southern Maine, w. to n. New Hampshire, and throughout Vermont, Connecticut, Rhode Island, and Massachusetts, except absent from Martha's Vineyard and Nantucket, Mass.

Status in New England: Locally common breeder, stable or increasing throughout much of its range.

Habitat: Prefers open, brushy fields, newly clear-cut areas, willow thickets along streams, edges of shrub wetlands, cattail-dominated wetlands, damp to dry upland fields, hedgerows, dense roadside growth, and orchards. In areas where range overlaps that of the alder flycatcher, prefers drier, more open shrubby habitat.

Special Habitat Requirements: Fairly open areas with scattered shrubs or forest edges.

Nesting: Egg dates: Early July, Massachusetts (Veit and Petersen 1993:298). June 11 to July 29, New York (Bull 1974:376). Clutch size: 3 to 5, typically 3 or 4. Incubation period: 12 to 15 days. Nestling period: 15 to 18 days. Broods per year: 1. Age at sexual maturity: 1 yr. Nest height: 3 to 25 ft (1.0 to 7.6 m). Typically 4 to 6 ft (1.2 to 1.8 m). Nest site: In a fork or saddled on a horizontal limb of a shrub, commonly willow, elder, viburnum, hawthorn, and others.

Territory Size: 0.8 to 2.9 acres (0.3 to 1.2 ha) (average 1.74 acres, 0.7 ha) for 73 territories in a dry marsh in Michigan (Walkinshaw 1966). Singing males of three separate populations had average territory sizes of 2.6, 3.2, and 4.5 acres (1.1, 1.3, and 1.8 ha) (Stein 1958).

Sample Densities: 25 to 30 pairs per square mile (10 to 11 pairs/km^2) in willow clump habitat in Illinois, 8 to 9 birds per 100 acres (40 ha) (Ford 1956 in Graber et al. 1974).

Food Habits: Flying insects.

Comments: Difficult to distinguish from alder flycatcher (both formerly named Traill's) even when in hand. Most widely accepted diagnostic characteristic is vocalization difference: willow, "fitz'-bew," alder, "fee-bee'-o."

Least Flycatcher

(*Empidonax minimus*)

Range: Breeds from s. Yukon and w.-central and s. Northwest Territories to s. Quebec and New Brunswick, s. to s. British Columbia and central Montana, e. to sw. Missouri, n. Ohio, and central New Jersey; in the Appalachians to nw. Georgia. Winters from central Mexico s. in Central America to Nicaragua.

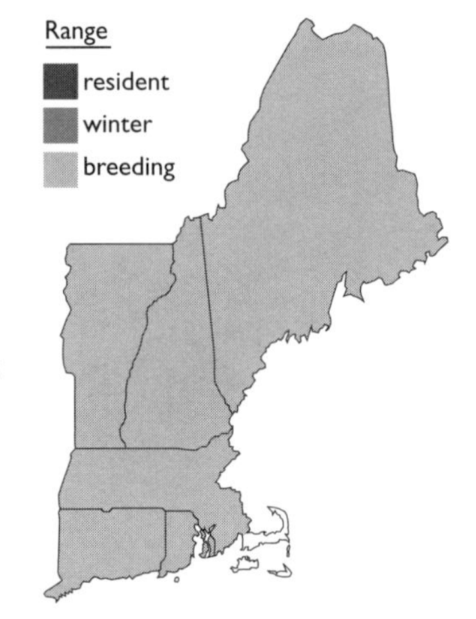

Distribution in New England: Throughout the region.

Status in New England: Fairly common breeder, except rare in coastal areas; absent from outer Cape Cod and the Islands. Fairly common migrant. Declining as both a breeder and a migrant (Holmes et al. 1986).

Habitat: Open mature (with sparse understory—Darveau et al. 1992) and second-growth deciduous and mixed forests (Hespenheide 1971) (occasionally northern conifer stands), forest edges and clearings, swamp and bog edges, shrubby fields, wooded residential areas, orchards. In northern hardwoods forests, habitat similar to that used by American redstart (Sherry and Holmes 1988).

Nesting: Egg dates: May 20 to late July (Veit and Petersen 1993:300). May 16 to July 28, New York (Bull 1974:377). Clutch size: 2 to 7, typically 4 (Briskie 1994). Incubation period: 12 to 15 days, typically 13 or 14 (Walkinshaw 1966, Harrison 1975, Briskie 1994). Nestling period: 12 to 17 days, typically 14 days. Broods per year: 1 or 2. Age at sexual maturity: 1 yr. Nest height: 2 to 60 ft (0.6 to 18.3 m), typically 10 to 20 ft (3.0 to 6.1 m). Nest site: In a crotch or on a limb of a deciduous or coniferous tree. Known to nest in apple, oak, pine, willow, sugar maple, and others.

Territory Size: 0.35 to 0.55 acre (0.1 to 0.2 ha) in oak-chestnut woodland in Virginia (Davis 1959). 0.03 to 0.5 acre (0.01 to 0.2 ha) (average 0.18 acre, 0.07 ha) in Michigan for 33 territories (MacQueen 1950). Usually less than 1 acre (0.4 ha) (Breckenridge 1956). 59 territories averaged 0.45 acres (0.18 ha) in New Hampshire (Sherry 1979). 10 territories in Ontario ranged from 0.08 to 0.95 acres (0.03 to 0.38 ha) and averaged 0.33 acres (0.13 ha) (Martin 1960).

Sample Densities: 2 nests per 27.67 acres (11.2 ha) of residential woodland in Illinois (Beecher 1942 in Graber et al. 1974). 9 nests per 19 acres (7.7 ha) in Virginia (Davis 1959)—oak-chestnut woodland. 2.7 pairs per acre (0.4 ha) in aspen-birch-maple habitat in Michigan (MacQueen 1950). 56 pairs per 100 acres (40 ha) in New Hampshire (Holmes and Robinson 1981), 60 pairs per 100 acres (40 ha) in Ontario (Kendeigh 1947).

Food Habits: Flying insects.

Comments: Least flycatchers have apparently varied widely in abundance since Colonial times—scarce and local before clearing, abundant around 1900 when much of the New England landscape was in old farmland, orchard, and woodlots, then less common as forests invaded fields. May compete with American redstarts for preferred habitat (Sherry 1979).

Eastern Phoebe

(Sayornis phoebe)

Range: Breeds from ne. British Columbia and w.-central and s. Northwest Territories to sw. Quebec and central New Brunswick, s. to s. Alberta, sw. South Dakota, central New Mexico, and central and ne. Texas across to Georgia and North Carolina. Winters primarily in the se. United States, n. along the Atlantic Coast to Virginia, and Mexico.

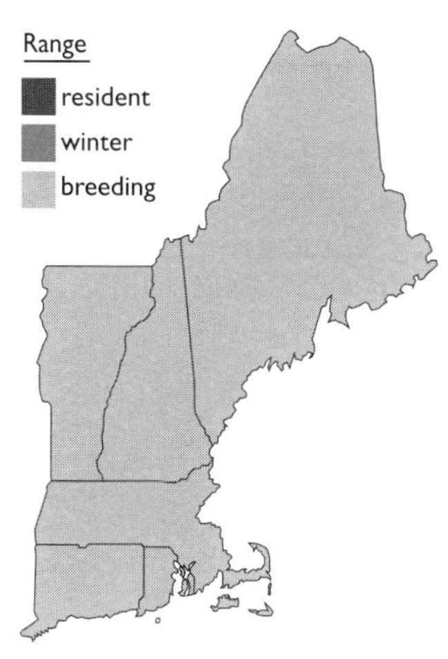

Distribution in New England: Throughout the region (except absent on Block Island). Winter: Occasionally overwinters in Massachusetts and Connecticut in very mild seasons (Veit and Petersen 1993:301).

Status in New England: Common and widespread breeder and migrant.

Habitat: Inhabits partially open woodlands, woodland edges, wooded ravines, agricultural and suburban areas, often near flowing water in forests.

Special Habitat Requirements: Sheltered ledges for nesting (Weeks 1994).

Nesting: Egg dates: April 27 to August 15, Massachusetts (Veit and Petersen 1993:301). April 20 to August

4, New York (Bull 1974:369). Clutch size: 3 to 8, typically 5 (Faanes 1980). Incubation period: Average 16 days (Smith 1942, Graber et al. 1974). Nestling period: 15 to 17 days. Broods per year: 2. Age at sexual maturity: 1 yr. Nest height: 2.5 to 20 ft (0.8 to 6.1 m), typically less than 15 ft (4.6 m). Nest site: Sheltered ledges (Hill and Gates 1988). Commonly nests on sheltered ledges of human structures including eaves of houses, bridges, and in culverts. Also under rock outcroppings and upturned tree roots.

Territory Size: 3.3 to 7.1 acres (1.3 and 2.9 ha) for 2 pairs nesting on buildings in Kansas (Fitch 1958). 0.7 acre (0.3 ha) in an Illinois floodplain forest (Fawver 1947 in Graber et al. 1974). 58 territorial males per 100 acres (40 ha) in birch-maple forests in Connecticut (Samson 1979).

Sample Densities: 6 nests per 30 acres (12.1 ha) in optimum habitat in Illinois (Graber et al. 1974). 7 pairs per 100 acres (40 ha) in mixed agricultural habitats in Maryland. 0.6 pairs per 100 acres (40 ha) in mixed forests and fields in Maryland (Stewart and Robbins 1958:201).

Food Habits: Flying insects, occasionally small fruits (Weeks 1994).

Comments: Phoebes usually choose one or more favorite perches from which to hawk insects. They are common victims of cowbird parasitism. Blocher (1936) reported parasitism in 50 percent of the nests observed in Illinois. This species benefits from forest management, moving into areas where cuttings have exposed ledge and rocks and created sunny forest openings in the vicinity of ledge.

Great Crested Flycatcher

(*Myiarchus crinitus*)

Range: Breeds from e.-central Alberta and central and se. Saskatchewan to sw. Quebec and central New Brunswick, s. to central and se. Texas, the Gulf Coast, and Florida, and w. to the e. Dakotas, w. Kansas, and w-central Oklahoma. Winters in central and s. Florida and from Mexico, s. to Colombia and n. Venezuela. A rare visitor in the Bahamas, Cuba, and Puerto Rico.

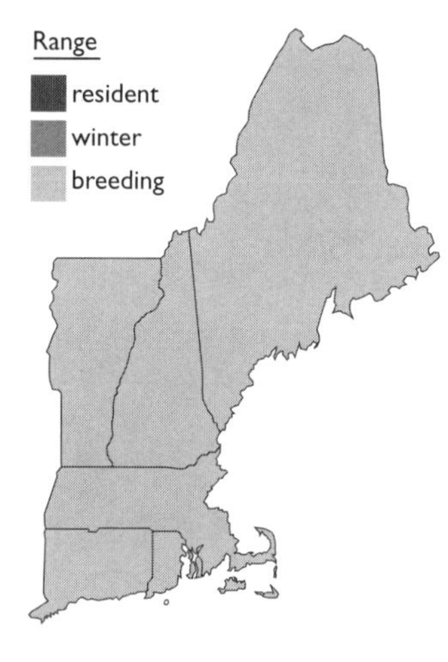

Distribution in New England: Throughout the region, except not reported from Block Island, R.I.

Status in New England: Common breeder in Maine, New Hampshire, Vermont, and Rhode Island; fairly common in Massachusetts (Veit and Petersen 1993:304), and uncommon in mature woodlands in Connecticut (Zeranski and Baptist 1990:175).

Habitat: Prefers hardwood woodlots, but is commonly found in old orchards, wooded residential areas, wooded swamps, and dry pitch-pine barrens. Favors woodlots with mature trees, but uses second-growth areas with scattered large cavity trees. Seldom found in deep forest, it has benefited from forest fragmentation and suburbanization (Lanyon 1997).

Special Habitat Requirements: Natural tree cavities or abandoned woodpecker holes for nesting; deciduous forest edge.

Nesting: Egg dates: May 27 to June 13, Massachusetts (Veit and Petersen 1993:305). May 22 to July 11, New York (Bull 1974:338). Clutch size: 4 to 8, typically 5 or 6. Incubation period: 13 to 15 days. Nestling period: 12 to 13 days. Broods per year: 1. Age at sexual maturity: 1 yr. Nest height: 3 to 75 ft (0.6 to 22.9 m), typically 10 to 20 ft (3.0 to 6.1 m). Nest site: In a cavity in a live or dead tree; accepts nest boxes of many types (see Lanyon 1997). Uses natural cavities or abandoned woodpecker holes. The female builds the nest (Lanyon 1997).

Territory Size: 0.6 to 4.6 acres (0.2 to 1.9 ha) (average 3.1 acres, 1.3 ha) for 26 territories in Illinois (Fawver 1947 in Graber et al. 1974). 4 to 8 acres (1.6 to 3.2 ha) (Stewart and Robbins 1958). 7.2, 6.6, 5.6 acres (2.9, 2.7, 2.3 ha) in forest-field edge habitat in Kansas (Fitch 1958).

Sample Densities: 50 birds per 100 acres (40 ha) in suburban habitats in Illinois (Ridgeway 1915). 8 pairs per 100 acres (40 ha) in mixed oak forest in Maryland. 7 pairs per 100 acres (40 ha) in dense second-growth oak-maple in Maryland. 4 pairs per 100 acres (40 ha) in hedgerows and active and abandoned farmland in Maryland (Stewart and Robbins 1958:200). 19.8 territorial males in red maple-birch forests in New York (Samson 1979).

Food Habits: Flying insects, insect larvae, fruits. Takes prey on the wing, but also hovers to glean from crevices in bark of trees, cracks in fallen logs, leaf surfaces. Birds spend much time foraging in forest canopy or edge of canopy.

Comments: Great crested flycatchers commonly incorporate cast snakeskins or bits of cellophane in their nests. Canopy foraging is more prevalent with interior woodland nesters than with edge nesters. Stomach analyses of 265 birds revealed a diet of 94 percent animal and 6 percent vegetable matter (Beal 1912 in Bent 1942:115). Originally a bird of forest interiors, this flycatcher has broadened its habitat to include more open areas and forest edges (Bent 1942).

Eastern Kingbird

(*Tyrannus tyrannus*)

Range: Breeding: From sw. and n.-central British Columbia, s. Northwest Territories, and central Manitoba, to s. Quebec and New Brunswick, s. to ne. California, n. Utah, nw. and central New Mexico, the Gulf Coast, and Florida. Migrates through Central America, Cuba, and the Bahamas. Winter: South to Colombia, Venezuela, Guyana, Brazil, Ecuador, Peru, and Bolivia.

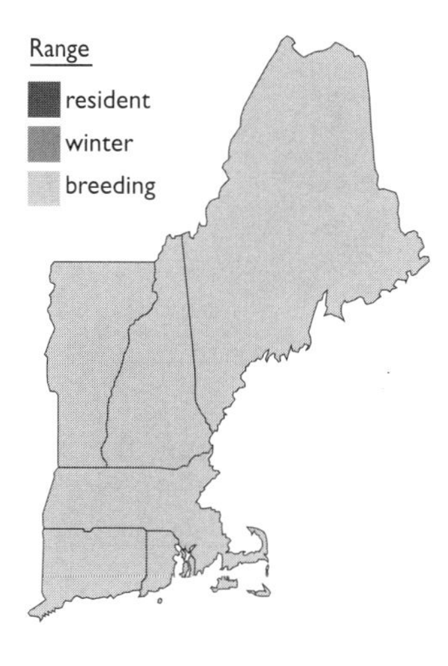

Distribution in New England: Throughout the region except at high elevation or in dense forests.

Status in New England: Common and widespread breeder, very common fall migrant.

Habitat: Frequents open areas with scattered trees or tall shrubs: orchards, forest edges or hedgerows along fields, roads, utility rights-of-way, golf courses, beaver impoundments, and wetlands, brushy areas along lakes, ponds, and streams. Occupies transitional vegetation in burned areas (Hamas 1983). Sometimes in open woodlands. Found at elevations up to 2,800 ft (850 m) in New Hampshire (Richards 1994:162 in Foss 1994).

Special Habitat Requirements: Open habitats with perches for aerial feeding.

Nesting: Egg dates: May 30 to July 4, Massachusetts (Veit and Petersen 1993:307). May 22 to July 16, New York (Bull 1974:364). Clutch size: 2 to 5, typically 3 to 4. Incubation period: 14 to 17 days (Murphy 1983b). Nestling period: 16 to 17 days (Murphy 1983b). Broods per year: 1. Age at sexual maturity: 1 yr. Nest height: 2 to 60 ft (0.6 to 18.3 m), typically 10 to 20 ft (3.0 to 6.1 m). Nest site: Often builds nest over water on a tree limb well away from the trunk or occasionally on top of a dead stub, or on a fence post if no trees are available. Frequently nests in hawthorn, apple, mulberry, or Norway spruce trees; also uses cottonwood, aspen, oak, willow, and maple (Mackenzie and Sealy 1981, Murphy 1996). Commonly places the nests 2.5 m (8 ft) above water (Blancher and Robertson 1985a).

Territory Size: 14 to 35 acres (5.7 to 14.2 ha) (4 pairs) (Odum and Kuenzler 1955). Territory size is probably more related to the open habitat used than to body size (Johnston 1971).

Sample Densities: Approximately 2 to 9 birds per 100 acres (40 ha) in suitable habitat in Illinois (Graber et al. 1974). 36 pairs per square mile (4 pairs/km^2) (maximum density) in North Dakota (Stewart and Kantrud 1972). 10 pairs per 100 acres (40 ha) in residential-orchard-lawn habitat in Maryland (Stewart and Robbins 1958:198).

Food Habits: Flying insects (staple), wild fruits; consumes over 200 kinds of insects and more than 40 kinds of fruits (Bent 1942). Birds seem to have favorite hawking perches. Commonly feeds over open land or water.

Comments: The kingbird is aggressive toward other birds, inhabits a wide array of habitats, and apparently has not changed in abundance in historic times. Smith (1966) noted kingbirds nesting in forested regions with internal clearings and extensive burned areas with standing trees.

Loggerhead Shrike

(*Lanius ludovicianus*)

Range: Breeding: From central Alberta, central Saskatchewan, s. Manitoba, Minnesota, central Wisconsin, central Michigan, and se. Ontario, s. to Mexico and the Gulf Coast. Very rare or absent from most of the Appalachians, Pennsylvania, New York, and New England. Winter: In the s. half of the United States, and s. to Oaxaca, Mexico.

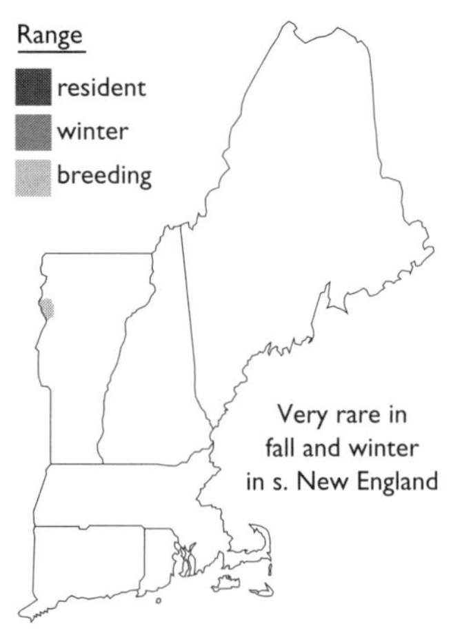

Distribution in New England: The Champlain Lowlands in Vermont (Kibbe 1985:260 in Laughlin and Kibbe 1985). Extirpated from Massachusetts and New Hampshire as a breeder.

Status in New England: Rare migrant and very rare winter resident. Virtually extirpated as a breeder (Bartgis 1992).

Habitat: Breeding: Inhabits open country with scattered shrubs or small trees such as abandoned pastures, fence rows, old orchards, agricultural fields, cemeteries, hedgerows, and mown roadsides.

Special Habitat Requirements: Open habitats with low vegetation and elevated perches.

Nesting: Egg dates: April 18 to June 28, New York (Bull 1974:453). Clutch size: 4 to 7, typically 5 to 6. Incubation period: 14 to 17 days. Nestling period: 16 to 20 days. Broods per year: 2. Age at sexual maturity: 1 yr. Nest height: 3 to 50 ft (0.9 to 15.2 m), typically not higher than 26 ft (8 m), in dense shrubs or a low tree

(Harrison 1975). Nest site: Builds in the dense foliage of a tree or shrub. Prefers to nest in thorny plants but is known to nest in oaks, pines, orchard trees, and grapevines.

Territory Size: 18.7 acres (7.6 ha) (Miller 1931). 23 acres (8.35 ha) in Florida (Yosef and Grubb 1993).

Sample Densities: 1.9 nests per mile (1.6 km) of hedge in Illinois (Graber et al. 1973).

Food Habits: Insects, small reptiles, amphibians, birds, and small mammals.

Comments: Loggerhead shrikes are one of the rarest breeding birds in Vermont. Habitat loss is partially responsible for the population decline of the loggerhead shrike; however, much research is needed, including survival on the wintering ground and use of pesticides (Yosef 1996, Brooks and Temple 1990, Gawlik and Bildstein 1990). Shrikes impale their prey in thorny trees or on barbed wire fences or hang the prey in the fork of a branch. Formerly nested in New Hampshire (C. Anderson pers. commun.) and Massachusetts (Veit and Petersen 1993:369).

Northern Shrike

(Lanius excubitor)

Range: Holarctic; breeds in North America in Alaska, the Yukon, sw. Mackenzie, and n. parts of Manitoba, Quebec, and Labrador. Winter: From s. Alaska and the s. half of Canada, s. to n. California, central Nevada, n. Arizona, n. New Mexico, Kansas, n. Missouri, central Illinois, Indiana, Ohio, Pennsylvania, and New Jersey.

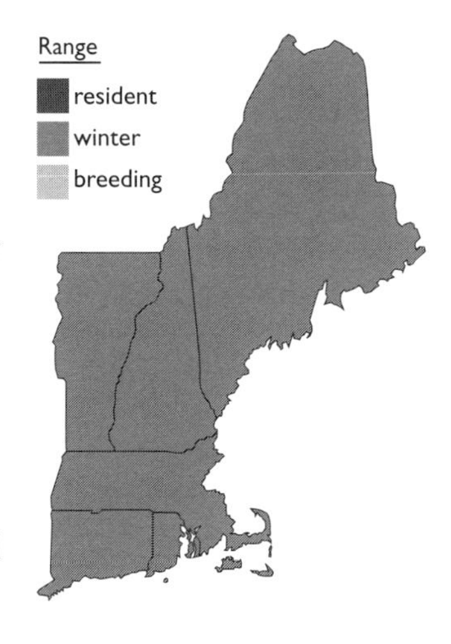

Distribution in New England: Irruptive winter resident throughout the region. Flight years are irregular, with "echo" flights following heavier flights, as in the winters of 1978–79 and 1979–80 (Veit and Petersen 1993:368).

Status in New England: Rare breeder (w. Vermont); rare migrant and winter resident (s. New England).

Habitat: Inhabits semiopen country with short grasses and scattered trees or shrubs for perches.

Special Habitat Requirements: Elevated perches (frequently uses fences and utility wires), short vegetation.

Food Habits: Rodents, especially mice, small birds (Cade 1967). Hunts widely over open fields with scattered perches—small trees or fences.

Comments: Northern shrikes are irruptive, and seen in New England in winter when boreal rodent and small bird populations are low—about every 4 yr (Pough 1949:133), but quite irregularly. Veit and Petersen (1993:368–369) summarize Brewster's (1906) account of a warden hired to shoot northern shrikes on Boston Common in order to protect house sparrows. The warden shot more than 50 shrikes in one winter (probably 1878–1879), an indication of the abundance of shrikes that winter and of the relative values ascribed to predators and introduced species at the time.

White-eyed Vireo

(Vireo griseus)

Range: Breeding: Southeastern Nebraska and central Iowa to s. Michigan, s. Ontario, and s. Massachusetts, s.

through e. Texas and Florida to San Luis Potosi, n. Hidalgo, n. Veracruz, and Tamaulipas. Winter: Southern Texas, s. to Honduras and e. across the Gulf Coast to Florida, n. to central coastal Northern Carolina. Also winters on Cuba and in the Bahamas.

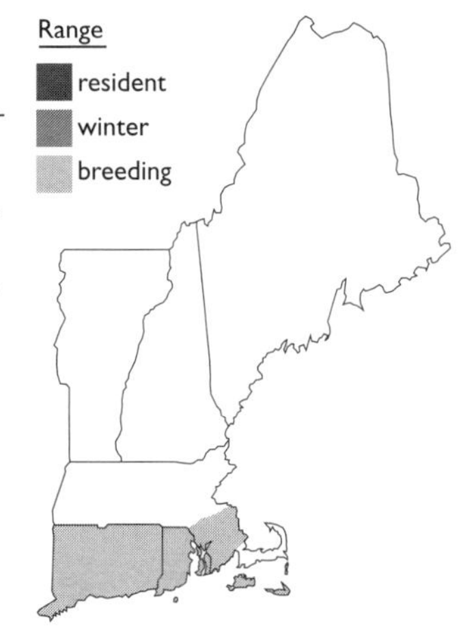

Distribution in New England: Breeds throughout Connecticut, although less common at higher elevations (Zeranski and Baptist 1990:208), Rhode Island, and in se. Massachusetts including Martha's Vineyard and Naushon Islands, but absent from outer Cape Cod (Veit and Petersen 1993:371), potentially breeding in the New Hampshire seacoast area (Foss 1994:228).

Status in New England: Locally common to common breeder in se. Massachusetts. Also occasional nesting pairs in e. Massachusetts n. to Plum Island.

Habitat: Breeding: Prefers dense secondary deciduous scrub, thick understory of bottomland forests, open woodlands or thickets near water, woodland margins, hedgerows, brambles, overgrown pastures, abandoned farmland. Deciduous thickets on low swampy or moist ground.

Special Habitat Requirements: Extensive low shrubby vegetation, brambles, saplings interspersed with taller trees on 10 to 20 percent of the area (Hopp et al. 1995).

Nesting: Egg dates: May 17 to July 17, New York (Bull 1974:454). Clutch size: 3 to 5, typically 4. Incubation period: 12 to 15 days (Bent 1950). Nestling period: 9 to 11 days (Hopp et al. 1995). Broods per year: 1 to 2. Age at sexual maturity: 1 yr. Nest height: 1 to 8 ft (0.3 to 2.4 m). Typically 2 to 6 ft (0.6 to 1.8 m). Nest site: A cone-shaped pendulous nest suspended from the fork of a low branch, usually well hidden by vegetation.

Territory Size: Approximately 2.5 to 3.5 acres (1.0 to 1.4 ha) per male (Stewart and Robbins 1958). Territories may be as small as 0.33 acre (0.13 ha) per male (Brewer 1955). 6.5 and 5.4 acres (2.6 and 2.2 ha) in stream and woodland edge thickets in Kansas (Fitch 1958). Average 3.3 acres (1.3 ha) in Florida (Bradley 1980). 0.35 acres (0.14 ha) in s. Illinois (Brewer 1955).

Sample Densities: West Virginia—10.0 territorial males per 100 acres (40 ha) in birch-oak forests (Samson 1979).

Food Habits: Beetles, moths, spiders, caterpillars, wild fruits. Feeds mainly in the inner canopy of low trees and scrub thickets. Seldom seen as it hunts through brush and vegetation.

Comments: In the late 1800s white-eyed vireos were fairly common and widespread breeders in s. New England, even (rarely) in New Hampshire, but between 1880 and 1930 they essentially disappeared except for local nesting along the Connecticut coast. Since about 1940, they have been gradually increasing in their former New England range (see Veit and Petersen 1993:371 and Zeranski and Baptist 1990:208 for historical review). Nearly half of white-eyed vireo nests are parasitized by brown-headed cowbirds (Hopp et al. 1995).

Yellow-throated Vireo

(*Vireo flavifrons*)

Range: Breeding: Southern Manitoba, Minnesota, s. Ontario, New Hampshire, and sw. Maine, s. to e. Texas, the Gulf Coast, and central Florida, and w. to the Dakotas, Nebraska, Kansas, Oklahoma, and w.-central Texas. Winter: Mainly in Mexico and Central and South America, from Colombia to French Guiana, including Trinidad and Tobago; a few winter in Florida.

Distribution in New England: Throughout Connecticut, Rhode Island, and Massachusetts except Cape Cod and the Islands. Absent from n. and e. Vermont and n. New Hampshire; locally in s. Maine.

Status in New England: Uncommon to local breeder, rare migrant.

Habitat: Edges of extensive tracts of mature, moist deciduous forests with partially open canopies; especially in bottomland forests and riparian woodlands (Rodewald and James 1996). Seldom in dense forests, rarely in conifers. Frequents roadsides, orchards, forest borders, and forested wetlands. Associated with north-facing slopes in s. parts of range. Estimated minimum forest area of 100 ha (250 acres) to sustain viable populations in central and e. Maryland (Robbins 1979). Most abundant in mature woods (Shugart et al. 1978).

Special Habitat Requirements: Mature, moist deciduous forests with partially open canopies.

Nesting: Egg dates: May 24 to June 17, Massachusetts (Veit and Petersen 1993:374). May 24 to June 18, New York (Bull 1974:457). Clutch size: 3 to 5, typically 4. Incubation period: 12 to 14 days. Nestling period: About 14 days. Broods per year: 1. Age at sexual maturity: 1 yr. Nest height: 1 to 20 m (3.1 to 66 ft) above the ground. Nest site: Suspended between a fork formed by a slender branch, usually near the trunk of a deciduous tree. Prefers maples, poplars, oaks, and elm trees for nesting in Ontario (Peck and James 1987).

Territory Size: About 10 acres (4.0 ha) (Robbins, unpublished data cited in Williamson 1971). 6.6 territorial males in birch-oak woods in West Virginia (Samson 1979).

Sample Densities: Maryland: 19 territorial males per 100 acres (40 ha) in virgin hardwood deciduous forest. 8 territorial males per 100 acres (40 ha) in second-growth river swamp. 7 territorial males per 100 acres (40 ha) in mixed oak forest. 3 territorial males per 100 acres (40 ha) in well-drained floodplain forest (Stewart and Robbins 1958:264).

Food Habits: Insects, especially adult and larval moths gleaned from foliage or bark. Also small amounts of fruit and, rarely, seeds (Rodewald and James 1996). Feeds in the upper part of the canopy (Williamson 1971).

Comments: Before 1900, yellow-throated vireos were widespread and common breeders in s. New England, especially in orchards (Merriam 1877) and shade trees (Sage et al. 1913). Pesticide spraying early in this century may have contributed to its decline (Zeranski and Baptist 1990:210). Yellow-throated vireos are well distributed in Vermont (Nichols 1985:266 in Laughlin and Kibbe 1985) and gradually increasing in New Hampshire (Robbins 1994a:242 in Foss 1994). Loss of streamside and lakeshore habitat and fragmentation of bottomland forests may threaten populations in breeding habitat (Robbins 1994a:242 in Foss 1994).

Blue-headed Vireo

(*Vireo solitarius*)

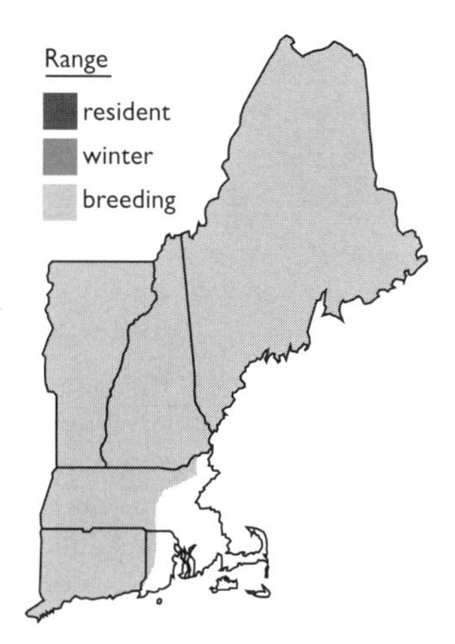

Range: Breeds from central British Columbia, e. through central Canada to n. Ontario and Newfoundland, sw. of and through the Rockies to s. California and w. Texas, s. through Mexico to Honduras, and e. of the Rockies to North Dakota, Illinois, and Massachusetts; in the Appalachian and Piedmont regions to e. Tennessee, Alabama, Georgia, South Carolina, North Carolina, Virginia, and Maryland. Winter: From s. California, central Texas, and n. portions of the Gulf States and North Carolina, s. to Costa Rica, and on Cuba and Jamaica. Occurs as a migrant in the Bahamas.

Distribution in New England: Throughout the region except for Cape Cod in Massachusetts and e. Rhode Island (Enser 1992).

Status in New England: Common (n. New England) to fairly common (w. Massachusetts) to uncommon (s. Connecticut) breeder.

Habitat: Breeds in extensive coniferous or mixed forests (preference for white pine, hemlock, or spruce) with a dense understory; tamarack or spruce swamps (James 1998). Occurs in closed-canopy forests and in sapling and pole-timber stands of mixed hardwoods in Grafton, Vt. (Thompson and Capen 1988). Associated with sawtimber stands of eastern hemlock in the White Mountain National Forest in New Hampshire and Maine (DeGraaf and Chadwick 1987). Also nests in mixed forests and deciduous forests that contain scattered conifers.

Nesting: Egg dates: May 14 to July 22, New York (Bull

1974:459). Clutch size: 3 to 5, typically 4. Incubation period: 13 to 15 days. Nestling period: 15 to 17 days. Broods per yr: 1 or 2. Age at sexual maturity: 1 yr. Nest height: 3.5 to 20 ft (1.1 to 6.1 m). Typically less than 10 feet (3.0 m). Nest site: Suspended from a forked horizontal branch, usually a conifer.

Territory Size: Averaged 12.5 acres (5 ha) and were often long and narrow, bordering along streams in the White Mountains (Sabo 1980).

Sample Densities: Average 29 birds per 100 acres (40 ha) in ponderosa pine forest in Colorado (Cruz 1975). 27 territorial males per 100 acres (40 ha) in virgin hemlock forest in Maryland. 17 territorial males per 100 acres (40 ha) in mature northern hardwood forest in Maryland (Stewart and Robbins 1958:265).

Food Habits: Insects, especially moths and caterpillars, small amounts of fruits. Chapin (1925 in Bent 1950:296) found the bulk of diet (306 stomachs) to be animal matter (96 percent) supplemented by small amounts of vegetable matter (4 percent). Feeds primarily in the lower and middle canopy.

Comments: The blue-headed vireo is the first vireo to arrive in spring and the last to leave in the fall. The blue-headed is the only vireo within its range that makes extensive use of coniferous forests, although it also occurs in deciduous habitats (James 1998). Clearing for settlement reduced its abundance; subsequent reversion to forest gradually increased it (Zeranski and Baptist 1990:209). The 1938 hurricane blew down so many pine stands that a reduction in solitary vireos was evident in the 1940s (Griscom and Snyder 1955).

Warbling Vireo

(*Vireo gilvus*)

Range: Breeds from se. Alaska, n. British Columbia, and s. Northwest Territories, se. to s. Ontario and New Brunswick, s. through Middle America and parts of n. and w. South America. Winters from central Mexico, s. through the breeding range. Resident populations in Middle and South America are considered to be a separate species, *Vireo leucophrys,* by some authors.

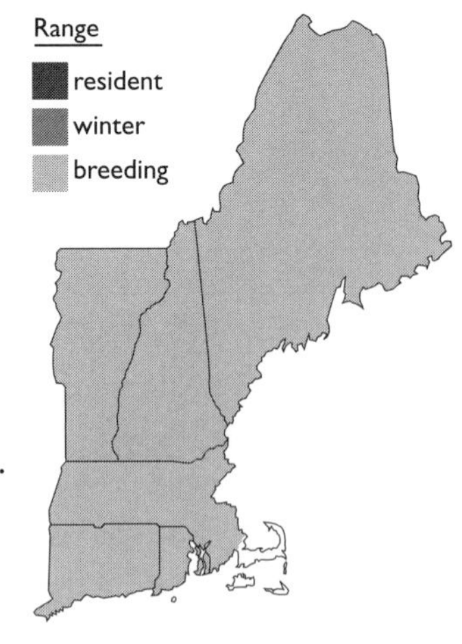

Distribution in New England: Breeds throughout the region, except absent from Cape Cod, Martha's Vineyard, and isolated nest records in 1974 from Nantucket, Mass., and Block Island, R.I.

Status in New England: Locally common and widespread breeder.

Habitat: Inhabits edges of mature riparian woodlands and bottomland forests, wooded farmsteads, shelterbelts, roadsides, shade trees in residential areas, cemeteries. Avoids extensive forests.

Special Habitat Requirements: Scattered mature deciduous trees or riparian forests.

Nesting: Egg dates: May 30 to 11 June, Massachusetts (Veit and Petersen 1993:375). May 16 to June 16, New York (Bull 1974:462). Clutch size: 3 to 5, typically 4. Incubation period: about 12 days. Nestling period: About 16 days. Broods per year: 1. Age at sexual maturity: 1 yr. Nest height: 20 to 90 ft (6.1 to 27.4 m). Nest site: In horizontal fork of a slender branch usually well away from trunk (Howes-Jones 1985). Typically protected by a canopy of leaves. Usually nests higher than other vireos.

Sample Densities: 10 territorial males per 100 acres (40 ha) in field with shrubs and stream-bordered trees in Maryland (Stewart and Robbins 1958:269). 29.6 territorial males per 100 acres (40 ha) in ash-basswood woodlands in New York (Samson 1979).

Food Habits: Insects, especially caterpillars, gleaned from leaf surfaces. Generally feeds in the middle and upper canopy of tall deciduous trees, largely on branch tips.

Comments: Warbling vireos were common breeding birds in orchards and residential shade trees in the 1800s. A long-term decline may have been due to pesticide spraying to control Dutch elm disease from about the turn of the century continuing into the 1960s (Zeranski and Baptist 1990:210). Veit and Petersen (1993:375) believe that habitat modification was responsible for the decline. Warbling vireos are gradually increasing in New England, occurring well into New Hampshire, even in the White Mountains at low elevations.

Philadelphia Vireo

(*Vireo philadelphicus*)

Range: Breeds from e.-central British Columbia to central Manitoba and sw. Newfoundland, s. to s. central Alberta, n.-central North Dakota, ne. Minnesota, s. Ontario, n. Vermont, n. New Hampshire, and Maine.

Distribution in New England: Breeding: Northern Vermont, n. New Hampshire, and central and n. Maine. Uncommon spring and fall migrant.

Status in New England: Absent from Connecticut, Massachusetts, and Rhode Island, rare to locally common breeder in Vermont, New Hampshire, and Maine.

Habitat: Deciduous, coniferous, or mixed forests, particularly early to mid-successional stages, woodland edges, clearings, and burned areas, neglected farmlands grown into small trees and tall shrubs interspersed with clearings, willow and alder thickets along streams.

Special Habitat Requirements: Early to mid-successional forest. Prefers yellow birch and ash as foraging substrates (Holmes and Robinson 1981).

Nesting: Egg dates: June 15 to July 15, Maine (Bent 1950:362). Clutch size: 3 to 5, typically 4. Incubation period: 11 to 14 days (Bent 1950, Peck and James 1987). Nestling period: 13 or 14 days (Moskoff and Robinson 1996). Broods per year: 1. Age at sexual maturity: 1 yr. Nest height: 10 to 40 ft (3.0 to 12.2 m). Nest site: Nest is hung in the fork of a slender horizontal twig of a deciduous tree or shrub.

Territory Size: 0.75 to 2.0 acres (0.3 to 0.8 ha) in Ontario (Rice 1978), 1.0 to 1.3 acres (0.4 to 0.5 ha) in mature deciduous forest in New Hampshire.

Food Habits: Insects, especially caterpillars; some wild fruits in autumn (less than 10 percent). Feeds in upper tree crowns, dense shrubs or forest regeneration.

Comments: Interspecifically territorial with the red-eyed vireo in some situations. May be adversely affected by selective removal of ash and yellow birch (Holmes and Robinson 1981).

Red-eyed Vireo

(*Vireo olivaceus*)

Range: Breeds from sw. British Columbia and s. Northwest Territories se. to central Ontario and the Maritime Provinces, s. to Oregon, Colorado, Oklahoma, Texas, the Gulf Coast, and Florida, s. through Middle America (except n. Mexico) and the n. two-thirds of South America. Winters in South America. Middle and South American breeding populations are recognized as separate species by some authors, *Vireo flavoviridis* and *Vireo chivi*, respectively.

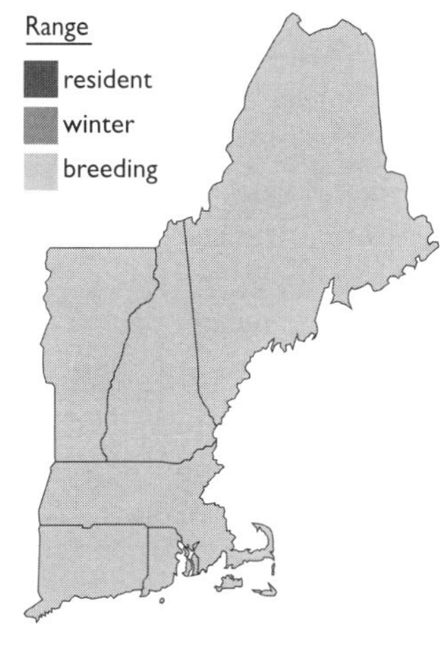

Distribution in New England: Widespread throughout the region.

Status in New England: Common breeder and migrant.

Habitat: Inhabits both upland and river-bottom forests of open deciduous and mixed forests with moderate to dense understories of saplings. Also in wooded clearings, borders of burns, orchards, and residential areas where shade trees provide a continuous canopy. Seldom found where conifers make up 75% or more of the basal area.

Special Habitat Requirements: Deciduous trees. A fairly continuous canopy rather than presence of an understory may be the chief habitat requirement (Lawrence 1953).

Nesting: Egg dates: May 25 to July 20, Massachusetts (Veit and Petersen 1993:377). May 13 to July 7, New York (Bull 1974:459). Clutch size: 2 to 5, typically 4. Incubation period: 12 to 14 days. Nestling period: 10 to 12 days. Broods per year: 1 or 2. Age at sexual maturity: 1 yr. Nest height: 2 to 60 ft (0.6 to 18.3 m). Typically 5 to 10 ft (1.5 to 3.0 m). DeGraaf et al. (1975) found red-eyed vireos (n = 20) nesting at an average height of 17 ft (5.2 m) in a nest site study in Massachusetts. Nest site: Suspended in the fork of a horizontal twig, often in a sapling, usually in a peripheral area of canopy.

Territory Size: 45 territories in Michigan averaged 1.7 acres (0.7 ha) per pair (Harrison 1975:172). 5 territories in mixed woods in Ontario ranged from 0.7 to 2.4 acres (0.3 to 1.2 ha) (average 1.4 acres, 0.6 ha) (Lawrence 1953).

Sample Densities: Maryland—60 territorial males per 100 acres (40 ha) in mature northern hardwood forest. 100 territorial males per 100 acres (40 ha) in virgin hardwood deciduous forest. 52 territorial males per 100 acres (40 ha) in dense second-growth forest. 34 territorial males per 100 acres (40 ha) in pine-oak forest. 10 territorial males per 100 acres (40 ha) in open slash area (Stewart and Robbins 1958:266).

Food Habits: Insects (more than 85 percent of diet), mainly caterpillars, moths, beetles, bugs, ants, gleaned from leaf surfaces, especially undersides. Glides rather than hops from branch to branch. Feeds in the uppermost branches of trees; most feeding occurs in the periphery of the middle and upper canopy.

Comments: Considered the most abundant bird in the e. deciduous forest (Robbins et al. 1966) and the most abundant vireo during migration.

Gray Jay

(Perisoreus canadensis)

Range: Breeding: From w. and central Alaska, n. Mackenzie, and sw. Keewatin, across to n. Quebec and n. Labrador, s. to n. California, central Idaho, e.-central Arizona, Black Hills of South Dakota, central Saskatchewan, n. Minnesota, s. Ontario, and n. New England. Winter: Generally throughout the breeding range.

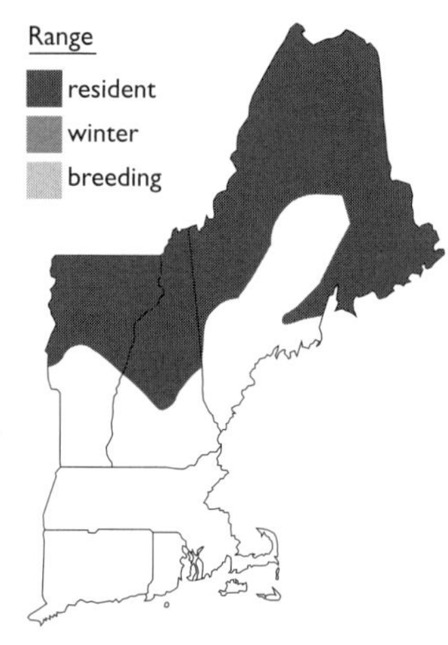

Distribution in New England: The Northeast Highlands and a section of the north central region in Vermont (Oatman 1985:202 in Laughlin and Kibbe 1985), n. New Hampshire, and n. and w. Maine (Adamus 1987).

Status in New England: Rare (Vermont) or locally common (New Hampshire and Maine) resident and breeder; very rare winter visitor to Massachusetts.

Habitat: Breeding: Inhabits northern coniferous (especially dense spruce-fir, black spruce, and pine forests, cedar bogs, and tamarack swamps) and nearby mixed forests. Rarely uses extensive deciduous forests. Winter: generally same as the breeding habitat.

Special Habitat Requirements: Coniferous forests.

Nesting: Egg dates: March to April (Goodwin 1976:250). Clutch size: 2 to 5, typically 3 or 4 (Strickland and Ouellet 1993). Incubation period: 16 to 18 days. Nestling period: About 15 days. Broods per year: 1. Age at sexual maturity: 1 yr. Nest height: 5 to 30 ft (1.5 to 9.1 m), typically 5 to 12 ft (1.5 to 3.7 m). Nest site: In solitary tree or clump of trees, usually conifers. Nest is often placed in crown of low tree or lower near trunk or branch tips and is usually well hidden. Nests in late winter in cold, apparently foodless, conditions, with eggs incubated at temperatures as low as −30°C (Strickland and Ouellet 1993).

Territory Size: Average 102.5 acres (41 ha) in the Yukon (Shank 1986).

Food Habits: Insects, carrion, berries, seeds, fledgling birds, small mammals, fungi, and lichens (Ouellet 1970).

Comments: Gray jays inhabit remote conifer forests, but are bold when encountered there, readily coming to camps to take bits of food. It was called "camp robber," "meat bird," and "whiskey jack," among other names, by woodsmen because it would eat any food left about camp. Gray jays are able to occupy permanent territories in climatically hostile regions by caching food for future use throughout the winter; gray jays have the largest salivary glands of any passerine bird; their sticky saliva binds food in a bolus for storage in conifer foliage (Strickland and Ouellet 1993).

Blue Jay

(*Cyanocitta cristata*)

Range: Resident from extreme e.-central British Columbia and central and se. Alberta to s. Quebec and Newfoundland, s. to central and se. Texas, the Gulf Coast, and s. Florida, and w. to e. Montana and e.-central New Mexico. Northern populations are partly migratory to the s. parts of the breeding range.

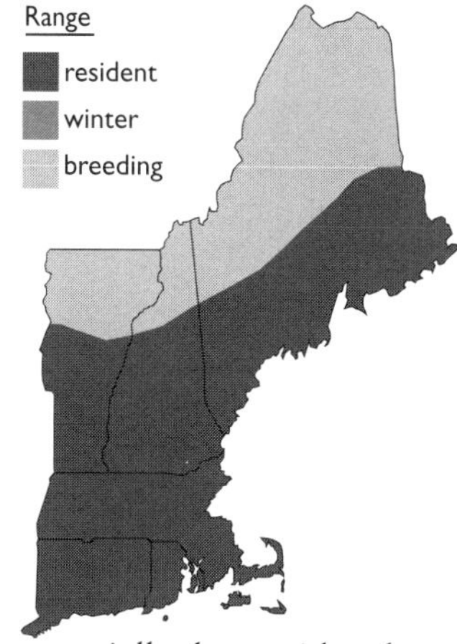

Distribution in New England: Throughout the region.

Status in New England: Common and widespread breeder; common to locally common winter resident.

Habitat: Inhabits deciduous and mixed woodlands, especially those with oak, beech, and hickory, but also occurs in coniferous forests. Also frequents wooded islands, farms, cities, suburbs, parks, roadside, and gardens—almost anywhere trees are found in grassy areas. Some n. birds move to the more s. parts of the breeding range (Goodwin 1976:262).

Nesting: Egg dates: April 28 to June 18, Massachusetts (Veit and Petersen 1993:322). April 28 to June 17, New York (Bull 1974:393). Clutch size: 3 to 6, typically 4 or 5. Incubation period: 17 to 18 days. Nestling period: 17 to 21 days. Broods per year: 1 north, 2 south. Age at sexual maturity: 2 yr (occasionally 1 yr). Nest height: 5 to 50 ft (1.5 to 15.2 m), typically 10 to 25 ft (3.0 to 7.6 m). Nest site: Prefers to nest in conifer thickets in mixed woodlands. Also builds in deciduous trees, shrubs, and shrubs overrun with vines. Nest may be close to trunk of tree or well out on a horizontal limb.

Territory Size: Territorial boundaries are not well defined (Goodwin 1976:267).

Sample Densities: 5 birds per 100 acres (40 ha) in well-defined floodplain forest in Maryland. 4 birds per 100 acres (40 ha) in mixed-oak forest in Maryland (Stewart and Robbins 1958). 5.0 territorial males per 100 acres (40 ha) in ash-maple woodlands in New York (Samson 1979).

Food Habits: Seeds, fruits, mast, occasionally takes insects, nestlings, young mice. Beal (1897 in Bent 1946:39) found that 76 percent of the annual diet (292 stomachs taken throughout the year) was vegetable matter and 24 percent was animal matter. Birds cache food in various places, possibly for winter use. Acorns are a staple food item throughout the year.

Comments: In the 1800s blue jays were primarily birds of rural areas, and only invaded cities and suburbs in the 1920s (Bagg and Eliot 1937).

American Crow

(*Corvus brachyrhynchos*)

Range: Breeds from n.-central British Columbia and sw. Mackenzie to central Quebec and s. Newfoundland, s. to Baja California, central Arizona, s. New Mexico, central and se. Texas, the Gulf Coast, and s. Florida. Northern populations are migratory, but winter within North America.

Distribution in New England: Throughout the region.

Status in New England: Common resident.

Habitat: Semiopen country of fields and woodlands; inhabits open deciduous, coniferous, and mixed forests, wooded river bottoms, orchards, woodlands adjacent to agricultural and other open land, parks, and suburban neighborhoods. In winter, individuals may move to lower elevations or to the coast.

Special Habitat Requirements: Open areas for foraging.

Nesting: Egg dates: May 5 to June 13, Massachusetts (Veit and Petersen 1993:323). March 30 to June 14, New York (Bull 1974:397). Clutch size: 3 to 8, typically 4 to 6. Incubation period: 18 days. Nestling period: About 25 days. Broods per year: 1 north, often 2 in south (Harrison 1975:139). Age at sexual maturity: 2 yr. Nest height: 10 to 70 ft (7.6 m). Nest site: Builds a large platform nest of sticks, usually on a horizontal branch or in a crotch of a tree near the trunk. Prefers to nest in conifers or oaks when available, but will nest on the ground, shrubs, or even on telephone pole crossbars.

Territory Size: Fitch (1958) found crows nonterritorial in Kansas and highly social in many activities. Average territory was 104 acres (42 ha), range 23 to 257 acres (9.4 to 104 ha) on Cape Cod (Chamberlain-Auger et al. 1990). Permanent residents defend a territory throughout the year (Chamberlain-Auger et al. 1990).

Sample Densities: 0.6 pair per 100 acres (40 ha) in mixed woodland and farmland habitat in Maryland (Stewart and Robbins 1958). 4 pairs nested within a distance of 100 yd (91.4 m) in Kansas (Fitch 1958). 8 pairs per square mile (3 pairs/km^2) in favorable habitat in North Dakota (Stewart and Kantrud 1972). 4.0 territorial males per 100 acres (40 ha) in second-growth hardwood forests in Connecticut (Samson 1979).

Food Habits: Omnivorous: Three-fourths of the diet is vegetable foods, including cultivated grains, seeds, wild and cultivated fruits, and nuts. Also eats small mammals, carrion, insects, eggs, young birds, millipedes, spiders, small crustaceans, frogs, and reptiles.

Comments: Will roost communally on cold winter nights, but returns to defend its territory at dawn (Chamberlain-Auger et al. 1990).

Fish Crow

(*Corvus ossifragus*)

Range: Breeds from New York and se. New Hampshire, s. along the Atlantic Coast to s. Florida, and w. to s. Texas; inland along major river systems to s. Illinois and e.-central Oklahoma.

Distribution in New England: Resident in se. New Hampshire along the coast, e. Massachusetts including Martha's Vineyard and Nantucket, e. Rhode Island,

and Connecticut along the coast and the Connecticut River Valley.

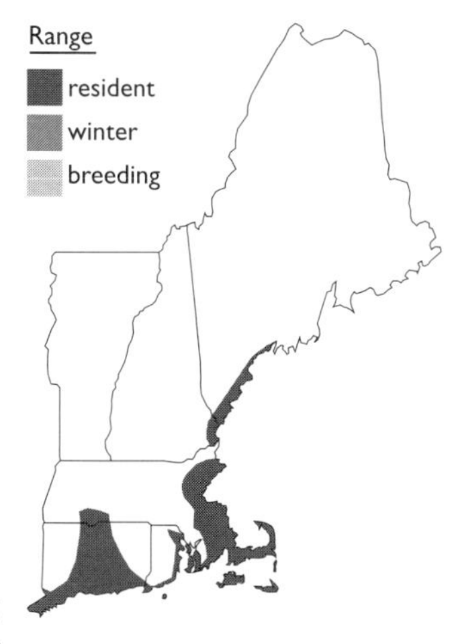

Status in New England: A rare resident in New Hampshire (Nevers 1994:184 in Foss 1994), locally common to fairly common along the coast and inland along the Connecticut River (Zeranski and Baptist 1990:185), uncommon to locally common along the coast and major river valleys in Massachusetts (Veit and Petersen 1993:324), and very uncommon to locally common in Rhode Island (Enser 1992:114).

Habitat: Inhabits low coastal areas, especially wooded marine shorelines, coastal marshes and beaches, estuaries, and forests near rivers and lakes. Occasionally in pine forests, orchards, old fields, and abandoned farmlands overgrown with pines and natural grasses.

Nesting: Egg dates: May 5 to June 6, Connecticut (Veit and Petersen 1993:325). March 20 to June 5, New York (Bull 1974:399). Clutch size: 3 to 5. Incubation period: 16 to 18. Nestling period: About 21 days. Broods per year: 1. Age at sexual maturity: 2 yr. Nest height: 20 to 80 ft (6.1 to 24.4 m), typically 50 ft (15.2 m). Nest site: Usually in small colonies in deciduous or coniferous trees.

Sample Densities: Colonies usually are made up of 2 to 4 pairs, each nesting in a separate tree.

Food Habits: Insects, grain, wild fruits, aquatic organisms, birds' eggs, carrion. Feeds on tidal flats, beaches, rookeries, banks of brackish rivers.

Comments: Fish crows have increased in southern New England since the 1950s. Also breeds in fertile farmland well inland from coast (100+ miles, 160+ km) in Pennsylvania and Maryland. Fish crows nested in Springfield, Mass., in 1986 (Veit and Petersen 1993:325). Fish crows often feed and roost amid flocks of common crows.

Common Raven

(*Corvus corax*)

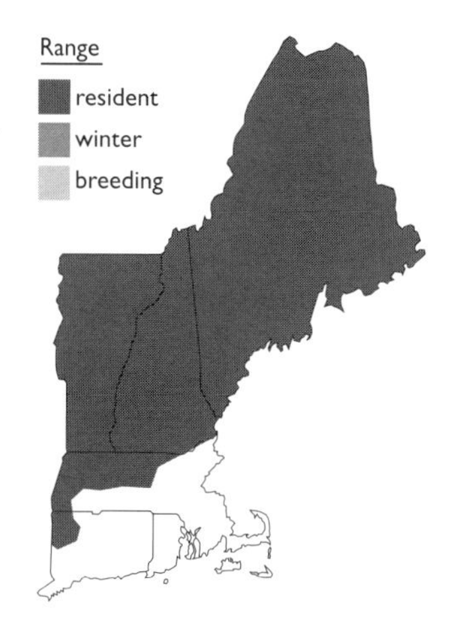

Range: Holarctic; in North America, breeds from Alaska and n. Canada, s. through the w. United States to Baja California and Mexico, and e. to the e. edge of the Rockies, w. Oklahoma, and central Texas; e. of the Rockies, s. to central Saskatchewan, n. Wisconsin, s. Ontario, Vermont, and Maine; also locally in the Appalachians to nw. Georgia. In the western part of its North American range, the raven is widespread, but in e. North America it is a bird of boreal forests and high mountains.

Distribution in New England: Breeds throughout Maine, New Hampshire, and Vermont, in n. and w. Massachusetts, and possibly nw. Connecticut (see Zeranski and Baptist 1990:186). Winter—uncommon (Massachusetts) to rare (Connecticut) winter visitor.

Status in New England: Locally uncommon breeder; reinvading its historic range in s. New England.

Habitat: Most common in open woodlands, clearings, open montane forests, and coastal regions. Generally avoids extensive, dense forests. Inhabits rocky seacoasts, steep canyons, boreal forests, foothills, mountains, tundra, and wooded marine islands.

Special Habitat Requirements: Cliffs or tall trees for nesting.

Nesting: Egg dates: March 26 to April 20, Massachusetts. March 24 to April 29, Maine (Bent 1946:214). Clutch size: 3 to 7, typically 4 to 6. Incubation period: 18 to 20 days. Nestling period: 35 to 42 days. Broods per year:

1. Age at sexual maturity: Probably 3 yr or more (Hooper and Dachelet 1976). Nest site: Usually high up in a tall tree or on a cliff ledge that is sheltered overhead and undercut or nearly vertical below. Nests may be used for successive years. Nests may be as close to 0.5 miles (0.8 km) to human dwellings (Hooper 1977).

Home Range: 2.6 to 4.2 square miles (6.7 to 10.9 km^2) (observed areas) in Wyoming (Craighead and Craighead 1969). Hooper (1977) observed ravens flying more than 1.2 miles (2 km) from nest sites.

Sample Densities: Nests are often spaced several miles apart (Harrison 1975). Hooper et al. (1975) found ravens nesting as close as 2.2 km (1.4 miles) and with an average distance of 4.3 km (1.7 miles) in Virginia.

Food Habits: Ravens are omnivorous, taking small mammals, carrion, birds, insects, and plant material (Harlow et al. 1975). Commonly feeds in seabird colonies (coast), garbage heaps; highways (road kills).

Comments: Forbush (1927) cites Wood (1634) and Josselyn (1674) stating that ravens were numerous on Cape Cod in early settlement times; since the 1960s ravens began to increase in Maine and n. New Hampshire and have been expanding their range southward; see Zeranski and Baptist (1990:186) and Veit and Petersen (1993:326) for a review of the historical trend. Harlow et al. (1975) found that medium to large mammals (apparently in the form of carrion from road kills or natural mortality) were the predominant food items taken in winter and spring in Virginia.

Horned Lark

(Eremophila alpestris)

Range: Breeding: In North America from w. and n. Alaska, the arctic coast of n. Canada, Prince Patrick, Devon, and Baffin Islands, and n. Labrador, s. to Oaxaca, sw. Louisiana, central Missouri, n. Alabama, and North Carolina; in South America, in the e. Andes of Colombia. Winter: Southern Canada, s. throughout the breeding range, s. to Baja California and Mexico except the Yucatan Peninsula, and locally or irregularly to the Gulf Coast.

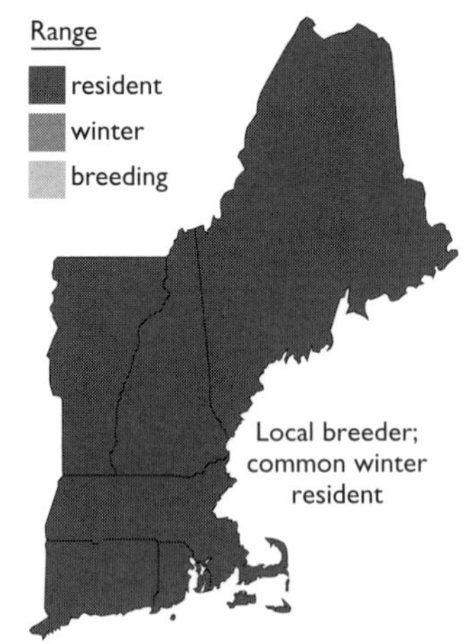

Distribution in New England: Locally throughout the region where open habitat with sparse vegetation occurs.

Status in New England: Locally uncommon breeder, generally declining with loss of agricultural habitat. Locally common winter resident.

Habitat: Breeding: Inhabits a wide variety of open treeless habitats, from coastal dunes and alpine tundra to prairies and deserts. Prefers sparse vegetation such as grazed fields, golf courses, airports, dunes, athletic fields, barnyards, and cemeteries. Habitats not associated with specific vegetation types (With and Webb 1993). More abundant in areas with barren ground present (Beason 1995). Avoids wooded areas. Found at elevations from sea level to >5,000 feet (Behle 1942). Winter: Similar to breeding habitats. Seeks bare cropland and exposed ground with short vegetation. Often concentrates along roadsides when ground is covered by snow (Beason 1995).

Special Habitat Requirements: Sparsely vegetated open areas.

Nesting: Egg dates: March 16 to July, Massachusetts (Veit and Petersen 1993:310). February 28 to July 31, New York (Bull 1974:381). Clutch size: 2 to 5, typically 4. Incubation period: 11 days. Nestling period: 8 to 10 days. Broods per year: 2. Age at sexual maturity: 1 yr. Nest site: A shallow depression on the ground (Beason 1974), nest is often paved with small pebbles along a portion of the rim. Nest placed in the open or next to a tuft of grass, a rock, or clump of sod.

Territory Size: Approximately 0.8 ha (2 acres) on burned-over grassland in Evanston, Ill. (1 pair); 5.0 ha (approximately) in garden and grainfield for 1 pair in Ithaca, N.Y. (Pickwell 1931:134). About 12 acres (4.9 ha) in field in Kansas (Fitch 1958).

Sample Densities: 6 pairs occupied 72 acres (29.1 ha) of field in Kansas (Fitch 1958). 160 pairs per square mile (6 pairs/km^2) (maximum density) in favorable habitat in North Dakota (Stewart and Kantrud 1972).

Food Habits: Summer—mainly weed and grass seeds, feeds insects to young (Beason 1995). Winter—mainly grass and weed seeds and waste grains. Feeds on snow-free barrens with abundant weed seeds in winter.

Comments: The horned lark is a prairie bird that extended its range eastward into New England in the 1880s as the land was cleared for agriculture. It is an early nesting species; early clutches are occasionally destroyed by snowstorms. McAtee (1905 in Pickwell 1931:31) found that the vegetable portion of the diet taken in a year accounted for 79.4 percent of total. Horned larks are gregarious in winter, and are commonly seen with snow buntings and Lapland longspurs on coastal dunes and beaches in winter.

Purple Martin

(Progne subis)

Range: Breeds from sw. British Columbia, s. to Baja California, from ne. and e.-central British Columbia and central Alberta to s. Ontario and New Brunswick and central Nova Scotia, s. through the Mexican highlands to Michoacán, and throughout the Caribbean, the Gulf Coast, and s. Florida. Local in the Rocky Mountains but avoids most other mountainous areas. Migrates through Central America and the Caribbean. Winters in South America from Colombia, Venezuela, and the Guianas s., e. of the Andes, to n. Bolivia and se. Brazil.

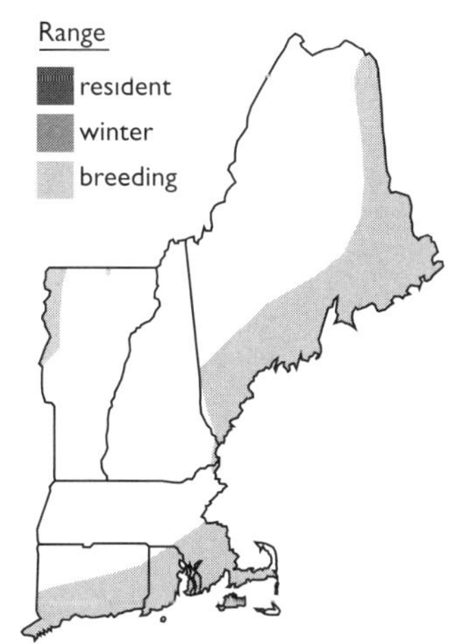

Distribution in New England: Breeds locally in s. and ne. Maine, s. New Hampshire, n. Champlain Lowlands and Lake Memphremagog in n. Vermont, ne. and se. Massachusetts, including Martha's Vineyard, Mass., s.-central Rhode Island, and s. Connecticut.

Status in New England: Uncommon to locally common breeder; colonies widely separated.

Habitat: Inhabits open country near water—fields, parks, suburban yards, farmland, meadows, coastal and freshwater marsh edges, open shores of lakes and ponds.

Special Habitat Requirements: Open habitat near water for foraging, and large, multiroomed nest boxes.

Nesting: Egg dates: May 21 to July 13, New York (Bull 1974:390). Clutch size: 3 to 8, typically 4 or 5. Incubation period: 15 to 18 days. Nestling period: 26 to 31 days. Broods per year: 1. Age at sexual maturity: 1 yr. Nest height: 15 to 20 ft (4.6 to 6.1 m). Nest site: Originally nested in abandoned woodpecker holes; today nests almost exclusively in nest boxes in eastern North America, preferably near water. Birds favor large, multiroomed birdhouses set on poles 15 to 20 ft (4.6 to 6.1 m) high. Easily driven out of nest sites by starlings and house sparrows.

Territory Size: Restricted to the nest cavity (Allen and Nice 1952) or the martin house compartment and the adjoining porch (Brown 1979).

Sample Densities: Colonies may consist of as many as 200 pairs (Bull and Farrand 1977).

Food Habits: Flying insects taken on the wing from the air or water surface.

Comments: Purple martins were probably uncommon nesting birds before New England was cleared for settlement, after which time they were common, even locally abundant in s. New England. The introduction of house sparrows and starlings in the late 1800s resulted in striking declines of purple martins as they were deprived of nest boxes and cavities (Zeranski and Baptist 1990:178–179). Martins in the East had inhabited ancestral tree cavities in forest edge and riparian areas until the late 1800s (Brown 1997), but since that time have been apparently totally dependent upon martin houses erected by people (Veit and Petersen 1993:312). In late summer, martins gather in large flocks, often roosting in urban areas (Robbins et al. 1966) prior to their southward migration. Dependence on insects for food makes them vulnerable to starvation during long periods of cold, wet weather.

Tree Swallow

(Tachycineta bicolor)

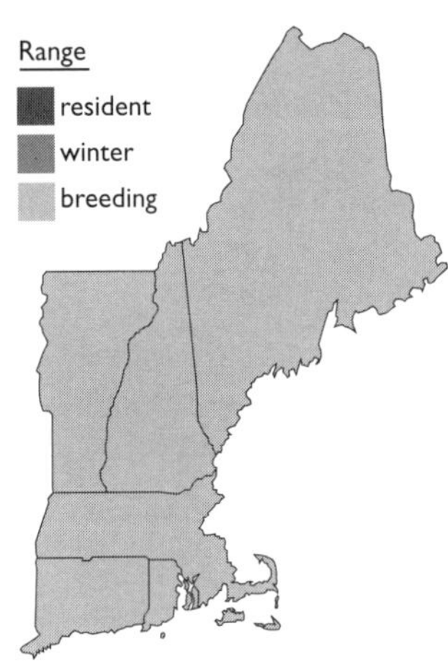

Range: Breeds from w. and central Alaska and central Yukon to n. Quebec and central Labrador, s. along the Pacific Coast to s. California and s.-central New Mexico; generally sporadic or irregular as a breeder e. of the Rocky Mountain States and s. of the upper Mississippi and Ohio Valleys, and along the Atlantic Coast s. of Massachusetts. Winters from s. California, sw. Arizona, Texas, the Gulf Coast, and the Atlantic Coast from New York s. to Nicaragua and in the Greater Antilles; casual or irregular to Costa Rica, Panama, Colombia, and Guyana.

Distribution in New England: Breeds throughout the region.

Status in New England: Common breeder, abundant fall migrant, especially along the coast. Rare in early winter along the coast.

Habitat: Open areas, usually near water: fields, pastures, beaver ponds, wooded swamps and marshes with dead standing trees containing old woodpecker holes. Also uses recent clear-cuts if snags with old woodpecker holes are present (DeGraaf 1991).

Special Habitat Requirements: Tree cavities in the open for nesting.

Nesting: Egg dates: April 19 to June 15, Massachusetts (Veit and Petersen 1993:314). May 5 to June 29, New York (Bull 1974:382). Clutch size: 2 to 8, typically 4 to 7 (Paynter 1954, Stutchberry and Robertson 1988). Incubation period: 13 to 16 days. Nestling period: 15 to 25 days (varies with food supply), typically 18 to 22 days (Paynter 1954, Burt 1977). Broods per year: 1, rarely 2 (Chapman 1955, Hussell 1983). Age at sexual maturity: 1 yr. Nest height: 4 to 15 ft (1.2 to 4.6 m). Nest site: Natural cavity or old woodpecker hole in a trunk or dead limb of dead or living tree, holes in buildings, nest boxes. Is usually a solitary nester but may nest in small colonies where suitable cavities abound and flying insects are abundant. Prefers to nest near a body of water.

Territory Size: 10 to 15 m around the nest site (Robertson et al. 1992).

Sample Densities: 40 occupied nest boxes in 28 acres (11.3 ha) of modified woodland in Illinois (Beecher 1942). Birds will nest within 7 ft (2.1 m) of each other in the presence of abundant food (Scott et al. 1977). Densities of up to 150 pairs per 0.3 ha (0.7 acres) are possible in nest boxes spaced no less than 2 m (6.6 ft) apart (Whittle 1926).

Food Habits: Flying insects (summer), berries, and seeds are taken to supplement the winter diet when insects are less abundant. Feeds over bodies of water and, in late summer, over fields.

Comments: The tree swallow's habit of eating bayberries apparently enables the tree swallow to return to the Northeast earlier in the spring, linger later into the fall, and winter farther north than other swallows. It commonly gathers along the coast in large flocks in autumn. Forest management in which dead snags are left standing is important to tree swallows and other birds that need existing cavities for nesting (Robertson et al. 1992).

Northern Rough-winged Swallow

(Stelgidopteryx serripennis)

Range: Breeding: From se. Alaska, central British Columbia, and s. Alberta to sw. Quebec and Maine, s. through Middle and South America to s. Chile and central Argentina. Winter: From Louisiana, Texas, and s. Florida, s. through the Greater Antilles, Middle America, and n. South America. Populations in Argentina, Urugay, and s. Brazil are austral migrants.

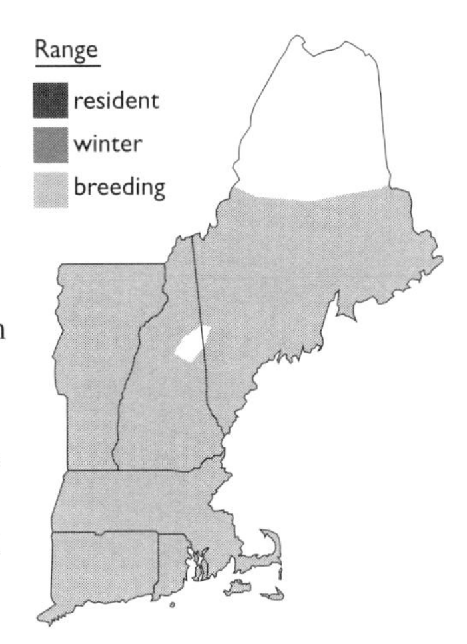

Distribution in New England: Breeds throughout the region, except absent from the White Mountains in New Hampshire (Elkins 1994:170 in Foss 1994), and n. Maine (Adamus 1988).

Status in New England: Uncommon breeder and migrant.

Habitat: Inhabits open country, wherever suitable nest sites exist within 0.6 miles (1 km) from water. Feeds over fresh water and saltwater, including marshes, scrub-shrub wetlands, coastal harbors, lakes, ponds, and reservoirs.

Special Habitat Requirements: Suitable nest sites within 0.6 mile (1 km) of water.

Nesting: Egg dates: May 7 to June 8, Massachusetts (Veit and Petersen 1993:316). May 19 to July 5, New York (Bull 1974:385). Clutch size: 4 to 8, typically 6 or 7. Incubation period: 16 days. Nestling period: 19 to 20 days. Broods per year: 1. Age at sexual maturity: 1 yr. Nest site: Commonly nests singly, or in small colonies of 2 to 25 pairs, frequently at the edge of a bank swallow colony (DeJong 1996). Suitable nest site habitat includes rocky gorges, shale banks, stony road cuts, railroad embankments, stream banks, gravel pits. Sometimes nests in culverts, drain pipes, crevices or holes in walls of buildings, bridges, and wharves. Often occupies abandoned bank swallow or kingfisher holes. Rarely nests in tree holes; will use artificial nest tubes placed in a cliff face or quarry bank (DeJong 1996).

Territory Size: Territory is limited to the immediate vicinity of the nest entrance (Lunk 1962:29).

Food Habits: Flying insects.

Comments: The northern rough-winged swallow has extended its range northward in historical times and was first reported in New England in 1851, and next in 1874, with the first nesting in Connecticut in 1876 (Zeranski and Baptist 1990:180). In 1895 it first nested in Massachusetts and in 1905 in Vermont (Norse 1985:194 in Laughlin and Kibbe 1985). The name "rough-winged" describes the serrations on the outer primary flight feather of the wings. Feeds over open water more than most swallow species.

Bank Swallow

(Riparia riparia)

Range: Holarctic; breeds in North America from w. and central Alaska and central Yukon to central Quebec and s. Labrador, s. to s. California, w. Nevada, s. New Mexico, s. Texas, n. Alabama, e. Virginia, and casually, nw. North Carolina and s.-central South Carolina. Migrates through Central America and the Caribbean. Winters in South America from Colombia, Venezuela, and the Guianas s., essentially e. of the Andes, to Peru, n. Argentina, and Paraguay.

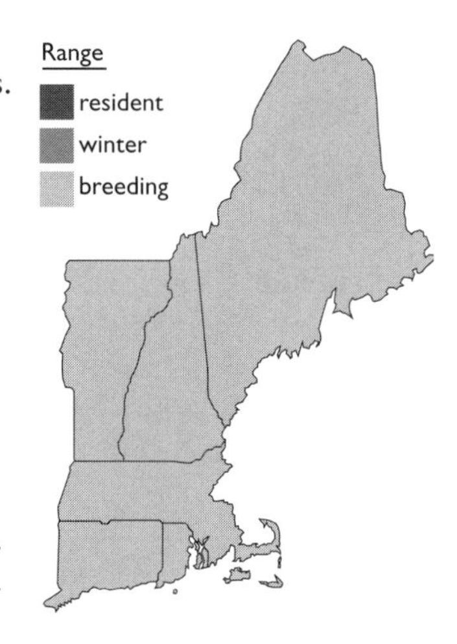

Distribution in New England: Nests throughout the region where suitable soft banks provide nest sites.

Status in New England: Locally common to abundant breeder at scattered riparian locations throughout the region.

Habitat: Inhabits grasslands, cultivated fields, or other open areas adjacent to water and suitable nest sites in river banks, ocean bluffs, gravel pits, road cuts, hardened sawdust piles, and other exposed bank surfaces in sand or clay.

Special Habitat Requirements: Open habitat near flowing water and steep but stabilized banks of erodable soils: sand, gravel or clay.

Nesting: Egg dates: May 28 to June 17, Massachusetts (Veit and Petersen 1993:316). May 15 to July 13 (second brood), New York (Bull 1974:383). Clutch size: 2 to 8, typically 5 (Hjertaas et al. 1988). Incubation period: 14 to 16 days. Nestling period: 18 to 22 days, to 30 days where food supply is limited. Broods per year: 1. Age at sexual maturity: 1 yr. Nest site: A burrow dug by both sexes usually near the top of a stable, grass-topped bank (Freer 1979). Depth varies from 9 in (22.9 cm) to 6 ft (1.6 m). Birds may restore existing burrows and form dense colonies where possible.

Territory Size: Territory is restricted to the area immediately surrounding the nest site.

Sample Densities: Minimum spacing of nest holes in a Wisconsin study was 4 in (10.2 cm). Most holes were 5 to 7 in (12.7 to 17.8 cm) apart (Petersen 1955).

Food Habits: Flying insects (nearly 100 percent of diet) taken from the air or by skimming the water surface. Also commonly feeds over grasslands, especially pastures.

Comments: The bank swallow is one of the few passerines with an almost cosmopolitan distribution; it is one of the most widely distributed swallows in the world. In the Old World, it is known as the sand martin (Garrison 1999). Colonial feeding may be an adaptation that allows for more effective discovery of insect swarms. Birds typically nest in dense colonies of 10 to more than 300 nests. Nesting is synchronized—more than 70 percent of the young leave the nest within a 6-day period (Emlen and Demong 1974).

Cliff Swallow

(Petrochelidon pyrrhonota)

Range: Breeds from w. and central Alaska and central Yukon to n. Ontario, s. Quebec, and New Brunswick, s. to Mexico, sw. Louisiana, s. to n. portion of the Gulf States and s. North Carolina; also in the Lake Okeechobee region of s. Florida. Migrates through Central America, the Caribbean, and n. South America. Winters in Brazil, Paraguay, Uruguay, and Argentina.

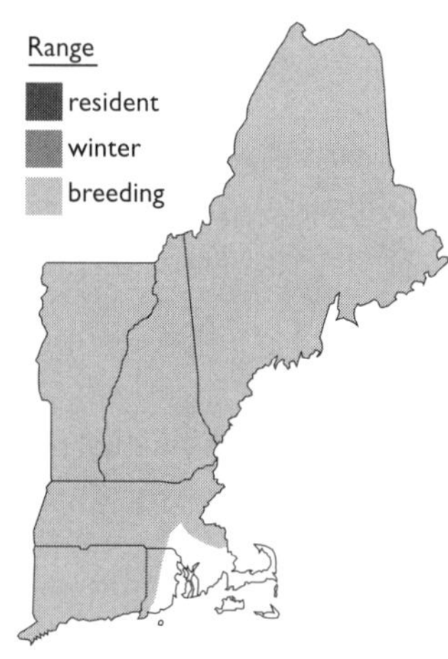

Distribution in New England: Breeds throughout Maine, New Hampshire, and Vermont, Plymouth County and w. Massachusetts, generally w. of the Connecticut River Valley in Connecticut, and in w. Rhode Island.

Status in New England: Common to locally common breeder in n. New England. Local and uncommon in Connecticut (Zeranski and Baptist 1990:181).

Habitat: Historically restricted to the vicinity of cliffs and banks; now occurs over open country, including grasslands, agricultural land, residential areas, bridges, dams, freeway overpasses, and other areas with mud supplies and potential nest sites. Apparently prefers to forage over grassy areas, plowed fields, and other open terrestrial areas (Brown and Brown 1995).

Special Habitat Requirements: A vertical substrate with an overhang for nesting, a mud supply for nest construction, fresh water with a smooth surface for drinking, and open foraging areas near the nest site (DeGraaf and Rappole 1995:353).

Nesting: Egg dates: May 25 to July 20, Massachusetts (Veit and Petersen 1993:318). May 9 to July 14, New York (Bull 1974:389). Clutch size: 3 to 6, typically 3 to 4 (Grant and Quay 1977). Incubation period: Average 13 to 15 days. Nestling period: About 24 days. Broods per year: 1. Age at sexual maturity: 1 yr. Nest site: Nests colonially under bridges or dams, eaves, and interior of barns and sheds. Solitary nesting occasionally occurs. Forms colonies of up to 3,500 active nests (Brown and Brown 1995).

Territory Size: Restricted to the distance the bird can reach with bill from rim of nest (Emlen 1952).

Sample Densities: More than 100 nests have been counted at a single barn (Bull 1974:389).

Food Habits: Flying insects make up nearly 100 percent of diet (Bent 1942:476). Often feed high in the sky (in excess of 100 ft) (30.5 m). Birds have been seen feeding up to 4 miles (6.4 km) from nest site (Emlen 1954).

Comments: Cliff and barn swallows may nest in the same barn, but competition is minimal because cliff swallows build nests on the outside or near the entrance and barn swallows nest deeper in the interior (Samuel 1971). Much of cliff swallow habitat has been usurped by house sparrows, and they are declining as agricultural land gives way to suburbs and forests.

Barn Swallow

(*Hirundo rustica*)

Range: Cosmopolitan; breeds in North America from s.-coastal and se. Alaska and s. Yukon across to central Manitoba, n. Ontario, and s. Quebec, s. to Mexico, the Gulf Coast, n.-central Florida, and s. North Carolina. Migrates through the Caribbean. Winters throughout Central and South America, casually n. to the sw. United States and throughout s. Florida.

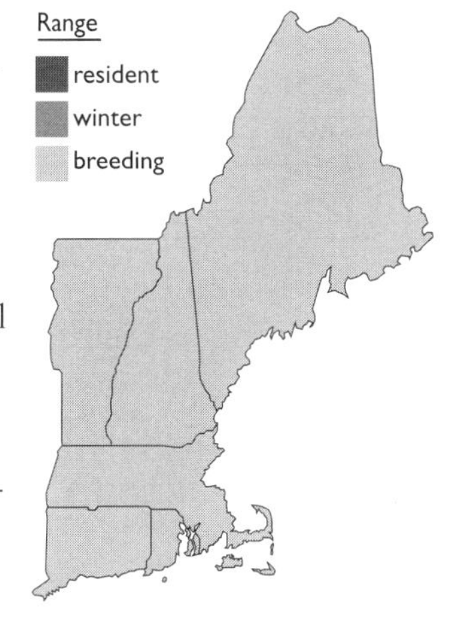

Distribution in New England: Breeds throughout the region.

Status in New England: Common to abundant breeder and migrant, especially in fall.

Habitat: Nests wherever suitable nest sites can be found (barns, under bridges, culverts, sheds), but favors those near farmlands, marshes, lakes, rural and suburban areas.

Special Habitat Requirements: Nest sites, especially buildings, near open habitats.

Nesting: Egg dates: May 18 to July 12, Massachusetts (Veit and Petersen 1993:320). May 11 to August 3, New York (Bull 1974:366). Clutch size: 4 to 6, typically 4 or 5. Incubation period: About 15 days. Nestling period: 16 to 23 days. Broods per year: 1 or 2 (at warmer latitudes).

Age at sexual maturity: 1 yr. Nest site: Nests inside sheds and barns (often in colonies—Snap 1976), under bridges, culverts. Formerly nested on cliffs, in caves, and in niches in rocks.

Territory Size: Probably restricted to the nest site.

Sample Densities: Usually 6 to 8 nests per site is maximum, but as many as 55 nests have been reported in a single barn (Harrison 1975:132) and 63 at a Lunenburg, Mass., barn (B. Blodget pers. commun.). 20 pairs per square mile (8 pairs/km^2) in favorable habitat in North Dakota (Stewart and Kantrud 1972). 11 pairs per 100 acres (40 ha) in mixed agricultural and residential habitats including buildings (Stewart and Robbins 1958:214).

Food Habits: Flying insects, occasionally takes fruits. Commonly feeds over ponds, lakes, rivers, and fields, seldom feeds more than 0.5 mile (0.8 km) from nest site (Samuel 1971).

Comments: Barn swallows have adapted almost exclusively to nesting in man-made structures (Davis 1937). The diet consists almost entirely of animal matter (Bent 1942:450). Nearly all the food is taken on the wing. Swallows in Illinois spent much time feeding over edge shrub areas. Feeding densities averaged 26 birds per 100 acres (40 ha) (Graber et al. 1972).

Black-capped Chickadee

(Poecile atricapillus)

Range: Breeds from w. and central Alaska, Saskatchewan, s. Quebec, and Newfoundland, s. to nw. California, ne. Nevada, central New Mexico, ne. Oklahoma, central Indiana, and n. New Jersey, and in the Appalachians at higher elevations (DeGraaf et al. 1991:313).

Distribution in New England: Throughout the region.

Status in New England: Common resident.

Habitat: Breeding: Deciduous, coniferous, or mixed woodlands (mixed preferred). Frequents dense woodlands to thickets, orchards, parks, and urban areas, wherever suitable nesting cavities exist or can be excavated.

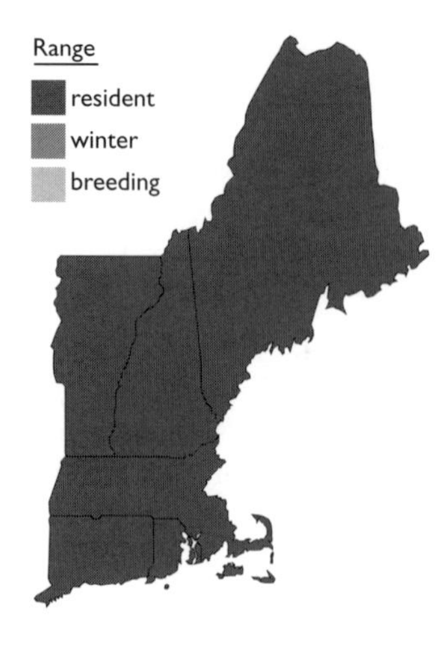

Special Habitat Requirements: Comparatively open sites near deep woods, and dead standing trees larger than 4 in (10.2 cm) diameter breast height for nesting and feeding.

Nesting: Egg dates: May 4 to July 12, Massachusetts (Veit and Petersen 1993:327). April 29 to July 15, New York (Bull 1974:401). Clutch size: 5 to 10, typically 6 to 8. Incubation period: 12 to 13 days. Nestling period: About 16 days (Smith 1993). Broods per year: 1 or 2. Age at sexual maturity: 1 yr. Nest height: 1 to 50 ft (0.3 to 15.2 m), typically 4 to 10 ft (1.2 to 3.0 m). Nest site: Usually excavates the nest hole in soft decayed wood of a dead tree or branch; will also use existing cavities of other birds or nest boxes. Both sexes share cavity excavation, but the female builds the nest within (Smith 1993). Prefers to nest in aspen, paper birch, yellow birch, willow, basswood, maple, and white ash. Favors open forest or forest edges adjacent to open areas for foraging. Associated with intermediate canopy cover (76.8 percent) and a dense understory in pole-sized hardwoods in Vermont (Thompson and Capen 1988).

Territory Size: Sizes ranged from 8.4 to 17.1 acres (3.4 to 6.9 ha) (average 13.2 acres, 5.3 ha) in different habitats (Odum 1941). 2.3 acres (0.9 ha) for 1 pair in Kansas (Fitch 1958).

Sample Densities: Average 1 pair per 22 acres (8.9 ha) in suitable habitat (Odum 1941). 24.6 territorial males per 100 acres (40 ha) in ash-basswood woodlands in New York (Samson 1979). Winter: 1 bird per 2.66 acres (1.1 ha) in bottomland woods in New York (Butts 1931).

Food Habits: About 50 percent animal (mostly insects and spiders) and 50 percent plant (Smith 1993). Insects, seeds, fruits. Chickadees glean from bark crevices, leaf, branch, and twig surfaces. Kluyver (1961) found that birds fed more often in pine groves with abundant caterpillars than in adjacent oak woods.

Comments: Black-capped chickadees in New England irregularly show massive southward fall migrations of mostly juvenile birds in response to lower food availability (low cone crops) in northern forests (Bagg

1969). Often nests in open woods or forest edges and feeds in deep woods (Odum 1941). Generally roosts in cavities in cold winter weather. The black-capped chickadee is the state bird of Maine and Massachusetts.

Boreal Chickadee

(*Poecile hudsonicus*)

Range: Largely resident in boreal forests of w. and central Alaska and central Yukon, e. to n. Ontario and Labrador, s. to extreme n.-central Washington, nw. Montana, n. Minnesota, n. Michigan, e. to ne. Vermont, n. New Hampshire, and e.-central and w.-central Maine.

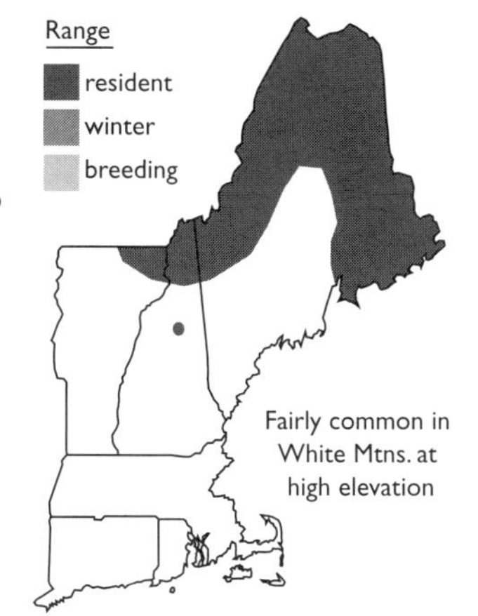

Distribution in New England: Northeastern Vermont (Oatman 1985:212 in Laughlin and Kibbe 1985), n. New Hampshire (Smith 1994:190 in Foss 1994), and e.-central and w.-central Maine (Ficken et al. 1996). Rare irregular winter visitor in Massachusetts (Veit and Petersen 1993:329), very rare in Connecticut (Zeranski and Baptist 1990:187).

Status in New England: Fairly common local resident in n. New England. Uncommon to rare winter visitor in s. New England.

Habitat: Northern coniferous forests, where it inhabits spruce, balsam, and dense pine forests, and white cedar and hemlock swamps, tamarack bogs, and occasionally birch and streamside willows. Common from 2,500 to 4,800 ft (760 to 1,460 m) in the White Mountains in New Hampshire (Smith 1994:190 in Foss 1994). Inhabits both young and mature conifer forests (Erskine 1977).

Special Habitat Requirements: Conifer forests with available cavities, or decaying trees.

Nesting: Egg dates: Late May, June, early July, New Hampshire (Smith 1994:190 in Foss 1994). June 11 to July 17 (late), New York (Bull 1974:402). Clutch size: 4 to 9, typically 6 or 7. Incubation period: 12 to 13 days (Harrison 1975). Nestling period: About 18 days. Broods per year: 1. Age at sexual maturity: 1 yr. Nest height: 1 to 10 ft (0.3 to 3.0 m). Nest site: Decaying stub or tree, preferably with firm exterior and soft interior. Bird may excavate several holes before choosing one for nest. Sometimes uses natural cavities or old woodpecker holes.

Territory Size: Larger than 5 ha (12.4 acres) in spruce-fir forest (McLaren 1975).

Food Habits: Arthropods and seeds of conifers and birches (Oatman 1985:212 in Laughlin and Kibbe 1985) gleaned from bark crevices, leaves, twigs, branches while spiraling head-downward down tree boles.

Comments: Nest site selection seems to be influenced more by the softness of the wood than by species of tree (McLaren 1975).

Tufted Titmouse

(*Baeolophus bicolor*)

Range: Largely resident from ne. Nebraska, central and e. Iowa, s. Wisconsin, n. Ohio, s. Ontario, central New York, s. Vermont, s. New Hampshire, and s. Maine, s.

to Connecticut, w. Texas, the Gulf Coast, and s. Florida and w. to central Kansas, e. Oklahoma, and e. Mexico. Recently expanding range northward.

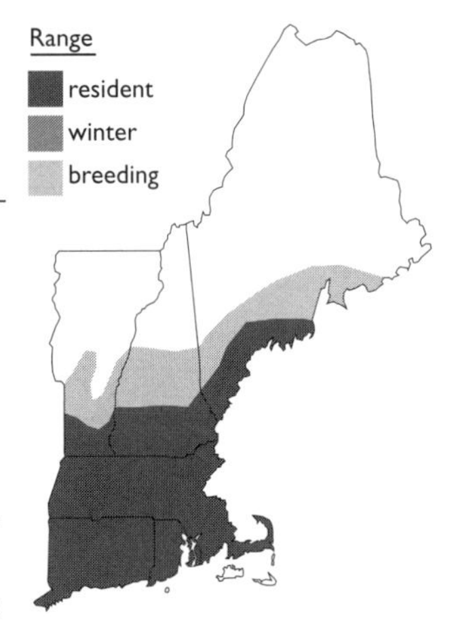

Distribution in New England: Southern Maine, s. New Hampshire, s. Vermont, and throughout Massachusetts, Connecticut, and Rhode Island, except absent from the islands.

Status in New England: Common resident.

Habitat: Breeding: Deciduous and mixed woods, deciduous wooded swamps, and river bottoms. Also occurs in orchards, city parks, and suburban areas. Usually does not occur at elevations above 2,000 ft (610 m) (Root 1988). Winter: Same as breeding habitat; common at feeding stations.

Special Habitat Requirements: Nesting cavities in deciduous or mixed woods.

Nesting: Egg dates: May and June, Massachusetts (Veit and Petersen 1993:331). April 29 to May 27, New York (Bull 1974:404). Clutch size: 3 to 9, typically 5 or 6 (Grubb and Pravosudov 1994). Incubation period: 12 to 14 days. Nestling period: 15 to 18 days. Broods per year: 1. Age at sexual maturity: 1 yr. Nest height: 3 to 90 ft (0.9 to 27.4 m). Nest site: Usually nests in natural tree cavities or abandoned woodpecker holes; occasionally nests in bird boxes.

Territory Size: 2.9 acres (1.2 ha) in oak-hickory-elm habitat in Kansas (Fitch 1958). 13.5 acres (5.4 ha) in Missouri (Brawn and Samson 1983). 5 birds were repeatedly recorded year-round within a 0.6-mile (0.9-km) radius from a banding station in Michigan (Van Tyne 1948). Average sizes of minimum home ranges in winter ranged from 10.4 to 19.7 acres (4.2 to 8.0 ha) in Kansas (Fitch 1958).

Sample Densities: Maryland—13 pairs per 100 acres (40 ha) in well-drained floodplain forest. 13 pairs per 100 acres (40 ha) in upland oak forest. 11 pairs per 100 acres (40 ha) in second-growth river swamp. 6 pairs per 100 acres (40 ha) in pine-oak forest (Stewart and Robbins 1958:226). 10.8 pairs per 100 acres (40 ha) in birch-basswood forests in Pennsylvania (Samson 1979).

Food Habits: Insects, spiders, snails, berries, seeds of sumac, yellow poplar, alder, poison ivy, and bayberry; and some mast from branch and leaf surfaces (spring and summer), branch surfaces (winter) on the ground, especially exposed soil (Fitch 1958). Generally feeds in the canopy, but often near the ground when not disturbed by observers. Forages on the trunks and major limbs of trees.

Comments: Tufted titmice are relatively recent arrivals in New England, with only five records for Connecticut until 1934; by 1949 they nested in Connecticut (Zeranski and Baptist 1990:188). They first nested in Massachusetts in 1958 (Kricher 1981, Veit and Petersen 1993:330), extending their range in the Northeast, likely due to suburban feeding stations (Boyd 1962).

Red-breasted Nuthatch

(Sitta canadensis)

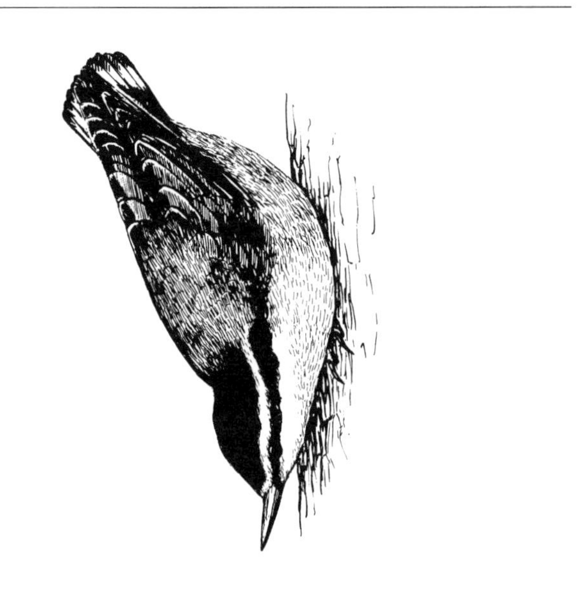

Range: Breeds from s.-coastal and se. Alaska, s. Yukon, central Manitoba, and Newfoundland, s. to s. California, central and se. Arizona, central Colorado, Wyoming, sw. North Dakota, s. Manitoba, s. Michigan, and n.-central Ohio; in the Appalachian Mountains to e. Tennessee and w. North Carolina; and s. to se. Pennsylvania, s. New Jersey, and s. New York. Winters throughout most of the breeding range except at the higher latitudes and elevations, irregularly s. to Baja California, s. Arizona, s. Texas, and central Florida.

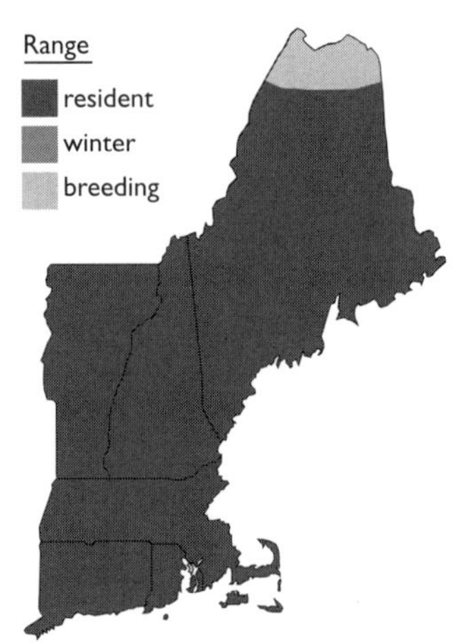

Distribution in New England: Throughout the region, except absent from Block Island in Rhode Island.

Status in New England: Common breeder in Maine, New Hampshire, and Vermont; an uncommon and local breeder in Massachusetts (Veit and Petersen 1993:332), and a very local breeder in the Northwest and Northeast Hills in Connecticut (Zeranski and Baptist 1990:188); uncommon and localized in coniferous forests in Rhode Island (Enser 1992:117). Common to uncommon winter resident and migrant.

Habitat: Breeding: Prefers coniferous forests, but sometimes inhabits mixed woodlands. Also uses coniferous forested wetlands. Winter: Mainly coniferous forests, but also frequents mixed woodlands with cone-bearing trees.

Special Habitat Requirements: Decaying trees or live trees with natural cavities.

Nesting: Egg dates: May 18 to June 10, Massachusetts (Veit and Petersen 1993:332). April 30 to June 17. Clutch size: 4 to 7, typically 5 or 6. Incubation period: 12 days. Nestling period: 18 to 21 days. Broods per year: 1. Age at sexual maturity: 1 yr. Nest height: 5 to 40 ft (1.5 to 12.2 m), typically 15 ft (4.6 m). Nest site: Generally uses natural cavities or old woodpecker holes for nesting, but can excavate its own cavity in soft, decaying wood of rotted stubs or dead branches. Smears pitch from coniferous trees below or around the entrance hole, even when the nest is in a deciduous tree or nest box. The pitch may serve as protection against predators; adults avoid contact with pitch by flying directly into the nest cavity (Kilham 1972).

Sample Densities: 0.4 territorial males per 100 acres (40 ha) in mixed mesophytic forests in Pennsylvania (Samson 1979).

Food Habits: Seeds—especially those of spruce, fir, and pine. Also feeds on spiders and some insects, including spruce budworms (Crawford et al. 1990). In August 1985, red-breasted nuthatches were seen sallying to feed aerially on moths flying from the tips of pine boughs (R. DeGraaf and P. Lyons, pers. obs.).

Comments: Red-breasted nuthatches are far more common in coniferous forests than in mixed forests.

White-breasted Nuthatch

(*Sitta carolinensis*)

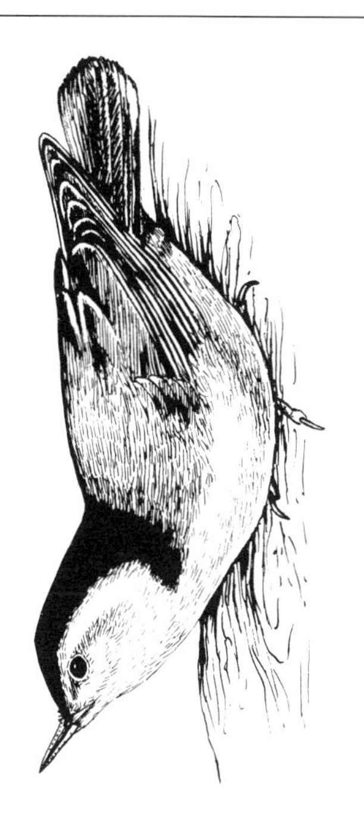

Range: Resident from nw. Washington, s. British Columbia, central Montana, s. Manitoba, n. Minnesota, n. Michigan, New Brunswick, and Nova Scotia, s. to Baja California, s. Nevada, central and se. Arizona, the highlands of Mexico, w. and e.-central Texas, and n. Florida. Absent from most of the Great Plains.

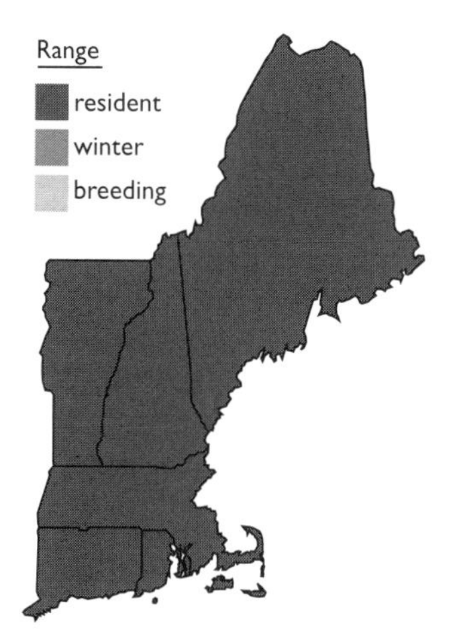

Distribution in New England: Resident throughout the region, except no evidence of nesting on Nantucket in Massachusetts or on Block Island, R.I. (Enser 1992:118).

Status in New England: Common resident.

Habitat: Mature deciduous forests and mixed forests, usually avoids boreal coniferous forests (Peck and James 1987, Andrle and Carroll 1988). Prefers forest edges near open areas such as fields, parks, roads, lakes, and ponds (Peck and James 1987). Also in residential neighborhoods and open woodlands.

Special Habitat Requirements: Natural tree cavities for nesting, preferably in trees with a diameter breast height of 12 in (30.5 cm) (Thomas et al. 1979).

Nesting: Egg dates: April 3 to May, Massachusetts (Veit and Petersen 1993:333). April 3 to June 6, New York (Bull 1974:407). Clutch size: 5 to 9, typically 8. Incubation period: 12 to 14 days (Allen 1929, Ritchison 1981). Nestling period: 14 to 26 days (Bent 1948, Ritchison 1981). Broods per year: 1. Age at sexual maturity: 1 yr. Nest height: 15 to 50 ft (4.6 to 15.2 m). Nest site: Prefers natural cavities in living trees with an opening twice as wide as its body (Pravosudov and Grubb 1993), but will use cavities in dying trees or old woodpecker holes. Rarely, if ever, excavates its own cavity.

Territory Size: Winter territories ranged from 25 to 37.5 acres (10 to 15 ha) per pair in woodlands and about 50 acres (20.2 ha) per pair in semiwooded country (Butts 1931).

Sample Densities: 1 pair per 24 acres (9.6 ha) in New York (Butts 1931). 6 pairs per 100 acres (40 ha) in oak/tulip-poplar forest in Maryland. 5 pairs per 100 acres (40 ha) in semiopen floodplain forest (sycamore, ash, elm in Maryland) (Stewart and Robbins 1958:228). 3.6 territorial males per 100 acres (40 ha) in birch-basswood in Pennsylvania (Samson 1979).

Food Habits: Insects, seeds, fruits, mast. The main summer diet consists of gypsy moth larvae and tent caterpillars, beetles, spiders, and ants (Hardin and Evans 1977) on the trunks and larger branches of trees and in bark crevices. The spring diet consists of more than 79 percent insects. The winter diet contains 26 percent animal and 67 percent vegetable (Forbush 1929). Mast from oak, beech, and hickory may provide important food sources (Ellison 1985:218 in Laughlin and Kibbe 1985). Often uses feeding stations, especially in winter.

Comments: White-breasted nuthatches move head downward as they forage on tree trunks. Pairs commonly store large amounts of food in crevices of bark throughout their territory (Kilham 1974). Much competition for natural cavities occurs between white-breasted nuthatches and gray and red squirrels (Kilham 1968b). The bird's name is probably derived from its habit of using crevices to hold fast large seeds and nuts which it then hacks open—*hatch* being a corruption of *hack* (Pravosudov and Grubb 1993).

Brown Creeper

(*Certhia americana*)

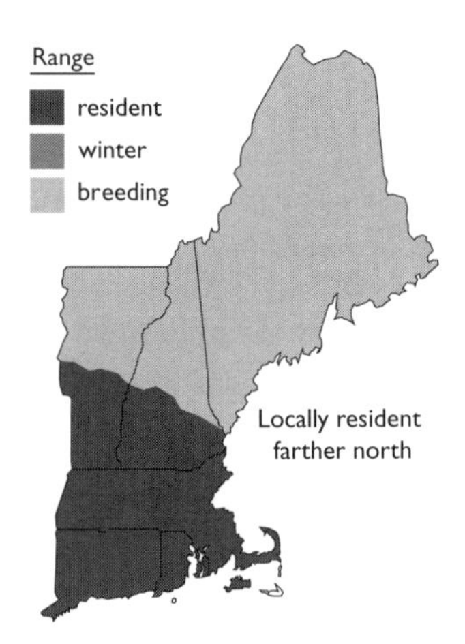

Range: Resident from sw., central, and se. Alaska, central Alberta, central Manitoba, and Newfoundland, s. to s. California, across to extreme w. Texas, se. Nebraska, se. Missouri, s. Ontario, e. Ohio, and West Virginia; in the Appalachians to e. Tennessee and w. North Carolina; and to the lowlands of Virginia, Maryland, and Delaware. Breeds also through Mexico into Central America. Winters generally through the breeding range, withdrawing from the higher latitudes and elevations and s. throughout the e. United States and s. Texas, the Gulf Coast, and central Florida.

Distribution in New England: Generally throughout the region, except Block Island, R.I., and Nantucket, Mass. Uncommon and local in the Northwest Hills and scarce along the coast in Connecticut (Zeranski and Baptist 1990:199).

Status in New England: Common breeder in Maine, New Hampshire, Vermont, and Rhode Island; uncommon in Massachusetts and Connecticut (Veit and Petersen 1993:334, Zeranski and Baptist 1990:199). Uncommon to common migrant and winter resident.

Habitat: Inhabits dense, mature coniferous, deciduous, and mixed woodlands and forested wetlands with standing trees and loose bark. Also occurs in upland forests within 197 ft (60 m) from water (Davis 1978). Associated with a closed canopy (87.4 percent) mixed deciduous-coniferous forest in Vermont (Thompson and Capen 1988). In winter found in open woodlands, scrub forests, parks, and suburban areas.

Special Habitat Requirements: Standing dead trees with loose bark.

Nesting: Egg dates: May 6 to 23, Massachusetts (Veit and Petersen 1993:334). April 24 to June 30, New York (Bull 1974:412). Clutch size: 5 to 9, typically 5 or 6. Incubation period: 14 to 15 days. Nestling period: 13 to 15 days. Broods per year: 1 or 2. Age at sexual maturity: 1 yr. Nest height: 5 to 15 ft (1.5 to 4.6 m). Nest site: Constructs nest under a loose bark flap of a live, dead, or dying tree. Less often uses a rotted knothole or abandoned woodpecker cavity.

Food Habits: Insects; a small amount of vegetable material, mainly mast. Explores bark crevices, trunk, sides and undersides of limbs while moving upward on trees. Sometimes attracted to suet at feeding stations in winter.

Comments: The availability of nest sites is a critical factor in whether brown creepers nest in a given area. They need trees in a very specific condition—recently dead with loose bark flaps. The nest is attached to the trunk under the bark flap with cocoons and spider egg cases.

Carolina Wren

(Thryothorus ludovicianus)

Range: Resident from e. Nebraska, Iowa, and se. Minnesota across to s. Ontario, extreme sw. Quebec, and s. New England, s. to Mexico, and Gulf Coast and Florida.

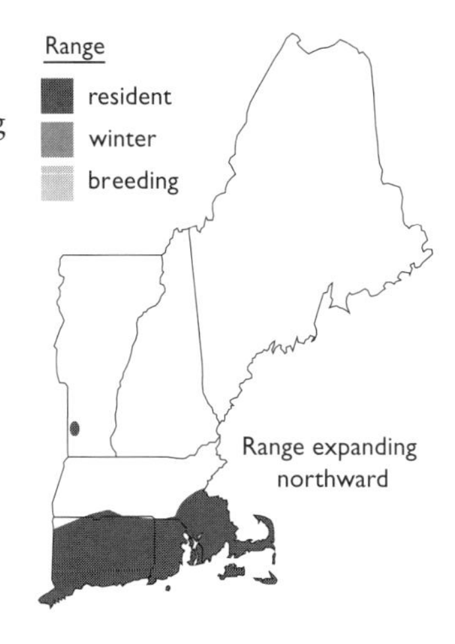

Distribution in New England: Throughout se. Massachusetts including Martha's Vineyard and Nantucket, expanding northward, now locally at lower elevations in large river valleys (Veit and Petersen 1993:335, Petersen in Haggerty and Morton 1995), throughout Rhode Island, and Connecticut, and nesting reported from sw. Vermont (Ellison 1985:222 in Laughlin and Kibbe 1985).

Status in New England: Fairly common local resident, but subject to fluctuations from year to year depending upon winter severity.

Habitat: Breeds in a variety of habitats from lowland streambank tangles to upland brushy slopes and woodland edges, cutover forests, slash piles, old fields, gardens. Prefers edges of moist forests with a dense, shrubby understory. Winter: Low, flat area in large river valleys, near tidal streams, in coastal thickets, and deep ravines.

Special Habitat Requirements: Low, brushy vegetation.

Nesting: Egg dates: Mid June, Massachusetts (Veit and Petersen 1993:337). April 15 to August 15, birds in s. New York have at least 2 broods (Bull 1974:417). Clutch size: 4 to 8, typically 5 or 6. Incubation period: 14 days. Nestling period: 13 to 14 days. Broods per year: 2. Age at sexual maturity: 1 yr. Nest height: To 10 ft (0.3 m), typically less than 10 ft (3 m). Nest site: Typically nests in natural tree cavities, woodpecker holes, overturned root cavities, bird houses, and various man-made structures.

Territory Size: Wrens occupied a wooded ravine-pond habitat in Kansas for 4 yr and defended the following areas: 5.8, 9.2, 3.9, and 7.6 acres (2.3, 3.7, 1.6, and 3.1 ha) (Fitch 1958). Average 0.3 acre (0.1 ha) in a swamp-thicket in Illinois (Brewer 1955). Average territory size of 10.25 acres (4.1 ha) in Tennessee (Strain and Mumme 1988). Average size of 2.5 acres (1 ha) in North Carolina (Simpson 1984).

Sample Densities: Maryland: 11 territorial males per 100 acres (40 ha) in hardwood forest (oak/tulip-poplar) with scattered pine. 6 territorial males per 100 acres (40 ha) in well-drained flood-plain forest (Stewart and Robbins 1958:235). 7.2 territorial males per 100 acres (40 ha) in birch-basswood forests in Pennsylvania (Samson 1979).

Food Habits: Insects, occasionally takes wild fruits.

Gleans over the trunks of trees, branches of shrubs, leaf surfaces, ground litter.

Comments: Carolina wrens are fairly recent breeding birds in New England, first reported in Connecticut in 1878 and first nesting there in 1895 (Zeranski and Baptist 1990:191). Northern limit of this species varies with the degree of winter severity (expands in mild winters and recedes with harsh weather). The contents of 291 stomachs collected throughout the year held 94 percent animal and 6 percent vegetable matter (Beal et al. 1916 in Bent 1948:209).

House Wren

(*Troglodytes aedon*)

Range: Breeds from s. and w.-central British Columbia and n. Alberta, e. to sw. Quebec and New Brunswick, and s. to n. Baja California, s. Arizona and New Mexico, w. and n. Texas, central Arkansas, s. Tennessee, and North Carolina. Winter: From s. California to n. Texas, the n. portion of the Gulf States, and coastal Maryland s. to s. Baja California, the Gulf Coast, Florida, and throughout Mexico to Oaxaca and Veracruz. Resident from Oaxaca through Central America to s. South America.

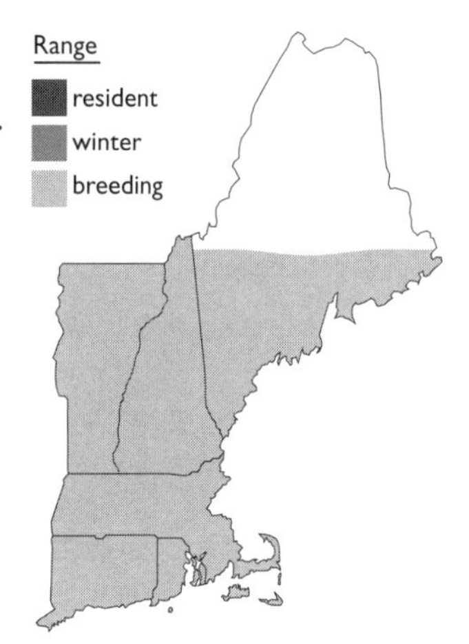

Distribution in New England: Breeds in s. Maine and throughout New Hampshire, Vermont, and Massachusetts.

Status in New England: Common inland, but uncommon on outer Cape Cod and the Islands.

Habitat: Historically associated with deciduous forest edges and open, shrubby woods, but has adapted to suburban backyards, woodlot edges, clearings, city parks, farmland, orchards, gardens, and swampy woodlands with dense thickets. Avoids high elevations in the East.

Special Habitat Requirements: Thickets and cavities for nesting in trees with a minimum diameter breast height of 10 in (25.4 cm) (Thomas et al. 1979).

Nesting: Egg dates: May 25 to August 1, Massachusetts (Veit and Petersen 1993:338). May 15 to July 31 (second brood), New York (Bull 1974:413). Clutch size: 5 to 8, typically 6 to 7. Incubation period: 12 to 15 days. Nestling period: 12 to 18 days. Broods per year: 2. Age at sexual maturity: 1 yr. Nest height: To 10 ft (3 m), but typically less. Nest site: Uses almost any cavity as a nest site, including cavities in trees, fenceposts, tin cans, eaves of buildings, nest boxes, large skulls, and even laundry hanging to dry on the line.

Territory Size: 178 territories ranged from 0.25 to 2.75 acres (0.1 to 1.1 ha) in forest edge and shrubby pasture habitat in Ohio (Kendeigh 1941b).

Sample Densities: 40 pairs per square mile (15 pairs/km^2) (maximum density) in favorable habitat in North Dakota (Stewart and Kantrud 1972). 100 territorial males per 100 acres (40 ha) in farmyard and orchard in Maryland. 50 territorial males per 100 acres (40 ha) in damp deciduous scrub with standing snags in Maryland. 14 territorial males per 100 acres (40 ha) in unsprayed orchard in Maryland (Stewart and Robbins 1958:232).

Food Habits: Small terrestrial insects, beetles, caterpillars, and bugs in low woody vegetation (Guinan and Sealy 1989).

Comments: House wren populations were severely depressed in New England following the introduction of house sparrows in Boston in 1868 (Forbush 1929), and did not regain their former abundance until the 1930s when house sparrow populations declined. House wrens destroy clutches (punctures eggs) of their own species or other passerines near the nest (Belles-Isles and Picman 1986). Readily nesting in small woodlots, forest edges, and suburban areas, house wrens have benefited from the breaking up of extensive forest areas (Johnson 1998).

Winter Wren

(*Troglodytes troglodytes*)

Range: Holarctic; breeds in e. North America. Breeds from coastal s. and se. and n. British Columbia to central Quebec and s. Labrador, s. to central California, central Idaho, se. Manitoba, s. Wisconsin, and se. New York, and in the Appalachians to ne. Georgia. Winters from s. Alaska and British Columbia, e. to ne. Colorado, central Iowa, s. Michigan, and Massachusetts, s. to s. California, s. Texas, the Gulf Coast, and central Florida.

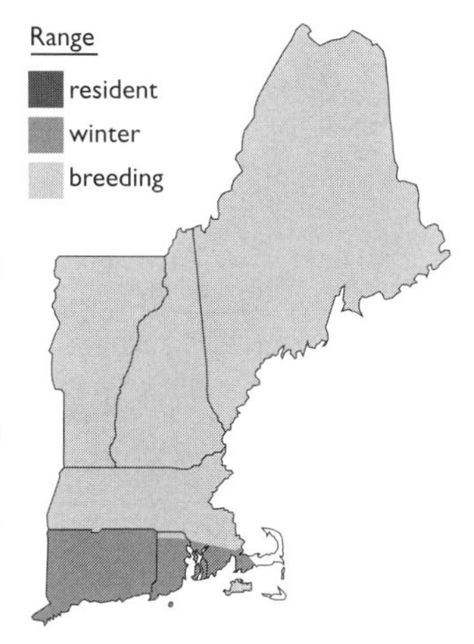

Distribution in New England: Breeds throughout Maine, New Hampshire, Vermont, w. and n. Massachusetts, although occasionally nests in cool cedar and maple swamps in the eastern part of the state (Veit and Petersen 1993:339); s. to nw. Connecticut. Winters throughout Connecticut and Rhode Island, n. to s.-central and e. Massachusetts, except absent from outer Cape Cod and Nantucket.

Status in New England: An uncommon and local breeder in w. Massachusetts. In winter uncommon and irregular in e. Massachusetts (Veit and Petersen 1993:339), s. to Connecticut in the Northwest and Northeast Hills, and rarely near the coast or in the Connecticut River Valley (Zeranski and Baptist 1990:191), and very uncommon in Rhode Island (Enser 1992:122).

Habitat: Breeding: Inhabits dense undergrowth of coniferous (especially spruce and fir) or mixed forests, generally near water. Frequents thickets near woodland streams, boreal bogs, white cedar swamps, banks of marshy ditches, slash piles, downed trees, cool ravines, and edges of clear cuts. Winter: Prefers moist coniferous and deciduous woodlands with a dense understory.

Special Habitat Requirements: Moist coniferous woodlands with low woody vegetation or low-lying cold bogs or swamps.

Nesting: Egg dates: May 20 to August 8, Maine (Veit and Petersen 1993:339). May 22 to July 7, New York (Bull 1974:415). Clutch size: 4 to 7, typically 5 or 6. Incubation period: 14 to 16 days. Nestling period: Probably about 2 weeks (Harrison 1975). Broods per year: 2. Age at sexual maturity: 1 yr. Nest site: Usually nests in a hollow base of a stump or upturned tree roots, in a hollow log, brush heap or rocky crevice, or mossy hummock. Occasionally will use man-made structures or an abandoned, low woodpecker hole.

Territory Size: Approximately 1 to 7 acres (0.4 to 2.8 ha) (average 2 to 3 acres, 0.8 to 1.2 ha) in garden-woodland areas (Armstrong 1956:430). 3 acres (1.2 ha) in subalpine habitat in the White Mountains in New Hampshire (Sabo 1980).

Food Habits: Insects gleaned from the ground.

Comments: The winter wren has been a regular migrant and winter resident in Connecticut throughout the historical period. Despite the name, winter wrens are not especially hardy, and do not winter north of Massachusetts.

Sedge Wren

(*Cistothorus platensis*)

Range: Breeds in North America, from extreme e.-central Alberta and central Saskatchewan e. to n. Michigan and

s. New Brunswick, s. to e.-central Arkansas, central Kentucky, and se. Virginia, and w. to central North Dakota and e. Kansas. Resident locally in Middle America and in South America locally in the Andes from Colombia to Chile and from Brazil to Tierra del Fuego. Winter: From w. Tennessee and Maryland to ne. Mexico, Texas, the Gulf Coast, and Florida.

Distribution in New England: Rare and local breeder and migrant.

Status in New England: Very rare and irregular breeder; most commonly reported as a fall migrant.

Habitat: Breeds in fresh or brackish sedge meadows and shallow sedge marshes with scattered shrubs and little or no standing water. Also uses damp upper margins of coastal marshes, ponds, or wetlands. Sites will be abandoned if they are too dry from ditching or drought, or too wet from flooding (Gibbs and Melvin 1992c).

Special Habitat Requirements: Wet meadows or drier margins of marshes.

Nesting: Egg dates: May 25 to July 7, Massachusetts (Veit and Petersen 1993:341). May 28 to July 30, New York (Bull 1974:419). Clutch size: 4 to 8, typically 7 (Walkinshaw 1935). Incubation period: 12 to 14 days. Nestling period: 12 to 14 days. Broods per year: 1 to 3. Age at sexual maturity: 1 yr. Nest height: 1 to 3 ft (0.3 to 0.9 m), typically 2 to 3 ft (0.6 to 0.9 m). Nest site: Builds nest over land or water in dense vegetation; usually places nest, interwoven with live grasses, less than 3 ft (1 m) above the substrate. Males build multiple domed nests in their territory that are used for nesting, as dormitories, and possibly as decoys for predators (Verner 1965, Burns 1982). Sedge wrens are nomadic breeders; breeding areas shift from year to year.

Sample Densities: 35 to 40 singing males were counted in a 10-acre (4-ha) marsh (Harrison 1975:152). 10 territorial males per 100 acres (40 ha) in switchgrass marsh-meadow in Maryland (Stewart and Robbins 1958:238).

Food Habits: Insects, spiders gleaned from the ground, marsh vegetation.

Comments: The secretive habits of this species have made it difficult to study. Seldom found in same area 2 yr in succession in New Hampshire (C. Anderson pers. commun.). Habitat loss is recognized as an important factor in the decline of sedge wrens. Apparently in long-term decline in Connecticut (Zeranski and Baptist 1990:192), and among the rarest nesting passerines in Massachusetts (Veit and Petersen 1990:340). The decline of agriculture and loss of wet pasture may be a factor in the rarity of this species.

Marsh Wren

(*Cistothorus palustris*)

Range: Breeds from sw. and e.-central British Columbia and n. Alberta, east to n. Michigan and e. New Brunswick, s. to Baja California, sw. Arizona, extreme w. and s. Texas, the Gulf Coast, and e.-central Florida. Generally very local in interior North America. Winter: In coastal areas throughout the breeding range, and in the interior from the s. United States to Mexico.

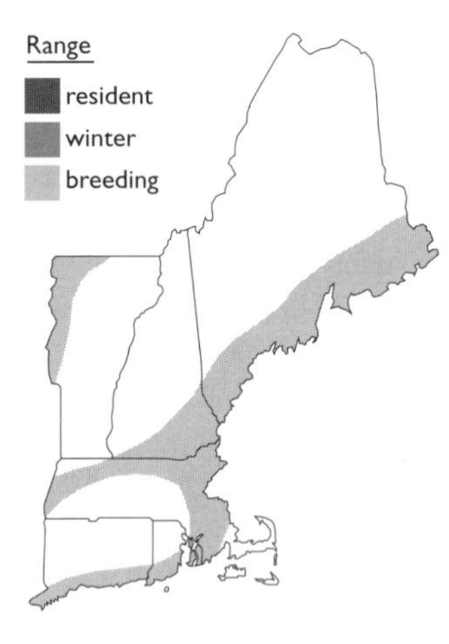

Distribution in New England: Breeds locally through s. Maine, s. New Hampshire, the Champlain Valley and s. Vermont; Massachusetts except outer Cape Cod, Martha's Vineyard, and Nantucket; s. throughout Connecticut and Rhode Island.

Status in New England: Locally common breeder, but has been declining in the East.

Habitat: Breeds in large fresh, brackish, or salt marshes with an abundance of tall emergent vegetation such as cattails, loosestrife, sedges, or rushes. Also uses emergent vegetation along the margins of slow-moving rivers, ponds, and lakes. Wetlands smaller than 0.4 ha are rarely used. Prefers large marshes with narrow-leaved cattails (Kroodsma and Verner 1997).

Special Habitat Requirements: Marshes with tall emergent vegetation.

Nesting: Egg dates: May 27 to August 1, Massachusetts (Veit and Petersen 1993:342). May 22 to August 7, New York (Bull 1974:419). Clutch size: 3 to 8, typically 5. Incubation period: 12 to 16 days. Nestling period: 14 to 16 days. Independence when about 23 days old (Verner 1965). Broods per year: 2. Age at sexual maturity: 1 yr. Nest height: 3 to 6 ft (1 to 3 m) above the marsh substrate. Nest site: Males build domed elliptical nests, usually in tall marsh vegetation such as cattails or sedges growing in shallow water, less often in small bushes or trees. Male constructs many dummy nests (Metz 1991) and uses some for roosting (Bull 1974:419). As in sedge wrens, the number of nests built strongly correlates with number of females to whom males are mated (Verner 1965).

Territory Size: 11 territories averaged 60 m² (71.8 square yards) in tall marsh grasses interspersed with shorter grasses along a river in Georgia. 22 territories averaged 85.3 m² (102 square yards) in *Spartina* spp. along a creek in Georgia (Kale 1965). Territories ranged from 2,600 square feet (241.5 m²) to 38,700 square feet (3,595 m²) (average 15,000 square feet, 1,393.3 m²) in cattails with scattered stands of bulrush. Territories ranged from 1,800 to 9,600 square ft (167.2 to 891.7 m²) (average 5,000 square feet, 464.4 m²) in narrow strips and patches of bulrush in Washington (Verner 1965). The territories of monogamous males ranged from 13,000 to 15,000 square feet (1,207.6 to 1,393.3 m²) in cattail-sedge association and 30,000 square feet (2,786.7 m²) in grasses in New York (Welter 1935).

Sample Densities: Maryland—104 territorial males per 100 acres (40 ha) in uniform needlerush marsh. 36 territorial males per 100 acres (40 ha) in cattail marsh (Stewart and Robbins 1958:236).

Food Habits: Insects, spiders, aquatic insects gleaned from the stems and leaves of marsh vegetation, water.

Comments: Evidently, the denser the emergent vegetation, the greater the number of young fledged (Leonard and Picman 1987). The best-known features of the marsh wren are its polygynous mating system and habit of building multiple nests, its habit of destroying eggs and nests of other marsh wrens as well as those of other species, and its large song repertoire. Males place multiple nests higher as the vegetation grows. Males escort prospective females attracted to the nests; if a female accepts one, she lines it with soft materials before laying eggs (Kroodsma and Verner 1997).

Golden-crowned Kinglet

(*Regulus satrapa*)

Range: Breeds from s. Alaska to n. Alberta, s. Quebec, and Newfoundland, s. in the coastal and interior mountains to s. and e. California, s. Utah, s.-central New Mexico, Mexico, Guatemala, and e. of the Rockies to s. Manitoba, n.-central Michigan, New York, e. Tennessee, w. North Carolina, n. New Jersey, and s. Maine. Winter: From s.-coastal Alaska and s. Canada, s. to n. Baja California, through the breeding range to Guatemala, the Gulf Coast, and central Florida.

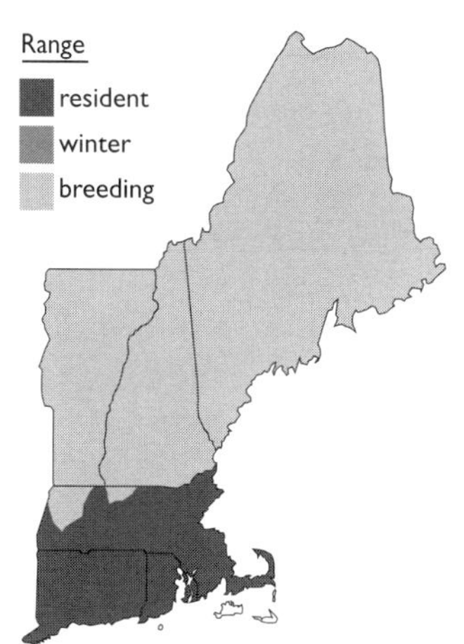

Distribution in New England: Breeding: Throughout Maine, Vermont, New Hampshire, Connecticut, Rhode Island, and n. and sw. Massachusetts and Plymouth County, absent from outer Cape Cod and the islands. Winter: Throughout Connecticut and Rhode Island, except absent from Block Island, north to s. and e. Massachusetts, except rare on outer Cape Cod and the Islands.

Status in New England: Uncommon to local breeder. A rare to local nester in Connecticut, primarily in the Northwestern and Northeastern Hills (Zeranski and

Baptist 1990:194); uncommon and local breeder in Massachusetts, primarily in the Berkshires but locally in n.-central and e. parts of the state (Veit and Petersen 1993:343); and most abundant in the Green Mountains and Northeastern Highlands in Vermont (Ellison 1985:233 in Laughlin and Kibbe 1985), n. New Hampshire, and Maine. Winter: Uncommon to fairly common in Connecticut (Zeranski and Baptist 1990:194), and uncommon to very common in Massachusetts (Veit and Petersen 1993:343).

Habitat: Breeding: Dense coniferous forests and plantings (especially spruce), but will nest in pine, fir, hemlock, and tamarack woods and cedar bogs. Spruce plantations as small as 1 ha (Andrle and Carroll 1988) provide suitable habitat in central and w. New York (Bull 1974:443) and e. Massachusetts (Veit and Petersen 1993:342). Winter: Moist coniferous and occasionally mixed forests with low tangles of weedy growth (Lepthien and Bock 1976, Pough 1949:126).

Nesting: Egg dates: June 5 to August 1 (for 2 broods), Massachusetts (Veit and Petersen 1993:342). May 28 to June 26 (1st brood), New York (Bull 1974:444). Clutch size: 5 to 11, typically 8 or 9 (Ingold and Galati 1997). Incubation period: 15 days—only female incubates. Broods per year: Commonly 2. The span of time covering both nests is about 94 days. The second nest is generally started before the nestlings in the first nest fledge. Male feeds the first brood while female constructs the second nest and lays and incubates the second clutch (Ingold and Galati 1997). Nest height: 6 to 60 ft (1.8 to 18.2 m), typically 30 to 60 ft (9.1 to 18.2 m). Nest site: Usually woven into twigs of a horizontal limb of a conifer. Nests in New England mountains from 300 to 1,600 m elevation.

Territory Size: 1.5 acres (0.6 ha) in mature spruce forests in subalpine areas of the White Mountains (Sabo 1980).

Sample Densities: 1 pair per 2 acres (0.8 ha) in Adirondack coniferous forest (Andrle 1971). 32 pairs per 100 acres in virgin spruce-hemlock bog forest in Maryland (Stewart and Robbins 1958:255).

Food Habits: Mostly small, soft-bodied arthropods: spiders, mites, insects on leaves, branches, and twigs, trunks (bark crevices).

Comments: Golden-crowned kinglets have been extending their breeding range southward in New York and s. New England and in the Midwest (Ingold and Galati 1997) by nesting in plantations of spruce (native and exotic) with a minimum diameter breast height of 6 in (15 cm) and dense, closed canopies. In n. New England this species is primarily found in boreal forest habitats.

Ruby-crowned Kinglet

(*Regulus calendula*)

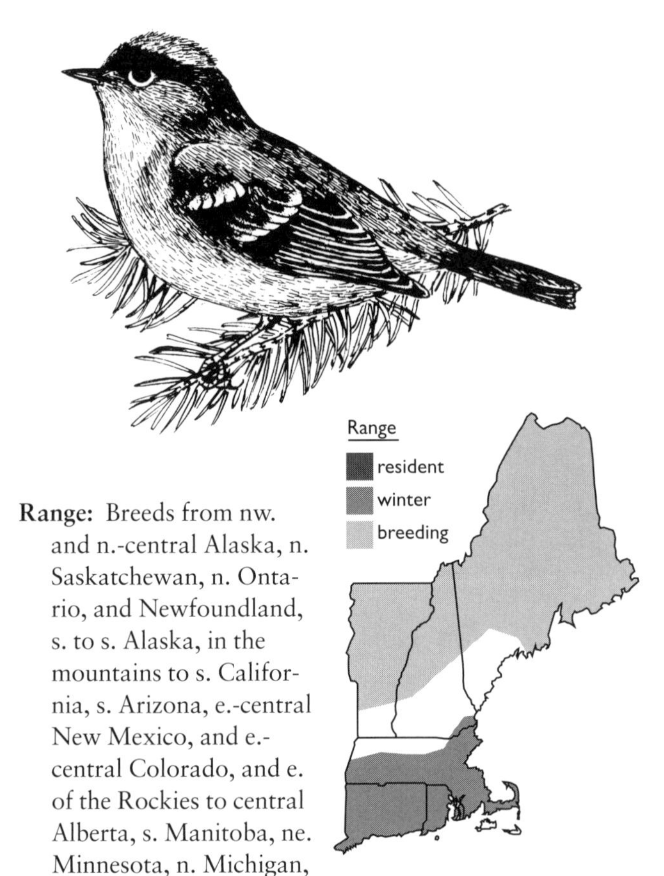

Range: Breeds from nw. and n.-central Alaska, n. Saskatchewan, n. Ontario, and Newfoundland, s. to s. Alaska, in the mountains to s. California, s. Arizona, e.-central New Mexico, and e.-central Colorado, and e. of the Rockies to central Alberta, s. Manitoba, ne. Minnesota, n. Michigan, n. New York, n. Maine, and Nova Scotia. Winters from s. British Columbia, Idaho, n. Arizona, Nebraska, s. Ontario, and New Jersey, s. to Baja California, s. Texas, s. Florida, and through Mexico to Guatemala.

Distribution in New England: Breeding: Northern New Hampshire, Vermont, and Maine (except absent from southernmost portions of Maine) and w. Massachusetts (2 records). Winter: Coastal Massachusetts, Connecticut, and Rhode Island.

Status in New England: Fairly common breeder in northern New England, common to occasionally abundant migrant, especially in spring; rare and irregular in winter.

Habitat: Coniferous forests in pure or mixed stands of spruce, fir, tamarack or pine, forest edges, and bogs, from 1,000 ft to 4,500 ft (300 to 1,400 m) elevation (Richards 1994:210 in Foss 1994).

Nesting: Egg dates: May 12 (Ingold and Wallace 1994). Clutch size: 5 to 11, typically 7 to 9. Incubation period: 12 to 14 days (Bent 1949, Peck and James 1987). Nestling period: Little known: thought to be 12 days (Bent 1949). One nest in Ontario fledged after 16 days

(Ingold and Wallace 1994). Broods per year: 1. Age at sexual maturity: 1 yr. Nest height: 2 to 100 ft (0.6 to 30.5 m), typically 15 to 60 ft (4.6 to 18.2 m). Nest site: Nest is usually well concealed in coniferous shrub or tree at tip of horizontal branch, typically in spruce.

Territory Size: Territories in two Ontario studies averaged 7.3 acres (2.9 ha) and ranged from 2.8 to 15 acres (1.1 to 6.0 ha, $n = 9$) (Ingold and Wallace 1994), and 6.3 acres (2.5 ha, $n = 2$) in Ontario (Kendeigh 1947).

Sample Densities: 12.4 and 32.9 pairs per 100 acres (40 ha) in spruce-fir in Maine (Stewart and Aldrich 1952), 3 pairs per 100 acres (40 ha) of spruce-tamarack (Gillespie and Kendeigh 1982). 32 and 48 pairs per 100 acres (40 ha) in northern hardwoods in New Hampshire (Sabo and Holmes 1983). 80 pairs per 100 acres (40 ha) in Ontario (Martin 1960).

Food Habits: In summer, insects on the needles and upper twigs of tree crowns. In winter, insects supplemented with seeds and fruits; forages lower and also uses leaves and stalks of herbaceous plants and saplings in thickets and deciduous woods.

Comments: Ruby-crowned kinglets are present most of the year in s. New England. They remain until December and are among the earliest spring arrivals, but apparently do not remain in the coldest part of winter.

Blue-gray Gnatcatcher

(*Polioptila caerulea*)

Range: Breeds from s. Oregon, n. California, s. Idaho, central Utah, Colorado, Nebraska, w. Iowa, se. Minnesota, Michigan, sw. Quebec, central New York, central Vermont, and s. Maine, south to Baja California, se. Texas, the Gulf Coast, and s. Florida, throughout Mexico to Central America. Winters from s. California, s. Nevada, w. and central Arizona, central Texas, the s. portions of the Gulf States, and on the Atlantic Coast from Virginia, s. to Mexico, Honduras, and in the Bahamas and Cuba.

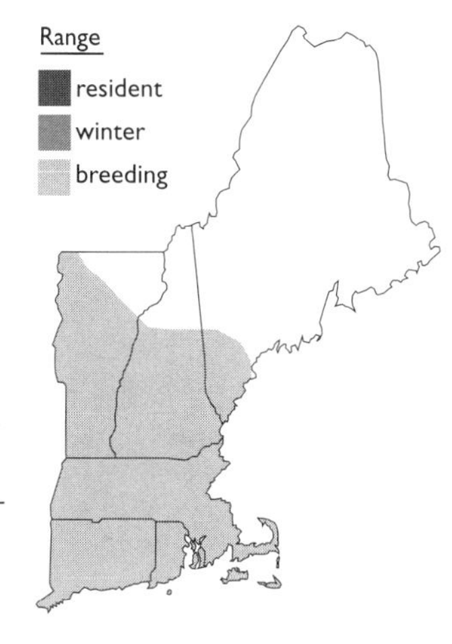

Distribution in New England: Fairly common throughout Massachusetts and Connecticut, absent from coast in Rhode Island, rare but increasing in Vermont (mostly in Connecticut River drainage and Champlain Lowlands), sparse in s. and central New Hampshire, rare in southernmost Maine.

Status in New England: Fairly common (Massachusetts) to rare (Maine) breeder; increasingly widespread.

Habitat: Nests in open, moist woodlands interspersed with brushy clearings, bottomland forests with closed canopies, floodplains, wooded swamps, streamside thickets. Prefer broad-leaved forest but also use pine or mixed woods.

Special Habitat Requirements: An abundant supply of arthropods (Root 1967).

Nesting: Egg dates: Probably mid May to mid June, Massachusetts (Veit and Petersen 1993:346). May 14 to June 17, New York (Bull 1974:441). May 16 to 26 June in Vermont (Ellison 1992). Clutch size: 3 to 5, typically 4 or 5 (Nice 1932). Incubation period: Average 13 days (range: 11 to 15 days; Ellison 1992). Nestling period: 10 to 13 days. Broods per year: 1 or 2. Age at sexual maturity: 1 yr. Nest height: 2 to 85 ft (0.6 to 26 m). Nest site: Usually near water, high in a deciduous or coniferous tree saddled on a horizontal limb or in a fork. Nests in a variety of trees—limb size and shape seem to be more important than tree species (Bent 1949).

Territory Size: 9 territories ranged from 2.2 to 7.4 acres (0.9 to 3.0 ha), average 4.6 acres (1.8 ha), in oak woodland and chaparral in California (Root 1970). 1 territory covered 2.2 acres (0.9 ha) along a wooded ravine and grove of trees in Kansas (Fitch 1958). If nest site abandoned, gnatcatcher commonly reuses material from old nest to build a new one.

Sample Densities: Maryland—7 pairs per 100 acres (40 ha) in semiopen floodplain forest. 6 pairs per 100 acres (40 ha) in unsprayed orchard (Stewart and Robbins

1958:254). Reaches high densities in the South and Southeast. 132 territories per 100 acres (40 ha) in hardwood swamps in South Carolina (Strom 1983), 116 per 100 acres (40 ha) in unmanaged riparian swamp forest in Arkansas (Christman 1983).

Food Habits: Apparently feeds exclusively on small insects and spiders (Root 1967) at tips of branches, leaf surfaces, bark. Prefers to feed in the high canopy of forest trees. In Vermont, half of identifiable prey were adult and larval moths (Ellison 1991).

Comments: Before 1900, blue-gray gnatcatchers were rare stragglers from the South. In the 1940s the bird expanded its range northeastward to Connecticut, and first nested there in 1950 (Bull 1964). It now nests throughout Massachusetts, and has nested as far north as Ossipee in central New Hampshire.

Eastern Bluebird

(Sialia sialis)

Range: Breeds from s. Saskatchewan, s. Quebec, and w. Nova Scotia, s. to s. Texas and s. Florida, and w. to the Dakotas, w. Kansas, Texas, and se. New Mexico; also in se. Arizona and through the highlands of Mexico to Nicaragua. Winters from the middle portions of the e. United States, s. throughout the breeding range, and on Cuba. Individuals from Guatemala s. are resident.

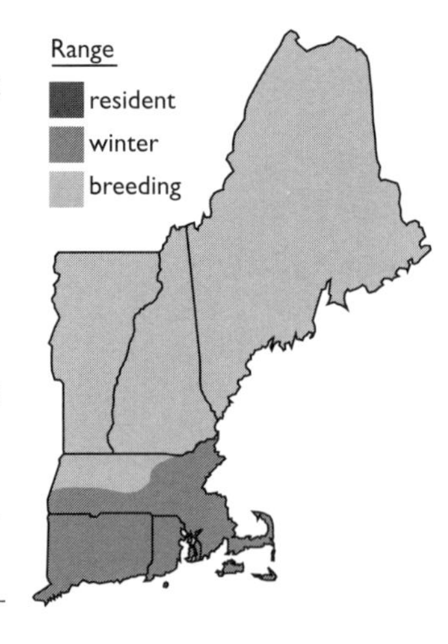

Distribution in New England: Breeds throughout the region, except not reported from Block Island, R.I., or Nantucket, Mass. Winter resident in e. Massachusetts, Connecticut, and Rhode Island, except absent from Block Island.

Status in New England: Uncommon, but increasing breeder and migrant; uncommon (Connecticut) to rare (Massachusetts) winter resident.

Habitat: Breeding: Inhabits open areas such as fields, orchards, agricultural land, shelterbelts, riparian woodlands, pitch-pine/oak forests, residential areas, beaver ponds (Zeleny 1976), and clearings created by fire or logging (Conner and Adkisson 1974). Winter: Forest edges, shrub areas, and fields or open grassy areas.

Special Habitat Requirements: Low cavities for nesting and perches for foraging (Pinkowski 1977).

Nesting: Egg dates: April 1 to August 18, New York (Bull 1974:438). Clutch size: 3 to 8, typically 4 or 5 (Peakall 1970). Incubation period: 13 to 15 days. Nestling period: 15 to 18 days (Hartshorne 1962). Broods per year: 2 or 3. Age at sexual maturity: 1 yr. Nest height: 5 to 20 ft (1.5 to 6.1 m), typically 5 to 12 ft (1.5 to 3.7 m). Nest site: Natural cavities, old woodpecker holes, or nest boxes in open areas or at the edge of a forest.

Territory Size: 5.4, 8.6, and 7.0 acres (2.2, 3.5, and 2.8 ha) for 3 territories in Kansas (Fitch 1958). 2.25 acres (1.0 ha) in Arkansas (Thomas 1946).

Home Range: Pinkowski (1977) found bluebirds foraging on areas ranging in size from 4.5 to 38.9 ha (11.1 to 96.1 acres) during nestling periods.

Sample Densities: 30 birds per 100 acres (40 ha) in orchard in Illinois. 34 birds per 100 acres (40 ha) in edge shrubbery in Illinois. 25 birds per 100 acres (40 ha) in residential habitat in Illinois. 13 birds per 100 acres (40 ha) in second-growth or cutover woods in Illinois (Graber et al. 1971). 9.2 territorial males per 100 acres (40 ha) in red maple woodlands in New York (Samson 1979).

Food Habits: Insects, especially grasshoppers, crickets, beetles, and caterpillars, make up about 68 percent of diet; fruit represents about 32 percent of diet (Bent 1949:247). Prefers to feed in areas with bare soil or sparse ground cover (Pinkowski 1977).

Comments: The eastern bluebird was likely uncommon in interior New England before extensive clearing in the eighteenth century; it was common by the mid nineteenth century in s. New England (Merriam 1877). Bluebirds declined after introduction of the house sparrow in the late 1800s and especially after the European starling reached New England in the 1920s—both competed for nest sites. The use of metal fenceposts, pesticides, and the decline of agriculture all contributed to bluebird decline. Two factors that enhanced the population were the spread of beaver in remote areas, providing many nest sites in standing dead trees, and the nest-box programs in rural and suburban areas. In n. New England, clear-cuts with standing, cavity-bearing snags provide bluebird nesting habitat for several years following cutting (DeGraaf 1991). Eastern bluebird populations are affected in the short term by climate: Severe winters or severe spring storms have been shown to lower the population for several years (Sauer and Droege 1990).

Veery

(*Catharus fuscescens*)

Range: Breeds from s.-central and se. British Columbia to New Brunswick and sw. Newfoundland, s. to central Oregon, s. Idaho, ne. South Dakota, n. Illinois, and n. Ohio, in the mountains through West Virginia, w. and central Maryland, e. Kentucky, w. and central Virginia, e. Tennessee, and w. North Carolina to nw. Georgia, and in the Atlantic region to e. Pennsylvania, central New Jersey, and the District of Columbia. Also in e.-central Arizona.

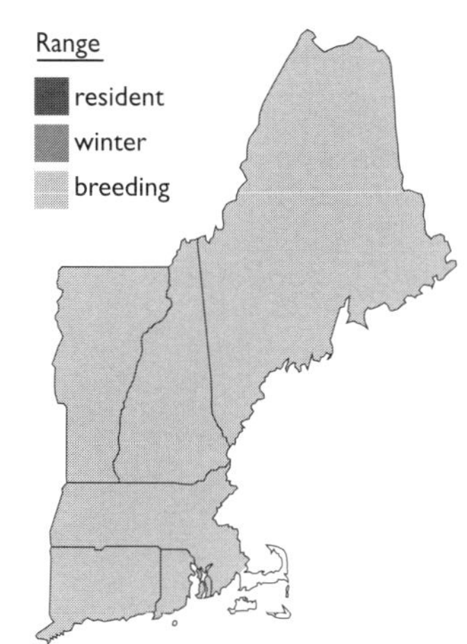

Distribution in New England: Breeds throughout the region, except absent from Block Island in Rhode Island, and outer Cape Cod, Martha's Vineyard, and Nantucket in Massachusetts.

Status in New England: Common breeder throughout the region, except rare on Cape Cod, and absent on the islands of Massachusetts, with the exception of Naushon, where it is common (Veit and Petersen 1993:350). Common spring, uncommon fall migrant.

Habitat: Nests in moist deciduous or mixed woods with intermediate canopy, bottomland forests, wooded swamps, and damp ravines. Also in regenerating clear-cuts (DeGraaf 1991). Prefers thickets of early deciduous second growth and open woods with dense understory of ferns, shrubs, and saplings. Absent from high elevations.

Special Habitat Requirements: Moist woodlands with thick understory of low trees and shrubs.

Nesting: Egg dates: May 20 to June 30, Massachusetts (Veit and Petersen 1993:350). May 16 to July 23 (Kibbe 1985:240 in Laughlin and Kibbe 1985). Clutch size: 3 to 5, typically 4. Incubation period: 10 to 12 days. Nestling period: 10 to 12 days (Bent 1949). Broods per year: 1 or 2. Age at sexual maturity: 1 yr. Nest height: To 3 ft (0.9 m), typically very low. Nest site: On the ground or low in a shrub, tree, or brush pile, often well hidden on a tussock of ferns or other groundcover.

Territory Size: 0.25 to 7.5 acres in Connecticut (0.1 to ≥ 3.0 ha) (Bertin 1975 in Moskoff 1994). Sixty-one territories in Ontario averaged 0.63 acres (0.25 ha) (Martin 1960).

Sample Densities: 12 pairs on a 3-acre (1.2-ha) plot of lakeside forest with laurel understory in New Hampshire (Harding 1925). 8 territorial males per 100 acres (40 ha) in virgin hemlock stand (Stewart and Robbins 1958:251). 40 territorial males per 100 acres (40 ha) in ash-basswood forest in New York (Samson 1979).

Food Habits: Mainly insects (about 60 percent) and wild fruits and seeds (40 percent) gathered on the forest floor (leaf litter), leaf and branch surfaces in lower canopy.

Comments: Veery populations likely peaked in New England after clearing of the white pine around 1910 produced extensive young hardwood forest.

Bicknell's Thrush

(*Catharus bicknelli*)

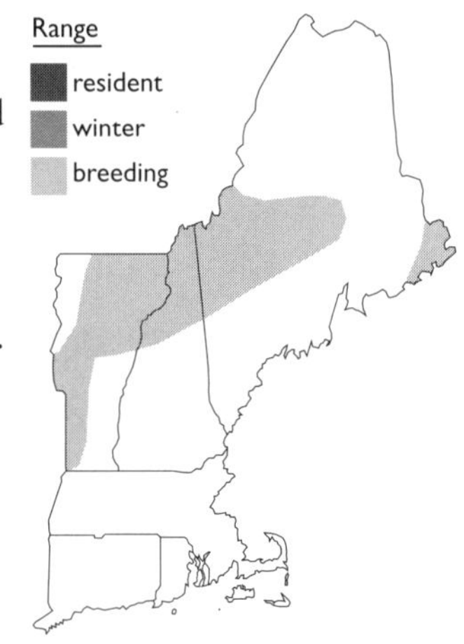

Range: Breeds from the e. north shore of the Gulf of St. Lawrence and the Gaspé Peninsula locally, especially on coasts s. through the Maritimes (excluding Newfoundland), s. to ne. New York (Adirondack Mountains and Catskills), Vermont, n. New Hampshire, and central Maine (Ouellet 1993). Winter: In the West Indies, especially Hispaniola (Veit and Petersen 1993:351), but also Cuba, Puerto Rico, and St. Croix (Ouellet 1993).

Distribution in New England: Breeds in Vermont, restricted to the highest peaks of the Taconics and Green Mountains (Kibbe 1985:242 in Laughlin and Kibbe 1985); White Mountains, n. New Hampshire; and central Maine. Has disappeared from Mt. Greylock in Massachusetts, where 5 to 10 pairs occurred in the early 1900s (Atwood et al. 1996) until 1973 (Veit and Petersen 1993:351).

Status in New England: Locally common to uncommon breeder, uncommon migrant.

Habitat: Nests primarily in dense, stunted forests of balsam fir and red spruce near treeline. In New England, breeds regularly only at the higher elevations of Maine (>915 m) (Boone and Krohn 1996:110), New Hampshire (1,070 to 1,370 m) (Richards 1994:218 in Foss 1994), and Vermont (820 to 1,250 m) (Atwood et al. 1996).

Because of its obligate subalpine spruce-fir habitat (Atwood et al. 1996, Wallace 1939), Bicknell's thrush is generally restricted to mountaintops in New England, rendering its habitat distribution very patchy in the region. Of the high-elevation (>915 m) areas known to be occupied, few exceed 1,000 ha in area (Atwood et al. 1996). Bicknell's thrush has not been reported in low-elevation regenerating clear-cuts in New England, as has been reported in e. Canada (Ouellet 1993). In winter, not restricted to mountain forests, but apparently restricted to mature, moist to wet tropical forest habitats (Rimmer et al. 1998).

Special Habitat Requirements: In New England, high-elevation forests consisting primarily of dense, stunted conifers, especially spruce and balsam fir.

Nesting: The following is for gray-cheeked thrush (*C. minimus*), which until 1995 included Bicknell's thrush. Egg dates: June 12 to June 27, New York (Bull 1974:435). Clutch size: 3 to 5, typically 4. Incubation period: 13 to 14 days. Nestling period: 11 to 13 days. Broods per year: 1. Age at sexual maturity: 1 yr. Nest height: 3 to 12 ft (6.1 m), typically about 6 ft (1.8 m) in conifers. Nest site: Usually in a bush or the fork of a low conifer limb, occasionally in a birch, sometimes on the ground under a low-hanging limb.

Territory Size: (Gray-cheeked thrush as explained above): 5.3 acres (2.1 ha) in the White Mountains of New Hampshire (Sabo 1980).

Sample Densities: (Gray-cheeked thrush as explained above): 9± territorial males/km^2 in a virgin spruce grove in the White Mountains in n. New Hampshire (Sabo 1980).

Food Habits: Insects, wild fruits.

Comments: Bicknell's thrush has not been reported recently at some historic breeding sites, including Mt. Greylock and Saddleball Mountain in Massachusetts, Glebe Mountain, Green Peak, Mt. Ascutney, and Molly Stark Mountain in Vermont, and Mt. Pemigewasset, Mt. Monadnock, Mt. Sunapee, and North Goat Mountain in New Hampshire (Atwood et al. 1996). It is uncertain whether Bicknell's thrush breeds in coastal Maine. It occurs in various low-elevation coastal areas in Quebec (Ouellet 1993, Gauthier and Aubrey 1995), New Brunswick and Nova Scotia, and Prince Edward Island (Erskine 1992), but could not be confirmed in multiple searches by experienced observers (Atwood et al. 1996). It is not known whether or not Bicknell's thrush is declining; its breeding habitat in New England is in small patches at high elevation, habitat that has experienced high levels of tree mortality (Miller-Weeks and Smorank 1993). Furthermore, the Caribbean wintering grounds of the species have been heavily deforested (Arendt 1992); such habitat effects indicate that Bicknell's thrush populations are precarious, and warrant careful monitoring (Atwood et al. 1996).

Swainson's Thrush

(*Catharus ustulatus*)

Range: Breeds from w. and central Alaska, n. Saskatchewan, central Quebec, and Newfoundland, s. to s. Alaska, s. and e.-central California, central Utah, n.-central New Mexico, extreme n. Nebraska, e. Montana, s. Manitoba, n. Minnesota, s. Ontario, n. Pennsylvania, and s. Maine. Also in e. West Virginia, w. Virginia, and w. Maryland. Winter: Nayarit and southern Tamaulipas, s. through Middle America to Guyana, w. Brazil, Peru, Bolivia, and n. Argentina.

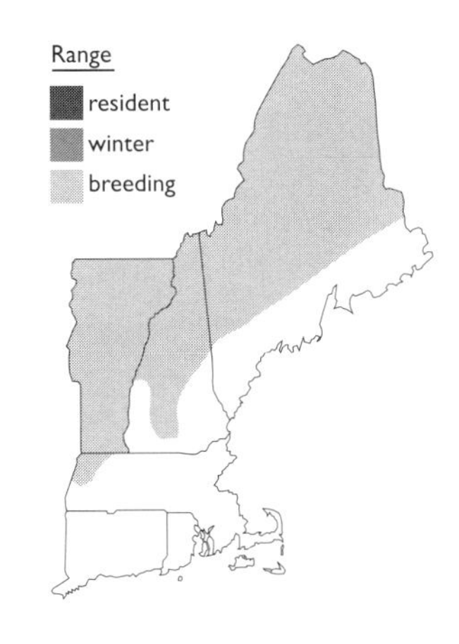

Distribution in New England: Northwestern Connecticut (rare), w. Massachusetts in the Berkshires, throughout Vermont and New Hampshire, except absent from the most sw. and se. regions of the state, throughout Maine, where common along the coast. Absent from Rhode Island.

Status in New England: Rare to locally common breeder; declining in the Berkshires of Massachusetts (Veit and Petersen 1993:352).

Habitat: Dense coniferous forests, especially spruce-fir forests in low damp areas or near water, sometimes in mixed or deciduous forests (hemlock and northern hardwoods) at lower elevations. Occurs in recent clearcuts, riparian thickets, dense, tall deciduous or mixed understory, young forest stands, or mature forests. Prefers forest interiors to edge. Breeds in mixed forest at high elevations in both undisturbed and selectively logged habitat in Vermont (Kibbe 1985:244 in Laughlin and Kibbe 1985). Rarely below 1,000 feet (300 m) or above 4,500 feet (1,370 m) in New Hampshire (Elkins 1994:220 in Foss 1994). First appear in clear-cut areas 2 yr after harvest, become common after 4 yr, and begin to decline after 15 yr, and again in moist mixed old-growth forests (DeGraaf 1991), as in the West (Mannan and Meslow 1984).

Special Habitat Requirements: Coniferous or mixed forests, especially adjacent to water or in low damp areas.

Nesting: Egg dates: June 10 to July 11, New York (Bull 1974:433). Clutch size: 3 to 5, typically 3 to 4. Incubation period: 10 to 14 days. Nestling period: 10 to 12 days. Broods per year: 1. Age at sexual maturity: 1 yr. Nest height: 2 to 20 ft (0.6 to 6.1 m), typically 4 to 8 ft (1.2 to 2.4 m). Nest site: Usually in a crotch close to the trunk or on a horizontal limb of a spruce or fir tree, often in dense stands of small coniferous trees.

Food Habits: Beetles, ants, weevils, bees, caterpillars, spruce-bud moth, spiders, and wild fruits. Beal (1915 in Bent 1949:181) found that the March to November diet of 403 birds consisted of 64 percent animal and 36 percent vegetable matter. Feeds on the forest floor, foliage and branch surfaces, often high in trees.

Comments: Swainson's thrushes nest very consistently between 700 and 1,100 m (2,300 to 3,600 ft) elevation in Vermont.

Hermit Thrush

(*Catharus guttatus*)

Range: Breeds from w. and central Alaska, n. Saskatchewan, and Newfoundland, s. to s. Alaska, in the mountains to s. California, s. Nevada, s. New Mexico, and w. Texas, and e. of the Rockies to central Alberta, central Wisconsin, s. Ontario, central Pennsylvania, w. Virginia, w. Maryland, s. New York, and in the Black Hills in sw. South Dakota. Winter: Southern British Columbia and the ne. United States, s. through Mexico (except Yucatan Peninsula) to Guatemala, s. Texas, and Florida.

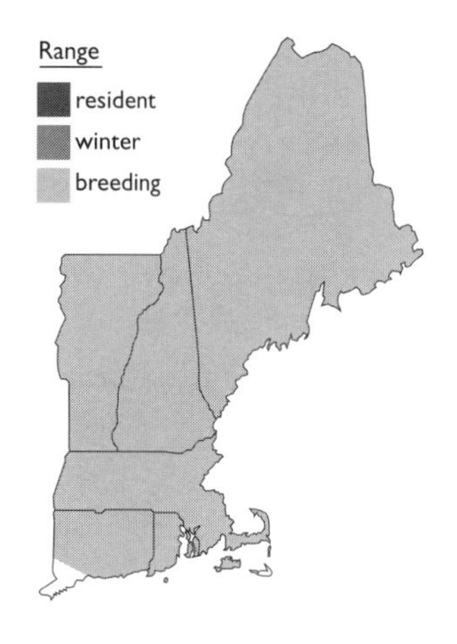

Distribution in New England: Breeding: Throughout the region except the spruce-covered outer islands of the Maine coast (Morse 1972). Winter: Southeastern Massachusetts, s. Rhode Island, and s. Connecticut.

Status in New England: Common breeder (Maine) to uncommon (se. Massachusetts and Connecticut); fairly common migrant; rare to uncommon, but regular in all but the most severe winters in s. New England (Veit and Petersen 1993:354).

Habitat: Generally inhabits dense coniferous or mixed forests with a nearly closed or closed canopy, or forest edge habitat. Occupies dry mixed forests, especially pine-oak, in Maine (Morse 1971); dry forests with a moderate understory away from streams in New Hampshire (Holmes and Robinson 1988), and mid-successional sapling to saw-timber forests stands in Vermont (Noon 1981). Restricted to coniferous forests (hemlock and white pine) in the interior of Massachusetts, and pine-oak forests in the se. part of the state (Veit and Petersen 1993:354). Also inhabits forest edge habitat such as margins of ponds and lakes, meadows, roads, utility cuts, mountain bogs, clear-cuts (Dilger 1956, Peck and James 1987, Jones and Donovan 1996). Not sensitive to intensive forest management in New England, as it is in se. Wyoming (Keller and Anderson 1992).

Special Habitat Requirements: Relatively extensive, rather dense coniferous or mixed forests.

Nesting: Egg dates: May 17 to June 14, Massachusetts (Veit and Petersen 1993:354). May 12 to August 24, New York (Bull 1974:431). Clutch size: 3 to 6, typically 3 or 4. Incubation period: 11 to 13 days, typically 12. Nestling period: 12 days. Broods per year: 2. Age at sexual maturity: 1 yr. Nest height: typically on the ground. Nest site: Usually under rock ledges, or under low overhanging limbs. Occasionally locates nest at the base or low to the ground in small trees or shrubs. Generally nests in undisturbed forests, occasionally nests in disturbed habitat such as clear cuts, old fields, or second growth areas (Jones and Donovan 1996).

Territory Size: 1.8 acres, range 0.15 to 8.35 acres (0.72 ha, range 0.06 to 3.34 ha) in Algonquin Provincial Park, Ontario (Martin 1960).

Sample Densities: 40 to 93 birds per 100 acres (40 ha) in second-growth or cutover woods (Fawks 1937, 1938). 20 birds per 100 acres (40 ha) in bottomland forest in Illinois (Karr 1968). 6 to 10 birds per 100 acres (40 ha) in upland forest in Illinois (Weise 1951 in Graber et al. 1971). 15.2 individuals per 100 acres (40 ha) in New Hampshire (Holmes et al. 1986). 5 to 8 pairs per 100 acres (40 ha) in Vermont (Carpenter 1973).

Food Habits: Insects, worms, snails, amphibians, reptiles, fruits. Beal (1915 in Bent 1949:153) found the stomach contents of 551 hermit thrushes contained 65 percent animal and 35 percent vegetable matter. Birds switch to mainly vegetable matter in fall and winter (berries and buds). Hermit thrushes explore leaf litter, bark while gleaning (Holmes and Robinson 1988).

Comments: The hermit thrush has expanded its range southward into Connecticut and Rhode Island in historic times (Zeranski and Baptist 1990:200). The hermit thrush is the state bird of Vermont.

Wood Thrush

(*Hylocichla mustelina*)

Range: Breeds from se. North Dakota, n. Michigan, n. Vermont, Maine, and Nova Scotia, s. to e.-central

Texas, the Gulf Coast, and n. Florida, and w. to e. South Dakota, central Kansas, and e. Oklahoma. Migrates through Cuba and the Bahamas. Winter: From s. Texas south to Panama.

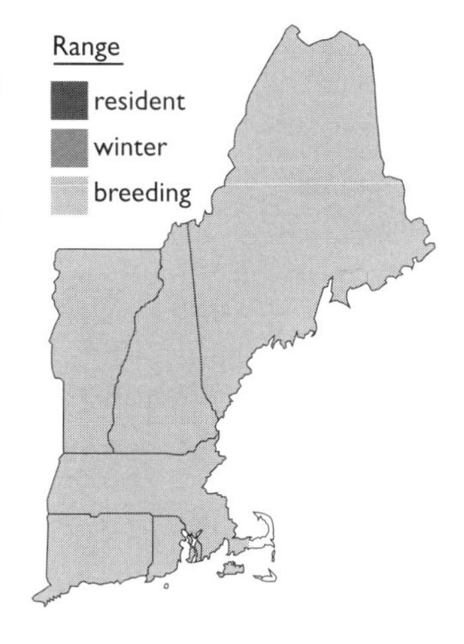

Distribution in New England: Breeds throughout the region, except absent from Nantucket and rare on outer Cape Cod.

Status in New England: Common, but declining across the breeding range.

Habitat: Nests in both the interiors and edges of mature, deciduous or mixed forests, particularly damp woodlands near swamps or water. Primary habitat features include trees taller than 53 ft (16 m), a shrub-subcanopy layer, shade, moist soil, and leaf litter (Roth et al. 1996). In New England, found on wooded slopes up to 2,000 ft (606 m) elevation (Terres 1980). Also nests in wooded residential areas and city parks. May be somewhat tolerant of forest fragmentation on breeding grounds, commonly found in woodlots 2.5 to 12.5 acres (1 to 5 ha in size) (Whitcomb et al. 1981, Roth and Johnson 1993). First appear in clear-cut areas after 10 yr, and become common after 15 yr (DeGraaf 1991). Habitat preference overlaps those of the veery and hermit thrush, but the wood thrush prefers more mature forest than do the others.

Special Habitat Requirements: Mature, moist deciduous or mixed forests with closed canopies.

Nesting: Egg dates: May 25 to June 26, Massachusetts (Veit and Petersen 1993:356). May 17 to July 7, New York (Bull 1974:429). Clutch size: 2 to 5, typically 3 or 4. Incubation period: 13 to 14 days. Nestling period: 12 to 13 days. Broods per year: 2. Age at sexual maturity: 1 yr. Nest height: Most nests located below 20 ft (6 m) in shrubs and small trees, average nest height 10 ft (3 m) above ground, occasionally located to 60 ft (20 m) in trees (Harrison 1975, Roth et al. 1996). Nest site: Builds a compact cup nest on a horizontal limb, in a fork of a sapling or tree, or well hidden in dense shrubbery. Hemlock is the preferred nest tree in Vermont (Kibbe 1985:248 in Laughlin and Kibbe 1985).

Territory Size: 0.2 to 7 acres (0.08 to 2.8 ha) in forested habitat (Weaver 1939, Graber et al. 1971). 1.4 acres (0.6 ha) in woodland edge in Kansas (Fitch 1958).

Sample Densities: Maryland—40 territorial males per 100 acres (40 ha) in virgin hardwood deciduous forest. 16 territorial males per 100 (40 ha) in shrub swamp. 11 territorial males per 100 acres (40 ha) in mature northern hardwood forest. 10 territorial males per 100 acres (40 ha) in mixed oak forest. 4 territorial males per 100 acres (40 ha) in pine-oak forest (Stewart and Robbins 1958:246). 40 territorial males per 100 acres (40 ha) in ash-basswood forests in New York (Samson 1979). 9.2 ± 5.6 territorial males per 100 acres (40 ha) in New Hampshire (Holmes and Sherry 1988). 15 ± 2.8 territorial males in a 37.5-acre (15-ha) suburban woods in Delaware (Roth and Johnson 1993). 30 to 37 territorial males per 100 acres (40 ha) in a mixed forest-homestead site in Vermont, but 5 to 26 territorial males per 100 acres (40 ha) considered more typical (Nicholson 1975, 1978, Kibbe 1985:248 in Laughlin and Kibbe 1985).

Food Habits: Insects, fruits. Gleans from leaf litter on the ground and from understory vegetation. Like the veery, turns leaves over with bill while foraging in the leaf litter.

Comments: The wood thrush is the largest woodland thrush in New England, common in s. New England since the 1800s. The reversion of former agricultural land to forest greatly increased the amount of breeding habitat for the wood thrush, but its wintering grounds continue to be cleared in Mexico.

American Robin

(*Turdus migratorius*)

Range: Breeds from w. and n. Alaska, se. Northwest Territories, n. Quebec, Labrador, and Newfoundland, s. to

s. California, central Arizona, Mexico, s. Texas, and central Florida. Winter: From s. Alaska, the n. United States, and Newfoundland, s. to Baja California, s. Texas, and s. Florida, throughout Mexico to Guatemala, the Bahamas, and Cuba.

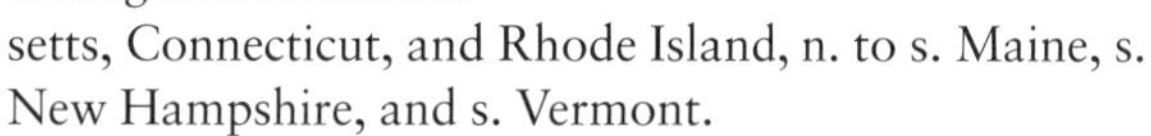

Distribution in New England: Breeds throughout the region. Resident throughout Massachusetts, Connecticut, and Rhode Island, n. to s. Maine, s. New Hampshire, and s. Vermont.

Status in New England: Abundant breeder and migrant, locally common winter resident. Probably the most abundant and widespread inland New England breeding bird in settled areas, it is scarce in heavily forested parts of the region.

Habitat: Ubiquitous in open deciduous and mixed woodlands and woodland edges, fields, orchards, farmlands, hedges, city parks, and suburban yards. Resident where trees and shrubs with persistent fruits and berries provide food and sheltered wooded areas provide roosting cover.

Special Habitat Requirements: Conifers for early nests.

Nesting: Egg dates: April 12 to July 25, Massachusetts (Veit and Petersen 1993:358). March 23 to July 19, New York (Bull 1974:428). Clutch size: 2 to 7 (Howard 1967), typically 3 or 4 (Young 1955). Incubation period: 12 to 14 days (Young 1955). Nestling period: 14 to 16 days. Broods per year: 2 or 3. Age at sexual maturity: 1 yr. Nest height: To 70 ft (21.3 m), typically 5 to 15 ft (1.5 to 4.5 m). Nest site: The first nest of the season is usually in a conifer, but will use deciduous trees for successive nests (Howell 1942). Late-season nest success is higher in deciduous trees in full leaf than in conifers (Knupp et al. 1977). White pine, maple, and apple trees are preferred nest trees (DeGraaf et al. 1975). Nests are mud-lined and are usually built on a horizontal branch or in a fork of a tree (Nickell 1944), but commonly uses shrubs and the ledges of buildings.

Territory Size: 0.30 to 0.75 acre (0.1 to 0.3 ha) (Collins and Boyajian 1965:133). 0.11 to 0.60 acres (0.4 to 0.24 ha) (average 0.30 acres, 0.1 ha) in Wisconsin (Young 1951). 41 pairs per 100 acres (40 ha) in Vermont (Nicholson 1978).

Sample Densities: 132 birds per 100 acres (40 ha) in urban residential areas in Illinois (Graber et al. 1971). 56 birds per 100 acres (40 ha) in edge shrubbery in central Illinois (Graber et al. 1971). 14 birds per 100 acres (40 ha) in second-growth or cutover woods in Illinois (Fawks 1937, 1938).

Food Habits: Wild and cultivated fruits, earthworms, insects. Foraging behavior related to grass length, and mowing (Eiserer 1980). Feeds in rich loamy soil, on fruit-bearing trees, shrubs, vines, and lawns. About 60 percent of the diet is vegetable matter and 40 percent animal matter (Bent 1949:25).

Comments: The robin is the state bird of Connecticut, and has probably always been common in the region. A study of banding returns by Hickey (1943) indicated that almost three-fourths of the young robins that survived their first winter returned to nest within 16 km (10 miles) of their birthplaces. Although suburban areas appear to be ideal robin habitat, productivity there is low due to predation by crows on eggs and nestlings and by cats on fledglings (Howard 1974).

Gray Catbird

(*Dumetella carolinensis*)

Range: Breeds from s. British Columbia, s. Ontario, and Nova Scotia, s. to central New Mexico and n. Florida, and w. to n. and s.-central Washington, s.-central and e. Oregon, n.-central Utah, and central and ne. Arizona. Winter: From n.-central and e. Texas, the central portions of the Gulf States, and the Atlantic Coast lowlands from se. Massachusetts, s. along the Gulf-Caribbean slope of Central America to Panama. Also winters in the Bahamas, Cuba, Jamaica, and Hispaniola.

Distribution in New England: Throughout the region. Some winter in se. Massachusetts, s. Rhode Island, and s. Connecticut.

Status in New England: Common breeder throughout the region, uncommon in winter along s. New England coasts.

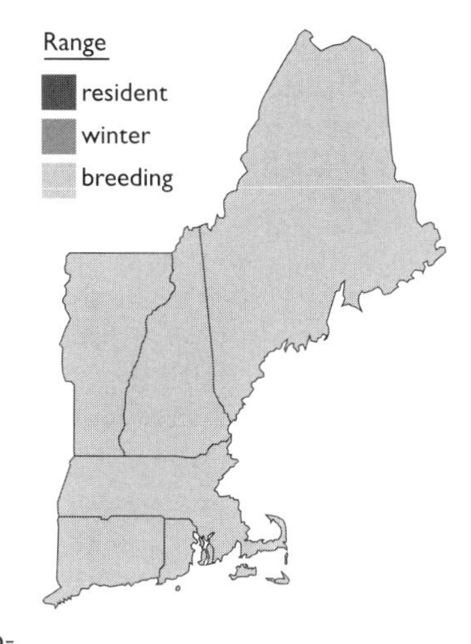

Habitat: Commonly in dense shrubby thickets, briars, vines along woodland borders, regenerating hardwoods, roadside shrubs, old house sites, abandoned farmlands and home sites. Rarely found in coniferous forests. Most abundant in coastal lowland shrublands and dense shrub-swamps in Massachusetts and Rhode Island (Veit and Petersen 1993:360, Enser 1992:131), associated with moist thickets along streams or woodland margins in Massachusetts and New Hampshire (Veit and Petersen 1993:360, Robbins 1994:228 in Foss 1994). One of seven species found most often in deciduous wetlands in central Massachusetts (Swift et al. 1984). In drier parts of the West, uses low moist thickets of forest edges or open places, riparian thickets, and aspen forest with dense understories. Winters primarily in forested habitats in the tropics (Cimprich and Moore 1995). Inhabits dense shrubby habitats in mature forests (Hamel 1992b); rain forest, gallery forest, pine plantations, and shaded pastures (Rappole and Warner 1980, Rappole et al. 1983, Rappole et al. 1992).

Special Habitat Requirements: Low, dense shrubby vegetation.

Nesting: Egg dates: May 22 to August 10, Massachusetts (Veit and Petersen 1993:360). May 5 to June 13, New York (Bull 1974:422). Clutch size: 3 to 5, typically 4 (Harrison 1975). Incubation period: 12 to 15 days. Nestling period: 9 to 15 days, typically 11. Broods per year: 2. Age at sexual maturity: 1 yr. Nest height: 2 to 10 ft (1 to 3 m), typically usually within 6 ft (2 m) of the ground (Cimprich and Moore 1995). Nest site: Builds in dense thickets of briars, vines, or shrubs such as multiflora rose, barberry, hedges, or occasionally in an evergreen shrub (Nickell 1965). Nests are typically well hidden by foliage (Stauffer and Best 1986, DeGraaf and Rappole 1995:381).

Territory Size: 0.16 to 0.36 acres (0.06 to 0.1 ha), average 0.26 acres (0.1 ha) in swamp-thicket in Illinois (Brewer 1955). 0.8 acres (0.32 ha) in s. Ontario (Darly et al. 1971).

Sample Densities: 10 territorial males per 100 acres (40 ha) in streamside habitat in Vermont (Farrar 1973). 1 nest per 8 acres (3.2 ha) in mixed shrub-small tree habitat in beech-maple-hemlock community in New York (Kendeigh 1946). 80 territorial males per 100 acres (40 ha) in scrub-shrub wetland, 35 territorial males per 100 acres (40 ha) in brushy abandoned farmland in Maryland (Stewart and Robbins 1958:241). 5 territorial males per 100 acres (40 ha) in ash-basswood habitat in New York (Samson 1979).

Food Habits: Small fruits, insects. Feeds in leaf litter on the ground, in the understory, and in the canopy (Cimprich and Moore 1995).

Comments: Catbirds use a wide variety of habitats and have been common in s. New England throughout the historical period. Connecticut had the highest counts of catbirds on the continent during the Breeding Bird Survey 1965–1979 (Robbins et al. 1986).

Northern Mockingbird

(*Mimus polyglottos*)

Range: Resident regularly from n. California, e. Oregon, nw. Nevada, n. Utah, se. Wyoming, sw. South Dakota, e. Nebraska, e. to s. New York and central New England, sporadically or locally n. to s. Prairie Provinces, central and ne. Minnesota, s. Wisconsin, s. Michigan, s. Ontario, sw. Quebec, Nova Scotia and Newfoundland, s. to s. Baja California, through Mexico to Oaxaca and Veracruz, and to se. Texas, the Gulf Coast, s. Florida (including the Florida Keys), the Bahama Islands and

Greater Antilles (e. to Anegada in the Virgin Islands, and recently on Little Cayman in the Cayman Islands). Northern populations are partially migratory. Casual n. to British Columbia, sw. Keewatin, n. Ontario, and Prince Edward Island.

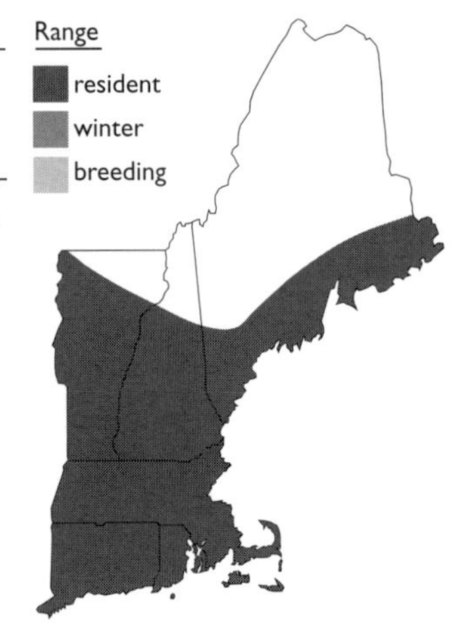

Distribution in New England: Breeds throughout the region, except ne. Vermont, n. New Hampshire, and n. Maine. Winters from s. Vermont, New Hampshire, and coastal Maine s.

Status in New England: Common resident.

Habitat: A wide variety of open to partly open habitats: woodland edges, pastures with scattered fruit-bearing shrubs or small trees, cultivated areas, gardens, parks, often in cities and suburban areas. Wintering habitat similar to breeding habitat; among thickets that bear persistent fruits, especially multiflora rose. In New England, almost always associated with multiflora rose in winter.

Special Habitat Requirements: Low, dense woody vegetation, elevated perches, a variety of persistent edible fruits.

Nesting: Egg dates: April 27 to July 21, New York (Bull 1974:425). June 3 to August 19 in Massachusetts (Veit and Petersen 1993:362). Clutch size: 3 to 6, typically 4 or 5 (Derrickson and Breitwisch 1992). Incubation period: 11 to 13 days (Peck and James 1987). Nestling period: 12 days (Laskey 1962). Broods per year: 2 or more (Laskey 1962). Age at sexual maturity: 1 yr. Nest height: 1.5 to 10 ft (0.5 to 3.0 m). Typically 3 to 10 ft (0.9 to 3.0 m). Nest site: Usually in a thicket or vines or in a dense tree (often an evergreen). Prefers sites near houses, especially porch vines, garden, lawn and foundation plantings. Prefers to nest in multiflora rose (DeGraaf et al. 1975).

Territory Size: Territories of 5 pairs of mockingbirds ranged from 26,650 to 60,000 square feet (2,475 to 5,573.3 m^2) (Michener and Michener 1935). Winter: Four females defended areas that ranged from 3,750 to 20,000 square feet (348.3 to 1857.8 m^2) (Michener and Michener 1935).

Sample Densities: Maryland: 15 territorial males per 100 acres (40 ha) in suburban-residential habitat. 2 territorial males per 100 acres (40 ha) in mixed agricultural habitats (Stewart and Robbins 1958:239).

Food Habits: Wild or cultivated fruits, seeds, insects. Gleans from ground litter and grasses, shrubs, trees.

Comments: The mockingbird has been spreading northward into New England since the 1870s. It was an erratic visitor from 1870 to the 1950s and a very local rare breeder in Massachusetts (Howe and Allen 1901). In the 1960s and 1970s, mockingbirds became widespread residents in s. New England, but had already nested in se. New Hampshire by 1957 (Robbins 1994:230 in Foss 1994). The expansion of mockingbirds northward has been widely attributed to the planting and subsequent spread of multiflora rose, which provides ideal nest sites and persistent fruit through the winter. Both male and female sing; unmated males sing at night. A male's repertoire often contains more than 150 song types, acquired through imitating other birds, vocalizations of non-avian species, and mechanical sounds (Derrickson and Breitwisch 1992).

Brown Thrasher

(*Toxostoma rufum*)

Range: Breeding: From se. Alberta, e. to New England and s. to Colorado, n. and e. Texas, the Gulf Coast, and s. Florida. Winter: In the south from Texas e., ranging n. in the Mississippi Valley to Illinois and along the Atlantic Coast to Massachusetts.

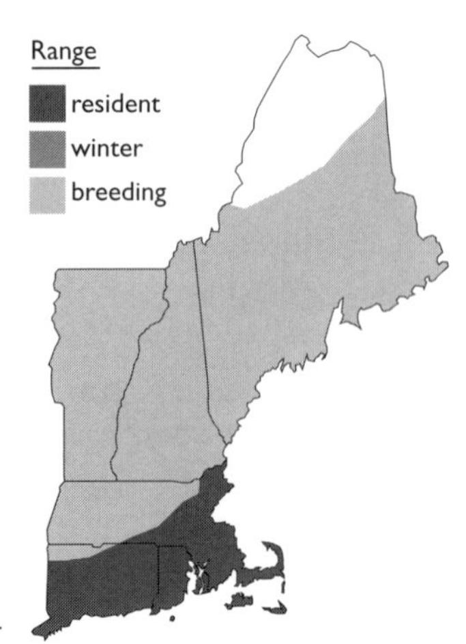

Distribution in New England: Breeds throughout the region, except absent from n. Maine.

Winter resident in e. Massachusetts and the Connecticut River Valley, Connecticut, and Rhode Island.

Status in New England: Common to fairly common breeder. Rare but regular in winter e. of the Berkshires and most numerous on the coastal plain in Massachusetts (Veit and Petersen 1993:363), and rare along the coast and very rare inland in Connecticut (Zeranski and Baptist 1990:204).

Habitat: Commonly inhabits dry thickets in wooded areas, pitch-pine/scrub-oak barrens, second growth, power-line rights-of-way, brushy fields, hedgerows, briar patches, roadsides, forest edges and clearings. Absent from higher mountains in New England. Resident primarily in coastal areas where the climate is mild and sparse snow cover allows birds to find persistent fruits and berries.

Special Habitat Requirements: Low, dense, woody vegetation for nesting and cover (Graber et al. 1970).

Nesting: Egg dates: May 9 to June 21, Massachusetts (Veit and Petersen 1993:362). May 6 to June 26, New York (Bull 1974:423). Clutch size: 2 to 5, typically 4. Incubation period: 12 to 13 days. Nestling period: 12 to 14 days. Broods per year: 1. Age at sexual maturity: 1 yr. Nest height: To 14 ft (4.3 m), typically 2 to 7 ft (0.6 to 2.1 m). Nest site: On the ground (rarely) or low in dense cover of shrubs or vines; early nests are generally built lower than later nests (Erwin 1935).

Territory Size: Average 1.6 acres (0.6 ha) in forest edge in Illinois (Graber et al. 1970).

Sample Densities: 3 pairs per 100 acres (40 ha) in forest edge (Holmes 1950 in Graber et al. 1970). 189 birds per 100 acres (40 ha) in hedgerows and 76 birds per 100 acres (40 ha) in edge shrubbery (Graber and Graber 1963). 86 birds per 100 acres (40 ha) in second-growth or cutover woods (Fawks 1937). 1.4 territorial males in upland and swamp thickets in Massachusetts (Samson 1979).

Food Habits: Insects (about 66 percent); berries, mast (acorns), and grain (about 33 percent) (Pough 1949:110).

Comments: Brown thrashers were likely abundant in the late 1800s in New England following farmland abandonment, and are rapidly declining as forests invade remaining old fields in s. New England today (Robbins et al. 1986). Brown thrashers are less inclined to nest near human habitations than the other mimic thrushes, gray catbird and mockingbird, preferring pastures or brushy hillsides.

European Starling

(Sturnus vulgaris)

Range: Originally palearctic; breeds in North America from se. Alaska, e. across the s. half of Canada to s. Labrador, and s. throughout most of the United States except in extreme sw., though now invading the n. Sonoran Desert. Winters throughout the United States.

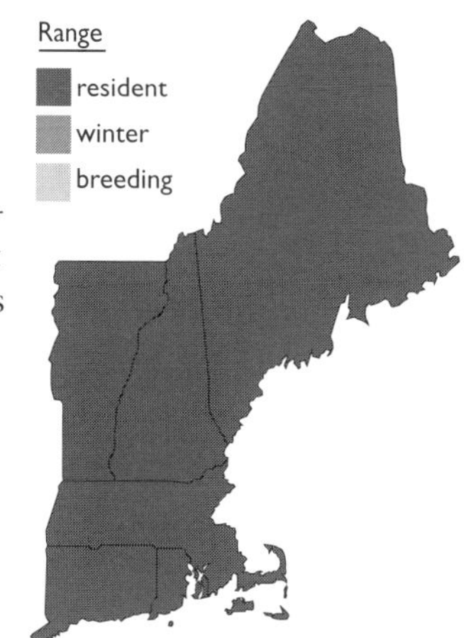

Distribution in New England: Throughout the region.

Status in New England: Common and widespread breeder in s. New England, local in n. forested country; abundant migrant, and winter resident from Massachusetts s., primarily in urban areas.

Habitat: Farms, cities, orchards, gardens, parks. Prefers rural areas with pastures, cultivated fields, and hayfields. Avoids extensive forests. In winter, roosts in dense vegetation or on buildings in villages and cities.

Special Habitat Requirements: Cavities for nesting. Minimum diameter breast height of trees suitable for nesting is about 10 in (25.4 cm).

Nesting: Egg dates: Early April to July, Massachusetts (Veit and Petersen 1993:370). April 10 to June 15, New York (Bull 1974:541). Clutch size: 2 to 7, typically 4 to 5 (Cabe 1993). Incubation period: 11 to 12 days (Cabe 1993). Nesting period: 18 to 22 days. Broods per year: 1 or 2. Probably single-brooded north of 48° latitude (Kessel 1953). Age at sexual maturity: 1 yr. Nest height:

2 to 60 ft (0.6 to 18.3 m). Typically 10 to 25 ft (3.0 to 7.6 m). Nest site: A cavity almost anywhere (selected by the male), often in natural or old woodpecker (especially flicker) holes, but commonly in barns and other buildings, drain pipes, and cupolas (Feare 1984).

Territory Size: Birds defended a 10- to 20-in (25.4- to 50.8-cm) radius around nest holes (Kessel 1957).

Sample Densities: 78 breeding females per square mile (20 breeding females/km^2) (some in nest boxes) on a farm in Scotland (Dunnet 1955).

Food Habits: Extremely diverse diet: insects, seeds, fruits, cultivated grains. Feeds on the soil surface and subsurface to depths not exceeding length of bill (Williamson 1975). Preferred feeding habitat: Lawns, meadows, grazed fields; starlings prefer to forage in low vegetation; feed up to three-fourths of a mile from nest site (Kessel 1957).

Comments: Introduced from Europe, starlings are highly adaptable and compete successfully with other birds for nest cavities. The entire North American population of starlings is believed to have descended from two different flocks released in Central Park in New York City: 20 birds were released on March 16, 1890, and 40 more were released on April 25, 1891 (Zeranski and Baptist 1990:207). By 1900 they had nested in Connecticut (Bent 1950), and they became abundant and widespread by 1920. Today the North American population is more than 200 million, and probably the most successful avian introduction to this continent (Cabe 1993). An intense competitor for nest sites, the starling has had a detrimental effect on many native cavity-nesting birds (Cabe 1993).

American Pipit

(Anthus rubescens)

Range: Holarctic; breeds in North America throughout Alaska and the e. Aleutians and from n. Yukon, s. through British Columbia, se. Alberta, Washington, and w. Montana, locally on mountaintops from Oregon, Utah, and Colorado, s. to the Sierra Nevada and on Mt. Gorgonio in California, the San Francisco and White Mountains in Arizona, central New Mexico, and from the Canadian Arctic s. to s. Mackenzie, s. Keewatin, n. Manitoba, s. Labrador, Newfoundland, and, locally, the Gaspé Peninsula, s. Quebec, n. Maine (Mt. Katahdin), and central New Hampshire (Mt. Washington). Winters in North America from the s. United States primarily in coastal areas n. to s. British Columbia and New York, casually in the interior and ne. to s. Canada, s. to Guatemala and El Salvador, s. Mexico (including the Yucatan Peninsula).

Distribution in New England: Breeds on Mount Katahdin and recently on Mount Washington (1 nest in 1991 described in Foss 1994:383). Migrates regularly through the region between the breeding and wintering seasons. Winters occasionally n. to Massachusetts (Veit and Petersen 1993:364).

Status in New England: Rare breeder (see earlier discussion); locally common to very common migrant, more numerous in fall than in spring. Rare and irregular in winter.

Habitat: Nests on Arctic tundra, alpine tundra, sedge meadows, and moss-covered rocky coastal hills along Arctic coasts that include eroded turf, tussocks, or tilted rocks to provide snow-free sites early in the nesting season (Verbeek 1970). Winter: Bare fields and grasslands, lakeshores, pastures, and other completely open areas (DeGraaf and Rappole 1995: 384). In fall migration, mostly in coastal salt marshes, mud flats, and along beaches, also along tidal creeks and open fields (Bent 1950:34).

Special Habitat Requirements: In migration: Bare fields, open shores. Breeding: Snow-free sites provided by tussocks or rocks at high elevations or high latitudes (Verbeek 1994).

Nesting: Egg dates: From about 4 days after a rise in temperature, sometimes only 4 days after the nest site is free of snow, and can extend into early July (Foss 1994:383). Clutch size: 3 to 7, typically 4 or 5. Incubation period: About 14 days. Nestling period: 14 or 15 days; adults feed young for 14 days after fledging. Nest: Constructed of grass, on the ground in or at the base of a tussock, under a rock edge or in a rock pile, or in a bare spot often partly recessed and facing away from prevailing winds and drifting snow (Verbeek 1970).

Food Habits: Insects gleaned from the ground and low vegetation while walking through vegetation, or gleans wind-blown insects on alpine snowfields (Verbeek 1970). Also small seeds and berries.

Comments: Nesting pipits were first found on Mt. Washington on June 29, 1991, by C. C. Rimmer in a sedge meadow just above 5,400 ft (1,650 m) elevation (Foss 1994:384). Several nesting pairs have been observed since (C. C. Rimmer pers. commun.). Pipits have long nested in a restricted area of alpine meadows near the summit of Mt. Katahdin; the population has been stable since 1985 (Boone and Krohn 1996:134).

Bohemian Waxwing

(Bombycilla garrulus)

Range: Holarctic; breeds in North America from central Alaska, Yukon, sw. Mackenzie, and n. Manitoba, s. to n. parts of Washington, Idaho, and Montana, central Saskatchewan, and central Manitoba. Winters s. to Washington, Colorado, the Great Lakes, and Maine, e. to Ontario, s. Quebec, Nova Scotia, and the n. tier of states, irregularly to California, Arizona, n. New Mexico, and n. Texas.

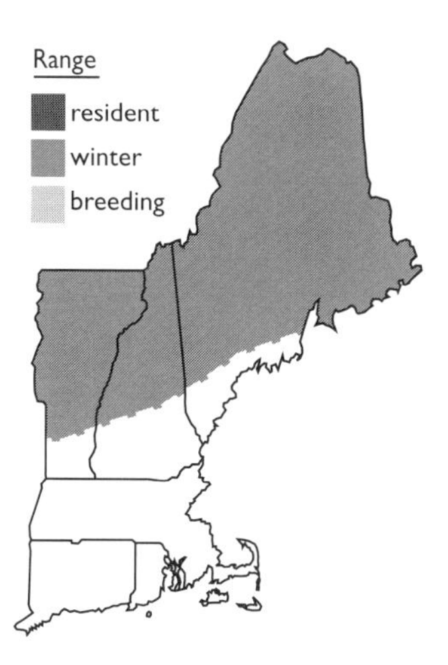

Distribution in New England: Irregular winter visitor throughout n. New England, very rare and erratic in Massachusetts and Connecticut.

Status in New England: Irregular to rare.

Habitat: Wherever persistent berries or fruits are available in boreal and temperate areas.

Food Habits: Berries in winter.

Comments: Bohemian waxwings were rarer in New England before 1960 than they are at present, with a definite increase since 1955 in Massachusetts (Veit and Petersen 1993:366).

Cedar Waxwing

(*Bombycilla cedrorum*)

Range: Breeds from se. Alaska, central British Columbia, Alberta, Saskatchewan, n. Manitoba, Ontario, central Quebec, and Newfoundland, s. to n. California, Nevada, Utah, Colorado, South Dakota, central Missouri, Illinois, Indiana, n. Georgia, w. North Carolina, and Virginia. Winters from s. British Columbia, Montana, Saskatchewan, Manitoba, Ontario, New York, New England, s. to Panama, and throughout the Greater and Lesser Antilles.

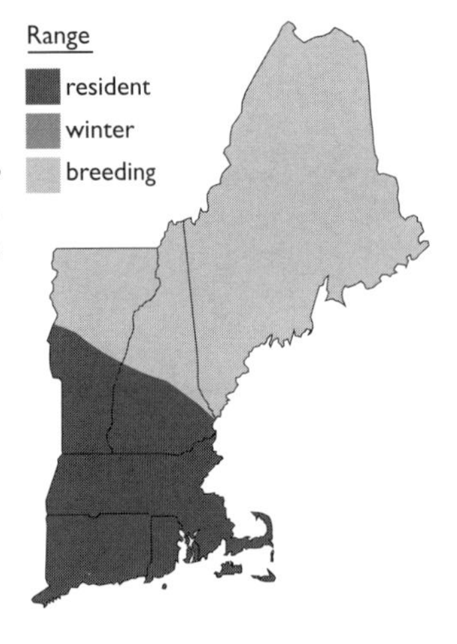

Distribution in New England: Breeding: Throughout the region. Winter: Southern Vermont, s. New Hampshire, and throughout Massachusetts, Connecticut, and Rhode Island.

Status in New England: Locally common breeder, generally increasing. Erratic migrant; rare to locally common in winter, especially near the coast.

Habitat: Inhabits a wide variety of habitats that provide fruits and berries: old fields, open deciduous and coniferous forests (avoids dense forests), forest edges, deciduous forested wetland edges, agricultural land, orchards, and residential areas (Lea 1942). Frequently nests in riparian areas that provide nesting shrubs and trees, fruits, and emerging aquatic insects (Witmer et al. 1997).

Special Habitat Requirements: Trees and shrubs that produce fruit and berries.

Nesting: Egg dates: May 30 to late August, Massachusetts (Veit and Petersen 1993:367). June 5 to September, New York (Bull 1974:450). Clutch size: 2 to 6, typically 4 or 5. Incubation period: 12 to 16 days. Nestling period: 12 to 18 days. Broods per year: 1 or 2. Age at sexual maturity: 1 yr. Nest height: 4 to 50 ft (1.2 to 15.2 m). Typically 6 to 20 ft (1.8 to 6.1 m). Nest site: Prefers dense coniferous thickets (often cedar), but nests in a variety of deciduous trees and shrubs. Nest is placed on a horizontal limb, often in a crotch next to main trunk.

Territory Size: 3 territories on an island in Lake Erie had the following areas: 0.06, 0.5, and 0.23 acre (0.02, 0.2, and 0.09 ha) (Putnam 1949).

Sample Densities: 20 nests were found in a 2.3-acre (0.9-ha) white pine plantation in Michigan (Rothstein 1971). 11 nests were located within a radius of 25 ft (7.6 m) in Ontario (Harrison 1975:166). 16 pairs per 100 acres (40 ha) in open hemlock-spruce bog in Maryland (Stewart and Robbins 1958). 4.8 territorial males per 100 acres (40 ha) in ash-basswood forests in Pennsylvania (Samson 1979).

Food Habits: Fresh and dried sugary fruits and flowers and insects taken on the wing.

Comments: Cedar waxwings are one of New England's most common breeding birds. Waxwings tend to nest late in summer when there is an abundant supply of wild fruits. Birds nest singly or in loose colonies. A second nest is often begun and eggs laid before the young in the first nest have fledged. During most of the year they nomadically roam the countryside in small to large flocks, behavior that is characteristic of species that specialize on patchily distributed foods (Witmer et al. 1997).

Blue-winged Warbler

(*Vermivora pinus*)

Range: Breeds from e. Nebraska and se. Minnesota, e. to s. Vermont and s. Maine, s. to Arkansas, n. Alabama, n. Georgia, w. South Carolina, and Delaware. Range is expanding in the Northeast. Winter: Winters in Mexico and n. Central America, casually to Panama.

Range
resident
winter
breeding

Distribution in New England: Southern Vermont, s. New Hampshire, and the s. tip of Maine, throughout Connecticut, Rhode Island, and Massachusetts except absent on outer Cape Cod and Nantucket. Most abundant in the Connecticut River Valley.

Status in New England: Uncommon to locally common breeder and increasing; uncommon migrant away from breeding localities (Veit and Petersen 1993:378).

Habitat: Nests in brushy growth near the borders of swamps or streams, forest edges, abandoned fields and pastures, thickets, and second-growth woods. Prefers brushy old pastures and old fields with saplings more than 10 ft (3 m) tall (Robbins et al. 1966:254). Prefers moister habitat than the closely related golden-winged warbler. In a regenerating hardwood forest, blue-winged warblers were abundant in regeneration-age stands (1 to 9 yr) and common in sapling-age stands (10 to 20 yr), but absent in older stands in Texas (Dickson et al. 1993).

Special Habitat Requirements: Old fields with scattered shrubs and small trees, commonly near water.

Nesting: Egg dates: May 29 to June 6, Massachusetts (Veit and Petersen 1993:378). May 18 to June 17, New York (Bull 1974:468). Clutch size: 4 to 7, typically 5. Incubation period: 10 to 12 days. Nestling period: 8 to 10 days. Broods per year: 1. Age at sexual maturity: 1 yr. Nest site: On the ground on a foundation of dry leaves, surrounded by bushes or tangles of vines and grasses.

Territory Size: Less than 1 acre (0.4 ha) to almost 2 acres (0.8 ha) per pair (New York) (Ficken and Ficken 1968). Burke Lake, Mich., 2.0 ha (5 acres) per pair—habitat: An extensive tamarack swamp surrounded by higher, drier, oak-hickory woods. Island Lake, Mich., 1.3 ha (4.6 acres) per pair—habitat: A low swamp of tamarack, poison sumac, red osier, and gray dogwoods and poplars (Murray and Gill 1976).

Food Habits: Caterpillars, beetles, ants, spiders from twigs of shrubs and tops of small trees.

Comments: Blue-winged warblers historically occurred w. of the Appalachians (Short 1962) and expanded their range into New England through Connecticut about 1880. The blue-winged is expanding its range potentially at the expense of the closely related golden-winged warbler, which is declining, although the two use quite similar habitats. Blue-winged warblers hybridize with golden-winged warblers. There are two general hybrid types, although the intergrades are continuously variable. "Brewster's" is most common; cross-breeding with the golden-winged parental form produces the much rarer "Lawrence's." The blue-winged warbler replaces the golden-winged warbler within 50 yr of initial hybridization (Confer and Knapp 1981).

Golden-winged Warbler

(*Vermivora chrysoptera*)

Range: Breeds from s. Manitoba, central Minnesota, and n. Wisconsin, e. to w. Vermont, s. New Hampshire, w. Massachusetts, and s. to sw. Iowa, s. Ohio, and s. Connecticut; in the Appalachian Mountains, s. to n. Georgia. Winters in Central America, Colombia, and Venezuela.

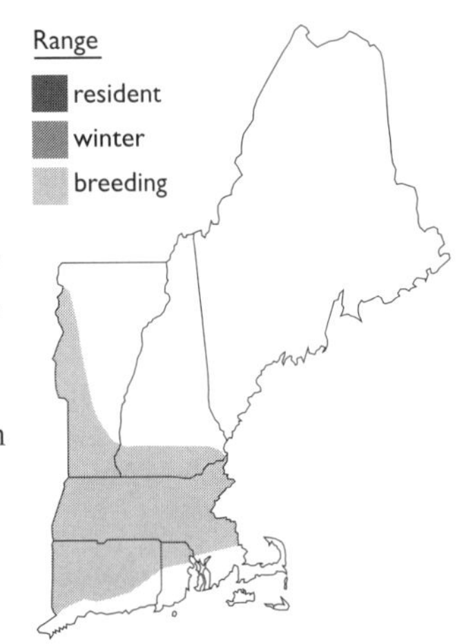

Distribution in New England: Western and s. Vermont, s. New Hampshire with most sightings along the coastal plain (Gavutis 1994:252 in Foss 1994), throughout Massachusetts except the se., Cape Cod, n. Rhode Island, and n. Connecticut. Does not occur in Maine.

Status in New England: Very uncommon, local, and decreasing breeder. Rare migrant away from known breeding areas (Veit and Petersen 1993:379).

Habitat: Early-successional openings in deciduous forests that follow fire or logging, second-growth woods, especially gray birch stands; dense scrubby thickets, and brush-bordered lowland areas. Also inhabits old fields or overgrown pastures with few trees and a dense understory of forbs, grasses, or ferns. Habitat is similar to that used by the invasive blue-winged warbler. Sometimes found on higher ground but avoids mountains.

Special Habitat Requirements: Brushy open areas, especially clearings in deciduous woodlands with saplings, forbs, grasses.

Nesting: Egg dates: May 20 to June 20, Massachusetts (Veit and Petersen 1993:380). May 18 to June 16, New York (Bull 1974:469). Clutch size: 4 to 6, typically 4 to 5. Incubation period: 10 to 12 days. Nestling period: 8 to 10 days. Broods per year: 1. Age at sexual maturity: 1 yr. Nest site: On or near the ground, built of dried grasses and bark. Generally supported by weed stalks, or at the base of shrubs or trees.

Territory Size: Less than 1 acre (0.4 ha) to almost 2 acres (0.8 ha) per pair (New York) (Ficken and Ficken 1968). Territories usually consisted of overgrown fields with many shrubs and small trees (under 20 ft) (6.0 m), bordered by taller deciduous trees. Burke Lake, Mich., 2.7 ha (6.7 acres) per pair—habitat: An extensive tamarack swamp surrounded by higher, drier oak-hickory woods. Island Lake, Mich., 1.9 ha (4.7 acres) per pair—habitat: A low swamp of tamarack, poison sumac, red osier, and gray dogwood and poplars (Murray and Gill 1976).

Sample Density: 17 territorial males per 100 acres (40 ha) in dense second-growth forest in Maryland (Stewart and Robbins 1958:276).

Food Habits: Small insects and larvae, canker-worms, spiders at the terminal twigs of high branches in tall trees.

Comments: Like the blue-winged warbler, golden-winged warblers expanded their breeding range into New England in the late 1800s. The golden-winged warbler arrived sooner and was a common breeder before the blue-winged warbler, but is declining in s. New England and expanding in the n. part of its range. See Zeranski and Baptist (1990:214) for a historical review. Occupies similar habitat as the blue-winged warbler, with which it interbreeds; the fertile hybrids are often called "Brewster's" and "Lawrence's" warblers (Confer 1992b). Golden-winged warblers may breed at higher elevations and slightly farther north than blue-winged warblers (Bull 1974:469). The rapid decline of golden-winged warbler in s. New England since the 1970s may be due to displacement by the blue-winged warbler and a loss of early successional habitat. It may be possible to create habitat for the golden-winged warbler by logging, burning, and intermittent farming (Confer 1992a). Golden-winged warblers are occasionally encountered on breeding bird surveys in regenerating hardwood clear-cuts as far n. as the White Mountains (DeGraaf pers. observ.).

Tennessee Warbler

(*Vermivora peregrina*)

Range: Breeds from se. Alaska and s. Yukon across Canada to n.-central Quebec and s. Labrador, and s. to s.-central British Columbia, nw. Montana, n. Minnesota, ne. New York, and s. Maine. Winter: Mexico to Colombia and Venezuela.

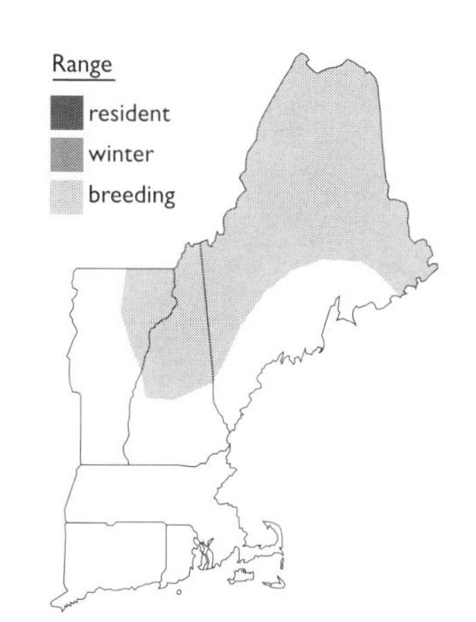

Distribution in New England: Northern New Hampshire and Vermont with two confirmed breeding locations (Winhall and Rutland) in s. Vermont (Oatman 1985:274 in Laughlin and Kibbe 1985). Throughout Maine except the s.-coastal region. Absent as a breeder from Massachusetts, Rhode Island, and Connecticut.

Status in New England: Common breeder in Maine (Boone and Krohn 1996), and n. New Hampshire; common to uncommon migrant in spring, rare in fall.

Habitat: Northern coniferous and mixed forests. Inhabits forest openings with grass, dense shrubs, and scattered clumps of young trees (Pough 1949:156). Found in tamarack or white-cedar bogs with abundant sphagnum, and muskeg borders with grass and low bushes present in Michigan (Baker 1979). Occasionally in dry pine lands, or in recent clear-cuts where small trees have been left (Bowdish and Philipp 1916). In Vermont, favors brushy areas scattered with small conifers at the edges of bogs, streams and other damp places (Oatman 1985 in Laughlin and Kibbe 1985). Generally inhabits northern tamarack swamps at elevations between 1,000 to 2,000 ft (300 to 610 m) and spruce-fir forests at elevations higher than 4,000 ft (1,220 m) in New Hampshire (Richards 1994:254 in Foss 1994). Occasionally found in regenerating clear-cuts in northern hardwoods in the White Mountains (DeGraaf pers. observ.). Abundant in spruce budworm outbreak areas and absent from non-budworm areas in n. New England (Crawford and Titterington 1979).

Special Habitat Requirements: Brushy, semiopen habitat in coniferous or mixed forests.

Nesting: Egg dates: June 10 to July 10. Peak: June 17 to June 26, New Brunswick (Bent 1953:89). Clutch size: 4 to 7, typically 6. Incubation period: 11 to 12 days. Broods per year: 1. Nest site: A cup nest on the ground, typically in sphagnum moss or grass, usually well concealed by overtopping vegetation.

Food Habits: Almost entirely insectivorous. Populations fluctuate markedly from year to year in response to periodic outbreaks of spruce budworm caterpillars on which the species is a specialist (Rimmer and McFarland 1998). Also takes weevils, flies, plantlice, grasshoppers, caterpillars, grubs, beetles, spiders, some fruit. Feeds in the terminal foliage of trees, generally feeding to 40 ft (12.2 m) high (MacArthur 1958).

Comments: A spruce budworm specialist, populations fluctuate with spruce budworm outbreaks (Richards 1994:254 in Foss 1994).

Nashville Warbler

(*Vermivora ruficapilla*)

Range: Breeds in s. interior British Columbia and nw. Montana, s. to nw. and s.-central California and extreme w.-central Nevada; and from central Saskatchewan to s. Quebec, Nova Scotia, and New Brunswick, s. to s. Manitoba, s. Wisconsin, s. Michigan, n. New Jersey, and Rhode Island. Winter: Southern Texas, s. to Mexico, Guatemala, and Belize, rarely in California and s. Florida.

Distribution in New England: Breeds throughout the region, except absent from Cape Cod and Nantucket, Mass.

Status in New England: Common breeder in n. New England, fairly common to uncommon in Massachusetts, rare in Connecticut. Uncommon migrant.

Habitat: Inhabits a variety of habitats, from cut-over second-growth, deciduous or mixed forests with an open canopy and shrubby undergrowth (Dobkin 1994), to tamarack and spruce bogs with sphagnum. Prefers an understory of twigs, leaves, and branches (Sodhi and Paszkowski 1995). Also occurs in regenerating areas that have been burned or cut, abandoned pastures and fields, woodland edges, swales, and shallow to dry bedrock openings in forests, especially those containing aspen or birch (DeGraaf and Rappole 1995). In the East, sometimes inhabits mountain slopes as high as 4,620 ft (1,400 m), but not found above timberline (Williams 1996a). In boreal Canada, primarily in black spruce woods, also in birch and poplar (Williams 1996a). Associated with habitat with a dense ground cover (70 percent) and a moderate canopy cover (65.9 percent) in west-central Maine (Collins 1983b). Occurs in open canopy (50 percent) forests, especially regenerating clear-cut areas in Grafton, Vt. (Thompson and Capen 1988). Uncommon in fields and disturbed seral forests in Vermont (Kibbe 1985:280 in Laughlin and Kibbe 1985).

Special Habitat Requirements: Second-growth deciduous forests with scattered trees interspersed with brush.

Nesting: Egg dates: May 21 to June 21, Massachusetts (Veit and Petersen 1993:383). May 19 to June 10, New York (Bull 1974:477). Clutch size: 3 to 5. Incubation period: 11 days. Nestling period: 11 to 12 days. Broods per year: 1. Age at sexual maturity: 1 yr. Nest site: Depression in moss or beneath canopy of dried, dead bracken fern; well hidden.

Territory Size: About 1/2 acre (0.2 ha) per pair in Ontario (Lawrence 1948). 1.1 ha in the White Mountains in New Hampshire (Sabo 1980).

Sample Densities: 39 territorial males per 100 acres (40 ha) in scrub spruce bog in Maryland. 21 territorial males per 100 acres (40 ha) in open hemlock-spruce bog in Maryland (Stewart and Robbins 1958:280). 9 territorial males/km^2 in mixed forest, increased to 24 pair/km^2 in a virgin spruce forest in the White Mountains in New Hampshire (Sabo 1980). 5 to 15 territorial males per 100 acres (40 ha) in Vermont (Kibbe 1985:280 in Laughlin and Kibbe 1985).

Food Habits: Adults, larvae, and eggs of various insects including small grasshoppers, plantlice, caterpillars, and beetles. Hops from the bottom to the top of a tree, hawking insects encountered.

Comments: The Nashville warbler, so named by Alexander Wilson, who first saw it in 1811 near Nashville, Tennessee (Williams 1996a), expanded its range into New England from w. of the Appalachians in historical times, beginning about 1830 in Connecticut (Bagg and Eliot 1937). Declining after 1900, its breeding range retreated northward and to higher elevations (Griscom and Snyder 1955). Flying insects sometimes taken in flycatcher fashion; mostly insectivorous (Griscom and Sprunt 1957:83). Clearing land for farming may have increased the population by creating second-growth habitat (Williams 1996), but the Nashville warbler did not greatly increase in response to forest clearing as did the chestnut-sided warbler.

Northern Parula

(*Parula americana*)

Range: Breeds from se. Manitoba and central Ontario, e. to New Brunswick, s. to s.-central and s. Texas, the Gulf Coast, and Florida, and w. to the e. edge of the Great Plains states. Winter: Florida and the West Indies and from Veracruz and Oaxaca, s. through Mexico to Costa Rica.

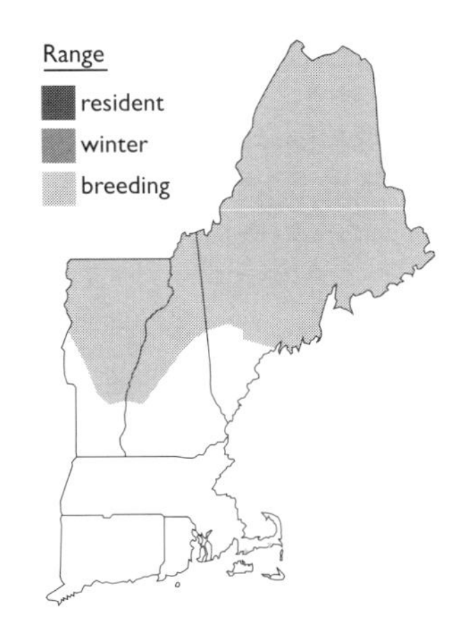

Distribution in New England: Breeds throughout n. and central Vermont, to n. and central New Hampshire, throughout Maine, except absent s. of Portland; records of nesting from 1974 to 1979 on Cape Cod, Naushon Island, and Martha's Vineyard in Massachusetts (Veit and Petersen 1993:385), nesting is rare and irregular in Connecticut (Zeranski and Baptist 1990:217).

Status in New England: Common breeder in areas where the lichen *Usnea* is found.

Habitat: Mature, moist forests and wooded riparian habitat, especially in spruce, hemlock, and balsam fir woodlands where mosslike lichens are found (*Usnea* in the North, Spanish moss in the South). Most abundant in woodlands consisting of sugar maple, red maple, paper birch, and yellow birch, 40 to 70 yr of age in Nova Scotia (Morgan and Freedman 1986, Freedman et al. 1981). Occur along forest edges along wetlands, streams, or ponds. Most common in spruce woods along forest or forest/shore edges in Maine (Morse 1977, 1980). Associated with closed-canopy forests, variable conifer cover, and trees in the smaller size classes in west-central Maine (Collins 1983b). Tolerates areas with moderate levels of timber harvest activity (Johnson and Brown 1990), but absent from clear-cut and strip-cut areas in Nova Scotia (Freedman et al. 1981). Seldom in deep woods (Smith 1994:258 in Foss 1994). Sensitive to forest fragmentation; requires approximately 250 acres (100 ha) to sustain breeding populations, although populations have been documented in smaller woods (Robbins 1989).

Special Habitat Requirements: Prefers to nest in, or use as nesting material, the lichen *Usnea*.

Nesting: Egg dates: May 17 to June 27, New York (Bull 1974:480). Clutch size: 2 to 7, typically 3 or 5. Incubation period: 12 to 14 days. Nestling period: 10 to 11 days. Broods per year: 1. Age at sexual maturity: 1 yr. Nest height: 6 to 100 ft (1.8 to 30.5 m). Nest site: Usually hanging near the end of a twig that is covered with *Usnea* (Graber and Graber 1951).

Territory Size: Range from 0.75 to 1.3 acres (0.75 to 1.3 ha) in mainland spruce forests, but occur on offshore islands 0.16 ha in size in Maine (Morse 1977).

Sample Densities: Maryland—47 territorial males per 100 acres (40 ha) in well-drained floodplain forest. 29 territorial males per 100 acres (40 ha) in poorly drained floodplain forest. 19 territorial males per 100 acres (40 ha) in second-growth river swamp. 12 territorial males per 100 acres (40 ha) in pine-oak forest (Stewart and Robbins 1958:281). 2.4 territorial males per 100 acres (40 ha) in birch-basswood forests in Pennsylvania (Samson 1979). 6.8 territorial males per mile (1.6 km) of shoreline in Maine (Morse 1967). Up to 18 territorial males per 100 acres (40 ha) in Vermont (Kibbe 1985:282 in Laughlin and Kibbe 1985).

Food Habits: Beetles, plantlice, inchworms, small hairy caterpillars, spiders from branches, twigs, and leaves of trees. Often hangs upside down, chickadee fashion.

Comments: Composition of diet is 98 percent animal, 2 percent vegetable (Wetmore 1916 in Bent 1953:143). Has disappeared from many forests in the Northeast with the decline of *Usnea*, a lichen sensitive to air pollution. *Usnea* lichen was extirpated in Connecticut in the 1920s, resulting in the virtual extirpation of northern parulas.

Yellow Warbler

(Dendroica petechia)

Range: Breeds from nw. and n.-central Alaska and n. Yukon to n. Ontario, central Quebec, and s. Labrador, s. through Mexico and Central America to n. South America, central and ne. Texas, n. Arkansas, central Georgia, and central South Carolina. Migrates and winters throughout the Caribbean. Winters from s. California, sw. Arizona, Mexico, and s. Florida, s. to

Venezuela, Brazil, Peru, and the Galapagos. Also resident in s. Florida, the Greater Antilles, the Virgin Islands, and other islands in the West Indies; also in Central America, coastal Colombia, Ecuador, and Venezuela.

Distribution in New England: Throughout the region.

Status in New England: Common breeder and migrant.

Habitat: Prefers open, often moist, habitats such as willow and alder thickets along stream and lake borders, swamp edges, marshes, brushy bottomlands, deciduous forested wetlands, alder, and thickets. Also inhabits forest edges, shrubby growth in suburban areas, orchards, old fields, agricultural areas, gardens, and roadside thickets. Uses only deciduous habitats except on islands off the coast of Maine where it occurs in coniferous forests (Morse 1989:110). Generally absent at elevations >1,000 ft (300 m) in New Hampshire (Staats 1994:260 in Foss 1994). Associated with open habitats characterized by few tall trees (canopy cover 27 percent, diameter breast height >7.5 and <23 cm) and a high percent ground cover in w.-central Maine (Collins 1983b).

Special Habitat Requirements: Dense deciduous thickets with few taller trees, commonly near water.

Nesting: Egg dates: May 20 to June 20, Massachusetts (Veit and Petersen 1993:385). May 15 to July 3 (Bull 1974:481). Clutch size: 3 to 6, typically 4 or 5. Incubation period: 11 to 12 days. Nestling period: 9 to 12 days. Broods per year: 1. Age at sexual maturity: 1 yr. Nest height: 2 to 15 ft (0.6 to 4.6 m) (Schrantz 1943). Typically 3 to 8 ft (0.9 to 2.4 m). DeGraaf et al. (1975) found 19 yellow warbler nests at an average height of 22 ft (6.6 m) in five habitat types ranging from rural to urban in Massachusetts. Nest site: Placed in a fork or crotch of a shrub, sapling, or tree; sometimes located in multiflora rose. Adaptable in choice of nest site.

Territory Size: Individual territories as small as 0.4 acres (0.16 ha) (Harrison 1975).

Sample Densities: 68 pairs per square mile (26 pairs/km^2) in favorable habitat in North Dakota (Stewart and Kantrud 1972). 63 territorial males per 100 acres (40 ha) in shrubby field with stream-bordered trees in Maryland. 5 territorial males per 100 acres (40 ha) in field and edge habitat in Maryland (Stewart and Robbins 1958:282). 59.2 territorial males per 100 acres (40 ha) in red maple-birch habitat in New York.

Food Habits: Insects—caterpillars of gypsy moth and brown-tail and tent caterpillars, cankerworms, beetles, weevils, plantlice, and grasshoppers. Also takes spiders. Food composition is 94 percent animal, 6 percent vegetable (Forbes 1883 in Bent 1953:171). Feeds on small tree limbs generally 4 to 40 ft (1.2 to 12.2 m) high (MacArthur 1958), branch hopping, gleaning, and hawking.

Comments: Yellow warblers have one of the widest ranges of any North American passerine, and are one of the most heavily parasitized by cowbirds. Yellow warblers are second only to chestnut-sided warblers in abundance in Vermont (Kibbe 1985:284 in Laughlin and Kibbe 1985), but are apparently declining in Connecticut due to suburban sprawl, excessive use of pesticides, and brood parasitism by cowbirds (Zeranski and Baptist 1990:217).

Chestnut-sided Warbler

(Dendroica pensylvanica)

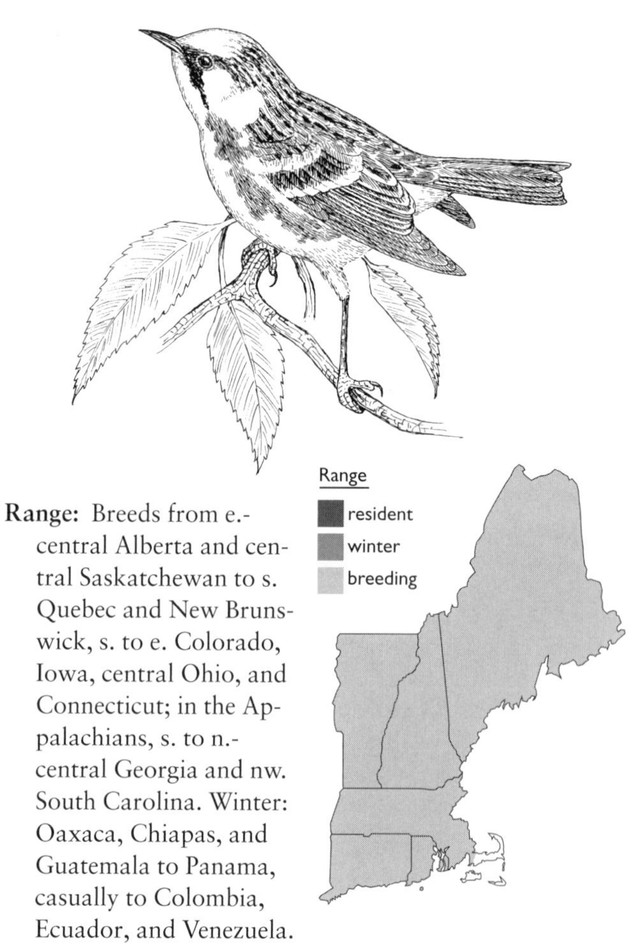

Range: Breeds from e.-central Alberta and central Saskatchewan to s. Quebec and New Brunswick, s. to e. Colorado, Iowa, central Ohio, and Connecticut; in the Appalachians, s. to n.-central Georgia and nw. South Carolina. Winter: Oaxaca, Chiapas, and Guatemala to Panama, casually to Colombia, Ecuador, and Venezuela. Also in the Greater Antilles, e. to the Virgin Islands.

Distribution in New England: Throughout the region, except absent on outer Cape Cod and the Islands in Massachusetts.

Status in New England: Abundant and widespread breeder in n. New England, fairly common in s. New England.

Habitat: Shrubby, open and dry areas with deciduous and mixed small trees, prefers second-growth areas of clear-cut forests or regenerating shelterwood cuts. Also abandoned pastures and fields, forest edges and clearings, low shrubs, briar thickets, brushy hillsides and brooksides, and roadside thickets, woodland clearings, and burns (DeGraaf and Rappole 1995:417–418). Inhabits forest with a mean canopy cover of 35.8 percent, and canopy height (9.5 m) in Grafton, Vt. (Thompson and Capen 1988). Inhabits regenerating clear-cuts 2 yr after harvest, becomes common in cuts after 4 yr, and numbers begin to decline when cuts are 10 yr old in New Hampshire (DeGraaf 1991). Associated with open habitats with few trees and dense ground cover in w.-central Maine (Collins 1983).

Special Habitat Requirements: Early second-growth deciduous woodlands with dense vegetation 1 to 3 m tall for nesting and foraging.

Nesting: Egg dates: May 22 to June 5, Massachusetts (Veit and Petersen 1993:387). May 20 to July 25, New York (Bull 1974:495). Clutch size: 4 to 5, typically 4. Incubation period: 11 to 12 days. Nestling period: 10 to 11 days. Broods per year: Usually 1, but known to have second brood in New Hampshire (D. King pers. commun.). Age at sexual maturity: 1 yr. Nest height: 1 to 4 ft (0.3 to 1.2 m), typically 2 ft (0.6 m). Nest site: Well concealed in low bush, sapling, briars, or vines.

Territory Size: Prior to mating, 4 territories measured 1.2, 1.3, 1.3, and 2.5 acres (0.5, 0.5, 0.5, and 1.0 ha). During incubation, males increased territory size by 200 to 700 ft (61 to 213.4 m) to encompass 2 to 12 acres (0.8 to 4.9 ha) in New York (Kendeigh 1945b).

Sample Densities: 79 territorial males per 100 acres (40 ha) in dense second growth in Maryland. 67 territorial males per 100 acres (40 ha) in open slash oak-maple area (Stewart and Robbins 1958:294).

Food Habits: Almost completely insectivorous, mainly larvae of lepidoptera and diptera and some adults; also spiders. Berries consumed when insects sparse (Richardson and Brauning 1995).

Comments: In the early 1800s, the chestnut-sided warbler was apparently a rare migrant in New England—Audubon considered it one of the rarest birds in North America (Bent 1953). The clearing of the forests for settlement greatly benefited the chestnut-sided warbler; it was common in New England by the mid 1800s (Minot 1895). Chestnut-sided warblers are usually absent from extensive mature forests in Vermont (Kibbe 1985:286 in Laughlin and Kibbe1985).

Magnolia Warbler

(Dendroica magnolia)

Range: Breeds from ne. British Columbia, w.-central and s. Northwest Territories, e. to n.-central Manitoba and s.-central and e. Quebec, and s. to s.-central British Columbia, central Saskatchewan, ne. Minnesota, central Michigan, w. Virginia, nw. New Jersey, and Connecticut. Winter: Oaxaca and central Veracruz to Panama, occasionally along the Gulf Coast and West Indies, e. to the Virgin Islands.

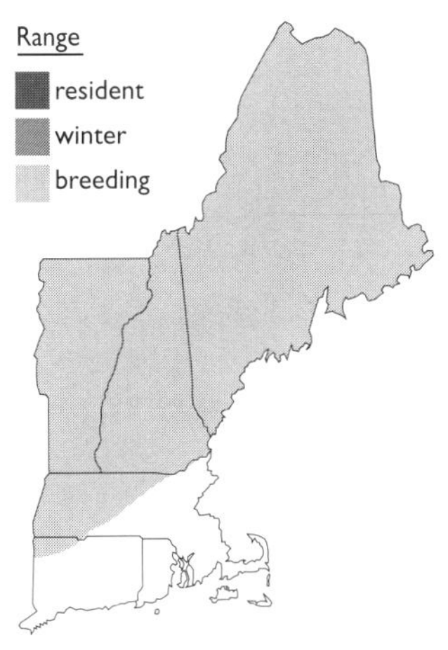

Distribution in New England: Throughout Maine, New Hampshire, Vermont, w. and central Massachusetts, and nw. Connecticut.

Status in New England: Locally common breeder, but rare in nw. Connecticut; common spring migrant.

Habitat: Breeding: Coniferous and mixed coniferous-deciduous forests, coniferous forested wetlands, woodland clearings with small conifers, successional forests, forest edges, and thickets along roadsides (DeGraaf et al. 1991). Prefers dense stands of young spruce and fir. Generally occurs in forests with a 50 percent canopy cover and an average canopy height of 37 ft (11.2 m) in Grafton, Vt. (Thompson and Capen 1988); in forests with a 64.2 percent canopy cover and canopy height of 48.6 ft (17.7 m) in w.-central Maine (Collins 1983). Nests at elevations > 1,000 ft (303 m) in the Northwest Hills of Connecticut (Zeranski and Baptist 1990:218). Inhabits white pine and deciduous saplings that recolonize abandoned pastures in Vermont (Kibbe 1985:288 in Laughlin and Kibbe 1985) and old pastures invaded by spruce in n. New Hampshire and Maine (DeGraaf pers. observ.).

Special Habitat Requirements: Stands of young conifers.

Nesting: Egg dates: June 15 to 24, Massachusetts (Veit and Petersen 1993:388). May 25 to July 11, New York (Bull 1974:482). Clutch size: 3 to 5, typically 4. Incubation period: 11 to 13 days (Harrison 1975). Nestling period: 8 to 10 days. Broods per year: 2. Age at sexual maturity: 1 yr. Nest height: 1 to 15 ft (0.3 to 4.5 m). Typically 1 to 10 ft (0.3 to 3.0 m). Nest site: Prefers young spruce, fir, and hemlock, but may use hardwoods. Well-concealed nest located on horizontal branch near the trunk (Hall 1994).

Territory Size: 20 males had territories that averaged 1.8 acres (0.7 ha) in size—habitat: Hemlock, beech in New York (Kendeigh 1945b). Differences in breeding territory size occurred in different forest types: aspen (average) 1.8 acres (0.7 ha); conifer-birch (average) 2.2 acres (0.9 ha); mixed (average) 2.4 acres (1.0 ha); maple (average) 3.3 acres (1.3 ha) (Stenger and Falls 1959). Average territory 1.8 acres (0.72 ha) in size in coniferous forests on small islands in Maine (Morse 1977).

Sample Densities: 22 pairs per 40 ha (100 acres), Loud's Island, Me.—forest is 83 percent red spruce, 14 percent white spruce; 15 pairs per 40 ha (100 acres), Marsh Island, Me.—100 percent white spruce; 42 pairs per 40 ha (100 acres), Harbor Island, Me.—100 percent white spruce (Morse 1976). 80 males per 100 acres (40 ha) in virgin hemlock forest in Maryland. 63 males per 100 acres (40 ha) in open hemlock-spruce bog in Maryland. 33 males per 100 acres (40 ha) in scrub spruce bog (Stewart and Robbins 1958:283). Greater densities are associated with habitat and stage of succession, conifer abundance, and density of understory (Hall 1994).

Food Habits: Weevils, leaf-beetles, leaf hoppers, plantlice, scale insects, ants, caterpillars, moths from branches of small trees or shrubs.

Comments: The magnolia warbler expanded its breeding range southward into Massachusetts and nw. Connecticut early in this century with the regrowth of high-elevation coniferous forests (Bagg and Eliot 1937). Magnolia warblers have long been considered common in n. New Hampshire, and logging in n. New England will continue to provide habitat for this striking bird.

Cape May Warbler

(Dendroica tigrina)

Range: Breeds from sw. and s.-central Northwest Territories and ne. British Columbia, e. to central Ontario, s. Quebec, and New Brunswick, s. to central Alberta, se. Manitoba, n. Wisconsin, s. Ontario, ne. New York, ne. Vermont, n. New Hampshire, and e.-central Maine. Winter: Central and s. Florida and the West Indies, and casually from the Yucatan Peninsula to Panama.

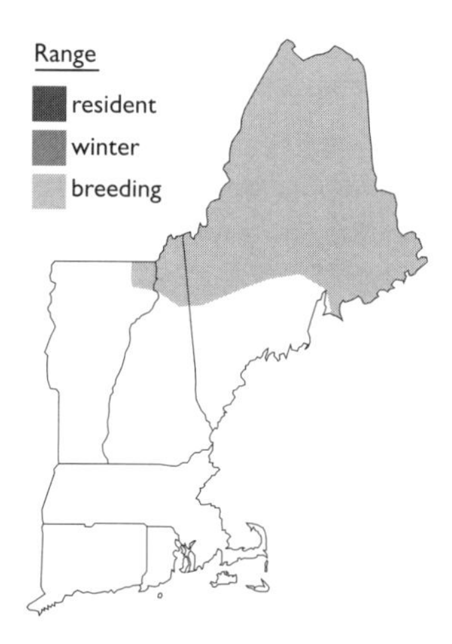

Distribution in New England: A scarce to uncommon breeder in the Northeast Highlands and the n.-central region of Vermont (Oatman 1985:290 in Laughlin and Kibbe 1985). Breeds in n. New Hampshire, and throughout central and n. Maine.

Status in New England: Common (Maine) to uncommon (Vermont) breeder; variously common to very uncommon spring and fall migrant in s. New England, but increasing since 1955 (Veit and Petersen 1993:389).

Habitat: Inhabits fairly open boreal forests, especially those with a high percentage of mature spruce. Also inhabits dense stands of spruce forests with a scattering of taller crowns that spire above the general canopy level, open land with smaller trees, and coniferous forest edges, especially if birches or hemlocks are present

(DeGraaf et al. 1991). In Vermont, prefers dense stands of fairly small black spruce or second-growth stages of other spruce species growing in damp areas (Oatman 1985:290 in Laughlin and Kibbe 1985).

Special Habitat Requirements: Tall stands of conifers, especially spruce, for nesting.

Nesting: Egg dates: June 10 to June 29. Peak: June 12 to June 20, New Brunswick (Bent 1953:224). Clutch size: 4 to 9, typically 6 or 7, larger than most wood-warblers. Broods per year: 1. Age at sexual maturity: 1 yr. Nest height: Typically 12 to 15 m. Nest site: Usually near the top of a coniferous tree close to the trunk (MacArthur 1958). Typically made of sphagnum moss, fine twigs, grasses, and lined with feathers (Harrison 1975).

Sample Densities: 28 pairs per 100 acres (40 ha) spruce-fir forest near Lake Nipigon, Ontario (Kendeigh 1947 in Griscom and Sprunt 1957:118). Densities fluctuate greatly with spruce budworm (*Choristoneura fumiferana*) outbreak cycles.

Food Habits: A spruce budworm specialist (Baltz and Latta 1998). Also insects, ants, small adults and larvae of moths, flies, beetles, small crickets, termites, larvae of dragonflies, and spiders. Feeds at the tips of dense branches and new buds of firs and spruces near tops of trees.

Comments: "Cape May" is an odd name for a bird that breeds in n. New England; Alexander Wilson (1766–1813) named it from a specimen shot at Cape May, N.J., and given to him in 1811 (Oatman 1985:290 in Laughlin and Kibbe 1985). Cape May Warblers are probably dependent on sporadic outbreaks of insects such as the spruce budworm (Morse 1978) that result in superabundant food supplies (MacArthur 1958). With their larger-than-average clutch sizes, Cape May warblers respond quickly to such food bonanzas, and then their population declines between outbreaks (Baltz and Latta 1998).

Black-throated Blue Warbler

(*Dendroica caerulescens*)

Range: Breeding: Western and central Ontario to New Brunswick, s. to ne. Minnesota, central Michigan, ne. Pennsylvania, and s. New England, and in the Appalachians to ne. Georgia and nw. South Carolina. Winter: Primarily in the Caribbean: From s. Florida to the West Indies, casually in Central America and Colombia.

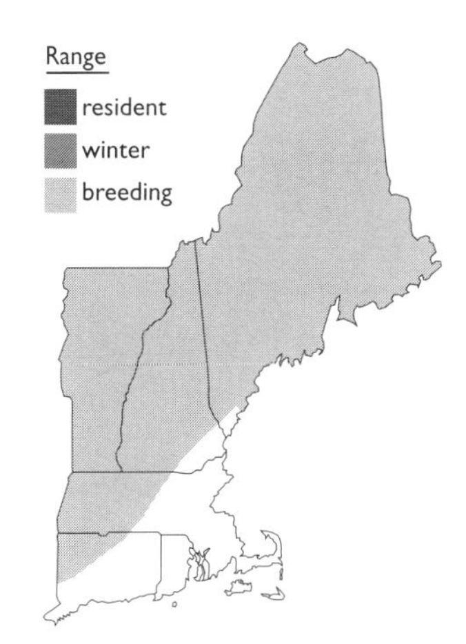

Distribution in New England: Throughout New England, except absent in e. Massachusetts, Rhode Island, and se. Connecticut. In Connecticut, increasing in the Northwest Hills, most abundant on Mt. Riga in Salisbury (Zeranski and Baptist 1990).

Status in New England: Common (n. New England) to fairly common (s. New England) breeder; fairly common spring and fall migrant.

Habitat: Most commonly found in large, more or less continuous tracts of northern hardwood forests with a dense understory of coniferous or deciduous shrubs or saplings (Darveau et al. 1992), also at edges of woodland clearings, generally in moist places. Found in closed-canopy (88 percent), pole-sized stands in Grafton, Vt. (Thompson and Capen 1988), and where canopy cover was 84 percent, canopy height 46.2 ft

(14.0 m) in pole stands in w.-central Maine (Collins 1983). Occurs predominantly in thinned saw-timber stands of maple, beech, and yellow birch in the White Mountain National Forest in New Hampshire and Maine (DeGraaf and Chadwick 1987). Most abundant at elevations of 400–700 m in the White Mountains of New Hampshire (Holmes 1994). Associated with increased timber harvest levels in e. Maine (Johnson and Brown 1990).

Special Habitat Requirements: Deciduous and mixed woodlands with thick understory of shrubs and saplings (Pough 1949:164).

Nesting: Egg dates: May 23 to June 10, Massachusetts (Veit and Petersen 1993:392). May 29 to July 16, New York (Bull 1974:484). Clutch size: Typically 4, 2 to 5 in New Hampshire (Holmes and Sherry 1992). Incubation period: 12 to 13 days. Nestling period: 10 days (Holmes 1994). Broods per year: 1 or 2, sometimes 3 (Holmes et al. 1992). Age at sexual maturity: 1 yr. Nest height: 4 in (10 cm) to 20 ft (6.1 m), New York (Bull 1974:484), commonly to 2 ft in deciduous shrubs. Nest site: In a fork of coniferous or deciduous saplings, or in low shrubs.

Territory Size: 5 acres (2 ha) in New Hampshire (Holmes and Sherry 1992).

Sample Densities: Maryland— 58 territorial males per 100 acres (40 ha) in virgin hemlock forest; 48 territorial males per 100 acres (40 ha) in young second growth; 17 territorial males per 100 acres (40 ha) in scrub spruce bog (Stewart and Robbins 1958:286).

Food Habits: Insects—mainly lepidopteran larvae (Holmes 1994), crane-flies, mosquitoes, plantlice. Feeds in upper branches, hawking, branch and twig gleaning in shrub and subcanopy (Holmes and Sherry 1994:268 in Foss 1994).

Comments: Considered a deep-woods warbler, but occupies clear-cuts in northern hardwoods as soon as 15 yr after harvest (DeGraaf 1991). The black-throated blue warbler is the only common forest warbler that forages extensively in the understory. It is also the New England warbler that exhibits the most striking sexual dimorphism in plumage, so much so that the two sexes were initially thought to be different species (see Holmes 1994).

Yellow-rumped Warbler

(*Dendroica coronata*)

Range: The eastern race (*coronata* group) breeds from w. Alaska and central Northwest Territories to n.-central Labrador, s. to n. British Columbia, se. Saskatchewan, central Michigan, and Massachusetts, and in the Appalachians to e. West Virginia. The western race (*auduboni* group) breeds from central British Columbia and sw. Saskatchewan, s. to s. California and n. Baja California, e. to w. Texas and s. to w. Chihuahua. Winter: The *coronata* group—from sw. British Columbia through the Pacific States, s. Arizona, and Colorado, and from Kansas e. across the central United States to New England, s. to Panama, and throughout the West Indies, where it is one of the last warblers to arrive in the Caribbean. The *auduboni* group winters from sw. British Columbia and Idaho to s. Baja California, through Mexico to Guatemala and w. Honduras.

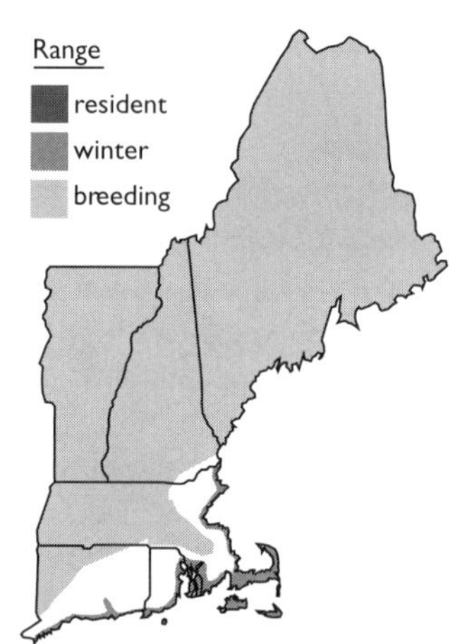

Distribution in New England: Throughout Vermont, New Hampshire except along the coast, and Maine; Connecticut, rare and local in the Northwest and Northeast Hills, with scattered reports of nesting elsewhere (Zeranski and Baptist 1990:220); to w. Rhode Island, and

Massachusetts, mainly at higher elevations in the central and w. part of the state, and locally in white pines near the se. coastal plain, but absent from Cape Cod and the islands (Veit and Petersen 1993:391). Winter: Connecticut, along the coast and Connecticut River Valley, occasionally inland (Zeranski and Baptist 1990:220), Rhode Island, Massachusetts from Nantucket and Cape Cod w. to Berkshire County (Veit and Petersen 1993:392), and along the coast of Maine.

Status in New England: Common breeder at higher elevations, abundant spring and fall migrant.

Habitat: Conifer and mixed forests and young coniferous growth near forest edges. Prefers spruce-fir woodlands, occasionally found in mature white pine forests near the se. coastal plain in Massachusetts (Veit and Petersen 1993:391). Nests almost exclusively in mature white pine forests in Rhode Island (Enser 1992:147). Associated with forests with a high percent canopy cover (84.3 percent), variable conifer cover, and trees in the smaller size classes (diameter breast height 15.1 to 23 cm) in w.-central Maine (Collins 1983). Occur most often in pole-size white pine and spruce-fir stands in the White Mountain National Forest in New Hampshire and Maine (DeGraaf and Chadwick 1987). Responded negatively to increasing levels of timber harvest in e. Maine (Johnson and Brown 1990).

Special Habitat Requirements: Coniferous trees for nesting.

Nesting: Egg dates: May 23 to June 5, Massachusetts (Veit and Petersen 1993:392). Clutch size: 3 to 5, typically 4 (Harrison 1975). Incubation period: 12 to 13 days. Nestling period: 12 to 14 days. Broods per year: 1 or 2. Age at sexual maturity: 1 yr. Nest height: 4 to 50 ft (1.2 to 15.2 m). Typically 15 to 20 ft (4.6 to 6.1 m). Nest site: Usually in a small coniferous tree, typically saddled on a branch of spruce, hemlock, or cedar. Sometimes in a deciduous tree such as maple or birch.

Sample Densities: 30 pairs per 40 ha (100 acres), Loud's Island, Me., 83 percent red spruce and 14 percent white spruce (Morse 1976). 39 pairs per 40 ha (100 acres), Marsh Island, Me., 100 percent white spruce (Morse 1976). 31 pairs per 40 ha (100 acres), Harbor Island, Me., 100 percent white spruce (Morse 1976).

Food Habits: Insects in summer—plantlice, caterpillars, small grubs, ants, and leaf beetles. In winter—eggs and larvae of some insects, bayberries, berries of red cedar, woodbine, viburnums, honeysuckle, mountain ash, and poison ivy. Feeds over the trunks and branches from tops of trees to ground level, and in the air.

Comments: The yellow-rumped warbler winters farther north than any other wood-warbler; it is able to shift its diet from insects to bayberries during winter, and is able to winter along the coasts from s. Maine s. The yellow-rumped is one of the most common warblers in North America, and one of the most ecologically generalized, foraging in a broad range of microhabits and using a range of feeding techniques from flycatching to leaf-gleaning. In the nonbreeding season found in a wide range of habitats where it consumes a substantial amount of fruit (Hunt and Flaspohler 1998).

Black-throated Green Warbler

(*Dendroica virens*)

Range: Breeds in e.-central British Columbia and n. Alberta to central Ontario and Newfoundland, s. to central Alberta, s. Manitoba, Pennsylvania, and n. New Jersey, and s. in the Appalachians to n. Alabama and Georgia. Winters: South Texas and the Greater Antilles, s. through e. Mexico and Central America to n. South America (Colombia and Venezuela).

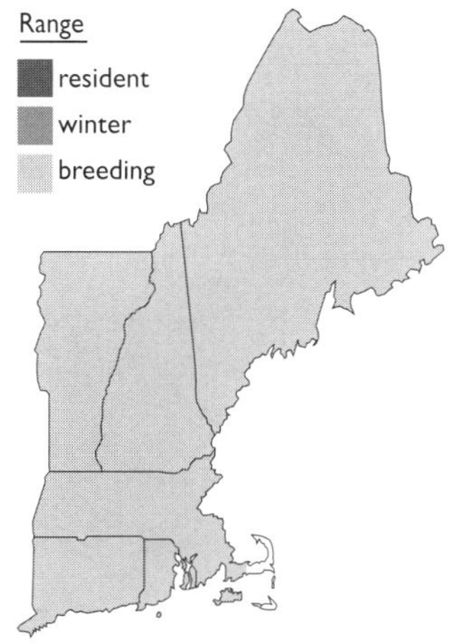

Distribution in New England: Throughout New England from Massachusetts n., except outer Cape Cod and Nantucket in Massachusetts; locally in much of Connecticut.

Habitat: Inhabits mature mixed woodlands (especially hardwood-hemlock stands and northern hardwood-spruce), coniferous forest with large trees, and larch bogs (DeGraaf and Rappole 1995). In New Hampshire, occurs in pine, spruce-fir, northern hardwood–spruce, and birch-beech-maple forests (Collins 1983a). In Vermont, generally found in northern hardwood forest that includes hemlock or red spruce (Ellison 1985:296 in Laughlin and Kibbe 1985). White pine and mixed forest frequented in Massachusetts (Bent 1953). Occasionally inhabits second-growth hardwoods and abandoned fields. Sensitive to logging activity; percentage of stands occupied decreased with increasing harvest activity (Johnson and Brown 1990).

Special Habitat Requirements: Mature coniferous or mixed woodlands.

Nesting: Egg dates: May 30 to June 18, Massachusetts (Veit and Petersen 1993:394). May 24 to July 2, New York (Bull 1974:489). Clutch size: 4 to 5, typically 4. Incubation period: 12 days. Nestling period: 8 to 10 days. Broods per year: 1, occasionally 2 (Harrison 1975). Age at sexual maturity: 1 yr. Nest height: 3 to 80 ft (0.9 to 24.4 m). Typically 15 to 20 ft (4.6 to 6.1 m). Nest site: Usually on a horizontal or drooping branch, usually in conifers, but occasionally in hardwoods.

Territory Size: 21 territories ranged from 0.6 to 2.5 acres (0.2 to 1.0 ha); average size 1.6 acres (0.6 ha) (Kendeigh 1945a) in New York: hemlock-beech forest. Territories are smaller in preferred conifer forests than in mixed forests (Morse 1993).

Sample Densities: Sometimes the most common breeding species in ne. coniferous forests (Morse 1993). 71 pairs per 40 ha (100 acres), Loud's Island, Me., 83 percent red spruce, 14 percent white spruce (Morse 1976). 61 pairs per 40 ha (100 acres) Marsh Island, Me., 100 percent white spruce (Morse 1976). 83 pairs per 40 ha (100 acres), Harbor Island, Me., 100 percent white spruce (Morse 1976). 36 pairs per 40 ha (100 acres) in n. Maine; primarily deciduous forest (Holmes et al. 1986). 9 territorial males per 100 acres (40 ha) in mature northern hardwood forest (Stewart and Robbins 1958:288).

Food Habits: Insects—small, cryptic leaf rollers, leaf-eating caterpillars, beetles, flies, gnats, and plantlice. Also takes mites, cankerworms, spiders, some berries. Gleans over the limbs and foliage of evergreens 10 to 50 ft (3.0 to 15.2 m) above ground, hopping or hovering followed by gleaning or occasional hawking.

Comments: Strongly associated with hemlocks. Populations may be at risk due to loss of winter habitat and spraying of pesticides in northern forests to control spruce budworm (Morse 1993).

Blackburnian Warbler

(Dendroica fusca)

Range: Breeds from central Alberta, e. to s. Quebec and Nova Scotia, s. to s. Manitoba, ne. Ohio, Pennsylvania, and se. New York, and in the Appalachians to South Carolina and n. Georgia. Winter: From mid Central America to Venezuela, Colombia, Ecuador, Peru, and Bolivia.

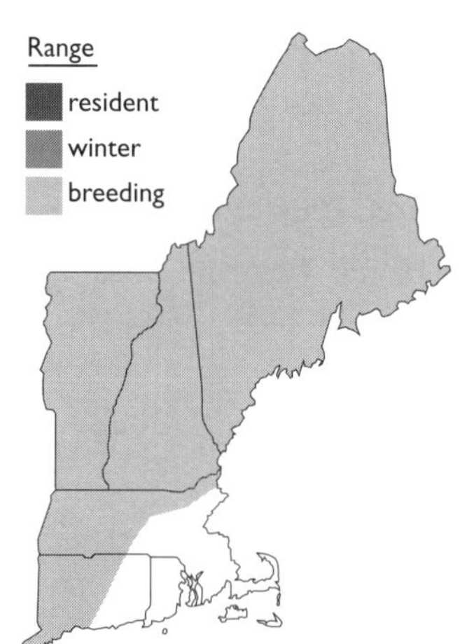

Distribution in New England: Breeds throughout Maine, Vermont, New Hampshire, and w. Massachusetts and Connecticut; not found in Rhode Island or in e. Massachusetts.

Status in New England: Common breeder in n. New England, uncommon in s. New England in higher elevations; increasing in Connecticut (Zeranski and Baptist 1990:222).

Habitat: Prefers mature conifer forests of hemlock, pines, fir, spruce. Also found in mixed forests or moist forests where spruces are thickly draped with bearded lichen (*Usnea*) (Morse 1994). In the Appalachians, inhabits oak forests along ridges. Nests in conifers, but will use deciduous trees in the southern part of the range or where conifers are scarce. Strong affinity for saw-timber-size spruce and fir in n. New Hampshire and w. Maine (DeGraaf and Chadwick 1987). Inhabits forests with high canopy cover (84 percent), variable coniferous cover, and many trees in the smaller class sizes: >3 to <9.1 inches (>7.5 cm to <23 cm) diameter breast height) in north-central Maine (Collins 1983). In Vermont, prefers old-growth stands of coniferous or mixed forests, but known to occupy deciduous forests containing even a single tall red spruce or hemlock (Ellison 1985:298 in Laughlin and Kibbe 1985).

Special Habitat Requirements: Mature coniferous and mixed forests.

Nesting: Egg dates: June 6 to June 26, Massachusetts (Veit and Petersen 1993:396). June 1 to June 24, New York (Bull 1974:492). Clutch size: 3 to 5, typically 4. Incubation period: Unknown, probably 11 to 12 days (Harrison 1975). Nestling period: Unknown. Broods per year: 1. Age at sexual maturity: 1 yr. Nest site: High up in a tree (usually a spruce or hemlock), situated well away from the trunk or in a small fork near the top of the tree.

Territory Size: 9 territories averaged 1.3 acres (0.5 ha) in size per pair in New York; habitat hemlock-beech (Kendeigh 1945a). Ranged from 0.4 to 0.6 ha (1 to 1.5 acres) along the coast of Maine; habitat both red and white spruce (Morse 1976). Average territory 1.1 ha (2.8 acres) in New Hampshire; habitat deciduous forest with patchily distributed conifers (Holmes and Sherry 1992).

Sample Densities: 26 pairs per 40 ha (100 acres), Loud's Island, Me., 83 percent red spruce, 14 percent white spruce (Morse 1976). 17 pairs per 40 ha (100 acres), Marsh Island, Me., 100 percent white spruce (Morse 1976). 12 pairs per 40 ha (100 acres) in n. Maine in bottomland spruce-fir forests and 36 pairs per 40 ha (100 acres) in adjacent higher elevation mixed forest (Morse 1976).

Food Habits: Almost entirely insects such as beetles, caterpillars, ants, crane-flies. Forages high in trees with rapid limb-to-limb gleaning, occasionally hovering or hawking.

Comments: Nests and forages in the tops of northern conifers, especially spruces.

Pine Warbler

(*Dendroica pinus*)

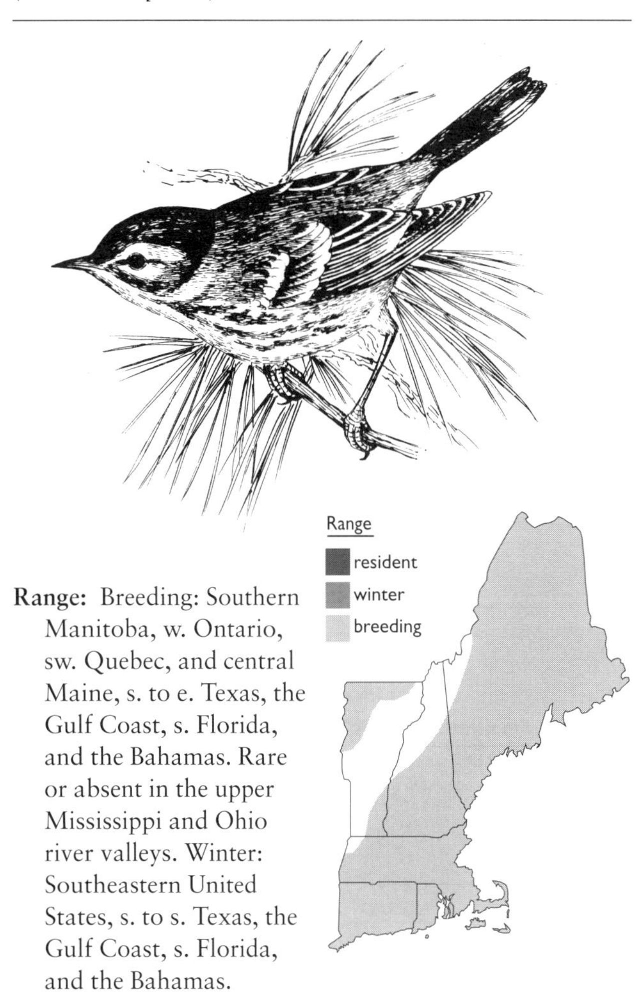

Range: Breeding: Southern Manitoba, w. Ontario, sw. Quebec, and central Maine, s. to e. Texas, the Gulf Coast, s. Florida, and the Bahamas. Rare or absent in the upper Mississippi and Ohio river valleys. Winter: Southeastern United States, s. to s. Texas, the Gulf Coast, s. Florida, and the Bahamas.

Distribution in New England: Throughout Rhode Island, especially in areas with pitch-pine forests (Enser 1992:150). In Connecticut, nests locally in the Northwest and Northeast Hills, and in coastal pitch-pine stands (Zeranski and Baptist 1990:223). In Massachusetts, considered a common breeder on the se. coastal plain and Martha's Vineyard and Nantucket, but uncommon to local on the mainland (Veit and Petersen 1993:397), absent from Mt. Greylock. Found locally throughout Vermont, in s. New Hampshire, and Maine, except absent from mountainous regions.

Status in New England: Locally common to rare breeder, uncommon migrant away from breeding areas.

Habitat: Inhabits pine forests, favoring open pitch-pine woods with tall trees. Frequents coastal pine barrens, less common inland. Nests in red pine and white pine, but shows a preference for pitch pine in New Hampshire (Robbins 1994:276 in Foss 1994). Inhabits tall stands of pine greater than 70 ft (21.3 m) with little undergrowth in New Hampshire; at lower elevations found in open stands of white pine (Ellison 1985:300 in Laughlin and Kibbe 1985).

Special Habitat Requirements: Open pine forests. Pitch pine is preferred, but other species of pine are used as well.

Nesting: Egg dates: May 8 to late July, Massachusetts (Veit and Petersen 1993:397). May 4 to June 6, New York (Bull 1974:502). Clutch size: 3 to 5, typically 4. Incubation period: Probably 12 to 13 days (Harrison 1975). Broods per year: 1. Age at sexual maturity: 1 yr. Nest height: 8 to 80 ft (2.4 to 24.4 m). Nest site: Saddled on a horizontal branch well out from the trunk; sometimes situated among the small twigs toward the end of a limb, obscured from below by a cluster of pine needles.

Sample Densities: Maryland—76 territorial males per 100 acres (40 ha) in immature loblolly-shortleaf pine stand. 20 territorial males per 100 acres (40 ha) in mature scrub pine (Stewart and Robbins 1958:297).

Food Habits: Insects—adult and larvae of beetles, ants, grasshoppers, moths, bugs, flies, and scale insects. Also takes spiders and small amounts of pine and birch seeds, berries of wax myrtle.

Comments: Generally associated with pines, especially pitch pine, where they occur within its breeding range in New England. Pine warblers are still one of the most common summer songbirds in the pine barrens of se. Massachusetts and Cape Cod, but have been declining since the 1940s and 1950s due to succession to oak through fire control.

Prairie Warbler

(*Dendroica discolor*)

Range: Breeds from e. Nebraska, central Missouri, n. Illinois, central Wisconsin, n. Michigan, s. Ontario, se. New York, and New Hampshire, s. to e. Texas, the Gulf Coast, and s. Florida. Winter: Central Florida, s. to the West Indies, and Central America and occasionally to n. South America.

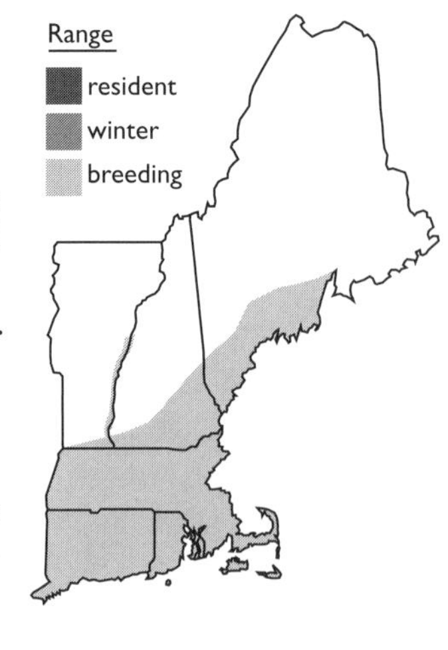

Distribution in New England: Throughout Connecticut, Rhode Island, Massachusetts, the Eastern Foothills in Vermont, s. New Hampshire, and s. Maine.

Status in New England: Common (Connecticut and se. Massachusetts) to uncommon and local (Vermont, se. New Hampshire) breeder; population declining in parts of breeding range as forests reclaim abandoned fields. Uncommon migrant.

Habitat: Open forests with little undergrowth, forest edges, shrubby second-growth areas, dry fields, road cuts, utility corridors, brushy gravel pits, and Christmas tree plantations. Avoids dense forests. Abundant in pitch-pine/scrub-oak barrens, especially those that are periodically burned (Veit and Petersen 1993:399). Almost exclusively in abandoned fields and dry hillsides in Vermont (Clark 1984:302 in Laughlin and Kibbe 1985). Young stands of pine 10 to 30 ft (3 to 9 m) tall (Robbins et al. 1966:268) and deciduous saplings. Logging and burning create favorable habitat.

Special Habitat Requirements: Dry areas with low trees and shrubs, favors areas with some coniferous cover.

Nesting: Egg dates: May 28 to June 18, Massachusetts (Veit and Petersen 1993:399). May 25 to June 29, New York (Bull 1974:505). Clutch size: 3 to 5, typically 4. Incubation period: 10 to 14, typically about 12 days (Nolan 1978:235). Nestling period: 8 to 11 days, typically 9 days. Broods per year: 1. Age at sexual maturity: 1 yr. Nest height: Less than 1 ft (0.3 m) to 45 ft (13.7 m). Typically 3.3 to 6.6 ft (1.0 to 2.0 m) (Nolan 1978:127). Nest site: Usually well hidden in upright fork of sapling or shrub, less frequently in vines. American elm, sugar maple, hawthorn, scrub oak, and bayberry are important nest plants (Nolan 1978:133, Bent 1953:431).

Territory Size: Average territory size is 4.3 acres (1.7 ha) in Indiana (Nolan 1978). Singing and surveillance perches are required for a territory to be suitable. Nolan (1978) found that nearly 75 percent of male prairie warblers reclaimed territories in successive years.

Food Habits: Larvae and adults of beetles, bugs, butterflies and moths, wasps, bees, flies, grasshoppers, and spiders. Gleans and flycatches in trees, saplings, shrubs, herbaceous vegetation, and in the air.

Comments: Breeding habitat influenced to a great extent by availability of Christmas tree plantations (Harrison 1975:196) as old fields revert to forest. Prairie warblers have declined greatly as fall migrants over the last half century (Veit and Petersen 1993:400).

Palm Warbler

(*Dendroica palmarum*)

Range: Breeds from s. Northwest Territories and n. Alberta to central Quebec and s. Newfoundland, s. to ne. British Columbia, central Alberta, n. Minnesota, s. Quebec, Maine, and Nova Scotia. Winter: From n.-central Texas to North Carolina, s. to s. Texas, the Gulf Coast, southern Florida, and the West Indies.

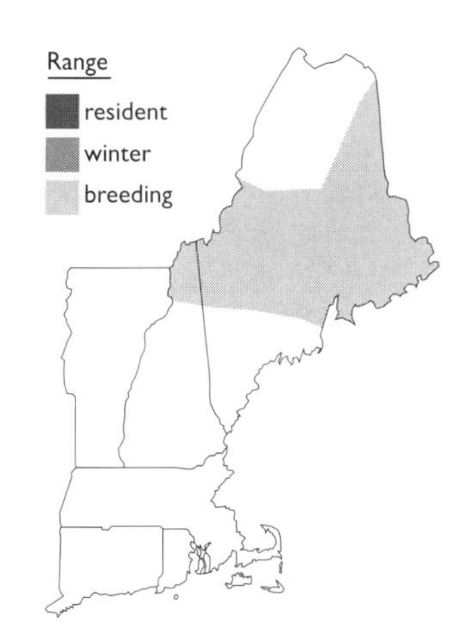

Distribution in New England: Central and n. Maine and northernmost New Hampshire.

Status in New England: Uncommon to local breeder, fairly common to common spring migrant, rare in winter along the Massachusetts coast.

Habitat: Breeds in open bogs in boreal forests (especially black spruce, tamarack or white cedar with sphagnum), dry heaths, dry open spruce or jack pine forests.

Special Habitat Requirements: Open bogs with sphagnum.

Nesting: Egg dates: May 18 to June 8, Nova Scotia (Bent 1953:449). Clutch size: 4 to 5. Incubation period: 12 days. Nestling period: 12 days. Broods per year: 1, possibly 2. Age at sexual maturity: 1 yr. Nest site: Usually on the ground, deep in sphagnum moss hummocks, may nest in the low branches of conifer saplings (DeGraaf and Rappole 1995:348).

Food Habits: Mainly insects such as beetles, ants, caterpillars, grasshoppers, gnats, mosquitoes, flies, and mayflies. Vegetable matter, especially barberries during the winter months. Gleans on the ground, in low shrubs.

Comments: Despite its name, the palm warbler is one of the northernmost of the *Dendroica* warblers; it was described by J. P. Gmelin from a wintering specimen on Hispaniola (Wilson 1996). The palm warbler is at the se. end of its breeding range in New England, and is a very rare breeder in northernmost New Hampshire. Like the pine warbler, it arrives very early in spring, and is one of the last warblers to leave in fall.

Bay-breasted Warbler

(*Dendroica castanea*)

Range: Breeds from sw. Northwest Territories to n.-central Saskatchewan and Newfoundland, s. to ne. British Columbia, s. Manitoba, and ne. Minnesota to s. Maine. Migrates through Central America and the Caribbean. Winter: Northern South America—n. Colombia and w. Venezuela—and occasionally s. to Peru and Brazil.

Distribution in New England: Breeds in ne. Vermont, based on a single confirmed breeding record (Oatman 1985:304 in Laughlin and Kibbe 1985), n. New Hampshire, and much of central and n. Maine; absent from s. Maine.

Status in New England: Fairly common to rare breeder. Has declined since the early 1980s in the e. part of the breeding range (DeGraaf and Rappole 1995:439). Uncommon spring and fall migrant outside of budworm outbreak years.

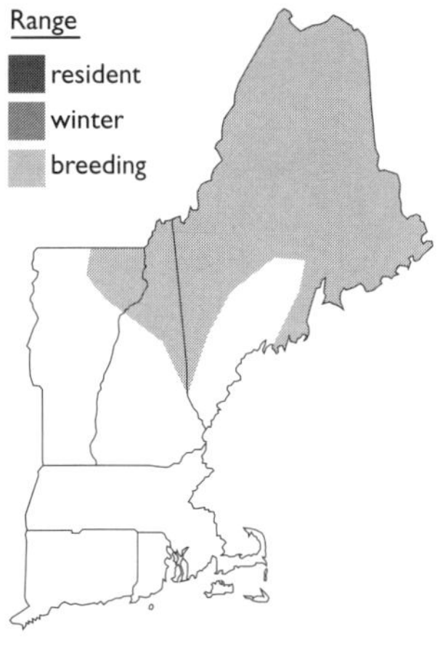

Habitat: Boreal forests, mature northern coniferous or mixed forests, especially balsam fir. Uses small forest openings or edges adjacent to small clearings, fence rows, highways, bogs, or streams. Occasionally in mixed forest adjacent to ponds. Breeds at elevations up to 4,000 ft (1,220 m) in coniferous or mixed woods in New Hampshire (Andrews 1994:304 in Foss 1994). Population densities increase with spruce budworm outbreaks, then decline or become scarce or absent once the outbreak is over (Morse 1989:99). Found in broad-leaf foliage in the White Mountains in New Hampshire (Sabo 1980).

Special Habitat Requirements: Second-growth boreal forests with trees 6 to 10 ft (1.8 to 3.4 m) tall (Pough 1949:174).

Nesting: Egg dates: June 5 to July 2, peak: June 17 to June 25, New Brunswick (Bent 1953:389). Clutch size: 4 to 7, typically 4 or 5 (Mendall 1937). Incubation period: 12 to 13 days. Nestling period: 11 to 12 days. Broods per year: 1. Age at sexual maturity: 1 yr. Nest height: 4 to 40 ft (1.2 to 12.2 m). Typically 15 to 25 ft (4.6 to 7.6 m). Nest site: Along horizontal branch of a conifer, usually spruce (preferably black spruce) or balsam fir, occasionally a pine or hardwood (Williams 1996b). Usually 5 to 10 ft (1.5 to 3.0 m) out from the trunk.

Territory Size: 1.5 ha (3.7 acres) in subalpine forests with mixed vegetation in the White Mountains of New Hampshire (Sabo 1980).

Sample Densities: During spruce budworm outbreaks, reports of 160 territorial males per 40 ha (100 acres) in the 1950s and 1960s (Erskine 1992), and 80 territorial males per 40 ha (100 acres) in n. Maine (Morse 1980).

Food Habits: Caterpillars, ants, beetles, leafhoppers, houseflies, spiders. Searches the foliage of trees (especially red spruce) at all heights but mainly in interior of tree crowns with slow, deliberate movements, often spending much time in the same tree (MacArthur 1958). Occasionally hangs upside down while gleaning prey.

Comments: Bay-breasted warblers may depend on periodic outbreaks of abundant insects such as the spruce budworm (MacArthur 1958).

Blackpoll Warbler

(*Dendroica striata*)

Range: Breeds from w. and n.-central Alaska throughout most of central Canada around the lower part of Hudson Bay to the Atlantic Coast, coincidental with boreal forests. Migrates s. across the e. United States and through the Caribbean. Winter: South America, s. to Peru and Brazil. May stray to Costa Rica when migrating through the Caribbean.

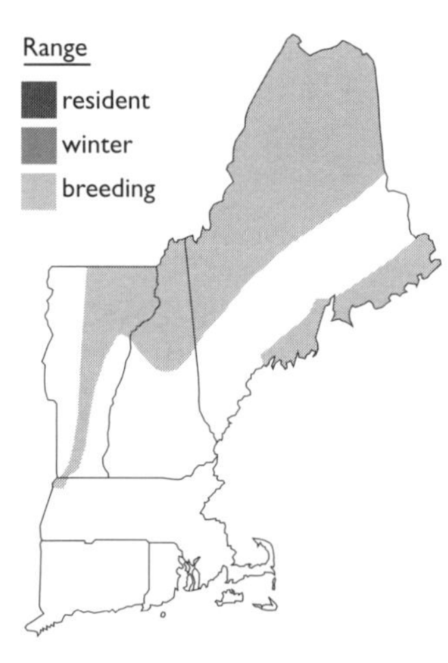

Distribution in New England: Breeds at Mt. Greylock in nw. Massachusetts (Veit and Petersen 1993:402), throughout Vermont (particularly the Green Mountains, the Northeast Highlands, the Taconic Mountains, and the n.-central region), to n. New Hampshire, and n. Maine and coastal areas n. of Casco Bay. Not reported from Connecticut and Rhode Island.

Status in New England: Uncommon breeder throughout the boreal forests of northern New England, uncommon and local on Mt. Greylock, Mass. Common to abundant spring and fall migrant.

Habitat: Generally inhabits northern coniferous forest,

especially spruce and fir forests, favors small (stunted, young) or medium-size conifers, and forest edges. Often in swampy fit flats and in stunted spruce on the upper slopes of mountains. Prefers nesting at elevations between 4,000 and 4,500 ft (1,220 and 1,370 m) in forests dominated by balsam fir interspersed with red and black spruce and heart-leaved paper birch in New Hampshire (Richards 1994:282 in Foss 1994). Associated with forests with moderate canopy cover (65.9 percent), high percent ground cover (70 percent), and trees in the smaller class sizes (diameter breast height 7.5 to 23 cm) in w.-central Maine (Collins 1983).

Special Habitat Requirements: Low northern coniferous forests, prefers spruce.

Nesting: Egg dates: June 15 to early July, Mt. Greylock, Mass. (Veit and Petersen 1993:402); June 15 to June 26, Vermont (Kibbe 1985:306 in Laughlin and Kibbe 1985). June 5 to July 10, New York (Bull 1974:499). Clutch size: 3 to 5, typically 4 or 5. Incubation period: 11 to 12 days. Nestling period: 10 to 12 days. Broods per year: 1, occasionally 2 (Ehrlich et al. 1988). Age at sexual maturity: 1 yr. Nest height: 2 to 7 ft (0.6 to 2.1 m). Nest site: Usually low in a spruce or other conifer, against the trunk of the tree supported by one or two horizontal branches; rarely on the ground.

Sample Densities: 12 pairs per 40 ha (100 acres) in Vermont (Metcalf 1977).

Food Habits: Insects such as spruce-gall lice, cankerworms, mosquitoes, fall webworms, locusts, ants, gnats; some seeds and berries.

Comments: Has the longest migration of any New World land bird, annually making a round trip of 2,500 to 5,000 miles (4,000 to 8,000 km) between the breeding and wintering grounds, much of it over water. It also has the most northerly breeding range and southerly wintering ground of any North American nesting warbler (Morse 1989:27).

Cerulean Warbler

(Dendroica cerulea)

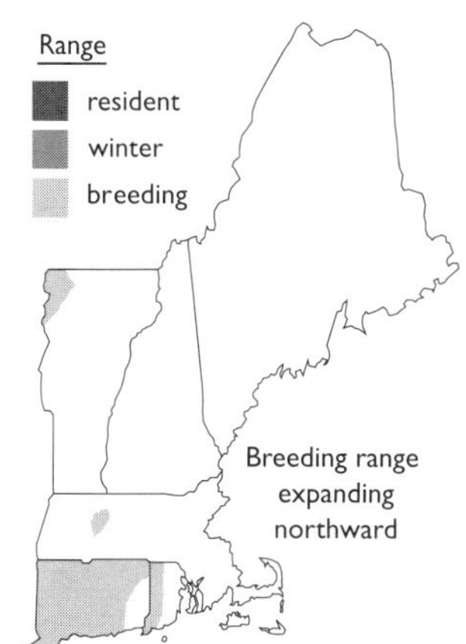

Range: Breeds from se. Minnesota, s. Wisconsin, s. Michigan, Ontario, New York, and w. New England, s. to ne. Texas, se. Louisiana, Mississippi, Alabama, and central North Carolina. Winter: Colombia, Venezuela, Ecuador, Peru, and Bolivia.

Distribution in New England: In Vermont in the Champlain Valley (Ellison 1985:308 in Laughlin and Kibbe 1985), in central Massachusetts in the Quabbin Reservoir (Veit and Petersen 1993:403), in Connecticut in the Northwestern Hills along the Housatonic River and along the Connecticut River in Haddam and East Haddam (Zeranski and Baptist 1990), and in w. Rhode Island. Not found in Maine or New Hampshire.

Status in New England: Rare and local breeder, rapidly declining as a breeder throughout much of its breeding range (Robbins et al. 1992) but gradually increasing in the Northeast (Veit and Petersen 1993:403) as a result of forest maturation (Hamel 1992a). Rare migrant.

Habitat: Prefers mature deciduous forests in floodplain, bottomland, or other mesic conditions. Prefers forests with tall trees, a semiopen canopy, and sparse understory (Robbins et al. 1992). Generally occupies upper canopy. Hamel (1992a) states that tree size is more important than species in defining suitable breeding habitat.

Special Habitat Requirements: Extensive (>250 ha) mature deciduous forests with sparse understories and closed or semiopen canopies.

Nesting: Egg dates: May 19 to June 23, New York (Bull 1974:490). Clutch size: 3 to 5, typically 4. Incubation period: 12 to 13 days. Nestling period: 9 to 10 days. Broods per year: 1. Nest site: Usually in the fork of a tall tree, some distance from the trunk with an open area below. Elm was a common nest tree in New York State (Bull 1974:490). Oak was a common nest tree in Pennsylvania (Harrison 1984). Also known to nest in maples, basswood, and yellow poplar.

Territory Size: In Tennessee, territories defended in forest stands with large trees but absent in stands of small trees (Hamel 1992a).

Sample Densities: 4.8 territorial males per 20 ha (50 acres) in birch-basswood habitat in Pennsylvania (Van Velzen 1977).

Food Habits: Mainly insects such as wasps, ants, bees, beetles, weevils, and caterpillars. Hawks and foliage-gleans.

Comments: The cerulean warbler has one of the most restricted winter ranges among Neotropical migrants: strictly in the primary humid evergreen forest along a narrow elevational zone between 620 and 1,300 m in the Andean foothills, one of the most intensively logged and cultivated regions in the Neotropics (Robbins et al. 1992).

Black-and-white Warbler

(Mniotilta varia)

Range: Breeds from w.-central Northwest Territories, n. Alberta, and central Saskatchewan, to s. Quebec and Newfoundland, s. to e. Montana, central Texas, Louisiana, Alabama, Georgia, and North Carolina. Winter: From s. Texas and Florida through Mexico and Central America to n. South America. Also migrates and winters throughout the Caribbean.

Range
resident
winter
breeding

Distribution in New England: Throughout the region.

Status in New England: Common to uncommon widespread breeder, common inland migrant, very rare in winter.

Habitat: Mature and somewhat open or second-growth deciduous or mixed forests, especially those containing saplings or scrubby trees. May have a preference for moist forests (Kricher 1995). Usually not abundant in coniferous forests unless hardwood understory is present. Moderately closed canopy (70.5 percent), canopy height 51.2 ft (15.5 m) in pole-timber stands in Grafton, Vt. (Thompson and Capen 1988); fairly closed canopy cover (84.3 percent), canopy height 46.2 ft (14.0 m) in pole stands in w.-central Maine (Collins 1983). Known to nest and forage in regenerating (<6 yr old) clear-cuts in New Hampshire (D. King pers. commun.). Reported reoccupying cutover forest 7 to 12 yr after harvest in spruce-fir forest in Maine (Titterington et al. 1979). First appears in clear-cuts 3 yr after harvest, becoming common after 10 yr, and remains present until the next cutting cycle (White Mountains in New Hampshire) (DeGraaf 1991). Absent in clear-cut and thinned plots but present in strip-cut plots in Nova Scotia (Freedman et al. 1981). Also in wooded city parks, cemeteries, and wooded roadsides.

Special Habitat Requirements: Deciduous and mixed forests and woodlands, especially those composed of immature or scrubby trees.

Nesting: Egg dates: May 18 to 30, Massachusetts (Veit and Petersen 1993:405); May 10 to June 30, New York (Bull 1974:463); May 26 to June 19, Vermont (Ellison 1985:310 in Laughlin and Kibbe 1985). Clutch size: 4 to 5, typically 5. Incubation period: 10 to 13 days. Nesting period: 8 to 12 days (Kricher 1995). Broods per year: 1, 2 suspected (Peck and James 1983). Age at sexual maturity: 1 yr. Nest site: A depression in the ground at the base of a tree, stump, overturned roots, or rock, or in the shelter of a log usually hidden from above. Occasionally in depression at top of stump (Kricher 1995). Constructed of dry leaves and grass.

Sample Densities: Maryland—21 territorial males per 100 acres (40 ha) in dense second growth. 13 territorial males per 100 acres (40 ha) in open slash area. 11 territorial males per 100 acres (40 ha) in virgin hardwood forest (Stewart and Robbins 1958:270).

Food Habits: Gleans wood-boring insects, click beetles, plantlice, small caterpillars, moths, spiders, egg masses, and pupae from bark crevices of tree trunks and main branches, generally to 35 ft (10.7 m) high (MacArthur 1958).

Comments: In the Connecticut Valley in central Massachusetts, black-and-white warblers were one of seven breeding bird species that accounted for 72 percent of the bird observations in deciduous forested wetlands (Swift et al. 1984). Commonly parasitized by brown-headed cowbirds. Black-and-white warblers are one of the earliest arriving warblers in spring, and are the only New England wood-warbler that forages along the trunk of trees like a nuthatch or brown creeper.

American Redstart

(Setophaga ruticilla)

Range: Breeds from se. Alaska, e. to Labrador and Newfoundland, s. to Utah, se. Oklahoma, and e. Texas, e. to South Carolina. Absent as a breeding bird through most of the Great Plains region. Migrates through Central America and the Caribbean. Winter: Baja California, s. Texas, and central Florida, s. to Brazil. Also throughout the Caribbean.

Distribution in New England: Throughout the region, although nearly absent along the coasts of Massachusetts, Rhode Island, and Connecticut (Veit and Petersen 1993:405, Enser 1992, Zeranski and Baptist 1990:227).

Status in New England: Common breeder, common to very common migrant.

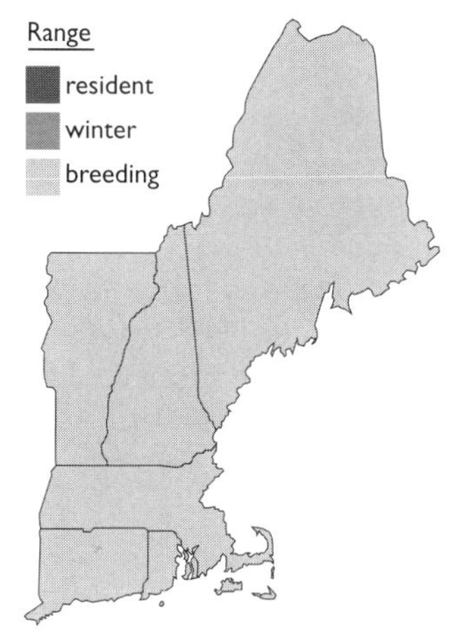

Habitat: Occupies a wide variety of open wood habitats in summer (Sherry and Holmes 1997). Prefers moist, early-successional deciduous forest edge habitat (Hunt 1996) and deciduous forests with a thick understory of saplings. Occasionally in conifer or mixed forests (Sabo 1980). Inhabits forest edges, deciduous saplings bordering pastures, suburban shade trees and shrubs, willow and alder thickets adjacent to streams and wooded swamps. Most abundant in pole stands of red maple and northern hardwoods in the White Mountains in New Hampshire (DeGraaf and Chadwick 1987). More abundant in forests with a mixture of stand ages than in mature forests in the White Mountains in New Hampshire (Welsh and Healy 1993). Associated with forests with a partly open canopy cover, few conifers, and trees in the smaller class sizes in Vermont (Thompson and Capen 1988). Occurs in forests with 84 percent canopy cover, variable conifer cover, and trees in the smaller size classes in w.-central Maine (Collins 1983). Positively associated with high levels of logging that produce hardwood regeneration in e. Maine (Johnson and Brown 1990). Does not avoid edges in New Hampshire (Sherry and Holmes 1997).

Special Habitat Requirements: Early successional deciduous habitats.

Nesting: Egg dates: May 29 to June 21, Massachusetts (Veit and Petersen 1993:405); May 27 to July 6, Vermont (Ellison 1985:312 in Laughlin and Kibbe 1985). May 22 to July 16, New York (Bull 1974:522). Clutch size: 3 to 5, typically 4. Incubation period: 11 to 12 days. Broods per year: 1. Age at sexual maturity: 1 yr. Nest height: 4 to 30 ft (1.2 to 9.1 m). Nest site: In upright crotch of a tree or on a horizontal limb, sapling, or shrub.

Territory Size: 0.43 acres (0.17 ha) in watersheds with 8-yr-old early-successional habitat or 22-yr-old mid-successional habitat, and 1.35 acres (0.54 ha) in mature forests >80 yr of age in the White Mountains in New Hampshire (Hunt 1996). Slightly less than 1 acre (0.4 ha) per pair (Griscom and Sprunt 1957:241); 1 acre (0.4 ha) or less per pair (Hickey, in Bent 1953); 0.8 acres (0.3 ha) per male (Ficken 1962); 6 territories on 1.4 acres (0.6 ha) (average 0.24 acres, 0.1 ha).

Sample Densities: Highest densities in early-successional hardwood habitats and lowest in mature coniferous forests in the White Mountains in New Hampshire (Hunt 1996). 7 males were sighted 10 to 20 m (33 to 66 ft) apart in area 100 m² (120 square yards) (Baker 1944) in a thick stand of young sugar maples (saplings) with a scattering of large deciduous trees. 36 pairs per 40 ha (100 acres) Harbor Island, Me., in white spruce (Morse 1976). 51 territorial males per 100 acres (40 ha) in well-drained floodplain forest in Maryland, and 91 territorial males per 100 acres (40 ha) in second-growth river swamp (Stewart and Robbins 1958:316).

Food Habits: Insects such as caterpillars, bugs, flies, moths, small grasshoppers, beetles, and wasps. Also takes spiders and small amounts of fruit. Generally feeds at heights between 5 and 50 ft (1.5 and 15.2 m) (MacArthur 1958) while branch and twig gleaning, hawking.

Comments: The American redstart is one of the most common and widespread warblers in New England, both as a breeding bird and migrant, and is probably the most common spring migrant.

Worm-eating Warbler

(Helmitheros vermivorus)

Range: Breeds from se. Nebraska, se. Iowa to s. and e. central Ohio, and se. Massachusetts, s. to se. Oklahoma and ne. Texas to nw. Florida. Winter: The West Indies, Veracruz, Chiapas, and the Yucatan Peninsula, s. through Central America (primarily on the Caribbean slope) to Panama. A few winter in s. Florida.

Distribution in New England: Local throughout Connecticut and Rhode Island, less numerous along the coast. Occurs in s. Massachusetts in Berkshire and Norfolk counties and on Mt. Tom (Veit and Petersen 1993:408). Possibly in the New Hampshire seacoast area (Gavutis 1994:288 in Foss 1994).

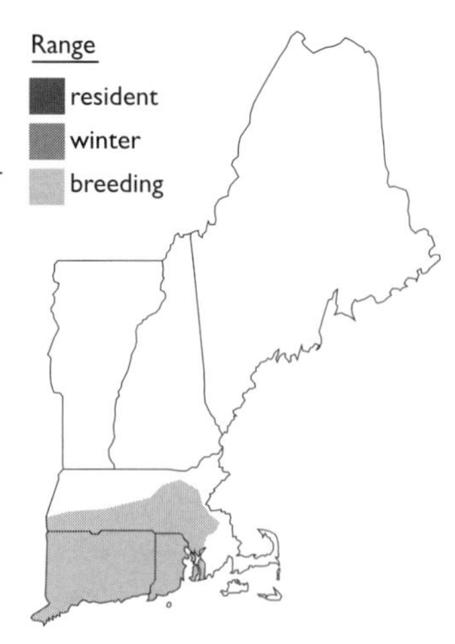

Status in New England: Locally common to rare breeder; rare but regular migrant.

Habitat: Nests in dense understory vegetation in ravines, dry wooded hillsides, extensive stands of mature deciduous forests with thick understory, and edges of streams or swamps rimmed by shrubs or vines.

Special Habitat Requirements: Deciduous or mixed mature forests that include ravines or hillsides with dense undergrowth (Hanners and Patton 1998).

Nesting: Egg dates: Mid June, Massachusetts (Veit and Petersen 1993:408). May 24 to June 18, New York (Bull 1974:466). Clutch size: 3 to 6, typically 4 to 5. Incubation period: 13 days. Nestling period: 10 days. Broods per year: 1. Age at sexual maturity: 1 yr. Nest site: In hollow on the ground often at the base of a tree, rock, or log. Frequently on hillsides or sides of ravines.

Food Habits: Mainly caterpillars (once known as "worms"). Gleans through the leaf litter of the forest floor and probes into suspended dead leaves in early spring (Hanners and Patton 1998).

Comments: The population of worm-eating warblers increased with the reversion of farm land to forest. Exhibiting a northward range expansion; first nested in Connecticut in 1879, in Massachusetts in 1949. Since that time, breeding has occurred both eastward and northward (Zeranski and Baptist 1990:229, Veit and Petersen 1993:408). Very sensitive to forest fragmentation.

Ovenbird

(*Seiurus aurocapillus*)

Range: Breeds from ne. British Columbia, s. Northwest Territories, n. Alberta, across s. Canada to Newfoundland, s. to e. Colorado, e. Oklahoma, n. Arkansas, and the Middle Atlantic States to n. Georgia. Winter: Coastal South Carolina, Florida, the Gulf States, coastal Texas, the West Indies, Mexico, s. to Central America, Venezuela, Colombia, Ecuador, and Peru.

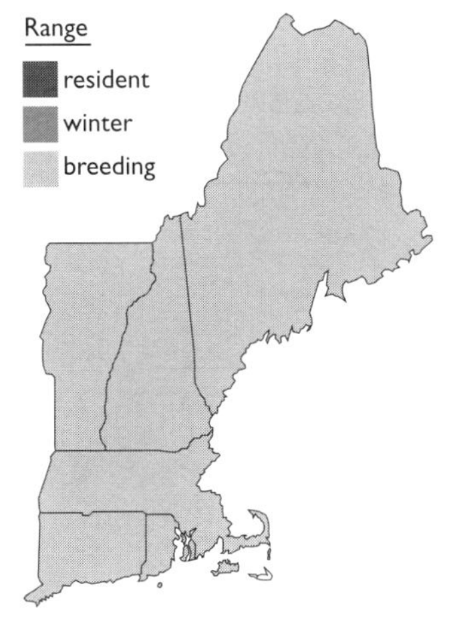

Distribution in New England: Throughout the region, except absent from Nantucket and the outer islands of Massachusetts.

Status in New England: Common breeder; common to uncommon migrant.

Habitat: Usually in open, mature, dry deciduous or mixed forest with little understory and an abundance of fallen leaves and logs. Sometimes associated with undisturbed forest (Freedman et al. 1981), but breeds in a wide range of forest conditions and sizes. Occasionally inhabits floodplain or swamp forests, in the north, inhabits jack pine and spruce forests; in Vermont, occupies northern hardwood forests comprised of beech, sugar maple, yellow birch, and hemlock; least common in extensive spruce-fir forests (Turner 1985:290 in Laughlin and Kibbe 1985). Thinning may reduce ovenbird abundance until the canopy closes (Johnston 1970). Inhabits forests with a mean canopy height 22.7 m and 87.3 percent canopy closure in Vermont (Thompson and Capen 1988). Reports of decreased breeding success in forest fragments in Pennsylvania (Porneluzi et al. 1993), Illinois (Robinson 1992), and Missouri (Faaborg 1990), but fragmentation is not a problem in New England.

Special Habitat Requirements: Large area of contiguous mature deciduous or mixed forest interior habitat (Van Horn and Donovan 1994).

Nesting: Egg dates: May 1 to July 8, Massachusetts (Veit and Petersen 1993:409). May 17 to July 22, New York (Bull 1974:507). Clutch size: 3 to 6, typically 4 or 5. Incubation period: 12 days. Nestling period: 8 to 10 days. Broods per year: generally 1, occasionally 2 if first attempt fails (Bent 1953). Age at sexual maturity: 1 yr. Nest site: On ground (sloping or level), in natural or excavated depression of dead leaves, sometimes at base of tree or log. Primarily leaves, grass, supple plant stems, hair, used to construct a domed nest resembling a clay oven with an entrance hole close to the ground.

Territory Size: 0.5 to 4.5 acres (0.2 to 1.8 ha) per pair (Hann 1937); 21 territories averaged 1.6 acres (0.6 ha), range 0.25 to nearly 3 acres (0.1 to 1.2 ha)—habitat: hemlock-beech (Kendeigh 1945a).

Sample Densities: Maryland—40 territorial males per 100 acres (40 ha) in mixed oak forest. 26 territorial males per 100 acres (40 ha) in dense second growth. 24 territorial males per 100 acres (40 ha) in young second growth (resulting from cutting). 17 territorial males per 100 acres (40 ha) in pine-oak forest (Stewart and Robbins 1958:302). In New Hampshire, the number of territorial males ranges from 10 to 28 per 100 acres (40 ha) in mature northern hardwoods (King 1995).

Food Habits: Insects such as plantlice, caterpillars (hairy and hairless), other larvae, moths, butterflies, grasshoppers, and crickets. Also consumes small snails, slugs, myriapods, earthworms, and spiders from the leaf litter, debris of forest floor, and herbaceous vegetation.

Comments: In New Hampshire, nest predation is greater near edges (0 to 200 m from edge) than in forest interior (200 to 400 m from edge) (King et al. 1996). Ovenbirds are reported to be sensitive to logging activity in Minnesota (Collins et al. 1982), e. Maine (Johnson and Brown 1990), and Nova Scotia (Freedman et al. 1981), but in extensive forests in the White Mountains of New Hampshire, matedness and fledging success are similar near clear-cut edges and in forest interiors (King et al. 1996). Pairs may mate in successive years (Bent 1953:458). Despite the substantial changes in New England's forests in historic times, ovenbirds have long remained fairly common (Linsley 1843, Merriam 1877, Sage et al. 1913, Zeranski and Baptist 1990:230).

Northern Waterthrush

(*Seiurus noveboracensis*)

Range: Breeds from Alaska and s. Northwest Territories, across Canada to central Labrador and Newfoundland, s. to nw. Washington, and e. to central Michigan, ne. Ohio, se. West Virginia, Pennsylvania, New York, and Massachusetts.

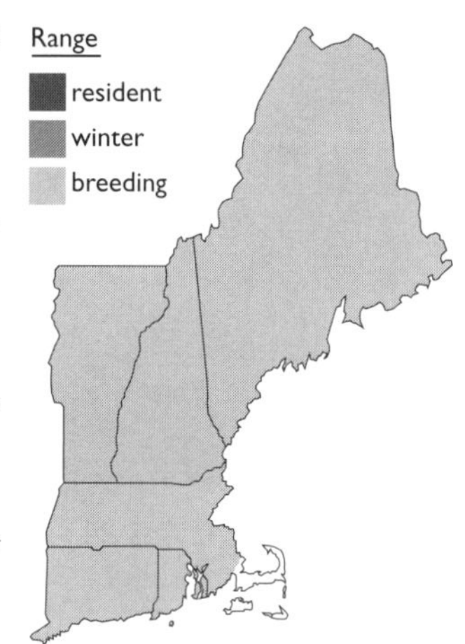

Distribution in New England: Throughout New England, except absent from Cape Cod and the Islands in Massachusetts, and unlikely along the coast and the southernmost portions of the Connecticut River Valley in Connecticut.

Status in New England: Uncommon to locally common breeder, common migrant.

Habitat: Favors dense thickets along edges of deciduous or coniferous forested wetlands, ponds, and montane bogs with pools of standing water containing hummocks (Craig 1985). Rarely inhabits riparian swamps and streams with numerous fallen trees. Frequents red maple and white cedar swamps in Massachusetts. Intolerant of disturbance from timber harvest in Maine (Johnson and Brown 1990). Commonly breeds up to 2,500 ft (850 m) in elevation.

Special Habitat Requirements: Cool, shady, wet brushy areas with open pools of water and hummocks.

Nesting: Egg dates: May 21 to June 15, Massachusetts (Veit and Petersen 1993:411). May 10 to June 28, New York (Bull 1974:509). Clutch size: 3 to 6, average 4 (Peck and James 1987). Incubation period: 12 to 14 days. Nestling period: 9 days. Broods per year: 1. Age at sexual maturity: 1 yr. Nest site: In cavities among roots of fallen trees and in rotten stumps, also at bases of standing trees or stumps, on sides of fern clumps, under cover along banks of wooded ponds or streams.

Territory Size: Eleven territories averaged 2.4 acres (1.0 ha) and ranged between 1.9 to 3.7 acres (0.8 to 1.5 ha) in New York. Ten territories averaged 1.2 acres (0.47 ha) in Connecticut (Eaton 1957).

Sample Densities: Maryland—84 territorial males per 100 acres (40 ha) in open hemlock-spruce bog. 33 territorial males per 100 acres (40 ha) in scrub spruce bog (Stewart and Robbins 1958:303). 12 territorial males per 100 acres (40 ha) in coniferous forest in Vermont (Metcalf 1977). One of the seven most abundant species in deciduous forested wetlands (Swift et al. 1984).

Food Habits: Aquatic insects, beetle larvae, moths, mosquitoes, ants. Also takes small crustaceans, mollusks, and worms. Primarily forages on ground along water's edge, also gleans from foliage (Eaton 1995).

Comments: Numbers have increased north of Connecticut since the turn of the century. First nested in Massachusetts in 1905, and now occupies most nesting habitat in the state (Veit and Petersen 1993:411). The northern waterthrush is more common than the Louisiana waterthrush in Vermont and New Hampshire.

Louisiana Waterthrush

(*Seiurus motacilla*)

Range: Breeds from e. Nebraska, central Iowa, and e.-central Minnesota to central New York and New

England, s. to e. Texas, central Louisiana, central Georgia, and the Carolinas. Winter: The West Indies, Mexico, Central America, Colombia, and Venezuela.

Distribution in New England: Breeds throughout Connecticut, central and w. Massachusetts, s. Vermont, s. New Hampshire, and s. Maine.

Status in New England: Uncommon but increasing (Connecticut) to local breeder (Massachusetts n.); uncommon migrant.

Habitat: Inhabits extensive deciduous and mixed bottomland forests with moss-covered logs and a thick understory along swiftly flowing streams, occasionally in wooded swamps that contain water channels. Sometimes inhabits shrub-swamps, bogs, or areas near swamp pools or lake edges. Considered an area-sensitive species, requiring a minimum of 250 acres (100 ha) to sustain viable breeding populations in Maryland (Robbins 1979). Avoids edge habitat (Chasko and Gates 1982).

Special Habitat Requirements: Woodlands with flowing water (Craig 1985).

Nesting: Egg dates: April 25 to June 20, New York (Bull 1974:510). Clutch size: 4 to 6, typically 5. Incubation period: 12 to 14 days. Nestling period: 9 to 10 days. Broods per year: 1. Age at sexual maturity: 1 yr. Nest height: 1 to 6 ft (0.3 to 1.8 m). Typically on ground. Nest site: In a cavity under steep bank along streams or in upturned roots of a fallen tree over or near water. A path of leaves may lead to the nest (Harrison 1984).

Territory Size: Generally linear in shape, running along streams (Robinson 1995). 1,312 ft in length (400 m) in New York (Eaton 1958). 358 m (range 188 to 538 m) in Connecticut (Craig 1981 in Robinson 1995). 930 m (range 375 to 1,200 m) in Illinois (Robinson 1990 in Robinson 1995).

Sample Densities: 4 territorial males per mile of stream in Vermont (Kibbe 1985:318 in Laughlin and Kibbe 1985). 4.8 territorial males per 100 acres (40 ha) in birch-basswood forests in Pennsylvania (Samson 1979).

Food Habits: Aquatic insects and invertebrates, beetles, ants, caterpillars, scale insects, spiders, and mollusks. Feeds along the sandy margins of streams, ground gleaning, leaf-pulling and gleaning, hawking, and hovering (Robinson 1990).

Comments: The Louisiana waterthrush has expanded its range e. and n. since 1900 into New England.

Mourning Warbler

(*Oporornis philadelphia*)

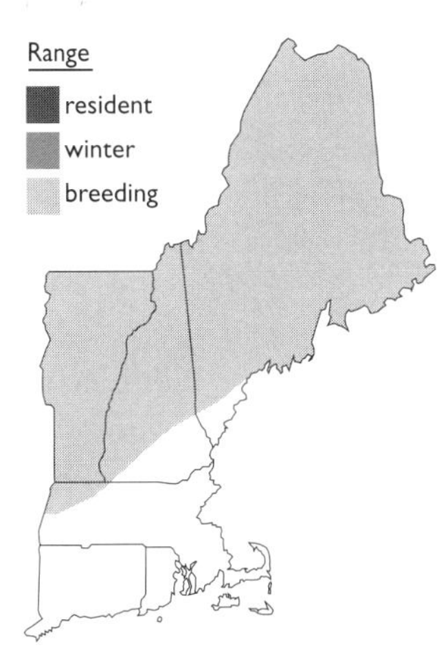

Range: Breeds e. of the Rocky Mountains from central Alberta across Canada and the n. United States from ne. British Columbia to Newfoundland, s. to North Dakota, n. Wisconsin, central Michigan, and w. Massachusetts, and through the Appalachian Mountains to Virginia. Winter: Southern Nicaragua to Venezuela, Colombia, and Ecuador.

Distribution in New England: Throughout Maine except the most s. coastal region, n. New Hampshire, Vermont, and in w. Massachusetts (Berkshire and nw. Franklin counties).

Status in New England: Locally common to uncommon breeder, uncommon migrant.

Habitat: Disturbed second growth: deciduous, shrubby early successional growth, dense undergrowth in open woods or shelterwood cuts, and occasionally in partially open coniferous and deciduous woodlands with herb and shrub understories (Pitocchelli 1993). Inhabits brushy hillsides, shrubby margins of utility corridors, overgrown fields, and lowland swamps or bogs. Occurs in small clearings or burned areas with blackberry and pin cherry thickets (Nichols 1984:320 in Laughlin and Kibbe 1985). Inhabits logged areas with living and dead trees present, including aspen, balsam, pine, and spruce (Niemi and Hanowski 1984). Positively associated with high levels of timber harvest in e. Maine, likely due to presence of brushy deciduous edge vegetation (Johnson and Brown 1990). First appears in clear-cut areas after

2 yr, becomes abundant in 5 yr, and begins to decline after 7 to 10 yr in the White Mountains, New Hampshire (DeGraaf 1991). Associated with open habitat with high ground cover (88.2 percent) and few trees in Maine (Collins 1983b). Frequently observed in the brushy edges of ski trails in Vermont (Nichols 1984:320 in Laughlin and Kibbe 1985).

Special Habitat Requirements: Stands of dense saplings, shrubs (Pough 1949:185).

Nesting: Egg dates: May 28 to July 7, New York (Bull 1974:514). Clutch size: 3 to 5, typically 4. Incubation period: 12 days (Cox 1960). Nestling period: 8 to 9 days (Cox 1960). Broods per year: 1. Age at sexual maturity: 1 yr. Nest height: To 2 ft (0.6 m) (rarely). Nest site: Typically on ground in tangles of briars, weeds, or grasses, commonly near clearings, bogs, or logging roads (Cox 1960).

Territories: 0.61 ha in New Hampshire (Wallace 1949).

Sample Densities: 10 territorial males per 100 acres (40 ha) in dense second growth in Maryland (Stewart and Robbins 1958:308). 51.8 territorial males per 100 acres (40 ha) in ash-basswood forests in New York (Samson 1979).

Food Habits: Beetles, lepidopteran larvae, spiders in thick underbrush within 2 m of the ground (Pitocchelli 1993).

Comments: Rarely encountered by early ornithologists, the mourning warbler is a beneficiary of the clearing of forests in the nineteenth century and subsequent regrowth of shrublands and young forest (Zeranski and Baptist 1990:233). Mourning warblers depend on clearcut logging, utility corridors, and other activities that create early successional habitat.

Common Yellowthroat

(*Geothlypis trichas*)

Range: Breeds from se. Alaska to n. Alberta and Newfoundland, s. to n. Baja California, in Mexico to Oaxaca and Veracruz, and to s. Texas, the Gulf Coast, and s. Florida. Winter: Along the Pacific Coast from n. California across s. Arizona, s. New Mexico, s. Texas, the Gulf States, and South Carolina, and along the Atlantic Coast from New Jersey, Virginia, and Delaware to Florida; also throughout the West Indies, Mexico, Central America, and casually to n. Colombia.

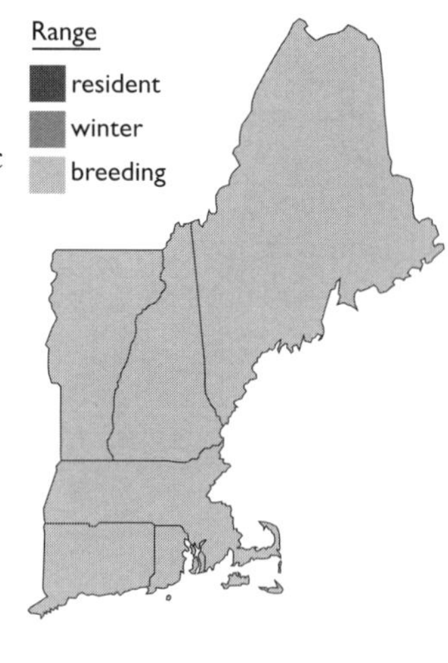

Distribution in New England: Throughout the region.

Status in New England: Common breeder and migrant.

Habitat: Dense brushy habitat with scattered shrubs and trees, most often in damp or wet habitats: beaver meadows, riparian thickets, scrub marshes, and swampy areas at the edge of damp woods, streams, and ponds. Occasionally in dry thickets or dense undergrowth in open woodland. Associated with open areas with little canopy cover (31 percent) and dense ground cover (86 percent) in Grafton, Vt. (Thompson and Capen 1988); 27 percent canopy cover and 88 percent dense ground cover in w.-central Maine (Collins 1983). First appear in clear-cuts 2 yr after harvest, become common after 6 yr, and decline after 10 yr in White Mountain National Forest in New Hampshire and Maine (DeGraaf 1991). Positively associated with areas of high harvest activity—likely due to presence of regenerating hardwood stands (Johnson and Brown 1990). Abundant in deciduous forested wetlands in the Connecticut River Valley region in Massachusetts (Swift et al. 1984).

Special Habitat Requirements: Moist areas with dense, herbaceous vegetation mixed with shrubs and small trees.

Nesting: Egg dates: May 24 to June 17, Massachusetts (Veit and Petersen 1993:417). May 15 to July 12, New York (Bull 1974:515). Clutch size: 3 to 6, typically 4. Incubation period: 12 to 13 days. Nestling period: 8 to 10 days. Broods per year: 1, 2 likely in the South (Harrison 1975). Age at sexual maturity: 1 yr. Nest height: To 3 ft (0.9 m). Typically on ground. Nest site: Among weeds, sedges or shrubs, in grassy tussocks, sometimes among ferns, or higher in low shrubs or tangles of briars.

Territory Size: 0.8 to 1.8 acres (0.3 to 0.7 ha) per pair in the Geddes Marsh area (Michigan) (Stewart 1953); 7 pairs in 5 or 6 acres (2.0 to 2.4 ha), averaging less than

1 acre (0.4 ha) per pair in shrubby field habitat in New York (Kendeigh 1945a).

Sample Densities: 9.7 males per square mile (4 males/km²) (Stewart 1953). 69 males per 100 acres (40 ha) (Stewart 1953). 1 pair per 2 acres (0.8 ha) (Hofslund 1957). 111 territorial pairs per 100 acres (40 ha) in hedgerow bordering brook in Maryland (Stewart and Robbins 1958:309). 17.3 males per 50 acres (20 ha) in red maple-birch forests in New York (Samson 1979).

Food Habits: Cankerworms, fall webworms, gypsy moths, caterpillars, grasshoppers, leafhoppers, plantlice, spiders. Feeds on the ground (in grasses); low shrubs. Males and females forage at different heights (males higher) but do not differ in use of plants (Morimoto and Wasserman 1991b).

Comments: The common yellowthroat occurs from low to high (3,600 ft) elevations in New England, and has been common throughout historic times. It is one of the most abundant breeding warblers in New England, and is frequently parasitized by cowbirds.

Hooded Warbler

(*Wilsonia citrina*)

Range: Breeds from se. Iowa. n. Illinois, extreme s. Michigan and Ontario, s. New York, and s. New England, s. to e. Texas, the Gulf of Mexico, and n. Florida. Migrates through the Greater Antilles and Central America. Winter: Mexico to Panama. Fairly common in the Virgin Islands; uncommon in Puerto Rico (DeGraaf and Rappole 1995:460).

Distribution in New England: Southern Connecticut and s. Rhode Island.

Status in New England: Locally uncommon breeder, primarily along the coast, rare migrant.

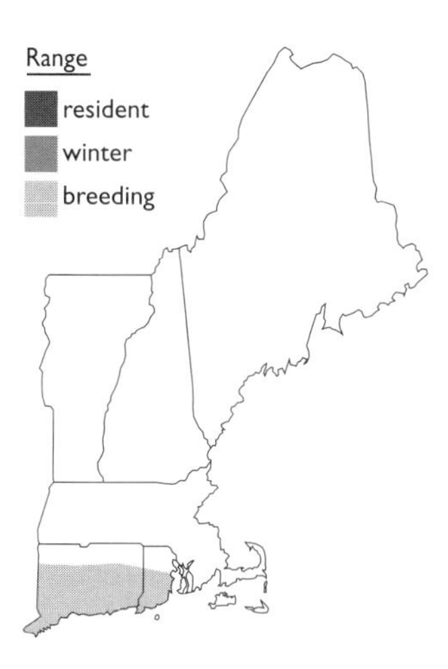

Habitat: Extensive deciduous or mixed forests of maple, beech, oak, with a dense understory of blackberry, cherry, gooseberry, maple-leaf viburnum, and spice bush (Gartshore 1988). Also inhabits wooded swamps with a dense understory. Prefers mature stands with forest openings, but will inhabit younger stands. Invades selectively logged deciduous forests 1 to 5 yr after harvest, where it remains as long as a shrubby understory exists (Evans and Stutchbury 1994). May benefit from openings created by patch cutting (McComb 1985). An area-sensitive species that requires 80 to 250 acres (30 to 100 ha) of forest to sustain a viable breeding population (Robbins 1979). Inhabits cypress-gum swamps in the Southeast.

Special Habitat Requirements: Low, dense, deciduous woody vegetation.

Nesting: Egg dates: May 29 to June 24, Massachusetts (Veit and Petersen 1993:418). May 25 to July 10, New York (Bull 1974:518). Clutch size: 3 to 5. Incubation period: 12 days. Nestling period: 8 to 9 days. Broods per year: 1 or 2. Age at sexual maturity: 1 yr. Nest height: 1 to 6 ft (0.3 to 1.8 m). Typically 2 to 3 ft (0.6 to 0.9 m). Nest site: In a horizontal fork of a bush, sapling, or in herbaceous vegetation.

Territory Size: Range from 1.3 to 1.9 acres (0.5 to 0.75 ha), males usually occupy the same territories in subsequent years (Evans and Stutchbury 1994). 1.0 territorial males in 40 ha (100 acres) in hickory-oak woods in Maryland (Samson 1979). Territories usually include small clearings with a shrub understory for nest sites (Bent 1953).

Sample Densities: 8 territorial males per 40 ha (100 acres) in Crawford County, Pa. (Mark and Stutchbury 1994). Maryland—32 territorial males per 100 acres (40 ha) in second-growth river swamp. 32 territorial males per 100 acres (40 ha) in young second growth. 17 territorial males per 100 acres (40 ha) in open slash area. 8 territorial males per 100 acres (40 ha) in upland oak forest (Stewart and Robbins 1958:313).

Food Habits: Grasshoppers, locusts, caterpillars, plantlice, wasps, ants, moths, beetles, flies, bugs, caddisflies. Hawks and flycatches prey.

Comments: Hooded warblers are at the northern limit of their breeding range in s. New England; they have nested a few times in Massachusetts, last confirmed in 1968 (Veit and Petersen 1993:418). Among warblers of the South, the hooded warbler most frequently occurs in New England.

Wilson's Warbler

(*Wilsonia pusilla*)

Range: Breeds from n. Alaska, n. Yukon, n. Ontario, se. Labrador, and Newfoundland, s. to s. California, central Nevada, n. Utah, n. New Mexico, central Ontario, n. New England (n. White Mountains), and Nova Scotia. Winter: Southern California and s. Texas to Panama.

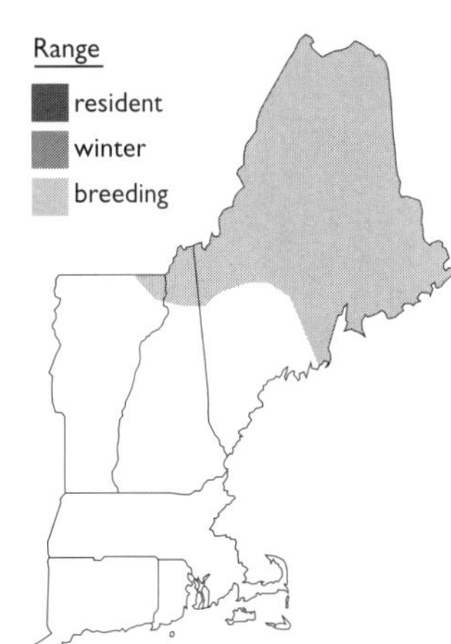

Distribution in New England: Breeds throughout much of Maine (except southern Maine), the Northeast Highlands of Vermont (Oatman 1985:324 in Laughlin and Kibbe 1985), and the n. White Mountains in New Hampshire.

Status in New England: Uncommon and local breeder, fairly common spring and fall migrant.

Habitat: Prefers open, wet areas in early stages of regeneration. Inhabits tamarack bogs, willow and alder scrub-shrub swamps, riparian thickets, edges of beaver ponds, and sphagnum bogs. Occasionally occurs in drier habitats in thickets or areas in early successional stages such as regenerating forests or old fields.

Special Habitat Requirements: Open, wet areas in early stages of regeneration.

Nesting: Egg dates: June 6 to June 21, New Brunswick (Bent 1953:639). Clutch size: 4 to 6, typically 5. Incubation period: 11 to 12 days. Nestling period: 10 to 11 days. Broods per year: 1. Age at sexual maturity: 1 yr. Nest site: On ground or sunken in moss or sedges, usually at base of alders or smaller shrubs, or at base of a sapling. May nest in loose colonies in favorable habitat.

Territory Size: Mean 0.57 ha (1.4 acres), range 0.2 to 1.3 ha (0.5 to 3.2 acres); mean 0.48 ha (1.2 acres), range 0.3 to 1.0 ha (0.7 to 2.47 acres), in California (Stewart 1973).

Sample Densities: Only four breeding records exist for Vermont (Oatman 1985:324 in Laughlin and Kibbe 1985), and it has never been recorded outside of Coos County, N.H. DeGraaf (pers. observ.) found a small colony (5+ pairs) at the edge of a shrub swamp abutting a fir stand in the Kilkenny area (New Hampshire) in 1979 and 1980.

Food Habits: Flies, gnats, plantlice, small caterpillars, other larvae, small grasshoppers, spiders, occasionally berries. Gleans from twigs to 25 ft (7.6 m) above ground (MacArthur 1959).

Comments: Named for Alexander Wilson (1766–1813), the father of American ornithology. Wilson's warbler appears to be more closely associated with wetlands in New England than elsewhere in its range.

Canada Warbler

(*Wilsonia canadensis*)

Range: Breeds from central Alberta, e. to s. Quebec and Nova Scotia, s. to s. Manitoba, central Minnesota, central Michigan, and through the Appalachian Mountains to n. Georgia. Migrates through Central America. Winters in Venezuela, Colombia, Ecuador, Peru, and Brazil.

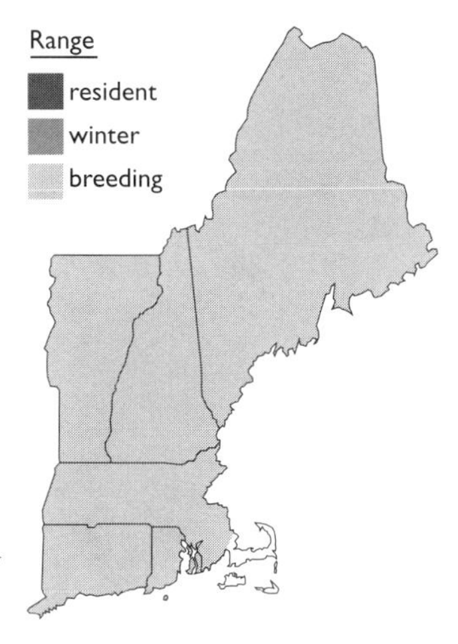

Distribution in New England: Throughout the region, except absent from Cape Cod and the Islands in Massachusetts. Local in Connecticut coastal areas and the Connecticut River Valley (Zeranski and Baptist 1990:236).

Status in New England: Fairly common breeder; common migrant.

Habitat: Occupies a variety of habitat types from upland to lowlands. Favors deciduous forested swamps, dense undergrowth in cool, moist, mature woodlands and along streams and swamps, and cedar bogs. In northern hardwoods with softwood understory in White Mountain National Forest in New Hampshire and Maine (DeGraaf and Chadwick 1987). Inhabits forests with a high percent shrub cover (70 percent), moderate canopy cover (64 percent), and few conifers in the canopy in w.-central Maine (Collins 1983). Common in Massachusetts oak stands with laurel (*Kalmia*) understories. First appear in clear-cuts 5 yr after harvest, become common after 15 yr, and remain abundant until the next cutting cycle (DeGraaf 1991). May prefer deciduous forested wetlands in the Connecticut River Valley region of Massachusetts; it was one of seven species accounting for 72 percent of breeding bird observations in this habitat (Swift et al. 1984). Generally not found at higher elevations in New Hampshire due to absence of dense deciduous growth (Sabo 1980).

Special Habitat Requirements: Forest with dense understory, especially along streams, bogs, swamps, or moist areas.

Nesting: Egg dates: May 31 to June 30, New York (Bull 1974:521). Clutch size: 3 to 5, 4 to 5 laid with equal frequency in New Hampshire (Ellison 1985:326 in Laughlin and Kibbe 1985). Incubation period: Probably 12 days. Nestling period: Unknown, estimated to be 12 days (Harrison 1975). Broods per year: 1. Age at sexual maturity: 1 yr. Nest site: On or near the ground, atop mossy logs or stumps, cavities in banks or amid roots of windthrows, among fern stands. Nests are usually in the vicinity of a stream, pond, or other body of water.

Territory Size: One male occupied a singing area of 0.6 acre (0.2 ha) until nesting began, at which time he expanded his movements to 2 acres (0.8 ha). Another male roamed 3 acres (1.2 ha) after nesting began (New York) (Kendeigh 1945b). Habitat: Hemlock-beech. 0.6 acres (0.2 ha) before nesting, 2 to 3 acres (0.8 to 1.2 ha) after nesting (Kendeigh 1945b).

Sample Densities: Maryland—45 territorial males per 100 acres (40 ha) in dense oak-maple second growth. 32 territorial males per 100 acres (40 ha) in young second growth (after cutting). 21 territorial males per 100 acres (40 ha) in open hemlock-spruce bog (Stewart and Robbins 1958:315). 1.2 territorial males in dry mixed forests in Ohio (Samson 1979). Mt. Greylock, Mass.: 28 counted June 6, 1952 (Veit and Petersen 1993:419).

Food Habits: Mosquitoes, flies, moths, beetles, small hairless caterpillars, spiders. Hawks prey and ground gleans in the leaf litter.

Comments: The Canada warbler spends relatively little time on its breeding grounds; it is usually one of the last warblers to arrive in spring and one of the first to depart after the nesting cycle (Conway 1999). Diet consists wholly of insects and spiders. Declining since the 1980s in the eastern part of its range (DeGraaf and Rappole 1995). May be vulnerable to forest fragmentation.

Yellow-breasted Chat

(*Icteria virens*)

Range: Breeds in s. British Columbia, North Dakota, s. Minnesota, s. Ontario, Vermont, and New Hampshire, s. to s.-central Baja California, the Gulf Coast, n.-central Florida, and Mexico. Winter: Southern Texas and s. Florida, s. through Central America and w. Panama.

Distribution in New England: Southern Connecticut, no longer breeds in Rhode Island. Winter: Southeastern

Massachusetts, except not reported from Martha's Vineyard, w. to s. Rhode Island and s. Connecticut.

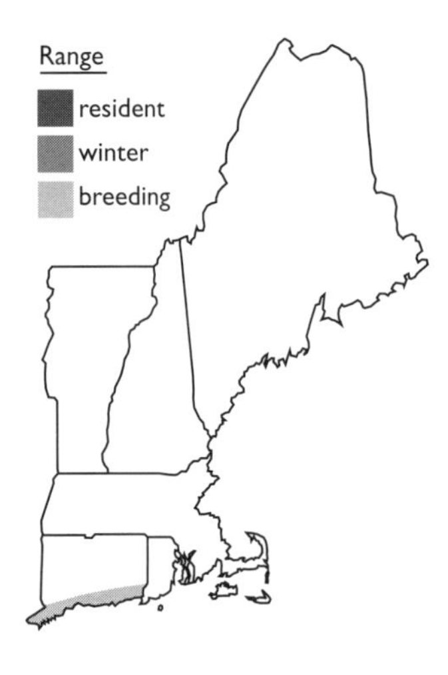

Status in New England: Rare, local, and erratic breeder in Massachusetts (Veit and Petersen 1993:421). Very rare in Connecticut; the last confirmed nest was in Greenwich in 1985 (Zeranski and Baptist 1990:236).

Habitat: Breeding: Extensive open areas of dense, shrubby habitat, especially along streams. Preference for open fields with scattered shrubs and thickets for nesting. Inhabits briar thickets, vines, small trees, and tall shrubs. Also occurs in forest edges, hedgerows, and extensive areas of hardwood regenerating forest. Winter: Scrub thickets, young second growth, and woodland edges.

Special Habitat Requirements: Dense shrubs and vines with no overtopping trees, often near water.

Nesting: Egg dates: May 25 to July 13, New York (Bull 1974:516). Clutch size: 3 to 5, typically 4. Incubation period: 11 to 12 days. Nestling period: 8 to 11 days. Broods per year: 1, occasionally 2. Age at sexual maturity: 1 yr. Nest height: 2 to 8 ft (0.6 to 2.4 m) above ground. Nest site: In a bush, small sapling, or tangle of grapevines, catbrier, brambles, and so on, occasionally on the ground.

Territory Size: 1.2 to 2.5 acres (0.5 to 1.0 ha) per pair, though individuals may roam well into a neighboring territory. Habitat: grown-over abandoned fields in northern Virginia (Dennis 1958).

Sample Densities: Maryland—36 territorial males per 100 acres (40 ha) in shrubby field with trees and stream. 28 territorial males per 100 acres (40 ha) in damp deciduous scrub with snags. 15 territorial males per 100 acres (40 ha) in dry deciduous scrub resulting from fire (Stewart and Robbins 1958:311).

Food Habits: Beetles, bugs, ants, wasps, weevils, mayflies, various caterpillars including tent caterpillars and grasshoppers, true bugs, raspberries, wild strawberries, blackberries, wild grapes.

Comments: Historically, the yellow-breasted chat occupied areas cleared of forest and declined as the area reverted to forest. It was likely common after disturbances such as fire. It is declining in the Northeast, due largely to reforestation of agricultural land and the almost total absence of shrublands.

Scarlet Tanager

(*Piranga olivacea*)

Range: Breeds from s. Manitoba, w. Ontario, s. Quebec, and New Brunswick, s. to e. North Dakota, central Nebraska, s. Kansas, e. Oklahoma, central Arkansas, n. Alabama, and n. Georgia; casually in the w. United States. Winter: Panama and Colombia s., e. of the Andes, through e. Ecuador and Peru to nw. Bolivia and South America.

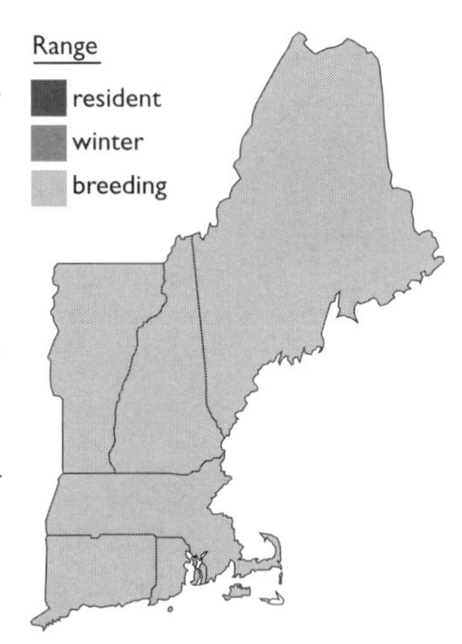

Distribution in New England: Throughout the region, except absent from Nantucket in Massachusetts.

Status in New England: Common and widespread breeder; common to uncommon migrant.

Habitat: Mature dry deciduous and mixed woodlands, forested deciduous wetlands, and bottomland or moist upland forests. Inhabits a wide range of forest stages, but most abundant in mature forests. Also in wooded parks, orchards, large suburban shade trees, roadsides, and powerline cuts. Associated with closed-canopy (79 percent) deciduous forests in Grafton, Vt. (Thompson and Capen 1988). Associated with saw-timber stands of northern hardwood and red maple (DeGraaf and Chadwick 1987); first appear in clear-cut areas 25 yr after harvest in White Mountain National Forest in New

Hampshire and Maine (DeGraaf 1991). 68 percent of 28 nests in oak-hickory woods and tamarack swamps were in trees with a diameter breast height greater than or equal to 9 in (23 cm) (Prescott 1965:21).

Special Habitat Requirements: Mature or pole-sized deciduous or mixed woodlands.

Nesting: Egg dates: May 20 to July 23, New York (Bull 1974:544). Clutch size: 3 to 5, usually 4. Incubation period: 13 to 14 days. Nestling period: About 15 days. Broods per year: 1. Age at sexual maturity: 1 yr. Nest height: Typically 20 to 35 ft (6.1 to 10.7 m). Average 33 ft (10 m) above ground in Vermont (Ellison 1985:328 in Laughlin and Kibbe 1985). Nest site: Usually placed well out from trunk on a horizontal branch in a leaf cluster or position where it is shaded from above and open to the ground below. It is usually placed where it can be approached by unobstructed flight from adjacent trees (Prescott 1965:20). Sometimes the eggs can be seen through the flimsy nest from below.

Territory Size: 1.2 territorial males in mixed mesophytic forests in Ohio (Samson 1979).

Sample Densities: Maryland—26 territorial males per 100 acres (40 ha) in virgin central hardwood deciduous forest. 15 territorial males per 100 acres (40 ha) in mature hardwood forest. 14 territorial males per 100 acres (40 ha) in dense second-growth (oak-maple) forest (Stewart and Robbins 1958:331).

Food Habits: Insects, fruits. Seven-eighths of the diet is animal, and one-eighth is vegetable (McAtee 1929 in Bent 1958:485). Gleans prey from leaves and twigs of outer tips of limbs, dead branches. Commonly forages in the open canopy of mixed and deciduous forests, often oaks or beeches (Holmes and Robinson 1981).

Comments: Historically, scarlet tanagers probably declined throughout the 1800s as forests were cleared for agriculture (Forbush 1929), but since the 1920s they have again become common with the return of deciduous woodlands (Bagg and Eliot 1937, Zeranski and Baptist 1990:238). During exclusively cold or rainy weather in spring, scarlet tanagers may feed on the ground, gleaning insects washed from the trees (Zumeta and Holmes 1978).

Eastern Towhee

(Pipilo erythrophthalmus)

Range: Breeds from s. British Columbia to s. Maine, s. to s. Baja California, Guatemala, n. Oklahoma, e. Louisiana, and s. Florida. Winter: From s. British Columbia, Utah, Colorado, the s. Great Lakes area, and along the Atlantic Coast, s. throughout the breeding range.

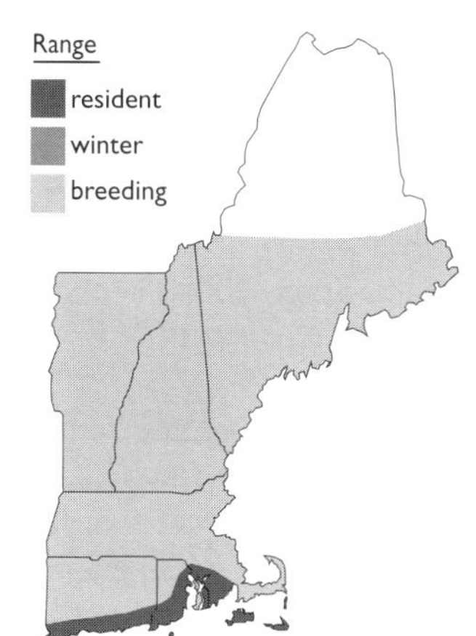

Distribution in New England: Southern Maine, throughout New Hampshire and Vermont, although most abundant in the s. parts of these states (Janeway 1994:312 in Foss 1994, Ellison 1985:336 in Laughlin and Kibbe 1985), throughout Massachusetts, Connecticut, and Rhode Island. Winter: Southeastern Massachusetts on Cape Cod and the Islands, s. Rhode Island, and Connecticut along the coast and at feeding stations (Veit and Petersen 1993:432, Ellison 1985:336 in Laughlin and Kibbe 1985, Enser 1992:167).

Status in New England: Common and widespread breeder in suitable habitat, but rapidly declining because of forest succession; rare winter resident along the Connecticut coast (Zeranski and Baptist 1990:244).

Habitat: Generally inhabits dense, brushy fields and dry pastures, forest edges, brushy areas along road edges and utility rights-of-way, pitch-pine/scrub-oak forests, clearings, and mid to late stages of secondary succession in burned-over areas, regenerating clear-cuts, and open selective and shelterwood or seed-tree cuts (Ellison 1985:336 in Laughlin and Kibbe 1985, Greenlaw 1996, Morimoto and Wasserman 1991a).

Special Habitat Requirements: Dense brushy dry cover.

Nesting: Egg dates: May 26 to June 19, Vermont (four records) (Ellison 1985:336 in Laughlin and Kibbe 1985). May 15 to August 4, New York (Bull 1974:570). Clutch size: 2 to 6, depending on food supply (Greenlaw 1978) typically 3 or 5. Incubation period: 12 to 13 days. Nestling period: 10 to 12 days. Broods per year: 2. Age at sexual maturity: 2nd year (Greenlaw 1996). Nest height: To 5 ft (1.5 m), typically on ground. Nest site: On or near ground in brushy cover or low in a shrub. Commonly parasitized by cowbirds.

Territory Size: Territory size varies throughout the breeding season in Massachusetts (Wasserman 1983). Mean territory size was 0.65 acres (0.26 ha) in pine barrens in se. Massachusetts (Morimoto and Wasserman 1991a). Mean territory size was 4 acres (1.6 ha), range 1.6 to 6.1 acres (0.64 to 2.44 ha), in mesic oak forests, and 3 acres (1.2 ha), range 1.8 to 4.1 acres (0.71 to 1.65 ha), in xeric pine-oak forests in New Jersey (Greenlaw 1996).

Sample Densities: 21 territorial males per 100 acres (40 ha) in a mesic oak forest and 32 territorial males per 100 acres (40 ha) in a xeric pine-oak forest in New Jersey (Greenlaw 1969 in Greenlaw 1996). 104 pairs per square mile (40 pairs/km^2) in favorable habitat in North Dakota (Stewart and Kantrud 1972). 50 territorial males per 100 acres (40 ha) in dry deciduous scrub; 33 territorial males per 100 acres (40 ha) in open slash area; 32 territorial males per 100 acres (40 ha) in young second growth (following cutting); 6 territorial males per 100 acres (40 ha) in young second growth (following cutting); 6 territorial males per 100 acres (40 ha) in pine-oak forest (Stewart and Robbins 1958:348).

Food Habits: Insects, seeds, fruits, snails. Scratches through the leaf litter of forest floor, scattering leaves with its beak, and occasionally gleans insects from foliage and fruits in low tree canopies (Greenlaw 1996).

Comments: Eastern towhees are large sparrows that scratch noisily in the leaf litter for seeds and insects. A bird of disturbed forests and dry forest edges, the eastern towhee is rapidly declining in the Northeast as agricultural land and brushy edges revert to forest or are lost to development.

American Tree Sparrow

(Spizella arborea)

Range: Breeds from Alaska and n. Yukon e. to n. Quebec and Labrador, and s. to nw. British Columbia, n. Saskatchewan, and central Quebec. Winters from s. Canada s. to s. U.S.

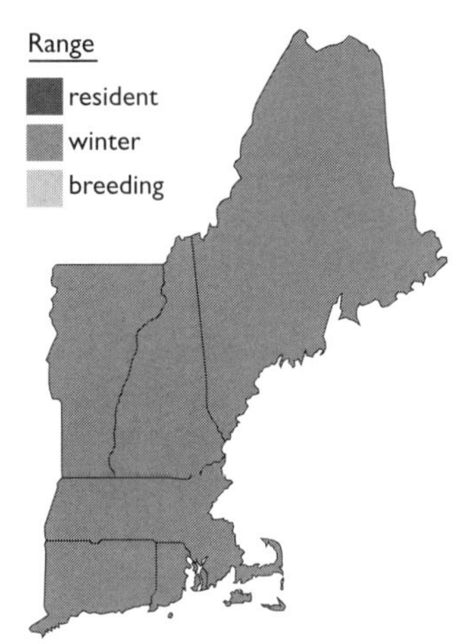

Distribution in New England: Winters throughout the region.

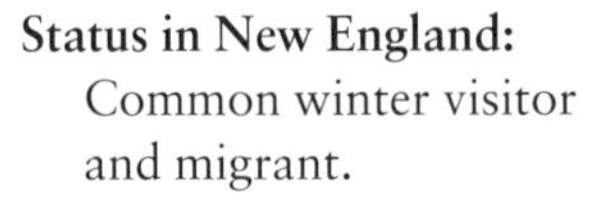

Status in New England: Common winter visitor and migrant.

Habitat: Open country, brushy edges of fields, weedy pastures, marshes, hedgerows, farmland (Naugler 1993). When deep snow covers old fields, tree sparrows move into swampy woodlands.

Territory Size: Nonterritorial in winter, but occupy winter ranges 300 to 1,200 m in diameter (Heydweiller 1935). May exhibit winter site fidelity (Brooks 1985). Individuals may wander up to 10 km from winter range.

Food Habits: Small grass and weed seeds, e.g., crabgrass, panicgrass, pigweed, ragweed, goldenrod, aster. Frequents feeding stations, may eat snow to obtain water (Bent 1968).

Comments: "American tree sparrow" is a misnomer for this bird; most individuals breed in northern Canada in areas north of the tree line and only winter in old-field habitats here (Naugler 1993). Tree sparrows have been common winter visitors since the early 1800s (Merriam 1877, Sage et al. 1913), but have probably declined, at least in s. New England as old fields have been lost to development or reverted to forest (Zeranski and Baptist

1990:245). Individuals may wander several miles from winter range in search of food (Sargent 1959).

Chipping Sparrow

(*Spizella passerina*)

Range: Breeds from e.-central and se. Alaska and central Yukon to n. Manitoba, s. Quebec, and sw. Newfoundland, s. to sw. and e.-central California, central and e. Texas, the Gulf Coast, and nw. Florida, through the highlands of Mexico to Nicaragua. Winters from central California, n. Texas, Tennessee, and Maryland, s. in Mexico throughout the breeding range.

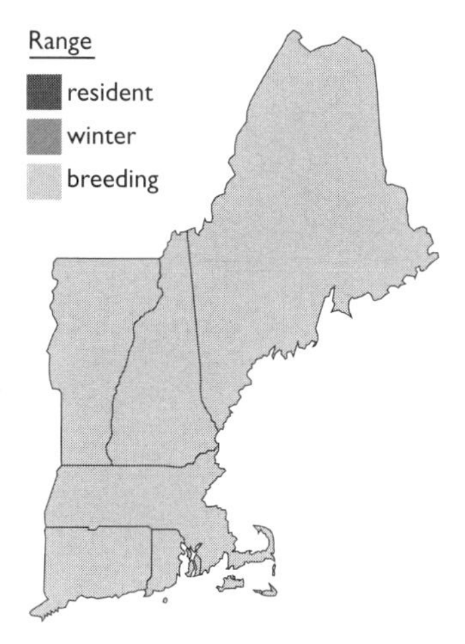

Distribution in New England: Widespread throughout the region. A few winter in se. Massachusetts, mid Cape Cod, and Martha's Vineyard (Veit and Petersen 1993:434).

Status in New England: Common breeder and migrant; rare but increasing early winter resident in s. New England.

Habitat: Breeds in suburban residential areas, farms, orchards, open mixed woodlands, clearings in forests and woodland edges, borders of lakes and streams (Middletown 1998). Avoids extensive forests. Winter: Coastal areas in s. New England.

Nesting: Egg dates: May 12 to June 26, Massachusetts (Veit and Petersen 1993:434). May 2 to July 19, New York (Bull 1974:583). Clutch size: 2 to 5, typically 3 or 4. Incubation period: 11 to 14 days. Nestling period: 10 to 14 days, typically 9 or 10 (Bent 1968). Broods per year: 2. Age at sexual maturity: 1 yr. Nest height: 1 to 25 ft (0.3 to 7.6 m). Typically 3 to 10 ft (0.9 to 3.0 m). Nest site: In a tree, shrub or vine; rarely on ground. Nest is often low in ornamental evergreens, Christmas tree plantations (Buech 1982), or Japanese barberry (*Berberis*), typically well concealed. First nests are generally placed lower than second nests (Walkinshaw 1944).

Territory Size: 1 to 1.5 acres (0.4 to 0.6 ha) in residential area in Michigan (Walkinshaw 1944). Two-thirds of an acre (0.3 ha) in Michigan (Sutton 1960). 7.6 acres (3.1 ha) in South Carolina (Odum and Kuenzler 1955).

Sample Densities: Maryland—90 territorial males per 100 acres (40 ha) in suburban residential area with orchard and lawn. 48 territorial males per 100 acres (40 ha) in unsprayed apple orchard. 18 territorial males per 100 acres (40 ha) in mixed agricultural habitats, including hedgerows and wood margins (Stewart and Robbins 1958:363).

Food Habits: Insects, seeds. March through November diet: 38 percent animal, 62 percent vegetable (Judd 1900 in Bent 1968:1175). Commonly feeds in open areas: bare or sparsely vegetated ground.

Comments: Chipping sparrows are among the few North American birds that have adapted well to human uses of the land. They are conspicuous in settled areas—they sing incessantly all day and readily nest in ornamental plantings.

Field Sparrow

(*Spizella pusilla*)

Range: Breeds from nw. and se. Montana and n. North Dakota to sw. Quebec and s. New Brunswick, s. to w.

Kansas, s. Texas, the Gulf Coast, and s. Georgia. Winter: From Kansas to Massachusetts, and s. to Mexico, the Gulf Coast, and s. Florida.

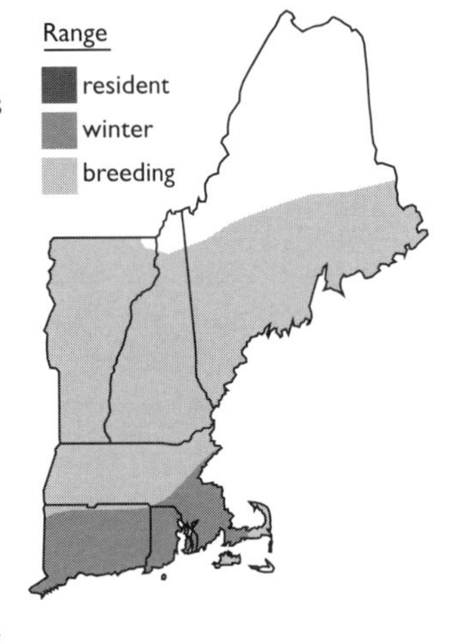

Distribution in New England: Breeds throughout Connecticut (Zeranski and Baptist 1990:246), Massachusetts except on Nantucket (Veit and Petersen 1993:436), Rhode Island except absent from Block Island (Enser 1992:169), Vermont (Ellison 1985:340 in Laughlin and Kibbe 1985), New Hampshire (Janeway 1994:316 in Foss 1994), and central and s. Maine (Boone and Krohn 1996:262). Winters throughout Connecticut (Zeranski and Baptist 1990:246) and se. Massachusetts, except outer Cape Cod and the islands (Veit and Petersen 1993:436); absent from Vermont, New Hampshire, and Maine (Carey et al. 1994).

Status in New England: Fairly common breeder in Connecticut (Zeranski and Baptist 1990:246), Massachusetts (Veit and Petersen 1993:436), Rhode Island (Enser 1992:169), and s. and central Vermont (Ellison 1985:340 in Laughlin and Kibbe 1985) and s. New Hampshire (Janeway 1994:316 in Foss 1994). Uncommon in n. Vermont, n. New Hampshire, and s. and central Maine. Common to uncommon from Connecticut (Zeranski and Baptist 1990:246) to se. Massachusetts (Veit and Petersen 1993:436).

Habitat: Breeds in old fields in early stages of succession with scattered woody vegetation (Evans 1978), lightly overgrown pastures, abandoned hayfields, unplowed agricultural fields, powerline corridors, overgrown gravel pits, briar thickets, woodland edges. Also cutover pine forests (DeGraaf et al. 1991), occasionally Christmas tree farms, orchards and tree nurseries (Peterjon and Rice 1991). Typically absent from suburban habitats (Carey et al. 1994). Winters in abandoned agricultural fields, pastures, forest edges, and fencerows (Allaire and Fisher 1975).

Special Habitat Requirements: Breeding: Open grassy areas with low shrubs or trees (Fretwell 1968). Winter: Forb- or grass-dominated habitats with brushy areas for cover from predators (Pearson 1991).

Nesting: Egg dates: May 8 to June 12, Massachusetts (Veit and Petersen 1993:437). May 16 to August 17 (second brood), New York (Bull 1974:586). Clutch size: 2 to 5, typically 3 or 4. Incubation period: 10 to 12 days (Best 1978, Walkinshaw 1968). Nestling period: 7 or 8 days (Walkinshaw 1968). Broods per year: 2 (rarely 3). Age at sexual maturity: 1 yr. Nest height: To 4 ft (1.2 m), typically on ground. Nest site: Early nests are usually on or near the ground in a tuft of grass. Later nests may be up to 4 ft high (1.2 m) in shrubs or trees.

Territory Size: 0.31 to 1.62 ha (0.8 to 4 acres) in shrub-grassland habitat in Illinois (Best 1977). 0.75 to 2.0 acres (0.3 to 0.8 ha) (average 1.3 acres, 0.5 ha) on semi-wooded hillsides or idle prairie grass pasture in Iowa (Crooks and Hendrickson 1953). Less than 2 acres to 5 or 6 (0.8 to 2 or 2.4 ha) in various habitats (Bent 1968:1220); 2 to 6 acres (0.8 to 2.4 ha) (Walkinshaw 1968).

Sample Densities: 8 pairs per 10 acres (4 ha) in fallow field in Michigan (Berger in Bent 1968). 1 pair per 3 acres (1.2 ha) in suitable habitat in Michigan (Walkinshaw 1939b). 80 males per 100 acres (40 ha) of unmowed apple orchard (Stewart and Robbins 1958:364).

Food Habits: In summer, insects >50 percent of diet, the rest, seeds of weeds and grasses (Evans 1964, Allaire and Fisher 1975). In winter, >90 percent seeds (Pulliam and Enders 1971, Allaire and Fisher 1975). Gleans from ground, also flies up and pulls stems to ground (Carey et al. 1994).

Comments: Field sparrows are at the northern limits of their breeding range in New England; they were abundant nesters before about 1915, and are declining with forest succession and development.

Vesper Sparrow

(*Pooecetes gramineus*)

Range: Breeds from s. Northwest Territories and central Saskatchewan to s. Quebec and Nova Scotia, s. to e.

and s. California, central New Mexico, Kansas, and North Carolina. Winters from central California, central Texas, s. Illinois, and Connecticut, s. to Mexico and Guatemala, the Gulf Coast, and central Florida.

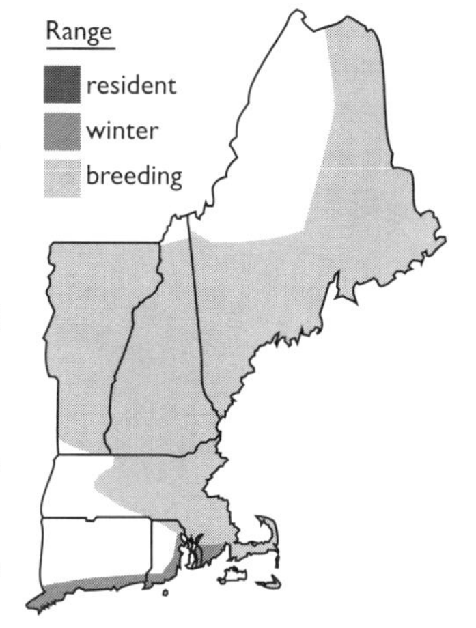

Distribution in New England: Breeding: Locally throughout ne. and s. Maine, New Hampshire, Vermont, and central and e. Massachusetts, except absent from Martha's Vineyard, Connecticut and s. Rhode Island, and Block Island. Winter: Coastal Connecticut, Rhode Island, and se. Massachusetts, depending on winter severity.

Status in New England: Uncommon and local breeder (rare in Connecticut); populations are declining in eastern North America with the disappearance of specialized breeding habitat. Uncommon migrant; rare in winter.

Habitat: Inhabits sparsely vegetated dry uplands such as short-grass meadows, grazed pastures, hayfields, blueberry barrens, cultivated grain fields, dry open uplands such as potato field edges, burned and cutover areas, coastal beach grass, airfields, and open-country roadsides.

Special Habitat Requirements: Dry upland areas with short, sparse herbaceous vegetation and conspicuous singing perches.

Nesting: Egg dates: April 15 to August 11, Massachusetts (Veit and Petersen 1993:438). May 5 to August 16, New York (Bull 1974:581). Clutch size: 3 to 6, typically 4 or 5. Incubation period: 12 to 13 days. Nestling period: 9 to 14 days. Broods per year: 2. Age at sexual maturity: 1 yr. Nest site: In or at base of grass tussock in depression in ground. Early nests may be completely exposed from above until concealed by surrounding growing vegetation.

Territory Size: 1.2 to 1.8 acres (0.5 to 0.7 ha) per pair in uncultivated field in Michigan (Bent 1968:869). 1.5 to 2.7 acres (0.6 to 1.1 ha) (average 2.2 acres, 0.9 ha) for 5 territories in Wisconsin grasslands (Wiens 1969:35).

Sample Densities: 3 pairs per 10 acres (4 ha) in a fallow field bordered by woods in Michigan (Bent 1968:869). Range of 8 to 12 pairs annually in a 14-acre (5.7-ha) uncultivated field in Michigan (Bent 1968:869). 40 pairs per square mile (15 pairs/km^2) in favorable habitat in North Dakota (Stewart and Kantrud 1972). 5 males per 80 acres (32.4 ha) in grassland in Wisconsin (Wiens 1969:53).

Food Habits: Insects and other small invertebrates (33 percent), weed seeds (66 percent) (Bent 1968:875). Feeds in grasses and weeds, sparsely vegetated ground.

Comments: The vesper sparrow is in severe decline with the loss of open habitats. A century ago it was an abundant breeding bird in the rural New England landscape; even a few decades ago it was common in agricultural areas. With losses of the last bits of cropland, the species has little breeding habitat left in New England. It was likely rare in New England before extensive clearing in the 1700s, and thrived briefly during the peak of agriculture.

Savannah Sparrow

(*Passerculus sandwichensis*)

Range: Breeds from Alaska and n. Yukon to n. Labrador and Newfoundland, s. in coastal regions to w.-central California, and in the interior to central California, n. New Mexico, Mexico, Nebraska, Kentucky, and New Jersey. Winters from s. British Columbia, Oklahoma, s. Illinois, and s. New England, s. to Central America and Cuba.

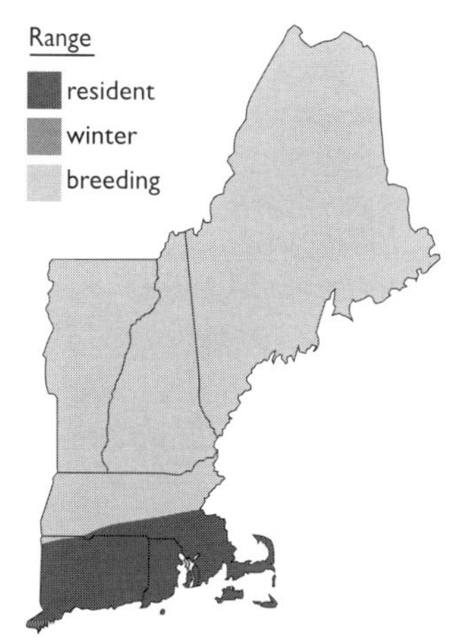

Distribution in New England: Breeds throughout the region in suitable habitat. Primarily s. coasts in winter.

Status in New England: Common (Cape Cod moorlands, Champlain Lowlands, and n.-central Vermont) to uncommon (New Hampshire) breeder. Appears to be declining with habitat succession. Common spring migrant, especially along the coast, rare in winter in s. New England.

Habitat: Hayfields, meadows, lightly grazed pastures, tundra, salt marshes, open moorlands, sand dunes, agricultural fields (especially alfalfa), airports.

Special Habitat Requirements: Large areas of grassland of intermediate height.

Nesting: Egg dates: May 21 to June 29, Massachusetts (Veit and Petersen 1993:440). May 11 to July 16, New York (Bull 1974:572). Clutch size: 2 to 6, typically 4 to 5 (Dixon 1978). Incubation period: 10 days (Welsh 1975). Average 10 to 13.2 days (Wheelwright and Rising 1993). Nestling period: 8 to 14 days (Bent 1968, Potter 1974, Welsh 1975). Broods per year: 2 or 3 (Dixon 1978). Age at sexual maturity: 1 yr. Nest site: In hollow on ground, typically hidden by a canopy of surrounding vegetation, often in grass tufts.

Territory Size: 0.16 to 1.09 ha (0.4 to 2.7 acres), average 1.4 acre (0.57 ha), for 16 territories in grasslands (Wiens 1973). 0.4 to 4.3 acres (0.2 to 1.7 ha), average 1.7 acres (0.7 ha), for 91 territories in grasslands in Wisconsin (Wiens 1969:35). 99 territories ranged from 0.21 to 1.91 ha (0.5 to 4.7 acres) in Nova Scotia (Stobo and McLaren 1975:32). Average 0.28 acres (0.11 ha) in Michigan (Potter 1974), 0.43 acres (0.17 ha) in Nova Scotia (Welsh 1975); territories on Kent Island, N.B., ranged from 0.13 to 0.75 acres (0.05 to 0.30 ha) (Wheelwright and Rising 1993).

Sample Densities: 115.9 pairs/km² (301 pairs per square mile) in grassland in Wisconsin (Wiens 1973). 37 territorial males per 80 acres (32.4 ha) in grasslands in Wisconsin (Wiens 1969:53). 120 pairs per square mile (46 pairs/km²) in favorable habitat in North Dakota (Stewart and Kantrud 1972). 50 territorial males per 100 acres (40 ha) in lightly grazed pasture in Maryland (Stewart and Robbins 1958:351). Greatest on islands: up to 8 pairs per acre (20 per ha) on Kent Island, N.B. (Wheelwright and Rising 1993), 0.6 pairs per acre (1.4 per ha) in Michigan (Potter 1974), 0.9 pairs per acre (2.3 per ha) in Quebec (Bedard and Meunier 1983).

Food Habits: Insects, especially beetles and grasshoppers, seeds of grasses and weeds.

Comments: Savannah sparrows probably increased with forest clearing in the settlement period, and have gradually declined inland with forest succession. The "Ipswich" savannah sparrow, *P. s. princeps,* breeds on Sable Island, Nova Scotia, and winters along the Atlantic Coast from Massachusetts to Georgia. It is a locally common winter resident restricted to sand dunes, beaches, and salt marshes on the immediate coast (Veit and Petersen 1993:441). This species was named by Alexander Wilson for the town of Savannah, Georgia, where the type specimen was collected (Wheelwright and Rising 1993).

Grasshopper Sparrow

(*Ammodramus savannarum*)

Range: Breeds from s. interior British Columbia and s. Alberta to sw. Quebec and s. Maine, s. to s. California, central Colorado, n. and s.-central Texas, central Georgia, and central North Carolina. There are also local resident populations throughout Central America, s. to Colombia and Ecuador, and in the West Indies on Jamaica, Hispaniola, and Puerto Rico. Winters from central California (rare) and s. Arizona to Tennessee and North Carolina, s. to Central America. Birds from North America winter from the s. United States to Costa Rica.

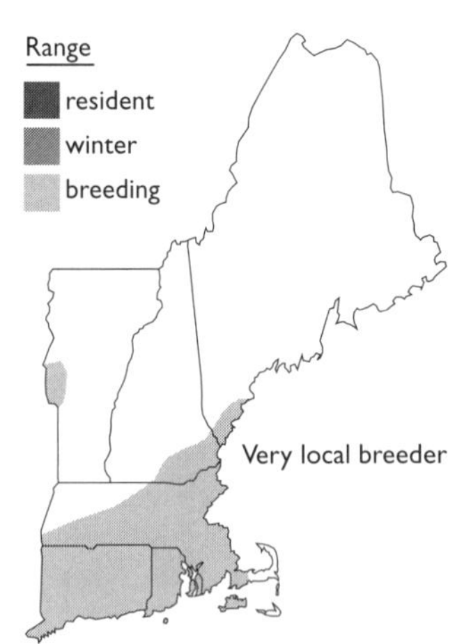

Distribution in New England: Patchy distribution where suitable habitat exists in s. Maine, s. New Hampshire, s. Vermont, including the s. Champlain Lowlands,

Connecticut, Rhode Island, and central and e. Massachusetts, except outer Cape Cod.

Status in New England: Rare to uncommon and local breeder, has been declining as remaining agricultural land is developed or reverts to forest. Rare fall migrant, very rare in winter.

Habitat: Generally prefers moderately open grasslands with patchy bare ground: dry hayfields, especially those with alfalfa and red clover, weedy fallow fields, prairies, even coastal dunes in Massachusetts (Vickery 1996), and airfields generally greater than 25 acres (10 ha) in size. Avoids fields with greater than 30 percent shrub cover (Johnston and Odum 1956). Reportedly occupying capped landfills in s. New England (Brad Blodgett, pers. commun). Habitat similar to that of Savanah sparrow (Vickery 1996).

Special Habitat Requirements: Dry grassy areas with conspicuous song perches.

Nesting: Egg dates: May 30 to August 6 (three records), Vermont (Ellison 1985:346 in Laughlin and Kibbe 1985). May 27 to August 6, New York (Bull 1974:580). Clutch size: 3 to 6, typically 4 or 5. Incubation period: 12 to 13 days. Nestling period: 9 days. Broods per year: 2. Age at sexual maturity: 1 yr. Nest site: In a depression on the ground, usually well hidden by surrounding weeds and grasses. Prefers orchard grass, alfalfa, and clover. Birds are solitary or nest in small colonies.

Territory Size: 6 territories averaged 3.4 acres (1.4 ha) in Iowa prairie (Kendeigh 1941a). 0.32 to 1.34 ha (0.8 to 3.3 acres), average 0.73 ha (1.8 acres), for 16 territories in grasslands (Wiens 1973). 0.8 to 4.3 acres (0.3 to 1.7 ha), average 2.1 acres (0.8 ha), in grasslands in Wisconsin (Wiens 1969:35). 1.2 to 3.3 acres (0.5 to 1.3 ha), average 2.0 acres (0.8 ha), for 22 territories on a farm in West Virginia (Smith 1963).

Sample Densities: 92 pairs/km^2 (239 pairs per square mile) on grassland in Wisconsin (Wiens 1973). 30 territorial males per 80 acres (32.4 ha) in grasslands in Wisconsin (Wiens 1969:53). 60 pairs per square mile (23 pairs/km^2) in favorable habitat in North Dakota (Stewart and Kantrud 1972). 77 territorial males per 100 acres (40 ha) in weedy fallow field in Maryland. 32 territorial males per 100 acres (40 ha) in weedy pasture in Maryland (Stewart and Robbins 1958:352).

Food Habits: Insects, seeds of annual weeds and grasses.

Comments: The grasshopper sparrow is in decline as farmlands revert to shrubby fields. Like other grassland sparrows, it probably increased during the settlement period—it was an "abundant" nester in the mid 1800s in Connecticut (Merriam 1877), although it apparently has always been considered rare and local in New Hampshire (Forbush 1929).

Fox Sparrow

(*Passerella iliaca*)

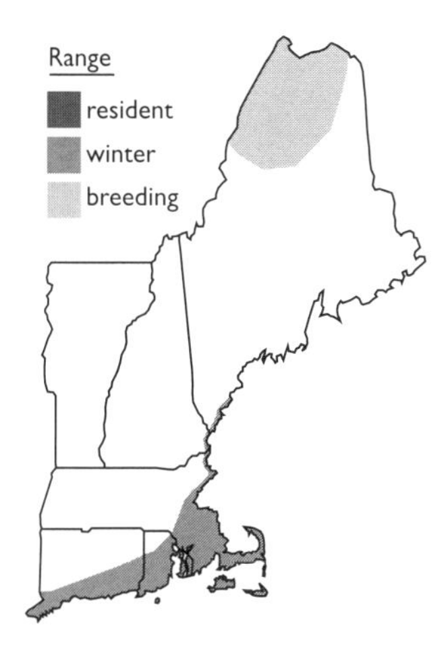

Range: Breeds from Alaska and the Yukon to n. Quebec and n. Labrador, s. to nw. Washington, in the w. mountains to s. California and central Colorado, and e. of the Rockies, across central Canada to s. Quebec. Winters from s. Alaska and s. British Columbia, s. through the Pacific States, and from central Arizona, Kansas, and New Brunswick, s. to Mexico and central Florida (DeGraaf et al. 1991:481).

Distribution in New England: Breeding: Northern Maine. Winter: Coastal areas from se. Massachusetts (Veit and Petersen 1993:448).

Status in New England: Uncommon breeder (n. Maine). Common inland spring migrant, uncommon in winter (Connecticut, occasionally Massachusetts).

Habitat: Breeding: Forests with dense understory, dense thickets of moist deciduous shrubs, areas of forest regenerating after timber harvest, krummholz, forest openings, margins of streams and ponds.

Special Habitat Requirements: Dense shrubby undergrowth.

Nesting: Clutch size: 4 to 5. Incubation period: 12 to 14 days. Nestling period: 9 to 11 days. Broods per year: 2. Nest height: up to 20 ft (6 m). Nest site: Typically on ground, but will nest in bushes or trees. Prefers conifers.

Food Habits: Insects, weed seeds, fruits.

Comments: Most common during spring migration, fox sparrows prefer thickets and shrubbery, not weedy fields like most other sparrows. Veit and Petersen (1993:448) state that fox sparrows have one of the most precise migration windows of any Massachusetts passerine; significant flights seem to be associated with weather conditions that ground large numbers of migrants. Fox sparrows vary greatly in abundance from year to year.

Song Sparrow

(*Melospiza melodia*)

Range: Breeds from Alaska across Canada, s. of Hudson Bay, to Newfoundland and s. across the n. United States, along the Pacific Coast, and into Mexico. Resident throughout the n. United States and the Pacific Coast. Winters along coasts from Alaska and Newfoundland, s. throughout the s. half of the breeding range.

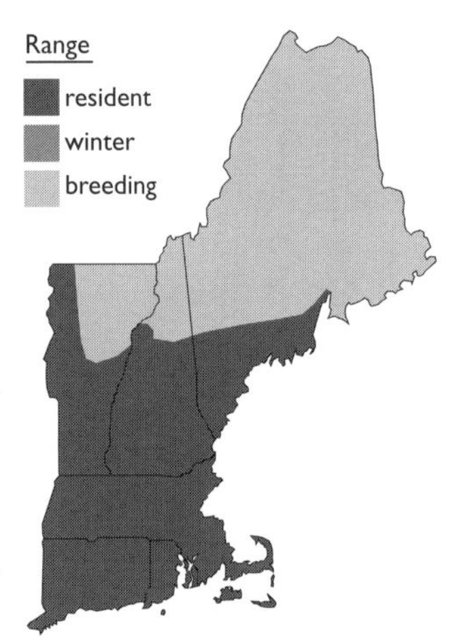

Distribution in New England: Breeding: Throughout the region. Winter: Southern Maine, central and s. New Hampshire, w. and s. Vermont, and throughout Massachusetts, Connecticut, and Rhode Island.

Status in New England: Very common breeder and migrant, common winter resident from Massachusetts s.

Habitat: Particularly common along the margins of waterways, marshy areas along the coast, and forest edges adjacent to bogs and clearings, also common in brushy or weedy fields, swamps, forest edges, roadsides, hedgerows, farms, and residential areas.

Special Habitat Requirements: Moist areas with brushy vegetation.

Nesting: Egg dates: May 1 to August 29, Vermont (Ellison 1985:348 in Laughlin and Kibbe 1985); April 30 to July 29, Massachusetts (Veit and Petersen 1993:448). April 17 to August 11, New York (Bull 1974:600). Clutch size: 3 to 6, typically 3 to 5. Incubation period: 12 to 13 days. Nestling period: 10 to 14 days. Broods per year: 2 or 3. Age at sexual maturity: 1 yr. Nest height: To 12 ft (3.7 m), typically 0 to 4 ft (0 to 1.2 m). Nest site: Early nests are usually on the ground and are typically well hidden in grasses or weeds or concealed under a bush or brush pile. Subsequent nests may be on the ground or elevated in a shrub. May raise height of successive nests with the growth of herbaceous vegetation. *Rosa multiflora* and *Rubus* spp. are preferred nest site vegetation (DeGraaf et al. 1975).

Territory Size: Ranges from 0.5 to 1.5 acres (0.2 to 0.6 ha) in favorable habitat (Nice 1937:74). From 167 to 822 m^2 (0.1 to 0.6 acres) on an island off British Columbia (Tompa 1962).

Home Range: Resident birds in winter may range over an area 6 to 10 times as large as breeding territory (Nice 1937:63).

Sample Densities: Maryland—21 territorial males per 19.2 acres (7.8 ha) in shrubby field. 3 territorial males per 9.5 acres (3.8 ha) in open hemlock-spruce bog. 4.5 territorial males per 20.5 acres (8.3 ha) in infrequently mowed apple orchard (Stewart and Robbins 1958).

Food Habits: Insects, weed seeds, fruits.

Comments: Virtually all records show that song sparrows have long been common summer residents in northern (Allen 1909) and southern (Bagg and Eliot 1937) New England. Nests in a wide variety of habitats. Prefers wet lowland situations with low, irregular plant growth and abundant sunlight.

Lincoln's Sparrow

(*Melospiza lincolnii*)

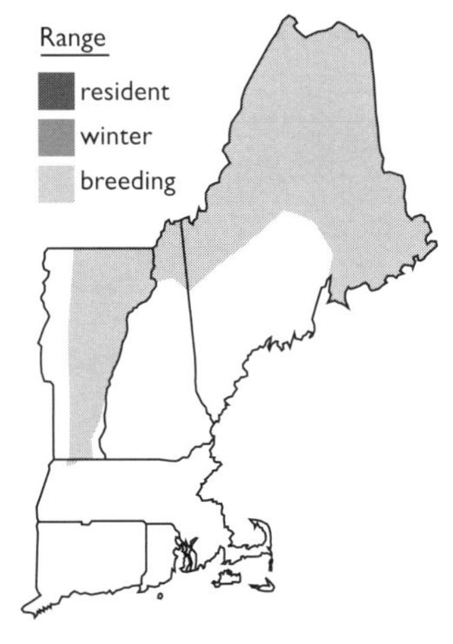

Range: Breeds from w.-central Alaska across most of Canada, s. along the Pacific Coast and the Rocky Mountains in s. California and n. New Mexico, and into the n. Great Lakes states and n. New England. Winter: From s. California, s. Arizona, Texas, and New Mexico, s. throughout Mexico to Costa Rica. Migrates throughout continental North America between its breeding and wintering ranges.

Distribution in New England: Massachusetts, one nesting record in town of Florida, Berkshire County (Veit and Petersen 1993:449); s. and ne. Vermont, n. New Hampshire, and n. Maine.

Status in New England: Fairly common breeder (n. Maine, ne. Vermont, n. New Hampshire), local in s. Vermont. Uncommon migrant.

Habitat: Prefers bogs, wet meadows, subalpine and montane willow-sedge and moss-dominated habitats. At lower elevations, inhabits forest edges and moist shrubby (especially willow) areas, stands of aspen, cottonwood, and willow; spruce-tamarack bogs, and drier sites including regrowth in clearings and openings created by logging and burning. Inhabits cutover spruce-fir flats with scattered shrubs and saplings in New Hampshire (Richards 1994:330 in Foss 1994). Avoids densely forested riparian areas.

Special Habitat Requirements: Low, brushy thickets along edges of fields, waterways, or in wet shrubby meadows.

Nesting: Egg dates: June 10 to June 28, New York (Bull 1974:599). Clutch size: 3 to 5, typically 4 (Ammon 1995). Incubation period: About 13 or 14 days. Nestling period: 10 to 11 days (Ammon 1995). Broods per year: 1 or 2. Age at sexual maturity: 1 yr. Nest site: Often on tussock of grass or sedge or in mosses and lichens. Usually well hidden by surrounding vegetation.

Territory Size: About 1 acre (0.4 ha) in forest edge habitat in Ontario (Bent 1968:1440).

Sample Densities: 20 to 28 pairs per 100 acres (40 ha) in Ontario (Speirs and Speirs 1968).

Food Habits: Insects (more than 60 percent in summer); weed seeds, grain. 31 birds taken in Massachusetts and New York in February, April, May, September, and October had consumed 42 percent animal and 58 percent vegetable material (Judd 1901 in Bent 1968:1451).

Comments: Lincoln's sparrow is a northern species that only occurred in New Hampshire in this century and first nested there in 1947, was "well established" there in 1956, and may be undergoing a minor increase in population and breeding range extension southward (Richards 1985:330 in Foss 1994).

Swamp Sparrow

(*Melospiza georgiana*)

Range: Breeds from w.-central and s. Northwest Territories and n. Manitoba across to s. Labrador, and s. to ne. and e.-central British Columbia, the Dakotas, n. Illinois, and Maryland. Winters from e. Nebraska through the Great Lakes region to Massachusetts, s. to Texas, the Gulf Coast, and Florida, and w. across New Mexico to se. Arizona, s. to Jalisco, Mexico.

Distribution in New England: Breeds throughout the region except Martha's Vineyard or Nantucket, Mass., or Block Island, R.I. Winter: Southern Massachusetts, except outer Cape Cod and the Islands, throughout Rhode Island except absent from Block Island, and throughout Connecticut, although most abundant along the coast and in the lower Connecticut River Valley.

Status in New England: Common to fairly common breeder, common migrant, uncommon in winter.

Habitat: A wide variety of wetland types: Emergent and shrub-scrub freshwater wetlands, especially those with rank marsh grasses, sedges, reeds, and cattails. Also brushy wet meadows, sloughs, bogs, and shrubby shorelines of lakes and ponds. Rarely in brackish meadows but has adapted to tidal marshes (Greenberg and Droege 1990); avoids heavily wooded wetlands (Mobray 1997).

Special Habitat Requirements: Marshes with open shallow water, low rank vegetation, and elevated songposts (Mobray 1997, Reinert and Golet 1979), but sometimes nests in rank grasses in wet northern clearcuts.

Nesting: Egg dates: May 13 to July 14, Massachusetts (Veit and Petersen 1993:451). May 15 to July 22, New York (Bull 1974:599). Clutch size: 3 to 6, typically 4 or 5. Incubation period: 12 to 15 days. Nestling period: 9 to 13 days. Broods per year: 1 or 2. Age at sexual maturity: 1 yr. Nest site: Often directly above water, on bent down grasses among cattails, or in a low bush. Frequently builds nest over water 0.5 to 2 ft (0.2 to 0.6 m) deep or more. Sutton (1960) found that birds preferred to nest in mixed vegetation (cattail, spirea, sedge, dwarf birch, and tamarack saplings) rather than in pure cattails.

Sample Densities: 21 birds per 100 acres (40 ha) in open hemlock-spruce bog in Maryland (Robbins 1949 in Bent 1968).

Food Habits: Insects (more than 80 percent in spring and early summer), weed seeds (90 percent in late summer and fall). Feeds in shallow water, on marsh vegetation.

Comments: The breeding range of the swamp sparrow extended southward in Massachusetts and Connecticut in the mid 1800s (Bagg and Eliot 1937). The swamp sparrow is the most insectivorous member of the genus *Melospiza* (Bent 1968).

White-throated Sparrow

(*Zonotrichia albicollis*)

Range: Breeds from se. Yukon and w.-central and s. Mackenzie to s. Labrador and Newfoundland, s. to central interior British Columbia, n.-central North Dakota, n. Wisconsin, and n. New Jersey. Winter: From se. Iowa and s. Wisconsin, e. to Massachusetts, s. to Mexico, the Gulf Coast, and Florida, and w. across Texas to California.

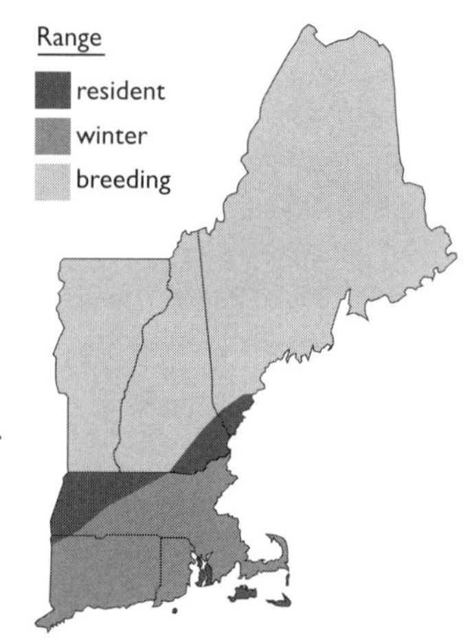

Distribution in New England: Breeds throughout the region. Winter resident primarily along the s. New England coast.

Status in New England: Common breeder throughout Maine, New Hampshire, Vermont, w. Massachusetts, and ne. and nw. Connecticut. Rare in e. Massachusetts, absent from Cape Cod and the islands, e. Connecticut, s. Connecticut River Valley, and Rhode Island. In winter uncommon in Vermont and inland Massachusetts and Connecticut, fairly common in coastal Massachusetts and Connecticut, and the s. Connecticut River Valley. Abundant spring and fall migrant in s. New England.

Habitat: Breeds in open deciduous, mixed, or coniferous forests with brushy clearings, open stunted tree growth at higher elevations, borders of bogs and beaver ponds, forest with extensive fire or insect damage, also cutover and open second-growth woodlands (Thompson and Capen 1988, DeGraaf 1991). Uses hedgerows, dense thickets in ravines or adjacent to watercourses or swamps, weedy fields, urban parks in winter.

Special Habitat Requirements: Breeding: Open forest with brushy understory. Winter: weedy fields, dense thickets for cover from predation (Schneider 1984).

Nesting: Egg dates: May 20 to July 14, Massachusetts (Veit and Petersen 1993:452). May 27 to August 1 (Ellison 1985:354 in Laughlin and Kibbe 1985). Clutch size: 3 to 7, typically 4 (Falls and Kopachena 1994). Incubation period: 11 to 14 days, typically 12. Nestling period: 7 to 12 days, typically 8 or 9 (Lowther and Falls 1968). Broods per yr: 1 or 2. Age at sexual maturity: 1 yr. Nest height: To 3 ft (0.9 m), typically on ground. Nest site: On or close to ground, in brush pile, under fallen limb, in grass hummock or mat of dead grasses or bracken fern. Typically located at edge of a clearing and well concealed by ground vegetation.

Territory Size: 110 territories ranged in size from 0.5 to 2.7 acres (0.2 to 1.1 ha) (average 0.52 acres, 0.3 ha) in Algonquin Provincial Park in Ontario (Martin 1960). 90 territories averaged 2.5 acres (0.99 ha) in Algonquin Provincial Park, Ontario (range 0.55 to 8.2 acres, 0.22 to 3.3 ha) (Falls and Kopachena 1994). Winter: Foraging ranges 0.25 to 1.0 acres (0.1 to 0.4 ha) (Piper and Wily 1990).

Sample Densities: Martin (1960) found densities varied from no birds in bog and hardwood forest to 56 territorial males per 100 acres (40 ha) in balsam fir and white spruce. Winter: Up to 55 birds per acre (22 birds per ha) (Root 1988).

Food Habits: Breeding: Mostly insects in summer, some greens and fruits. Winter: Seeds, fruits, also buds and catkins. Common visitor at feeding stations. Feeds by ground gleaning primarily, also branch and leaf gleaning.

Comments: Before 1900 the white-throated sparrow was a "rare" breeder in w. Massachusetts, and its range extended s. into Connecticut after 1915 (Bagg and Eliot 1937). White-throated sparrows also extended their range into lower elevations fairly recently; they were unknown at elevations below 1,000 ft early in this century (Forbush 1929), but since 1950 breed in low-elevation sites. The species consists of two plumage morphs that differ in habitat use. Pairs with males with white eye-stripes use open forest and forest edge. Pairs with males with tan eye-stripes use both open forest and denser forest (Knapton and Falls 1982).

Dark-eyed Junco

(Junco hyemalis)

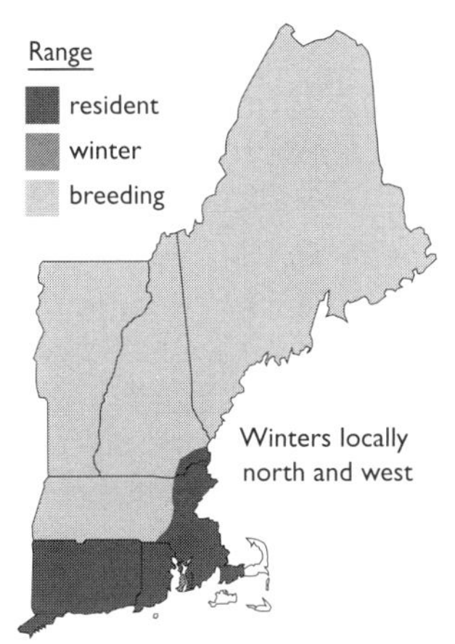

Range: Breeds from Alaska and central Yukon to Labrador and Newfoundland, s. to central coastal California, in the mountains to e. California, central Arizona, and w. Texas, s. Alberta, n. and e.-central Minnesota, central Michigan, s. New England, and in the Appalachian Mountains to n. Georgia and nw. South Carolina; also in the Black Hills. Winters from central and s.-coastal Alaska, coastal British Columbia, and across s. Canada, s. to Mexico, the Gulf Coast, and n. Florida.

Distribution in New England: Breeds throughout Maine, Vermont, New Hampshire, w. and central Massachusetts, and Connecticut, but rare and local in e. Massachusetts, absent from the se. coastal plain, and Cape Cod and the Islands (Veit and Petersen 1993:455). Resident: Southern Maine, except absent from the s. coastal region, s. New Hampshire, e. and central Massachusetts, although less abundant or rare in the se. and absent from the Islands, s. and e. Connecticut, and throughout Rhode Island.

Status in New England: Common breeder in n. New England, w. Massachusetts, uncommon in the Northwest Hills of Connecticut. Common migrant and winter resident.

Habitat: Breeding: Inhabits edges and small openings in coniferous and mixed forests, woodland clearings, wooded stream borders, open woodlands, overgrown fields, brushy cover bordering mountain meadows, logging roads, and old burns. Avoids deep forest interiors. Winter: Prefers weedy fields but also inhabits open woodlands, hedgerows, farmyards, and feeding stations in residential areas.

Special Habitat Requirements: Openings and edges in wooded habitats that have dense herbaceous vegetation.

Nesting: Egg dates: Late May to mid June, Massachusetts (Veit and Petersen 1993:455). May 5 to August 7, Vermont (Ellison 1985:356 in Laughlin and Kibbe 1985). April 28 to August 15, New York (Bull 1974:588). Clutch size: 3 to 6, typically 4 or 5. Incubation period: 12 to 13 days. Nestling period: 9 to 12 days. Broods per year: 2. Age at sexual maturity: 1 yr. Nest site: Often on ground under weeds and grasses, on slope, under fallen log or at base of tree or roadbank in cavity formed by roads. Occasionally nests low in shrub or tree.

Home Range: 27, 33, and 17 ha (66.7, 81.5, and 42 acres)—winter for two flocks (one flock used two home ranges) (Gottfried and Franks 1975).

Food Habits: Insects, wild fruits, weed seeds. Commonly feeds in weed patches, hedgerows in winter.

Comments: Juncos are one of the most abundant breeding birds in New Hampshire and nest at the highest elevations in the state. They have been a common breeder in northern New England for at least a century. Juncos feed on the ground in all seasons except when deep snow forces them to search in shrubs and forbs. Formerly called the slate-colored junco.

Lapland Longspur

(*Calcarius lapponicus*)

Range: Holarctic; in North America breeds from w. and n. Alaska and n. Yukon across n. Canada, including many islands, to Labrador, and s. to s.-coastal Alaska and s. Keewatin. Winter: From coastal s. Alaska and s. British Columbia across the n. United States to Nova Scotia, and s. to se. California, Colorado, Arkansas, and Maryland; casually farther s.

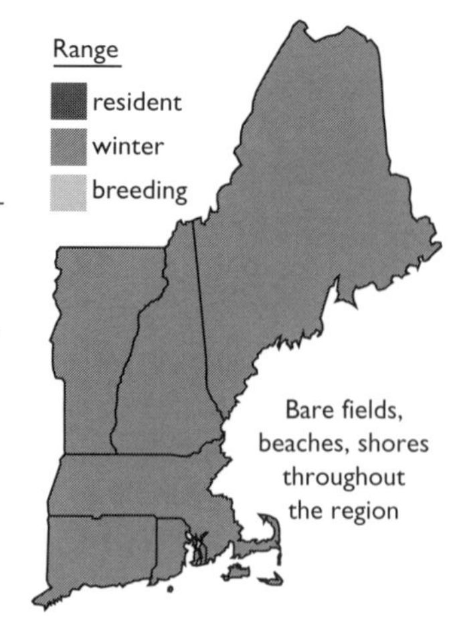

Distribution in New England: Winter resident throughout the region, although most common in coastal areas.

Status in New England: Locally common migrant and winter resident s. to Massachusetts, rare in Connecticut (Zeranski and Baptist 1990:256).

Habitat: Winter: Extensive open areas such as stubble fields, barren ground, beaches, soft marshes, and open weedy meadows.

Food Habits: Seeds of weeds and grasses. Feeds in short grasses, on bare earth: stubble fields, coastal sandy areas where vegetation is sparse.

Comments: Lapland longspurs are commonly seen with snow buntings and horned larks, primarily along coasts and in the agricultural areas of the Connecticut River Valley.

Snow Bunting

(*Plectrophenax nivalis*)

Range: Holarctic; breeds in North America on the Arctic tundra from n. Alaska to Prince Patrick and n. Ellesmere Island, s. to extreme nw. British Columbia, e.-central Mackenzie, and n. Labrador. Winter: West-central and s. Alaska and s. Canada, s. to California, Colorado, Missouri, and North Carolina, casually farther s.

Range
resident
winter
breeding
Snow-free fields, shores throughout the region

Distribution in New England: Throughout the region in winter.

Status in New England: Common to locally abundant migrant and winter resident.

Habitat: Breeds in rocky areas with or near vegetated tundra or sedge meadows for feeding (Lyon and Montgomerie 1995). In winter, open, seedy and barren fields, lake shores, beaches, sand dunes, salt marshes, and road sides.

Special Habitat Requirements: In winter, barren fields and beaches.

Nesting: Egg dates: May–July. Clutch size: 3 to 9, usually 4 to 7 (Terres 1980). Incubation period: 10 to 15 days, average 12 days (Parmelee 1968, Lyon and Montgomerie 1995). Nestling period: 9 to 15 days, average 10 to 13. Broods per year: 1, occasionally 2 (Smith and Marquiss 1995). Age at sexual maturity: 1 yr. Nest site: On the ground. Rock cavities, deep cracks and fissures, under loose rocks on the ground, will use artificial nest sites (such as barrels, metal cans and boxes, rubble, skulls), rarely exposed on open ground (Lyon and Montgomerie 1995).

Food Habits: Seeds of grasses and weeds, also some buds and invertebrates during summer (Parmelee 1968). Seeds gleaned from tips of weeds and grasses, invertebrates occasionally taken by flycatching close to the ground. Will jump against stems of weeds and grasses to scatter seeds (Cramp and Perrins 1994), tears young leaves off grasses and forbs (Bazely 1987).

Comments: Snow buntings are most abundant in late fall/early winter and gradually become less numerous during the course of the winter in s. New England (Veit and Petersen 1993:458). Most common along the coast and in the agricultural Connecticut River Valley in association with horned larks.

Northern Cardinal

(*Cardinalis cardinalis*)

Range: Resident from se. South Dakota, central Minnesota, n. Wisconsin, and s. Ontario to Nova Scotia and Massachusetts, s. to the Gulf Coast and s. Florida. Local in sw. Texas, New Mexico, s. Arizona, and se. California. Throughout the Mexican lowlands.

Range
resident
winter
breeding

Distribution in New England: Throughout Connecticut, Rhode Island, Massachusetts, and s. and central Vermont and the Champlain Lowlands (Pilcher 1985:330 in Laughlin and Kibbe 1985). Absent from northernmost New Hampshire and n. Maine.

Status in New England: Common and widespread resident; range expanding northward.

Habitat: Forest edges, open woodlands, shrub swamps, thickets, parks, suburban yards, open swamps, residential areas, less common in forests unless thickets are present. Easily attracted to feeding stations with sunflower seeds.

Special Habitat Requirements: Thick underbrush or shrubs.

Nesting: Egg dates: Late April to late June, Massachusetts (Veit and Petersen 1993:425). April 10 to September 9, New York (Bull 1974:548). Clutch size: 2 to 5, typically 3 to 4. Incubation period: 12 to 13 days. Nestling period: 9 to 11 days. Broods per year: 2 or 3, occasionally 4. Age at sexual maturity: 1 yr. Nest height: 2 to 12 ft

(0.6 to 3.6 m), usually 4 to 5 ft (1.2 to 1.5 m). Nest site: In dense shrubs, small deciduous or coniferous trees, tangles of vines, thickets, briars.

Territory Size: 0.51 to 2.32 ha (1.3 to 5.7 acres), average 1.18 ha (2.9 acres), in Tennessee, but 10.97 to 23.24 ha (27.1 to 57.4 acres), average 18.81 ha (46.5 acres), in Ontario (Dow 1969). 0.31 to 0.45 acres (0.1 to 0.2 ha), average 0.37 acre (0.1 ha), in swamp thicket in Illinois (Brewer 1955). Cardinals range no further than a few miles from their territory during their lifetime (Laskey 1944).

Sample Densities: 30 males per 100 acres (40 ha) in oak-hickory forests with clearings and hedgerows in Tennessee. 0.48 males per 100 acres (40 ha) in beech-maple woodlots in Ontario (Dow 1969). 23 territorial males per 100 acres (40 ha) in semiopen floodplain forest. 5 territorial males per 100 acres (40 ha) in field and edge (Stewart and Robbins 1958:333).

Food Habits: Seeds and fruits, waste grains, insects, spiders, centipedes, snails, slugs (Austin 1968).

Comments: Northern cardinals are recent residents of New England. Very accidental in Connecticut in the mid 1800s (Linsley 1843), they increased somewhat after 1930 (Griscom and Snyder 1955), first nested in Massachusetts in 1961, and by 1970 nested throughout the state (Veit and Petersen 1993:425). They expanded their range into Vermont and New Hampshire along the major river valleys. Both males and females sing.

Rose-breasted Grosbeak

(Pheucticus ludovicianus)

Range: Breeds from s. Northwest Territories across s. Canada to Nova Scotia, s. to n.-central North Dakota and Kansas, central Oklahoma, s. Missouri, central Indiana, and Ohio to central New Jersey and s. along the Appalachians to n. Georgia. Migrates through the West Indies, but a rare visitor to Puerto Rico and the Virgin Islands. Winters: Central Mexico to Venezuela, Colombia, Ecuador, and Peru; rarely in the sw. United States.

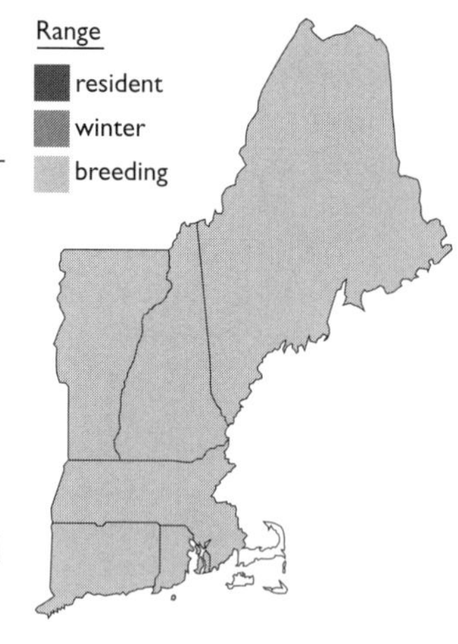

Distribution in New England: Throughout Maine, New Hampshire, Vermont, Connecticut, Massachusetts, although a rare breeder on Cape Cod and absent from the Islands (Veit and Petersen 1993:426), throughout Rhode Island, except absent along the outer coastal zone, and relatively rare in pine-dominated w. portions of the state (Enser 1992:165).

Status in New England: Fairly common breeder but declining throughout much of the breeding range since the early 1980s. Uncommon fall migrant.

Habitat: Edges of mature moist deciduous or mixed forests, openings, second-growth deciduous or mixed woods, borders of swamps and streams, wooded swamps, thick shrubs or brush, shrubs bordering pastures, suburban areas with large trees, and old orchards. Associated with pole stands of hardwoods with a canopy closure of 83 percent and canopy height of 71 ft (21.5 m) in Grafton, Vt. (Thompson and Capen 1988). Found only in undisturbed hardwood forests in Nova Scotia (Freedman et al. 1981), yet tolerates moderate levels of timber harvest activity in e. Maine (Johnson and Brown 1990). First appear in clear-cut areas 3 yr after harvest, become common after 15 yr, and will inhabit forests until the next clear-cut in the White Mountain National Forest in New Hampshire and Maine (DeGraaf 1991).

Special Habitat Requirements: Edges of mature deciduous forest with dense brush or sapling stands.

Nesting: Egg dates: May 23 to June 15, Massachusetts (Veit and Petersen 1993:426). May 6 to July 19, New York (Bull 1974:549). Clutch size: 3 to 6, typically 4. Incubation period: 12 to 14 days. Nestling period: 9 to 12 days. Broods per year: 1, occasionally 2. Age at sexual maturity: 1 yr. Nest height: 6 to 26 ft (1.8 to 7.9 m). Typically 10 to 12 ft (3.0 to 4.6 m). Nest site: Usually built in the fork of a deciduous tree. Less commonly placed in a deciduous or evergreen shrub. Equal occurrence of nests in yellow birch, sugar maple, and American beech with an average height of 49.2 ft (15.0 m) in

White Mountains, N.H. (Sherry and Holmes 1994:308 in Foss 1994).

Food Habits: Insects and spiders (about 50 percent); the balance of diet is seeds, fruits.

Comments: Rose-breasted grosbeaks were apparently very scarce in New England before the 1840s. They began to increase in the 1870s to about 1940, the time of extensive young forest in New England. It seems strange today, but rose-breasted grosbeaks, like cardinals, were commonly kept as cage birds from the South around the turn of the century.

Indigo Bunting

(*Passerina cyanea*)

Range: Breeds from se. Saskatchewan and n. Minnesota to central and s. New Brunswick, s. to s. New Mexico, central and se. Texas, the Gulf Coast, and central Florida; locally in central Colorado, sw. Utah, central Arizona, and s. California. Winter: Southern Mexico, the Greater Antilles (except Hispaniola), and the Virgin Islands, s. to Panama; also in s. Florida and infrequently in coastal Texas and elsewhere in the southern part of the breeding range.

Distribution in New England: Breeds throughout Connecticut, Rhode Island, Massachusetts, except on Nantucket and outer Cape Cod (Veit and Petersen 1993:428); throughout Vermont, New Hampshire, and Maine, except absent in nw. Maine.

Status in New England: Common breeder and migrant, but declining over much of its New England breeding range as old fields revert to forest.

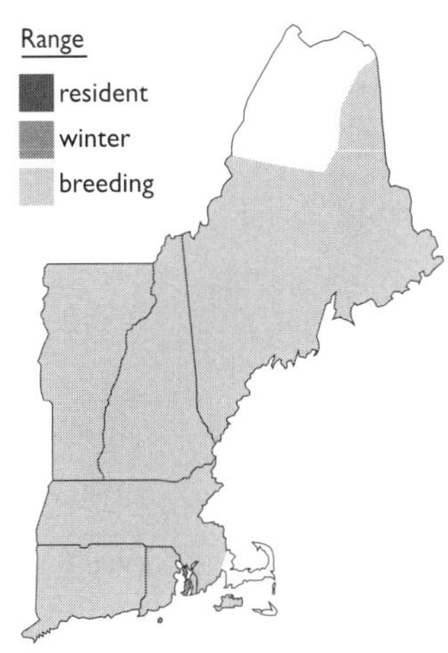

Habitat: Edges of woods, old burns, open brushy fields, roadside thickets, open deciduous woodlands, brushy ravines, early- to mid-successional stages of reforestation, ski slopes, orchards, utility rights-of-way. The presence of tall trees and high perches may be a habitat requirement (Herbert and Elkins 1994:310 in Foss 1994). Tends to be more numerous along streams. Partial or improvement cuts in mixed mesophytic forests of the Southeast are beneficial to the indigo bunting (McComb et al. 1989). Associated with open-canopy (13.9 percent) habitat with dense vegetation <10 ft (<3 m) tall in Vermont (Thompson and Capen 1988). Scarce to absent in deep woods, urban areas, or intensively cultivated lands (Payne 1992).

Special Habitat Requirements: High song perches, thick weeds or shrubs, open areas at forest edges, old fields.

Nesting: Egg dates: June 3 to June 22, Massachusetts (Veit and Petersen 1993:429). May 26 to August 3, New York (Bull 1974:551). Clutch size: 2 to 6, typically 3 or 4. Incubation period: only females brood—12 to 13 days (Payne 1992). Nestling period: 8 to 10 days (Payne 1992). Broods per year: 1 or 2, probably single-brooded in New England. Age at sexual maturity: 1 yr. Nest height: 2 to 12 ft (0.6 to 3.7 m), typically 3 ft (0.9 m). Nest site: In dense cover, usually in weeds or brambles, or in the fork of a shrub or low tree (Bradley 1948).

Territory Size: 2.7 acres (1.1 ha) in sapling, shrub and vine habitat in Kansas (Fitch 1958). Average 0.26 acre (0.1 ha) in swamp-thicket in Illinois (Brewer 1955).

Sample Densities: 5 nests per 7.08 acres (2.9 ha) in thickets (Beecher 1942). 9 to 18 pairs per mile (1.6 km) in forest edge (Johnston 1947). 13 pairs per 25 acres (10.1 ha) in apple orchard (Stewart and Robbins 1958:337). 9.8 territorial males per 100 acres (40 ha) in ash-basswood in New York (Samson 1979). 15.2 territorial males per 100 acres (40 ha) in West Virginia (Hall 1983).

Food Habits: Insects, spiders, berries (Payne 1992), also seeds of grasses and herbs. Gleans from branches, leaf surfaces, bare soil. Commonly feeds in cornfields in late summer.

Comments: Indigo buntings were abundant in New England 100 years ago, even as far n. as Franconia, N.H. They depend on forest disturbance to provide forest edge and weedy old fields; they are common breeders in ski areas, nesting up to 3,500 ft in the mountains of Vermont and New Hampshire.

Bobolink

(*Dolichonyx oryzivorus*)

Range: Breeds from s. interior British Columbia across s. Canada and central Ontario, s. to e. Oregon, central Colorado, central Illinois, and central New Jersey. Migrates through Central America, the Caribbean, and n. South America. Winter: Southern South America (mostly east of the Andes) from Peru and central Brazil s. to n. Argentina.

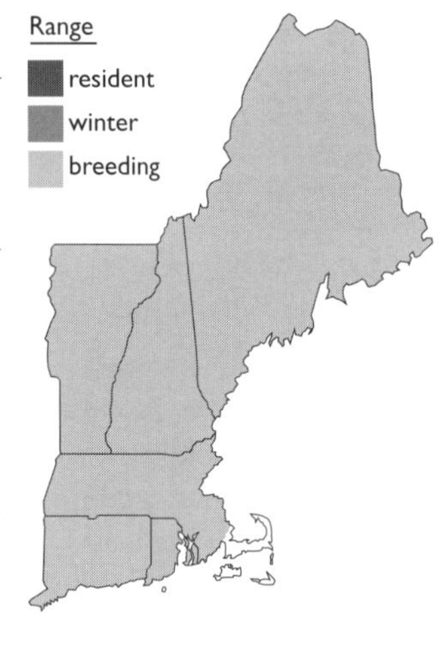

Distribution in New England: Breeds throughout Maine; New Hampshire except the "core of the White Mountains and much of the North Country where suitable habitat is lacking" (Steele 1994:338 in Foss 1994); Vermont except generally absent at elevations above 2,500 ft (765 m) and from heavily forested areas (Ellison 1985:358 in Laughlin and Kibbe 1985); Massachusetts except outer Cape Cod, Nantucket, and Martha's Vineyard, although a few pair regularly attempt to nest there (Veit and Petersen: 458); Rhode Island except absent from Block Island (Enser 1992:179); and Connecticut, although more abundant in the northern part of the state (Zeranski and Baptist 1990:258).

Status in New England: Locally common breeder; numbers have declined since the early 1990s with the loss of grasslands and agricultural fields. Common spring and fall migrant.

Habitat: Generally restricted to large open fields of tall grass, hay fields, grain crops, fields with a mix of grass and broad-leaved forbs (red clover, dandelion), but also inhabits wet meadows, fallow fields, ungrazed to lightly grazed mixed-grass prairies, floodplain fields (Martin and Gavin 1995), and historically, large beaver meadows (Belknap 1792). More abundant (67 percent) in large open hay fields $\geq$8 yr old since the last time of plowing and seeding, with low amounts of total vegetation cover, low alfalfa and legume cover, high proportions of grass cover, and high litter cover in New York (Bollinger and Gavin 1992).

Special Habitat Requirements: Large expanses of grassland or older ($\geq$8 yr) hay fields with little or no alfalfa, high litter cover and scattered broad-leafed forbs for nest-site cover (Bollinger and Gavin 1992).

Nesting: Egg dates: May 18 to June 20, New York (Bull 1974:523). Clutch size: 4 to 7, typically 5 or 6. Incubation period: 11 to 13 days, typically 11. Nestling period: 10 to 11 days (Martin and Gavin 1995). Broods per year: 1, occasionally 2 (30 percent in New York) (Gavin 1984). Age at sexual maturity: 1 yr. Nest site: Always on the ground, usually in a hollow scrape in the ground or in a natural depression, rarely above ground attached to plant stems. Nests in dense stands of vegetation such as hay, clover, or weeds such as dandelions. Often nests in wet habitats, transitional between areas with drier soils and those with poor drainage (Willenberger 1980, Martin and Gavin 1995).

Territory Size: 22 territories ranged from 2.7 to 12.1 acres (1.1 to 4.9 ha), average 6.3 acres (2.6 ha), in grasslands in Wisconsin (Wiens 1969:35). 1.2 acres (0.49 ha) in New York (Bollinger 1988 in Martin and Gavin 1995). Territories center around the nest and tend to be smaller in high-quality habitat (Martin and Gavin 1995).

Sample Densities: 9 territorial males per 80 acres (32.4 ha) in grasslands in Wisconsin (Wiens 1969:53). 100 pairs per square mile (39 pairs/km^2) in favorable habitat in North Dakota (Stewart and Kantrud 1972).

Food Habits: Insects, weed and grass seeds; in New England, the summer diet consists of 70 to 90 percent insects, replaced almost entirely by grain (90 percent) in September (Forbush 1929 V. 2:404). Commonly feeds in cultivated grain fields in fall.

Comments: Changes in haying practices (earlier and more frequent mowing and modern mowing and raking equipment) and loss of agricultural land to development have contributed to the bobolink's decline in the Northeast (Wiens 1969:41, Bollinger et al. 1990, Bollinger and Gavin 1992). Conservation efforts should include creation of large patches (at least 10–15 ha) of sparse grass-dominated vegetation resembling old hayfields, and hay-cropping the patches after mid July every 2–3 years to prevent excessive encroachment of woody vegetation (Bollinger and Gavin 1992).

Red-winged Blackbird

(*Agelaius phoeniceus*)

Range: Breeds from the s. tip of Alaska, and Yukon down to n. Washington, across n. Canada and the United States, including Idaho, Montana, Wyoming, the Great Lakes states, and New England. Resident in the rest of the United States s. through Mexico to Costa Rica. Northern birds winter in much of the s. portion of the breeding range as far s. as Mexico. Also resident in the Bahamas and Cuba.

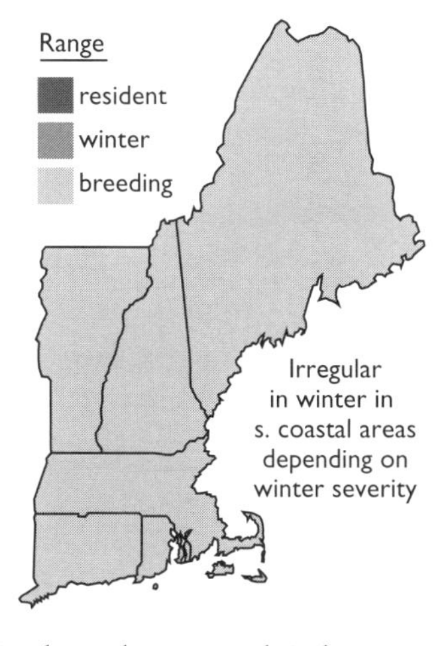

Distribution in New England: Breeding: Widespread throughout the region. In mild winters, coastal New Hampshire and s. throughout coastal Massachusetts, Connecticut, and Rhode Island.

Status in New England: Locally abundant breeder, but declining with the loss of fields and wetland habitats. Abundant migrant, uncommon and irregular winter resident in s. New England.

Habitat: Breeding: Marshes and agricultural areas near wetlands; also in shrub-swamps, marshy ditches along roadsides, old fields, pastures, and wetlands in urban parks. Prefers areas with trees nearby and where open habitat edges, especially marsh/field edges, are highly interspersed (DeGraaf and Rappole 1995:507). Winter: Coastal marshes, roosts characterized by dense cover, deciduous thickets, or coniferous stands (Yasukawa and Searcy 1995).

Special Habitat Requirements: Emergent vegetation adjacent to open areas with scattered tall shrubs and trees.

Nesting: Egg dates: May 10 to June 18, Massachusetts (Veit and Petersen 1993:461). April 26 to July 9, New York (Bull 1974:526). Clutch size: 3 to 5, typically 3 or 4. Incubation period: 11 to 13 days. Broods per year: 2 or 3. Age at sexual maturity: 1 to 3 yr: females—2 years, males—probably 3 yr (Orians and Beletsky 1989). Nest height: 3 inches to 14 ft (7.6 cm to 4.3 m). Nest site: Nest is a narrow, deep cup of grass, reeds, and weed rootlets, usually attached to emergent vegetation (particularly cattails). Also nests in upland habitat such as alfalfa and grain fields, pastures, and areas of mixed shrubs and grasses.

Territory Size: Territories are smaller in marsh habitats than those in uplands; mean marsh territory size 1,625 m^2, mean upland territory 2,895 m^2 (average computed by Searcy and Yasukawa 1995). Average sizes ranged from 200 to 600 m^2 (0.05 to 0.15 acres) in bulrush with a little cattail in Washington (Holm 1973). Average size of 21 territories in cattail clumps surrounded by grassland was 2,512 square feet (233 m^2). Average size of 22 territories in main area of cattail marsh including central and peripheral territories was 10,653 square feet (990 m^2) (Orians 1961). 51 marsh territories averaged 0.17 acre (0.07 ha) (range 0.06 to 1.12 acres, 0.02 to 0.5 ha), upland territories averaged 0.54 acre (0.2 ha) (range 0.33 to 0.99 acres, 0.07 to 0.4 ha) (Case and Hewitt 1963).

Sample Densities: 16 pairs per 100 acres (40 ha) in marsh, 11 pairs per 100 acres (40 ha) in uplands (Case and Hewitt 1963). 164 pairs per square mile (63 pairs/km^2) in favorable habitat in North Dakota (Stewart and Kantrud 1972). 73 territorial males per 100 acres (40 ha) in cattail marsh, 36 territorial males per 100 acres (40 ha) in shrubby field with stream-bordered trees (Stewart and Robbins 1958:322).

Food Habits: Primarily plant matter during the

nonbreeding season (weed seeds, grains), primarily animal matter during breeding season (insects), but varies with sex, time of year, and access to plant matter (Yasukawa and Searcy 1995).

Comments: In the fall, redwings congregate in large flocks. While they have been at least locally common throughout the historic period, they increased greatly in recent decades, but are now declining. The fall diet consists almost entirely of weed seeds.

Eastern Meadowlark

(*Sturnella magna*)

Range: Breeds from n. Minnesota and s. Ontario across to s. New Brunswick, s. through the e. United States to Texas, the Gulf Coast, and Florida, w. to sw. South Dakota, central Nebraska, and central Arizona, s. throughout Central America and also to South America, as far as Colombia in the w. and Brazil in the e. Resident on Cuba. Winters from central Arizona, Kansas, central Wisconsin, s. New England, and Nova Scotia, s. throughout the breeding range, casually farther n. Northern populations winter as far south as Mexico.

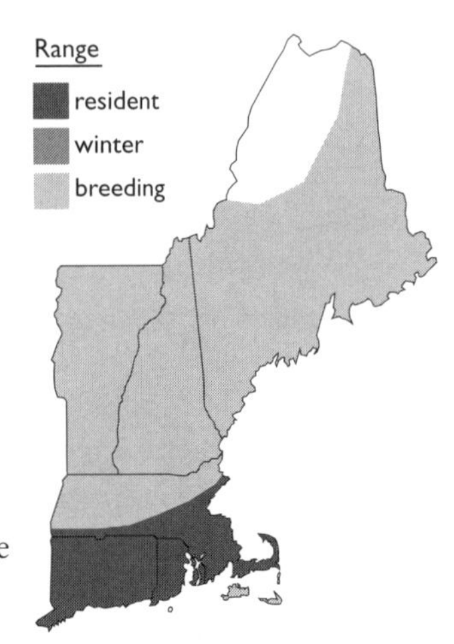

Distribution in New England: Occurs throughout the region, except absent from nw. Maine and Block Island in Rhode Island. Erratic in winter in Massachusetts, most common along the se. coastal region (Veit and Petersen 1993:462), uncommon along the coast in Connecticut.

Status in New England: Fairly common to locally common breeder and migrant; erratic in winter in s. New England; a decline noted in the central and e. United States due to loss of grassland habitat.

Habitat: Large grassy fields of intermediate height and density but also uses grassy meadows, hay fields, tallgrass prairies, agricultural fields of alfalfa, and clover, and open weedy orchards. Bobolinks use fields that are more lush, vesper sparrows less so.

Special Habitat Requirements: Extensive open grassland with elevated song perches.

Nesting: Egg dates: April 21 to June 28, Massachusetts (Veit and Petersen 1993:462). May 9 to August 4, New York (Bull 1974:524). Clutch size: 2 to 6, typically 3 to 5. Incubation period: 13 to 15 days. Nestling period: 10 to 12 days. Broods per year: 2. Age at sexual maturity: 1 yr. Nest site: On ground in a natural depression or one scraped by female, partially or entirely domed with nest materials and adjacent vegetation and opening on the side (Lanyon 1995). Prefers to nest in cover 10 to 20 in (25 to 50 cm) high.

Territory Size: 3 to 15 acres (1.2 to 6.1 ha) in moist lowlands in Wisconsin (Lanyon 1957). 18 territories ranged from 4.3 to 7.9 acres (1.7 to 3.2 ha), average 5.8 acres (2.3 ha), in grasslands in Wisconsin (Wiens 1969:35).

Home Range: 2.8 acres (1.1 ha) in a field of brome grass in Kansas (Fitch 1958).

Sample Densities: 20.9 nests per 100 acres (40 ha) in pasture, 12.6 nests per 100 acres (40 ha) in hayfield in Illinois. Ungrazed pasture had more nests than grazed pasture (Roseberry and Klimstra 1970). 12 territorial males per 80 acres (32.4 ha) in grasslands in Wisconsin (Wiens 1969:53).

Food Habits: Insects, especially beetles and grasshoppers, weed seeds, grass seeds, waste grain seed.

Comments: The meadowlark was most abundant in New England in the nineteenth century, during the peak of agriculture and dairying—like the bobolink, meadowlarks have gradually declined with the loss of grasslands. Belknap (1792) called the meadowlark the "marsh lark"; it likely inhabited extensive beaver meadows in pre-Colonial times (Steele 1985:342 in Foss 1994).

Rusty Blackbird

(*Euphagus carolinus*)

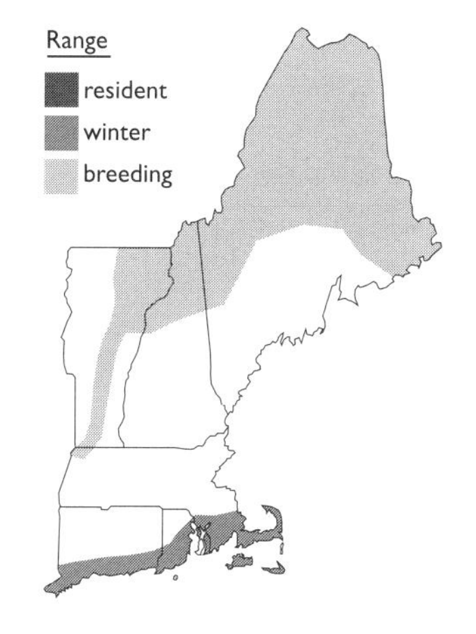

Range: Breeds from w. and n.-central Alaska and n. Yukon to s. Keewatin, n. Quebec and central Labrador, s. to sw. and s.-coastal Alaska, central interior British Columbia, central Manitoba, n. New England, and ne. New York. Winter: South-coastal Alaska and se. British Columbia to s. Ontario and s. New England, s. to Texas, the Gulf Coast, and n. Florida, and w. to central Colorado.

Distribution in New England: Breeding: Western Massachusetts (Berkshire and Franklin counties) at higher elevations (Veit and Petersen 1993:464), n. to Vermont in the Green Mountains and the Northeast and North-Central Highlands (Nichols 1985:364 in Laughlin and Kibbe 1985), n. New Hampshire, and n. and central Maine.

Status in New England: Rare and local breeder, common to abundant inland migrant, rare but regular in winter in s. New England (Veit and Petersen 1993:464).

Habitat: Wooded swamps, tree-bordered marshes, beaver ponds, muskegs, boreal bogs and stream borders with alder and willow thickets, wooded islands in lakes (Avery 1995). Frequently nests in thickets of spruce and balsam fir 2 to 10 ft (0.6 to 3.0 m) above water, occasionally in deciduous shrubs such as willows, alders, buttonbush, or sweetgale in Vermont (Nichols 1985:364 in Laughlin and Kibbe 1985). Generally found at elevations between 1,000 to 4,000 ft (300 to 1,200 m) in New Hampshire (Richards 1994:344 in Foss 1994). In winter, also in open woodland, scrub, pastures, fields.

Special Habitat Requirements: Boreal wooded wetlands.

Nesting: Egg dates: May 19 to May 29, Massachusetts (Veit and Petersen 1993:465). May 7 to June 15, New York (Bull 1974:533). Clutch size: 3 to 6, typically 4 or 5. Incubation period: 14 days. Nestling period: About 13 days. Broods per year: 1. Age at sexual maturity: 1 yr. Nest height: 2 to 20 ft (0.6 to 6.1 m), typically less than 10 ft (3.0 m). Nest site: Solitary nester. Nest is often in dense foliage against the trunk of young conifers, especially balsam and spruce, often at the edge of or over water (Kennard 1920). Also builds in deciduous shrubs in marshes such as sweet gale and buttonbush.

Territory Size: Breeding territories are sometimes large. Nests may be 0.5 mile (0.8 km) or more apart (Harrison 1975:217).

Food Habits: Insects, seeds of weeds, grains, wild fruit, occasionally preys on other birds (Messerly 1979, Campbell 1974, Woodruff and Woodruff 1991). The stomach contents of 132 birds taken in all months of the year except June and July contained 53 percent animal and 47 percent vegetable matter (Beal in Bent 1958:288).

Comments: Migratory rusty blackbirds are most common in wooded swamps, where they gather in large flocks. They are recent nesters in New England, first nesting in New Hampshire in 1953 (Richards 1985:344 in Foss 1994) and in the late 1970s in Massachusetts (Veit and Petersen 1993:465). The name comes from the winter plumage.

Common Grackle

(*Quiscalus quiscula*)

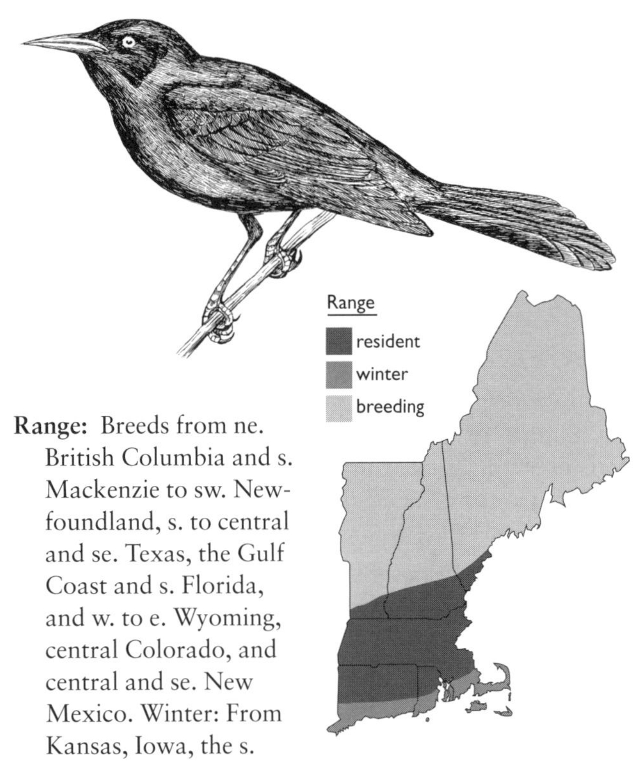

Range: Breeds from ne. British Columbia and s. Mackenzie to sw. Newfoundland, s. to central and se. Texas, the Gulf Coast and s. Florida, and w. to e. Wyoming, central Colorado, and central and se. New Mexico. Winter: From Kansas, Iowa, the s. Great Lakes region, New England, and Nova Scotia, s. to se. New Mexico, s. Texas, the Gulf Coast, and Florida.

Distribution in New England: Breeds throughout the region. Resident in southernmost Maine, Vermont, and New Hampshire, Massachusetts, n. Connecticut, and n. Rhode Island. In winter, occasionally in s. Vermont and New Hampshire (Ellison 1985:366 in Laughlin and Kibbe 1985, Smith 1994:346 in Foss 1994), local and uncommon in Massachusetts (Veit and Petersen 1993:467), uncommon to common along the coast in Connecticut (Zeranski and Baptist 1990:262), and Rhode Island.

Status in New England: Common breeder and migrant; local and uncommon in winter in s. New England.

Habitat: Prefers to nest in open habitats, especially agricultural areas, suburbs, city parks, forest edges, open margins of lakes, ponds, and rivers, scrub swamps, and forested wetlands. In winter, agricultural areas, forests (especially coniferous), and city parks, often near water.

Special Habitat Requirements: Open areas with open water for foraging adjacent to graves or woodlots for nesting and roosting.

Nesting: Egg dates: May 2 to June 1, Massachusetts (Veit and Petersen 1993:467). April 12 to June 4, New York (Bull 1974:536). Clutch size: 3 to 7, typically 4 to 5. Incubation period: Average 13.5 days (Peer and Bollinger 1997). Nestling period: 12 to 15 days (Maxwell and Putnam 1972). Nest height: 1 to 60 ft (0.3 to 18.3 m). Typically 10 to 20 ft (3.0 to 6.1 m). Solitary or colonial nesters. Usually nests in small colonies of 20 to 30 pairs. Nest site: Prefers conifers but uses deciduous trees and shrubs.

Territory Size: Both male and female defend only a small area surrounding nest (Ficken 1963).

Sample Densities: 92 pairs per square mile (35 pairs/km^2) in favorable habitat in North Dakota (Stewart and Kantrud 1972). 34.6 per 100 acres (40 ha) in red maple-birch in New York (Samson 1979).

Food Habits: Ground-dwelling insects, fruits, mast, waste grains, small quantities of fish, crustaceans, amphibians, nesting birds, and eggs. Feeds in open fields, shores of ponds, lawns.

Comments: Grackles are gregarious in all seasons, and are highly adaptable, nesting almost anywhere. Ornamental evergreens are commonly used for nesting, especially stands or groves of pines or spruces. Maxwell et al. (1976) found 24 percent of 2,601 nests located in red cedar. Records of grackles using cavities and birdboxes, even a mailbox, indicate an ability to utilize marginal habitats. Grackles have been common in New England throughout historic times, so ubiquitous in early Colonial times that bounties were paid for their destruction (Zeranski and Baptist 1990:262).

Brown-headed Cowbird

(*Molothrus ater*)

Range: Breeds from se. Alaska, n. British Columbia, and s. Northwest Territories, e. to s. Quebec and s. Newfoundland, and s. to n. Baja California, Guerrero, Puebla, and Veracruz,Mexico, the Gulf Coast, and central Florida. Winter: From n. California, central Arizona, the Great Lakes region, and New England, s. to the Isthmus of Tehuantepec in Mexico, the Gulf Coast, and s. Florida.

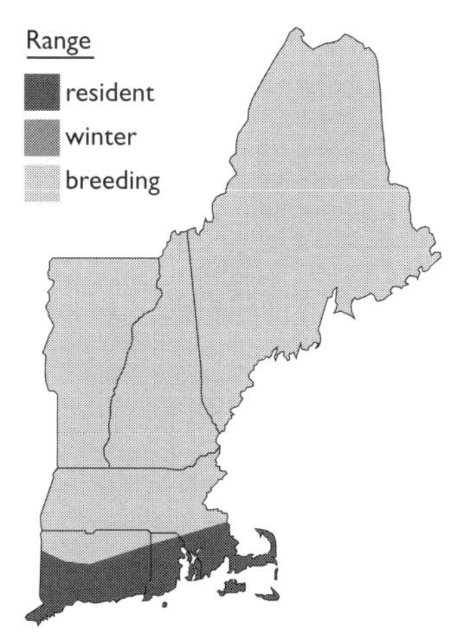

Distribution in New England: Breeding: Throughout the region. Winter: Southeastern Massachusetts including the Islands, s. Connecticut, and s. Rhode Island.

Status in New England: Common and widespread breeder and abundant migrant, but declining in e. North America. Locally fairly common in winter.

Habitat: Prefers breeding habitat with low or scattered trees that are interspersed with short or mowed grassland (Lowther 1993). Also inhabits open coniferous and deciduous woodlands, forest edges, brushy thickets, fields, pastures, agricultural land, orchards, and suburban areas. May prefer field-forest ecotones based on high number of parasitized nest in this habitat (Brittingham and Temple 1983, Johnson and Temple 1990). Primarily winters in agricultural land and feeding stations.

Special Habitat Requirements: Open fields, actively grazed pasture lands, mowed grassy areas.

Nesting: Egg dates: May 15 to July 1, Massachusetts (Veit and Petersen 1993:469). April 23 to July 31, New York (Bull 1974:539). Clutch size: 1 to 6 (Payne 1965), typically 3 (usually lays only one egg per nest). Female cowbirds wander widely, overlap their breeding ranges, and can lay up to 40 eggs per season (Lowther 1993). Nestling period: 8 to 13 days (Woodward and Woodward 1979). Broods per year: 3 or 4. Age at sexual maturity: 1 yr. Nest site: Builds no nest; lays its eggs in the nests of other birds (over 220 species, of which 144 have raised young cowbirds successfully) (Friedmann 1929, 1963, Friedmann et al. 1977). Song sparrows, yellow warblers, red-eyed vireos, and chipping sparrows are the most common hosts (Harrison 1975, Lowther 1993), but many warblers and the wood thrushes are frequent hosts. In Vermont, 22 species have been hosts to cowbird parasitism (Ellison 1985:369 in Laughlin and Kibbe 1985).

Territory Size: Apparently does not defend an area but has a fixed breeding area in which female lays eggs (Friedmann 1929). Females in the Northeast defend an area averaging 50 acres (20 ha) against other females (Dufty 1982).

Home Range: About 20 to 30 acres (8.1 to 12.1 ha) in floodplain habitat (open weedy fields with scattered trees) in Ohio (Nice 1937:154).

Sample Densities: 152 pairs per square mile (59 pairs/km^2) in favorable habitat in North Dakota (Stewart and Kantrud 1972). 10.6 territorial males per 100 acres (40 ha) in ash-hickory-oak forest in Missouri (Samson 1979).

Food Habits: Seeds of weeds, grasses, grains, insects. Commonly feeds in grain fields, pastures where they often feed among cattle.

Comments: It is generally believed that cowbirds moved east from the Great Plains as e. North America was cleared and farm animals introduced. Belknap (1792) did not list the cowbird as occurring in New Hampshire, but by the early 1800s it nested in Burlington, Vt. Females may lay up to 40 eggs during the breeding season (Lowther 1993). Birds are often seen feeding in mixed flocks with red-winged blackbirds or common grackles. Both sexes may flock in all seasons.

Orchard Oriole

(*Icterus spurius*)

Range: Breeds from se. Saskatchewan, s. Manitoba, and central Minnesota, e. to s. Vermont, se. New Hampshire, s. to Michoacán, Mexico, Texas, the Gulf Coast, and central Florida, and w. to e. Colorado. Migrates through Cuba. Winters from s. Mexico (including the Yucatan Peninsula) to Columbia and Venezuela, casually to s. Texas, rarely in coastal California.

Distribution in New England: Breeds in Massachusetts e. of the Taconic Mountains, throughout Rhode Island and Connecticut, except absent in the Northwest Hills; a few records for s. Maine, s.-central Vermont and se. New Hampshire.

Status in New England: Local and uncommon breeder in central and e. Massachusetts and Connecticut; rare in Rhode Island, although appears to be increasing since 1975. Uncommon spring, rare fall migrant (Veit and Petersen 1993:469).

Habitat: Common in orchards, farmland, gardens, shade trees in suburbs and along rural roads, riparian zones, floodplains (Bull 1964, Scharf and Kren 1996), primarily occupies lower elevations, avoids heavy forest (Godfrey 1986).

Special Habitat Requirements: Open woodlands or open areas with scattered trees.

Nesting: Egg dates: May 28 to June 25, Massachusetts (Veit and Petersen 1993:470). May 18 to June 22, New York (Bull 1974:530). Clutch size: 3 to 7, typically 4 or 5. Incubation period: 12 to 14 days. Nestling period: 11 to 14 days. Broods per year: 1. Age at sexual maturity: 1 yr. Nest height: 4 to 70 ft (1.2 to 21.3 m). Typically 10 to 20 ft (3.0 to 6.1 m). Nest site: Nest is suspended in a forked twig in a tree or shrub and is well concealed by dense foliage.

Sample Densities: Maryland—29 territorial males per 100 acres (40 ha) in farmyards. 15 territorial males per 100 acres (40 ha) in suburban residential area. 10 territorial males per 100 acres (40 ha) in shrubby field with stream-bordered trees (Stewart and Robbins 1958:323). 37 territorial males per 100 ha in overgrown pasture and 25 territorial males per 100 ha of brushy field in West Virginia (Hall 1983).

Food Habits: Insects gleaned from foliage represent more than 90 percent of diet, wild fruits form the remainder. Stomachs of 11 birds taken in May and June in Maryland contained 91 percent animal and 9 percent vegetable material (Judd 1902 in Bent 1958:200).

Comments: The orchard oriole is one of the earliest departing breeding birds, departing Massachusetts by the end of July. Orchard orioles commonly nest in small colonies of several nests, often quite close together, and have been reported to exhibit cooperative breeding (Scharf and Kren 1996). Formerly quite abundant in s. New England, orchard orioles were more abundant than Baltimore orioles on Long Island in the 1840s (Bull 1964).

Baltimore Oriole

(*Icterus galbula*)

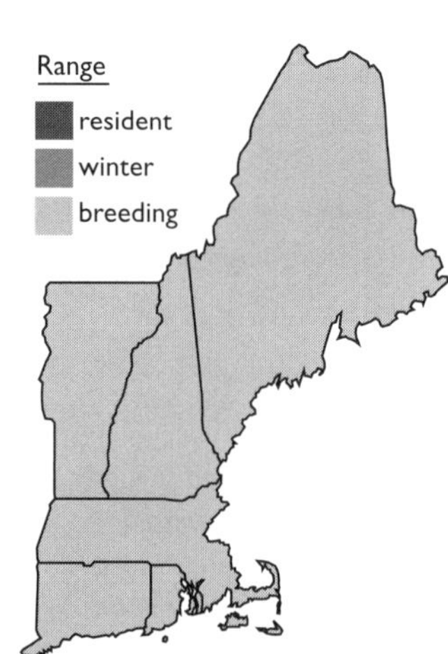

Range: Breeds from s. interior British Columbia and central Alberta to central Maine and central Nova Scotia, s. to s. Texas, Mexico, the central Gulf States, central North Carolina, and Delaware. Winter: From Mexico, to Colombia and Venezuela. Also in Cuba, Jamaica, and Puerto Rico.

Distribution in New England: Throughout the region, except absent from Block Island, R.I. Considered scarce on Nantucket, Mass. (Veit and Petersen 1993:471).

Status in New England: Fairly common and widespread breeder, common spring migrant, rare in s. New England in winter.

Habitat: Inhabits open and semiopen areas: roadside shade trees, orchards, cemeteries, parks, deciduous forest edges, wooded river bottoms, and shelter belts. Avoids dense forests.

Special Habitat Requirements: Tall deciduous trees for nesting.

Nesting: Egg dates: May 24 to July 4, Massachusetts (Veit and Petersen 1993:471). May 29 to July 3 (11 records),

Vermont (Ellison 1985:372 in Laughlin and Kibbe 1985). May 15 to June 13, New York (Bull 1974:530). Clutch size: 4 to 6, typically 4 or 5. Incubation period: 12 to 14 days. Nestling period: 11 to 14 days. Broods per year: 1. Age at sexual maturity: 1 yr. Nest height: 6 to 60 ft (1.8 to 18.3 m). Typically 25 to 30 ft (7.6 to 9.1 m). Nest site: Usually attaches a pendant nest by the rim to the tip of a long drooping branch. Most frequently uses large deciduous trees, especially elms, cottonwoods, and sugar maples growing in the open, but will use a wide variety of deciduous trees.

Sample Densities: 20 pairs per square mile (8 pairs/km^2) in favorable habitat in North Dakota (Stewart and Kantrud 1972). 10 territorial males per 100 acres (40 ha) in shrubby field with stream-bordered trees in Maryland (Stewart and Robbins 1958:324).

Food Habits: Insects, fruit. The diet is mainly animal material (83 percent) and is supplemented by lesser amounts of vegetable material (17 percent), mostly fruits (Forbush 1913:226).

Comments: Baltimore orioles, named for Lord Baltimore's colors, arrive in New England with the blossoming of apple and cherry trees. The birds preferentially nested in elms (Graf and Greeley 1976), and it was thought that the great losses of elms in the 1960s would lead to a decline in orioles, but Dutch elm disease had no effect on oriole numbers. The Baltimore oriole has benefited from settlement, being a bird of semiopen habitats.

Pine Grosbeak

(*Pinicola enucleator*)

Range: Holarctic; resident in North America from w. and central Alaska and n. Yukon, e. to n. Quebec and n. Labrador, s. to central California and n. New Mexico, and, e. of the Rockies, to n. Alberta, s. Ontario, and central Maine. Winters from central Alaska and s. Mackenzie, e. to Labrador and s. throughout the breeding range; in invasion years, occurs farther.

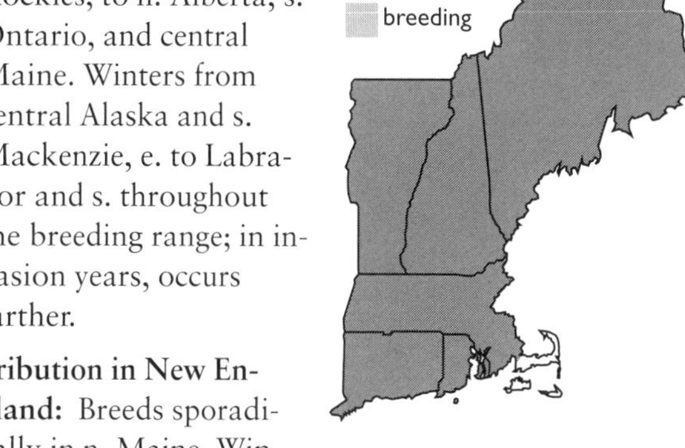

Distribution in New England: Breeds sporadically in n. Maine. Winters throughout the region except absent from Cape Cod, Martha's Vineyard and Nantucket, Mass., and Block Island, R.I.

Status in New England: Irregular and uncommon breeder (n. Maine). An erratic winter visitor throughout the region.

Habitat: Inhabits northern spruce-fir forests, typically at high elevations, usually at the edge of a forest opening. Also occurs in spruce stands bordering bogs or barren areas with clumps of dwarf spruce and tamarack. Prefers stands with large trees and low to intermediate canopy cover. May remain in the breeding range or move s. to open coniferous forests, residential areas with feeders, orchards, or roadsides in winter.

Special Habitat Requirements: Coniferous forests of spruce-fir or pine.

Nesting: Clutch size: 2 to 5, typically 4. Incubation period: 13 to 14 days. Broods per year: 1. Age at sexual maturity: 1 yr. Nest site: Low in coniferous tree (often spruce) or shrub.

Food Habits: Buds, seeds, some insects in spring and summer. Winter diet: 99.1 percent vegetable (commonly buds), 0.9 percent animal (365 stomachs). Summer diet: 84 percent vegetable, 16 percent animal (29 stomachs) (Gabrielson 1924 in Bent 1968:330).

Comments: Major flights of pine grosbeaks occur at irregular intervals; large flights have occurred in this century in the winters of 1903–1904, 1929–1930, 1951–1952, 1961–1962, and 1977–1978 (Zeranski and Baptist 1990:265). Flocks congregate to feed on roadside seeds, especially of white ash, in winter.

Purple Finch

(*Carpodacus purpureus*)

Range: Breeds from n. British Columbia, s. Yukon, and n. and central Alberta, e. to central Ontario and Newfoundland, s. to Baja California, and e. of the Great Plains, to central Minnesota, n. Ohio, West Virginia, and se. New York. Winter: From sw. British Columbia, s. to Baja California, and from s. Manitoba, e. to Newfoundland, s. to Texas, the Gulf Coast, and Florida.

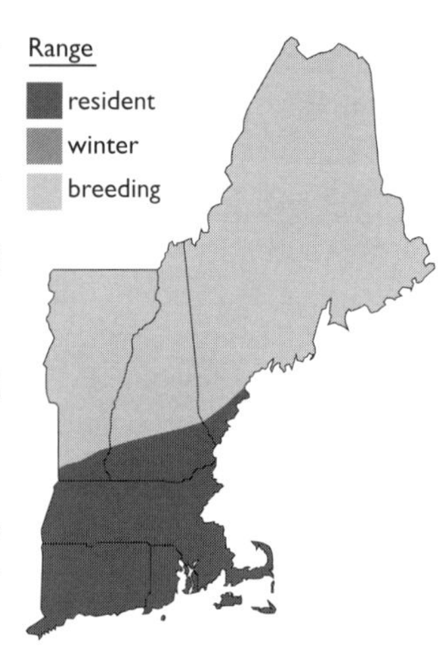

Distribution in New England: Throughout the region. Resident throughout much of the region, except absent from n. Maine in some years.

Status in New England: Common (n. New England, Massachusetts) to uncommon (Connecticut) resident in s. New England. Widespread uncommon to common breeder in n. New England—may be resident or semi-migratory in the north, depending on food supply.

Habitat: Primarily inhabits moist or cool coniferous forest edges, open mixed woodlands, and ornamental coniferous plantations. Also occurs in clearings associated with bogs, mixed coniferous-deciduous forests, riparian edges, pastures, and lawns scattered with conifers and shrubs. Most common at elevations below 2,000 ft (610 m), but observed at elevations up to 5,000 ft (1,520 m) in New Hampshire (McDermott 1994:354 in Foss 1994). In winter, commonly moves to lower elevations, frequently congregates around houses with feeding stations. Roosts in dense evergreens or thickets.

Special Habitat Requirements: Coniferous trees.

Nesting: Egg dates: May 10 to June 19, Massachusetts (Veit and Petersen 1993:474). May 13 to July 16, New York (Bull 1974:556). Clutch size: 2 to 8, typically 3 to 5 (Wootton 1996). Incubation period: 13 days. Nestling period: 13 to 16 days. Broods per year: 1 or 2. Nest height: 5 to 60 ft (1.5 to 18.3 m). Nest site: In the East, typically on a horizontal branch of a conifer, commonly spruce, but occasionally nests in a deciduous tree or shrub.

Territory Size: Males probably defend territories, but more research needed.

Sample Densities: Averages 8 pairs per 100 acres (40 ha): 2.3 pairs per 100 acres (40 ha) in mixed hardwood forests in Connecticut; 3.8 pairs per 100 acres (40 ha) in hemlock and white pine forest in Connecticut; 5.6 pairs per 100 acres (40 ha) in spruce-fir woods in Maine; 7.0 pairs per 100 acres (40 ha) in mixed forest in Maine; 12.0 pairs per 100 acres (40 ha) in fir and birch forest in Maine; 19.5 pairs per 100 acres (40 ha) in a spruce plantation in Maine (Wootton 1996).

Food Habits: Buds and seeds, blossoms, fruit, and occasionally insects.

Comments: The purple finch is the state bird of New Hampshire. Suburban ornamental conifers and Christmas tree plantations have influenced the southward range expansion of this species (Harrison 1975:230).

House Finch

(*Carpodacus mexicanus*)

Range: Breeds from central British Columbia, n.-central Wyoming, and w. Nebraska, s. to central Mexico. Introduced to New York in 1940. Now breeds also from

Illinois, s. Ontario, s. Quebec, Vermont, and s. Maine, s. to Missouri, Georgia, and South Carolina (Veit and Petersen 1993:475). Winter: Generally throughout the breeding range.

Distribution in New England: Throughout New England, n. to s. Maine. Semimigratory depending upon food supply.

Status in New England: Common to locally common resident.

Habitat: Almost exclusively in settled areas such as rural, suburban, and urban yards, parks, and farms—areas with buildings and lawns (Hill 1993).

Special Habitat Requirements: Developed areas with open ground.

Nesting: Egg dates: Late March to early July (Veit and Petersen 1993:476). April 11 to July 20, New York (Bull 1974:560). Clutch size: 2 to 6, typically 4 or 5. Incubation period: 12 to 16 days, typically 13 to 14 days (Hill 1993). Nestling period: Average 15 days, range 11 to 19 days (Evenden 1957). Broods per year: 2 or more. Age at sexual maturity: 1 yr. Nest height: 3 to 20 ft (0.9 to 6.1 m). Nest site: Uses a variety of sites including buildings, ledges, tree cavities, bird houses, vines (especially ivy) on buildings, in porch light fixtures, even forgotten Christmas wreaths. In the e. United States, birds seem to be associated with conifers, especially cultivated varieties such as arborvitae and hedges (Elliott and Arbib 1953). Gregarious year-round, house finches are semicolonial nesters.

Food Habits: Seeds, buds, wild and cultivated fruits, insects.

Comments: The western house finch was introduced as the "Hollywood finch" to the Northeast on Long Island, N.Y., in 1940 when dealers released the illegally sold birds. Its range has expanded to include much of the Northeast, first nesting in Connecticut at Fairfield in 1957 (Zeranski and Baptist 1990:266), in Massachusetts (Ashley Falls) in 1958 (Veit and Petersen 1993:476), in Vermont (Bennington) in 1976 (Nichols 1985:376 in Laughlin and Kibbe 1985), and in New Hampshire (Hampton and Manchester) in 1975 (Janeway 1994:356 in Foss 1994). Its spread throughout eastern North America has been one of the most notable ornithological events of the twentieth century (Hill 1993).

Red Crossbill

(Loxia curvirostra)

Range: Holarctic; resident in North America from s. Alaska to Newfoundland, through s. Canada, and s. to the n. Great Lakes area, e. to s. New Hampshire. Also from the Pacific states e. to the Rocky Mountains, and s. into parts of Mexico. Wanders irregularly during the nonbreeding season.

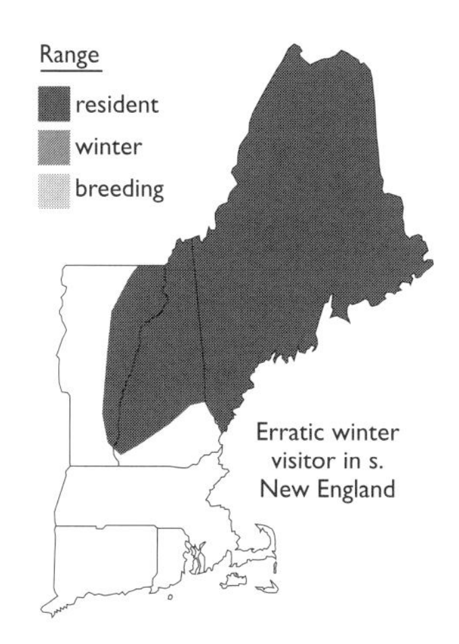

Distribution in New England: An erratic breeder in s. Vermont (Fichtel 1984:406 in Laughlin and Kibbe 1985), s. New Hampshire (Staats 1994:358 in Foss 1994), and e. Maine (Adamus 1988); rare in Massachusetts (Veit and Petersen 1993:477) after major flights. Wanders considerable distances in the nonbreeding season during major irruptions (Griscom 1937). Erratic winter visitor throughout the region.

Status in New England: Locally common to rare breeder in coniferous forests. Very erratic winter visitor to s. New England.

Habitat: Inhabits coniferous and mixed coniferous-deciduous forests from sea level to high mountains. In winter, only in cone-bearing coniferous forests and groves.

Special Habitat Requirements: Conifers with abundant seed production.

Nesting: Egg dates: Late March to April 16, Massachusetts (Veit and Petersen 1993:477). Mid-January to July, Maine (Forbush 1929). March 30 to April 30, New York (Bull 1974:566). Clutch size: 2 to 5, typically 3 or 4. Incubation period: 12 to 16 days. Nestling period: 15 to 25 days. Age at sexual maturity: 1 yr. Nest height: 5

to 80 ft (1.5 to 24.4 m). Typically 10 to 40 ft (3.0 to 12.2 m). Nest site: On horizontal branch of conifer, usually hidden in a tuft of needles, well out from trunk.

Territory Size: Little data available; appears to defend a small area around the nest (Lawrence 1949, Adkisson 1996).

Food Habits: Seeds of conifers, particularly spruce, pine, Douglas fir, and hemlock. Occasionally eats seeds and buds of hardwoods (birch, alder, boxelder), berries, and insects. Feeds on cones at tips of branches of trees.

Comments: Breeding periods are irregular, with nesting primarily in late winter and spring but reported in all months of the year; may be governed by availability of conifer seeds (Benkman 1990). Populations sporadically travel great distances from their northern core range and nest in suitable habitat thousands of miles from their usual breeding range, and then disappear (Dickerman 1986). Conifer seed abundance and availability determine where and when red crossbills nest. Most commonly seen along the coast in winter in s. New England.

White-winged Crossbill

(*Loxia leucoptera*)

Range: Holarctic; resident in North America from ne. Alaska, s. to the Pacific Northwest, throughout most of s. Canada, the e. coast of Canada, and n. New England.

Distribution in New England: Northern and central Maine, n. New Hampshire, and ne. Vermont.

Status in New England: Uncommon to rare breeder, erratic winter visitor in s. New England.

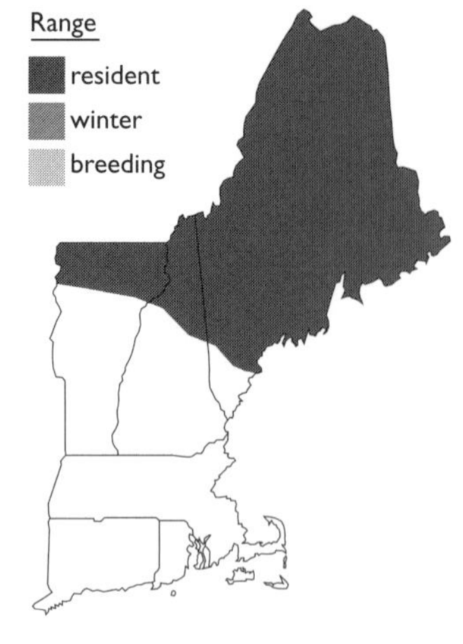

Habitat: Breeding: Coniferous forests, especially spruce or tamarack. In winter, cone-bearing conifer groves.

Special Habitat Requirements: Cone-bearing mature coniferous forests (Benkman 1993b).

Nesting: Clutch size: 2 to 5, typically 3 to 4. Incubation period: Probably 12 to 14 days. Nestling period: Unknown. Age at sexual maturity: <1 yr, known to nest at 5 months (Benkman 1989). Nest height: 3 to 40 ft (1 to 12 m) above ground (Taber 1968). Nest site: Nests in conifers.

Territory Size: Male defends an area of 1 to 2 acres (0.4 to 0.8 ha) around the nest tree in Utah (Smith 1978).

Food Habits: Primarily seeds of conifers; occasionally seeds of hardwood trees, especially birch and alder, or seeds of weeds, fruit, or small amounts of insects (Benkman 1987). Extracts seeds from conifer cones. Forage in flocks; individuals can eat up to 3,000 conifer seeds per day (Benkman 1992).

Comments: May breed during any month of the year, particularly from January to May (Benkman 1992). Breeding habits are little known. White-winged—are more common than red crossbills. Both species, like other winter finches, were apparently more abundant prior to 1955 than at any time since. More common inland than along the coast in s. New England (Veit and Petersen 1993:478). Crossbills are among the most specialized birds, adapted for extracting the seeds from conifer cones (Benkman 1993a). Crossbills are nomadic, moving their nesting and foraging sites in response to conifer seed availability, which varies over time and from region to region.

Common Redpoll

(*Carduelis flammea*)

Range: Holarctic; breeds in North America in the n. Arctic region from n. Alaska to Baffin Island, and s. to n. British Columbia, e. to Newfoundland. Resident in s. part of the breeding range. Winters in Alaska and n. Saskatchewan, e. to Newfoundland and s. to about mid United States.

Distribution in New England: In winter, throughout Maine, New Hampshire, Vermont, Massachusetts except absent from Martha's Vineyard and Nantucket, Rhode Island except absent from Block Island, and Connecticut, most often inland (Zeranski and Baptist 1990:268).

Status in New England: Erratic winter visitor and resident. Rare or absent in a given year. Occasionally abundant in Massachusetts (Veit and Petersen 1993:479).

Habitat: Open country: weedy fields with scattered small conifers or shrubs, weedy hollows among coastal dunes, thickets of alders and birches, which are an important winter staple, also at feeding stations in open country.

Food Habits: Seeds of weeds, grasses, conifers, birches, and alders. Adept at opening seed heads.

Comments: Like other winter finches, common redpolls have not appeared in large flocks in recent decades; flights of 2,000 to 2,500 were reported in both the Connecticut River Valley and in e. Massachusetts in the 1940s (Griscom and Snyder 1955); the largest recent winter flight in 1973–1974 contained more than 1,000 birds (Veit and Petersen 1993:479).

Hoary Redpoll

(*Carduelis hornemanni*)

Range: Holarctic; breeds in North America in w. and n. Alaska, n. Yukon, n. and e.-central Mackenzie, s. Victoria Island, Keewatin, ne. Manitoba, Southhampton Island, and n. Quebec. Also on Ellesmere, Bylot, and n. Baffin Islands, and in n. Greenland. Winters in the breeding range (except extreme n. areas) and s., irregularly, to s. Canada, and Montana to n. Illinois, New York, and New England.

Distribution in New England: In winter, erratic throughout the region, except not reported from Block Island in Rhode Island, or Martha's Vineyard or Nantucket in Massachusetts.

Status in New England: Rare and erratic winter visitor.

Habitat: Old fields, pastures, and birch or alder swamps or swales.

Food Habits: Seeds of birches, alders, and common grasses. Commonly feeds by opening seed heads.

Comments: Hoary redpolls probably occur regularly in New England during major flights of common redpolls (Veit and Petersen 1993:480).

Pine Siskin

(*Carduelis pinus*)

Range: Breeds from central Alaska s. through the w. United States and e. across Canada to Newfoundland. Winters throughout e. and s. United States from New England, the Great Lakes, and the n. Plains States, s. into Mexico. Resident in the s. Rocky Mountains, sw. deserts, and the Pacific Coast.

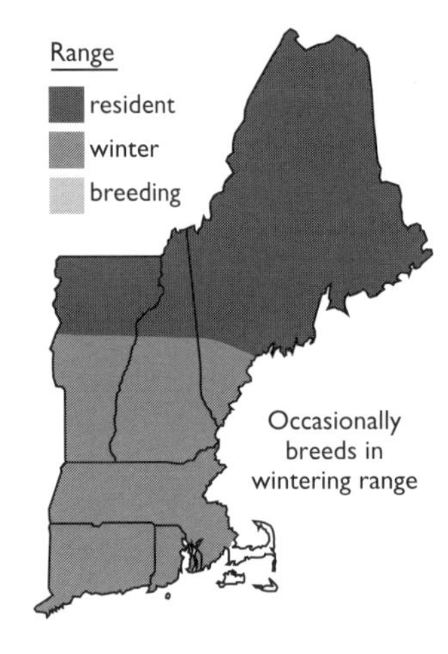

Distribution in New England: Throughout the region except absent from outer Cape Cod and Martha's Vineyard in Massachusetts and Block Island in Rhode Island; considered a rare and irregular breeder in the Northwest Hills region in Connecticut (Zeranski and Baptist 1990:268). Resident in n. Maine, New Hampshire, and Vermont. Winter: Massachusetts, reported from Provincetown, Nantucket, and Athol (Veit and Petersen 1993:482), and usually fairly common in Connecticut (Zeranski and Baptist 1990:268).

Status in New England: Uncommon and unpredictable breeder in n. New England; occasional breeder in Massachusetts, rare and irregular in the Northwest Hills of Connecticut. Abundant to rare (very irregular) in winter (Dawson 1997).

Habitat: Primarily inhabits coniferous forests, natural conifer stands or evergreen plantations, preferring those with low to intermediate canopy cover. Less numerous in second-growth alders, aspen, and broadleaf trees along coniferous forest edges.

Special Habitat Requirements: Coniferous forests.

Nesting: Egg dates: Probably late March to April but uncertain, Massachusetts (Veit and Petersen 1993:481). Mid March to April, New Hampshire (MacLeod 1985:367 in Foss 1994). April 25 to May 25, New York (Bull 1974:564). Clutch size: 2 to 6, typically 3 or 4. Incubation period: 13 days. Nestling period: About 15 days. Broods per year: 1, possibly 2. Age at sexual maturity: 1 yr. Nest height: 6 to 35 ft (1.8 to 10.7 m). Typically 20 ft (6.1 m). Nest site: Usually nests in loose colonies. Nest is a flat, compact cup, usually on a horizontal branch of a conifer and well out from the trunk. Used nests are rimmed with excreta, as are those of the American goldfinch. Nests exclusively in conifers.

Territory Size: Small area 3 to 6 ft (0.9 to 1.8 m) in diameter surrounding nest (Weaver and West 1943).

Food Habits: Summer—insects, spiders, buds, seeds, tender buds and leaves. Winter—seeds of annual weeds, conifers, birches, and alders.

Comments: Siskins usually breed at elevations of 3,000 ft (914 m) or more in New York, Vermont, and New Hampshire; lower in Maine. Birds feed in flocks in all seasons of the year (Rogers 1937). Numbers seem to fluctuate with cone crops, and they nest regularly in Massachusetts and Connecticut after pronounced southward winter flights (Veit and Petersen 1993:481).

American Goldfinch

(*Carduelis tristis*)

Range: Breeding: From s. British Columbia and n.-central Alberta, e. to central Ontario and sw. Newfoundland, and s. to California, s. Colorado, ne. Texas, central

Alabama, and South Carolina. Winter: From s. British Columbia, the n. United States, s. Ontario, and Nova Scotia, s. to Veracruz, Mexico, the Gulf Coast, and s. Florida.

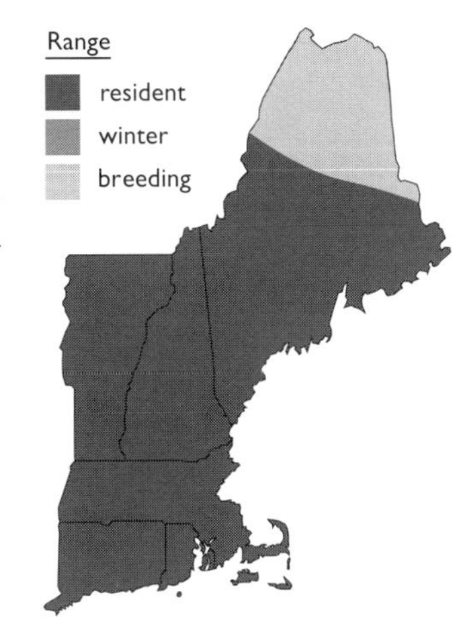

Distribution in New England: Throughout the region, northernmost birds migrate to s. New England. Resident in s. New England.

Status in New England: Common breeder, migrant, and winter resident.

Habitat: In summer, frequents open weedy fields, pastures with scattered trees, farmyards, marshes, forest edges, regenerating clear-cuts, wooded residential areas, fencerows and thickets (Middleton 1993). In the West, inhabits riparian areas, especially those with willow thickets along streams, ditches, and ponds, and wooded swamps. Winter: Old fields, woodlands, swamps and marshes, and residential areas with feeding stations.

Special Habitat Requirements: Open weedy fields and marshes with thistle and other composites, or cattails, and scattered woody growth for nesting.

Nesting: Egg dates: July 10 to early September (commonly mid August), Massachusetts (Veit and Petersen 1993:482). July 3 to September 16, New York (Bull 1974:563). Clutch size: 4 to 6, typically 5 (Walkinshaw 1938a). Incubation period: 12 to 14 days. Nestling period: 11 to 15 days. Broods per year: 1 or 2 (Walkinshaw 1939a). Age at sexual maturity: 1 yr. Nest height: 1 to 33 ft (0.3 to 10.1 m), typically 5 to 16.5 ft (1.5 to 5 m), sometimes 60 ft (18 m) in pine trees (Nickell 1951). The compact cup is rimmed with excreta after use. Nest site: Usually in a cluster of upright branches or in a fork of a horizontal limb; commonly near water.

Territory Size: Goldfinches do not always show strong territorial behavior (Nickell 1951), but defend the immediate vicinity of the nest (Drum 1939). Average territory size of 38 pairs was an area 95 ft (20 m) in diameter in a dry marsh in Wisconsin (Stokes 1950), defined by the song flights of the male.

Sample Densities: 38 pairs per 6.4 acres (2.6 ha) of dry marsh in Wisconsin (Stokes 1950). 40 pairs per square mile (15 pairs per km^2) in favorable habitat in North Dakota (Stewart and Kantrud 1972). 21 territorial males per 100 acres (40 ha) in shrubby field with stream-bordered trees in Maryland (Stewart and Robbins 1958:345).

Food Habits: Almost exclusively granivorous: primarily seeds of weeds and birches, alders, conifers (in winter), also succulent vegetation, buds, and some berries. Feeds at tips of weed stalks, fruit-bearing branches of trees and shrubs, breaking open seed heads.

Comments: American goldfinches are among the latest nesting birds in New England; their nesting coincides with seed production of thistles. The Canada thistle and dandelion are important for food and nesting material (Nickell 1951). Seasonally dimorphic, American goldfinches display the greatest difference between breeding and wintering plumage of all *Carduelis* finches (Middleton 1993).

Evening Grosbeak

(*Coccothraustes vespertinus*)

Range: Breeds from sw. and n.-central British Columbia, e. across n. Alberta, s. Manitoba, and s. Quebec to Nova Scotia and s. in the mountains, to central California, w.-central and e. Nevada, central and se. Arizona, s. New Mexico, the Mexico highlands to Michoacán, Puebla, and w.-central Veracruz, and, e. of the Rocky Mountains, to Minnesota, Michigan, and Massachusetts. Winters

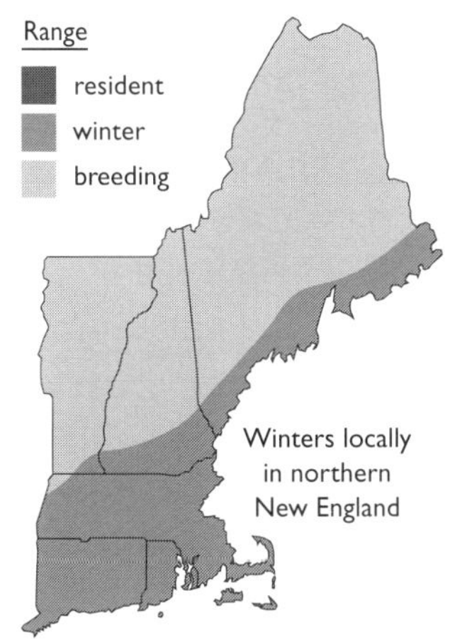

throughout the breeding range and s., sporadically, to s. California, s. Arizona, Oaxaca, the Gulf Coast, and central Florida. Mexican populations are resident.

Distribution in New England: Breeds throughout Vermont, Maine except along the s. coastal region, and New Hampshire except absent from the coastal region; sw. to Berkshire County in w. Massachusetts. Winters in coastal Maine and New Hampshire, Massachusetts including Cape Cod and the Islands, s. to Rhode Island and Connecticut.

Status In New England : Uncommon to very common migrant and winter resident, uncommon but increasing breeder in Vermont, n. New Hampshire, and Maine, two records for Massachusetts.

Habitat: Favors conifers (especially spruces and firs) throughout most of its range. Uses mixed forests for nesting in area of extended range with predominantly mixed forests or where trees are sparse. In winter forms large flocks and may move to lower elevations or to residential areas with feeding stations.

Special Habitat Requirements: Conifers for nesting.

Nesting: Egg dates: May 19 to June 4, New York (Bull 1974:553). Clutch size: 3 to 5, typically 3 or 4. Incubation period: 12 to 14 days. Nestling period: 13 to 14 days. Broods per year: Possibly 2 (Bull 1974:553). Nest height: 5 to 60 ft (6.1 to 18.3 m). Nest site: Usually in a conifer, occasionally in a deciduous tree. Nest built in dense foliage (Bekoff et al. 1987).

Food Habits: Buds, fruits, seeds, insects.

Comments: Evening grosbeaks historically bred only in nw. North America, and gradually spread southward and eastward from 1854 to 1887. The planting and self-seeding of boxelder (*Acer negundo*) across the prairies has been credited with this range extension (Bent 1968). Boxelder seeds are persistent and are preferred by evening grosbeaks to any other except possibly sunflower. The large flight of 1890–1891 accounted for most of New England's first records for this species (Zeranski and Baptist 1990:269), which first nested in Connecticut in 1962 (Bull 1974), in Massachusetts in 1980 (Veit and Petersen 1993:484), Vermont in 1926 (Marble 1926), and in New Hampshire in 1953 (Wallace 1953).

House Sparrow

(*Passer domesticus*)

Range: Essentially cosmopolitan due to introductions; resident throughout the inhabited portions of North America.

Range
resident
winter
breeding

Distribution in New England: Resident throughout urbanized and farming areas of the region.

Status in New England: Common to abundant resident.

Habitat: Villages, farms, cities, parks. Avoids heavily forested areas.

Nesting: Egg dates: February to September, Massachusetts (Veit and Petersen 1993:486). March 23 to July 16, New York (Bull 1974:542). Clutch size: 3 to 7, typically 5. Incubation period: About 14 days (Lowter and Cink 1992). Nestling period: 13 to 18 days (Weaver 1942). Broods per year: 2 or 3. Age at sexual maturity: 1 yr. Nest height: 10 to 50 ft (3.0 to 15.2 m). Nest site: Cavities, crevices in buildings, trees, billboards, bird houses, cupolas, rafters, dense ivy on buildings (Summers-Smith 1958). Pairs may be faithful to nest sites for life (Summers-Smith 1988:143).

Territory Size: Defense is limited to the nest site.

Sample Densities: C. A. North (1972) had 3.4 breeding birds per acre (0.4 ha) on his 160-acre (64.8-ha) study area. 80 pairs per square mile (30 pairs/km^2) in favorable habitat in North Dakota (Stewart and Kantrud 1972). A social species, the house sparrow tends to breed in small colonies of 10 to 20 pairs in clumped, rather than uniformly spaced, distributions (Summers-Smith 1988:139).

Food Habits: Insects, vegetables, fruits and seeds (summer), weed seeds and waste grains (winter), garbage.

Feeds on sparsely vegetated or bare earth, pavement, gleaning food from the ground.

Comments: House sparrows were first accidentally introduced to Boston in 1858 and then intentionally in 1868 and 1869, reportedly to help control gypsy moths and other insect pests (Veit and Petersen 1993:486). Peak populations occurred in 1890 to 1915, to the great detriment of nesting bluebirds, tree swallows, and house wrens, but declined with the replacement of the horse by the automobile. Birds are gregarious when feeding and roosting. The house sparrow competes successfully for nesting cavities and often usurps them from more desirable species of birds. A pair that has bred usually keeps the same nest site for life. Exceptions occur where sites are plentiful.

Literature Cited

Adamus, P. R. 1988. *Atlas of breeding birds in Maine, 1978–1983*. Augusta, Me.: Maine Department of Inland Fisheries and Wildlife.

Adkisson, C. A. 1996. Red crossbill (*Loxia curvirostra*). In: Poole, A.; Gill, F. (editors). *The birds of North America*. No. 256. Philadelphia: Academy of Natural Sciences; Washington, D.C.: American Ornithologists' Union.

Ailes, I. W. 1980. Breeding biology and habitat use of the upland sandpiper in central Wisconsin. *Passenger Pigeon* 42:53–63.

Aldrich, J. W. 1947. The Hungarian and Chukar partridges in America. Wildl. Leafl. 292. Washington, D.C.: U.S. Fish and Wildlife Services.

Allaire, R. J., Jr.; Fisher, C. D. 1975. Feeding ecology of three resident sympatric sparrows in eastern Texas. *Auk* 92:260–269.

Allan, T. A. 1985. Seasonal changes in habitat use by Maine spruce grouse. *Can. J. Zool.* 63:2738–2742.

Allen, A. A. 1929. Nuthatch. *Bird-Lore* 31:423–432.

Allen, A. A. 1939.*The golden plover and other birds*. American Bird Biographies. Second series. Ithaca, N.Y.: Comstock. 324 pp.

Allen, D. (editor). 1956. *Pheasants in North America*. Harrisburg, Pa.: Stackpole Books; Washington, D.C.: Wildlife Management Institute. 490 pp.

Allen, F. H. 1909. *Fauna of New England: Aves*. Vol. 7 of Occasional Papers of the Boston Society of Natural History. Boston: Boston Society of Natural History.

Allen, R. W.; Nice, M. M. 1952. A study of the breeding biology of the purple martin (*Progne subis*). *Am. Midl. Nat.* 47(3): 606–665.

Ammon, E. M. 1995. Lincoln's sparrow (*Melospiza lincolnii*). In: Poole, A.; Gill, F. (editors). *The birds of North America*. No. 191. Philadelphia: Academy of Natural Sciences; Washington, D.C.: American Ornithologists' Union.

Andrews, R. 1994. Acadian flycatcher. Pages 150–151. In: Foss, C. R. (editor). *Atlas of breeding birds in New Hampshire*. Dover, N.H.: Arcadia. 414 pp.

Andrle, R. F. 1971. Range extension of the golden-crowned kinglet in New York. *Wilson Bull.* 83: 313–316.

Andrle, R. F.; Carroll, J. R. 1988. *The atlas of breeding birds in New York State*. Ithaca, N.Y.: Cornell University Press.

Apfelbaum, S. I.; Seelbach, P. 1983. Nest tree, habitat selection and productivity of seven North American raptor species based on Cornell University nest record card program. *Raptor Res.* 17:97–113.

Arbib, R. 1982. The Blue List for 1982. *Am. Birds* 36:129–135.

Arendt, W. J., with collaborators. 1992. Status of North American migrant landbirds in the Caribbean region: A summary. Pages 143–147. In: Hagan, J. M., III; Johnston, D. W. (editors). *Ecology and conservation of neotropical migrant landbirds*. Washington, D.C.: Smithsonian Institution Press. 609 pp.

Armstrong, E. A. 1956. Territory in the wren (*Troglodytes troglodytes*). *Ibis* 98:430–437.

Armstrong, E.; Euler, D. 1983. Habitat use of two woodland buteo species in central Ontario. *Can. Field-Nat.* 97:200–207.

Armstrong, J. T. 1965. Breeding home range in the nighthawk and other birds; Its evolutionary and ecological significance. *Ecology* 46:619–629.

Armstrong, W. H. 1958. Nesting and food habits of the long-eared owl in Michigan. *Mich. State Univ. Biol. Ser.* 1:63–96.

Atwater, S.; Schnell, J. 1989. *Ruffed grouse*. Harrisburg, Pa.: Stackpole Books. 370 pp.

Atwood, J. L.; Rimmer, C. C.; McFarland, K. P.; Tsai, S. H.; Nagy, L. R. 1996. Distribution of Bicknell's thrush in New England and New York. *Wilson Bull.* 108:650–661.

Austin, J. E.; Miller, M. R. 1995. Northern pintail (*Anas acuta*). In: Poole, A.; Gill, F. (editors). *The birds of North America*. No. 163. Philadelphia: Academy of Natural Sciences; Washington, D.C.: American Ornithologists' Union.

Austin, O. L., Jr. (editor). 1968. *Life histories of North American cardinals, grosbeaks, buntings, towhees, finches, sparrows, and their allies*. U.S. Natl. Mus. Bull. 237. Washington, D.C.: Smithsonian Institution.

Averill, C. K., Jr. 1892. *Birds found in the vicinity of Bridgeport, Connecticut*. Bridgeport, Conn.: Bridgeport Scientific Society.

Avery, M. L. 1995. Rusty blackbird (*Euphagus carolinus*). In: Poole, A.; Gill, F. (editors). *The birds of North America*. No. 200. Philadelphia: Academy of Natural Sciences; Washington, D.C.: American Ornithologists' Union.

Bagg, A. C.; Parker, H. M. 1951. The turkey vulture in New England and eastern Canada up to 1950. *Auk* 68:315–333.

Bagg, A. M. 1969. A summary of the fall migration season, 1968, with special attention to the movements of black-capped chickadees. *Audubon Field Notes* 23:4–12.

Bagg, A. C.; Eliot, S. A., Jr. 1937. *Birds of the Connecticut Valley in Massachusetts*. Northampton: Hampshire Bookshop. 813 pp.

Bailey, W. 1955. *Birds in Massachusetts—When and where to find them*. South Lancaster: College Press. 234 pp.

Baker, B. W. 1944. Nesting of the American redstart. *Wilson Bull.* 56:83–90.

Baker, D. E. 1979. Tennessee warbler nesting in Chippewa County. *Jack-Pine Warbler* 57:25.

Baldassarre, G. A.; Bolen, E. G. 1994. *Waterfowl ecology and management*. New York: John Wiley and Sons.

Ballam, J. M. 1984. The use of soaring by the red-tailed hawk (*Buteo jamaicensis*). *Auk* 101:519–524.

Bartgis, R. 1992. Loggerhead shrike. Pages 281–297. In: Schneider, K. J.; Pence, D. M. (editors). *Migratory nongame birds of management concern in the Northeast*. Hadley, Mass.: U.S. Department of the Interior Fish and Wildlife Service.

Bateman, H., Jr. 1977. King rail (*Rallus elegans*). Pages 93–104. In: Sanderson, G. C. (editor). *Management of migratory shore and upland game birds in North America.* Washington, D.C.: International Association of Fish and Wildlife Agencies.
Baumgartner, F. M. 1939. Territory and population in the great horned owl. *Auk* 56:274–282.
Bazely, D. 1987. Snow buntings feeding on leaves of saltmarsh grass during spring migration. *Condor* 89:190–192.
Beard, E. B. 1964. Duck brood behavior at the Seney National Wildlife Refuge. *J. Wildl. Manage.* 28:492–521.
Beason, R. C. 1974. Breeding behavior of the horned lark. *Auk* 91:65–74.
Beason, R. C. 1995. Horned lark (*Eremophila alpestris*). In: Poole, A.; Gill, F. (editors). *The birds of North America.* No. 195. Philadelphia: Academy of Natural Sciences; Washington, D.C.: American Ornithologists' Union.
Becker, D. M.; Sieg, C. H. 1987. Home range and habitat utilization of breeding male merlins, *Falco columbarius*, in southern Montana. *Can. Field-Nat.* 101:398–403.
Bedard, J.; Meunier, M. 1983. Parental care in the savannah sparrow. *Can. J. Zool.* 61:2836–2843.
Beecham, J. J.; Kochert, M.N. 1975. Breeding biology of the golden eagle in southwestern Idaho. *Wilson Bull.* 87: 506–513.
Beecher, W. J. 1942. *Nesting birds and the vegetation substrate.* Chicago: Chicago Ornithological Society. 69 pp.
Behle, W. H. 1942. Distribution and variation of the horned larks of western North America. *Univ. Calif. Publ. Zool.* 46:205–316.
Bekoff, M.; Scott, A. C.; Conner, D. A. 1987. Nonrandom nest-site selection in evening grosbeaks. *Condor* 89:819–829.
Belanger, L.; Couture, R. 1988. Use of man-made ponds by dabbling duck broods. *J. Wildl. Manage.* 52(4):718–723.
Belant, J. L.; Seamans, T. W.; Gabrey, S. W.; Ickes, S. K. 1993. Importance of landfills to nesting herring gulls. *Condor* 95:817–830.
Belknap, J. 1792. *The history of New Hampshire.* Boston: Bradford and Read.
Belles-Isles, J. C.; Picman, J. 1986. House wren nest-destroying behavior. *Condor* 88:190–193.
Bellrose, F. C. 1976. *Ducks, geese and swans of North America.* Harrisburg, Pa.: Stackpole Books. 544 pp.
Bellrose, F. C. 1980. *Ducks, geese and swans of North America.* 2nd ed. Harrisburg, Pa.: Stackpole Books. 544 pp.
Bellrose, F. C.; Holm, D. J. 1994. *Ecology and management of the wood duck.* Harrisburg, Pa.: Stackpole Books.
Bellrose, F. C.; Johnson, K. L.; Meyers, T. U. 1964. Relative value of natural cavities and nesting houses for wood ducks. *J. Wildl. Manage.* 28:661–676.
Bengtson, S. 1970. Location of nest-sites of ducks in Lake Myvatn area, northeast Iceland. *Oikos* 21:218–229.
Benkman, C. W. 1987. Food profitability and the foraging ecology of crossbills. *Ecol. Monogr.* 57:251–267.
Benkman, C. W. 1989. Breeding opportunities, foraging rates, and parental care in white-winged crossbills. *Auk* 106:483–485.
Benkman, C. W. 1990. Foraging rates and the timing of crossbill reproduction. *Auk* 107:376–386.
Benkman, C. W. 1992. White-winged crossbill (*Loxia leucoptera*). In: Poole, A. F.; Stettenheim, P.; Gill, F. B. (editors). *The birds of North America.* No. 27. Philadelphia: Academy of Natural Sciences; Washington, D.C.: American Ornithologists' Union.
Benkman, C. W. 1993a. Adaptation to single resources and the evolution of crossbills (*Loxia*) diversity. *Ecol. Monogr.* 63:305–325.
Benkman, C. W. 1993b. Logging, conifers, and the conservation of crossbills. *Conserv. Biol.* 7:473–479.
Bent, A. C. 1921. *Life histories of North American gulls and terns.* U.S. Natl. Mus. Bull. 113. Washington, D.C.: Smithsonian Institution. 345 pp.
Bent, A. C. 1923. *Life histories of North American wild fowl.* Part I. U.S. Natl. Mus. Bull. 126. Washington, D.C.: Smithsonian Institution. 244 pp.
Bent, A. C. 1926. *Life histories of North American marsh birds.* U.S. Natl. Mus. Bull. 135. Washington, D.C.: Smithsonian Institution. 490 pp.
Bent, A. C. 1929. *Life histories of North American shore birds.* Part II. U.S. Natl. Mus. Bull. 146. Washington, D.C.: Smithsonian Institution. 412 pp.
Bent, A. C. 1932. *Life histories of North American gallinaceous birds.* U.S. Natl. Mus. Bull. 162. Washington, D.C.: Smithsonian Institution. 490 pp.
Bent, A. C. 1937. *Life histories of North American birds of prey.* Part I. U.S. Natl. Mus. Bull. 167. Washington, D.C.: Smithsonian Institution. 409 pp.
Bent, A. C. 1938. *Life histories of North American birds of prey.* Part II. U.S. Natl. Mus. Bull. 170. Washington, D.C.: Smithsonian Institution. 495 pp.
Bent, A. C. 1939. *Life histories of North American woodpeckers.* U.S. Natl. Mus. Bull. 174. Washington, D.C.: Smithsonian Institution. 334 pp.
Bent, A. C. 1940. *Life histories of North American cuckoos, goatsuckers, hummingbirds, and their allies.* U.S. Natl. Mus. Bull. 176. Washington, D.C.: Smithsonian Institution. 506 pp.
Bent, A. C. 1942. *Life histories of North American flycatchers, larks, swallows, and their allies.* U.S. Natl. Mus. Bull. 179. Washington, D.C.: Smithsonian Institution. 555 pp.
Bent, A. C. 1946. *Life histories of North American jays, crows, titmice.* U.S. Natl. Mus. Bull. 191. Washington, D.C.: Smithsonian Institution. 495 pp.
Bent, A. C. 1948. *Life histories of North American nuthatches, wrens, thrashers, and their allies.* U.S. Natl. Mus. Bull. 195. Washington, D.C.: Smithsonian Institution. 475 pp.
Bent, A. C. 1949. *Life histories of North American thrushes, kinglets and their allies.* U.S. Natl. Mus. Bull. 196. Washington, D.C.: Smithsonian Institution. 454 pp.
Bent, A. C. 1950. *Life histories of North American wagtails, shrikes, vireos, and their allies.* U.S. Natl. Mus. Bull. 197. Washington, D.C.: Smithsonian Institution. 411 pp.
Bent, A. C. 1953. *Life histories of North American wood warblers.* Parts I and II. U.S. Natl. Mus. Bull. 203. Washington, D.C.: Smithsonian Institution. 734 pp.
Bent, A. C. 1958. *Life histories of North American blackbirds, orioles, tanagers, and allies.* U.S. Natl. Mus. Bull. 211. Washington, D.C.: Smithsonian Institution. 549 pp.
Bent, A. C. 1968. *Life histories of North American cardinals, grosbeaks, buntings, towhees, finches, sparrows, and allies.* Parts I, II, and III. U.S. Natl. Mus. Bull. 237. Washington, D.C.: Smithsonian Institution. 1889 pp.
Berger, A. J. 1951. Nesting density of Virginia and sora rails in Michigan. *Condor* 53:202.

Bergman, R. D.; Swain, P.; Weller, M. W. 1970. A comparative study of nesting Forster's and black terns. *Wilson Bull.* 82:435–444.

Bertin, R .J. 1977. Breeding habitats of the wood thrush and veery. *Condor* 79(3):303–311.

Best, L. B. 1977. Territory quality and mating success in the field sparrow (*Spizella pusilla*). *Condor* 79:192–203.

Best, L.B. 1978. Field sparrow reproductive success and nesting ecology. *Auk* 95:9–22.

Best, L. B.; Stauffer, D. F. 1986. Factors confounding evaluation of bird-habitat relationships. Pages 209–216. In: Verner, J.; Morrison, M. L.; Ralph, C. J. (editors). *Wildlife 2000: Modeling habitat relationships of terrestrial vertebrates.* Madison: University of Wisconsin Press.

Bird, T. J.; Bird, D. M. 1990. Peregrine falcons, *Falco peregrinus*, nesting in urban environments: a review. *Can. Field-Nat.* 104(2):209–218.

Blackwell, B. F.; Sinclair, J. A. 1995. Evidence of secondary consumption of fish by double-crested cormorants. *Mar. Ecol. Prog. Ser.* 123:1–4.

Blancher, P. J.; Robertson, R. J. 1985. A comparison of eastern kingbird breeding biology in lakeshore and upland habitats. *Can. J. Zool.* 63:2305–2312.

Blocher, A. 1936. Cowbirds. *Oologist* 53(10):131–133.

Blodgett, B. G. 1988. The half-century battle for gull control. *Mass. Wildl.* 38(2):4–10.

Boag, D. A; Schroeder, M. A. 1992. Spruce grouse (*Falcipennis canadensis*). In: Poole, A. F.; Stetteneheim, P.; Gill, F. (editors). *The birds of North America.* No. 5. Philadelphia: Academy of Natural Sciences; Washington, D.C.: American Ornithologists' Union.

Boag, D. A.; Sumanik, K. M. 1969. Characteristics of drumming sites selected by ruffed grouse in Alberta. *J. Wildl. Manage.* 33:621–628.

Bollinger, E. K. 1988. Breeding dispersion and reproductive success of bobolinks in an agricultural landscape. Ph.D. dissertation; Cornell University, Ithaca, N.Y.

Bollinger, E. K.; Bollinger, P. B.; Gavin, T. A. 1990. Effects of hay-cropping on eastern populations of the bobolink. *Wildl. Soc. Bull.* 18:142–150.

Bollinger, E. K., Gavin, T. A. 1992. Eastern bobolink populations: Ecology and conservation in an agricultural landscape. Pages 497–506. In: Hagan, J. M.; Johnston, D. M. (editors). *Ecology and conservation of Neotropical migrant landbirds.* Washington, D.C.: Smithsonian Institution Press.

Bond, R. R. 1957. Ecological distribution of breeding birds in the upland forests of southern Wisconsin. *Ecol. Monogr.* 27:351–384.

Boone, R. B.; Krohn, W. B. 1996. Maine State-Wide GAP Analysis. Draft report. Maine Cooperative Fish and Wildlife Research Unit. Orono: University of Maine.

Borror, A. C. 1994a. Double-crested cormorant. Pages 6–7. In: Foss, C. R. (editor). *Atlas of breeding birds in New Hampshire.* Dover, N.H.: Arcadia. 414 pp.

Borror, A. C. 1994b. Glossy ibis (*Plegadis falcinellus*). Pages 372–373. In: Foss, C. R. (editor). *Atlas of breeding birds in New Hampshire.* Dover, N.H.: Audubon Society of New Hampshire. 414 pp.

Bortolotti, G. R. 1994. Effect of nest-box size on nest-site preference and reproduction in American kestrels. *J. Raptor Res.* 28(3):127–133.

Bosakowski, T. 1984. Roost selection and behavior of the long-eared owl (*Asio otus*) wintering in New Jersey. *Raptor Res.* 18:137–142.

Bosakowski, T.; Speiser, R. E. 1994. Macrohabitat selection by nesting northern goshawks: Implications for managing eastern forests. *Stud. Avian Biol.* 16:46–49.

Bowdish, B. S.; Philipp, P. B. 1916. The Tennessee warbler in New Brunswick. *Auk* 33: 1–8.

Boyd, E. M. 1962. A half-century's changes in the bird-life around Springfield, Massachusetts. *Bird-Banding* 33:137–148.

Bradley, H. L. 1948. A life history study of the indigo bunting. *Jack-Pine Warbler* 26:103–113.

Bradley, R. A. 1980. Vocal and territorial behavior in the white-eyed vireo. *Wilson Bull.* 92:302–311.

Brauning, D. W. 1992. *Atlas of breeding birds in Pennsylvania.* Pittsburgh, Pa.: University of Pittsburgh Press.

Brawn, J. D.; Samson, F. B. 1983. Winter behavior of tufted titmice. *Wilson Bull.* 95:222–232.

Breckenridge, W. J. 1956. Measurements of the habitat niche of the least flycatcher. *Wilson Bull.* 68:47–51.

Brennan, L. A. 1999. Northern bobwhite (*Colinus virginianus*). In: Poole, A.; Gill, F. (editors). *The birds of North America.* No. 397. Philadelphia: Academy of Natural Sciences; Washington, D.C.: American Ornithologists' Union.

Brewer, R. 1955. Size of home range in eight bird species in a southern Illinois swamp-thicket. *Wilson Bull.* 67:140–141.

Brewster, W. 1906. *The birds of the Cambridge region of Massachusetts.* Memoirs of the Nuttall Ornithological Club. No. IV. Cambridge, Mass.: Nuttall Ornithological Club. 426 pp.

Brewster, W. 1924. *The birds of the Lake Umbagog region of Maine (Parts 1 and 2).* Bull. Mus. Comp. Zool. 66. Cambridge, Mass.: Harvard College.

Brigham, R. M. 1989. Roost and nest sites of the common nighthawk: Are gravel roofs important? *Condor* 91:722–724.

Briskie, J. V. 1994. Least flycatcher (*Empidonax minimus*). In: Poole, A.; Gill, F. (editors). *The birds of North America.* No. 99. Philadelphia: Academy of Natural Sciences; Washington, D.C.: American Ornithologists' Union.

Brittingham, M. C.; Temple, S. A. 1983. Have cowbirds caused forest birds to decline? *Bioscience* 33:31–35.

Broley, C. L. 1947. Migration and nesting of Florida bald eagles. *Wilson Bull.* 59:3–20.

Brooks, B. L.; Temple, S. A. 1990. Dynamics of a loggerhead shrike population in Minnesota. *Wilson Bull.* 102:441–450.

Brooks, E. W. 1985. Fidelity of an American tree sparrow to a wintering area. *J. Field Ornithol.* 56:406–407.

Brooks, R. P.; Davis, W. J. 1987. Habitat selection by breeding belted kingfishers (*Ceryle alcyon*). *Am. Midl. Nat.* 117:63–70.

Brown, C. R. 1979. Territoriality in the purple martin. *Wilson Bull.* 91:583–591.

Brown, C. R. 1997. Purple martin (*Progne subis*). In: Poole, A.; Gill, F. (editors). *The birds of North America.* No. 287. Philadelphia: Academy of Natural Sciences; Washington, D.C.: American Ornithologists' Union.

Brown, C. R.; Brown, M. B. 1995. Cliff swallow (*Hirundo pyrrhonota*). In: Poole, A.; Gill, F. (editors). *The birds of North America.* No. 149. Philadelphia: Academy of Natural Sciences; Washington, D.C.: American Ornithologists' Union.

Brown, L.; Amadon, D. 1968. *Eagles, hawks, and falcons of the world.* 2 vols. New York: McGraw-Hill.

Buckley, P. A.; Paxton, R. P.; Cutler, D. 1976. The nesting season: Hudson-Delaware region. *Am. Birds* 30(5):932–938.
Buech, R. R. 1982. Nesting ecology and cowbird parasitism of clay-colored, chipping, and field sparrows in a Christmas tree plantation. *J. Field Ornithol.* 53:363–369.
Bull, E. 1975. Habitat utilization of the pileated woodpecker, Blue Mountains, Oregon. M.S. thesis, Oregon State University, Corvallis.
Bull, E. L.; Duncan, J. R. 1993. Great gray owl (*Strix nebulosa*). In: Poole, A.; Gill, F. (editors). *The birds of North America.* No. 41. Philadelphia: Academy of Natural Sciences; Washington, D.C.: American Ornithologists' Union.
Bull, E. L.; Holthausen, R. S. 1993. Habitat use and management of pileated woodpeckers in northeastern Oregon. *J. Wildl. Manage.* 57:335–345.
Bull, E. L.; Jackson, J. E. 1995. Pileated woodpecker (*Dryocopus pileatus*). In: Poole, A.; Gill, F. (editors). *The birds of North America.* No. 148. Philadelphia: Academy of Natural Sciences; Washington, D.C.: American Ornithologists' Union.
Bull, E. L.; Meslow, C. 1977. Habitat requirements of the pileated woodpecker in northeastern Oregon. *J. Forestry* 77:335–337.
Bull, J. 1964. *Birds of the New York area.* New York: Harper & Row. 540 pp.
Bull, J. 1974. *Birds of New York State.* Garden City, N.Y.: Doubleday Natural History Press. 655 pp.
Bull, J.; Farrand, J., Jr. 1977. *The Audubon Society field guide to North American birds: Eastern region.* New York: Alfred A. Knopf. 775 pp.
Bump, G.; Darrow, R. W.; Edminster, F. C.; Crissy, W. F. 1947. *The ruffed grouse: Life history, propagation, management.* Albany: New York Conservation Department. 896 pp.
Burger, J. 1978. Competition between cattle egrets and native North American herons, egrets, and ibises. *Condor* 80:15–23.
Burger, J. 1987. Physical and social determinants of nest site selection in the piping plover in southern New Jersey. *Condor* 89:881–918.
Burger, J.: Lesser, F. 1978. Determinants of colony site selection in common terns (*Sterna hirundo*). *Colon. Waterbirds* 1:118–127.
Burns, F. L. 1911. A monograph of the broad-winged hawk (*Buteo platyperus*). *Wilson Bull.* 23:139–320.
Burns, J. T. 1982. Nests, territories, and reproduction of sedge wrens (*Cistothorus platensis*). *Wilson Bull.* 94:338–349.
Burt, E. H., Jr. 1977. Some factors in the timing of parent chick recognition in swallows. *Anim. Behav.* 25:231–239.
Buss, I. O.; Hawkins, A. S. 1939. The upland plover at Faville Grove, Wisconsin. *Wilson Bull.* 51:202–220.
Butler, R. W. 1992. Great blue heron (*Ardea herodia*). In: Poole, A. F.; Stettenheim, P.; Gill, F. B. (editors). *The birds of North America.* No. 25. Philadelphia: Academy of Natural Sciences; Washington, D.C.: American Ornithologists' Union.
Butts, W. K. 1931. A study of the chickadee and white-breasted nuthatch by means of marked individuals. *Bird-Banding* 2:1–26.
Cabe, P. R. 1993. European starling (*Sturnus vulgaris*). In: Poole, A.; Gill, F. (editors). *The birds of North America.* No. 48. Philadelphia: Academy of Natural Sciences; Washington, D.C.: American Ornithologists' Union.
Cade, T. C. 1967. Ecological and behavior aspects of predation by the northern shrike. *Living Bird* 6:43–86.
Cairns, W. E. 1982. Biology and behavior of breeding piping plovers. *Wilson Bull.* 94:531–545.
Campbell, R. W. 1974. Rusty blackbirds prey on sparrows. *Wilson Bull.* 86(3):291–293.
Cannings, R. J. 1987. The breeding biology of northern saw-whet owls in southern British Columbia. Pages 193–198. In: Nero, R. W.; Clark, R. J.; Knapton, R. J.; Hamre, R. H. (editors). *Biology and conservation of northern forest owls.* Gen. Tech. Rep. RM-142. Ft. Collins, Colo.: U.S. Forest Service.
Cannings, R. J. 1993. Northern saw-whet owl (*Aegolius acadicus*). In: Poole, A.; Gill, F., eds. *The birds of North America.* No. 42. Philadelphia: Academy of Natural Sciences: Washington, D.C.: American Ornithologists' Union.
Carey, M.; Burhans, D. E.; Nelson, D. A. 1994. Field sparrow (*Spizella pusilla*). In: Poole, A.; Gill, F. (editors). *The birds of North America.* No. 103. Philadelphia: Academy of Natural Sciences; Washington, D.C.: American Ornithologists' Union.
Carpenter, B. 1973. Deciduous-coniferous second-growth north woods. *Am. Birds* 27:95.
Carroll, J. P. 1993. Gray partridge (*Perdix perdix*). In: Poole, A.; Gill, F. (editors). *The birds of North America.* No. 57. Philadelphia: Academy of Natural Sciences; Washington, D.C.: American Ornithologists' Union.
Carter, B. C. 1958. The American goldeneye in central New Brunswick. *Can. Wildl. Serv. Wildl. Manage. Bull. Ser.* 2(9):1–47.
Carter, J. W. 1992. Upland sandpiper (*Bartramia longicauda*). Pages 235–251. In: Schneider, K. J.; Pence, D. M. (editors). *Migratory nongame birds of management concern in the Northeast.* Newton Corner, Mass.: U.S. Fish & Wildlife Service, Region 5.
Case, N. A.; Hewitt, O. H. 1963. Nesting and productivity of the red-winged blackbird in relation to habitat. *Living Bird* 2:7–20.
Catling, P. M. 1972. A study of the boreal owl in southern Ontario with particular reference to the irruption of 1968–69. *Can. Field-Nat.* 86:223–232.
Chamberlain-Auger, J. A.; Auger, P. A.; Strauss, E. G. 1990. Breeding biology of American crows. *Wilson Bull.* 102:615–622.
Chapman, L. B. 1955. Studies of a tree swallow colony. *Bird-banding* 26:45–70.
Chasko, G. G.; Gates, J. E. 1982. Avian habitat suitability along a transmission-line corridor in an oak-hickory forest region. *Wildl. Monogr.* 82. 41 pp.
Christman, S. P. 1983. Mississippi Delta bottomland hardwoods (unmanaged). *Am. Birds* 37:67.
Ciaranca, M. A.; Allin, C. C.; Jones, G. S. 1997. Mute swan (*Cygnus olor*). In: Poole, A.; Gill, F. (editors). *The birds of North America.* No. 273. Philadelphia: Academy of Natural Sciences; Washington, D.C.: American Ornithologists' Union.
Cimprich, D. A.; Moore, F. R. 1995. Gray catbird (*Dumetella carolinensis*). In: Poole, A.; Gill, F. (editors). *The birds of North America.* No. 167. Philadelphia; Academy of Natural Sciences; Washington, D.C.: American Ornithologists' Union.
Clark, R. J. 1975. A field study of the short-eared owl, *Asio flammeus* (Pontoppidan), in North America. *Wildl. Monogr.* 47:1–67.
Coleman, J. S.; Fraser, D. J. 1989. Habitat use and home ranges of black and turkey vultures. *J. Wildl. Manage.* 53:782–792.
Collins, H. H., Jr.; Boyajian, N. R. 1965. Familiar garden birds of America. New York: Harper and Row. 309 pp.

Collins, S. L. 1983a. Geographic variation in habitat structure of the black-throated green warbler (*Dendroica virens*). *Auk* 100:382–389.

Collins, S. L. 1983b. Geographic variation in habitat structure for the wood warblers in Maine and Minnesota. *Oecologia* 59:246–252.

Collins, S. L.; James, F. C.; Risser, P. G. 1982. Habitat relationships of wood warblers (Parulidae) in northern central Minnesota. *Oikos* 39:50–58.

Collopy, M. W. 1984. Parental care and feeding ecology of golden eagle nestlings. *Auk* 101:753–760.

Colvin, B. A. 1985. Common barn-owl population declines in Ohio and the relationship to agricultural trends. *J. Field Ornithol.* 56:224–235.

Confer, J. L. 1992a. Golden-winged warbler (*Vermivora chrysoptera*). Pages 369–383. In: Schneider, K. J.; Pence, D. M. (editors). *Migratory nongame birds of management concern in the Northeast.* Hadley, Mass.: U.S. Department of the Interior, Fish and Wildlife Service.

Confer, J. L. 1992b. Golden-winged warbler (*Vermivora chrysoptera*). In: Poole, A. F.; Stettenheim, P.; Gill, F. B. (editors). *The birds of North America.* No. 20. Philadephia: Academy of Natural Sciences; Washington, D.C.: American Ornithologists' Union.

Confer, J. L.; Knapp, K. 1981. Golden-winged warblers and blue-winged warblers: The relative success of a habitat specialist and generalist. *Auk* 98:108–114.

Conner, R. N. 1976. Nesting habitat for red-headed woodpeckers in southwestern Virginia. *Bird-Banding* 47:40–43.

Conner, R. N.; Adkisson, C. S. 1974. Eastern bluebirds nesting in clearcuts. *J. Wildl. Manage.* 38:934–935.

Conner, R. N.; Adkisson, C. S. 1977. Principal component analysis of woodpecker nesting habitat. *Wilson Bull.* 89:122–129.

Conner, R. N.; Hooper, R. G.; Crawford, H. S.; Mosby, H. S. 1975. Woodpecker nesting habitat in cut and uncut woodlands in Virginia. *J. Wildl. Manage.* 39:144–150.

Conway, C. J. 1995. Virginia rail (*Rallus limicola*). In: Poole, A.; Gill, F. (editors). *The birds of North America.* No. 173. Philadelphia: Academy of Natural Sciences; Washington, D.C.: American Ornithologists' Union.

Conway, C. J. 1999. Canada warbler (*Wilsonia canadensis*). In: Poole, A.; Gill, F. (editors). *The birds of North America.* No. 421. Philadelphia: Academy of Natural Sciences; Washington, D.C.: American Ornithologists' Union.

Cornwell, G. W. 1963. Observations on the breeding biology and behavior of a nesting population of belted kingfishers. *Condor* 65:426–431.

Cottrille, W. P.; Cottrille, B. D. 1958. Great blue heron: Behavior at the nest. *Misc. Publ. Mus. Zool. Univ. Mich.* 102:1–15.

Coulter, M. W. 1957. Food of wood ducks in Maine. *J. Wildl. Manage.* 21:235–236.

Coulter, M. W.; Mendall, H. L. 1968. Northeastern states. Pages 90–101. In: Barske, P. (editor). *Black duck; Evaluation, management, and research: A symposium.* Bolton, Mass.: Atlantic Waterfowl Council and Wildlife Management Institute.

Coulter, M. W.; Miller, W. R. 1968. Nesting biology of black ducks and mallards in northern New England. *Vt. Fish and Game Dept. Bull.* 68(2):1–74.

Cox, G. W. 1960. A life history of the mourning warbler. *Wilson Bull.* 75:5–28.

Craig, E. H.; Craig, T. H.; Powers, L. R. 1988. Activity patterns and home-range use of nesting long-eared owls. *Wilson Bull.* 100:204–213.

Craig, R. J. 1985. Comparative habitat use by Northern and Louisiana Waterthrushes. *Wilson Bull.* 97:347–355.

Craighead, J.; Craighead, F. 1969. *Hawks, owls and wildlife.* New York: Dover. 443 pp.

Cramp, S.; Perrins, C. M. 1994. *The birds of the western paleartic.* Oxford: Oxford University Press.

Crawford, H. S.; Jennings, D. T.; Stone, T. L. 1990. Red-breasted nuthatches detect early increases in spruce budworm populations. *North. J. Appl. Forestry* 7:81–83.

Crawford, H. S.; Titterington, R. W. 1979. Effects of silvicultural practices on bird communities in upland spruce-fir stands. Pages 110–119. In: DeGraaf, R. M.; Evans, K. E. (editors). *Management of north central and northeastern forest for nongame birds.* Gen. Tech. Rep. NC-51. St. Paul, Minn.: U.S. Department of Agriculture, Forest Service.

Crocoll, S. T. 1994. Red-shouldered hawk (*Buteo lineatus*). In: Poole, A.; Gill, F. (editors). *The birds of North America.* No. 107. Philadelphia: Academy of Natural Sciences; Washington, D.C.: American Ornithologists' Union.

Crocoll, S. T.; Parker, J. W. 1989. The breeding biology of broad-winged and red-shouldered hawks in western New York. *Raptor Res.* 23:125–139.

Crooks, M. P.; Hendrickson, G. O. 1953. Field sparrow life history in central Iowa. *Iowa Bird Life* 23:10–13.

Crosby, M. S. 1922. Mute swans on the Hudson. *Auk* 39:100.

Crowley, S. C. 1994. Habitat use and population monitoring of secretive waterbirds in Massachusetts. M.S. thesis, University of Massachusetts, Amherst.

Cruickshank, A. D. 1942. *Birds around New York City.* New York: American Museum of Natural History.

Cruz, A. 1975. Ecology and breeding biology of the solitary vireo. *J. Colo.-Wyo. Acad. Sci.* 7(6):36–37.

Cuthbert, N. L. 1954. A nesting study of the black tern in Michigan. *Auk* 71:36–63.

Darley, J. A.; Scott, D. M.; Taylor, N. K. 1971. Territorial fidelity of catbirds. *Can. J. Zool.* 49:1465–1478.

Darveau, M.; DesGranges, J. L.; Gauthier, G. 1992. Habitat use by three breeding insectivorous birds in declining maple forests. *Condor* 94:72–82.

Darveau, M.; Gauthier, G.; Desgranges, J. L.; Mauffette, Y. 1993. Nesting success, nest sites, and parental care of the least flycatcher in declining maple forests. *Can. J. Zool.* 71:1592–1601.

Davis, C. M. 1978. A nesting study of the brown creeper. *Living Bird* 17:237–263.

Davis, D. W. 1959. Observations on territorial behavior of least flycatchers. *Wilson Bull.* 71:73–85.

Davis, E. M. 1937. Observations on nesting barn swallows. *Bird-Banding* 8:66–73.

Davis, W. E. 1986. Effects of old nests on nest-site selection in black-crowned night-herons and snowy egrets. *Wilson Bull.* 98:300–303.

Davis, W. E., Jr. 1993. Black-crowned night heron (*Nycticorax nycticorax*). In: Poole, A.; Gill, F. (editors). *The birds of North America.* No. 74. Philadelphia: Academy of Natural Sciences; Washington, D.C.: American Ornithologists' Union.

Davis, W. E.; Kushlan, J. A. 1994. Green heron (*Butorides virescens*). In: Poole, A.; Gill, F. (editors). *The birds of North America.* No. 129. Philadelphia: Academy of Natural Sciences;Washington, D.C.: American Ornithologists' Union.

Davis, W. J. 1982. Territory size of *Megaceryle alcyon* along a stream habitat. *Auk* 99:353–362.
Dawson, W. R. 1997. Pine Siskin (*Carduelis pinus*). In: Poole, A.; Gill, F. (editors). *The birds of North America*. No. 280. Philadelphia: Academy of Natural Sciences; Washington, D.C.: American Ornithologists' Union.
DeGraaf, R. M. 1975. Suburban Habitat Associations of Birds. Ph.D. dissertation, University of Massachusetts, Amherst. 295 pp.
DeGraaf, R. M. 1985. Breeding bird assemblages in New England northern hardwoods. Pages 5–22. In: Regan, R. J.; Capen, D. E. (editors). *The impact of timber management practices on nongame birds in Vermont*. Conf. Proceedings. Montpelier, Vt.
DeGraaf, R. M. 1991. Breeding bird assemblages in managed northern hardwood forests in New England. Pages 154–171. In: Rodiek, J. E.; Bolen, E. G. (editors). *Wildlife and habitats in managed landscapes*. Washington, D.C.: Island Press.
DeGraaf, R. M.; Chadwick, N. L. 1987. Forest type, timber size class, and New England breeding birds. *J. Wildl. Manage.* 51:212–217.
DeGraaf, R. M.; Pywell, H. R.; Thomas, J. W. 1975. Relationships between nest height, vegetation and housing density in New England suburbs. *Trans. Northeast Wildl. Conf.* 32:130–150.
DeGraaf, R. M.; Rappole, J. H. 1995. *Neotropical migratory birds: Natural history, distribution, and population change*. Ithaca, N.Y.: Cornell University Press. 491 pp.
DeGraaf, R. M.; Scott, V. E.; Hamre, R. H.; Ernst, L.; Anderson, S. H. 1991. *Forest and rangeland birds of the United States: Natural history and habitat use*. Agric. Handbook 688. Washington, D.C.: U.S. Forest Service. 625 pp.
DeJong, M. J. 1996. Northern rough-winged swallow (*Stelgidopteryx serripennis*). In: Poole, A.; Gill, F. (editors). *The birds of North America*. No. 234. Philadelphia: Academy of Natural Sciences; Washington, D.C.: American Ornithologists' Union.
Dennis, J. V. 1958. Some aspects of the breeding ecology of the yellow-breasted chat (*Icteria virens*). *Bird-Banding* 29:169–183.
Dennis, J. V. 1969. The yellow-shafted flicker (*Coloptes auratus*) on Nantucket Island, Massachusetts. *Bird-Banding* 40(4):290–308.
Derrickson, K. C.; Breitwisch, R. 1992. Mockingbird (*Mimus polyglottos*). In: Poole, A. F.; Stettenheim, P.; Gill, F. (editors). *The birds of North America*. No. 7. Philadelphia: Academy of Natural Sciences; Washington, D.C.: American Ornithologists' Union.
Dexter, R. W. 1977. Synopsis of the 1976 season for chimney swifts at Kent State University. *Bird-Banding* 48(1):73–74.
Dexter, R. W. 1981. Nesting success of chimney swifts related to age and the number of adults at the nest, and the subsequent fate of the visitors. *J. Field Ornithol.* 52:228–232.
Dickerman, R. W. 1986. A review of the red crossbill in New York. Pt. 2. Identification of specimens from New York. *Kingbird* 36:127–134.
Dickson, J. G.; Thompson F. R., III; Conner, R. N.; Franzreb, K. E. 1993. Effects of silviculture on Neotropical migratory birds in central and southeastern oak pine forests. Pages 374–385. In: *Status and management of Neotropical migratory birds*. Gen. Tech. Rep RM-229. Fort Collins, Colo.: U.S. Department of Agriculture, Forest Service, Rocky Mountain Forest and Range Exp. Sta.
Dijak, W. D.; Tannenbaum, B.; Parker, M. A. 1990. Nest-site characteristics affecting success and reuse of red-shouldered hawk nests. *Wilson Bull.* 102:480–486.
Dilger, W. C. 1956. Adaptive modifications and ecological isolating mechanisms in the thrush genera *Catharus* and *Hylocichla*. *Wilson Bull.* 68:171–199.
Dixon, C. L. 1978. Breeding biology of the savannah sparrow on Kent Island. *Auk* 95:235–246.
Dobkin, D. S. 1992. Neotropical Migrant Landbirds in the Northern Rockies and Great Plains. Publ. R1–93–34. Missoula, Mont.: U.S. Department of Agriculture, Forest Service, Northern Region.
Dobkin, D. S. 1994. *Conservation and management of Neotropical migrant landbirds*. Moscow: University of Idaho Press.
Dow, D. D. 1969. Home range and habitat of the cardinal in peripheral and central populations. *Can. J. Zool.* 47:103–115.
Drewien, R. C.; Springer, P. F. 1969. Ecological relationships of breeding blue-winged teal to prairie potholes. Pages 102–115. In: *Saskatoon Wetlands Seminar*. Report Series No. 6. Ottawa, Ontario: Canadian Wildlife Service.
Drum, M. 1939. Territorial studies on the eastern goldfinch. *Wilson Bull.* 51:69–77.
Drury, W. H. 1973. Population changes in New England seabirds. *Bird-Banding* 44:267–313.
Drury, W. H.; Nisbet, I. C. T. 1972. Movement of herring gulls in New England. *Wildl. Res. Rep.* 2:173–212.
DuBowy, P. J. 1996. Northern shoveler (*Anas clypeata*). In: Poole, A.; Gill, F. (editors). *The birds of North America*. No. 217. Philadelphia. Academy of Natural Sciences; Washington, D.C.: American Ornithologists' Union.
Duebbert, H. F.; Lokemoen, J. T. 1977. Upland nesting of American bittern, marsh hawks, and short-eared owls. *Prairie Nat.* 9:33–39.
Dufty, A. M. 1982. Movements and activities of radio-tracked brown-headed cowbirds. *Auk* 99:316–327.
Dugger, B. D.; Dugger, K. M.; Fredrickson, L. H. 1994. Hooded merganser (*Lophodytes cucullatus*). In: Poole, A.; Gill, F. (editors). *The birds of North America*. No. 98. Philadelphia: Academy of Natural Sciences; Washington, D.C.: American Ornithologists' Union.
Duncan, J. R.; Duncan, P. A. 1998. Northern hawk owl. In: Poole, A.; Gill, F. (editors). *The birds of North America*. No. 356. Philadelphia: Academy of Natural Sciences; Washington, D.C.: American Ornithologists' Union.
Dunn, E. H. 1979. Nesting biology and development of young in Ontario black terns. *Can. Field-Nat.* 93:276–281.
Dunn, E. H.; Agro, D. J. 1995. Black tern (*Chlidonias niger*). In: Poole, A.; Gill, F. (editors). *The birds of North America*. No. 147. Philadelphia: Academy of Natural Sciences; Washington, D.C.: American Ornithologists' Union.
Dunnett, G. M. 1955. The breeding of the starling (*Sturnus vulgaris*) in relation to its food supply. *Ibis* 97:619–662.
Eadie, J. M.; Keast, A. 1982. Do goldeneye and perch compete for food? *Oecologia* 55:225–230.
Eadie, J. M.; Mallory, M. L.; Lumsden, H. G. 1995. Common goldeneye (*Bucephala clangula*). In: Poole, A.; Gill, F. (editors). *The birds of North America*. No. 170. Philadelphia: Academy of Natural Sciences; Washington, D.C.: American Ornithologists' Union.

Eaton, E. H. 1914. *Birds of New York*. Albany: State University of New York.
Eaton, S. W. 1957. A life history study of *Seiurus noveboracensis*. *Sci. Stud. St. Bonaventure U.* 19:7–36.
Eaton, S. W. 1958. A life history of the Louisiana waterthrush. *Wilson Bull.* 70:211–236.
Eaton, S. W. 1992. Wild turkey (*Meleagris gallopavo*). In: Poole, A. F.; Stettenheim, P.; Gill, F. B. (editors). *The birds of North America*. No. 22. Philadelphia: Academy of Natural Sciences; Washington, D.C.: American Ornithologists' Union.
Eaton, S. W. 1995. Northern waterthrush (*Seiurus noveboracensis*). In: Poole, A.; Gill, F. (editors). *The birds of North America*. No. 182. Philadelphia: Academy of Natural Sciences; Washington, D.C.: American Ornithologists' Union.
Eddleman, W. R.; Knopf, F. L.; Meanley, B. F.; Reid F. A.; Zembal, R. 1988. Conservation of North American rallids. *Wilson Bull.* 100:458–475.
Edminster, F. C. 1954. *American game birds of field and forest*. New York: Charles Scribner's Sons. 490 pp.
Eiserer, L. A. 1980. Effects of grass length and mowing on foraging behavior of the American robin (*Turdus migratorius*). *Auk* 97(3):576–580.
Elkins, K. C. 1978. The fall migration, August 1–November 30, 1977. *N.H. Audubon* 14(2):2–6.
Elkins, K. C. 1994. Northern rough-winged swallow. Pages 170–171. In: Foss, C. R. (editor). *The atlas of breeding birds in New Hampshire*. Dover, N.H.: Arcadia. 456 pp.
Elliott, J. J.; Arbib, R. S., Jr. 1953. Origin and status of the house finch in the eastern United States. *Auk* 70:31–37.
Ellison, L. N. 1971. Territoriality in Alaskan spruce grouse. *Auk* 88:652–664.
Ellison, L. N. 1973. Seasonal social organization and movements of spruce grouse. *Condor* 75:375–385.
Ellison, W. G. 1985a. Blackburnian warbler. Pages 298–299. In: Laughlin, S. B.; Kibbe, D. P. (editors). *The atlas of breeding birds of Vermont*. Hanover, N.H.: University Press of New England. 656 pp.
Ellison, W. G. 1985b. Black-throated green warbler (*Dendroica virens*). Pages 296–297. In: Laughlin, S. B.; Kibbe, D. P. (editors). *The atlas of breeding birds of Vermont*. Hanover, N.H.: University Press of New England. 656 pp.
Ellison, W. G. 1985c. Cattle egret. Pages 40–41. In: Laughlin, S. B.; Kibbe, S. P. (editors). *The atlas of breeding birds of Vermont*. Hanover, N.H.: University Press of New England. 656 pp.
Ellison, W. G. 1985d. Eastern wood-pewee. Pages 172–173. In: Laughlin, S. B.; Kibbe, D. P. (editors). *The atlas of breeding birds of Vermont*. Hanover, N.H.: University Press of New England. 656 pp.
Ellison, W. G. 1985e. Red-tailed hawk. Pages 84–85. In: Laughlin, S. B.; Kibbe, D. P. (editors). *The atlas of breeding birds of Vermont*. Hanover, N.H.: University Press of New England. 656 pp.
Elllison, W. G. 1991. The mechanism and ecology of range expansion by the blue-gray gnatcatcher. M.S. Thesis, University of Connecticut, Storrs.
Ellison, W. G. 1992. Blue-gray gnatcatcher (*Polioptila caerulea*). In: Poole, A. F.; Stettenheim, P.; Gill, F. B. (editors). *The birds of North America*. No. 23. Philadelphia: Academy of Natural Sciences; Washington, D.C.: American Ornithologists' Union.
Emlen, J. T., Jr. 1952. Social behavior in nesting cliff swallows. *Condor* 54:117–199.
Emlen, J. T. 1954. Territory, nest building and pair formation in the cliff swallow. *Auk* 71:16–35.
Emlen, S.; Demong, N. J. 1974. Adaptive significance of synchronized breeding in the bank swallow. Abstract, 92nd meeting, American Ornithologists' Union, Norman, Okla.; p.8.
England, E. G. 1940. A nest of the arctic three-toed woodpecker. *Condor* 42:242–245.
Enser, R. W. 1992. *The atlas of breeding birds in Rhode Island*. Providence: Rhode Island Department of Environmental Management. 206 pp.
Erickson, A. B. 1941. A study of Wilson's snipe. *Wilson Bull.* 53:62.
Eriksson, M. O. G. 1978. Lake selection of goldeneye ducklings in relation to the abundance of food. *Wildfowl* 29:81–85.
Eriksson, M. O. G. 1983. The role of fish in the selection of lakes by nonpiscivorous ducks: Mallard, teal, and goldeneye. *Wildfowl* 34:27–32.
Ernst, S. G. 1945. The food of the red-shouldered hawk in New York State. *Auk* 62:452–453.
Erskine, A. J. 1971. *Buffleheads*. Monograph No. 4. Ottawa: Canadian Wildlife Service. 240 pp.
Erskine, A. J. 1977. *Birds in boreal Canada: Communities, densities, and adaptations*. Monograph No. 41. Ottawa: Canadian Wildlife Service.
Erskine, A. J. 1992. *Atlas of breeding birds of the Maritime provinces*. Halifax: Nimbus Publishing and Nova Scotia Museum.
Erwin, W. G. 1935. Some nesting habits of the brown thrasher. *J. Tenn. Acad. Sci.* 10:179–204.
Evans, E. W. 1978. Nesting responses of field sparrows (*Spizella pusilla*) to plant succession in a Michigan old field. *Condor* 80:34–40.
Evans, F. C. 1964. The food of vesper, field and chipping sparrows nesting in an old field in southern Michigan. *Am. Midl. Nat.* 72:57–75.
Evans, Ogden, L. J.; Stutchbury, B. J. 1994. Hooded warbler (*Wilsonia citrina*). In: Poole, A.; Gill, F. (editors). *The birds of North America*. No. 110. Philadelphia: Academy of Natural Sciences; Washington, D.C.: American Ornithologists' Union.
Evenden, F. G., Jr. 1957. Observations on nesting behavior of the house finch. *Condor* 59:112–117.
Evrard, J. O.; Bacon, B. R.; Grunewald, T. R. 1987. Unusual upland nests of the ring-necked duck. *J. Field Ornithol.* 58 (1):31.
Faaborg, J. 1976. Habitat selection and territorial behavior of the small grebes of North Dakota. *Wilson Bull.* 88:390–399.
Faanes, C. A. 1980. Breeding biology of eastern phoebes in northern Wisconsin. *Wilson Bull.* 92:107–110.
Falls, J. B.; Kopachena, J. G. 1994. White-throated sparrow (*Zonotrichia albicollis*). In: Poole, A.; Gill, F. (editors). *The birds of North America*. No. 128. Philadelphia: Academy of Natural Sciences; Washington, D.C.: American Ornithologists' Union.
Farrar, R. B. 1973. Rural stream border. *Am. Birds* 27:982–983.
Fawks, E. 1937. Bird-Lore's first breeding-bird census. Second-growth hardwood. *Bird-Lore* 39(5):380.
Fawks, E. 1938. Bird-Lore's second breeding-bird census. Second-growth hardwood. *Bird-Lore* 40(5):359.
Fawver, B. J. 1947. Bird copulation of an Illinois flood plain forest. *Ill. Acad. Sci. Trans.* 40:178–189.
Feare, C. J. 1984. *The starling*. Oxford: Oxford University Press.

Fichtel, C. 1985. Common loon. Pages 30–31. In: Laughlin, S. B.; Kibbe, D. P. (editors). *The atlas of breeding birds of Vermont.* Hanover, N.H.: University Press of New England.

Ficken, M. S. 1962. Agonistic behavior and territory in the American redstart. *Auk* 79:607–632.

Ficken, M. S.; Ficken, R. W. 1968. Territorial relationships of blue-winged warblers, golden-winged warblers, and their hybrids. *Wilson Bull.* 80:442–451.

Ficken, M. S.; McLaren, M. A.; Hailman, J. P. 1996. Boreal chickadee (*Parus hudsonicus*). In: Poole, A.; Gill, F. (editors). *The birds of North America.* No. 254. Philadelphia: Academy of Natural Sciences; Washington, D.C.: American Ornithologists' Union.

Ficken, R. W. 1963. Courtship and aggressive behavior of the common grackle (*Quiscalus quiscula*). *Auk* 80:5272.

Finch, D. M. 1990. Effects of predation and competitor interference on nesting success of house wrens and tree swallows. *Condor* 92:674–687.

Fischer, R. B. 1958. *The breeding biology of the chimney swift (Chaetura pelagica* L.). New York State Mus. Sci. Serv. Bull. 368. Albany: New York State Museum. 141 pp.

Fitch, H. S. 1958. Home ranges, territories and seasonal movements of vertebrates of the Natural History Reservation. *Univ. Kans. Publ. Mus. Nat. Hist.* 11:63–326.

Fitch, H. S.; Swenson, F.; Tillotson, D. F. 1946. Behavior and food habits of the red-tailed hawk. *Condor* 48:205–237.

Fogarty, M. J.; Arnold, K. A. 1977. Common snipe. Pages 189–209. In: Sanderson, Glen C. (editor). *Management of migratory shore and upland game birds in North America.* Washington, D.C.: International Association of Fish and Wildlife Agencies.

Foote, L. E. 1946. *A history of wild game in Vermont.* Montpelier: Vermont Fish and Game Department.

Forbes, J. E.; Warner, D. W. 1974. Behavior of a radio-tagged saw-whet owl. *Auk* 91:783–795.

Forbush, E. H. 1913. *Useful birds and their protection.* Boston: Massachusetts State Board of Agriculture. 451 pp.

Forbush, E. H. 1916. *A history of gamebirds, wildfowl, and shore birds of Massachusetts and adjacent states.* 2nd ed. Boston: Massachusetts State Board of Agriculture.

Forbush, E. H. 1925. *Birds of Massachusetts and other New England states.* Vol. 1. Boston: Massachusetts Department of Agriculture.

Forbush, E. H. 1927. *Birds of Massachusetts and other New England states.* Vol. 2. Boston: Massachusetts Department of Agriculture. 461 pp.

Forbush, E. H. 1929. *Birds of Massachusetts and other New England states.* 3 vols. Boston: Massachusetts Department of Agriculture.

Forbush, E. H.; May, J. B. 1939. *Natural history of the birds of eastern and central North America.* Boston: Houghton Mifflin. 554 pp.

Foss, C. R. 1994. Turkey vulture. Pages 40–41. In: Foss, C. R. (editor). *Atlas of breeding birds in New Hampshire.* Dover, N.H.: Arcadia. 414 pp.

Franks, E. C.; Martin, W. 1967. Thirty-first breeding-bird census. Upland oak-hickory forest. *Audubon Field Notes* 21(6):615.

Fredrickson, L. H. 1970. Breeding biology of American coots in Iowa. *Wilson Bull.* 82:445–457.

Fredrickson, L. H. 1971. Common gallinule breeding biology and development. *Auk* 88:914–919.

Fredrickson, L. H. 1977. American coot (*Fulica americana*). Pages 123–147. In: Sanderson, G. C. (editor). *Management of migratory shore and upland game birds in North America.* Washington, D.C.: International Association of Fish and Wildlife Agencies.

Freedman, B.; Beauchamp, C.; McLaren, I. A; Tingley, S. I. 1981. Forestry management practices and populations of breeding birds in a hardwood forest in Nova Scotia. *Can. Field-Nat.* 95:307–311.

Freer, V. M. 1979. Factors affecting site tenacity in New York bank swallows. *Bird-Banding* 50:349–357.

Fretwell, S. 1968. Habitat distribution and survival in the field sparrow (*Spizella pusilla*). *Bird-Banding* 39:293–306.

Friedmann, H. 1929. *The cowbirds.* Springfield, Ill.: Charles C. Thomas. 421 pp.

Friedmann, H. 1963. *Host relations of the parasitic cowbirds.* U.S. Nat. Mus. Bull. 233. Washington, D.C.: Smithsonian Institution.

Friedmann, H.; Kiff, L. F.; Rothstein, S. I. 1977. A further contribution to knowledge of the host relations of the parasitic cowbirds. *Smithsonian Contrib. Zool.* 235:1–75.

Friley, C. E.; Bennett, L. J.; Hendrickson, G. O. 1938. The American coot in Iowa. *Wilson Bull.* 50:81–86.

Garrison, B. A. 1999. Bank swallow (*Riparia riparia*). In: Poole, A.; Gill, F. (editors). *The birds of North America.* No. 414. Philadelphia: Academy of Natural Sciences; Washington, D.C.: American Ornithologists' Union.

Gartshore, M. E. 1988. A summary of the breeding status of hooded warblers in Ontario. *Ontario Birds* 6:84–99.

Gates, J. M. 1972. Red-tailed hawk populations and ecology in east-central Wisconsin. *Wilson Bull.* 84:421–433.

Gauthier, G. 1993. Bufflehead (*Bucephala albeola*). In: Poole, A.; Gill, F. (editors). *The birds of North America.* No. 67. Philadelphia: Academy of Natural Sciences; Washington, D.C.: American Ornithologists' Union.

Gauthier, J.; Aubry, Y. 1993. *Atlas of the breeding birds of southern Quebec.* St Foy, Que.: Canadian Wildlife Service.

Gavin, T. A. 1984. Broodedness in bobolinks. *Auk* 101:179–181.

Gavutis, G. W., Jr. 1994a. Golden-winged warbler. Pages 252–253. In: Foss, C. R. (editor). *Atlas of breeding birds in New Hampshire.* Dover, N.H.: Arcadia. 414 pp.

Gavutis, G. W., Jr. 1994b. Willet. Pages 82–83. In: Foss, C. R. (editor). *Atlas of breeding birds in New Hampshire.* Dover, N.H.: Arcadia. 414 pp.

Gawlik, D. E.; Bildstein, K. L. 1990. Reproductive success and nesting habitat of loggerhead shrikes in north-central South Carolina. *Wilson Bull.* 102:37–38.

Gelbach, F. R. 1994. *The eastern screech-owl: Life history, ecology, and behavior in suburbia and the countryside.* College Station: Texas A&M University Press.

Gelbach, F. R. 1995. Eastern screech-owl (*Otus asio*). In: Poole, A.; Gill, F. (editors). *The birds of North America.* No. 165. Philadelphia: Academy of Natural Sciences; Washington, D.C.: American Ornithologists' Union.

Gibbs, J. P.; Faaborg, J. 1990. Estimating the viability of ovenbird and Kentucky warbler populations in forest fragments. *Conserv. Biol.* 4(2):193–196.

Gibbs, J. P.; Longcore, J. R.; McAuley, D. G.; Ringelman, J. K. 1991. Use of wetland habitats by selected nongame water birds in Maine. Fish and Wildl. Res. Rep. 9. Washington, D.C.: U.S. Fish and Wildlife Service.

Gibbs, J. P.; Melvin, S. M. 1990. An assessment of wading birds and other wetlands avifauna and their habitats in Maine. Unpubl. rep. Bangor: Maine Department of Inland Fisheries and Wildlife.
Gibbs, J. P.; Melvin, S. M. 1992a. American bittern. In: Schneider, D. J.; Pence, D. M. (editors). *Migratory nongame birds of management concern in the Northeast.* Newton Corner, Mass.: U.S. Fish and Wildlife Service, Region 5. 400 p.
Gibbs, J. P.; Melvin, S. M. 1992b. Pied-billed grebe, *Podilymbus podiceps*. Pages 31–49. In: Schneider, D. J.; Pence, D. M. (editors). *Migratory nongame birds of management concern in the Northeast.* Newton Corner, Mass.: U.S. Fish and Wildlife Service, Region 5. 400 pp.
Gibbs, J. P.; Melvin, S. M. 1992c. Sedge wren (*Cistothorus platensis*). Pages 191–209. In: Schneider, D. J.; Pence, D. M. (editors). *Migratory nongame birds of management concern in the Northeast.* Newton Corner, Mass.: U.S. Fish and Wildlife Service, Region 5.
Gibbs, J. P.; Melvin, S. M.; Reid, F. A. 1992a. Least Bittern. In: Poole, A. F.; Stettenheim, P.; Gill, F. B. (editors). *The birds of North America.* No. 17. Philadelphia: Academy of Natural Sciences; Washington, D.C.: American Ornithologists' Union.
Gibbs, J. P.; Melvin, S.; Reid, F. A. 1992b. American bittern (*Botaurus lentiginosus*). In: Poole, A. F.; Stettenheim, P.; Gill, F. B. (editors). *The birds of North America.* No. 17. Philadelphia: Academy of Natural Sciences; Washington, D.C.: American Ornithologists' Union.
Gillespie, W. L.; Kendeigh, S. C. 1982. Breeding bird populations in northern Manitoba. *Can. Field-Nat.* 96:272–281.
Glahn, J. F. 1974. Study of breeding rails with recorded calls in north-central Colorado. *Wilson Bull.* 86:206–214.
Glover, F. A. 1953. Nesting ecology of the pied-billed grebe in northwestern Iowa. *Wilson Bull.* 65:32–39.
Godfrey, W. E. 1979. *The birds of Canada.* Ottawa: National Museum of Natural Sciences. 428 pp.
Godfrey, W. E. 1986. *The birds of Canada.* Natl. Mus. Can. Bull. 203. 423 pp.
Good, T. P. 1998. Great black-backed gull (*Larus marinus*). In: Poole, A.; Gill, F. (editors). *The birds of North America.* No. 330. Philadelphia: Academy of Natural Sciences; Washington, D.C.: American Ornithologists' Union.
Goodrich, L. J.; Crocoll, S. C.; Senner, S. E. 1996. Broad-winged hawk (*Buteo platypterus*). In Poole, A.; Gill, F. (editors). *The birds of North America.* No. 218. Philadelphia: Academy of Natural Sciences; Washington, D.C.: American Ornithologists' Union.
Goodwin, D. 1976. *Crows of the world.* Ithaca, N.Y.: Cornell University Press. 359 pp.
Goodwin, D. 1977. *Pigeons and doves of the world.* 2nd ed. Ithaca, N.Y.: Cornell University Press. 446 pp.
Gottfried, B. M.; Franks, E. C. 1975. Habitat use and flock activity of dark-eyed juncos in winter. *Wilson Bull.* 87:374–383.
Graber, J. W.; Graber, R. R.; Kirk, E. L. 1977. Illinois birds: Picidae. *Ill. Nat. Hist. Surv. Biol. Notes* 102:1–73.
Graber, R. R.; Graber, J. W. 1951. Nesting of the parula warbler in Michigan. *Wilson Bull.* 63:75–83.
Graber, R. R.; Graber, J. W. 1963. A comparative study of bird populations in Illinois, 1906–1909 and 1956–1958. *Ill. Nat. Hist. Surv. Bull.* 28(3):383–528.
Graber, R. R.; Graber, J. W.; Kirk, E. L. 1970. Illinois birds: Mimidae. *Ill. Nat. Hist. Surv. Biol. Notes* 68:1–38.
Graber, R. R.; Graber, J. W.; Kirk, E. L. 1971. Illinois birds: Turididae. *Ill. Nat. Hist. Surv. Biol. Notes* 75:1–44.
Graber, R. R.; Graber, J. W.; Kirk, E. L. 1972. Illinois birds: Hirundinidae. *Ill. Nat. Hist. Surv. Biol. Notes* 80:1–36.
Graber, R. R.; Graber, J. W.; Kirk, E. L. 1973. Illinois birds: Laniidae. *Ill. Nat. Hist. Surv. Biol. Notes* 83:1–18.
Graber, R. R.; Graber, J. W.; Kirk, E. L. 1974. Illinois birds: Tyrannidae. *Ill. Nat. Hist. Surv. Biol. Notes* 86:1–56.
Graf, R. L.; Greeley, F. 1976. The nesting site of the northern oriole in Amherst, Massachusetts. *Wilson Bull.* 88:359–360.
Grant, G. S.; Quay, T. L. 1977. Breeding biology of cliff swallows in Virginia. *Wilson Bull.* 89:286–290.
Greenberg, R.; Droege, S. 1990. Adaptations to tidal marshes in breeding populations of the swamp sparrow. *Condor* 92:393–404.
Greenlaw, J. S. 1978. The relation of breeding schedule and clutch size to food supply in the rufous-sided towhee. *Condor* 80:24–33.
Greenlaw, J. S. 1996. Eastern towhee (*Pipilo erythrophthalmus*). In: Poole, A.; Gill, F. (editors). *The birds of North America.* No. 262. Philadelphia: Academy of Natural Sciences; Washington, D.C.: American Ornithologists' Union.
Greij, E. D. 1994. Common moorhen. Pages 145–157. In: Tracha, T. C.; Braun, C. E. (editors). *Migratory shore and upland game bird management in North America.* Kingville, Texas: Caesar Kleberg Wildlife Research Institute.
Grice, D.; Rogers, J. P. 1965. The wood duck in Massachusetts. Final Rep., Federal Aid in Wildlife Restoration Project W19-R. Boston: Division of Fisheries and Game. 96 pp.
Griese, H. J.; Ryder, R. A.; Braun, C. E. 1980. Spatial and temporal distribution of rails in Colorado. *Wilson Bull.* 92:96–102.
Griscom, L. 1937. A monographic study of the red crossbill. *Proc. Boston Soc. Nat. Hist.* 41:77–210.
Griscom, L. 1949. *The birds of Concord.* Cambridge, Mass.: Harvard University Press. 340 pp.
Griscom, L.; Snyder, D. E. 1955. *The birds of Massachusetts—An annotated and revised check list.* Salem, Mass.: Peabody Museum. 295 pp.
Griscom, L.; Sprunt, A., Jr. 1957. *The warblers of America.* New York: Devin-Adair Co. 356 pp.
Gross, A. O. 1923. The black-crowned night heron (*Nycticorax nycticorax noevius*) of Sandy Neck. *Auk* 40:1–30, 191–214.
Gross, A. O. 1940. Eastern nighthawk. Pages 206–234. In: Bent, A. C. *Life histories of North American cuckoos, goat suckers, hummingbirds, and their allies.* U.S. Natl. Mus. Bull. 176. Washington, D.C.: U.S. National Museum.
Grubb, T. C., Jr. 1995. *Constructing bald eagle nests with natural materials.* Research Note. RM-RN-535. Ft. Collins, Colo.: U.S. Forest Service.
Grubb, T. C., Jr.; Pravosudov, V. V. 1994. Tufted titmouse (*Parus bicolor*). In: Poole, A.; Gill, F. (editors). *The birds of North America.* No. 86. Philadelphia: Academy of Natural Sciences; Washington, D.C.: American Ornithologists' Union.
Guinan, D. M.; Sealy, S. G. 1989. Foraging-substrate use by house wrens nesting in natural cavities in a riparian habitat. *Can. J. Zool.* 67:61–67.
Gullion, G. W. 1953. Territorial behavior of the American coot. *Condor* 55:169–186.
Gullion, G. W. 1972. Improving your forested lands for ruffed grouse. *Misc. J. Ser. Minn. Agric. Exp. Sta. Publ.* 1439:1–34.

Hagar, D. C., Jr. 1957. Nesting populations of red-tailed hawks and horned owls in central New York. *Wilson Bull.* 69:263–272.

Hagar, J. A. 1941. Little blue heron nesting in Massachusetts. *Auk* 58:568–569.

Hagar, J. A. 1969. History of the Massachusetts peregrine falcon population, 1935–1957. Pages 123–131. In: Hickey, J. J. (editor), *Peregrine falcon populations, their biology and decline*. Madison: University of Wisconsin Press. 596 pp.

Haggerty, T. M.; Morton, E. S. 1995. Carolina wren (*Thryothorus ludovicianus*). In: Poole, A.; Gill, F. (editors). *The birds of North America*. No. 188. Philadelphia: Academy of Natural Sciences; Washington, D.C.: American Ornithologists' Union.

Hall, G. A. 1983. *West Virginia birds*. Spec. Publ. No. 7. Pittsburgh: Carnegie Museum of Natural History.

Hall, G. A. 1994. Magnolia warbler (*Dendroica magnolia*). In: Poole, A.; Gill, F. (editors). *The birds of North America*. No. 136. Philadelphia: Academy of Natural Sciences; Washington, D.C.: American Ornithologists' Union.

Hamas, M. J. 1983. Nest-site selection by eastern kingbirds in a burned forest. *Wilson Bull.* 95:476–477.

Hamas, M. J. 1994. Belted kingfisher (*Ceryle alcyon*). In: Poole, A.; Gill, F. (editors). *The birds of North America*. No. 84. Philadelphia: Academy of Natural Sciences; Washington, D.C.: American Ornithologists' Union.

Hamel, P. B. 1992a. Cerulean warbler *(Dendroica cerulea)*. Pages 385–400. In: Schneider, D. J.; Pence, D. M. (editors). *Migratory nongame birds of management concern in the Northeast*. Newton Corner, Mass.: U.S. Fish and Wildlife Service, Region 5. 400 pp.

Hamel, P. B. 1992b. *The land manager's guide to the birds of the south*. Chapel Hill, N.C.: The Nature Conservancy.

Hamerstrom, F. N.; Hamerstrom, F.; Hart, J. 1973. Nest boxes: An effective management tool for kestrels. *J. Wildl. Manage.* 37:400–403.

Hammond, D. E.; Wood, R. L. 1977. New Hampshire and the disappearing loon. Meredith, N.H.: Loon Preservation Committee. 16 pp.

Hann, H. W. 1937. Life history of the ovenbird in southern Michigan. *Wilson Bull.* 49:145–237.

Hanners, L. A.; Patton, S. R. 1998. Worm-eating warbler (*Helmitheros vermivorus*). In: Poole, A.; Gill, F. (editors). *The birds of North America*. No. 367. Philadelphia: Academy of Natural Sciences; Washington, D.C.: American Ornithologists' Union.

Hardin, K. I.; Evans, D. E. 1977. Cavity nesting bird habitat in the oak-hickory forests—a review. Gen. Tech. Rep. NC-30. St. Paul, Minn.: U.S. Department of Agriculture, Forest Service, North Central Forest Experiment Station. 23 pp.

Harding, K. C. 1925. Semi-colonization of veeries. *Northeastern Bird-Banding Assoc.* 1(1):4–7.

Hardy, R. 1939. Nesting habits of the western red-tailed hawk. *Condor* 41:79–80.

Harlow, R. C. 1922. The breeding habits of the northern raven in Pennsylvania. *Auk* 39:399–410.

Harlow, R. F.; Hooper, R. G.; Chamberlain, D. R.; Crawford, H. S. 1975. Some winter and nesting season foods of the common raven in Virginia. *Auk* 92:298–306.

Harrison, H. 1975. *A field guide to birds' nests* (*in the United States east of the Mississippi River*). Boston: Houghton Mifflin. 350 pp.

Harrison, H. H. 1984. *Wood warblers' world*. Simon and Schuster: New York. 335 pp.

Hartshorne, J. M. 1962. Behavior of the eastern bluebird at the nest. *Living Bird* 1:131–149.

Hatch, J. 1982. The cormorants of Boston harbor and Massachusetts bay. *Bird Observer Eastern Mass.* 10:65–73.

Hayward, G. D.; Hayward, P. H. 1993. Boreal owl (*Aegolius funereus*). In: Poole, A.; Gill, F. (editors). *The birds of North America*. No. 63. Philadelphia: Academy of Natural Sciences; Washington, D.C.: American Ornithologists' Union.

Healy, W. M. 1992a. Population influences: environment. Pages 129–143. In: Dickson, J. G. (editor). *The wild turkey: Biology and management*. Harrisburg, Pa.: Stackpole. 463 pp.

Healy, W. M. 1992b. Behavior. Pages 46–65. In: Dickson, J. G. (editor). *The wild turkey: Biology and management*. Harrisburg, Pa.: Stackpole. 463 pp.

Hecht, W. R. 1951. Nesting of the marsh hawk at Delta, Manitoba. *Wilson Bull.* 63:167–176.

Heightmeyer, M. E.; Fredrickson, L. H.; Krause, G. 1989. Water and habitat dynamics of the Mingo Swamp in southeastern Missouri. *U.S. Dept. Interior, Fish Wildl. Serv., Fish Wildl. Res.* 6:1–26.

Henny, D. J.; Schmid, F. C.; Martin, E. M.; Hood, L. L. 1973. Territorial behavior, pesticides, and the population ecology of red-shouldered hawks in central Maryland 1943–1971. *Ecology* 54:545–554.

Hepp, G. R.; Bellrose, F. C. 1995. Wood duck (*Aix sponsa*). In: Poole, A.; Gill, F. (editors). *The birds of North America*. No. 169. Philadelphia: Academy of Natural Sciences; Washington, D.C.: American Ornithologists' Union.

Hepp, G. R.; Hair, J. D. 1977. Wood duck brood mobility and utilization of beaver pond habitats. *Proc. Annu. Conf. Southeast. Assoc. Fish Wildl. Agencies* 31:216–225.

Hespenheide, H. A. 1971. Flycatcher habitat selection in the eastern deciduous forest. *Auk* 88:61–74.

Heydweiller, A. M. 1935. A comparison of winter and summer territories and seasonal variations of the tree sparrow (*Spizella a. arborea*). *Bird-Banding* 6:1–11.

Hickey, J. J. 1942. Eastern population of the duck hawk. *Auk* 59:176–204.

Hickey, J. J. 1943. *A guide to bird watching*. New York: Oxford University Press. 262 pp.

Hickey, J. J. (editor). 1969. *Peregrine falcon populations: Their biology and decline*. Milwaukee: University of Wisconsin Press. 596 pp.

Hickey, J. J.; Anderson, D. W. 1969. The peregrine falcon: Life history and population literature. Pages 3–42. In: Hickey, J. J. (editor). *Peregrine falcon populations: Their biology and decline*. Milwaukee: University of Wisconsin Press.

Hilden, O. 1964. Ecology of duck populations in the island group of Valassaaret, Gulf of Bothnia. *Ann. Zool. Fenn.* 1:1–279.

Hill, G. E. 1993. House finch (*Carpodacus mexicanus*). In: Poole, A.; Gill, F. (editors). *The birds of North America*. No. 146. Philadelphia: Academy of Natural Sciences; Washington, D.C.: American Ornithologists' Union.

Hill, N. P. 1965. *The birds of Cape Cod, Massachusetts*. New York: William Morrow.

Hill, S. R.; Gates, J. E. 1988. Nesting ecology and microhabitat of the eastern phoebe in the central Appalachians. *Am. Midl. Nat.* 120:313–324.

Hills, C. F., III. 1978. The Birds of Fairfield County, Connecticut and Vicinity. Unpublished manuscript.

Hjertaas, D. G.; Hjertaas, P.; Maher, W. J. 1988. Colony size and reproductive biology of the bank swallow (*Riparia riparia*) in Saskatchewan, Canada. *Can. Field-Nat.* 102:465–470.

Hochbaum, H. A. 1944. *The canvasback on a prairie marsh.* Washington, D.C.: Wildlife Management Institute. 201 p.

Hofslund, P. B. 1957. Cowbird parasitism of the northern yellowthroat. *Auk* 74:42–48.

Hohman, W. L.; Pritchert, R. D.; Pace, R. M., III; Wodington, D. W.; Helm, R. 1990. Influence of ingested lead on body mass of wintering canvasbacks. *J. Wildl. Manage.* 54(2):211–215.

Holm, C. H. 1973. Breeding sex ratios, territoriality, and reproductive success in the red-winged blackbird (*Agelaius phoeniceus*). *Ecology* 54:356–365.

Holmes, R. T. 1994. Black-throated blue warbler. In: Poole, A.; Gill, F. (editors). *The birds of North America.* No. 87. Philadelphia: Academy of Natural Sciences; Washington, D.C.: American Ornithologists' Union.

Holmes, R. T.; Robinson, S. K. 1981. Tree species preferences of foraging, insectivorous birds in a northern hardwood forest. *Oecologia* 48:31–35.

Holmes, R. T.; Robinson, S. K. 1988. Spatial patterns, foraging tactics, and diets of ground-foraging birds in a northern hardwoods forest. *Wilson Bull.* 100:377–394.

Holmes, R. T.; Sherry, T. W. 1988. Assessing population trends of New Hampshire forest birds: Local vs. regional patterns. *Auk* 105:756–768.

Holmes, R. T.; Sherry, T. W. 1992. Site fidelity of migratory warblers in temperate breeding and Neotropical wintering areas: Implications for population dynamics, habitat selection, and conservation. Pages 563–575. In: Hagan, J. M.; Johnston, D. W. (editors). *Ecology and conservation of Neotropical migrant landbirds.* Washington, D.C.: Smithsonian Institution Press.

Holmes, R. T.; Sherry, T. W.; Marra, P. P.; Petit, K. E. 1992. Multiple brooding, nesting success, and annual productivity of a neotropical migrant, the black-throated blue warbler (*Dendroica caerulescens*), in an unfragmented temperate forest. *Auk* 109:321–333.

Holmes, R. T.; Sherry, T. W.; Sturges, F. W. 1986. Bird community dynamics in a temperate deciduous forest: Long-term trends at Hubbard Brook. *Ecol. Monogr.* 56:201–220.

Holt, D. W.; Andrews, E.; Claflin, N. 1991. Nonbreeding season diet of northern saw-whet owls, *Aegolius acadicus*, on Nantucket Island, Massachusetts. *Can. Field-Nat.* 105:382–385.

Holt, D. W.; Childs, N. N. 1991. Non-breeding season diet of long-eared owls in Massachusetts. *J. Raptor Res.* 25:23–24.

Holt, D. W.; Leasure, S. M. 1993. Short-eared owl (*Asio flammeus*). In: Poole, A.; Gill, F. (editors). *The birds of North America.* No. 62. Philadelphia: Academy of Natural Sciences; Washington, D.C.: American Ornithologists' Union.

Hooper, R. G. 1977. Nesting habitat of common ravens in Virginia. *Wilson Bull.* 89:233–242.

Hooper, R. G.; Crawford, H. S.; Chamberlain, D. R.; Harlow, R. F. 1975. Nesting density of common ravens in the ridge-valley region of Virginia. *Am. Birds* 29:931–935.

Hooper, R. G.; Dachelet, C. A. 1976. Flocks of non-breeding common ravens in Virginia. *Raven* 47(1):23–24.

Hopp, S. L.; Kirby, A.; Boone, C. A. 1995. White-eyed vireo (*Vireo griseus*). In: Poole, A.; Gill, F. (editors). *The birds of North America.* No. 168. Philadelphia: Academy of Natural Sciences; Washington, D.C.: American Ornithologists' Union.

Horak, G. J. 1970. A comparative study of the foods of the sora and Virginia rail. *Wilson Bull.* 89:373–379.

Houston, C. S.; Smith, D. G.; Rohner, C. 1998. Great horned owl (*Bubo virginianus*). In: Poole, A.; Gill, F. (editors). *The birds of North America.* No. 372. Philadelphia: Academy of Natural Sciences; Washington, D.C.: American Ornithologists' Union.

Howard, D. V. 1967. Variation in the breeding season and clutch size of the robin in the northeastern United States and maritime provinces of Canada. *Wilson Bull.* 79:432–440.

Howard, D. V. 1974. Urban robins: A population study. Pages 67–75. In: Noyes, J. H.; Progulske, D. R. (editors). *Wildlife in an urbanizing environment.* Plan. Res. Div. Ser. 28. Amherst: Holdsworth Natural Resources Center, University of Massachusetts.

Howe, R. H., Jr.; Allen, G. M. 1901. *The birds of Massachusetts.* Cambridge, Mass.: Nuttall Ornithological Club. 154 pp.

Howell, J. C. 1942. Notes on the nesting habits of the American robin. *Am. Midl. Nat.* 28:529–603.

Howell, T. R. 1952. Natural history and differentiation in the yellow-bellied sapsucker. *Condor* 54:237–282.

Howes-Jones, D. 1985. Nesting habits and activity patterns of warbling vireo (*Vireo gilvus*) in southern Ontario. *Can. Field-Nat.* 99:484–489.

Hoyt, S. F. 1957. The ecology of the pileated woodpecker. *Ecology* 38:246–256.

Hughes, J. M. 1999. Yellow-billed cuckoo (*Coccyzus americanus*). In: Poole, A.; Gill, F. (editors). *The birds of North America.* No. 418. Philadelphia: Academy of Natural Sciences; Washington, D.C.: American Ornithologists' Union.

Hunt, P. D. 1996. Habitat selection by American redstarts along a successional gradient in northern hardwoods forest: Evaluation of habitat quality. *Auk* 113(4):875–888.

Hunt, P. D.; Flaspohler, D. J. 1998. Yellow-rumped warbler (*Dendroica coronata*). In: Poole, A.; Gill, F. (editors). *The birds of North America.* No. 376. Philadelphia: Academy of Natural Sciences; Washington, D.C.: American Ornithologists' Union.

Hussell, D. J. T. 1983. Tree swallow pairs raise two broods per season. *Wilson Bull.* 95:470–471.

Ingold, J. L.; Galati, R. 1997. Golden-crowned kinglet (*Regulus satrapa*). In: Poole, A.; Gill, F. (editors). *The birds of North America.* No. 301. Philadelphia: Academy of Natural Sciences; Washington, D.C.: American Ornithologists' Union.

Ingold, J. L.; Wallace, G. E. 1994. Ruby-crowned kinglet (*Regulus calendula*). In: Poole, A.; Gill, F. (editors). *The birds of North America.* No. 119. Philadelphia: Academy of Natural Sciences; Washington, D.C.: American Ornithologists' Union.

Jackson, J. A. 1970. A quantitative study of the foraging ecology of downy woodpeckers. *Ecology* 51:318–323.

Jahn, L. R.; Hunt, R. A. 1964. Duck and coot ecology and management in Wisconsin. *Wisc. Conserv. Dept. Tech. Bull.* 33:1–212.

James, R. D. 1998. Blue-headed vireo (*Vireo solitarius*). In: Poole, A.; Gill, F. (editors). *The birds of North America.* No. 379. Philadelphia: Academy of Natural Sciences; Washington, D.C.: American Ornithologists' Union.

Janeway, E. C. 1994. Yellow-bellied flycatcher. Pages 148–149. In: Foss, C. R. (editor). *Atlas of breeding birds in New Hampshire.* Dover, N.H.: Arcadia. 414 pp.

Jarvis, W. L.; Southern, W. E. 1976. Food habits of ring-billed gulls breeding in the Great Lakes region. *Wilson Bull.* 88:62–75.

Jenni, D. A. 1969. A study of the ecology of four species of herons during the breeding season at Lake Alice, Alachua County, Florida. *Ecol. Monogr.* 39:245–270.

Johnsgard, P. A. 1973. *Grouse and quails of North America.* Lincoln: University of Nebraska Press. 553 pp.

Johnsgard, P. A. 1975. *Waterfowl of North America.* Bloomington: Indiana University Press. 575 pp.

Johnsgard, P. A. 1988. *North American owls: Biology and natural history.* Washington, D.C.: Smithsonian Institution Press.

Johnson, K. 1995. Green-winged teal (*Anas crecca*). In: Poole, A.; Gill, F. (editors). *The birds of North America.* No. 193. Philadelphia: Academy of Natural Sciences; Washington, D.C.: American Ornithologists' Union.

Johnson, L. S. 1998. House wren (*Troglodytes aedon*). In: Poole, A.; Gill, F. (editors). *The birds of North America.* No. 380. Philadelphia: Academy of Natural Sciences; Washington, D.C.: American Ornithologists' Union.

Johnson, R. G.; Temple, S. A. 1990. Nest predation and brood parasitism of tall grass prairie birds. *J. Wildl. Manage.* 54:106–111.

Johnson, R. R.; Dinsmore, J. J. 1985. Brood-rearing and post-breeding habitat use by Virginia rails and soras. *Wilson Bull.* 97:551–554.

Johnson, R. R.; Dinsmore, J. J. 1986. Habitat use by breeding Virgina rails and soras. *J. Wild. Manage.* 50:387–392.

Johnson, W. N., Jr.; Brown, P. W. 1990. Avian use of a lakeshore buffer strip and an undisturbed lakeshore in Maine. *No. J. Appl. Forestry* 7:114–117.

Johnston, D. W. 1970. High density of birds breeding in a modified deciduous forest. *Wilson Bull.* 82:79–82.

Johnston, D. W. 1971. Niche relationships among some deciduous forest flycatchers. *Auk* 88:796–804.

Johnston, D. W.; Odum, E. 1956. Breeding bird populations in relation to plant succession on the Piedmont of Georgia. *Ecology* 37:50–62.

Johnston, R. F. 1992. Rock dove. In: Poole, A. F.; Stettenheim, P.; Gill, F. B. (editors). *The birds of North America.* No. 13. Philadelphia: Academy of Natural Sciences; Washington, D.C.: American Ornithologists' Union.

Johnston, V. R. 1947. Breeding birds of the forest edge in Illinois. *Condor* 49:45–53.

Jones, P. W.; Donovan, T. M. 1996. Hermit thrush (*Catharus guttatus*). In: Poole, A.; Gill, F. (editors). *The birds of North America.* No. 261. Philadelphia: Academy of Natural Sciences; Washington, D.C.: American Ornithologists' Union.

Josselyn, J. [1672] 1833. An account of two voyages to New England. *Mass. Hist. Soc. Coll., 3rd series* 3. 273 pp.

Judd, S. D. 1901. *The relation of sparrows to agriculture.* Bur. Biol. Surv. Bull. 15. Washington, D.C.: U.S. Department of Agriculture.

Jurek, R. M.; Leach, H. R. 1977. Shorebirds. Pages 301–320. In: Sanderson, G. C. (editor). *Management of migratory shore and upland game birds in North America.* Washington, D.C.: International Association of Fish and Wildlife Agencies.

Kale, H. W., II. 1965. *Ecology and bioenergetics of the long-billed marsh wren in Georgia salt marshes.* Publ. No. 5. Cambridge, Mass.: Nuttall Ornithological Club. 142 pp.

Karr, J. R. 1968. Habitat and avian diversity on strip-mined land in east central illinois. *Condor* 70:348–357.

Kaufmann, G. W. 1987. Growth and development of Sora and Virginia rail chicks. *Wilson Bull.* 99:432–440.

Kaufmann, G. W. 1989. Breeding ecology of the sora, *Porzana carolina*, and the Virginia rail, *Rallus limicola*. *Can. Field-Nat.* 103:270–282.

Keeler, J. E. 1977. Mourning dove (*Zenaida macroura*). Pages 275–298. In: Sanderson, G. C. (editor). *Management of migratory shore and upland game birds in North America.* Washington, D.C.: International Association of Fish and Wildlife Agencies.

Keith, A. R. 1964. A thirty-year summary of the nesting of the barn owl on Martha's Vineyard, Massachusetts. *Bird-Banding* 35:22–31.

Keith, L.B. 1961. A study of waterfowl ecology on small impoundments in southeastern Alberta. *Wildl. Monogr.* 6:1–88.

Keller, M. E.; Anderson, S. H. 1992. Avian use of habitat configurations created by forest cutting in southeastern Wyoming. *Condor* 94:55–65.

Kendeigh, S. C. 1941a. Birds of a prairie community. *Condor* 43:165–174.

Kendeigh, S. C. 1941b. Territorial and mating behavior of the house wren. *Univ. Ill. Biol. Monogr.* 18(3):1–120.

Kendeigh, S. C. 1945a. Community selection of birds on the Helderberg Plateau of New York. *Auk* 62:418–436.

Kendeigh, S. C. 1945b. Nesting behavior of wood warblers. *Wilson Bull.* 57:145–164.

Kendeigh, S. C. 1946. Breeding birds of the beech-maple-hemlock community. *Ecology* 27:226–245.

Kendeigh, S. C. 1947. Bird population studies in the coniferous forest biome during a spruce budworm outbreak. *Ontario Dept. Lands Forests, Biol. Bull.* 1:1–100.

Kennamer, J. E.; Kennamer, M.; Brenneman, R. 1992. Chapter 2—History. Pages 6–17. In: Dickson, J. G. (editor). *The wild turkey: Biology and management.* Harrisburg, Pa.: Stackpole. 463 pp.

Kennamer, R. A.; Hepp, G. R. 1987. Frequency and timing of second broods in wood ducks. *Wilson Bull.* 99:655–662.

Kennard, F. H. 1920. Notes on the breeding habits of the rusty blackbird in northern New England. *Auk* 37:412–422.

Keppie, D. M.; Whiting, R. M., Jr. 1994. American woodcock (*Scolopax minor*). In: Poole, A.; Gill, F. (editors). *The birds of North America.* No. 100. Philadelphia: Academy of Natural Sciences; Washington, D.C.: American Ornithologists' Union.

Kessel, B. 1953. Second broods in the European starling in North America. *Auk* 70:479–483.

Kessel, B. 1957. A study of the breeding biology of the European starling (*Sturnus vulgaris* L.) in North America. *Am. Midl. Nat.* 58(2):257–331.

Kibbe, D. P. 1985a. American bittern. Page 35. In: Laughlin, S. B.; Kibbe, D. P. (editors). *The atlas of breeding birds of Vermont.* Hanover, N.H.: University Press of New England. 656 pp.

Kibbe, D. P. 1985b. American coot. Pages 102–103. In: Laughlin, S. B.; Kibbe, D. P. (editors). *The atlas of breeding birds of Vermont.* Hanover, N.H.: University Press of New England. 656 pp.

Kibbe, D. P. 1985c. Gray partridge. Pages 88–89. In: Laughlin, S. B.; Kibbe, D. P. (editors). *The atlas of breeding birds of Vermont.* Hanover, N.H.: University Press of New England. 656 pp.

Kibbe, D. P.; Laughlin, S. B. 1985. Double-crested cormorant. Pages 391–392. In: Laughlin, S. B.; Kibbe, D. P. (editors). *The*

atlas of breeding birds of Vermont. Hanover, N.H.: University Press of New England. 656 pp.
Kilham L. 1958a. Pair formation, mutual tapping, and nest hole selection of red-bellied woodpeckers. *Auk* 75:318–329.
Kilham, L. 1958b. Territorial behavior of wintering red-headed woodpeckers. *Wilson Bull.* 70:347–358.
Kilham, L. 1962. Breeding behavior of yellow-bellied sapsuckers. *Auk* 79:31–43.
Kilham, L. 1963. Food storing of red-bellied woodpeckers. *Wilson Bull.* 75:227–234.
Kilham, L. 1965. Differences in the feeding behavior of male and female hairy woodpeckers. *Wilson Bull.* 77:134–143.
Kilham, L. 1968a. Reproductive behavior of hairy woodpeckers. II. Nesting and habitat. *Wilson Bull.* 80:286–305.
Kilham, L. 1968b. Reproductive behavior of white-breasted nuthatches. I. Distraction display, bill-sweeping and nest hole defense. *Auk* 85:477–492.
Kilham, L. 1971. Reproductive behavior of yellow-bellied sapsuckers. Pt. 1. Preference for nesting in *Fomes*-infected aspens and nest hole interrelations with flying squirrels, raccoons, and other animals. *Wilson Bull.* 83:159–171.
Kilham, L. 1972. Death of red-breasted nuthatch from pitch around nest hole. *Auk* 89:451–452.
Kilham, L. 1973. Reproductive behavior of the red-breasted nuthatch: I. Courtship. *Auk* 90:597–609.
Kilham, L. 1974. Covering of stores by white-breasted and red-breasted nuthatches. *Condor* 76:108–109.
Kilham, L. 1983. *Life history studies of the woodpeckers of eastern North America*. Publ. 20. Cambridge, Mass.: Nuttall Ornithological Club.
King, D. I.; Griffin, C. R.; DeGraaf, R. M. 1996. Effects of clearcutting on habitat use and reproductive success of ovenbirds (*Seiurus aurocapillus*) in an extensively forested landscape. *Conserv. Biol.* 10:1380–1386.
King, J. R. 1955. Notes on the life history of Traill's flycatcher. *Auk* 72:148–173.
Kirk, D. A.; Mossman, M. J. 1998. Turkey vulture (*Cathartes aura*). In: Poole, A.; Gill, F. (editors). *The birds of North America*. No. 339. Philadelphia: Academy of Natural Sciences; Washington, D.C.: American Ornithologists' Union.
Kitchen, D. W.; Hunt, G. S. 1969. Brood habitat of the hooded merganser. *J. Wildl. Manage.* 33:605–609.
Kluyver, H. M. 1961. Food consumption in relation to habitat in breeding chickadees. *Auk* 78:532–550.
Knapton, R. W.; Falls, J. B. 1982. Polymorphism in the white-crowned sparrow: Habitat occupancy and nest-site selection. *Can. J. Zool.* 60:452–459.
Knupp, D. M.; Owen, R. B., Jr.; Diamond, J. B. 1977. Reproductive biology of American robins in northern Maine. *Auk* 94(1):80–85.
Korschgen, L. J. 1967. Feeding habits and foods. Pages 137–198. In: Hewitt, O. H. (editor). *The wild turkey and its management*. Washington, D.C.: Wildlife Society.
Kricher, J. C. 1981. Range expansion of the tufted titmouse (*Parus bicolor*), in Massachusetts. *Am. Birds* 35:750–753.
Kricher, J. C. 1995. Black-and-white warbler (*Mniotilta varia*). In: Poole, A.; Gill, F. (editors). *The birds of North America*. No. 158. Philadelphia: Academy of Natural Sciences; Washington, D.C.: American Ornithologists' Union.
Krohn, W. B.; Allen, R. B.; Moring, J. R.; Hutchinson, A. E. 1995. Double-crested cormorants in New England: Population and management histories. *Colonial Waterbirds* 18:99–109.
Kroodsma, D. E.; Verner, J. 1997. Marsh wren (*Cistothorus palustris*). In: Poole, A.; Gill, F. (editors). *The birds of North America*. No. 308. Philadelphia: Academy of Natural Sciences; Washington, D.C.: American Ornithologists' Union.
Kushlan, J. A. 1976. Feeding behavior of North American herons. *Auk* 93:86–94.
Lanyon, W. E. 1957. The comparative biology of the meadowlarks *Sturnella* in Wisconsin. *Publ. Nuttall Ornithol. Club* 1:1–67.
Lanyon, W. E. 1995. Eastern meadowlark (*Sturnella magna*). In: Poole, A.; Gill, F. (editors). *The birds of North America*. No. 160. Philadelphia: Academy of Natural Sciences; Washington, D.C.: American Ornithologists' Union.
Lanyon, W. E. 1997. Great crested flycatcher (*Myiarchus crinitis*). In: Poole, A.; Gill, F. (editors). *The birds of North America*. No. 300. Philadelphia: Academy of Natural Sciences; Washington, D.C.: American Ornithologists' Union.
Laskey, A.R. 1944. A study of the cardinal in Tennessee. *Wilson Bull.* 56:27–44.
Laskey, A. R. 1962. The breeding biology of mockingbirds. *Auk* 79:596–606.
Laubhan, M. K.; Reid, F. A. 1991. Characteristics of yellow-crowned night-heron nests in lowland hardwood forests of Missouri. *Wilson Bull.* 103(3):486–491.
Laubhan, M. K.; Rundle, W. D.; Swartz, B. I.; Reid, F. A. 1991. Diurnal activity patterns and foraging success of yellow-crowned night-herons in seasonally flooded wetlands. *Wilson Bull.* 103(2):272–277.
Laughlin, S. B. 1985a. Eastern screech-owl. Pages 132–133. In: Laughlin, S. B.; Kibbe, D. P. (editors). *The atlas of breeding birds of Vermont*. Hanover, N.H.: University Press of New England. 656 pp.
Laughlin, S. B. 1985b. Great horned owl. Pages 134–135. In: Laughlin, S. B.; Kibbe, D. P. (editors). *The atlas of breeding birds of Vermont*. Hanover, N.H.: University Press of New England. 656 pp.
Laughlin, S. B.; Kibbe, D. P. (editors). 1985. *The atlas of breeding birds of Vermont*. Hanover, N.H.: University Press of New England. 656 pp.
Lawrence, L. de K. 1947. Five days with a pair of nesting Canada jays. *Can. Field-Nat.* 61:1–12.
Lawrence, L. de K. 1948. Comparative study of the nesting behavior of chestnut-sided and Nashville warblers. *Auk* 65:204–219.
Lawrence, L. de K. 1949. The red crossbill at Pimisi Bay, Ontario. *Can. Field-Nat.* 63:147–160.
Lawrence, L. de K. 1953. Nesting life and behavior of the red-eyed vireo. *Can. Field-Nat.* 67(2):47–77.
Lawrence, L. de K. 1967. *A comparative life history study of four species of woodpeckers*. Ornithol. Monogr. No. 5. Lawrence, Kans.: American Ornithologists' Union. 156 pp.
Lea, R. B. 1942. A study of the nesting habits of the cedar waxwing. *Wilson Bull.* 54:225–237.
Lehner, P. N. 1965. Some observations on the ecology of the mourning dove in New York. *N.Y. Fish Game J.* 12:147–169.
Leonard, M. L.; Picman, J. 1987. Nesting mortality and habitat selection by marsh wrens. *Auk* 104:491–495.
Lepthien, L. W.; Bock, C. E. 1976. Winter abundance patterns of North American kinglets. *Wilson Bull.* 88:483–485.

Linehan, J. T. 1973. Nest records of cerulean warblers in Delaware. *Wilson Bull.* 85:482–483.
Linsley, Rev. J. H. 1843. A catalogue of the birds of Connecticut, arranged according to their natural families. *Am. J. Sci. Arts* 46:51.
Looman, S. J.; Shirley, D. L.; White, C. M. 1996. Productivity, food habits, and associated variables of barn owls utilizing nest boxes in central Utah. *Great Basin Nat.* 56(1):73–84.
Loranger, A. J. 1980. Fall foods of the Ring-necked Pheasant in Connecticut, 1963–1979. Master's thesis, University of Connecticut, Storrs.
Low, G.; Mansell, W. 1983. *North American marsh birds.* New York: Harper & Row. 192 pp.
Lowther, J. K.; Falls, L. B. 1968. White-throated sparrow. Pages 1364–1392. In: Bent, A. C. *Life histories of North American cardinals, grosbeaks, buntings, towhees, finches, sparrows and allies*, pt. 3. Austin, O. L., Jr. (editor). U.S. Natl. Mus. Bull. 237. Washington, D.C.: U.S. National Museum.
Lowther, P. E. 1993. Brown-headed cowbird (*Molothrus ater*). In: Poole, A.; Gill, F. (editors). *The birds of North America.* No. 47. Philadelphia: Academy of Natural Sciences; Washington, D.C.: American Ornithologists' Union.
Lowther, P. E.; Cink, C. L. 1992. House sparrow (*Passer domesticus*). In: Poole, A. F.; Stettenheim, P.; Gill, F. B. (editors). *The birds of North America.* No. 12. Philadelphia: Academy of Natural Sciences; Washington, D.C.: American Ornithologists' Union.
Lumsden, H. G.; Robinson, J.; Hartford, R. 1986. Choice of nest boxes by common goldeneyes in Ontario. *Wilson Bull.* 92:497–505.
Lunk, W. A. 1962. *The rough-winged swallow: A study of its breeding biology in Michigan.* Publ. No. 4. Cambridge, Mass.: Nuttall Ornithological Club. 155 pp.
Luoma, J. R. 1987. Black duck declines: An acid rain link. *Audubon* May:19–24.
Lyon, B.; Montgomerie, R. 1995. Snow bunting and McKay's bunting (*Plectrophenax nivalis* and *Plectrophenax hyperboreus*). In: Poole, A.; Gill, F. (editors). *The birds of North America.* No. 198–199. Philadelphia: Academy of Natural Sciences; Washington, D.C.: American Ornithologists' Union.
MacArthur, R. H. 1958. Population ecology of some warblers of northeastern coniferous forests. *Ecology* 39:599–619.
MacArthur, R. H. 1959. On the breeding distribution pattern of North American migrant birds. *Auk* 76:318–325.
Mackenzie, D. E.; Sealy, S. G. 1981. Nest site selection in eastern and western kingbirds: A multivariate approach. *Condor* 83:310–312.
MacLeod, I. C. 1994. Pine siskin. Pages 362–363. In: Foss, C. R. (editor). *Atlas of breeding birds in New Hampshire.* Dover, N.H.: Arcadia. 414 pp.
MacQueen, P. M. 1950. Territory and song in the least flycatcher. *Wilson Bull.* 62:194–205.
MacWhirter, R. B.; Bildstein, K. L. 1996. Northern harrier (*Circus cyaneus*). In: Poole, A.; Gill, F. (editors). *The birds of North America.* No. 210. Philadelphia: Academy of Natural Sciences; Washington, D.C.: American Ornithologists' Union.
Maddock, M. 1989. Color and first age of breeding in cattle egrets as determined from wing-tagged birds. *Corella* 13:1–8.
Mallory, M. L.; McNicol, D. K.; Weatherhead, P. J. 1994. Habitat quality and reproductive effort of common goldeneyes nesting near Sudbury, Canada. *J. Wildl. Manage.* 58:552–560.
Manci, K. M.; Rusch, D. H. 1988. Indices to distribution and abundance of some inconspicuous waterbirds at Horicon Marsh. *J. Field Ornithol.* 59:67–75.
Mannan, R. W.; Meslow, E. C. 1984. Bird populations and vegetation characteristics in managed and old-growth forests, northeastern Oregon. *J. Wildl. Manage.* 48:1219–1238.
Marble, R. M. 1926. Nesting of evening grosbeak at Woodstock, Vermont. *Auk* 43:549.
Markley, M. H. 1967. Limiting factors. Pages 199–243. In: Hewitt, O. H. (editor). *The wild turkey and its management.* Washington, D.C.: Wildlife Society.
Mark, D.; Stutchbury, B. S. 1994. Response of a forest-interior songbird to the threat of cowbird parasitism. *Anim. Behav.* 47:275–280.
Marks, J. S. 1985. Yearling male long-eared owls breed near natal nest. *J. Field Ornithol.* 56:181–182.
Marks, J. S. 1986. Nest site characteristics and reproductive success of long-eared owls in southwestern Idaho. *Wilson Bull.* 98:547–560.
Marks, J. S.; Evans, D. L.; Holt, D. W. 1994. Long-eared owl (*Asio otus*). In: Poole, A.; Gill, F. (editors). *The birds of North America.* No. 133. Philadelphia: Academy of Natural Sciences; Washington, D.C.: American Ornithologists' Union.
Marti, D. D. 1992. Common barn owl. In: Poole, A. F.; Stettenheim, P.; Gill, F. (editors). *The birds of North America.* No. 1. Philadelphia: Academy of Natural Sciences; Washington, D.C.: American Ornithologists' Union.
Martin, J. W. 1987. Behavior and habitat use of breeding northern harriers in southwestern Idaho. *J. Raptor Res.* 21:57–66.
Martin, N. D. 1960. An analysis of bird populations in relation to forest succession. Algonquin Provincial Park, Ontario. *Ecology* 41:126–140.
Martin, S. G.; Gavin, T. A. 1995. Bobolink (*Dolichonyx oryzivorus*). In: Poole, A.; Gill, F. (editors). *The birds of North America.* No. 176. Philadelphia: Academy of Natural Sciences; Washington, D.C.: American Ornithologists' Union.
Marzilli, V. 1989. Up on the roof. *Maine Fish Wildl.* 31:25–29.
Massachusetts Audubon Society. 1977. Bay state herons. *Mass. Audubon* 17(3):8–9.
Matray, P. F. 1974. Broad-winged hawk nesting and ecology. *Auk* 91:307–324.
Maxwell, G. R., II; Putnam, L .S. 1972. Incubation care of young and nest success of the common grackle (*Quiscalus quiscula*) in northern Ohio. *Auk* 89:349–359.
Maxwell, G. R.; Nocilly, J. N.; Shearer, R. I. 1976. Observations at a cavity nest of the common grackle and an analysis of grackle nest sites. *Wilson Bull.* 88:505–507.
McCarty, J. P. 1996. Eastern wood-pewee (*Contopus virens*). In: Poole, A.; Gill, F. (editors). *The birds of North America.* No. 245. Philadelphia: Academy of Natural Sciences; Washington, D.C.: American Ornithologists' Union.
McComb, W. C. 1985. Habitat associations of birds and mammals in an Appalachian forest. *Proc. Annu. Conf. Southeast. Assoc. Fish Wildl. Agencies* 39:420–429.
McComb, W. C.; Groetsch, P. L.; Jacoby, G. E.; McPeek, G. A. 1989. Response of forest birds to an improvement cut in Kentucky. *Proc. Annu. Conf. Southeast. Assoc. Fish Wildl. Agencies* 43:313–325.
McGilvrey, F. B. 1968. A guide to wood duck production habitat requirements. *U.S. Bureau Sport Fisheries Wildl. Res. Publ.* 60:1–32.

McIntyre, J. W. 1988. *The common loon, spirit of northern lakes.* Minneapolis: University of Minnesota Press. 228 pp.
McIntyre, J. W.; Barr, J. F. 1997. Common loon (*Gavia immer*). In: Poole, A.; Gill, F. (editors). *The birds of North America.* No. 313. Philadelphia: Academy of Natural Sciences; Washington, D.C.: American Ornithologists' Union.
McLaren, M. A. 1975. Breeding biology of the boreal chickadee. *Wilson Bull.* 87:344–354.
McNeil, R.; Leger, C. 1987. Nest-site quality and reproductive success of early and late-nesting double-crested cormorants. *Wilson Bull.* 99:262–267.
Meanley, B. 1969. *Natural history of the king rail.* North Am. Fauna No. 67. Washington, D.C.: Bureau of Sport Fisheries and Wildlife. 108 pp.
Meanley, B. 1992. King rail. In: Poole, A. F.; Stettenheim, P.; Gill, F. (editors). *The birds of North America.* No. 3. Philadelphia: Academy of Natural Sciences; Washington, D.C.: American Ornithologists' Union.
Melvin, S. M.; Gibbs, J. P. 1996. Sora (*Porzana carolina*). In: Poole, A.; Gill, F. (editors). *The birds of North America.* No. 250. Philadelphia: Academy of Natural Sciences; Washington, D.C.: American Ornithologists' Union.
Mendall, H. L. 1937. Nesting of the bay-breasted warbler. *Auk* 54:429–439.
Mendall, H. L. 1958. *The ring-necked duck in the Northeast.* Orono: University of Maine Press. 317 pp.
Mendall, H. L.; Aldous, C. M. 1943. *The ecology and management of the American woodcock.* Orono: Maine Cooperative Wildlife Research Unit. 201 pp.
Meng, H. K.; Rosenfield, R. N. 1988. Cooper's hawk: Reproduction. Pages 331–349. In: Palmer, R. S. (editor). *Handbook of North American birds, vol. 4, part 1: Diurnal raptors.* New Haven, Conn.: Yale University Press.
Menkens, G. E., Jr.; Anderson, S. H. 1987. Nest site characteristics of a predominantly tree-nesting population of golden eagles. *J. Field Ornithol.* 58(1):22–25.
Merriam, C. H. 1877. A review of the birds of Connecticut. *Trans. Conn. Acad.* 4. New Haven: Connecticut Academy.
Messerly, E. H. 1979. Rusty blackbird feeds on American goldfinch. *Bull. Okla. Ornithol. Soc.* 12(1):6–7.
Metcalf, L. N. 1977. Fortieth breeding bird census; Coniferous forest. *Am. Birds* 31(1):53–54.
Metz, K. J. 1991. The enigma of multiple nest building by male marsh wrens. *Auk* 108:170–173.
Meyerriecks, A. J. 1962. Green heron. Pages 419–426. In: Palmer, R. S. (editor). *Handbook of North American birds.* Vol. 1. New Haven, Conn.: Yale University Press.
Michener, H.; Michener, J. R. 1935. Mockingbirds, their territories and individualities. *Condor* 37:97–140.
Middleton, A. L. A. 1993. American goldfinch (*Carduelis tristis*). In: Poole, A.; Gill, F. (editors). *The birds of North America.* No. 80. Philadelphia: Academy of Natural Sciences; Washington, D.C.: American Ornithologists' Union.
Middleton, A. L. A. 1998. Chipping sparrow (*Spizella passerina*). In: Poole, A.; Gill, F. (editors). *The birds of North America.* No. 334. Philadelphia: Academy of Natural Sciences; Washington, D.C.: American Ornithologists' Union.
Miller, A. B. 1934. Cuckoos and caterpillars. *Bird-Lore* 36:301.
Miller, A. H. 1931. Systematic revision and natural history of the American shrikes (*Lanius*). *Univ. Calif. Publ. Zool.* 38(2):11–242.
Miller, J. R.; Miller, J. T. 1948. Nesting of the spotted sandpiper at Detroit, Michigan. *Auk* 65:558–567.
Miller, R. S.; Nero, R. W. 1983. Hummingbird-sapsucker association in northern climates. *Can. J. Zool.* 61:1540–1546.
Miller-Weeks, M.; Smoronk, D. 1993. *Aerial assessment of red spruce and balsam fir condition in the Adirondack region of N.Y., the Green Mountains of Vt., the White Mountains of N.H., and the mountains of western Maine, 1985–1986.* Durham: U.S. Department of Agriculture, Forest Service. 141 pp.
Minor, W. F.; Minor, M.; Ingraldi, M. F. 1993. Nesting of red-tailed hawks and great horned owls in a central New York urban/suburban area. *J. Field Ornithol.* 64(4):433–439.
Minot, H. D. 1895. *The land-birds and game-birds of New England.* Cambridge, Mass.: Riverside Press.
Mirarchi, R. E.; Baskett, T. S. 1994. Mourning dove (*Zenaida macroura*). In: Poole, A.; Gill, F. (editors). *The birds of North America.* No. 117. Philadelphia: Academy of Natural Sciences; Washington, D.C.: American Ornithologists' Union.
Mobray, T. B. 1997. Swamp sparrow (*Melospiza georgiana*). In: Poole, A.; Gill, F. (editors). *The birds of North America.* No. 279. Philadelphia: Academy of Natural Sciences; Washington, D.C.: American Ornithologists' Union.
Moldenhauer, R. R.; Regelski, D. J. 1996. Northern parula (*Parula americana*). In: Poole, A.; Gill, F. (editors). *The birds of North America.* No. 215. Philadelphia: Academy of Natural Sciences; Washington, D.C.: American Ornithologists' Union.
Moore, W. S. 1995. Northern flicker (*Colaptes auratus*). In: Poole, A.; Gill, F. (editors). *The birds of North America.* No. 166. Philadelphia: Academy of Natural Sciences; Washington, D.C.: American Ornithologists' Union.
Morgan, K.; Freedman, B. 1986. Breeding bird communities in a northern hardwood forest succession in Nova Scotia. *Can. Field-Nat.* 100:506–519.
Morimoto, D. C.; Wasserman, F. E. 1991a. Dispersion patterns and habitat associations of rufous-sided towhees, common yellowthroats, and prairie warblers in the southeastern Massachusetts pine barrens. *Auk* 108:264–276.
Morimoto, D. C.; Wasserman, F. E. 1991b. Intersexual and interspecific differences in the foraging behavior of rufous-sided towhees, common yellowthroats, and prairie warblers in the pine barrens of southeastern Massachusetts. *J. Field Ornithol.* 62:436–449.
Morse, D. H. 1967. Competitive relationships between parula warblers and other species during breeding season. *Auk* 84(4):490–502.
Morse, D. H. 1971. Effects of the arrival of a new species upon habitat utilization by two forest thrushes in Maine. *Wilson Bull.* 83:57–65.
Morse, D. H. 1972. Habitat differences of Swainson's and hermit thrushes. *Wilson Bull.* 84:206–208.
Morse, D. H. 1976. Variables affecting the density and territory size of breeding spruce-woods warblers. *Ecology* 57:290–301.
Morse, D. H. 1977. The occupation of small islands by passerine birds. *Condor* 79:399–412.
Morse, D. H. 1978. Populations of bay-breasted and Cape May warblers during an outbreak of the spruce budworm. *Wilson Bull.* 90(3)404–413.
Morse, D. H. 1980. Foraging and coexistence of spruce-woods warblers. *Living Bird* 18:7–25.

Morse, D. H. 1989. *American warblers: An ecological and behavioral perspective.* Cambridge, Mass.: Harvard University Press. 406 pp.
Morse, D. H. 1993. Black-throated green warbler (*Dendroica virens*). In: Poole, A.; Gill, F. (editors). *The birds of North America.* No. 55. Philadelphia: Academy of Natural Sciences; Washington, D.C.: American Ornithologists' Union.
Morse, D. H. 1994. Blackburnian warbler (*Dendroica fusca*). In: Poole, A.; Gill, F. (editors). *The birds of North America.* No. 102. Philadelphia: Academy of Natural Sciences; Washington, D.C.: American Ornithologists' Union.
Morse, T. E.; Jakabosky, J. L.; McCrow, V. P. 1969. Some aspects of the breeding biology of the hooded merganser. *J. Wildl. Manage.* 33:596–604.
Moskoff, W. 1994. Veery (*Catharus fuscescens*). In: Poole, A.; Gill, F. (editors). *The birds of North America.* No. 142. Philadelphia: Academy of Natural Sciences; Washington, D.C.: American Ornithologists' Union.
Moskoff, W.; Robinson, S. K. 1996. Philadelphia vireo (*Vireo philadelphicus*). In: Poole, A.; Gill, F. (editors). *The birds of North America.* No. 214. Philadelphia: Academy of Natural Sciences; Washington, D.C.: American Ornithologists' Union.
Muller, M. J.; Storer, R. W. 1999. Pied-billed grebe (*Podilymbus podiceps*). In: Poole, A.; Gill, F. (editors). *The birds of North America.* No. 410. Philadelphia: Academy of Natural Sciences; Washington, D.C.: American Ornithologists' Union.
Mumford, R. E. 1964. The breeding biology of the Acadian flycatcher. *Univ. Mich. Mus. Zool. Misc. Publ.* 125:1–50.
Murphy, M. T. 1983a. Ecological aspects of the reproductive biology of eastern kingbirds: Geographic comparison. *Ecology* 64:914–928.
Murphy, M. T. 1983b. Nest success and nesting habitats of eastern kingbirds and other flycatchers. *Condor* 85:208–219.
Murphy, M. T. 1996. Eastern kingbird (*Tyrannus tyrannus*). In: Poole, A.; Gill, F. (editors). *The birds of North America.* No. 253. Philadelphia: Academy of Natural Sciences; Washington, D.C.: American Ornithologists' Union.
Murphy, R. D.; Gratson, M. W.; Rosenfield, R. N. 1988. Activity and habitat use by a breeding male Cooper's hawk in a suburban area. *J. Raptor Res.* 22:97–100.
Murray, B. G., Jr.; Gill, F. B. 1976. Behavioral interactions of blue-winged and golden-winged warblers. *Wilson Bull.* 88:231–254.
Murray, N.; Stauffer, D. 1995. Nongame bird use of habitat in central Appalachian riparian forests. *J. Wildl. Manage.* 59:78–88.
Nagy, A. C. 1963. Population density of sparrow hawks in eastern Pennsylvania. *Wilson Bull.* 75:93.
National Wild Turkey Federation. 1986. *Guide to the American wild turkey.* National Wild Turkey Federation, Edgefield, S.C. 189 pp.
Naugler, C. T. 1993. American tree sparrow (*Spizella arborea*). In: Poole, A. F.; Stettenheim, P.; Gill, F. B. (editors). *The birds of North America.* No. 37. Philadelphia: Academy of Natural Sciences; Washington, D.C.: American Ornithologists' Union.
Nelson, M. W. 1972. The incubation period in Peales falcons. *Raptor Res.* 6(1):11–15.
Nettleship, D. N.; Duffy, D. C. 1995. The double-crested cormorant: Biology, conservation and management. *Colonial Waterbirds Spec. Publ.* 1, Vol. 18. 256 pp.
Newman D. L. 1958. A nesting of the Acadian flycatcher. *Wilson Bull.* 70:130–144.
Nice, M. M. 1932. Observations on the nesting of the blue-gray gnatcatcher. *Condor* 34:18–22.
Nice, M. M. 1937. Studies in the life history of the song sparrow. 1. A population study of the song sparrow. *Tran. Linnaean Soc. N.Y.* 4:1–247.
Nice, M. M. 1954. Problems of incubation periods in North American birds. *Condor* 56(4):173–197.
Nicholls, T. H.; Warner, D. W. 1972. Barred owl habitat use as determined by radiotelemetry. *J. Wildl. Manage.* 36:213–224.
Nichols, W. 1985. Mourning warbler. Pages 320–321. In: Laughlin, S. B.; Kibbe, D. P. (editors). *The atlas of breeding birds of Vermont.* Hanover, N.H.: University Press of New England. 656 pp.
Nicholson, J. 1975. Thirty-ninth breeding bird census; mixed forest, old field and homesite. *Am. Birds* 29:1107–1108.
Nicholson, J. 1978. Forty-first breeding bird census; mixed forest, old field and homesite. *Am. Birds* 32:72–73.
Nickell, W. P. 1943. Observations on the nesting of the killdeer. *Wilson Bull.* 55:23–28.
Nickell, W. P. 1944. Studies of habitats, locations and structural materials of nests of the robin. *Jack-Pine Warbler* 22(2):48–64.
Nickell, W. P. 1951. Studies of habitats, territory, and nests of the eastern goldfinch. *Auk* 68:447–470.
Nickell, W. P. 1965. Habitats, territory and nesting of the catbird. *Am. Midl. Nat.* 73:433–478.
Niemi, G. J.; Hanowski, J. M. 1984. Relationships of breeding birds to habitat characteristics in logged areas. *J. Wildl. Manage.* 48:438–443.
Nisbet, I. C. T. 1994. Evidence for double-brooding in common terns. *Colon. Waterbirds* 17(1):95–96.
Nolan, V., Jr. 1978. The ecology and behavior of the prairie warbler, *Dendroica discolor. Ornithol. Monogr.* 26:1–595.
Noon, B. R. 1981. The distribution of an avian guild along a temperate elevational gradient: The importance and expression of competition. *Ecol. Monogr.* 51:105–124.
Noon, J. 1990. *The Squam Lakes and their loons.* Meredith, N.H.; Loon Preservation Committee.
Norse, W. J.; Fichtel, C. 1985. Red-bellied sapsucker. Pages 156–157. In: Laughlin, S. B.; Kibbe, D. P. (editors). *The atlas of breeding birds of Vermont.* Hanover, N.H.: University Press of New England. 656 pp.
Norse, W. J.; Laughlin, S. B. 1985. Ruby-throated hummingbird. Pages 150–157. In: Laughlin, S. B.; Kibbe, D. P. (editors). *The atlas of breeding birds of Vermont.* Hanover, N.H.: University Press of New England. 656 pp.
North, C. A. 1972. Population dynamics of the house sparrow, *Passer domesticus* (L.), in Wisconsin, USA. Pages 195–210. In: Kendeigh, S. C.; Pinowski, J. (editors). *Productivity, population dynamics, and systematics of granivorous birds.* Proc. of general meeting of the working group on granivorous birds, IBP, PT section; Sept. 6–8, 1970; The Hague, Holland.
Novak, P. G. 1992. Black tern (*Chlidonias niger*). Pages 149–167. In: Schneider, D. J.; Pence, D. M. (editors). *Migratory nongame birds of management concern in the Northeast.* Newton Corner, Mass.: U.S. Fish and Wildlife Service, Region 5.
Noyes, J. H.; Jarvis, R. L. 1985. Diet and nutrition of breeding female redhead and canvasback ducks in Nevada. *J. Wildl. Manage.* 49:203–211.
Nummi, P. 1993. Food-niche relationships of sympatric mallards and green-winged teals. *Can. J. Zool.* 71:49–55.

Oatman, G. F. 1985. Tennessee warbler. Pages 278–279. In: Laughlin, S. B.; Kibbe, D. P. (editors). *The atlas of breeding birds of Vermont*. Hanover, N.H.: University Press of New England. 656 pp.
Odum, E. P. 1941. Annual cycle of the black-capped chickadee. *Auk* 58:314–333, 518–535.
Odum, E. P. 1942. Annual cycle of the black-capped chickadee. *Auk* 59:499–531.
Odum, E. P.; Johnston, D. E. 1951. The house wren breeding in Georgia. *Auk* 68:357–366.
Odum, E. P.; Kuenzler, E. J. 1955. Measurement of territory and home range size in birds. *Auk*. 72:128–137.
Odum, R. R. 1977. Sora (*Porzana carolina*). Pages 57–65. In: Sanderson, G. C. (editor). *Management of migratory shore and upland game birds in North America*. Washington, D.C.: International Association of Fish and Wildlife Agencies.
Ogden, J. C. 1975. Effects of bald eagle territoriality on nesting ospreys. *Wilson Bull.* 87:496–505.
Oliphant, L. W. 1985. North American merlin breeding survey. *Raptor Res.* 19:37–41.
Olson, S.; Marshall, W. 1952. The common loon in Minnesota. Occas. Pap. No. 5. St. Paul: University of Minnesota. 77 pp.
Orians, G. H. 1961. The ecology of blackbird (*Agelaius*) social systems. *Ecol. Monogr.* 31:285–312.
Orians, G. H.; Beletsky, L. D. 1989. Red-winged blackbird. Pages 193–197. In: Newton, I. (editor). *Life reproduction in birds*. New York: Academic Press.
Oring, L. W.; Gray, E. M.; Reed, J. M. 1997. Spotted sandpiper (*Actitis macularia*). In: Poole, A.; Gill, F. (editors). *The birds of North America*. No. 289. Philadelphia: Academy of Natural Sciences; Washington, D.C.: American Ornithologists' Union.
Ouellet, H. 1970. Further observations on the food and predatory habits of the gray jay. *Can. J. Zool.* 48:327–330.
Ouellet, H. 1993. Bicknell's thrush: Taxonomic status and distribution. *Wilson Bull.* 105:545–572.
Owen, R. B., Jr. 1977. American woodcock (*Philohela minor, Scolopax minor* of Edwards 1974). Pages 149–186. In: Sanderson, G. C. (editor). *Management of migratory shore and upland game birds in North America*. Washington, D.C.: International Association of Fish and Wildlife Agencies.
Palmer, R. S. (ed.). 1962. *Handbook of North American birds*. Vol. 1. New Haven, Conn.: Yale University Press. 567 pp.
Palmer, R. S. (editor). 1976. *Handbook of North American birds*. Vols. 2 and 3. *Waterfowl*. New Haven, Conn.: Yale University Press. 600 pp.
Parmelee, D. F. 1954. Notes on the breeding of certain ducks and mergansers in Dickenson County, Michigan. *Jack-Pine Warbler* 32:110–118.
Parmelee, D. F. 1968. *Plectrophenax nivalis nivalis* (Linneaus): Snow bunting. In: Austin, O. L., Jr. (editor). Life histories of North American cardinals, grosbeaks, buntings, towhees, finches, sparrows, and allies. *U.S. Natl. Mus. Bull.* 237:1652–1675.
Parmelee, D. F. 1992. Snowy owl (*Nyctea scandiaca*). In: Poole, A. F.; Stettenheim, P.; Gill, F. (editor). *The birds of North America*. No. 10. Philadelphia: Academy of Natural Sciences; Washington, D.C.: American Ornithologists' Union.
Payne, R. B. 1965. Clutch size and numbers of eggs laid by brown-headed cowbirds. *Condor* 67:44–60.
Payne, R. B. 1992. Indigo bunting. In: Poole, A. F.; Stettenheim, P.; Gill, F. (editors). *The birds of North America*. No. 4. Philadelphia: Academy of Natural Sciences; Washington, D.C.: American Ornithologists' Union.
Paynter, R. A., Jr. 1954. Interrelation between clutch-size, brood-size, prefledging survival, and weight in Kent Island tree swallows. *Bird-Banding* 25(2):35–58.
Peakall, D. B. 1970. The eastern bluebird: Its breeding season, clutch size, and nesting success. *Living Bird* 9:239–255.
Pearson, S. M. 1991. Food patches and the spacing of individual foragers. *Auk* 108:355–362.
Peck, G. K.; James, R. D. 1987. *Breeding birds of Ontario: Nidology and distribution, Vol. 2. Passerines*. Toronto: Life Science Misc. Publ., Royal Ontario Museum.
Peer, B. D.; Bollinger, E. K. 1997. Common grackle (*Quiscalus quiscula*). In: Poole, A.; Gill, F. (editors). *The birds of North America*. No. 271. Philadelphia: Academy of Natural Sciences; Washington, D.C.: American Ornithologists' Union.
Pendergast, B. A.; Boag, D. A. 1970. Seasonal changes in diet of spruce grouse in central Alberta. *J. Wildl. Manage.* 34:605–611.
Perkins, G. H.; Howe, C. D., Jr. 1901. *A preliminary list of the birds found in Vermont. Twenty-First Vermont Agricultural Report*. Montpelier: Vermont Department of Agriculture.
Peterjon, B. G.; Rice, D. L. 1991. *The Ohio breeding bird atlas*. Columbus: Ohio Department of Natural Resources.
Petersen, A .J. 1955. The breeding cycle in the bank swallow. *Wilson Bull.* 67:235–286.
Peterson, J. M.; Crocoll, S. T. 1992. Red-shouldered hawk (*Buteo lineatus*). In: Schneider, D. J.; Pence, D. M. (editors). *Migratory nongame birds of management concern in the Northeast*. Newton Corner, Mass.: U.S. Fish and Wildlife Service, Region 5. 400 pp.
Peterson, J. M. C.; Fichtel, C. 1992. Olive-sided flycatcher (*Contopus borealis*). Pages 149–170. In: Schneider, D. J.; Pence, D. M. (editors). *Migratory nongame birds of management concern in the Northeast*. Newton Corner, Mass.: U.S. Fish and Wildlife Service, Region 5.
Pickwell, G. B. 1931. The prairie horned lark. *Trans. Acad. Sci. St. Louis* 27:1–153.
Pierotti, R.; Annett, C. A. 1987. Reproductive consequences of dietary specialization and switching in an ecological generalist. Pages 417–441. In: Kamil, A. C.; Krebs, J.; Pulliam, H. R. (editors). *Foraging behavior*. New York: Plenum.
Pierotti, R.; Annett, C. A. 1991. Diet choice in the herring gull: Constraints imposed by reproductive and ecological factors. *Ecology* 72(1):319–328.
Pierotti, R. J.; Good, T. P. 1994. Herring gull (*Larus argentatus*). In: Poole, A.; Gill, F. (editors). *The birds of North America*. No. 124. Philadelphia: Academy of Natural Sciences; Washington, D.C.: American Ornithologists' Union.
Pinkowski, B. C. 1977. Foraging behavior of the eastern bluebird. *Wilson Bull.* 89:414.
Piper, W. H.; Wiley, R. H. 1990. Correlates of range size in wintering white-throated sparrows (*Zonotrichia albicollis*). *Anim. Behav.* 40:545–552.
Pistorius, A. 1985. Long-eared owl. Pages 138–139. In: Laughlin, S. B.; Kibbe, D. P. (editors). *The atlas of breeding birds of Vermont*. Hanover, N.H.: University Press of New England. 656 pp.
Pitelka, F. A. 1942. High population of breeding birds within an artificial habitat. *Condor* 44:172–174.
Pitocchelli, J. 1993. Mourning warbler (*Oporornis philadelphia*). In: Poole, A.; Gill, F. (editors). *The birds of North America*.

No. 72. Philadelphia: Academy of Natural Sciences; Washington, D.C.: American Ornithologists' Union.

Platt, J. B. 1976. Sharp-shinned hawk nesting and nest site selection in Utah. *Condor* 78:102.

Ponton, D. A. 1983. Nest site selection by peregrine falcons. *Raptor Res.* 17(1):27–28.

Poole, A. F. 1989. *Ospreys: A natural and unnatural history.* Cambridge, U.K.: Cambridge University Press.

Porneluzi, P.; Bednarz, J. C.; Goodrich, L.; Zawada, N.; Hoover, J. 1993. Reproductive performance of territorial ovenbirds occupying forest fragments and a contiguous forest in Pennsylvania. *Conserv. Biol.* 7:618–622.

Portnoy, J. W.; Dodge, W. E. 1979. Red-shouldered hawk nesting ecology and behavior. *Wilson Bull.* 91:104–117.

Pospichal, L. B.; Marshall, W. H. 1954. A field study of sora rail and Virginia rail in central Minnesota. *Flicker* 26:2–32.

Post, W.; Enders, F. 1970. Notes on a salt marsh Virginia rail population. *Kingbird* 20:61–67.

Poston, H. J. 1969. Relationships between the shoveler and its breeding habitat at Strathmore, Alberta. *Saskatoon Wetlands Seminar, Can. Wildl. Serv. Rep. Ser.* 6:132–137.

Poston, H. J. 1974. Home range and breeding biology of the shoveler. *Can. Wildl. Serv. Rep. Ser.* 25. Ottawa.

Potter, E. F. 1980. Notes on nesting yellow-billed cuckoos (*Coccyzus americanus*). *J. Field Ornithol.* 51:17–29.

Pough, R. H. 1949. *Audubon bird guide: Eastern land birds.* New York: Doubleday. 312 pp.

Pough, R. H. 1951. *Audubon water bird guide.* New York: Doubleday. 352 pp.

Poulin, R. G.; Grindal, S. D.; Brigham, R. M. 1996. Common nighthawk (*Chordeiles minor*). In: Poole, A.; Gill, F. (editors). *The birds of North America.* No. 213. Philadelphia: Academy of Natural Sciences; Washington, D.C.: American Ornithologists' Union.

Pratt, H. M. 1970. Breeding biology of great blue herons and common egrets in central California. *Condor* 72:407–416.

Pratt, H. M.; Winkler, D. W. 1985. Clutch size, timing of laying, and reproductive success in a colony of great blue herons and great egrets. *Auk* 102:49–63.

Pravosudov, V. V.; Grubb, T. C., Jr. 1993. White-breasted nuthatch (*Sitta carolinensis*). In: Poole, A.; Gill, F. (editors). *The birds of North America.* No. 54. Philadelphia: Academy of Natural Sciences; Washington, D.C.: American Ornithologists' Union.

Preble, N. A. 1957. The nesting habits of the yellow-billed cuckoo. *Am. Midl. Nat.* 57:474–482.

Prescott, K. W. 1965. The scarlet tanager (*Pirango olivacea*). *N.J. State Mus. Invest.* 2:1–159.

Preston, C. R.; Beane, R. D. 1993. Red-tailed hawk (*Buteo jamaicensis*). In: Poole, A.; Gill, F. (editors). *The birds of North America.* No. 52. Philadelphia: Academy of Natural Sciences; Washington, D.C.: American Ornithologists' Union.

Prince, H. H. 1968. Nest sites used by wood ducks and common goldeneyes in New Brunswick. *J. Wildl. Manage.* 32:489–500.

Prior, K. A. 1990. Turkey vulture food habits in southern Ontario. *Wilson Bull.* 102:706–710.

Pulliam, H. R.; Enders, F. 1971. The feeding ecology of five sympatric finch species. *Ecology* 52:557–566.

Putnam, L. S. 1949. The life history of the cedar waxwing. *Wilson Bull.* 61:141–182.

Rappole, J. H.; Morton, E. S.; Lovejoy, T. E.; Ruos, J. L. 1983. *Neartic avian migrants in the Neotropics.* Washington, D.C.: U.S. Fish and Wildlife Service.

Rappole, J. H.; Morton, E. S.; Ramos, M. A. 1992. Density, philopatry, and population estimates for songbird migrants wintering in Veracruz. Pages 337–344. In: Hagan, J. M.; Johnston, D. W. (editors). *Ecology and conservation of Neotropical migrant landbirds.* Washington, D.C.: Smithsonian Institution Press. 609 pp.

Rappole, J. H.; Warner, D. W. 1980. Ecological aspects of migrant bird behavior in Veracruz, Mexico. Pages 353–393. In: Keast, A; Morton, E. S. (editors). *Migrant birds in the Neotropics: Ecology, behavior, distribution, and conservation.* Washington, D.C.: Smithsonian Institution Press.

Ratcliffe, D. 1980. *The peregrine falcon.* Vermillion, S.D.: Buteo Books. 416 pp.

Raynor, G. S. 1941. The nesting habits of the whip-poor-will. *Bird-Banding* 12:98–104.

Reinert, S.; Golet, F. 1979. Breeding ecology of the swamp sparrow in a southern Rhode Island peatland. *Trans. Northeast Sect. Wildl. Soc.* 36:1–13.

Reller, A. W. 1972. Aspects of behavior ecology of red-headed and red-bellied woodpeckers. *Am. Midl. Nat.* 88:270–290.

Reynolds, R. T.; Meslow, E. C.; Wight, H. M. 1982. Nesting habitat of coexisting *Accipiter* in Oregon. *J. Wildl. Manage.* 46:124–138.

Rice, J. C. 1978. Behavioral interactions of two interspecifically territorial vireos. I.: Song discrimination and natural interactions. *Anim. Behav.* 26:527–549.

Richards, T. 1994. Hooded merganser. Pages 64–65. In: Foss, C. R. (editor). *Atlas of breeding birds in New Hampshire.* Dover, N.H.: Arcadia. 414 pp.

Richards, T.; Staats, S. A. 1994. Common merganser. Pages 38–39. In: Foss, C. R. (editor). *Atlas of breeding birds in New Hampshire.* Dover, N.H.: Arcadia. 414 pp.

Richardson, M.; Brauning, D. W. 1995. Chestnut-sided warbler (*Dendroica pensylvanica*). In: Poole, A.; Gill, F. (editors). *The birds of North America.* No. 190. Philadelphia: Academy of Natural Sciences; Washington, D.C.: American Ornithologists' Union.

Ridgeway, R. 1915. Bird-life in southern Illinois. *Bird-Lore* 17(2):91–103.

Riegner, M. F. 1982. The diet of yellow-crowned night-herons in the eastern and southeastern United States. *Colonial Waterbirds* 5:173–176.

Rimmer, C. C. 1992. Common loon (*Gavia immer*). Pages 3–30. In: Schneider, D. J.; Pence, D. M. (editors). *Migratory nongame birds of management concern in the Northeast.* Newton Corner, Mass.: U.S. Fish and Wildlife Service, Region 5. 400 pp.

Rimmer, C. C.; McFarland, K. P. 1998. Tennessee warbler (*Vermivora peregrina*). In: Poole, A.; Gill, F. (editors). *The birds of North America.* No. 350. Philadelphia: Academy of Natural Sciences; Washington, D.C.: American Ornithologists' Union.

Rimmer, C.C., McFarland, K. P.; and Goetz, J. E. 1998. Distribution, habitat use, and conservation of Bicknell's thrush and other montane forest birds in the Dominican Republic. Unpubl. Report. Woodstock, Vt.: Vermont Institute of Natural Science.

Ritchison, G. 1981. Breeding biology of the white-breasted nuthatch. *Loon* 53:184–187.

Robbins, C. S. 1979. Effect of forest fragmentation on bird populations. Pages 198–212. In: DeGraaf, R. M.; Evans, K. E. (editors). *Management of north central and northeastern forest for nongame birds*. Gen. Tech. Rep NC-51. St. Paul, Minn.: U.S. Department of Agriculture, Forest Service.

Robbins, C. S. 1994a. Yellow-throated vireo. Pages 242–243. In: Foss, C. R. (editor). *Atlas of breeding birds in New Hampshire*. Dover, N.H.: Arcadia. 414 pp.

Robbins, C. S. 1994b. Whip-poor-will. Pages 120–121. In: Foss, C. R. (editor). *Atlas of breeding birds in New Hampshire*. Dover, N.H.: Arcadia. 414 pp.

Robbins, C. S.; Bruun, B.; Zim, H. S. 1966. *Birds of North America*. New York: Golden Press. 340 pp.

Robbins, C. S.; Bystrak, D.; Geissler, P. H. 1986. *The breeding bird survey: Its first fifteen years, 1965–1979*. Resource Publication No. 157. Washington, D.C.: U.S. Fish and Wildlife Service.

Robbins, C. S.; Dawson, D. K.; Dowell, B. A. 1989. Habitat area requirements of breeding birds of the middle Atlantic States. *Wildl. Monogr.* 103.

Robbins, C. S.; Fitzpatrick, J. W.; Hamel, P. B. 1992. A warbler in trouble: *Dendroica cerulea*. Pages 549–562. In: Hagan, J. M.; Johnston, D. W. (editors). *Ecology and conservation of Neotropical migrant landbirds*. Washington, D.C.: Smithsonian Institution Press. 609 pp.

Robertson, R. J.; Stutchberry, B. J.; Cohen, R. R. 1992. Tree swallow (*Tachycineta bicolor*). In: Poole, A. F.; Stettenheim, P.; Gill, F. (editors). *The birds of North America*. No. 11. Philadelphia: Academy of Natural Sciences; Washington, D.C.: American Ornithologists' Union.

Robinson, S. K. 1992. Population dynamics of breeding Neotropical migrants in a fragmented Illinois landscape. Pages 408–418. In: Hagan, J. M.; Johnston, D. W. (editors). *Ecology and conservation of Neotropical migrant landbirds*. Washington, D.C.: Smithsonian Institution Press.

Robinson, T. R.; Sargent, R. R.; Sargent, M. B. 1996. Ruby-throated hummingbird (*Archilochus colubris*). In: Poole, A.; Gill, F. (editors). *The birds of North America*. No. 204. Philadelphia: Academy of Natural Sciences; Washington, D.C.: American Ornithologists' Union.

Robinson, W. D. 1990. Louisiana waterthrush foraging behavior and microhabitat selection in southern Illinois. Master's thesis, Southern Illinois University, Carbondale.

Robinson, W. D. 1995. Louisiana waterthrush. In: Poole, A.; Gill, F. (editors). *The birds of North America*. No. 151. Philadelphia: Academy of Natural Sciences; Washington, D.C.: American Ornithologists' Union.

Rodewald, P. G.; James, R. D. 1996. Yellow-throated vireo (*Vireo flavifrons*). In: Poole, A.; Gill, F. (editors). *The birds of North America*. No. 247. Philadelphia: Academy of Natural Sciences; Washington, D.C.: American Ornithologists' Union.

Rodgers, J. A., Jr.; Smith, H. T. 1995. Little blue heron (*Egretta caerulea*). In: Poole, A.; Gill, F. (editors). *The birds of North America*. No. 145. Philadelphia: Academy of Natural Sciences; Washington, D.C.: American Ornithologists' Union.

Rogers, T. L. 1937. Behavior of the pine siskin. *Condor* 39:143–149.

Root, R. B. 1967. The niche exploitation pattern of the blue-gray gnatcatcher. *Ecol. Monogr.* 37:317–350.

Root, R. B. 1970. The behavior and reproductive success of the blue-gray gnatcatcher. *Condor* 71:16–31.

Root, T. 1988. *Atlas of wintering North American birds: An analysis of Christmas Bird Count* data. Chicago: University of Chicago Press.

Roseberry, J. L.; Klimstra, W. D. 1970. The nesting ecology and reproductive performance of the eastern meadowlark. *Wilson Bull.* 82:243–267.

Rosenburg, C. 1992. Barn owl, *Tyto alba*. Pages 253–279. In: Schneider, D. J.; Pence, D. M. (editors). *Migratory nongame birds of management concern in the Northeast*. Newton Corner, Mass.: U.S. Fish and Wildlife Service, Region 5. 400 pp.

Rosene, W. 1969. *The bobwhite quail: Its life and management*. New Brunswick, N.J.: Rutgers University Press. 418 pp.

Rosenfield, R. N; Bielefeldt, J. 1993. Cooper's hawk (*Accipiter cooperii*). In: Poole, A.; Gill, F. (editors). *The birds of North America*. No. 75. Philadelphia: Academy of Natural Sciences; Washington, D.C.: American Ornithologists' Union.

Rosenfield, R. N.; Morasky, C. M.; Bielefeldt, J.; Loope, W. L. 1992. Forest fragmentation and island biogeography, a summary and bibliography. Tech. Rep. 92/08. U.S. Department of the Interior, National Park Service.

Roth, R. R.; Johnson, M. S.; Underwood, T. J. 1996. Wood thrush (*Hylocichla mustelina*). In: Poole, A.; Gill, F. (editors). *The birds of North America*. No. 246. Philadelphia: Academy of Natural Sciences; Washington, D.C.: American Ornithologists' Union.

Roth, R. R.; Johnson, R. K. 1993. Long-term dynamics of a wood thrush population breeding in a forest fragment. *Auk* 110:37–48.

Rothstein, S. I. 1971. High nest density and non-random nest placement in the cedar waxwing. *Condor* 73:483–485.

Ryder, J. P. 1993. Ring-billed gull (*Larus delawarensis*). In: Poole, A. F.; Stettenheim, P.; Gill, F. B. (editors). *The birds of North America*. No. 33. Philadelphia: Academy of Natural Sciences; Washington, D.C.: American Ornithologists' Union.

Sabo, S. R. 1980. Niche and habitat relations in subalpine bird communities of the White Mountains of New Hampshire. *Ecol. Monogr.* 50:241–259.

Sabo, S. R.; Holmes, T. T. 1983. Foraging niches and the structure of forest bird communities in contrasting montane habitats. *Condor* 85:121–138.

Sage, J. H.; Bishop, C. B.; Bliss, W. P. 1913. The birds of Connecticut. Bull. No. 20. Hartford, Conn.: State Geological and Natural History Survey.

Samson, F. B. 1979. Lowland hardwood bird communities. Pages 49–66. In: DeGraaf, R. M.; Evans, K. E. (editors). *Management of north central and northeastern forest for nongame birds*. Gen. Tech. Rep NC-51. St. Paul, Minn.: U.S. Department of Agriculture, Forest Service.

Samuel, D. E. 1971. The breeding biology of barn and cliff swallows in West Virginia. *Wilson Bull.* 83:284–301.

Sargent, T. D. 1959. Winter studies on the tree sparrow (*Spizella arborea*). *Bird-Banding* 30:27–37.

Sauer, J. R.; Droege, S. 1990. Recent population trends of the eastern bluebird. *Wilson Bull.* 102:239–252.

Sayler, J. C.; Lagler, K. F. 1946. The eastern belted kingfisher (*Megaceryle alcyon alcyon*) in relation to fish management. *Trans. Am. Fisheries Soc.* 76th Annual Meeting:97–117.

Scharf, W. C.; Kren, J. 1996. Orchard oriole (*Icterus spurius*). In: Poole, A.; Gill, F. (editors). *The birds of North America*. No. 255. Philadelphia: Academy of Natural Sciences; Washington, D.C.: American Ornithologists' Union.

Schneider, K. C. J. 1984. Dominance, predation, and optimal

foraging in white-throated sparrow flocks. *Ecology* 65:1820–1827.

Schnell, G. D. 1969. Communal roosts of wintering rough-legged hawks (*Buteo lagopus*). *Auk* 86(4):682–690.

Schrantz, F. G. 1943. Nest life of the eastern yellow warbler. *Auk* 60:367–387.

Scott, V. E.; Evans, K. E.; Patton, D. R.; Stone, C. P. 1977. *Cavity-nesting birds of North American forests*. Agric. Handbook 511. Washington, D.C.: U.S. Department of Agriculture.

Sealy, S. G. 1978. Possible influence of food on egg-laying and clutch size in the black-billed cuckoo. *Condor* 80:103–104.

Sealey, S. G. 1989. Defense of nectar resources by migrating Cape May warblers. *J. Field Ornithol.* 60:89–93.

Searcy, W. A.; Yasukawa, K. 1995. *Polygyny and sexual selection in red-winged blackbirds*. Princeton, N.J.: Princeton University Press.

Sedgwick, J. A.; Knopf, F. I. 1990. Habitat relationships and nest-site characteristics of cavity-nesting birds in cottonwood floodplains. *J. Wildl. Manage.* 54:112–124.

Sepik, G. F.; Owen, R. B., Jr.; Coulter, M. W. 1981. A landowner's guide to woodcock management in the Northeast. Pages 1–23. Misc. Rep. 253. Washington, D.C.: U.S. Department of the Interior, Fish and Wildlife Service.

Serrentino, P. 1992. Northern harrier (*Circus cyaneus*). Pages 89–118. In: Schneider, D. J.; Pence, D. M. (editors). *Migratory nongame birds of management concern in the Northeast*. Newton Corner, Mass.: U.S. Fish and Wildlife Service, Region 5. 400 pp.

Shank, C. C. 1986. Territory size, energetics, and breeding strategy in the corvidae. *Am. Nat.* 128:642–652.

Sheldon, W. G. 1967. *The book of the American woodcock*. Amherst: University of Massachusetts Press. 227 pp.

Sherry, T. W. 1979. Competitive interactions and adaptive strategies of American redstarts and least flycatchers in a northern hardwoods forest. *Auk* 96:265–283.

Sherry, T. W.; Holmes, R. T. 1988. Habitat selection by breeding American redstarts and least flycatchers in a northern hardwoods forest. *Auk* 105:350–364.

Sherry, T. W.; Holmes, R. T. 1997. American redstart (*Setophaga ruticilla*). In: Poole, A.; Gill, F. (editors). *The birds of North America*. No. 277. Philadelphia: Academy of Natural Sciences; Washington, D.C.: American Ornithologists' Union.

Short, H. L.; Drew, L. C. 1962. Observations concerning behavior, feeding and pellets of short-eared owls. *Am. Midl. Nat.* 67:424–433.

Short, L. L., Jr. 1962. The blue-winged and golden-winged warbler in New York. *Kingbird* 12:59–67.

Short, L. L. 1974. Habits and interactions of North American three-toed woodpeckers (*Picoides arcticus* and *Picoides tridactylus*). *Am. Mus. Novitates* 2547:1–42.

Shugart, H. H.; Dueser, R. D.; Anderson, S. H. 1974. Influence of habitat alterations on bird and small mammal populations. Pages 92–96. In: *Timber-wildlife Management Symp. Proc. Occas. Pap. 3*. Columbia: Missouri Academy of Science.

Shugart, H. H.; Smith, T. M.; Kitchings, J. T.; Kroodsma, R. L. 1978. The relationship of nongame birds to southern forest types and successional stages. Pages 5–16. In: DeGraaf, R. M. (editor). *Management of southern forests for nongame birds*. Gen. Tech. Rep SE-14. Asheville, N.C.: U.S. Forest Service.

Siegfried, W. R. 1978. Habitat and modern range expansion of the cattle egret. Pages 315–325. In: Sprunt, A., IV; Ogden, J. C.; Winckler, S. (editors). *Wading Birds*. National Audubon Society Research Report No. 7. New York: National Audubon Society.

Simson, B. S. 1984. Tests of habituation to song repertoires by Carolina wrens. *Auk* 101:244–254.

Simpson, M. B., Jr. 1972. The saw-whet owl population of North Carolina's southern Great Balsam Mountains. *Chat* 36:39–47.

Singer, F. J. 1974. Status of the osprey, bald eagle and golden eagle in the Adirondacks. *N.Y. Fish Game J.* 21(1):18–31.

Sjolander, S.; Agren, G. 1972. Reproductive behavior of the common loon. *Wilson Bull.* 84:296–308.

Slate, D. 1994. Great black-backed gull. Pages 94–95. In: Foss, C. R. (editor). *Atlas of breeding birds in New Hampshire*. Dover, N.H.: Arcadia. 414 pp.

Smith, D. G.; Devine, A. 1993. Winter ecology of the long-eared owl in Connecticut. *Conn. Warbler* 13:44–53.

Smith, D. G.; Gilbert, R. 1984. Eastern screech-owl home range and use of suburban habitats in southern Connecticut. *J. Field Ornithol.* 55:322–329.

Smith, D. G.; Wilson, C. R.; Frost, H. H. 1972. The biology of the American kestrel in central Utah. *Southwest Nat.* 17(1):73–83.

Smith, K. G. 1978. White-winged crossbills breed in northern Utah. *Western Birds* 9:79–81.

Smith, L. M.; Hupp, J. W.; Ratti, J. T. 1982. Habitat use and home range of gray partridge in eastern South Dakota. *J. Wildl. Manage.* 46(3):580–587.

Smith, R. D.; Marquiss, M. 1995. Production and costs of nesting attempts in snow buntings *Plectrophenax nivalis*: Why do they attempt 2nd broods? *Ibis* 137(4):469–476.

Smith, R. L. 1963. Some ecological notes on the grasshopper sparrow. *Wilson Bull.* 75:159–165.

Smith, S. 1994. Northern parula. Pages 258–259. In: Foss, C. R. (editor). *Atlas of breeding birds in New Hampshire*. Dover, N.H.: Arcadia. 414 pp.

Smith, S. M. 1993. Black-capped chickadee (*Parus atricapillus*). In: Poole, A. F.; Stettenheim, P.; Gill, F. B. (editors). *The birds of North America*. No. 39. Philadelphia: Academy of Natural Sciences; Washington, D.C.: American Ornithologists' Union.

Smith, W. J. 1966. Communication and relationships in the genus *Tyrannus*. *Nuttall Ornithol. Club Publ.* 6:1–250.

Smith, W. P. 1942. Nesting habitats of the eastern phoebe. *Auk* 59:410–417.

Smith, W. P. 1950. Some additions to the Vermont Bird List. *Vt. Bot. and Bird Clubs Joint Bull.* 18:55–59.

Snap, B. D. 1976. Colonial breeding in the barn swallow (*Hirundo rustica*) and its adaptive significance. *Condor* 78:471–480.

Snyder, D.; Bonney, C.; Robertson, W. B. 1948. Twelfth breeding bird census. Deciduous flood-plain forest. *Audubon Field Notes* 2(6):237.

Sodhi, N. S.; James, P. C.; Warkentin, I. G.; Oliphant, L. W. 1992. Breeding ecology of urban merlins (*Falco columbarius*). *Can. J. Zool.* 70:1477–1483.

Sodhi, N. S.; Oliphant, L. W. 1992. Hunting ranges and habitat use and selection of urban-breeding merlins. *Condor* 94:743–749.

Sodhi, N. S.; Oliphant, L. W.; James, P. C.; Warkentin, I. G. 1993. Merlin (*Falco columbarius*). In: Poole, A.; Gill, F. (editors). *The birds of North America*. No. 44. Philadelphia: Academy of Natural Sciences; Washington, D.C.: American Ornithologists' Union.

Sodhi, N. S; Paszkowski, C. A. 1995. Habitat use and foraging behavior on four parulid warblers in a second-growth forest. *J. Field Ornithol.* 66(2):277–288.
Spear, R. N., Jr. 1976. *Birds of Vermont*. Rev. ed. Burlington, Vt.: Green Mountain Audubon Society.
Speirs, J. M.; Speirs, D. H. 1968. *Melospiza lincolnii lincolnii,* Lincoln's sparrow. In: Bent, A. C. (editor). Life histories of North American cardinals, grosbeaks, buntings, towhees, finches, sparrows, and allies. *U.S. Natl. Mus. Bull.* 237(pt. 3):1434–1436.
Speiser, R.; Bosakowski, T. 1987. Nest site selection by northern goshawks in northern New Jersey and southeastern New York. *Condor* 89:387–394.
Speiser, R.; Bosakowski, T. 1988. Nest site preferences of red-tailed hawks in the highlands of southeastern New York and New Jersey. *J. Field Ornithol.* 59(4):361–368.
Spencer, O. R. 1943. Nesting habits of the black-billed cuckoo. *Wilson Bull.* 55:11–22.
Sperry, C. C. 1940. Food habits of a group of shorebirds: Woodcock, snipe, knot, and dowitcher. Wildl. Res. Bull. 1. Washington, D.C.: U.S. Department of the Interior, Biological Survey. 37 pp.
Spofford, W. R. 1971. The breeding status of the golden eagle in the Appalachians. *Am. Birds* 25:3–7.
Squires, J. R.; Reynolds, R. T. 1997. Northern goshawk (*Accipter gentilis*). In: Poole, A.; Gill, F. (editors). *The birds of North America.* No. 298. Philadelphia: Academy of Natural Sciences; Washington, D.C.: American Ornithologists' Union.
Staats, S. A. 1994. Yellow warbler (*Dendroica petechia*). Pages 260–261. In: Foss, C. R. (editor). *Atlas of breeding birds in New Hampshire.* Dover, N.H.: Arcadia. 414 pp.
Stauffer, D. F.; Best, L. B. 1986. Nest-site characteristics of open-nesting birds in riparian habitats in Iowa. *Wilson Bull.* 98:231–242.
Stegeman, L. C. 1957. Winter food of the short-eared owl in central New York. *Am. Midl. Nat.* 57:120–124.
Stein, R. C. 1958. The behavioral, ecological and morphological characteristics of two populations of the alder flycatcher (*Empidonax traillii*). *N.Y. State Mus. Sci. Serv. Bull.* 371:1–63.
Stenger, J.; Falls, J. B. 1959. The utilized territory of the ovenbird. *Wilson Bull.* 71:125–140.
Stenzel, L. E.; Huber, H. R.; Page, G. W. 1976. Feeding behavior and diet of the long-billed curlew and willet. *Wilson Bull.* 88:314–332.
Stewart, P. A. 1952. Dispersed breeding behavior, and longevity of banded barn owls in North America. *Auk* 69:227–245.
Stewart, R. E. 1949. Ecology of a nesting red-shouldered hawk population. *Wilson Bull.* 61:26–35.
Stewart, R. E. 1953. A life history of the yellowthroat. *Wilson Bull.* 65:99–115.
Stewart, R. E. 1962. Waterfowl populations in the upper Chesapeake region. Spec. Sci. Rep. Wildl. No. 65. Pages 1–208. Washington, D.C.: U.S. Fish and Wildlife Service.
Stewart, R. E.; Aldrich, J. W. 1952. Ecological studies of breeding bird populations in northern Maine. *Ecology* 33:226–238.
Stewart, R. E.; Kantrud, H. A. 1972. Population estimates of breeding birds in North Dakota. *Auk* 89:766–788.
Stewart, R. E.; Robbins, C. S. 1958. Birds of Maryland and the District of Columbia. North American Fauna 62. Washington, D.C.: U.S. Department of the Interior, Fish and Wildlife Service.
Stewart, R. M. 1973. Breeding behavior and life history of the Wilson's warbler. *Wilson Bull.* 85:21–30.
Stobo, W. T.; McLaren, I. A. 1975. *The Ipswich sparrow.* Halifax: Nova Scotia Institute of Science. 105 pp.
Stockard, C. R. 1905. Nesting habits of birds in Mississippi. *Auk* 22:146–158.
Stoddard, H. L. 1931. *The bobwhite quail: Its habits, preservation and increase.* New York: C. Scribner's Sons. 559 pp.
Stokes, A. W. 1950. Breeding behavior of the goldfinch. *Wilson Bull.* 62:107–127.
Storm, G. L.; Scott, J. G. 1989. Nesting. Pages 131–133. In: Atwater, S.; Schnell, J. (editors). *Ruffed grouse.* Harrisburg, Pa.: Stackpole. 370 pp.
Stotts, V. 1957. The black duck (*Anas rubripes*) in the upper Chesapeake Bay. *Proc. 10th Annu. Conf. Southeastern Assoc. Game and Fish Commission* 234–242.
Strain, J. G.; Mumme, R. L. 1988. Effects of food supplementation, song playback, and temperature on vocal territorial behavior in Carolina wrens. *Auk* 105:11–16.
Strickland, D.; Ouellet, H. 1993. Gray jay (*Perisoreus canadensis*). In: Poole, A. F.; Stettenheim, P.; Gill, F. B. (editors). *The birds of North America.* No. 40. Philadelphia: Academy of Natural Sciences; Washington, D.C.: American Ornithologists' Union.
Strohmeyer, D. L. 1977. Common gallinule (*Gallinula chloropus*). In: Sanderson, G. C. (editor). *Management of migratory and upland game birds in North America.* Washington, D.C.: International Association of Fish and Wildlife Agencies. 358 pp.
Strom, K. 1983. Virgin hardwood swamp forest. *Am. Birds* 37:65.
Stutchberry, B. J.; Robertson, R. J. 1988. Within-season and age-related patterns of reproductive performance in female tree swallows (*Tachycineta bicolor*). *Can. J. Zool.* 68:827–834.
Summers-Smith, D. 1958. Nest-site selection, pair formation, and territory in the house-sparrow (*Passer domesticus*). *Ibis* 100:190–203.
Summers-Smith, J. D. 1988. *The sparrows: A study of the genus* Passer. Calton, England: T. & A. D. Poyser, Ltd.
Sutcliffe, S. 1980. Aspects of the nesting ecology of common loons in New Hampshire. M.S. thesis, University of New Hampshire, Durham.
Sutton, G. M. 1960. The nesting fringillids of the Edwin S. George Reserve, southeastern Michigan (parts 6 and 7). *Jack-Pine Warbler* 38:46–65, 125–139.
Svoboda, F. J.; Gullion, G. W. 1972. Preferential use of aspen by ruffed grouse in northern Minnesota. *J. Wildl. Manage.* 36:1166–1180.
Swift, B. L.; Larson, J. S.; DeGraaf, R. M. 1984. Relationship of breeding bird density and diversity to habitat variables in forested wetlands. *Wilson Bull.* 96:48–59.
Taber, W. 1968. White-winged crossbill. Pages 527–544. In: Bent, A. C. (editor). *Life histories of North American cardinals, grosbeaks, buntings, towhees, finches, sparrows, and allies.* U.S. Natl. Mus. Bull. 237 (pt. 1). Washington, D.C.: Smithsonian Institution.
Tanner, J. T. 1942. *The ivory-billed woodpecker.* New York: Dover. 111 pp.
Tanner, W. D., Jr.; Hendrickson, G. O. 1956. Ecology of the sora in Clay County, Iowa. *Iowa Bird Life* 26(4):78–81.
Tate, G. R. 1992. Short-eared owl (*Asio flammeus*). Pages 171–190. In: Schneider, K. J.; Pence, D. M. (editors). *Migratory*

nongame birds of management concern in the Northeast. Newton Corner, Mass.: U.S. Fish and Wildlife Service, Region 5.

Taylor, J. W. 1971. The Virginia rail. *Virginia Wildl.* 32(9):27.

Telfair, R. C., II. 1994. Cattle egret (*Bubulcus ibis*). In: Poole, A.; Gill, F. (editors). *The birds of North America.* No. 158. Philadelphia: Academy of Natural Sciences; Washington, D.C.: American Ornithologists' Union.

Terres, J. K. 1980. *The Audubon Society encyclopedia of North American birds.* New York: Alfred A. Knopf.

Terrill, L. M. 1931. Nesting of the saw-whet owl in the Montreal District. *Auk* 48:169–174.

Thomas, J. W.; Anderson, R.; Maser, C.; Bull, E. 1979. Snags. Pages 60–77. In: Thomas, J. W. (editor). *Wildlife habitats in managed forests–The Blue Mountains of Oregon and Washington.* Agric. Handbook 553. Washington, D.C.: U.S. Department of Agriculture. 512 pp.

Thomas, R. H. 1946. An orchard oriole colony in Arkansas. *Bird-Banding* 17:161–167.

Thompson, R. R., III; Capen, D. E. 1988. Avian assemblages in seral stages of a Vermont forest. *J. Wildl. Manage.* 52:771–777.

Thompson, W. L.; Yahner, R. H.; Storm, G. L. 1990. Winter use and habitat characteristics of vulture communal roosts. *J. Wildl. Manage.* 54:77–83.

Thompson, Z. 1853. *Natural history of Vermont.* Published by the author, Burlington, Vt. Reprinted 1972. Rutland, Vt.: Charles Tuttle Co.

Threlfall, W.; Blacquiere, J. R. 1982. Breeding biology of the fox sparrow in Newfoundland. *J. Field Ornithol.* 53:235–239.

Timken, R. L.; Anderson, B. L. 1969. Food habits of common mergansers in the north-central United States. *J. Wildl. Manage.* 33:87–91.

Titterington, R. W.; Crawford, H. S.; Burgason, B. N. 1979. Songbird responses to commercial clearcutting in Maine spruce-fir forests. *J. Wildl. Manage.* 43:602–609.

Titus, K.; Mosher, J. A. 1981. Nest-site habitat selected by woodland hawks in the central Appalachians. *Auk* 98:270–281.

Titus, K.; VanDruff, L. 1981. Response of the common loon to recreational pressure in the Boundary Waters Canoe Area, northeastern Minnesota. *Wildl. Monogr.* 79. 59 pp.

Toland, B. R. 1985. Double brooding by American kestrels in central Missouri. *Condor* 87:434–436.

Toland, B. R. 1986. Hunting success of some Missouri raptors. *Wilson Bull.* 98:116–125.

Tompa, F. S. 1962. Territorial behavior: The main controlling factor of a local song sparrow population. *Auk* 79: 687–697.

Townsend, C. W. 1929. Breeding of the yellow-crowned night heron (*Nyctanassa violacea*) at Ipswich. *Bull. Essex County Ornithol. Club* 11:27–30.

Tremblay, J.; Ellison, L. N. 1979. Effects of human disturbance on breeding of black-crowned night herons. *Auk* 96:364–369.

Tuck, L. M. 1972. The snipes: A study of the genus *Capella. Can. Wildl. Serv. Monogr.* 5:1–428.

Uhlig, H. G. 1950. Resurvey of flock distribution of the wild turkey: West Virginia. *Pittman-Robertson Q.* 10(3):371–372.

Van Camp, L. F.; Henny, C. J. 1975. *The screech owl: Its life history and population ecology in northern Ohio.* North American Fauna No. 71. Washington, D.C.: U.S. Fish and Wildlife Service. 65 pp.

Van Horn, M. A.; Donovan, T. 1994. Ovenbird (*Seiurus aurocapillus*). In: Poole, A.; Gill, F. (editors). *The birds of North America.* No. 88. Philadelphia: Academy of Natural Sciences; Washington, D.C.: American Ornithologists' Union.

Van Tyne, J. 1948. Home range and duration of family ties in the tufted titmouse. *Wilson Bull.* 60:121.

Van Velzen, W. T. 1977. Fortieth breeding bird census. *Am. Birds* 31:24–93.

Veit, R.; Petersen, W. R. 1993. *Birds of Massachusetts.* Lincoln: Massachusetts Audubon Society.

Verbeek, N. A. M. 1970. Breeding ecology of the water pipit. *Auk* 87:425–451.

Verbeek, N. A. M.; Hendricks, P. 1994. American pipit (*Anthus rubescens*). In: Poole, A.; Gill, F. (editors). *The birds of North America.* No. 95. Philadelphia: Academy of Natural Sciences; Washington, D.C.: American Ornithologists' Union.

Verner, J. 1965. Breeding biology of the long-billed marsh wren. *Condor* 67:6–30.

Vernon, R. C. 1994. Pied-billed grebe. Pages 4–5. In: Foss, C. R. (editor). *Atlas of breeding birds in New Hampshire.* Dover, N.H.: Arcadia. 414 pp.

Vesall, D. B. 1940. Notes on nesting habits of the American bittern. *Auk* 52:207–208.

Vickery, P. D. 1996. Grasshopper sparrow (*Ammodramus savannarum*). In: Poole, A.; Gill, F. (editors). *The birds of North America.* No. 239. Philadelphia: Academy of Natural Sciences; Washington, D.C.: American Ornithologists' Union.

Walkinshaw, L. H. 1935. Studies of the short-billed marsh wren (*Cistothorus stellaris*) in Michigan. *Auk* 52:362–369.

Walkinshaw, L. H. 1937. The Virginia rail in Michigan. *Auk* 54:464–475.

Walkinshaw, L. H. 1938. Life history studies of the eastern goldfinch. Part I. *Jack-Pine Warbler.* 16:3–11, 14–15.

Walkinshaw, L. H. 1939a. Life history studies of the eastern goldfinch. Part II. *Jack-Pine Warbler.* 17:3–12.

Walkinshaw, L. H. 1939b. Nesting of the field sparrow and survival of the young. *Bird-Banding* 10:107–114.

Walkinshaw, L. H. 1940. Summer life of the sora rail. *Auk* 67:153–168.

Walkinshaw, L. H. 1944. The eastern chipping sparrow in Michigan. *Wilson Bull.* 56:193–205.

Walkinshaw, L. H. 1957. Yellow-bellied flycatcher nesting in Michigan. *Auk* 74:293–304.

Walkinshaw, L. H. 1966. Summer biology of Traill's flycatcher. *Wilson Bull.* 78:31–46.

Walkinshaw, L. H. 1968. Eastern field sparrow. Pages 1217–1235. In: *Life histories of North American cardinals, grosbeaks, buntings, towhees, finches, sparrows and allies.* U.S. Nat. Mus. Bull. 237. Washington, D.C.: Smithsonian Institution.

Wallace, G. J. 1939. Bicknell's thrush: Its taxonomy, distribution and life history. *Proc. Boston Soc. Nat. Hist.* 41:211–402.

Wallace, J. G. 1948. The barn owl in Michigan, its distribution, natural history and food habits. *Mich. State Coll. Agric. Exp. Tech. Bull.* 208.

Wallace, V. 1949. Partially cut over northern hardwoods slope. *Audubon Field Notes.* 3:259.

Wallace, V. H. 1953. Nesting season. *N.H. Bird News* 6:17–23.

Warkentin, I. G.; James, P. C. 1988. Nest-site selection by urban merlins. *Condor* 90:734–738.

Wasserman, F. E. 1983. Territories of rufous-sided towhees contain more than minimal food resources. *Wilson Bull.* 95:664–667.

Watts, B. D. 1989. Nest-site characteristics of yellow-crowned night-herons in Virginia. *Condor* 91:979–983.
Watts, B. D. 1995. Yellow-crowned night-heron (*Nyctanassa violacea*). In: Poole, A.; Gill, F. (editors). *The birds of North America.* No. 161. Philadelphia: Academy of Natural Sciences; Washington, D.C.: American Ornithologists' Union.
Weaver, F. G. 1939. Studies in the life history of the wood thrush. *Bird-Banding* 10:16–23.
Weaver, R. L. 1942. Growth and development of English sparrows. *Wilson Bull.* 54:183–191.
Weaver, R. L.; West, F. H. 1943. Notes on the breeding of the pine siskin. *Auk* 60:492–504.
Webster, C. G. 1964. Fall foods of soras from two habitats in Connecticut. *J. Wildl. Manage.* 28:163–165.
Weeks, H. P., Jr. 1994. Eastern phoebe (*Sayornis phoebe*). In: Poole, A.; Gill, F. (editors). *The birds of North America.* No. 94. Philadelphia: Academy of Natural Sciences; Washington, D.C.: American Ornithologists' Union.
Weller, M. W. 1961. Breeding biology of the least bittern. *Wilson Bull.* 73:11–35.
Weller, M. W.; Fredrickson, L. H. 1973. Avian ecology of a managed glacial marsh. *Living Bird* 12:269–291.
Welsh, C. J. E.; Healy, W. M. 1993. Effect of even-aged timber management on bird species diversity and composition in northern hardwoods of New Hampshire. *Wildl. Soc. Bull.* 21:143–154.
Welsh, D. 1975. Savannah sparrow breeding and territoriality on a Nova Scotia dune beach. *Auk* 92:235–251.
Welter, W. A. 1935. The natural history of the long-billed marsh wren. *Wilson Bull.* 47:3–34.
Werschkul, D. F.; McMahon, E.; Leitschuh, M. 1976. Some effects of human activities on the great blue heron in Oregon. *Wilson Bull.* 88:660–666.
Wetmore, A. 1916. *Birds of Puerto Rico.* Tech. Bull. 326. Washington, D.C.: U.S. Department of Agriculture. 140 pp.
Wheeler, A. H. 1992. Reproductive parameters for free ranging American kestrels (*Falco sparverius*) using nest boxes in Montana and Wyoming. *J. Raptor Res.* 26(1):6–9.
Wheelwright, N. T.; Rising, J. D. 1993. Savannah sparrow (*Passerculus sandwichensis*). In: Poole, A.; Gill, F. (editors). *The birds of North America.* No. 45. Philadelphia: Academy of Natural Sciences; Washington, D.C.: American Ornithologists' Union.
Whitcomb, R. F.; Robbins, C. S.; Lynch, J. F.; Whitcomb, B. L.; Klimkiewicz, M. K.; Bystrak, D. 1981. Effects of forest fragmentation on the avifauna of the eastern deciduous forest. Pages 125–205. In: Burgess, R. L.; Sharpe, D. M. (editors). *Forest island dynamics in man-dominated landscapes.* New York: Springer-Verlag.
Whittle, C. L. 1926. Notes on the nesting habits of the tree swallow. *Auk* 43:247–248.
Wiens, J. A. 1969. *An approach to the study of ecological relationships among grassland birds.* Ornithological Monographs No. 8. Lawrence, Kans.: American Ornithologists' Union.
Wiens, J. A. 1973. Interterritorial habitat variation in grasshopper and savannah sparrows. *Ecology* 54:877–884.
Wilcox, L. 1980. Observations of the life history of willets on Long Island, New York. *Wilson Bull.* 92:253–258.
Willard, D. E. 1977. The feeding ecology and behavior of five species of herons in southeastern New Jersey. *Condor* 79:462–470.
Willenberger, J. F. 1980. Vegetation structure, food supply, and polygyny in bobolinks (*Dolichonyx oryzivorus*). *Ecology* 61:140–150.
Williams, J. B. 1975. Habitat utilization by four species of woodpeckers in a central Illinois woodland. *Am. Midl. Nat.* 93(2):354–367.
Williams, J. M. 1996a. Nashville warbler (*Vermivora ruficapilla*). In: Poole, A.; Gill, F. (editors). *The birds of North America.* No. 205. Philadelphia: Academy of Natural Sciences; Washington, D.C.: American Ornithologists' Union.
Williams, J. M. 1996b. Bay-breasted warbler (*Dendroica castanea*). In: Poole, A.; Gill, F. (editors). *The birds of North America.* No. 206. Philadelphia: Academy of Natural Sciences; Washington, D.C.: American Ornithologists' Union.
Williamson, P. 1971. Feeding ecology of the red-eyed vireo (*Vireo olivaceous*) and associated foliage-gleaning birds. *Ecol. Monogr.* 41:129–152.
Williamson, P.; Gray, L. 1975. Foraging behavior of the starling (*Sturnus vulgaris*) in Maryland. *Condor* 77:84–89.
Willson, M. F. 1970. Foraging behavior of some winter birds of deciduous woods. *Condor* 72:169–174.
Wilson, J. E. 1959. The status of the Hungarian partridge in New York. *Kingbird* 9:54–57.
Wilson, W. H., Jr. 1996. Palm warbler (*Dendroica palmarum*). In: Poole, A.; Gill, F. (editors). *The birds of North America.* No. 238. Philadelphia: Academy of Natural Sciences; Washington, D.C.: American Ornithologists' Union.
With, K. A.; Webb, D. R. 1993. Microclimate of ground nests: The relative importance of radiative cover and wind breaks for three grassland species. *Condor* 95:401–413.
Witmer, M. C.; Mountjoy, D. J.; Elliot, L. 1997. Cedar waxwing (*Bombycilla cedrorum*). In: Poole, A.; Gill, F. (editors). *The birds of North America.* No. 309. Philadelphia: Academy of Natural Sciences; Washington, D.C.: American Ornithologists' Union.
Wood, N. A. 1951. *The birds of Michigan.* University of Michigan Museum of Zoology Misc. Publ. No. 75. 559 pp.
Wood, R.; Gelston, W. L. 1972. Preliminary report: The mute swans of Michigan's Grand Traverse Bay region. Rep. 2683. Michigan Department of Natural Resources. 6 pp.
Woodruff, C.; Woodruff, A. 1991. Predation by rusty blackbirds on songbirds at a winter feeder. *Chat* 55:54–55.
Woodward, P. W.; Woodward, J. C. 1979. Survival of fledgling brown-headed cowbirds. *Bird-Banding* 50:66–68.
Wootton, J. T. 1996. Purple finch (*Carpodacus purpureus*). In: Poole, A.; Gill, F. (editors). *The birds of North America.* No. 208. Philadelphia: Academy of Natural Sciences; Washington, D.C.: American Ornithologists' Union.
Yasukawa, K.; Searcy, W. A. 1995. Red-winged blackbird (*Agelaius phoeniceus*). In: Poole, A.; Gill, F. (editors). *The birds of North America.* No. 184. Philadelphia: Academy of Natural Sciences; Washington, D.C.: American Ornithologists' Union.
Yeager, L. E. 1955. Two woodpecker populations in relation to environmental change. *Condor* 57:148–153.
Yeatter, R. E. 1934. The Hungarian partridge in the Great Lakes Region. Bull. No. 5. Ann Arbor: University of Michigan School of Forestry and Conservation. 92 pp.
Yosef, R.; Grubb, T .C. 1993. Effect of vegetation height on hunting behavior and diet of loggerhead shrikes. *Condor* 95:127–131.
Yosef, R. 1996. Loggerhead shrike (*Lanius ludovicianus*). In:

Poole, A.; Gill, F. (editors). *The birds of North America.* No. 231. Philadelphia: Academy of Natural Sciences; Washington, D.C.: American Ornithologists' Union.
Young, A. D.; Titman, R. D. 1986. Costs and benefits to red-breasted mergansers nesting in tern and gull colonies. *Can. J. Zool.* 64:2339–2343.
Young, H. 1951. Territorial behavior of the eastern robin. *Proc. Linnaean Soc. N.Y.* 5862:1–37.
Young, H. 1955. Breeding behavior and nesting of the eastern robin. *Am. Midl. Nat.* 53:329–352.
Zeleny, L. 1976. *The bluebird.* Bloomington: Indiana University Press.
Zeranski, J. D.; Baptist, T. T. 1990. *Common loon.* In: *Connecticut birds.* Hanover, N.H.: University Press of New England. 328 pp.
Zicus, M. C.; Hennes, S. K.; Riggs, M. R. 1995. Common goldeneye nest attendance patterns. *Condor* 97:461–472.
Zimmerman, J. L. 1977. Virginia rail (*Rallus limnicola*). Pages 46–56. In: Sanderson, G. C. (editor). *Management of migratory shore and upland game birds in North America.* Washington, D.C.: International Association of Fish and Wildlife Agencies. 358 pp.
Zumeta, D. C.; Holmes, R. T. 1978. Habitat shifts and roadside mortality of scarlet tanagers during a cold wet New England spring. *Wilson Bull.* 90:575–586.

Mammals

This section provides information on the life history, distribution, and habitat associations of currently occurring inland mammals in New England. Scientific nomenclature generally follows *Mammal Species of the World* (Wilson and Reeder 1993), the standard reference used by the American Society of Mammalogists. This classification differs for two New England mammals from the *Revised Checklist of North American Mammals North of Mexico, 1997* (Jones et al. 1997). Wilson and Reeder (1993) consider the northern long-eared myotis a subspecies of Keen's myotis (*Myotis keenii*). We consider it a distinct species, *Myotis septentrionalis*, as do Wilson and Ruff (1999). Also, we have retained the striped skunk (*Mephitis mephitis*) within the Mustelidae after Wilson and Reeder (1993); Jones et al. (1997) place the skunks in a separate family, Mephitidae.

The National Museum of Natural History, Smithsonian Institution, maintains an electronic database of the nomenclature found in Wilson and Reeder (1993) at the following website address: <http://nmnhgoph.si.edu/msw/>. Species are arranged in phylogenetic order.

Five mammal species, gray wolf (*Canis lupus*), wolverine (*Gulo gulo*), mountain lion (*Puma concolor*), American elk or wapiti (*Cervus elaphus*), and woodland caribou (*Rangifer tarandus*), apparently have been extirpated from New England (Godin 1977). One subspecies, the sea mink (*Mustela vison macrodon*), has been extinct since about 1900 (Cardoza in press). We have included the gray wolf and mountain lion because of recent wolf sightings in Maine and several confirmed signs of mountain lion in the eastern Canadian provinces and numerous reports of mountain lion sightings each year in the New England states (Gerson 1988; Cumberland and Dempsey 1994; Stocek 1995; E. Orff, pers. commun.) and because of the great public interest in these species. We have omitted the wolverine due to the lack of reliable reports (if indeed it ever occurred in New England). Huntable populations of elk occur in the Blue Mountain Forest Association in Sullivan County, N.H., but the American elk is not included here because there are no longer any free-ranging elk (which had been reintroduced to the wild in 1903 and 1933 in N.H.) in New England. Woodland caribou has also been omitted, although two unsuccessful reintroduction attempts were made on Mount Katahdin, Maine, the latest attempt occurring in the mid to late 1980s (McCollough et al. 1989).

We have included the introduced black-tailed jackrabbit (*Lepus californicus*), successfully introduced to Nantucket Island, Mass. We have omitted descriptions of the beach vole (*Microtus breweri*), which only inhabits Muskeget Island, Mass.; the European rabbit (*Oryctolagus cuniculus*) found on several islands in Boston Harbor (Cardoza 1998); escaped European wild boar (*Sus scrofa*) found rarely in western New Hampshire and eastern Vermont; and the introduced fallow deer (*Dama dama*), until about the late 1970s found on Martha's Vineyard, Mass. (J. Cardoza, pers. commun.). Two subspecies, Penobscot meadow vole (*Microtus pennsylvanicus shattucki*) and Block Island meadow vole (*M. p. provectus*), are included in Appendix A as species of special concern but are not given complete accounts. The marine mammals that inhabit the coastal and pelagic waters of the northeastern United States are listed in Appendix C.

The relationships of New England mammals to forest habitats are not as well understood as are those of birds. For some species, life history and distribution data are still lacking, particularly for some shrews, bats, and rodents. For such species, this compilation must be regarded as a starting point. For many others, this revision illustrates the wealth of life history and distribution data that have been acquired by the wildlife and zoological research and management communities over the last 15 years. Mammalian associations with habitat structure at a variety of scales continue to warrant further study.

Species' distributions are constantly changing, reflecting changes in land use and the values ascribed to wildlife. Twenty or thirty years ago, few would ever have imagined moose reoccupying southern New England or the possibility of wolves recurring naturally in northern New England. Change is the only constant in the dynamic New England landscape.

Literature Cited

Cardoza, J. E. Seamink. In press. In: French, T. W.; Cardoza, J. E. (editors). *Rare Vertebrates of Massachusetts*. Lincoln: Massachusetts Audubon Society.

Cardoza, J. E. 1998. The European rabbit: History of introductions, and its status, biology, and environmental effects, with special reference to Massachusetts. Report on file. Westboro: Massachusetts Division of Fisheries and Wildlife. 43 pp. + illus.

Cumberland, R. E.; Dempsey, J. A. 1994. Recent confirmation of a cougar, *Felis concolor*, in New Brunswick. *Can. Field-Nat.* 108:224–226.

Gerson, H. B. 1988. Cougar, *Felis concolor*, sightings in Ontario. *Can. Field-Nat.* 102:419–424.

Godin, A. J. 1977. *Wild mammals of New England*. Baltimore, Md.: Johns Hopkins University Press. 304 pp.

Jones, C.; Hoffmann, R. S.; Rice, D. W.; Engstrom, M. D.; Bradley, R. D.; Schmidly, D. J.; Jones, C. A.; Baker, R. J. 1997. Revised checklist of North American mammals north of Mexico, 1997. Occasional Papers. Museum Texas Tech. University. 173:1–19.

McCollough, M.; Murphy, K.; Newton, L. 1989. Are logging and caribou compatible? Miscellaneous report 336:259. Orono, Me.: Maine Agric. Exp. Sta. Abstract only.

Stocek, R. F. 1995. The cougar, *Felis concolor*, in the Maritime Provinces. *Can. Field-Nat.* 109:19–22.

Wilson, D. E.; Reeder, D. M. 1993. *Mammal species of the world: A taxonomic and geographic reference*. Washington, D.C.: Smithsonian Institution Press in association with American Society of Mammalogists. 1,206 pp.

Wilson, D. E.; Ruff, S. (editors). 1999. *The Smithsonian book of North American mammals*. Washington, D.C.: Smithsonian Institution Press in association with American Society of Mammalogists. 750 pp.

Sources used to prepare maps of species' distributions in addition to expert reviews include:

Wild Mammals of New England. 1977. A. J. Godin. Baltimore, Md.: Johns Hopkins University Press.

Mammals of the Eastern United States. 1998. J. O. Whitaker, Jr., and W. J. Hamilton, Jr. Ithaca, N.Y.: Cornell University Press. 583 pp.

Handbook of Canadian Mammals—Marsupials and Insectivores. 1983. C. G. van Zyll de Jong, Ottawa: Canadian National Museum of Natural Sciences. 210 pp.

Handbook of Canadian Mammals—Bats. 1985. C. G. van Zyll de Jong. Ottawa: Canadian National Museum of Natural Sciences. 212 pp.

Mammalian Species. Various volumes (n.d.). Shippensburg, Pa.: American Society of Mammalogists.

Massachusetts Wildlife's State Mammal List. 1999. J. E.Cardoza; G. S. Jones; and T. W. French. Massachusetts Division of Fisheries & Wildlife website document at: <http://www.magnet.state.ma.us/dfwele/dfw/dfwmam.htm>.

Mammal Species

The following lists the references on mammals for further reading on life histories and distribution.

Citations by Region

North America

Chapman, J. A.; Feldhamer, G. A. 1982. *Wild mammals of North America, biology, management, and economics*. Baltimore, Md.: Johns Hopkins University Press. 1,168 pp.

Hall, E. R.; Kelson, K. R. 1959. *The mammals of North America*. 2 volumes. New York: Ronald Press.

Whitaker, J. O., Jr.: Hamilton, W. J., Jr. 1998. *Mammals of the eastern*

United States. 3rd edition. Ithaca, N.Y.: Cornell University Press. 583 pp.
Wilson, D. E.; Ruff, S. (editors). 1999. *The Smithsonian book of North American mammals.* Washington, D.C.: Smithsonian Institution Press in association with American Society of Mammalogists. 750 pp.

Canada

Banfield, A. W. F. 1974. *Mammals of Canada.* Toronto: University of Toronto Press. 438 pp.
Peterson, R. L. 1966. *The mammals of eastern Canada.* Toronto: Oxford University Press. 465 pp.
van Zyll de Jong, C. G. 1983. *Handbook of Canadian mammals. Marsupials and insectivores.* Ottawa: National Museums of Canada. 210 pp.
van Zyll de Jong, C. G. 1985. *Handbook of Canadian mammals. Bats.* Ottawa: National Museums of Canada. 212 pp.

Northeastern United States

Godin, A. 1977. *Wild mammals of New England.* Baltimore, Md. Johns Hopkins University Press. 304 pp.
Saunders, D. A. 1988. *Adirondack mammals.* Syracuse, N.Y.: State University of New York, College of Environmental Science and Forestry. 216 pp.

Great Lakes Region

Baker, R. 1983. *Michigan mammals.* East Lansing: Michigan State University Press. 642 pp.
Burt, W. H. 1957. *Mammals of the Great Lakes Region.* Ann Arbor: University of Michigan Press. 246 pp.
Hazard, E. B. 1982. *Mammals of Minnesota.* St. Paul: University of Minnesota Press. 286 pp.
Jackson, H. H. T. 1961. *Mammals of Wisconsin.* Madison: University of Wisconsin Press. 518 pp.
Jones, J. K., Jr.: Birney, E. C. 1988. *Handbook of mammals of the north-central states.* St. Paul: University of Minnesota Press. 346 pp.
Kurta, A. 1995. *Mammals of the Great Lakes region.* Ann Arbor: University of Michigan Press. 246 pp.

Central States Region

Mumford, R. E.; Whitaker, J. O., Jr. 1982. *Mammals of Indiana.* Bloomington: Indiana University Press. 537 pp.
Schwartz, C. W.; Schwartz, E. R. 1981. *The wild mammals of Missouri.* Columbia: University of Missouri Press and Missouri Department of Conservation. 356 pp.

Virginia Opossum

(Didelphis virginiana)

Range: Throughout the e. United States and s. Ontario, w. along Lakes Erie and Ontario to the Great Plains, s. to middle America. Also occurs along the West Coast as a result of transplants (Godin 1977:16, Gardner 1982:3, van Zyll de Jong 1983:32, Seidensticker et al. 1987:249). Has expanded range northward and westward during the past century (Hamilton 1958).

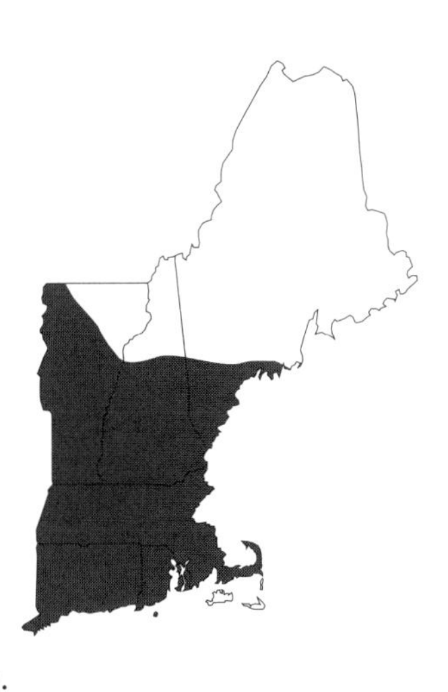

Distribution in New England: Opossums did not occur in New England before 1900. Now northern limits extend into s. Maine, central New Hampshire, and nw. Vermont. Winter severity limits northern distribution, which in Michigan is the −7°C January isotherm (Brocke 1970a, Tyndale-Biscoe 1973, van Zyll de Jong 1983:32).

Status in New England: Common to uncommon (Godin 1977:16).

Habitat: Dry to wet wooded areas; commonly found in wet woods near rivers and swamps, less often in wooded uplands or cultivated fields. Common near human habitation, where they are attracted to garbage.

Special Habitat Requirements: Dens—usually in abandoned burrow, tree cavity, hollow log, or brush pile; water (Llewellyn and Dale 1964, Hossler et al. 1994).

Reproduction: Age of sexual maturity: 8 to 12 months. Breeding period: Late January to early July in New York (Hamilton 1958). Gestation period: 13 days (Lay 1942). Young born: February to July in extremely undeveloped stage and remain in female's pouch for 60 days; weaned by days 96 to 106 (Gardner 1982:8). Litter size: 5 to 13, average 8 with larger litters in northern populations (Hartman 1923, Hartman 1952, Gardner 1982:8). Litters per year: 1 per year in north, 2 per year in s. United States, perhaps 3 per year further s. (Gardner 1982:8, Walker et al. 1975, McManus 1974).

Home Range: Recent radiotelemetry studies show discrete home ranges; males occupy larger areas averaging 270 acres (108 ha) that can overlap the home range areas of several females, which are smaller in size, averaging 127 acres (51 ha) and more stable (Fitch and Shirer 1970, Gillette 1980). Young begin to move on their own at 80 to 100 days, and gradually expand their range away from the maternal den (Gillette 1980). Juveniles and some females utilize sudden dispersal movements, whereas adult males make gradual shifts in their ranges (Gillette 1980).

Sample Densities: Density estimates are highly variable, ranging from lows of fewer than several opossums per square mile (less than 0.03/ha) in Iowa, Illinois, and Kansas (Wiseman and Hendrickson 1950, Verts 1963, Fitch and Sandidge 1953, Seidensticker et al. 1987:249), to highs of >150 opossums per square mile (>0.09/ha) in e. Texas, Illinois, and central New York (Lay 1942, Sanderson 1961, Holmes and Sanderson 1965, VanDruff 1971).

Food Habits: Insects, worms, fruits, nuts, carrion, and garbage; almost any vegetable or animal food (Lay 1942). Also preys on voles, shrews, and moles (Hamilton 1951, Taube 1947, Gardner 1982:22).

Comments: In winter, opossums become less active but do not hibernate (McManus 1971). Opossums are not active in air temperatures less than −12°C; survival limits are reached when air temperatures keep animals in underground dens more than 70 consecutive days (Brocke 1970b, van Zyll de Jong 1983:36). Individuals in the north often lack ears and tails due to frostbite. Avoids predators by feigning death and voiding noxious substances from its anal glands (Francq 1969), as well as by direct flight to a temporary refuge (e.g., grapevine tangles, trees, and ground holes) (Ladine and Kissell 1994). Gardner (1982:26) suggested that human activities—hunting, trapping, and roadkills—were the primary causes of opossum mortality. Seidensticker et al. (1987:249) reported that 35 percent of the animals equipped with radio transmitters or eartags were roadkilled, 4 percent were trapped, 17 percent died of excessive parasite loads, 9 percent died of winter exposure while in live traps, and 35 percent died of unknown causes. Opossums can form stable, hierarchical social relationships in captivity (Holmes 1991).

Masked Shrew

(*Sorex cinereus*)

Range: Throughout all but northernmost Canada and Alaska, s. through the Appalachians and central Washington to North Carolina and Tennessee, in the n. Rockies, s. to New Mexico and throughout the n. Great Lakes States (Godin 1977:22, Baker 1983:36, van Zyll de Jong 1983:67). The masked shrew is the most widely distributed shrew in North America (Godin 1977:22).

Distribution in New England: Widespread throughout New England in all terrestrial habitats.

Status in New England: Common to uncommon (Godin 1977:22).

Habitat: Damp deciduous and coniferous woodlands with grasses, rocks, logs, or stumps for cover; bogs and other moist swales. Less often in open country or in dry woods. In clear-cuts in West Virginia (Kirkland 1977b). Verme and Ozoga (1981) found shrew populations increased in clear-cuts and burned areas in swamp conifer cover in the Upper Peninsula of Michigan. Parker (1989) found masked shrews throughout young (2 to 17 yr old) conifer plantations in New Brunswick. In a 1-yr study, D'Anieri et al. (1987) found masked shrews a dominant component (48.6%) of the small mammal community in 4- to 5-yr-old glyphosate-treated clear-cuts in n. Maine. Vera and Servello (1994) found masked shrews a dominant component in regenerating spruce-fir stands treated with paper mill sludge in Maine. Occurred on 86 to 100 percent of forested plots in the White Mountains of New Hampshire and Maine (Yamasaki 1997).

Special Habitat Requirements: High humidity (moist sites) (Banfield 1974:9), ground cover especially leaves, rotten logs, herbaceous vegetation (Teferi and Herman 1995).

Reproduction: Age at sexual maturity: 4 months (Short 1961, Baker 1983:38, Teferi et al. 1992). Breeding period: Late April to late October or November (Banfield 1974:9, Teferi et al. 1992). Gestation period: Probably 18 days (Godin 1977:24, Peterson 1966:36, Baker 1983:38). Young born: Late April to September or October. Litter size: 4 to 10, average 7 (van Zyll de Jong 1983:71). Litters per year: 1 to 3 (Baker 1983:38, Kurta 1995:31).

Home Range: About 0.10 acre (0.04 ha) (Banfield 1974, Baker 1983:37).

Sample Densities: Densities of up to 9 individuals per acre (22 per ha) have been reported in favorable habitats (Banfield 1974:9). Density estimates vary from 4.9 to 74 shrews per acre (2 to 30 shrews per ha) (Buckner 1966). Year-to-year variation in summer populations can vary greatly from 9 to 93 shrews per acre (22.5 to 240 per ha) on mature hardwood sites at the McCormick Experimental Forest in Michigan's Upper Peninsula (Haverman 1973, Anderson 1977). Pitfall captures varied from 5.2 to 39.9 captures per 100 trap-nights in the White Mountains of New Hampshire and Maine (Yamasaki 1997). Captures per 100 trap-nights can fluctuate by a factor of 10 within a 5-yr period in New England forests (D. Snyder pers. commun., Yamasaki 1997).

Food Habits: Mainly insectivorous (lepidopteran larvae) and carnivorous (including young mice and salamanders) (Buckner 1964, Baker 1983:38, van Zyll de Jong 1983:70, Bellocq et al. 1994). Also consumes worms, spiders, snails, slugs, small amounts of vegetable matter, littoral amphipods and kelp flies (Hamilton 1930, Whitaker and French 1984). Seeds can be a winter food source (Criddle 1973). An important predator of sawfly cocoon populations (Buckner 1964). Considerable dietary overlap exists between masked and pygmy shrews in Michigan (Ryan 1986). Commonly feeds among litter on forest floor; known to forage in wrack deposits (decaying macroalgae along the strand line) in the upper intertidal zone in Nova Scotia (Stewart et al. 1989).

Comments: Young are weaned when approximately 20 days old (Forsyth 1976). Nests in grass, or under logs, rocks, or brush. Most active at night, especially on dark cloudy nights (van Zyll de Jong 1983:70, Teferi and Herman 1995). Precipitation also increases nocturnal activity (Buckner 1964, Doucet and Bider 1974, Vickery and Bider 1978). Adequate snow cover is essential to shrew survival as shrews remain active in subnivean habitats throughout winter (van Zyll de Jong 1983:70). Comprised 27.7 to 38.3 percent of small mammal samples on the White Mountains of New Hampshire and Maine (Yamasaki 1997).

Water Shrew

(Sorex palustris)

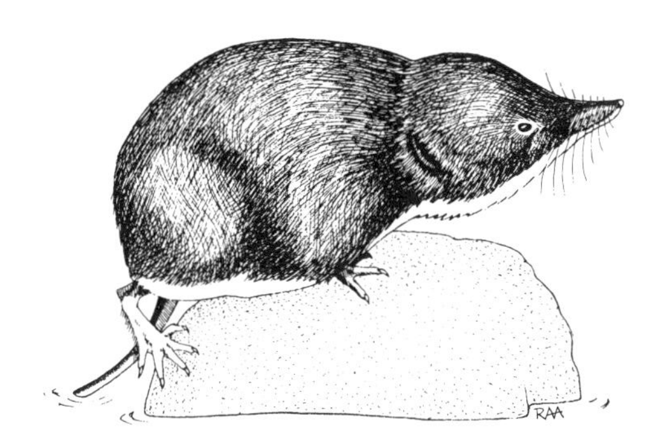

Range: Nova Scotia and s. Quebec, w. to British Columbia, s. through New England, much of New York, Pennsylvania, and the s. Appalachians. Also occurs in the Sierra Nevada and throughout the Rockies (Baker 1983:50, Beneski and Stinson 1987).

Distribution in New England: Throughout New England; near cold water bodies: swift flowing streams, beaver flowages, marshes, and bogs.

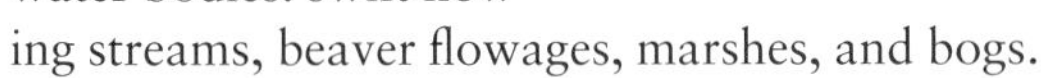

Status in New England: Uncommonly detected.

Habitat: Wet areas, especially grass-sedge marsh or shrub zones along ponds and streams in coniferous forest (Ozoga and Gaertner 1963, Rabe 1981, Timm 1975, Spencer and Pettus 1966, Wrigley et al. 1979). Also at wooded shores with favorable cover in the form of crevices beneath boulders, tree roots, or overhanging banks; can be found in areas away from water (van Zyll de Jong 1983:112). Occurred on 11 to 19 percent of forested plots in the White Mountains of New Hampshire and Maine (Yamasaki 1997).

Special Habitat Requirements: Herbaceous cover, cold water bodies (bog, stream, lake).

Reproduction: Age at sexual maturity: Possibly 9 months. Breeding period: Possibly February to August. Peak: Possibly March to July. Gestation period: Probably about 21 days (Conaway 1952, van Zyll de Jong 1983:114). Young born: Probably March to August. Litter size: 4 to 7. Litters per year: Possibly 2 to 3 are produced each year by mature females (Conaway 1952, Banfield 1974:14).

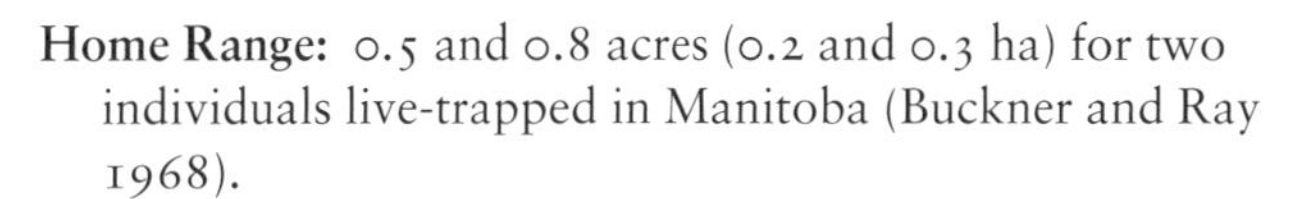

Home Range: 0.5 and 0.8 acres (0.2 and 0.3 ha) for two individuals live-trapped in Manitoba (Buckner and Ray 1968).

Sample Densities: Kirkland and Schmidt (1982) report low water shrew trapping success relative to other associated shrew species, which implies low densities (Beneski and Stinson 1987). Similar low trapping success (0.17 captures per 100 trap-nights) has been reported in the White Mountains (Yamasaki 1997).

Food Habits: Insectivorous—mainly eats mayfly, caddisfly, and stonefly larvae (Calder 1969, Conaway and Pfitzer 1952, Beneski and Stinson 1987, Whitaker and French 1984). Also takes snails, slugs, flatworms, small fish, and larval amphibians (Banfield 1974:14, Whitaker and Schmeltz 1973, Lampman 1947). Will cache surplus food and return later to finish the meal (Kurta 1995:42). Water shrew eats its weight in prey daily (Conaway 1952, Sorensen 1962).

Comments: Water shrews are solitary, active day and night, and throughout the winter (Sorensen 1962). Water shrews are known for being adept swimmers and divers (van Zyll de Jong 1983:113). Water shrews use tunnels dug by *Peromyscus*; dig their own short tunnels (less than 13 cm); and continually build and rebuild nests of small sticks and leaves (Sorenson 1962, van Zyll de Jong 1983:114). Little is known about the habits of this species; it has been found more than 100 m from streams in mature northern hardwood stands in northern New Hampshire (D. Rudis pers. commun.). Comprised 0.2 to 0.4 percent of small mammal samples on the White Mountains of New Hampshire and Maine (Yamasaki 1997).

Smoky Shrew

(Sorex fumeus)

Range: Maritime Provinces, s. Quebec and Ontario, w. to Lake Superior and s. Wisconsin; s. from n. New

England and New York through the Appalachian Mountains to ne. Georgia, Ohio, and Kentucky (Baker 1983:33, Owen 1984).

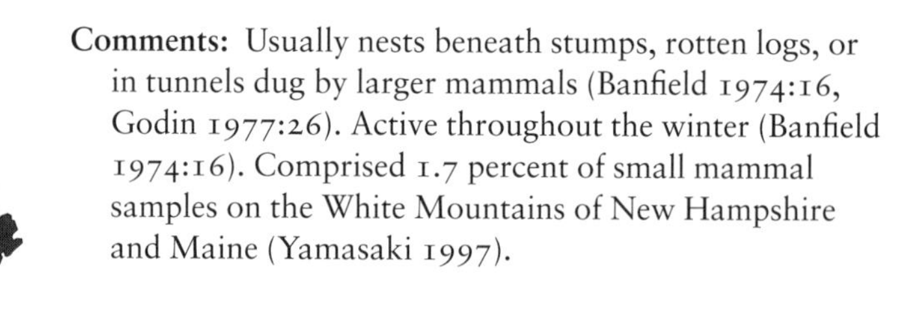

Distribution in New England: Throughout the region except absent from most of Rhode Island and se. Massachusetts.

Status in New England: Locally common to uncommon.

Habitat: Damp, boulder-strewn, upland deciduous and coniferous forests (often beech or maple, birch and hemlock) with thick leaf mold and old decaying logs (Owen 1984). DeGraaf et al. (1991) found significantly more smoky shrews in hardwoods than softwoods. Typically near streams with moss-covered banks (Burt and Grossenheider 1976, Godin 1977:26). Also uses early clear-cuts in coniferous woodlands (Kirkland 1977b), bogs and swamps (Owen 1984). Occurred on 32 to 40 percent of forested plots in the White Mountains of New Hampshire and Maine (Yamasaki 1997).

Special Habitat Requirements: Loose damp leaf litter—does not burrow, uses runways of other small mammals (Banfield 1974:16, Baker 1983:43) in shady wooded areas (Peterson 1966:38); open understory.

Reproduction: Age at sexual maturity: Spring following first winter (Owen 1984, van Zyll de Jong 1983:84). Breeding period: Late March to early August in New York (Godin 1977:27). Gestation period: About 20 days (Hamilton 1940). Young born: Mid April through August in New York (Hamilton 1940). Litter size: 5 to 6 (van Zyll de Jong 1983:84). Litters per year: Up to 3 (Hamilton 1940, Wrigley 1969).

Home Range: Baker (1983:43) suggests smoky shrews forage in a solitary manner in areas less than one acre (0.4 ha) in size.

Sample Densities: Ranged from 5 to 50 individuals per acre (12 to 123 per ha) in late summer in New York (Hamilton 1940). Richens (1974) trapped 0.1 to 0.92 animals per 100 trap-nights in nw. Maine. Pitfall captures varied from 0.75 to 3.97 captures per 100 trap-nights in the White Mountains of New Hampshire and Maine (Yamasaki 1997).

Food Habits: Mainly insectivorous, larvae and adults (80 percent) but also eats earthworms, spiders, snails, salamanders, small mammals, and birds (Hamilton 1940, Baker 1983:44, Owen 1984, Whitaker and French 1984).

Comments: Usually nests beneath stumps, rotten logs, or in tunnels dug by larger mammals (Banfield 1974:16, Godin 1977:26). Active throughout the winter (Banfield 1974:16). Comprised 1.7 percent of small mammal samples on the White Mountains of New Hampshire and Maine (Yamasaki 1997).

Long-tailed Shrew

(Sorex dispar)

Range: Nova Scotia, New Brunswick, central and w. Maine, and se. Quebec, s. in the Appalachians to North Carolina and Tennessee (Lincoln 1935, Martin 1966, Mather 1933, Osgood 1935, Starrett 1954, Kirkland et al. 1979, Kirkland 1981, Scott 1987).

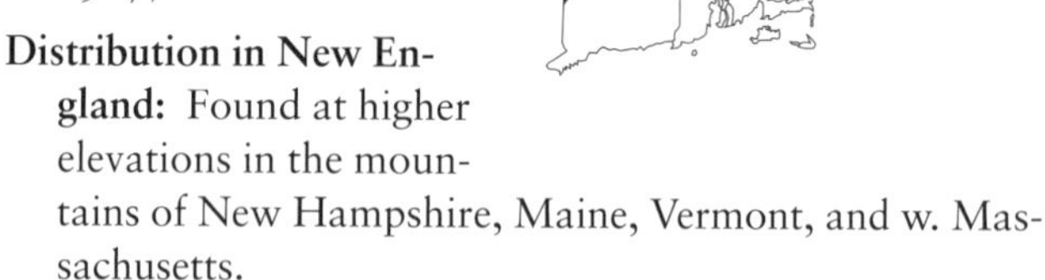

Distribution in New England: Found at higher elevations in the mountains of New Hampshire, Maine, Vermont, and w. Massachusetts.

Status in New England: Undetermined—possibly rare.

Habitat: Cold, damp coniferous forests, typically near moss-covered rocks and logs that provide shady protective crevices, or wooded talus slopes (Connor 1960, Richmond and Grimm 1950). Also found in deciduous and mixed forest (Burt and Grossenheider 1976). Occurred on 6 to 33 percent of forested plots in the White Mountains of New Hampshire and Maine (Yamasaki 1997). Found in a 1-yr-old red spruce clear-cut in West Virginia (Kirkland et al. 1976). Others have been taken in road construction rubble (Conaway and Pfitzer 1952).

Special Habitat Requirements: Rocky, wooded sites; dead woody debris; open understory; herbaceous vegetation.

Reproduction: Age at sexual maturity: Less than 1 yr. Breeding period: Possibly late April to August (Kirkland and Van Deusen 1979). Gestation period: Unknown. Young born: Probably May to August. Litter size: 2 to 5 reported; total of 4 records for litter size (Kurta 1995:33).

Home Range: Unreported.

Sample Densities: 7 individuals were trapped on 1 acre (0.4 ha) of talus in Pennsylvania (Richmond and Grimm 1950). Pitfall captures varied from 0.1 to 0.7 captures per 100 trap-nights in the White Mountains of New Hampshire and Maine (Yamasaki 1997).

Food Habits: Mainly insectivorous (Diptera, Coleoptera, and Orthoptera). Also eats centipedes and spiders (Connor 1960, Richmond and Grimm 1950, Conaway and Pfitzer 1952).

Comments: Little is known about this shrew. Probably is active during the day and at night (van Zyll de Jong 1983:91). Occasionally it is found in moderate numbers in favorable habitat and it is known to be partly subterranean. Composed 0.2 to 0.6 percent of small mammal samples on the White Mountains of New Hampshire and Maine (Yamasaki 1997). Also called the rock shrew.

Pygmy Shrew

(Sorex hoyi)

Range: Throughout Canada s. of the tundra to Alaska, s. to ne. Washington and n. Idaho; e. from ne. Montana to New England, and s. in the Appalachians to North Carolina (van Zyll de Jong 1983:121).

Distribution in New England: Northern New Hampshire, n. Vermont, and all but southernmost Maine.

Status in New England: Generally considered to be rare, but may be uncommon in n. New England.

Habitat: Found in a wide variety of deciduous and coniferous habitats, though some preference for mesic forest habitats has been noted; bogs, swamps, marshes, meadows, dry clearings, and savanna parkland (van Zyll de Jong 1983:123, Long 1972, Wrigley et al. 1979, Kirkland et al. 1987). Uses wet or moist habitats or less often dry areas close to water. Found in damp litter, especially near rotten stumps and logs in wooded areas. In New Hampshire, this species was more abundant in swamp hardwood than coniferous stands (DeGraaf et al. 1991). In a 1-yr study, D'Anieri et al. (1987) found pygmy shrews comprised 8.6 percent of the small mammals trapped in 4- to 5-yr-old glyphosate-treated clear-cuts in n. Maine. Occurred on 19 to 44 percent of forested plots in the White Mountains of New Hampshire and Maine (Yamasaki 1997).

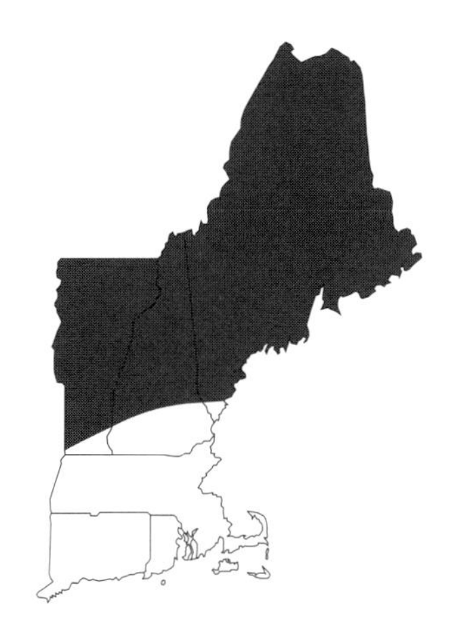

Special Habitat Requirements: Moist leafmold near water.

Reproduction: Age at sexual maturity: Spring following first winter (Kurta 1995:38). Breeding period: Uncertain—male lateral glands are visible from June to August and may indicate breeding period (van Zyll de Jong 1983:123, Long 1972). Gestation period: Probably 18 days (Kurta 1995:38). Young born: Uncertain—late June to early August. Litter size: Embryo counts from 3 to 8; average 5.75 (van Zyll de Jong 1983:123). Litters per year: May bear only 1 litter.

Home Range: Unreported.

Sample Densities: Estimates range from 0.52 per ha (Manville 1949) in the Upper Peninsula of Michigan to 0.7 to 1.2 per ha (Buckner 1964). Pitfall captures varied from 0.24 to 3.17 captures per 100 trap-nights in the White Mountains of New Hampshire and Maine (Yamasaki 1997). D'Anieri et al. (1987) trapped 0.59 animals per 100 trap-nights in 4- to 5-yr-old clear-cuts in n. Maine. Vera and Servello (1994) trapped 0.56 to 1.32 pygmy shrews per 100 trap-nights on paper mill sludge-treated clear-cuts in Maine.

Food Habits: Insectivorous, mainly hepialid larvae, grasshoppers, larch sawfly nymphs and adults, elaterid larvae and adults, large lepidopteran, and dipteran larvae, and spiders (van Zyll de Jong 1983:123, Long 1974, Whitaker and French 1984). Jack pine seeds were eaten in winter (Criddle 1973). Captive shrews known to scavenge rodent carcasses (Kurta 1995:38). Ants were avoided (Buckner 1964) in Manitoba; ants were a major dietary item in sympatric populations of both pygmy and masked shrews in Michigan (Ryan 1986).

Comments: Life history is little known. Captive individuals active for short periods throughout the day and night; peak activity occurs at night; behavior is similar to other *Sorex* species (Buckner 1964, Prince 1940, van Zyll de Jong 1983:123). Comprised 0.7 to 0.8 percent of small mammal samples on the White Mountains of New Hampshire and Maine (Yamasaki 1997). Use of pitfall traps increases the likelihood of observing pygmy shrews as well as the other smaller shrews (Prince 1941).

Northern Short-tailed Shrew

(*Blarina brevicauda*)

Range: Nova Scotia w. to Saskatchewan, s. to Nebraska and ne. Kansas, and s. Appalachians to n. Georgia (van Zyll de Jong 1983: 128, George et al. 1986).

Distribution in New England: Widespread throughout New England.

Status in New England: Common.

Habitat: Both forested and fairly open habitats (Miller and Getz 1977): deciduous, mixed, and less often coniferous forests with moist loose humus; especially common along stream banks and in meadows with tall rank grass or sedges, brush piles, and stone walls. Significantly more abundant in hardwoods than in softwoods (DeGraaf et al. 1991). Occurred on 28 to 92 percent of forested plots in the White Mountains of New Hampshire and Maine (Yamasaki 1997). Avoids dry, warm sites (Getz 1961a, Pruitt 1959). Favored grass-sedge marsh and willow-alder shrub zone in Manitoba (Wrigley et al. 1979).

Special Habitat Requirements: Low vegetation, loose leaf litter, high humidity.

Reproduction: Age at sexual maturity: Early females may mature in 6 weeks, but probably do not breed until a year after their birth. Breeding period: March to September. Gestation period: 17 to 22 days (van Zyll de Jong 1983:133, Kurta 1995:48). Young born: April to September. Litter size: 3 to 7, average 4.5 (van Zyll de Jong 1983:133). Litters per year: 3 or more (van Zyll de Jong 1983:133).

Home Range: Estimates average 2.5 ha (Blair 1940a, Blair 1941, Buckner 1966) with ranges overlapping other shrews (George et al. 1986).

Sample Densities: Densities up to 48 individuals per acre (119/ha) have been reported in good habitats (Banfield 1974:22). Pitfall captures varied from 0.2 to 2.3 captures per 100 trap-nights in the White Mountains of New Hampshire and Maine (Yamasaki 1997). D'Anieri et al. (1987) trapped 0.59 animals per 100 trap-nights in glyphosate-treated clear-cuts in n. Maine.

Food Habits: Earthworms (Oligochaeta), millipedes (Diplopoda), and insect larvae are major prey items (Mumford and Whitaker 1982, Linzey and Linzey 1973, Whitaker and French 1984). Voles and mice are uncommon prey items (Eadie 1944, 1948, 1949, 1952). Plants, sowbugs, snails, slugs, centipedes, and spiders are other common food items (Banfield 1974:23). May take nestlings from ground nests of small birds.

Comments: Active day and night throughout the year; the short-tailed shrew is one of the most abundant small mammals in New England (Godin 1977:30). More fossorial than other shrews; digs own tunnels and uses burrows of other vertebrate species, especially voles (Hamilton 1931a). Has a poisonous bite that allows for preying on larger mice and voles (Lawrence 1946), and immobilizes snails, earthworms, and beetles (Martin 1981). Comprised 1.0 to 6.7 percent of small mammal samples on the White Mountains of New Hampshire and Maine (Yamasaki 1997).

Least Shrew

(Cryptotis parva)

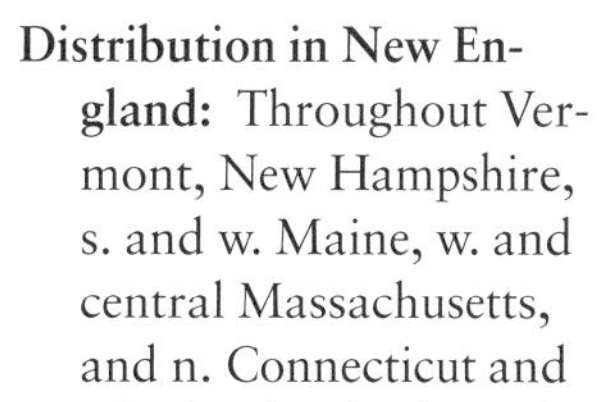

Range: Southwestern Connecticut, w. through central New York to South Dakota, s. through e. Texas and Florida to Central America (Jarrell 1965, Whitaker 1974).

Distribution in New England: Southwestern coastal Connecticut (Goodwin 1932, Goodwin 1942).

Status in New England: Undetermined—may be fairly common. Seldom caught in traps, but remains are often found in owl pellets (Banfield 1974:25).

Habitat: Open, dry grassy fields with or without scattered brush, also salt marshes, woodland edges (Banfield 1974:25, Godin 1977:34).

Special Habitat Requirements: Dry fields with loose soils for tunnels; often uses runways of larger mice and shrews (Davis and Joeris 1945).

Reproduction: Age at sexual maturity: About 40 days (Godin 1977:35). Breeding period: Early March to early November (at northern edge of range) (Hamilton 1944). Gestation period: 21 to 23 days (Mock 1970). Young born: Late March to late November. Litter size: 3 to 7, average 4.5 (Conaway 1958, Whitaker 1974, Kivett and Mock 1980). Litters per year: Probably 2 to 3.

Home Range: Estimated at 1,700 m^2 (0.41 acres) for one male and 2,800 m^2 for one female (Howell 1954, van Zyll de Jong 1983:139).

Sample Densities: Howell (1954) estimated densities of 1.73 to 4.95/ha (Whitaker 1974).

Food Habits: Insects (lepidopteran and coleopteran larvae and orthopterans), mollusks, centipedes, slugs, sow bugs, earthworms, frogs, lizards, *Endogone*, and vegetable matter (Hamilton 1944, Whitaker 1974, van Zyll de Jong 1983:139).

Comments: Rarely nests in burrows. More often uses hollows under stones, logs, or stumps. Highly social—31 individuals were found in one winter nest (McCarley 1959).

Hairy-tailed Mole

(Parascalops breweri)

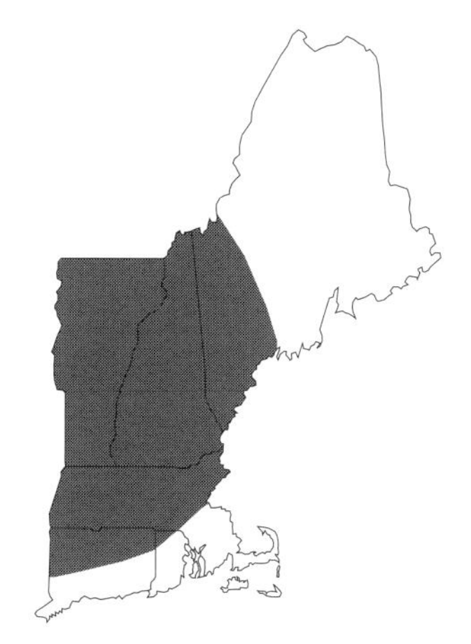

Range: New Brunswick, se. Canada, e. Ontario, s. through w. Maine and the Appalachian Mountains to e. Ohio and w. North Carolina (Hallett 1978, van Zyll de Jong 1983:159).

Distribution in New England: Throughout Vermont, New Hampshire, s. and w. Maine, w. and central Massachusetts, and n. Connecticut and Rhode Island. Throughout the light textured soils and mixedwood forests of New England.

Status in New England: Locally common.

Habitat: Open woods, grassy meadows, and ungrazed pastures with light, sandy loams and sand. Prefers areas with surface vegetative cover and sufficient moisture. Avoids heavy wet soils (Kurta 1995:51). Occurred on roughly 2 percent of forested plots in the White Mountains of New Hampshire and Maine (Yamasaki 1997).

Special Habitat Requirements: Loose, well-drained sandy loam soil.

Reproduction: Age at sexual maturity: 10 months (Eadie 1939). Breeding period: Late March and early April in New Hampshire (Eadie 1939). Gestation period: Four to six weeks (Hallett 1978). Young born: May and June (Hallett 1978). Litter size: 4 or 5, average 4. Litters per year: 1 (possibly 2).

Home Range: About 0.2 acre (0.1 ha) (Eadie 1939).

Sample Densities: An average density of 1.2 moles per acre (3 per ha) on 27 acres (11 ha) and a maximum density of 11 individuals per acre (27 per ha) has been reported in various habitats in New Hampshire (Eadie 1939). 10 to 12 moles per acre (25 to 30 per ha) have been reported in New York (Hamilton 1939a).

Food Habits: Earthworms, insects (especially Coleoptera, Diptera, Lepidoptera, and Hymenoptera), millipedes, centipedes, snails, slugs, and sowbugs (Eadie 1939, Hamilton 1941, Jensen 1986), forages on forest floor at night.

Comments: Fossorial; constructs two tunnel systems—one shallow (just below surface), the other deep (10 to 18 in, 25 to 46 cm) (Jensen 1986). Permanent deep tunnels are sites of breeding and winter nests (Eadie 1939, Gorman and Stone 1990) and may be used for several years.

Eastern Mole

(*Scalopus aquaticus*)

Range: Massachusetts w. to Wyoming, s. to central Texas and the Gulf of Mexico (Yates and Schmidly 1978).

Distribution in New England: Throughout Massachusetts, including Cape Cod, Connecticut, and Rhode Island (Yates and Schmidly 1978). The eastern mole is the most widely distributed mole in North America (Godin 1977:38).

Status in New England: Locally common.

Habitat: Pastures, meadows, lawns (less often in open woodland) in moist, loamy or sandy soils that permit easy digging (Arlton 1936). Often in moist (not wet) bottomlands where earthworms are plentiful; avoids heavy clay, stony, or gravely soils (Arlton 1936).

Special Habitat Requirements: Soft moist soils containing earthworms.

Reproduction: Age at sexual maturity: 1 yr. Breeding period: Peaks in late March and April (Conaway 1959). Gestation period: Four weeks to 45 days (van Zyll de Jong 1983:155, Kurta 1995:56). Young born: Late April or May. Litter size: 3 to 5, average 4 (van Zyll de Jong 1983:155). Litters per year: 1. Young are independent when about 1 month old (Godin 1977:39).

Home Range: Average area 0.74 acre (0.3 ha) for 7 moles in Kentucky, 4 males averaged 1.09 acre (0.4 ha), 3 females averaged 0.28 acre (0.1 ha) (Harvey 1976).

Sample Densities: Unknown.

Food Habits: Earthworms, insects (especially white grubs), and vegetable matter (Arlton 1936, Hisaw 1923b, Yates and Schmidly 1978, van Zyll de Jong 1983:154).

Comments: Fossorial; active throughout the year during hours of day and night except early morning and early evening (Harvey 1976). Digs tunnels just below surface for foraging and in dry or cold weather excavates deeper burrows 10 in or more deep as living quarters and thoroughfares to feeding areas (Hisaw 1923a, Yates and Schmidly 1978). Solitary except during breeding season (Leftwich 1972).

Star-nosed Mole

(*Condylura cristata*)

Range: Southern Labrador, w. to sw. Manitoba, s. through n. Indiana and Ohio, and the Appalachians through w. North Carolina and along the coast to the

ne. corner of North Carolina (Petersen and Yates 1980).

Distribution in New England: Throughout New England.

Status in New England: Common to uncommon.

Habitat: Occurs in low wet ground near bodies of water, swamps, wet meadows, occasionally wet spots in fields or low-lying woods. Also in mixed hardwood stands with dry soils but near water.

Special Habitat Requirements: Wet, mucky humus.

Reproduction: Age at sexual maturity: 10 months. Breeding period: April and May. Gestation period: About 45 days. Young born: May and June. Litter size: 3 to 7, average 5.4 (Davis and Peek 1970, Eadie and Hamilton 1956). Litters per year: 1.

Home Range: Probably about 1 acre (0.4 ha) (Banfield 1974:36).

Sample Densities: 25 and 41 moles/ha (10 and 17 moles per acre) have been reported in late winter in New York (Eadie and Hamilton 1956).

Food Habits: Aquatic annelids, leeches, aquatic insects (especially caddisfly and midge larvae, and plecopterans), crustaceans, mollusks, and small fish (occasionally). Terrestrial foods include earthworms, white grubs, isopods, and small amounts of vegetable material (Hamilton 1931b). Forages above ground at night.

Comments: An excellent swimmer, has been found swimming under the ice of streams and ponds (Hamilton 1931b). Spends much time foraging in bottom sediments of wetlands (van Zyll de Jong 1983:180). Gregarious, usually lives in small colonies (Eadie and Hamilton 1956). Much less fossorial than other moles (Petersen and Yates 1980, Hamilton 1931b). Star tentacle may be a sensitive tactile structure (Kurta 1995:58) or an electroreceptor, capable of detecting slight electrical fields emitted by aquatic prey (Gould et al. 1993).

Little Brown Myotis

(*Myotis lucifugus*)

Range: Labrador w. to s. Alaska across the limits of the boreal forest; s. throughout much of the 48 states from Georgia to Arkansas and s. California to the central highlands of Mexico (van Zyll de Jong 1985:70).

Distribution in New England: Throughout New England.

Status in New England: Common (Griffin 1940a, van Zyll de Jong 1985:70).

Habitat: Variable depending on season and setting. Three roost types can be described: hibernation (overwintering) sites, day, and night roosts (Fenton and Barclay 1980). Known hibernacula are usually caves and abandoned mines or hydroelectric dams (Griffin 1945, Fenton and Barclay 1980, Kurta and Teramino 1994). Hibernacula characteristics include temperatures between 2 and 5°C, and high humidity (>85 percent) (Fenton and Barclay 1980, Kurta 1995). Pregnant females seek warmer day roosts than males, using attics and barns in spring and early summer where they often form maternity colonies. Females evicted from maternity colonies in buildings were not seen in other nearby maternity colonies (Neilson and Fenton 1994) or in constructed bat shelters. Adult males and nonparous females use day roosts singly and in small groups, away from maternity colonies (Fenton and Barclay 1980). Communal night roosting is an important thermoregulatory mechanism for pregnant females in order to maintain high body temperatures for rapid embryo development and subsequent survival of overwintering young (Barclay

1982). Solitary night roosting is related to ambient night temperatures and may be an expression of the travel costs to a roost and the thermoregulatory roost benefits (Barclay 1982), as well as lower insect abundance at temperatures below 5°C (Anthony et al. 1981), or the need to digest a meal (Barclay 1982). Night roosts are rarely used when ambient temperatures exceed 15°C (Fenton and Barclay 1980).

Special Habitat Requirements: Females seek dark warm sites for maternity colonies due to increased energy costs of pregnancy and lactation (Kurta et al. 1987) in late April and disperse from July to mid September (Davis and Hitchcock 1965). Males use cooler daytime roosts, frequently in valleys near streams and marshes, and at higher slope positions in the mountains than do females (Krusic 1995, Sasse 1995). Males maintain high daytime body temperatures in order to be physiologically ready for the late-summer mating period, rather than entering energy-conserving torpor (reviewed in Kurta and Kunz 1988). Smaller wooden roost cavities provide more total insulation for solitary bats than larger cavities (Kurta 1985); clusters of individuals in roost cavities significantly increase the insulation effect (Kurta 1985).

Reproduction: Age at sexual maturity: About 4 to 9 months for females, longer at higher latitudes, at least 1 yr for males (van Zyll de Jong 1985:76, Baker 1983:100, Schowalter et al. 1978, Schowalter et al. 1979). Breeding period: Swarming behavior is a prelude to active phase breeding, which begins in August and continues through autumn; passive phase breeding with torpid females may occur in late fall (Fenton and Barclay 1980, Wai-Ping and Fenton 1988); ovulation and fertilization delayed until spring (Thomas et al. 1979); and mating is random and promiscuous (Wai-Ping and Fenton 1988). Gestation period: 50 to 60 days (estimate) (Wimsatt 1945), depending on the weather (Kurta 1995:68). Young born: Mid June to early July (Cagle and Cockrum 1943). Litter size: 1; twins are rare (van Zyll de Jong 1985:75). Litters per year: 1.

Home Range: Unknown.

Sample Densities: Summer density: Average 26 bats per square mile (10/km^2) over an 8,600-square-mile (22,274-km^2) area served by a cave in s. Vermont. Winter density: In caves in s. Vermont, 300,000 ± 30,000 (Davis and Hitchcock 1965). Little brown bats accounted for 71 percent of the mist net captures in 752 net hours at 18 sites [elevations range from 280 to 622 m (920 to 2040 ft)] in the White Mountains (Sasse 1995). Male little brown bats accounted for 62 percent of mist-net and harp-trap captures by Krusic (1995).

Food Habits: Congregate over water to drink and hawk flying insects (varying in size from 3 to 10 mm long), especially midges, mayflies, caddisflies, and mosquitoes, but also beetles, moths, stoneflies, true bugs, and termites (Anthony and Kunz 1977, van Zyll de Jong 1985:73). Also forages in areas of group selection softwood regeneration and along forest roads, trails, streams, ponds, and lakes in a forest-dominated landscape (Krusic et al. 1996). Nightly consumption of insects for pregnant females averages 2.5 g, lactating females 3.7 g, and juveniles 1.8 g (Anthony and Kunz 1977 in van Zyll de Jong 1985:73). Aerial foraging activities consume 61 to 66 percent of the average daily metabolized energy budget in females (Kurta et al. 1989). Typically forages in two nocturnal periods, from emergence after sunset to midnight and again just before dawn when returning to day roost locations (van Zyll de Jong 1985:73). Relatively low wing loading to aspect ratio; high maneuverability allows for foraging in forested habitats (Fenton 1990, Barclay 1991, Krusic and Neefus 1996). Experimental increases in canopy structure at forest clear-cut edges reduced foraging activity by small myotids (Brigham et al. 1997).

Comments: Breeding colonies of 12 to 1,200 little browns have been reported in Vermont. Females exhibit site fidelity to maternity colonies (Kurta 1995:68, Fenton and Barclay 1980). Longevity is high (banded males in se. Ontario in excess of 31 yr) (Keen and Hitchcock 1980, Fenton and Barclay 1980). Screech-owl predation can reduce bat use of nursery colonies (Barclay et al. 1982). Other bat predators include small carnivores, mice, raptors, and snakes (Fenton and Barclay 1980). However, maternity colonies in buildings are at risk from building modifications that potentially seal young inside buildings (Neilson and Fenton 1994). Repairs to exclude bats should be done in late fall/early winter in order to avoid this situation.

Northern Long-eared Bat, or Northern Myotis

(*Myotis septentrionalis*)

Range: Newfoundland and Nova Scotia, w. to Saskatchewan, s. to Wyoming and n. Florida (van Zyll de Jong 1985: 93, Kurta 1995:69).

Distribution in New England: Throughout New England, especially in the White Mountains of New Hampshire (Sasse 1995). Common during spring and summer on Martha's Vineyard (Buresch 1999).

Status in New England: Common to uncommon (Fitch and Shump 1979). Locally and irregularly distributed within its range.

Habitat: Known hibernacula are caves and abandoned mine shafts, or hydroelectric dams with temperatures between 2 and 7°C (Kurta 1995:69, Kurta and Teramino 1994), high humidity, and where the air is still (Fitch and Shump 1979). Hibernate singly or in small groups of two or three bats, tucked into small cracks and drill holes (Kurta 1995:69). Hibernation occurs in mixed species groups, especially with *Myotis lucifugus* and *Eptesicus fuscus* in the same location (Kurta and Teramino 1994). Females congregate in maternity colonies in groups of up to 60; males roost singly (Kurta 1995:69). Some use human dwellings (behind shutters and wooden shingles) and outbuildings for shelter; others use tree cavities for small nursery colonies (Sasse 1995) and loose bark crevices as roosts (Kurta 1995:69). Roost trees in forested landscapes tend to be clustered in stands of large trees: those with larger average live tree diameters, more total dead tree basal area, and more hardwood than softwood basal area than the surrounding forested area (Sasse 1995).

Special Habitat Requirements: A majority (66 percent) of roost trees used by radio-tagged females were in dead hardwoods (predominantly beech, sugar maple, and yellow birch) with cavities (Sasse and Pekins 1996); roosts in live silver maples with cavities, hollow green ash, and under the loose bark on dead trees were reported by Kurta (1995:69). On average, roost trees were larger [16.1 in (40.9 cm) diameter breast height], taller, and less decayed than the average available dead tree in the surrounding area (Sasse and Pekins 1996).

Reproduction: Age at sexual maturity: About 6 to 9 months for females, 1 yr for males. Breeding period: September to October with ovulation and fertilization delayed until spring. Gestation period: About 60 days (Kurta 1995:71). Young born: Mid June to early July (Kurta 1995:71, Sasse 1995). Litter size: 1. Litters per year: 1.

Home Range: Unknown.

Sample Densities: Several hundred individuals were observed hibernating in caves in Canada (Hitchcock 1949). Periodic hibernation counts in New Hampshire caves and abandoned mines range from high tens to low hundreds (New Hampshire Natural Heritage Inventory staff pers. commun.). Northern long-eared bats accounted for 27 percent of the mist-net captures in 752 net hours at 18 sites [elevations range from 280 to 622 m (920 to 2040 ft)] in the White Mountains (Sasse 1995). Northern long-eared bats accounted for 12 percent of mist-net and harp-trap captures by Krusic (1995).

Food Habits: Forages over ponds and clearings and within forests, below the tree canopy but above the shrub layer (Cowan and Guiguet 1965, Fitch and Shump 1979); and may glean stationary prey off of leaf surfaces (Faure et al. 1993). Stomachs of three individuals in Indiana contained assassin bugs, moths, butterflies, flies, leaf hoppers, and other unidentified insects (Whitaker 1972a); stone flies, caddisflies, weevils, scarab beetles, bark lice, and lacewings were also mentioned (van Zyll de Jong 1985:94, Griffith and Gates 1985, Kurta 1995:69). Lowest wing loading to aspect ratio of all New England bats; high maneuverability allows foraging in thick forest habitats (Fenton 1990, Barclay 1991, Krusic and Neefus 1996).

Comments: Almost undetectable using a broadband ultrasonic detector system (Krusic and Neefus 1996). Gating cave and mine entrances to maintain temperature and humidity patterns but limit human access to hibernacula sites is a practice used in many places (Kurta 1995:71). Current longevity record in the wild is 19 yr (Kurta 1995:71).

Indiana Myotis

(*Myotis sodalis*)

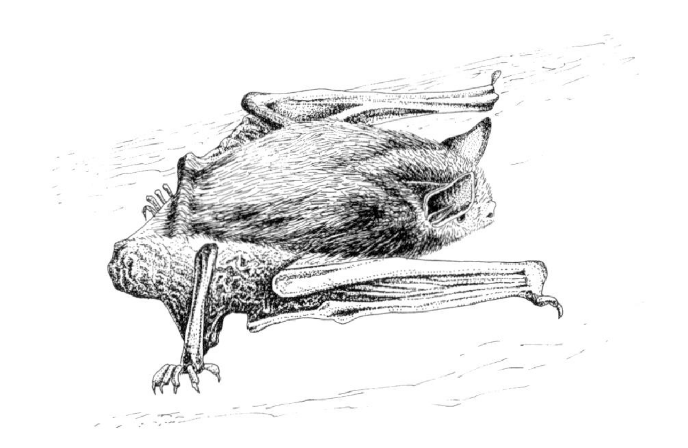

Range: Eastern New York, s. Vermont, New Hampshire, and Massachusetts; w. to s. Wisconsin and Michigan, s. to Georgia (Appalachians); nw. Arkansas and Oklahoma (Kurta et al. 1993a, Jones and Birney 1988, Thomson 1982, Kurta 1995:73, Brady et al. 1983).

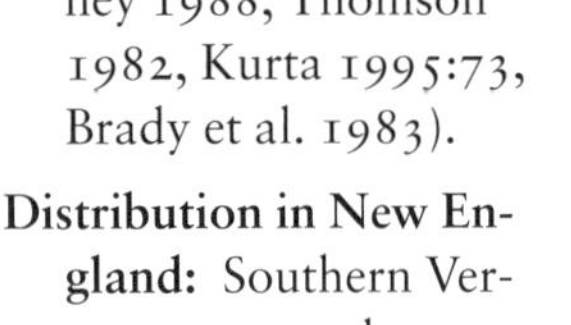

Distribution in New England: Southern Vermont, one undocumented report in the White Mountains of New Hampshire (Krusic 1995), w. Massachusetts, and nw. Connecticut. Closest known hibernacula just n. of Lake George in n. New York (A. Hicks, N.Y. DEC, pers. commun.) and Dorset, Vt. (C. Grove, USFS., pers. commun.).

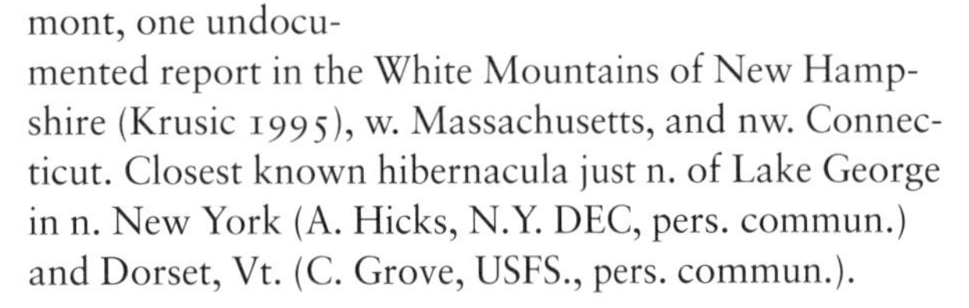

Status in New England: Rare and endangered.

Habitat: Favored hibernacula sites are limestone caves and mines that maintain midwinter ambient temperatures between 4 and 8°C and high humidity (87 percent) (Humphrey 1978, Thomson 1982). Commonly forages in dense floodplain forest just above or below the canopy (Kurta 1995:73). Maternity colonies found in upland and riparian forests, pastures, and open wetlands (Kurta 1995:73). Use of maternity colony roost trees shifts among the cluster [within 150 m (492 ft)] of suitable saw-timber-sized dead trees with sloughing bark (Kurta et al. 1996).

Special Habitat Requirements: Uses loose and peeling bark on dead or dying, large [40.9 cm (16.1 in) diameter breast height] hardwood trees in a seasonally flooded wetland, subjected to full sunlight, as maternity colony site (Kurta et al. 1993a, Kurta et al. 1996); uses loose bark on ash, hickories, elms, and sycamores (Humphrey et al. 1977, Kurta et al. 1993b). Individuals move among a cluster of roost trees throughout the summer, perhaps in response to changing microclimates or the ephemeral nature of loose and peeling bark (Gardner et al. 1991, Kurta et al. 1993a, Kurta et al. 1996).

Reproduction: Age at sexual maturity: About 6 months. Breeding period: Swarming and subsequent mating occur from mid August to late October (Thomson 1982). Hibernation period may last from mid September to early June (averages mid October to mid April). Gestation period: Probably 60 days, similar to the little brown bat (Kurta 1995:74). Young born: Late June. Litter size: 1.

Home Range: Pregnant females range over 52 ha (128 acres); lactating females range over 95 ha (232 acres) (Kurta 1995:74).

Sample Densities: Typically hibernates in dense, sizable single-species clusters; occasionally individuals roost within groups of little brown bats (Kurta and Teramino 1994, Mumford and Whitaker 1982). Occurs in tightly clustered groups in densities of 3,200 bats/m^2 (300 bats/ft^2) in hibernacula sites (Clawson et al. 1980). Exit counts of pregnant and lactating females and young at maternity colony roost trees ranged from 1 to 45, and the majority (89 percent) of counts ranged from 2 to 21 bats (Kurta et al. 1996).

Food Habits: Forages in the foliage of tree crowns 7 to 98 ft (2 to 30 m) tall along the shores of rivers and lakes and over floodplains (Humphrey et al. 1977). Four stomachs examined in Indiana contained ichneumons, leafhoppers, beetles, and unidentified wasps (Whitaker 1972a). Samples from 140 adults contained insect parts from the orders Lepidoptera, Coleoptera, Diptera, Trichoptera, Plecoptera, Homoptera, Hemiptera, and Ephemeroptera (Brack and LaVal 1985).

Comments: Maximum body temperatures measured at maternity colony roosts routinely exceeded 35°C to 40°C with no associated stress (Kurta et al. 1996); contrary to mortality noted at body temperatures >35°C (Henshaw and Folk 1966). Band recoveries revealed seasonal movements of up to 320 miles (512 km) (Hall 1960, Hall 1962). Seven caves in Indiana, Missouri, and Kentucky housed 85 percent of the total estimated population of Indiana bats in the 1970s (Brady et al. 1983). Populations had decreased 28 percent from 1960 to 1975 (Humphrey 1978). Gating hibernacula entrances to protect wintering environments from disturbance and vandalism can alter cave microclimates (Richter et al. 1993), requiring careful application of design standards (Brady et al. 1983). Observed longevity of 14 to 20 yr in the wild (LaVal and LaVal 1980, Thomson 1982, Kurta 1995:74).

Eastern Small-footed Myotis

(*Myotis leibii*)

Range: Southeastern Ontario, s. Quebec, and central Maine, s. along the Appalachians to n. Georgia and Alabama, and w. to Arkansas and Missouri (van Zyll de Jong 1985:118, Kurta 1995:65, Best and Jennings 1997).

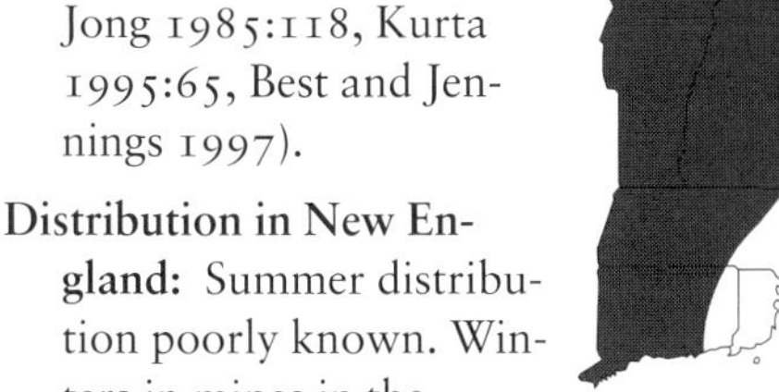

Distribution in New England: Summer distribution poorly known. Winters in mines in the White Mountains of New Hampshire (Krusic 1995) and the Adirondacks of New York (Whitaker and Hamilton 1998:90). Reported only from Hampden County, Mass. (Cardoza et al. 1999).

Status in New England: Uncommon.

Habitat: Favors dry, drafty, and cool (less than 4°C) hibernacula: near the entrances of mines, mine tunnels, and caves (Kurta 1995, Best and Jennings 1997). Will hibernate with *M. lucifugus, M. septentrionalis*, and *Eptesicus fuscus* (New Hampshire Natural Heritage Inventory staff pers. commun.). Females form small maternity colonies, often in rocky crevices on cliffs and sometimes in crevice-like places on buildings (Hitchcock 1955, Kurta 1995:65). Males are solitary.

Special Habitat Requirements: Tolerates cooler temperatures than most bat species, tends to enter hibernation last (late November) and leave first (early April) in e. Ontario (Fenton 1972), and similarly in Vermont (Mohr 1936). Tends to hibernate singly and in small groups, rather than in large clusters; hanging from the ceiling and walls, as well as occasionally under rocky rubble on the floor (Jones and Birney 1988, Kurta 1995:65). Warm weather day roosts include buildings, rock slabs, and under a stone (van Zyll de Jong 1985:119).

Reproduction: Age at sexual maturity: Unknown. Breeding period: Unknown. Gestation period: Unknown. Young born: Late May to early July (Quay 1948, Jones and Birney 1988, Kurta 1995:65). Litter size: Probably 1.

Home Range: Banding returns suggest small-footed bats travel fairly short distances [less than 40 km (25 mi)] between summer habitats and overwintering hibernacula (Hitchcock 1955, van Zyll de Jong 1985:119, Kurta 1995:65).

Sample Densities: Hibernacula counts average less than 20; but 400 individuals have hibernated in the same cave (Kurta 1995). Maternity colony numbers range from 12 to several dozen adult females and young (Jones and Birney 1988, Kurta 1995:65).

Food Habits: Unknown. Probably similar to other myotids. Flies low to the ground (1 to 3 m) and erratically (van Zyll de Jong 1985:119). Flies, bugs, beetles, leafhoppers, and flying ants found in stomachs of two specimens (Cockrum 1952, Kurta 1995:65).

Comments: Associated with caves in the foothills of mountains up to 2,000 ft (610 m) in coniferous woodlands (hemlock, spruce, white cedar) (Hitchcock 1949). Female survivorship is almost half that of males, perhaps due to greater reproductive effort expended, higher metabolic rates, and longer daytime activity periods through the summer (Hitchcock et al. 1984). Observed longevity of 12 yr in the wild has been reported (van Zyll de Jong 1985:119).

Silver-haired Bat

(*Lasionycteris noctivagans*)

Range: Southern Canada, w. to s. Alaska, s. to central California, central Texas, and South Carolina (van Zyll de Jong 1985:130, Kunz 1982).

Distribution in New England: Summer resident. Seen during migration foraging over marshes, lakes and ponds in the White Mountains (Krusic 1995). Detected during ultrasonic surveys on Martha's Vineyard (Buresch 1999).

Status in New England: Uncommon.

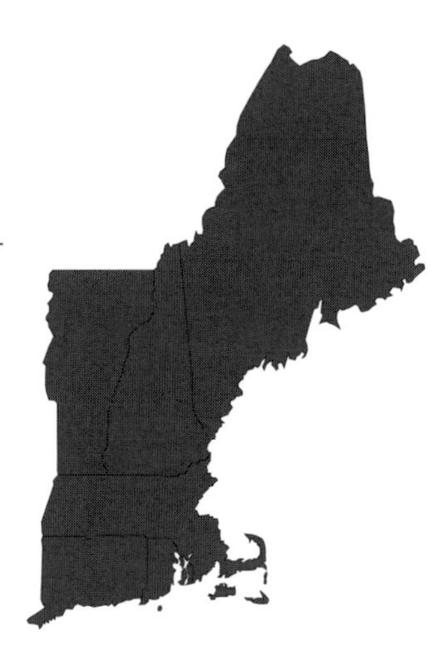

Habitat: Forages in hardwood clear-cuts (Krusic and Neefus 1996); coniferous and mixedwood forest areas near lakes, streams or ponds. Tends to forage in fairly open habitats, given the higher aspect ratio and high wing loading morphology required for long-distance flight (Krusic and Neefus 1996). Roosts in hollow trees and cavities, under loose bark, furrowed bark folds, and occasionally in buildings (Barclay et al. 1988, Kurta 1995:80).

Special Habitat Requirements: Roosts in *Salix* sp., *Acer* sp., and *Fraxinus* sp. between 1 and 5 m (3 to 16 ft) above ground in the Lake States (Kurta 1995:80). Dead trees with loose bark or cavities for summer roosting sites and nearby water courses.

Reproduction: Age at sexual maturity: First summer. Breeding period: Late September with delayed fertilization. Gestation period: 50 to 60 days (Druecker 1972). Young born: June or July. Litter size: 2 (occasionally 1) (Kurta and Stewart 1990). Litters per year: 1.

Home Range: Unknown.

Sample Densities: Exit counts from summer silver-haired bat roost trees ranged from 1 to 35 (average 12) in s. British Columbia (Vonhof 1996). Silver-haired bats accounted for less than 1 percent of the mist-net captures in 752 net hours at 18 sites [elevations range from 280 to 622 m (920 to 2040 ft)] in the White Mountains (Sasse 1995).

Food Habits: Forages near sunset over forest clearings, ponds, and streams; slow, acrobatic flier (Kurta 1995:80); often less than 20 ft above surface; may prefer emerging aquatic insects (Banfield 1974:54). An opportunistic feeder, consuming flies, beetles, moths, and swarms of anything else encountered (Barclay 1985, Kurta 1995:80); can pursue prey at close range using multiharmonic search-approach calls (Barclay 1986).

Comments: Migrates along well-defined coastal flyways in the Northeast to s. parts of range, probably in late October, and returns in April (Kurta 1995:80). Generally hibernates under loose bark or in tree cavities or buildings. Varies widely in abundance throughout its wide range. Silver-haired, hoary, and red bats appear to minimize foraging competition through several mechanisms: foraging at different times, using different types of search-approach calls for long- and short-range hunting, and selecting different prey (Reith 1980, Kunz 1973, Barclay 1985, 1986). Longevity in the wild is 12 yr, as estimated by tooth wear patterns (Kurta 1995:81). Both sexes tend toward solitary living (van Zyll de Jong 1985:131); sometimes females (up to a dozen) form maternity colonies (Kurta 1995:80).

Eastern Pipistrelle

(*Pipistrellus subflavus*)

Range: Southeastern Canada w. to ne. Minnesota, s. to e. Mexico and Honduras. Probably absent from n. Maine and much of lower Michigan and e. Upper Peninsula of Michigan (van Zyll de Jong 1985:124, Kurta 1995:85, Knowles 1992).

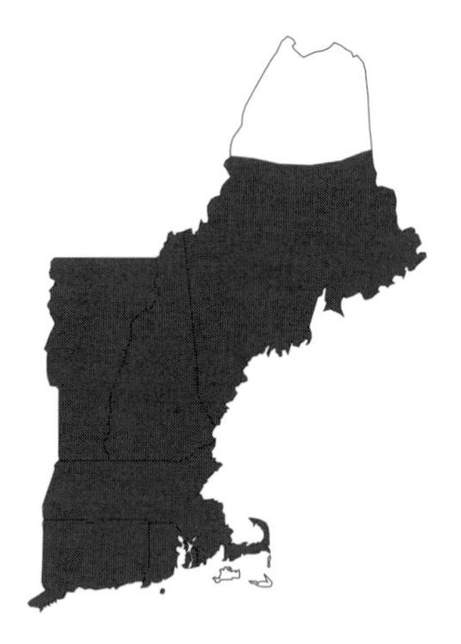

Distribution in New England: Summer resident, breeding distribution unknown. Winters in abandoned mines in central New Hampshire. Detected during ultrasonic surveys on Martha's Vineyard (Buresch 1999).

Status in New England: Uncommon to rare.

Habitat: Forages over water and along forest-field edges (Davis and Mumford 1962); forages over new clearcuts and group-cut openings in New Hampshire (Krusic and Neefus 1996), avoiding dense forests. Uses warm draft-free places for hibernation (mid October to May) (Banfield 1974:57) in caves, mines, rock crevices (Godin 1977:54), and occasionally hydroelectric dams

(Kurta and Teramino 1994). Pipistrelles in the northern latitudes roost in hollow trees (Kurta 1995:87), tree foliage (van Zyll de Jong 1985:125), and occasionally garages and barns (van Zyll de Jong 1985:125, Winchell and Kunz 1996); southern pipestrelles will use caves as maternity colony sites (Kurta 1995:87). Sometimes switches maternity colony roost sites (Whitaker 1998) like *M. sodalis* in Michigan (Kurta et al. 1996).

Special Habitat Requirements: Uses warmer draft-free hibernation sites than most bats (van Zyll de Jong 1985:125); open water over which to forage; roost trees.

Reproduction: Age of sexual maturity: Probably does not mate the first summer in northern populations (Fujita and Kunz 1984, van Zyll de Jong 1985:126). Breeding period: October to November and frequently in early spring. Gestation period: About 45 days (Hall 1956). Young born: Mid June to early July. Litter size: Usually 2. (Winchell and Kunz 1996).

Home Range: May feed within a radius extending at least 5 miles (8 km) from roosting site. Bats appear to migrate short distances [less than 50 km (30 mi)] from disbanding maternity colonies to hibernacula (Kurta 1995:87).

Sample Densities: Hibernacula counts range from one to several hundred (Kurta 1995:87), hanging singly from walls, generally not the ceiling. Males are solitary; females form small maternity colonies (less than 20 adults) in the summer (Kurta 1995:87).

Food Habits: Usually solitary feeder, flying slowly and erratically in small open areas, emerging from day roosts near sunset (van Zyll de Jong 1985:125). Prefers to feed over rivers, pastures (if large trees are nearby), and high in bordering trees in search of flies, beetles, ants, bugs, moths, wasps (Banfield 1974:57, Godin 1977:54, Griffith and Gates 1985). Leaf hoppers are an important food (Whitaker 1972a).

Comments: The eastern pipistrelle is the first species to enter hibernation in autumn and last species to leave in springtime (Kurta 1995:87). Reported longevity in the wild is 9 yr for females and 15 yr for males (van Zyll de Jong 1985:125). Griffin (1940b) notes successful homing from 85 miles. Call structure is indicative of open-area forager, with two strong harmonics at 20 kHz and 40 kHz, a call duration of 7.8 ms, and interpulse interval of 189 ms (MacDonald et al. 1994, Krusic and Neefus 1996).

Big Brown Bat

(*Eptesicus fuscus*)

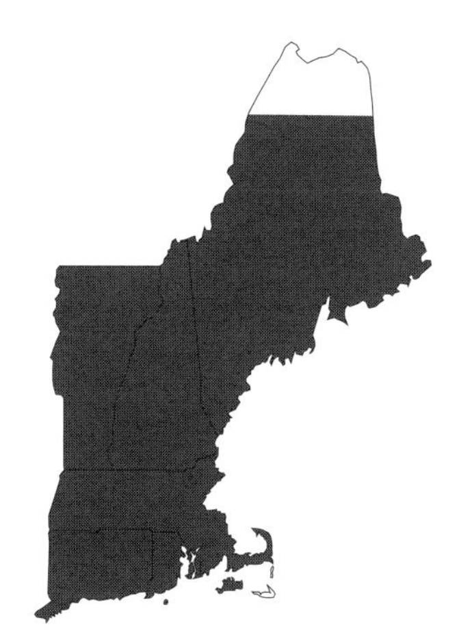

Range: Southern Canada w. to se. Alaska, s. to Venezuela and Colombia (Kurta and Baker 1990).

Distribution in New England: Throughout the region.

Status in New England: Common.

Habitat: Most abundant in agricultural landscapes, towns, and cities; least abundant in forest-dominated landscapes (Kurta 1995:88). Tends to be a habitat generalist, using a wide variety of hardwood and softwood forests and features, especially still water, roads and trails, regenerating, sapling/pole, and older age-classes in the White Mountains of New Hampshire (Krusic and Neefus 1996). Hibernacula usually are caves and mines, though buildings (van Zyll de Jong 1985:163, Whitaker and Gummer 1992) and sewers are sometimes used (Goehring 1972). Hibernacula are cold [around 0°C (32°F) and dry] (Hitchcock 1949, Kurta 1995:90). May hibernate with *Myotis lucifugus*, *M. septentrionalis,* and *M. leibii* in the same location (Kurta and Teramino 1994, New Hampshire Natural Heritage Inventory staff pers. commun.). Maternity colonies containing 20 to 70 adults are found in cool attics, eaves, and inside walls, barns, and tree cavities (van Zyll de Jong 1985:164, Kurta and Baker 1990); avoids hot attics. In British Columbia, maternity colonies are found in dead and hollow *Pinus ponderosa* (Brigham 1991); rock crevices are used singly. Males are solitary. Uses night roosts (barns, porch awning, or behind shutters) after foraging to rest and digest its meal (Kurta 1995:90).

Special Habitat Requirements: Uses cooler summer roosts

[less than 33 to 35°C (90°F)] than those used by little brown bats (Davis et al. 1968).

Reproduction: Age at sexual maturity: Females: 1 yr usually but sometimes earlier (Christian 1956, Brigham and Fenton 1986). Males: First autumn. Breeding period: September through March (Phillips 1966, Kurta and Baker 1990). Peak: September. Fertilization occurs in April. Gestation period: About 2 months. Young born: June. Litter size: Usually 1 in the West and 2 in the East. Litters per year: 1. Can relocate to other nearby maternity colony sites when evicted from buildings, unlike little brown bats, but with reduced reproductive success if evicted prior to parturition (Brigham and Fenton 1986).

Home Range: Travels short distances, usually no more than 48 to 80 km (30 to 50 mi), from maternity colony to hibernaculum site (Barbour and Davis 1969, Mills et al. 1975, Kurta 1995:88).

Sample Densities: Abundance decreases greatly going from the deciduous forest biome to the coniferous forest biome (Kurta et al. 1989). Maternity colony size varies from 5 to 700 individuals; generally between 25 to 75 individuals (Mills et al. 1975, Kurta and Baker 1990). Big brown bats accounted for 1 percent of the mist-net captures in 752 net hours at 18 sites [elevations range from 280 to 622 m (920 to 2040 ft)] in the White Mountains (Sasse 1995). Big brown bats accounted for almost 10 percent of mist-net and harp-trap captures by Krusic (1995).

Food Habits: Big brown bats tend to be coleopteran specialists (Whitaker 1972a, Whitaker 1995, Griffith and Gates 1985, Kurta 1995), but also capture a variety of other insect prey including hemipterans, hymenopterans, dipterans, plecopterans, neuropterans, and some lepidopterans (Hamilton 1933a, van Zyll de Jong 1985:163, Griffith and Gates 1985). Individuals may use the same feeding ground each night (Barbour and Davis 1969). Foraging habitat generalist (Humphrey 1982) flying over a range of heights above ground from 5 to 50 m, and 1 to 2 km from day roosts (Kurta and Baker 1990). Foraging activities were unaffected by experimental increased structure of forest–clear-cut edges at heights <18m (Brigham et al. 1997), as large bats were flying above cluttered edges.

Comments: Hibernation begins in November. Reproductive success may be negatively correlated with colony size (Mills et al. 1975). Maternity colonies found in buildings are at risk from building modifications to exclude bats: such activities can potentially seal young inside buildings (Neilson and Fenton 1994). Exclusion repairs should be done in late fall/early winter in order to avoid this situation. An important predator of agricultural pests (Whitaker 1995).

Red Bat

(*Lasiurus borealis*)

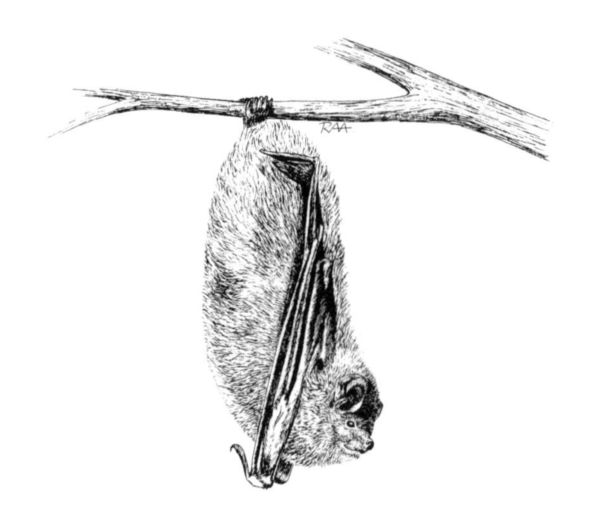

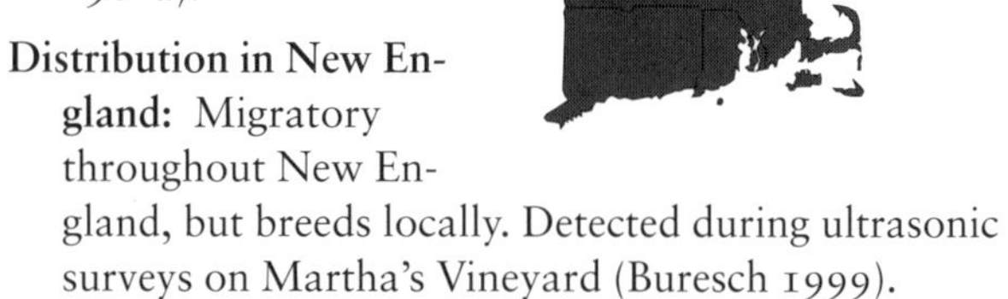

Range: Maritimes w. to Saskatchewan and the Rockies, s. through coastal Mexico and Central America to Chile and Argentina. Also along the Pacific Coast from w. Washington to Baja Peninsula (van Zyll de Jong 1985:137, Shump and Shump 1982a).

Distribution in New England: Migratory throughout New England, but breeds locally. Detected during ultrasonic surveys on Martha's Vineyard (Buresch 1999).

Status in New England: Uncommon to rare.

Habitat: Hibernates in much milder climates usually south of latitude 40°N (Davis and Lidicker 1956); uses tree foliage, hollow trees, woodpecker cavities, or under bark (van Zyll de Jong 1985:137). Summer roosts in the dense foliage of tree crowns (especially shade trees) or shrubs 1 to 6 m (3.5 to 20 ft) above the ground (McClure 1942, van Zyll de Jong 1985:137). Found in forests, open cultivated rural areas, and small towns (van Zyll de Jong 1985:137, Kurta 1995:76, Krusic and Neefus 1996). Uses a variety of hardwood and softwood habitats and features, especially still water, roads and trails, regenerating and older age classes in the White Mountains of New Hampshire (Krusic and Neefus 1996). Most active over water early in evening (Kunz 1973, Hart et al.

1993). Males are solitary; females with young in small family groups. Rarely found in buildings or caves except during migration. In Maryland, bats favored deciduous woodlands (Paradiso 1969). Greatest numbers were found along fence rows and forest edges (Constantine 1966).

Special Habitat Requirements: Roost sites provide dense leaf cover and shade above and to the sides, but open below (Shump and Shump 1982a). Uses a variety of hardwood shade trees (e.g., *Ulmus*, *Acer*, *Prunus*, *Celtis*, and *Juglans*); hangs from leaf petioles or small twigs; looks like a dead leaf (Constantine 1966, van Zyll de Jong 1985:138).

Reproduction: Age at sexual maturity: Second summer. Breeding period: August to September (Hamilton and Whitaker 1979), fertilization occurs in spring. Gestation period: 80 to 90 days (Jackson 1961). Young born: Late May to early July, mid June in Indiana (Whitaker and Mumford 1972) and Iowa (Kunz 1971). Litter size: 1 to 5, average 2.3. Litters per year: 1.

Home Range: Unknown; however, known to forage 600 to 1,000 yd (546 to 910 m) from day roosts (Jackson 1961).

Sample Densities: 1 individual per acre (2.4/ha) in Iowa (McClure 1942). Roosts singly or in a small family group (female and pups) up to 5 individuals at day roost sites (van Zyll de Jong 1985:138). Red bats accounted for less than 1 percent of the mist-net captures in 752 net hours at 18 sites [elevations range from 280 to 622 m (920 to 2040 ft)] in the White Mountains (Sasse 1995).

Food Habits: Begins foraging 1 to 2 hours after sunset (Kunz 1973), starting by flying slowly and erratically above tree crowns, then slowly descending to ground level, foraging in straight lines or large circles (Barbour and Davis 1969, Shump and Shump 1982a); forages routinely over the same areas. Eats moths, true bugs, beetles, hymenopterans, flies, crickets, cicadas, and other insects (Whitaker 1972a, Mumford 1973, Connor 1971); also will forage around barn or street lights (Shump and Shump 1982a). Red bats eavesdrop on other echolocating bats in order to locate vulnerable prey (Hickey and Fenton 1990). Forages in groups of up to 20 to 30 individuals (LaVal and LaVal 1979).

Comments: Migrates south in groups in autumn, wintering from Maryland and Washington, D.C., to the Gulf States (Banfield 1974:62, Paradiso 1969). Predators include sharp-shinned hawk, great horned owl, American kestrel, and blue jay (van Zyll de Jong 1985:138, Shump and Shump 1982a). Red bats are strong fliers, covering great distances over water (Godin 1977:59).

Hoary Bat

(*Lasiurus cinereus*)

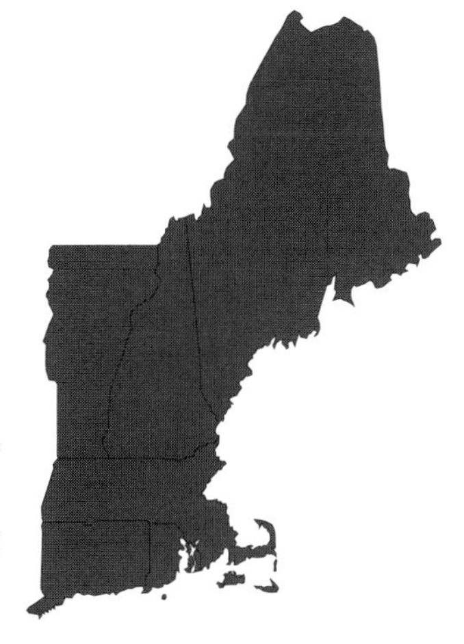

Range: Hudson Bay s. to central Florida and South America (Shump and Shump 1982b, Kurta 1995:77).

Distribution in New England: Migratory throughout the region.

Status in New England: Considered rare but may be more common than previously thought using ultrasonic detection devices (Krusic and Neefus 1996, Hickey and Neilson 1995).

Habitat: Found in forests, open cultivated rural areas, and small towns (van Zyll de Jong 1985:143, Kurta 1995:77, Krusic and Neefus 1996). Prefers coniferous forests but also uses deciduous woods and woodland edges, hedgerows, and trees in city parks (Godin 1977:60). Uses tree foliage, hollow trees, and woodpecker cavities (van Zyll de Jong 1985:143). Summer roosts in the dense foliage of tree crowns 3 to 12 m (10 to 39 ft) above the ground (Constantine 1966, van Zyll de Jong 1985:143). Uses a variety of hardwood and softwood habitats and features (including ponds, streams, and trails), especially regenerating and older age-classes, roads, and trails in the White Mountains of New Hampshire (Krusic and Neefus 1996), but at lower numbers compared with red bats. Majority of hoary bats hibernate in the much milder climates of the s. United States and Mexico (van Zyll de Jong 1985:144).

Special Habitat Requirements: Roost sites that provide dense leaf cover and shade above and to the sides, but open below (Shump and Shump 1982b). Uses a variety of hardwood shade trees and conifers (e.g., *Ulmus*, *Acer*, *Prunus*, *Maclura*, and *Picea*); hangs from small twigs and branches, blends well with lichen-encrusted branches of northern trees (van Zyll de Jong 1985:143, Kurta 1995:77).

Reproduction: Age at sexual maturity: Most become mature during first summer (Drueker 1972). Breeding period: September to November. Peak: Early September. Gestation period: Believed to be about 90 days (Jackson 1961). Young born: Late May to early June (Bogan 1972). Litter size: 1 to 4, average 2. Litters per year: 1.

Home Range: Feeding range may extend a mile (1.6 km) or more from roosting site (Paradiso 1969).

Sample Densities: Hoary bats accounted for less than 1 percent of the mist-net captures in 752 net hours at 18 sites [elevations range from 280 to 622 m (920 to 2040 ft)] in the White Mountains (Sasse 1995). Hoary bats accounted for 1 percent of mist-net and harp-trap captures by Krusic (1995) in the White Mountains.

Food Habits: Begins foraging in the fifth hour after sunset (Kurta 1995:79); flies fast with low maneuverability due to high wing loading and aspect ratios (Barclay 1985). Tends to forage directly in uncluttered air spaces at heights 7 to 15 m (23 to 49 ft) above the ground (Shump and Shump 1982b, Barclay 1984, Kurta 1995:79). Forages for large insects (moths, beetles, and dragonflies) over lakes and forest clearings; also forages around barn or street lights (Kurta 1995:79, Banfield 1974:64, Barclay 1985, Krusic and Neefus 1996). Out of 139 hoary bats examined in New Mexico, 136 contained moths, up to 25 individuals per bat (Ross 1967). Has been seen attacking pipistrelles in New York (Bishop 1947) and in California (Orr 1950). Uses single-harmonic search-approach calls at low frequencies for long-range detection of large insects in the open (Barclay 1986).

Comments: The hoary bat is the largest and most strikingly colored of eastern bats (Godin 1977:60). Migrates in waves to s. United States and Central America, but individuals have been found in the North during the winter months (Whitaker 1967). Males are solitary; females with young roost in small family groups.

Eastern Cottontail

(*Sylvilagus floridanus*)

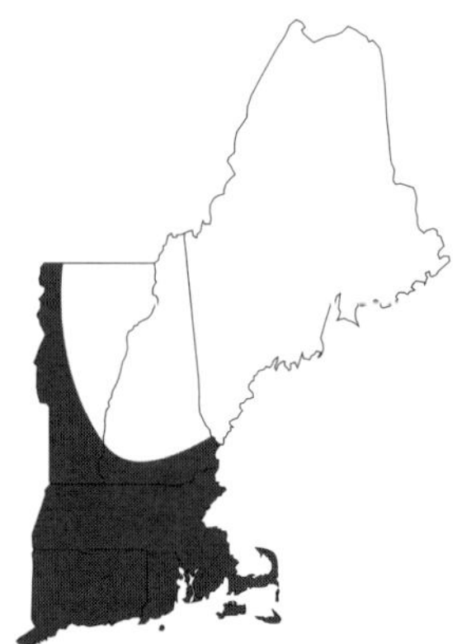

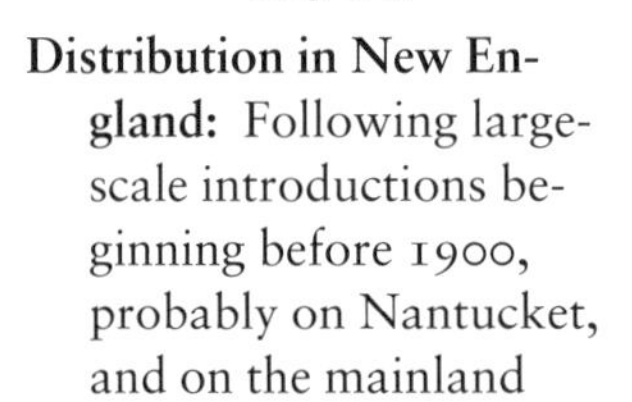

Range: East of the Rockies throughout the e. United States and extreme s. Canada and through e. Mexico and parts of Central America (Chapman et al. 1980, Kurta 1995:94).

Distribution in New England: Following large-scale introductions beginning before 1900, probably on Nantucket, and on the mainland early in the twentieth century, eastern cottontails have expanded their range in much of Vermont, s. New Hampshire, and s. New England (Jackson 1973, Johnston 1972, and summarized in Probert and Litvaitis 1996).

Status in New England: Common.

Habitat: Farmlands, pastures, fallow fields, open woodlands, thickets along fence rows and stone walls, edges of forests, swamps and marshes, suburban areas with adequate food and cover. Avoids dense woods.

Special Habitat Requirements: Brush piles, stone walls, dens or burrows for year-round protection from storms and cold weather. Interspersion of herbaceous and shrubby cover important (Allen 1984).

Reproduction: Age at sexual maturity: 2 to 3 months. Up to half of the year's crop breeds in the summer of their birth depending on geographic region (Chapman et al.

1980, Kurta 1995). Breeding period: March to September. Peak: April to August. Gestation period: 25 to 35 days, average 28 days (as summarized in Chapman et al. 1982:92). Young born: March to September. Young disperse from the nest at about 3 to 5 weeks (Godin 1977:67, Kurta 1995:96). Litter size: 3 to 8, typically 5 or 6. Litters per year: 3 to 4.

Home Range: Sizes range from about 0.5 to 40 acres (0.2 to 16.2 ha) or more (Dalke and Sime 1938, Godin 1977:68). Larger ranges (0.95 to 2.8 ha) for adult males and smaller ranges (0.95 to 1.2 ha) for adult females (McDonough 1960, Chapman et al. 1980). Usually less than 2 ha (5 acres) in the Great Lakes region (Kurta 1995); approximately 3.2 ha (8 acres) (Banfield 1974). Males have larger seasonal home ranges in spring and summer than in fall and winter (Althoff and Storm 1989).

Sample Densities: 0.46 animal per acre (1.1/ha) on 75-acre (30.4-ha) plot in Iowa during month of June, increasing to 1.65 per acre (4/ha) in August, followed by a drop to 0.89 per acre (2.2/ha) in October (Banfield 1974:77). Up to 8 rabbits per hectare (3/acre) in favorable habitat in the Great Lakes region (Trent and Rongstad 1974, Kurta 1995).

Food Habits: Crepuscular and nocturnal feeder. Most feeding takes place in the 2 to 3 hours after sunrise and within the hour following sunset. Summer foods: Tender parts of grasses and herbs. Winter foods: Bark, twigs, and buds of shrubs and young trees such as apple, maple, birch, and oak (Haugen 1942). Coprophagic (Eabry 1968, Chapman et al. 1980).

Comments: Female does not dig a burrow—uses abandoned woodchuck hole or digs a shallow nest in soft earth that is well concealed by surrounding vegetation. Probert and Litvaitis (1996) suggest that eastern cottontails occupy a wider range of habitats than New England cottontails because of their ability to utilize habitats with sparse understory vegetation. They further suggest that eastern cottontails can colonize disturbance patches sooner than New England cottontails and then maintain access to limited food resources by already being in place. Various predators include coyote, foxes, domestic dogs and cats, bobcats, hawks, and owls. Predators can be numerous on very diverse landscapes (Chapman et al. 1980, Oehler and Litvaitis 1996). Cottontails have smaller feet than snowshoe hares and are highly visible on snow because of their brown pelage. Predation can be high where snow cover is persistent and adequate escape cover (deep burrows) is lacking (Keith and Bloomer 1993).

New England Cottontail

(Sylvilagus transitionalis)

Range: Historic range is central and s. New England, s. through e. New York, e. Pennsylvania, New Jersey, and along the s. Appalachian Mountains [to slightly above 4,000 ft (1,220 m)]; to Alabama (Hall and Kelson 1959, Chapman 1975).

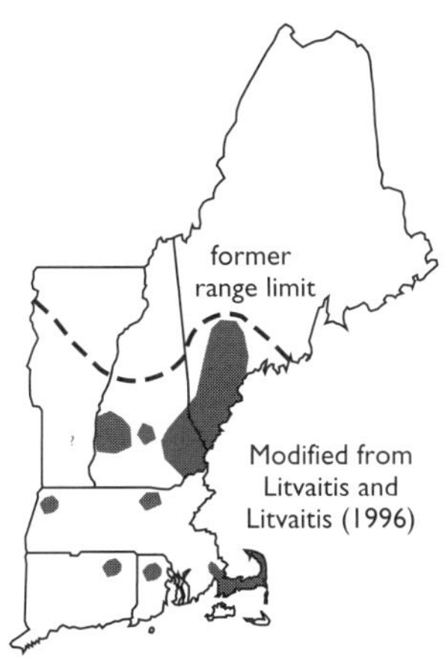

Distribution in New England: Current range reduced to isolated patches in Maine, New Hampshire, Massachusetts, Rhode Island, and Connecticut (Probert and Litvaitis 1996, Litvaitis and Litvaitis 1996). Suburbanization and forest maturation have restricted populations to small patches of early successional habitat in New England (Barbour and Litvaitis 1993, Litvaitis 1993, Brown and Litvaitis 1995, Litvaitis and Litvaitis 1996, Litvaitis et al. 1997), where they are vulnerable to local and regional extirpations (Litvaitis and Villafuerte 1996).

Status in New England: Uncommon or rare in Maine, New Hampshire, Connecticut, and Massachusetts; status unknown in Vermont and Rhode Island. Species of special concern in four of the six New England states (J. Litvaitis pers. commun.). See Appendix A.

Habitat: Brushy areas, open woodlands, swamps, mountains (Fay and Chandler 1955). Reported on beaches, salt marshes, and open land (Johnston 1972). Also in young forest clear-cuts, shrublands, and in hayfields or other grasslands. Dense cover and conifers are frequently components of habitats used by this species (Chapman et al. 1982:103). Occupied winter habitat patches ranged from 0.2 to 15 ha in size; dominated by shrubs, woody seedlings, and saplings (conditions commonly associated with secondary succession for 25 yr after disturbance); in association with abandoned or idle agricultural lands, power-line corridors, highway median strips, or cleared lands adjacent to industrial parks (Litvaitis 1993, Barbour and Litvaitis 1993, Brown and Litvaitis 1995, Probert and Litvaitis 1996, Litvaitis and Villafuerte 1996).

Special Habitat Requirements: Closely spaced patches of dense deciduous and coniferous shrubs, seedlings, and saplings at least 0.5 m tall and less than 7.5 cm (less than 3 in) in diameter (Barbour and Litvaitis 1993). Isolated patches of dense woody understory vegetation are much less frequently used (Barbour and Litvaitis 1993). In winter, New England cottontails occur much more frequently in patches greater than 10 ha (Litvaitis and Villafuerte 1996). Cottontails were in poorer physical condition and twice as vulnerable to predation in habitat patches less than 3 ha in size than in larger patches (Barbour and Litvaitis 1993, Brown and Litvaitis 1995, Litvaitis and Villafuerte 1996, Villafuerte et al. 1997). Increased proportions of agricultural and developed lands can be expected to produce more medium-sized carnivores and therefore increased predation rates (Oehler and Litvaitis 1996). Seldom ventures far from dense cover (Pringle 1960a).

Reproduction: Age at sexual maturity: Probably during second year (Dalke 1942). Breeding period: March to September. Peak: March to July (Chapman et al. 1982:92). Gestation period: 28 days (Dalke 1942). Young born: End of March to early April extending through July (Pringle 1960a). Litter size: 3 to 8, average 5. Litters per year: 2 or 3.

Home Range: 0.5 to 1.8 acres (0.2 to 0.7 ha) (Dalke 1937, McDonough 1960). Average 3 acres (1.2 ha) for 17 females and 8.3 acres (3.4 ha) for 10 males in swamp and upland woods (Dalke 1942).

Sample Densities: 0.3 to 7 rabbits per hectare (0.1 to 2.8 rabbits per acre) (Barbour and Litvaitis 1993); mean density differed between large (1.0/ha) and small (2.2/ha) patches, and density was correlated with patch size.

Food Habits: Summer: grasses and herbs. Winter: seedlings, bark, twigs, buds (prefers maple and oak) (Dalke and Sime 1941). Coprophagic (Eabry 1968, Chapman 1975). Foods consumed related to availability.

Comments: Adult females are slightly larger than males (Godin 1977:70). Crepuscular and nocturnal feeder. May be somewhat dominant over eastern cottontails; New England cottontails were able to maintain dominance in 50 percent of the behavioral trials testing a dominance hierarchy between the two cottontail species (Probert and Litvaitis 1996). New England cottontails continue to occupy stable shrub-dominated wetlands, especially if those habitats were occupied prior to eastern cottontail arrival (Hoff 1987). New England cottontail restoration efforts need to (1) manipulate habitat patches in areas still occupied by New England cottontails, (2) design vigorous vegetative management systems for habitat patches based on 2 or 3 distinct age classes of successional vegetation and a maximum rotation of 30 years in both forest-dominated and agriculture-dominated landscapes, and (3) recognize that size of the disturbance patch (15 to 75 ha [6 to 30 acres]) and distance (less than 500 m) to the next patch influence the relative success of management activities (Probert and Litvaitis 1996, Litvaitis and Villafuerte 1996).

Morphological characteristics of New England that distinguish them from eastern cottontails in winter include: shorter ear length, smaller body mass, presence of a black spot between the ears, lacking a white spot on forehead, and a black line on the anterior edge of the ear (Chapman et al. 1982:86, Litvaitis et al. 1991a). Chapman et al. (1982:83) suggests reclassifying *S. transitionalis* into two sister species, New England cottontail, occurring e. of the Hudson River, and Appalachian cottontail (*S. obscurus*), occurring w. and s. of the Hudson River. However, the limited variation in mitochondrial DNA sequences in *S. transitionalis* does not warrant reclassification into sister species (Litvaitis et al. 1997).

Snowshoe Hare

(*Lepus americanus*)

Range: Newfoundland, w. to Alaska, s. along the n. United States border, and s. in the Sierra Nevada, Rockies, and Appalachians (Bittner and Rongstad 1982:146).

Distribution in New England: Throughout the region.

Status in New England: Common in suitable habitat.

Habitat: Deciduous, coniferous, and mixed woodlands

(less often deciduous) with dense brushy understory (Litvaitis et al. 1985a, 1985b), coniferous swamps (Sievert and Keith 1985), cutover areas, burns in nearly all types of forests (Keith et al. 1984), and 7- to-15-yr-old commercially clear-cut softwood stands with dense hardwood and softwood regeneration (Monthey 1986). Favors second-growth aspen-birch in vicinity of conifers. In both e. Maine and Alaska, hare showed a significant shift from coniferous understory in winter to hardwood understory in summer (O'Donoghue 1983, Wolff 1980). Snowshoe hares use stands with very dense hardwood (Scott and Yahner 1989) and softwood understories during the winter (Litvaitis et al. 1985b). Pellet counts in Nova Scotia spruce-fir forest indicate lower hare use in stands with no understory cover, tree heights greater than 12 m, and canopy closures exceeding 60 percent (Orr and Dodds 1982).

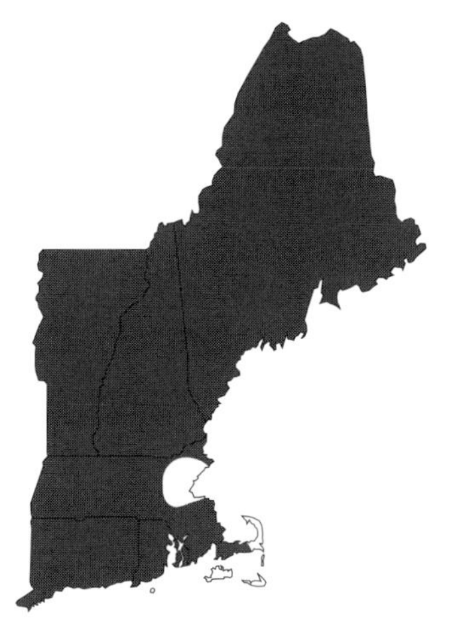

Special Habitat Requirements: Interspersion of dense shrubby cover for browse, and especially dense regenerating and sapling coniferous cover for protection from both terrestrial and avian predation (Brocke 1975, Carreker 1985).

Reproduction: Age at sexual maturity: In the spring following the year of birth. Breeding period: March to July (Bittner and Rongstad 1982:148). Gestation period: 34 to 40 days, average 37 (as summarized by Bittner and Rongstad 1982:148). Young born: May to August. Litter size: 1 to 6, average 3 (Aldous 1937, Dodds 1965). Litters per year: 1 to 3, occasionally 4 (Bittner and Rongstad 1982:148).

Home Range: Probably about 10 acres (4 ha) (Burt 1957). About 24 acres (10.1 ha) for adult males and 19 acres (7.7 ha) for adult females on an island in nw. Montana (Adams 1959). Daily ranges for both sexes were about 4 acres (1.6 ha) in mixed woodland-old field habitat in Canada (Bider 1961). Boutin (1984) demonstrated a decrease in late winter home range area when supplemental food was added to study areas in the Yukon Territory.

Sample Densities: Populations follow 10- or 11-yr cycles with densities ranging from 1 per square mile (259.2 ha) to several hundred per square mile (approximately 100 per km^2). Populations studied in Alberta demonstrated a 23-fold change from lowest to highest spring densities of 3,000 to 5,900 hares per square mile (259.2 ha) (Keith and Windberg 1978). Krebs et al. (1987) estimated 0 to 10 hares per hectare over a 6-yr period in the Yukon Territory. Litvaitis (1990) reported 0.4 to 1.5 hares per hectare in Maine.

Food Habits: Mainly crepuscular and nocturnal. Summer: Succulent vegetation such as clover, grasses, forbs, and ferns. Winter: Twigs, buds, and bark of small trees and seedlings such as alder, raspberry, mountain maple, and balsam fir. Hares maintain winter weight on woody forage characteristically less than 4 mm in diameter, though they will often clip stems up to 1.5 cm (Keith et al. 1984). Coprophagic (Bookhout 1959).

Comments: Pelage turns white in winter following fall molt and returns to brown after spring molt. Litvaitis (1991) found that winter pelage hares were trapped in sites with greater understory cover than brown pelage hares during periods with no snow cover, suggesting that habitat selection is a predator avoidance mechanism. Young are precocial, shortly after birth spending their days in separate hiding spots, and regrouping once a day to nurse (Rongstad and Tester 1971). Pellet counts on a series of long and narrow quadrats (5.08 cm × 305 cm [2 in × 10.1 ft]) provide an accurate extensive census for snowshoe hare (Krebs et al. 1987). Snowshoe hares seldom dig and normally do not enter abandoned dens or burrows of other animals as do cottontails (Godin 1977:74).

European Hare

(*Lepus europaeus*)

Range: Western Connecticut, e. New York, e. Pennsylvania, and w. New Jersey. Introduced to New York from

Europe between 1890 and 1910.

Distribution in New England: Locally in Litchfield and n. Fairfield counties, Connecticut (Bittner and Rongstad 1982, Paul Rego pers. commun.).

Status in New England: Probably uncommon.

Habitat: Open rolling country (mainly dairy land) with low vegetation and open views. Occasionally uses open woodlands with little ground cover.

Special Habitat Requirements: Open fields.

Reproduction: Age at sexual maturity: Probably first spring following birth. Breeding period: January to late July (Kurta 1995:104). Gestation period: About 42 days. Young born: March. Litter size: Average 4 (Schneider 1990). Litters per year: Generally 2 (Schneider 1990).

Home Range: 11 square miles (28.5 km^2) (Eabry 1970). About 12 acres (4.9 ha) (Banfield 1974).

Sample Densities: Average population density was about 25 hares per square mile (10/km^2) in Ontario with a potential density of 100 hares per square mile (39/km^2) under ideal conditions (Banfield 1974).

Food Habits: Summer: Grass, clover, corn, fruits such as raspberries, apples. Winter: Buds, bark, and twigs of young trees and seedlings. Coprophagic.

Comments: European hares are larger than snowshoe hares, with larger and wider ears, and more powerful hind legs and feet (Godin 1977:78). Builds no nest. Scrapes a hollow in ground near protective vegetation, rocks. Female caches leverets in separate forms; then tends young singly on a rotating basis (Kurta 1995:104). Introduced successfully from Hungary into Duchess County, N.Y. about 1893 as a game species for coursing (Whitaker and Hamilton 1998:193). Reintroduced at about 5-yr intervals until 1910 or 1911; hares gradually spread into s. Vermont, w. Massachusetts and Connecticut, central New Jersey, and easternmost Pennsylvania. At the turn of the century there were an estimated 20 to 40 hares per square mile in Sheffield, Berkshire County, Mass. (Godin 1977:79). European hares are also found in se. Ontario at higher densities than in introduced sites in the United States (Dean and DeVos 1965).

Black-tailed Jackrabbit

(*Lepus californicus*)

Range: Nebraska to Texas, w. to California and Oregon, s. into Mexico. Introduced as a game species to Nantucket and Martha's Vineyard, Mass., New Jersey, Maryland, Virginia, and s. Florida (Dunn et al. 1982, Best 1996).

Distribution in New England: Limited to Nantucket only (Godin 1977:77, Whitaker and Hamilton 1998:189).

Status in New England: Common on Nantucket.

Habitat: Stabilized beach dunes, as well as open fields and cultivated areas (Godin 1977:78).

Reproduction: Age at sexual maturity: Probably first spring following birth. Breeding period: Variable from February to May (in north) (Dunn et al. 1982). Gestation period: About 41 to 47 days. Young born: March to June. Litter size: 1 to 6 leverets (Godin 1977:78). Litters per year: 2 to 4 (Dunn et al. 1982).

Home Range: Home range size averages 16.2 to 20.2 ha and larger (Dunn et al. 1982, Harestad and Bunnell 1979) in w. United States and can overlap considerably (Whitaker and Hamilton 1998:191).

Sample Densities: Highly variable, from 0.2 to 0.9 per hectare in arid regions to 3 to 34.6 per hectare in more temperate regions (Dunn et al. 1982:130) in w. United States.

Food Habits: A variety of grasses, sedges, and forbs (Dunn et al. 1982:130); adults capable of consuming roughly 390 g of forage per day (Best 1996).

Comments: Introduced to Nantucket from Kansas as a game species in 1925 and again in the early 1940s as a substitute for the red fox for the traditional "ride to the hounds"; jackrabbits run in graceful bounds up to 35 mph over short distances (Godin 1977:77).

Eastern Chipmunk

(*Tamias striatus*)

Range: Quebec, w. to Manitoba, s. through most of e. United States to Louisiana and nw. Florida. Absent from most of Coastal Plain (Snyder 1982).

Distribution in New England: Throughout the region.

Status in New England: Common.

Habitat: Primarily deciduous forests, but also coniferous forests and brushy areas (Snyder 1982, Kurta 1995:114, King et al. 1998). Young as well as mature stands if sufficient understory or rocky cover is present.

Special Habitat Requirements: Cover in the form of decaying stumps and logs, rock piles and outcrops, and stone walls; elevated perches for observation and vocalization activities (Snyder 1982). Activity is centered around burrow locations (Bowers 1995).

Reproduction: Age at sexual maturity: Females—sometimes 3 months (usually 1 yr). Males—7 months to 1 yr (Yerger 1955). Breeding period: Late February to early April and late June to early July (Snyder 1982). Gestation period: About 31 days. Young born: Mid April to mid May and mid July to mid August.

Litter size: 1 to 8, average 4 or 5. Litters per year: Usually 2; spring litters survive better than late summer litters (Tryon and Snyder 1973).

Home Range: Varies temporally and geographically with mean minimum ranges from 0.03 to 0.6 ha and individual ranges varying from 100 m^2 to >1 ha (Elliott 1978, Snyder 1982, Bowers et al. 1990). Individual ranges overlap and fluctuate around defended core areas of about 20 percent of the range (Ickes 1974). Adult male ranges generally twice as large as females during breeding periods (Bowers et al. 1990). Home ranges expand in fall when foods are cached (Bowers et al. 1990). Seasonal availability of food (Mares et al. 1976) and water (Bowers et al. 1990) influence home range size.

Sample Densities: Two individuals per acre (5/ha) at onset of breeding season, increasing to 4 individuals or more per acre (10+/ha) at end of breeding season (Burt 1957). Elliott (1978) observed 6.3 individuals per acre (15.5/ha) in the Adirondacks. Varies geographically and temporally from 0.1 to 15.2 per acre (0.3 to 37.6/ha) (Yerger 1953). Four to 9 per acre (10 to 22/per ha) typical in Great Lakes region (Yerger 1953, Kurta 1995:114). Two- to fivefold annual fluctuation in population numbers is common in areas with lower densities (Snyder 1982).

Food Habits: Warm season foods include insects, fungi, birds and eggs, small mammals, and assorted amphibians and reptiles (Snyder 1982). Overwintering staples include a variety of seeds, nuts, and acorns that are cached underground (Forbes 1966, Snyder 1982) and sometimes last through the following spring and summer. Feeds during daylight hours.

Comments: Solitary and diurnal for most of the year (Yahner 1978). Enters torpid state during winter in underground burrows, but lacking large fat reserves, arouses frequently to feed on cached food reserves; sometimes forages aboveground during mild winter weather (Snyder 1982). Usually a terrestrial forager, will sometimes climb woody vegetation. Nests and separate food caches usually found in larger chambers of sometimes elaborate underground tunnel systems up to 10 m in length, usually less than 1 m from surface (summarized in Snyder 1982). Chipmunks are relatively long-lived (up to 8 yr), though most survive less than 2 yr (E. G. Allen 1938, Tryon and Snyder 1973, Smith and Smith 1972).

Woodchuck

(Marmota monax)

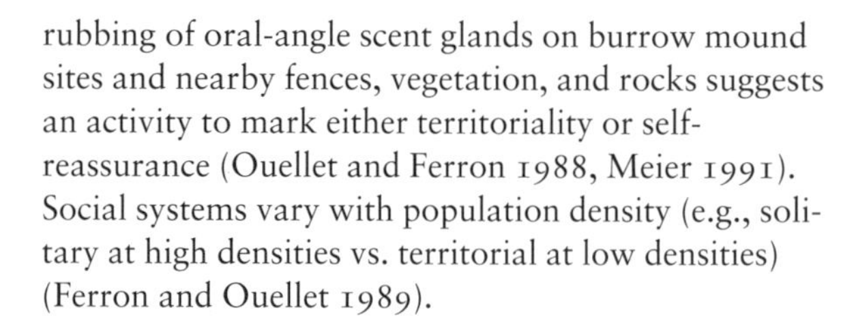

Range: Newfoundland w. to Alaska, s. through the e. United States to Arkansas and Alabama (Kwiecinski 1998).

Distribution in New England: Throughout the region.

Status in New England: Common.

Habitat: Open woodlands, rolling farmland, pastures, meadows, open brushy hillsides, grassy highway rights-of-way and utility corridors. Well adapted to human-dominated landscapes.

Special Habitat Requirements: Well-drained soils in which to burrow; herbaceous forage.

Reproduction: Age at sexual maturity: 10 to 25 percent may breed as yearlings, but commonly breeds during second year (Lee and Funderburg 1982:180). Breeding period: Early March to mid April. Gestation period: 31 to 32 days. Young born: Early April to mid May (Grizzell 1955). Litter size: 2 to 6, average 4 (Snyder and Christian 1960). Litters per year: 1.

Home Range: At high densities, home ranges overlap with neighbors and are maintained with dominant-subordinate hierarchy (Bronson 1963, 1964). At low densities, male ranges vary from 11.6 to 55.6 ha during the breeding season to 1.9 to 7.6 ha after the young disperse (Ferron and Ouellet 1989). Females have much smaller home ranges (0.1–0.6 ha) while accompanied by young and increase their home ranges (0.7–4.9 ha) after the young disperse (Ferron and Ouellet 1989).

Sample Densities: Range from 1 animal per 8.1 ha in Quebec (Ferron and Ouellet 1989) to highs of 1 animal per 1.1 ha in Pennsylvania (Bronson 1963, 1964). Muzzle rubbing of oral-angle scent glands on burrow mound sites and nearby fences, vegetation, and rocks suggests an activity to mark either territoriality or self-reassurance (Ouellet and Ferron 1988, Meier 1991). Social systems vary with population density (e.g., solitary at high densities vs. territorial at low densities) (Ferron and Ouellet 1989).

Food Habits: A diurnal feeder on succulent green vegetation such as dandelion, common plantain, red and white clover, spring grasses (low silica content), alfalfa, various herbs, and a variety of vegetable crops (Fall 1971, Grizzell 1955, Swihart 1990, Swihart 1991). Early spring foods include bark, buds, and twigs of black cherry, sumac, and dogwood (Kurta 1995:117). Occasionally eats small amounts of insects.

Comments: The woodchuck is the largest member of the squirrel family in New England. Fossorial except when feeding. Digs extensive system of burrows including a hibernation and nest chamber. Sometimes uses separate summer and winter dens. A solitary or territorial herbivore and true hibernator, the woodchuck does not cache food. Stores large amounts of body fat for winter dormancy; maintains a body temperature of 4°C (40°F) except during brief periods of arousal from October to March or April (Kurta 1995:117, Lee and Funderburg 1982:181). Scent-marking behaviors can have long-term localized effects on the distribution of woody plants in close proximity to burrows (Swihart and Picone 1991).

Gray Squirrel

(Sciurus carolinensis)

Range: Southern Quebec to Manitoba, s. to Texas and Florida (Flyger and Gates 1982a, Koprowski 1994).

Distribution in New England: Northward from s. New England into e. and central Maine, n. New Hampshire, and Vermont.

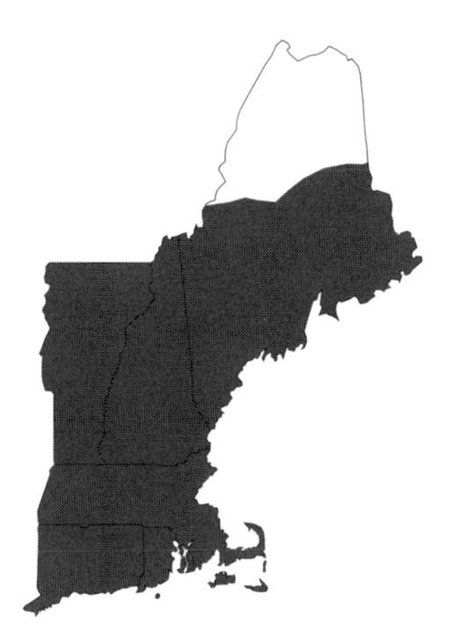

Status in New England: Common to abundant in s. and central New England; uncommon to common at the n. limits of range where it tends to inhabit towns and residential areas.

Habitat: Mature deciduous and mixed forests with hard mast-producing tree species, especially *Quercus* sp., *Carya* sp., and *Juglans* sp. Also found in forested bottomland, towns, suburban woodlots, and city parks (Baker 1983:204).

Special Habitat Requirements: Mast-producing tree species, cavity trees, tall trees for leaf nests [nests are usually 25 ft (7.6 m) or more above ground]; thick patchy understory vegetation and ground cover. At least 2 to 3 suitable cavity trees per acre (6 to 8 per ha) for optimum habitat (Nixon et al. 1980). Cavity openings are at least 3 in (8 cm) wide and 12 in (0.3 m) deep (Baker 1983:204).

Reproduction: Age at sexual maturity: About 3 months (Smith and Barkalow 1967) to 1 year (Allen 1954). Breeding period: January to February, and May to June (second litter). Gestation period: 44 days. Young born: March to April, and July to August (Kurta 1995:125). Litter size: 2 to 4 (Koprowski 1994). Litters per year: 2.

Home Range: Size varies from 0.5 to >20.2 ha (1.2 to 50 acres) throughout the year but is usually <5 ha (12.3 acres) (summarized in Koprowski 1994). Adult male home ranges > juveniles ≥ adult female home ranges (Koprowski 1994).

Sample Densities: <3/ha to >21/ha; dependent on tree seed availability and diversity of hard-mast-producing tree species (Koprowski 1994).

Food Habits: Diurnal feeder. Consumes nuts, buds, seeds and grains, fungi, fruits, bird eggs, inner bark of trees. Commonly caches food (scatterhoards) for future use. Will eat insects and pupae in spring and summer when preferred foods are scarce (H. Smith pers. commun.).

Comments: Arboreal, seldom wandering far from trees. In autumn, squirrels often move their home ranges short distances to areas with greater food supplies (Sharp 1960) and may occasionally migrate in large numbers over many miles (Larson 1962, Flyger 1969). Gray squirrel behavior suggests an age- and gender-based social hierarchy (Pack et al. 1967). Usually several squirrels share winter dens.

Red Squirrel

(*Tamiasciurus hudsonicus*)

Range: Boreal Canada w. to Alaska, s. in the Appalachians to Tennessee and in the Rockies to New Mexico (Flyger and Gates 1982b, Obbard 1987:265, Steele 1998).

Distribution in New England: Widely found throughout the region.

Status in New England: Common to uncommon.

Habitat: Coniferous, mixed and occasionally deciduous forests, rural woodlots (Hatt 1929, Layne 1954, King et al. 1998). Can be found using aspen stands with inclusions of conifers (Yahner 1987).

Special Habitat Requirements: Coniferous and mixed woodlands with mature trees that provide an adequate cone supply, and cavity tree dens (Hamilton 1939b, Hatt 1929, Layne 1954, Yahner 1980).

Reproduction: Age at sexual maturity: 10 to 12 months (Obbard 1987:265). Breeding period: January to September. Peak: Mid February to March and June to July (Klugh 1927, Layne 1954, Godin 1977:97, Lair 1985a, Obbard 1987). Gestation period: 31 to 35 days (Lair 1985a). Young born: March to May, August to September. Litter size: 1 to 7, typically 4 or 5. Litters per year: 1 or 2 (Lair 1985b).

Home Range: About 1 acre (0.4 ha) when food is plentiful (Hamilton 1939b). 2.73 to 6.03 acres (1.1 to 2.4 ha) (Banfield 1974:139). Less than 200 yd (182 m) in diameter (Burt and Grossenheider 1976). Steele (1998) summarizes mean territory size estimates of 0.35 to

0.66 ha in eastern conifer types circularly centered around midden and den sites.

Sample Densities: Obbard (1987:265) summarizes various density estimates; highest densities are in spruce forests, intermediate densities in mixed conifer and mixedwood forests, and the lowest densities in hardwood forests. Density estimates range from 250 to 400 squirrels/km² (647 to 1036 per square mile) in spruce forests to densities <100/km² (<259 per square mile) in hardwood forests.

Food Habits: Diurnal and crepuscular (Godin 1977:98). Feeds on conifer seeds, nuts, buds, sap, tender leaves, fruits, flowers, fungi, insects, bird eggs, and the young of small vertebrates (Hatt 1929, Klugh 1927, Layne 1954, Smith 1968). More carnivorous than gray squirrels. Caches food for winter use in large underground caches, not in scatter hoards as do gray squirrels (Godin 1977:98).

Comments: Red squirrels are a significant prey item for northern goshawk, fisher, and marten (Coulter 1966, Meng 1959, Powell 1981, Soutiere 1979). Ermine, long-tailed weasel, lynx, red-tailed hawk, and great horned owl are other red squirrel predators (Obbard 1987:265, Luttich et al. 1970, Rusch et al. 1972, Rusch and Reeder 1978, van Zyll de Jong 1966). Become inactive for short periods in winter to avoid cold and storms (Pauls 1978).

Southern Flying Squirrel

(Glaucomys volans)

Range: Eastern North America from s. Quebec and Nova Scotia w. to the Great Lakes and s. to e. Texas and s. Florida (Dolan and Carter 1977).

Distribution in New England: Throughout s. New England to the n. edge of their range in n. New Hampshire, central Maine (Dolan and Carter 1977), and along the Maine coast (Kurta 1995:134).

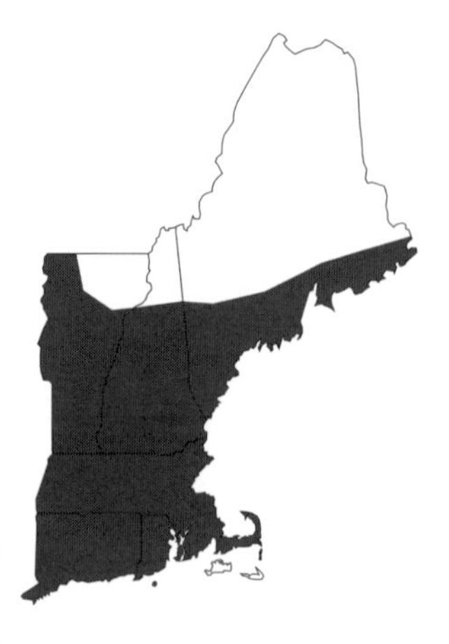

Status in New England: Common to uncommon.

Habitat: Mature deciduous and mixed forests, especially beech-maple, oak-hickory, and aspen.

Special Habitat Requirements: Several nest sites per individual (Muul 1968), mature woodland with cavity trees. Favors cavities with entrance diameters of 1.6 to 2 in (40 to 50 mm) (Dolan and Carter 1977).

Reproduction: Age at sexual maturity: About 1 yr for females (Jordan 1948). Breeding period: Late February to early March; June to July. Gestation period: About 40 days. Two mating periods. Young born: April and May, late July and August; peaks in April and August (Massachusetts). Litter size: 2 to 6, average 3 to 4. Litters per year: 2.

Home Range: Approximately 3.4 ha for females and 9.9 ha for males in New Hampshire at the n. edge of their distribution (Fridell and Litvaitis 1991); sometimes much larger than home ranges in more southern populations. Bendel and Gates (1987) found average minimum areas were 1.95 ha and 2.45 ha for males in Maryland. Average 0.41 per acre (0.17/ha) for females (may defend entire home range) and 0.53 per acre (0.21/ha) for males (no defense) in oak-maple habitat in New York (Madden 1974).

Sample Densities: Densities of up to 5 individuals per acre (12/ha) have been reported in woodland in New York (Sollberger 1943).

Food Habits: Hickory and beech nuts, acorns, seeds, fleshy fruits, and mushrooms. Highly carnivorous, readily takes moths, beetles, bird eggs and nestlings, and carrion (Kurta 1995:136). Stores food in den for winter use.

Comments: Nocturnal feeders (Sollberger 1940); solitary during the summer; groups up to 20 individuals will huddle in the same den, during winter (Kurta 1995:136). Favors abandoned woodpecker holes for den sites. Also builds dense summer nests of twigs in coniferous trees, especially in red cedar trees where forest is overtaking old fields. Frequently takes over bird boxes. Active throughout the year except during extreme winter cold. Interactions with cavity-dwelling birds over nest sites range from coexistence to ousting or killing cavity nesters, based on size (Stabb et al. 1989). Flying squirrels do not fly but rather volplane or glide to a landing site after leaping from a height. Depending on the air current, the squirrel may glide 150 ft (50 m) horizontal distance from a height of 60 ft (20 m) (Godin 1977:100). Females may carry young while gliding (Stack 1925).

Northern Flying Squirrel

(*Glaucomys sabrinus*)

Range: Canada w. to Alaska, s. in the e. United States to s. New England, and in the Appalachians s. to North Carolina. To the w. the range extends s. to North Dakota, Utah (Rocky Mountains), and n. California (Wells-Gosling and Heaney 1984).

Distribution in New England: Throughout n. New England, s. into Massachusetts except Plymouth County and Cape Cod and s. Connecticut (Wells-Gosling and Heaney 1984).

Status in New England: Common to uncommon.

Habitat: Mature mixedwood, deciduous, or coniferous forests. Favors cool, heavily wooded areas above 1,000 ft (305 m) in elevation.

Special Habitat Requirements: Hollow trees and tree cavities for winter dens. Builds spherical twig nests typically in conifers, close to the trunk for summer use (Cowan 1936). Arboreal lichens for winter food.

Reproduction: Age at sexual maturity: Probably 6 months to 1 yr. Breeding period: February to May and July. Gestation period: About 37 to 42 days (Kurta 1995:134). Young born: Late March to early July; late August or early September (Godin 1977:103). Litter size: 2 to 4. Litters per year: 1.

Home Range: Home range radii in Pennsylvania and North Carolina were 100 to 200 m, an estimated circular area of 0.7 to 3.1 ha (1.0 to 7.7 acres) (Weigl and Osgood 1974).

Sample Densities: Ranges from 0.3 to 10 per ha in favorable habitats (Wells-Gosling and Heaney 1984). Observations of 9 flying squirrels sharing the same winter nest cavity during cold weather (Wells-Gosling 1985).

Food Habits: Nocturnal feeder. Eats a variety of nuts, especially acorns; also takes conifer seeds, catkins, fruits, buds, insects, mushrooms, and bird eggs (Kurta 1995:134). Caches food in tree cavities.

Comments: Predators include a variety of hawks and owls, as well as marten, weasels, coyotes, and domestic cats. Active throughout the year except during extreme winter cold. Often highly social in winter dens.

Beaver

(*Castor canadensis*)

Range: Most of North America with the exception of the high arctic, parts of the sw. United States, Florida, and Mexico (Novak 1987:283).

Distribution in New England: Throughout the region.

Status in New England: Common.

Habitat: Small to large slowly flowing brooks, streams, rivers, or lakes that are usually, but not necessarily, bordered by woodland. Impoundments, reservoirs, and drainage ditches are readily occupied if food resources are nearby (Buech 1985, Broschart et al. 1989). Seasonally fluctuating or fast-moving waters tend to be avoided (Novak 1987:283).

Special Habitat Requirements: Optimum stream channel gradients are less than 3 percent (Hodgdon and Hunt 1953, Howard and Larson 1985); will use some stream channel gradients up to 12 percent (Olson and Hubert 1994). Wetlands and streams in wider valleys are more suitable than narrower or steeper valleys (Olson and Hubert 1994). A supply of young hardwoods, especially aspen and alder, within 100 ft of water (Belovsky 1984) for dam-building materials (Barnes and Mallik 1996, 1997) and forage, though foraging can occur several hundred feet beyond (Howard and Larson 1985).

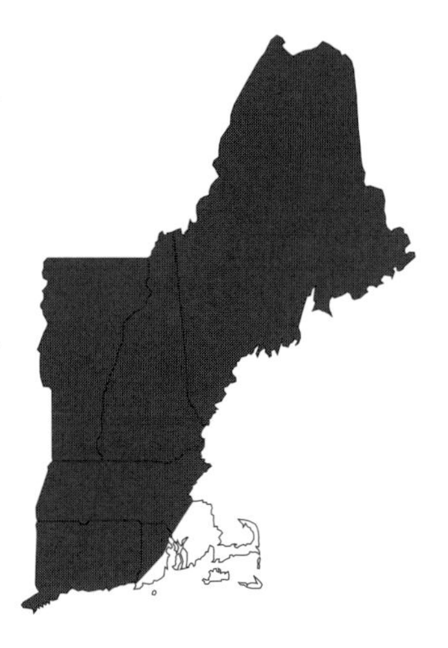

Reproduction: Age at sexual maturity: 1.5 to 2 yr (Larson 1967, Novak 1987:283). Breeding period: Mid January to mid March (Hodgdon and Hunt 1953). Peak: Mid February. Gestation period: About 106 days. Young born: Mid May to early June. Litter size: In ne. North America, average is 3.8, ranging from 2.4 to 5.5 as summarized by Novak (1987:283). Litter size may be related to type and amount of available food (Huey 1956). Litters per year: 1.

Home Range: Territorial activity focused around colony location (e.g., lodge and dam sites); a beaver colony can occupy 0.4 to 0.6 mi (0.6 to 0.9 km) of stream length (Novak 1987:283). Two-year-olds disperse from natal colony in late spring; capable of dispersing 5 to 10 miles and sometimes further (Olson and Hubert 1994).

Sample Densities: Family group or colony densities range from 0.4 to 12 per square mile (0.15 to 0.4 families/km^2) (summarized by Novak 1987:283). Family size varied from 3.2 to more than 8 individuals (Novak 1987:283).

Food Habits: Mainly a nocturnal feeder (Hodgdon and Larson 1973). Consumes bark of deciduous trees, especially aspens, balsam poplar, alder, willow, cottonwood, red-osier, birches, oaks, and maples. Beavers maximize food intake, taking the largest aspen stems possible depending on distance to water and the predation risk involved (Basey and Jenkins 1995). Additionally, summer diet includes substantial amounts of grasses, sedges, rushes, and herbaceous species, especially aquatic plants such as water lily and water arum (Northcott 1971, Belovsky 1984, Novak 1987:283). Caches woody twigs and branches near lodges for winter use under ice. Aspen and alder have more digestible energy and crude protein than red maple (Doucet and Ball 1994).

Comments: The beaver is the largest rodent in North America and the only species that creates its own habitat. The dam is the most obvious and impressive example of beaver activity; it provides the depth of water needed to float food and lodge-building material and provides protection. Dam building begins on the upstream side; the beavers lay sticks, leaves, sod, grass, mud, and even stones across a stream until the flow is checked and the water level rises. The height is extended from the pond side and the length extended to hold the impoundment at the desired level. The flatter the terrain, the longer the dam.

The lodge is usually started on an elevation on the pond bottom, and built up with logs and brush as the pond level rises with stream damming. The interior is kept hollow as the lodge is built, generally to a height of 5 to 6 ft above water and 20 to 30 ft across, although it may be higher and wider. The interior chamber is 6 to 8 ft wide and about 2 ft high, the floor just above the water level and bedded with dry sedges, grasses, or wood chips. It is accessed by one or more underwater tunnels (Godin 1977:106).

The monogamous pair bond between adults is long term (Novak 1987:283). Beaver are important prey items for wolf, coyote, fisher, and bobcat (Jenkins and Busher 1979, Samson and Crete 1997). Beaver induce long-term ecosystem and landscape-level changes by herbivory and physical habitat alteration. Beaver maintain diverse wetland systems and modify biogeochemical characteristics of drainages (Naiman et al. 1994, Naiman et al. 1986, Johnston and Naiman 1990a, 1990b, 1990c).

Deer Mouse

(*Peromyscus maniculatus*)

Range: Most of North America except n. Canada, w. Mexico, the se. United States, and the Atlantic Coastal Plain.

Distribution in New England: Throughout the region.

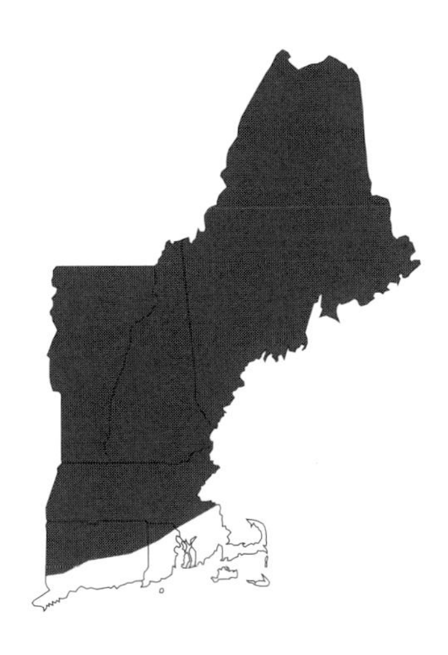

Status in New England: Common.

Habitat: Mainly occurs in coniferous and mixed forests, along field borders, stone walls, in outbuildings near areas with small trees and dense ground cover (Choate 1973, Godin 1977:111). Deer mice in s. Vermont used microhabitats with larger trees, higher canopy closure percentages, and more open ground cover than white-footed mice (Parren and Capen 1985). DeGraaf et al. (1991) found significantly more deer mice in softwoods than hardwoods. Uses recent forest clear-cuts (Kirkland 1977b), shrub openings, and recent burns (Kurta 1995:155). Monthey and Soutiere (1985) found deer mice more prevalent in uncut and partially cut softwoods rather than in *Rubus* or sapling stages. Occurred on 28 to 72 percent of forested plots in the White Mountains of New Hampshire and Maine (Yamasaki 1997).

Special Habitat Requirements: Down logs, rotting stumps, tree cavities, exposed rocks (stone walls, boulders and ledge) (Parren and Capen 1985).

Reproduction: Age at sexual maturity: Females 32 to 90 days (summarized by Millar 1989). Males about 60 days. Breeding period: April to August (Drickamer 1978). Gestation period: About 23 days. Young born: May to September. Litter size: 3 to 7, average 4. Litters per year: 3 or 4.

Home Range: Average 2.3 acres (0.9 ha) for adult males and 1.4 acres (0.6 ha) for adult females in virgin hardwood forest in Michigan (Blair 1942). 0.10 to 0.31 acre (0.04 to 0.13 ha) for adult males and 0.12 to 0.25 acre (0.05 to 0.10 ha) for adult females (Manville 1949). Wolff (1985) found means of 0.14 to 0.15 acre (0.06 ha) for both males and females in Virginia.

Sample Densities: Based on behavioral trials, territorial defense and aggressive behavior are density dependent and occur when populations exceed 25 to 30 mice per hectare (Wolff 1985). Snap-trap captures varied from 0.05 to 0.94 captures per 100 trap-nights in the White Mountains of New Hampshire and Maine (Yamasaki 1997).

Food Habits: Omnivorous; insects (including crickets, springtails, beetles, lepidopteran larvae, and grubs), earthworms, seeds (especially grasses and forbs), agricultural crops (e.g., corn, soybeans, and wheat), and woody seeds (e.g., nuts, conifer seeds, berries, and wild cherry). Animal matter can comprise 15 to 55 percent of the diet (Kurta 1995:156). In fall they cache seeds and nuts in hollow trees, logs, abandoned birds' nests, and in burrows near their nest chamber for winter use.

Comments: Deer mice are semiarboreal, readily climbing trees. Nocturnal (King 1968); usually more active during the later half of night (0100–0700 h) than earlier in the evening (1900–0100 h) (Drickamer 1987). Active throughout the year except during severe cold spells or winter storms. Nests in a variety of places including stone walls, buildings, old burrows of small mammals, under logs, or in tree cavities. Deer mice composed 1.1 to 6.3 percent of small mammal samples on the White Mountains of New Hampshire and Maine (Yamasaki 1997).

White-footed Mouse

(*Peromyscus leucopus*)

Range: Throughout most of the e. United States and Great Plains, w. to w. Montana and e. Arizona, and s. through e. Mexico and Yucatan, except n. Maine, n. Minnesota, n. Wisconsin, Florida, and coastal sections of the se. United States (Lackey et al. 1985).

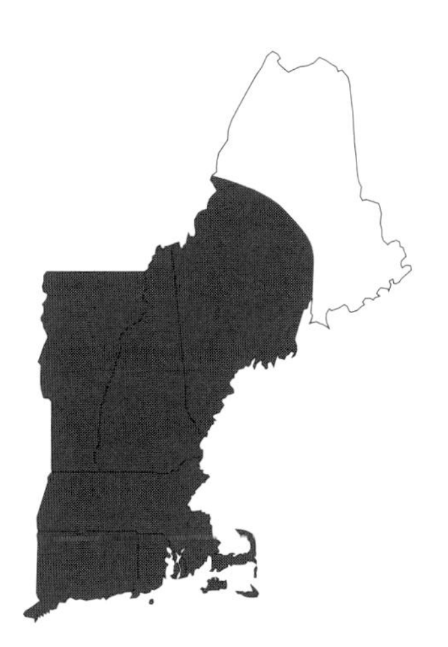

Distribution in New England: Throughout the region except n. and e. Maine.

Status in New England: Common.

Habitat: Interiors and edges of deciduous and mixed forests (Klein 1960, Garman et al. 1994) from sea level to above treeline. Clear-cuts, brushy fields and woodland clearings, pastures, streamside thickets, old buildings. Occurred on 36 to 72 percent of forested plots in the White Mountains of New Hampshire and Maine (Yamasaki 1997).

Special Habitat Requirements: Cover such as down logs (Planz and Kirkland 1992), rotting stumps, tree cavities, exposed rocks (stone walls, boulders and ledge).

Reproduction: Age at sexual maturity: Females: 38 to 51 days (summarized by Millar 1989). Breeding period: March to October (Drickamer 1978). Gestation period: 23 days (Svihla 1932, Millar 1989). Young born: April to November. Litter size: 1 to 7, typically 3 to 4. Litters per year: 2 to 4 (Snyder 1956, Millar 1989).

Home Range: Sizes ranged from 0.16 to 0.54 acre (0.06 to 0.22 ha) for adult males and 0.06 to 0.36 acre (0.02 to 0.15 ha) for adult females in mature oak-hickory in southern Michigan (Burt 1940). Wolff (1985) found means of 0.16 acre (0.06 ha) for males and 0.12 acre (0.05 ha) for females in Virginia.

Sample Densities: Based on behavioral trials, territorial defense and aggressive behavior are density dependent and occur when populations exceed 25 to 30 mice per hectare (Wolff 1985). Snap-trap captures varied from 0.09 to 1.77 captures per 100 trap-nights in the White Mountains of New Hampshire and Maine (Yamasaki 1997). Densities on a central Vermont mountain ranged from 3.7 mice per hectare to 93.4 mice per hectare (Brooks et al. 1998).

Food Habits: Omnivorous; animal matter (especially grubs and lepidopteran larvae) composes roughly 30 percent of the diet. Hard mast (hickory, beech, oak), maple and pine seeds, and cultivated grains and grass seeds are staple items. Soft mast (grapes, berries, cherries) is eaten seasonally, as well as small amounts of meat (carrion) (Lackey et al. 1985) or passerine bird eggs (DeGraaf and Maier 1996). Commonly caches food for future use.

Comments: Nests in a variety of places including stone walls, tree cavities, under stumps or logs, or in buildings. Nocturnal, usually more active during the earlier half of night (1900–0100 h) than later in the evening (0100–0700 h) (Drickamer 1987), and active in all seasons. Composed 2.3 to 11.7 percent of small mammal samples on the White Mountains of New Hampshire and Maine (Yamasaki 1997).

Southern Red-backed Vole

(Clethrionomys gapperi)

Range: Across Canada from British Columbia to Nova Scotia, s. through the Appalachians to n. Georgia, and in the Rockies s. to New Mexico. Also occurs across the n. border states (Merritt 1981).

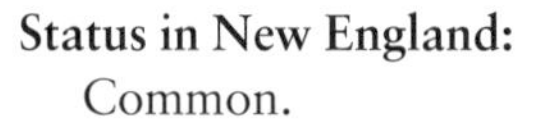

Distribution in New England: Throughout the region.

Status in New England: Common.

Habitat: Cool moist deciduous, mixed, or coniferous forests among mossy rocks, logs, tree roots, or other cover. Favors damp situations in coniferous or mixed woods. Highest densities found in subclimax communities (Cameron 1958). Less commonly found near stone walls at woodland edges or near talus slopes. Uses young clear-cuts in deciduous or coniferous woodlands (Miller and Getz 1972, Miller and Getz 1973, Kirkland 1978, Martell 1983, Monthey and Soutiere 1985, Vera and Servello 1994) and mixed forest (Lovejoy 1975). Occurred on 44 to 100 percent of forested plots in the White Mountains of New Hampshire and Maine (Yamasaki 1997).

Special Habitat Requirements: Water sources such as springs, brooks, or bogs, debris cover (fallen trees, stumps, rocks, slash) in cool coniferous or mixed forest.

Reproduction: Age at sexual maturity: Possibly 3 or 4 months (Blair 1941). Breeding period: Mid January to late November. Peak: February to October. Gestation period: 17 to 19 days (Svihla 1930). Young born: February to December. Litter size: 2 to 8, typically 4 to 6. Litters from higher latitudes and elevations tend to be larger (Innes 1978). Litters per year: 2 or 3 (Kurta 1995:162).

Home Range: Less than 0.5 ha (1.2 acres) according to

Merritt and Merritt (1978). Blair (1941) found 1 male using 3.6 acres (1.4 ha) in virgin northern hardwoods in n. Michigan.

Sample Densities: 4 to 40 per ha (2 to 16 per acre) summarized by Merritt (1981). Peak densities usually occur in late summer and early fall. Cyclic peaks in populations occur at 6- to 10-yr intervals (Patric 1962, Grant 1976). Pitfall captures varied from 0.8 to 11.42 captures per 100 trap-nights in the White Mountains of New Hampshire and Maine (Yamasaki 1997).

Food Habits: An omnivorous and opportunistic feeder throughout the year. Consumes vegetative shoots and leaf petioles in spring, ripe berries and fruit, and *Endogone* in summer, seeds and nuts in fall, seeds, tree roots, and deciduous tree bark in winter, and insects throughout the year (Merritt 1981, Kurta 1995:164).

Comments: Active at all hours but mainly nocturnal. Tends not to cache winter foods but actively forages throughout the winter under the snow. Commonly uses burrow systems of moles or large shrews (Criddle 1932). Nests under logs, stumps, and roots. Vole populations are vulnerable to freezing winter temperatures without an insulating blanket of snow on the ground and to spring meltwaters that flood underground tunnels (Kurta 1995:162). Major predators include broad-winged hawks, ermine, long-tailed shrews, foxes, bobcat, coyote, and domestic dogs and cats (Baker 1983:295, Kurta 1995:162). Composed 4.5 to 26.4 percent of small mammal samples on the White Mountains of New Hampshire and Maine (Yamasaki 1997).

Meadow Vole

(*Microtus pennsylvanicus*)

Range: Throughout Canada and Alaska, s. to Washington, n. New Mexico, and n. Georgia (Reich 1981).

Distribution in New England: Throughout the region.

Status in New England: Common to abundant.

Habitat: Fields, pastures, orchards, freshwater and saltwater marshes and meadows, borders of streams and lakes, open and wooded swamps, bogs; less commonly, open woods and clearcuts (Monthey and Soutiere 1985). Occurred on 6 to 19 percent of forested plots in the White Mountains of New Hampshire and Maine (Yamasaki 1997).

Special Habitat Requirements: Herbaceous vegetation, loose organic soils.

Reproduction: Age at sexual maturity: 4 to 5 weeks (Kurta 1995:172). Breeding period: Throughout the year, though reproductive activity sometimes ceases during midwinter (Kurta 1995:172). Peak: April to October. Gestation period: About 21 days. Young born: Throughout the year. Litter size: 1 to 9, typically 4 or 5. Litters per year: May produce 4 to 10 (Bailey 1924, Reich 1981). Known to produce 17 (Hamilton 1941).

Home Range: Averages less than 0.7 acre (0.3 ha) (Hamilton 1937); larger in summer than winter, and larger in marshes than meadows (Kurta 1995:170). Areas up to 40 m^2 surrounding nesting sites are aggressively defended (Kurta 1995:170). Seldom exceeded 0.06 acre (0.02 ha) in New York in good habitat (Hamilton 1937). Sizes may vary from 0.08 to 0.23 acre (0.03 to 0.09 ha) (Blair 1940b, Banfield 1974). Defensive behavior displayed during male encounters may indicate that they defend territories (Getz 1961b).

Sample Densities: There is a cyclic fluctuation in populations of about 4 yr (Hamilton 1937). Pitfall captures varied from 0.03 to 0.45 captures per 100 trap-nights in the White Mountains of New Hampshire and Maine (Yamasaki 1997).

Food Habits: Eats mainly vegetable material, especially tender grasses, bulbs, cambium of roots and stems, seeds, and grains. Occasionally caches food when supply is abundant and takes small amounts of meat when available.

Comments: The meadow vole is one of the most common mammals in New England. Active day and night throughout the year. Builds extensive tunnel and runway systems. Nests under boards, rocks, logs, and in other sheltered spots including tunnels. May damage nursery and orchard stock. Communal nesting of reproductively active females observed by McShea and Madison (1984). Composed 0.1 to 0.4 percent of small mammal samples on the White Mountains of New Hampshire and Maine (Yamasaki 1997). The larger beach vole, found only on Muskeget Island, Mass., was once considered a meadow vole. Burt and Grossenheider (1976) indicated that it might be a separate species, *M. breweri*. Jones et al. (1997) and Wilson and Reeder (1993) now recognize the beach vole as a separate species. Godin (1977:122) lists the Block Island meadow vole, *Microtus p. provectus*, as a subspecies restricted to Block Island, R.I.

Rock Vole

(*Microtus chrotorrhinus*)

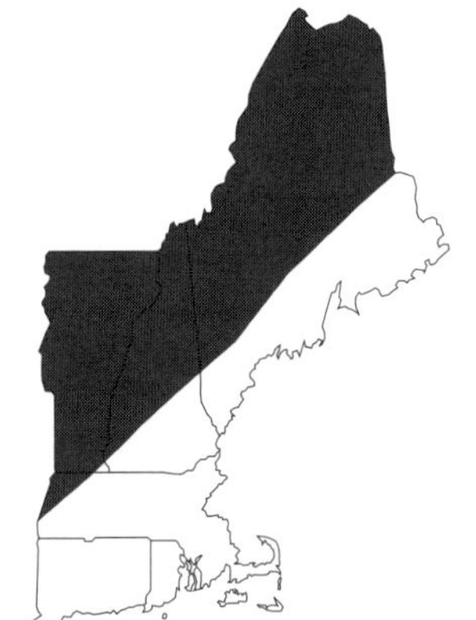

Range: Labrador and the Maritime Provinces (Cape Breton), w. to central Ontario and ne. Minnesota (Christian and Daniels 1985), s. through the mountains of n. New England and the Appalachians to North Carolina and Tennessee (Kirkland and Jannett 1982).

Distribution in New England: Throughout the mountains of n. and w. Maine, New Hampshire and Vermont.

Status in New England: Uncommon; may be locally common in appropriate habitat.

Habitat: Coniferous and mixed forests at higher elevations. Favors cool, damp, moss-covered rocks and talus slopes in vicinity of streams. Kirkland (1977a) captured rock voles in clear-cuts in West Virginia, habitat not previously reported for this species. Timm et al. (1977) found voles using edge between boulder field and mature forest in Minnesota. They also have been taken at lower elevations [1,312 to 1,509 ft (400 to 460 m)] in the Adirondacks and n. Maine (Kirkland and Knipe 1979, Clough 1987a). Occurred on 6 percent of forested plots in the White Mountains of New Hampshire and Maine (Yamasaki 1997).

Special Habitat Requirements: Cool, moist, rocky woodlands with herbaceous groundcover and flowing water.

Reproduction: Age at sexual maturity: Females and males are mature when body length exceeds 140 mm and 150 mm, respectively, and total body weight exceeds 30 g for both sexes (Martin 1971). Females born in late spring produce litters in first summer (Timm et al. 1977). Breeding period: Late March to mid October (Martin 1971). Gestation period: 19 to 21 days (Rosen 1978). Young born: Early spring to fall. Peak: June. Litter size: 1 to 7, typically 3 or 4. Litters per year: Up to 3.

Home Range: Unknown.

Sample Densities: 0.95 to 28.3 captures per 100 trap-nights in various geographic locations (summarized by Kirkland and Jannett 1982). Pitfall captures varied from 0.03 to 0.07 captures per 100 trap-nights in the White Mountains of New Hampshire and Maine (Yamasaki 1997).

Food Habits: Eats bunchberry, wavy-leafed thread moss, *Rubus* sp., *Vaccinium* sp., lepidopteran larvae, fungi including *Endogone*, bluebead lily, and Canada mayflower (Martin 1971, Kirkland and Jannett 1982). A captive subadult ate insects (Timm et al. 1977). Seems to be diurnal, with greatest feeding activity taking place in morning (Martin 1971).

Comments: First described on Mt. Washington in 1893 (Miller 1895). Occurs locally in small colonies throughout its range. Natural history information is incomplete for this species. Habitat preferences seem to vary geographically. Predators include bobcat, timber rattlesnake, copperhead, and short-tailed shrew (Kirkland and Jannett 1982). Composed 0.1 to 0.2 percent of small mammal samples on the White Mountains of New Hampshire and Maine (Yamasaki 1997). The rarity of this species in scientific collections may be due to the limited amount of trapping within its restricted habitat rather than actual scarcity of numbers (Godin 1977:126).

Woodland Vole

(*Microtus pinetorum*)

Range: North-central New England, w. to central Wisconsin, s. to e. Texas and n. Florida (Smolen 1981).

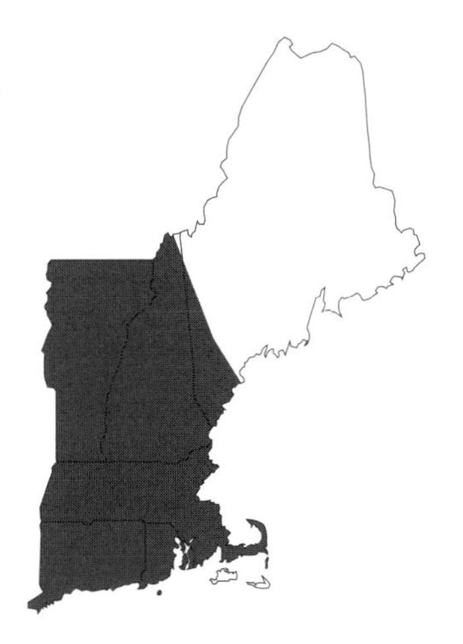

Distribution in New England: Throughout Vermont, Massachusetts, Connecticut, Rhode Island, and all but northernmost New Hampshire, and only in sw. Maine.

Status in New England: Common to uncommon.

Habitat: A wide range of habitats including deciduous forests, grasslands, meadows, and orchards (Hamilton 1938, Benton 1955). Favors well-drained uplands, but sometimes occurs in marshes and swamps. Occurred on less than 1 percent of forested plots in the White Mountains of New Hampshire and Maine (Yamasaki 1997).

Special Habitat Requirements: Uses variable depths of leaf litter or duff (Miller 1964) or grass; moist well-drained soils.

Reproduction: Age at sexual maturity: Females at 10 to 12 weeks; males later (Smolen 1981). Breeding period: Mid February through November in the Northeast; year-round farther s. (Smolen 1981). Gestation period: 20 to 24 days (Smolen 1981). Young born: Early March to early December. Litter size: 1 to 6, typically 2 to 4 (Smolen 1981). Litters per year: 1 to 4 (Goertz 1971).

Home Range: About 0.25 acre (0.10 ha) in oak-hickory woods in Michigan (Burt 1940). Home ranges of voles in Connecticut had average maximum diameters of 30.7 yd (33.7 m) for females and 30 yd (32.7 m) for males (Miller and Getz 1969).

Sample Densities: Densities in orchards usually greater than hardwood forest (Hamilton 1938) varying from 0 to 14.6 per hectare. Pitfall captures varied from 0.03 to 0.25 captures per 100 trap-nights in the White Mountains of New Hampshire and Maine (Yamasaki 1997).

Food Habits: Subterranean roots, tubers, and stems of forbs and grasses such as *Trifolium* sp., *Dicentra* sp., *Convolvulus* sp., *Viola* sp., *Agropyron* sp., and *Panicum* sp. are eaten in summer. *Endogone*, various berries, seeds, and nuts (including hickory, hazel, and oaks), *Rubus* and *Malus* bark, and lily bulbs are also eaten at different times of the year (summarized in Smolen 1981). Often caches large amounts of food in burrows.

Comments: Woodland voles are the most fossorial of New England microtine rodents, spending much time digging tunnel systems, foraging below ground, and moving along surface runways. Tunnels may be dug as deep as 12 in (30.5 cm) but are generally 3 or 4 in (7 to 10 cm) below ground surface. Nests are built under logs or rocks or in burrows well below ground. Active throughout the year. May be a severe pest species in nurseries, orchards, and potato farms in the Northeast (Anthony et al. 1986). Composed less than 0.1 percent of small mammal samples on the White Mountains of New Hampshire and Maine (Yamasaki 1997). Also called the pine vole. Important predators include snakes, numerous hawks and owls, foxes, raccoons, weasels, skunks, and opossums (Kurta 1995:175).

Muskrat

(*Ondatra zibethicus*)

Range: Throughout most of Canada except directly east and northwest of Hudson Bay. In most of the United States n. of Mexico, except parts of California, Texas, South Carolina, Georgia, and all of Florida (Boutin and Birkenholz 1987:314).

Distribution in New England: Widespread throughout the region (Willner et al. 1980).

Status in New England: Common to uncommon.

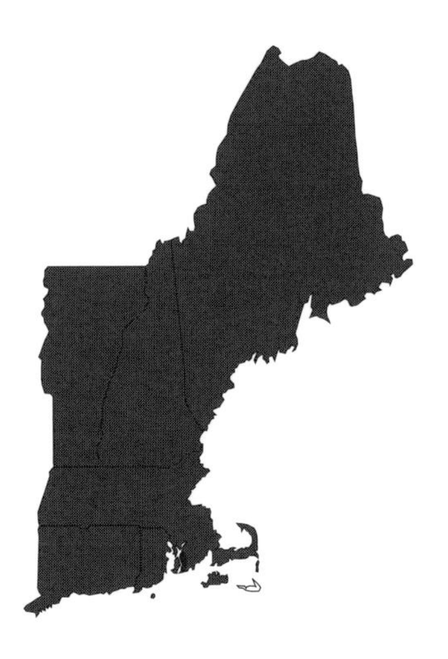

Habitat: Marshes, shallow portions of lakes, ponds, swamps, sluggish streams, drainage ditches (Johnson 1925). Most abundant in areas with cattails (Errington 1963) where the interspersion of emergent vegetation and water is at a 1:1 ratio (Weller 1978, Proulx and Gilbert 1983).

Special Habitat Requirements: Wetlands with dense emergent vegetation and stable water levels. Sufficient water depths (>15 cm or 6 in) needed to adequately cover house channels and provide access to houses beneath the water's surface (Proulx and Gilbert 1983). Muskrats use bank burrows either when dense emergent vegetation is absent or when water depths are less than 15 cm (Proulx and Gilbert 1983).

Reproduction: Age at sexual maturity: Possibly 6 months (H. Smith pers. commun.), and perhaps earlier in the s. United States (Boutin and Birkenholz 1987:314). Breeding period: Late February to August (as summarized in Boutin and Birkenholz 1987:314). Mid March to September (Chamberlain 1951) in Massachusetts. Gestation period: 28 or 30 days (Godin 1977). Young born: April or May and June or July. September and early October litters have been observed (H. Smith pers. commun.). Litter size: 5 to 8 in ne. United States and Canada (Clough 1987b), maximum of 14 embryos reported from Labrador (Chubbs and Phillips 1993a). Litters per year: 1.4 to 2.5 (Clough 1987b); generally 2 (Proulx and Gilbert 1983). Litter size positively correlated with latitude (Danell 1978), while number of litters per year is inversely related to latitude (Boyce 1977 cited in Perry 1982).

Home Range: Early work by Errington and Errington (1937) estimated that home ranges were usually contained within 200 yd (182 m) of den. Further work by Errington (1963) recorded an average home range diameter of 200 ft (61 m) in Iowa. Caley (1986) reported an average intercapture distance of 83 ft (25.3 m). Proulx and Gilbert (1983) calculated average summer home ranges of 5,210 to 11,970 ft^2 (484 to 1,112 m^2) in Ontario. Breeding muskrat pairs occupy home ranges exclusive of other pairs (Proulx and Gilbert 1983). Juveniles remain within parents' home range through autumn and often until spring (Boutin and Birkenholz 1987:314). Females with young will defend the nest site.

Sample Densities: Estimating densities usually involves counting the number of muskrat houses in an area multiplied by the number of muskrats per house (Boutin and Birkenholz 1987:314). Estimates of muskrats per house vary annually and seasonally from 2.8 to 5.0 (Lay 1945, Dozier et al. 1948, Dozier 1950, Proulx and Gilbert 1984). Bank-denning muskrats will be missed using this estimation technique (Boutin and Birkenholz 1987:314). Proulx and Gilbert (1983) estimated 18.1 to 22.6 animals per hectare on an Ontario marsh during a 2-yr study.

Food Habits: Consumes a variety of aquatic plants, especially cattails, reeds, pondweeds, bulrushes, and water lilies (summarized by Perry 1982:297, Proulx and Gilbert 1983). Freshwater clams and other small aquatic animals are also eaten (Schwartz and Schwartz 1981). Builds roofed feeding platforms near house. Most foraging is within 15 m (50 ft) of the primary lodge, and few movements exceeded 150 meters (500 ft.) (MacArthur 1978).

Comments: May construct a dome-shaped house of weeds over water (less than 2 ft [0.6 m] deep) for the nest or may dig a den in stream or ditch bank. Mainly nocturnal but often seen in daylight during the spring and autumn (Boutin and Birkenholz 1987:314). Active throughout the year. Muskrats using small ponds (less than 0.4 ha) are more likely to move between ponds than muskrats using larger ponds in Missouri (Shanks and Arthur 1952). Populations tend to follow a 10-yr cycle (Elton and Nicholson 1942). Low water levels in marshes increase mink predation on muskrats (Proulx et al. 1987). In winter, muskrats make fewer but longer foraging dives (MacArthur 1984). Muskrats move mainly in spring and fall, and may travel overland more than 20 miles in search of more favorable habitat (Errington 1939).

Southern Bog Lemming

(*Synaptomys cooperi*)

Range: Quebec w. to Manitoba, s. to Kansas, and e. to the Atlantic Coast of North Carolina (Baker 1983:337, Linzey 1983).

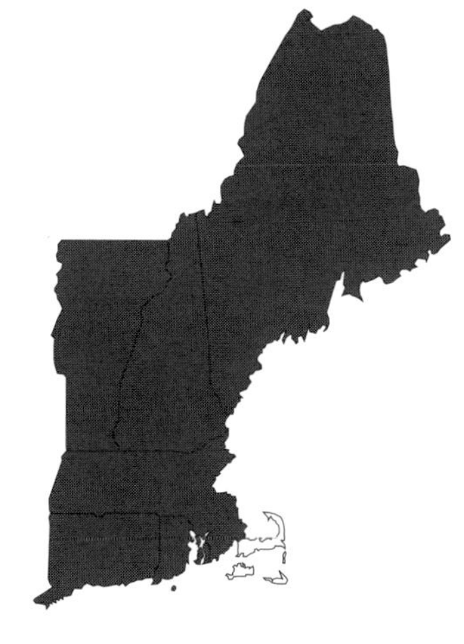

Distribution in New England: Throughout most of the region, but very locally distributed (Linzey 1983).

Status in New England: Uncommon; can be locally common.

Habitat: Uses a variety of open and forested habitats where shrublands and forests have an herbaceous ground cover; in both moist lowlands and well-drained and dry uplands (Baker 1983:338, Linzey and Cranford 1984). Uses wet grassy depressions and tamarack swamps (Pruitt 1959), open meadows and orchards (Gaines et al. 1977, Getz 1961c), moist deciduous and mixed forests (Kirkland 1977b, Timm 1975, Dice and Sherman 1922). Favors sphagnum bogs (Wrigley 1969, Robinson 1975) and deciduous woodlands with a thick layer of loose duff. Uses clear-cuts and other small forest openings with adequate ground cover (Kirkland 1977b, McKeever 1952). Occurred on 3 to 28 percent of forested plots in the White Mountains of New Hampshire and Maine (Yamasaki 1997). Tends to be outcompeted in more extensive and open grassy habitats by *Microtus pennsylvanicus* (Linzey and Cranford 1984).

Special Habitat Requirements: Moist soils, sphagnum bogs, areas with deep thick leaf mold (Godin 1977:134).

Reproduction: Age at sexual maturity: Probably 60 days (Baker 1983). Breeding period: Throughout the year. Peak: April to September. Gestation period: 23 days (Connor 1959). Young born: Throughout the year; most young are born between May and September. In New Jersey, females produced a litter every 67 days (average) in spring and summer (Connor 1959). Litter size: 1 to 8, typically 2 to 5. Litters per year: Probably 2 or 3 (Kurta 1995:183).

Home Range: 0.1 to 0.25 acre (0.04 to 0.1 ha) reported by Connor (1959) in New Jersey; 1 acre (0.40 ha) for 1 individual in sphagnum bog with tamarack and black spruce forming a dense canopy (Buckner 1957) in Manitoba; 0.20 to 0.50 acre (0.08 to 0.20 ha) (Banfield 1974:188). Getz (1960) calculated average home range areas for females at 0.06 ha; males at 0.04 ha. Females defend nest.

Sample Densities: Dramatic cyclic fluctuations of local southern bog lemming populations influence year-to-year presence in a single habitat (Beasley 1978). Estimates of bog lemmings at Rose Lake Wildlife Research Station in Michigan varied from 0.6 to 36 per acre (0.2 to 14/ha) (Blair 1948, Linduska 1950, Baker 1983:339). Connor (1959) estimated similar densities of 4 to 12 per hectare in New Jersey. Banfield (1974:188) reported densities of 2 to 14 per acre (0.8 to 5.6/ha). Pitfall captures varied from 0.03 to 0.69 captures per 100 trap-nights in the White Mountains of New Hampshire and Maine (Yamasaki 1997).

Food Habits: Tender succulent parts of herbaceous plants, especially leaves, stems, and seeds of grasses and sedges, fruits. Occasionally takes mosses, fungi, bark, roots, and some invertebrates (Linzey 1983, Kurta 1995:182).

Comments: Tunnel systems are deep, 6 to 12 in (15 to 30 cm) below ground, and complex with many side chambers for resting, feeding, and storing of food. Surface runways serve as travel lanes and usually are concealed from above by grassy cover (Baker 1983:340). Winter nest may be located in burrow, summer nest may be on surface in tuft of grass. Active during the day and night at all seasons of the year (Connor 1959). Linzey (1983) found *Synaptomys* to be mainly nocturnal. Preyed upon by weasels, owls (Linzey 1983), and other carnivores. Not considered colonial by Linzey (1983). Nonaggressive, tends to avoid interspecific encounters with *Microtus* (Rose and Spevak 1978, Gaines et al. 1979). Bright green fecal pellet piles are indicative of bog lemming activity, as are evenly sized herbaceous stem cuttings [1 to 3 in (3 to 8 cm) long] (Linzey 1983, Kurta 1995:182). Composed 0.1 to 0.4 percent of small mammal samples on the White Mountains of New Hampshire and Maine (Yamasaki 1997).

Northern Bog Lemming

(*Synaptomys borealis*)

Range: Alaska and British Columbia, s. to Washington and Idaho, e. to Wisconsin and Labrador (Hazard

1982, Clough and Albright 1987, Saunders 1988, Harper 1961, Anderson 1962, Wetzel and Gunderson 1949, Wright 1950).

Distribution in New England: The disjunct subspecies *Sphagnicola* is found in n. New Hampshire, n. to Mt. Katahdin, from Maine to New Brunswick and Quebec e. and s. of the St. Lawrence River (Godin 1977). Eight specimens recorded from five locations in Maine and New Hampshire over the last 100 years. The first U.S. specimen was collected near Fabyans at the w. base of Mt. Washington, Coos County, N.H. (Preble 1899); five specimens have come from Baxter State Park at 400 m elevation and on the alpine tableland on Mt. Katahdin, Piscataquis County, Me. (Dutcher 1903, Clough and Albright 1987); and one specimen each from Mt. Moosilauke, Grafton County, N.H. (Clough and Albright 1987), and Wild River drainage, Coos County, N.H. (M. Yamasaki, pers. obs.).

Status in New England: Rare and local.

Habitat: Found at elevations from 400 to 1375 m (Clough and Albright 1987) in mossy spruce woods, low elevation spruce-fir, hemlock and beech forests, sphagnum bogs, damp weedy meadows, and alpine sedge meadows.

Special Habitat Requirements: Moist to wet loose soils or leaf mold.

Reproduction: Age at sexual maturity: Unknown. Breeding period: May to August. Gestation period: Unknown. Young born: May to August. Litter size: 2 to 8, typically 4. Litters per year: Probably 2 or 3 (Kurta 1995:180).

Home Range: Unknown.

Sample Densities: Unknown.

Food Habits: Succulent parts of grasses and sedges, seeds, fungi.

Comments: Uses burrows several inches below ground and shallow runways on surface, similar to the southern bog lemming (Banfield 1974). Builds a spherical nest of dried grass in underground tunnels; winter nests are on the ground surface (Kurta 1995:180). Bright green fecal pellet piles are indicative of bog lemming activity, as are discarded herbaceous stem cuttings in lemming runways (Kurta 1995:180).

Norway Rat

(*Rattus norvegicus*)

Range: Throughout most of North America; numbers vary with climate and habitat (Jackson 1982, Baker 1983:346).

Distribution in New England: Localized around cities, towns, and farms.

Status in New England: Abundant.

Habitat: Concentrates in areas where food is abundant, such as waterfronts, farms, cities, and dumps. They also inhabit rural and suburban residences (Calhoun 1962).

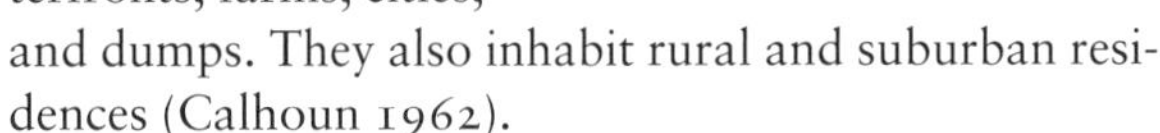

Special Habitat Requirements: Buildings, dumps, or loose soil for digging burrows near food supply.

Reproduction: Age at sexual maturity: 80 to 85 days. Breeding period: Throughout the year. Peaks: Spring and autumn. Gestation period: 21 to 22 days. Young born: In all seasons of year. Litter size: 8 to 12 (Jackson 1982:1080, Kurta 1995:188). Litters per year: 4 to 7 (Jackson 1982:1080).

Home Range: About 25 to 50 yd (23 to 46 m) in diameter (Banfield 1974:222). Movements were confined to an area 100 to 150 ft (30 to 46 m) in diameter both in residential and farm areas (Davis 1953); average radius of 30 to 50 m (98 to 164 ft) (Jackson 1982:1080).

Sample Densities: Urban densities vary from 25 to 150 per city block (Baker 1983:346).

Food Habits: Omnivorous, taking fruits, vegetables, grains, carrion and fresh meat, garbage.

Comments: Norway rats apparently are originally from Asia; they arrived in North America on the North Atlantic seaboard about 1775 (Godin 1977:138). Colonial and closely associated with humans. Probably the most adaptable and economically important rodent because of the damage it causes to buildings, stored food and grain, and the diseases it spreads to humans. Active mainly at night throughout the year. May dig extensive burrow systems for nesting and escaping predators. Most rats live less than 1 yr (Jackson 1982:1080).

The black rat (*Rattus rattus*) was introduced as early as 1609 with the early colonists. The species apparently thrived until the 1700s in most parts of its new range except in the northernmost parts where the larger, more aggressive Norway rat drove it out. Black rats were recorded from Vermont (Green 1936) to Rhode Island in 1903 (Cronan and Brooks 1968), but apparently no longer occur in New England. Newcomers may be temporarily found near seaports (Godin 1977:138), but none have been confirmed since the 1930s.

House Mouse

(*Mus musculus*)

Range: Throughout North America from coastal Alaska s. to Panama (Jackson 1982:1077, Baker 1983:352, Godin 1977:141).

Distribution in New England: Localized around cities, towns, and farms. Apparently now absent from Martha's Vineyard, Mass. (Cardoza et al. 1999).

Status in New England: Abundant.

Habitat: Buildings, fields, corncribs. Often burrows in fields and uses existing mouse runways during warm seasons of year, and moves indoors to escape winter cold.

Special Habitat Requirements: Buildings in winter.

Reproduction: Age at sexual maturity: 1.5 to 2 months (Jackson 1982:1080). Breeding period: Throughout the year. Peak: Early spring to late summer. Gestation period: 19 days (Jackson 1982:1080). Young born: Throughout the year. Litter size: 4 to 7 (Jackson 1982:1080). Litters per year: 5 to 8, typically 6.

Home Range: Average 1,560 ft^2 (145 m^2) for males and females (Lidicker 1966) in brush-grass habitat (on an island) with high population of *Microtus*. 3,925 ft^2 (365 m^2) in area with low (1 individual) *Microtus* population (Quadagno 1968). Males varied from 228.5 to 245.2 m^2; females varied from 220.3 to 238.8 m^2 in experimental dispersal trials in old field habitats (Maly et al. 1985). Average home range size varied from 250 to 1,700 m^2 (0.06 to 0.4 acre) in outdoor enclosures (Mikesic and Drickamer 1992).

Sample Densities: Densities of 300 or more mice per acre (741+/ha) were reported on an island (Lidicker 1966).

Food Habits: Fruits, grains, seeds, vegetables, plant roots, insects, almost any sweet or high protein food. Occasionally caches food.

Comments: This Old World species preceded the Norway rat into North America, probably at or soon after earliest settlement. Mainly nocturnal, active throughout the year. Colonial and highly social—may construct communal nests. Most house mice live less than 1.5 yr (Kurta 1995:186).

Meadow Jumping Mouse

(*Zapus hudsonius*)

Range: Most of Canada, Alaska, and the continental United States, s. to n. Georgia and Colorado (Whitaker 1972b).

Distribution in New England: Throughout the region.

Status in New England: Locally common.

Habitat: Occurs primarily in abandoned moist, grassy and brushy fields, marshes and meadows, willow-alder thickets along water courses, swamps and transition areas between lowlands and wooded uplands, and mixed occasionally-dry meadows. Seems to prefer areas with numerous shrubs and small trees (Sheldon 1934, Sheldon 1938). Occurred on 17 to 19 percent of forested plots in the White Mountains of New Hampshire and Maine (Yamasaki 1997).

Special Habitat Requirements: Thick herbaceous groundcover, moist loose soils for burrowing.

Reproduction: Age at sexual maturity: Less than 1 yr. Young females of first litter may breed during first year (Quimby 1951). Breeding period: Late April to early September. Peaks: Early June through August (Hamilton and Whitaker 1979). Gestation period: 18 days. Young born: May to early October. Litter size: 2 to 8, average 5 to 6. Litters per year: 2, possibly 3 (Whitaker 1963a).

Home Range: Average 0.38 acre (0.15 ha) for females and average 0.43 acre (0.17 ha) for males in Itasca Park in Minnesota (Quimby 1951). Approximately 0.89 acre (0.36 ha) average for males and 0.92 acre (0.37 ha) average for females in grassy area in Michigan (Blair 1940c).

Sample Densities: Can vary considerably from year to year. Estimates vary from 1.8 to 11.9 per acre (7.4 to 48.4 per ha) in Minnesota (Quimby 1951). Pitfall captures varied from 0.14 to 0.24 captures per 100 trapnights in the White Mountains of New Hampshire and Maine (Yamasaki 1997).

Food Habits: Animal matter (especially beetles and cutworms) and seeds compose 50 and 20 percent respectively of the posthibernation diet in spring (Whitaker 1972b). Berries, various fruits and nuts, *Endogone*, and insects (especially weevils and carabid beetles) add to the ripening grass and forb seeds that form the basic diet (Whitaker 1972b). Feeds on rootlets exposed to stream erosion (Cameron 1958). Does not store food (Whitaker 1972b).

Comments: The common name "jumping mouse" is a misnomer. They do not travel by jumping; they travel slowly by crawling through or under grass, or take short hops of 1 to 2 ft. When startled they may take a few jumps and then remain motionless (Godin 1977:144). Mainly nocturnal and solitary; activity increases with rainfall. Capable of climbing low shrubs (Sheldon 1938). Hibernate for longer periods in winter than most mammals (Godin 1977:144) in chambers 1 to 3 ft below ground, usually in a bank or hill (Banfield 1974:227). Composed 0.4 to 0.7 percent of small mammal samples on the White Mountains of New Hampshire and Maine (Yamasaki 1997). Key predators include red-tailed hawks, and barn and long-eared owls (Whitaker 1972b).

Woodland Jumping Mouse

(*Napaeozapus insignis*)

Range: Canadian Maritime Provinces and Newfoundland, w. to n. Minnesota and s. to e. Ohio and New Jersey and along Appalachians to n. Georgia (Whitaker and Wrigley 1972).

Distribution in New England: Throughout all but se. New England.

Status in New England: Locally common.

Habitat: Moist, cool woodlands with herbaceous groundcover and low woody plants in bordering streams, lakes, or ponds (Lovejoy 1973, Miller and Getz 1977). In New Hampshire significantly more abundant in hardwoods than softwoods (DeGraaf et al. 1991). Uses recent clear-cuts with herbaceous cover (Kirkland 1977b). Seldom ventures into bare open areas. Occurred on 56 to 94 percent of forested plots in the White Mountains of New Hampshire and Maine (Yamasaki 1997).

Special Habitat Requirements: Moist cool woodland, loose soils for burrowing, herbaceous cover (Whitaker and Wrigley 1972).

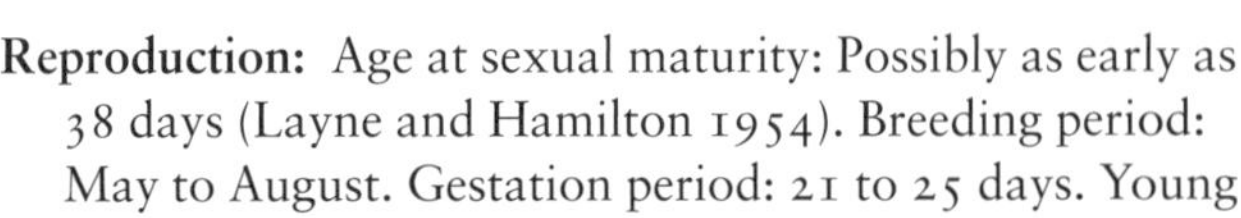

Reproduction: Age at sexual maturity: Possibly as early as 38 days (Layne and Hamilton 1954). Breeding period: May to August. Gestation period: 21 to 25 days. Young born: Late May to late August. Occasionally a second litter born in September (Godin 1977:148). Litter size: 2 to 7, typically 5. Litters per year: 1 or 2.

Home Range: Average 8.96 acres (3.63 ha) for an adult male and 6.55 acres (2.65 ha) for an adult female (Banfield 1974:230). 1.0 to 6.5 acres (0.40 to 2.63 ha) for females and 1.0 to 9.0 acres (0.40 to 3.64 ha) for males in virgin hardwood forest in Michigan (Blair 1941).

Sample Densities: Reported densities of 0.26 to 5.2 per acre (0.64 to 12.8/ha) were summarized by Whitaker and Wrigley (1972). Pitfall captures varied from 0.73 to 7.89 captures per 100 trap-nights in the White Mountains of New Hampshire and Maine (Yamasaki 1997).

Food Habits: Seeds, roots, and basal parts of herbaceous plants, fruits, subterranean fungus *Endogone*, and insects, especially lepidopteran larvae, craneflies, grubs, and centipedes (Hamilton 1935, Whitaker 1963b, Sheldon 1938). Does not cache food for winter survival; depends on accumulating sufficient body fat reserves for at least 6 months of profound hibernation.

Comments: Primarily nocturnal but active on cloudy days. Hibernates from October or November until April or May depending on the onset of cold weather (Preble 1956). Nest may be built in excavated chamber within burrow system, usually about 4 in (10 cm) below the surface of ground or under log or stump. Composed 5.2 to 58.2 percent of small mammal samples on the White Mountains of New Hampshire and Maine (Yamasaki 1997). Brower and Cade (1966) note the complementary local distribution of southern red-backed voles and woodland jumping mice; when red-backed voles are locally abundant, woodland jumping mice abundance tends to be very low. Wrigley (1972) suggests woodland jumping mice can survive 3 to 4 yr in the wild.

Woodland jumping mice walk on all four feet when moving slowly and take quadrupedal hops for greater speed. After several moderate hops, they tend to remain motionless under the nearest cover unless disturbed (Godin 1977:147). They can jump at least 6 ft in distance (Wrigley 1972); jumps of 2 to 3 m (6.6 to 9.8 ft) reported.

Porcupine

(Erethizon dorsatum)

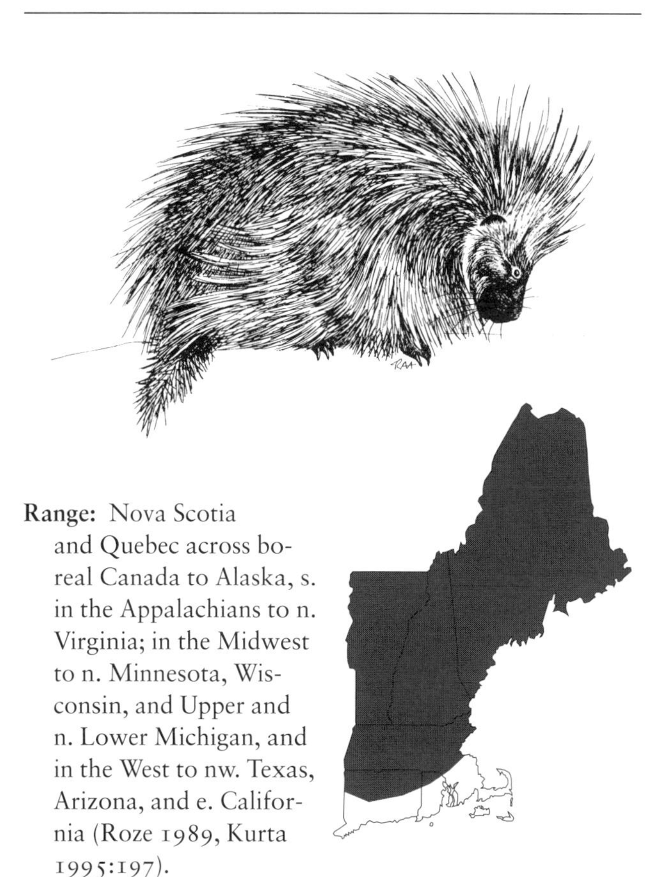

Range: Nova Scotia and Quebec across boreal Canada to Alaska, s. in the Appalachians to n. Virginia; in the Midwest to n. Minnesota, Wisconsin, and Upper and n. Lower Michigan, and in the West to nw. Texas, Arizona, and e. California (Roze 1989, Kurta 1995:197).

Distribution in New England: Throughout all but southernmost New England.

Status in New England: Common to uncommon.

Habitat: Mixed or coniferous forests with adequate denning sites, especially northern hardwood-hemlock. Not restricted to any plant or edaphic community (Dodge 1982:359).

Special Habitat Requirements: Den sites in rock ledges, trees, or other protected places.

Reproduction: Age at sexual maturity: 15 to 16 months. Breeding period: October through December, occasionally later. Gestation period: 205 to 217 days, average 210 days (Shadle 1951). Young born: April to June. Litter size: 1 (Hale and Fuller 1996). Litters per year: 1.

Home Range: Winter ranges averaged 6 acres (2.4 ha) in New Hampshire (Faulkner and Dodge 1962) and 13.3 acres (5.4 ha) in the Adirondacks of New York (Shapiro 1949). Spring and summer ranges ranged from 32 to 36 acres (13.0 to 14.6 ha) in conifer-hardwood forest in Minnesota (Marshall et al. 1962). Varies with climate and habitat (Dodge 1982:360).

Sample Densities: Reported densities range from 2 to 24.6

per square mile (1.2 to 9.5/km^2) as summarized in Woods (1973). Dodge (1967) reported as many as 100 porcupines in a 4- to 5-acre area in w. Massachusetts with many ledges, and six individuals in various parts of an abandoned house in New Hampshire.

Food Habits: Woody vegetation in winter, particularly inner cambium, buds, and needles of white pine and hemlock, and bark of sugar maple and birch. Summer foods include large quantities of basswood, aspen, elm, and birch leaves; grasses, flowers, fruits, seeds and nuts. Hemlock is a major winter food in the Northeast (Dodge 1967, Griesemer et al. 1994).

Comments: Mainly nocturnal, remaining active throughout the year. The den may be in a rocky cavern of ledge, in a hollow log, an abandoned building, or an abandoned fox or beaver den. Usually is solitary throughout the year but in winter several occasionally den together (Dodge 1982:359, Griesemer et al. 1996). Den sites, especially long-used sites, are recognizable by large piles of droppings; also uses "station trees" in summer, usually a hemlock or white spruce (Curtis and Kozicky 1944). May damage commercially grown trees or buildings. The fisher is the main natural predator of porcupines. Porcupines crave salt, and readily gnaw shed antlers, as well as axe handles and other tools left in the woods.

Coyote

(*Canis latrans*)

Range: Throughout North and Central America, s. of the tundra region (Bekoff 1977, Bekoff 1978, Voigt and Berg 1987:344, Parker 1995).

Distribution in New England: Throughout n. New England (Richens and Hugie 1974, Lawrence and Bossert 1969, Hilton 1978, Voigt and Berg 1987:344) and s. New England (Pringle 1960b, Parker 1995). Distribution limited by snow depths, prey size and density, and competition or lack thereof with larger predators like wolf or mountain lion (Todd et al. 1981, Voigt and Berg 1987:344).

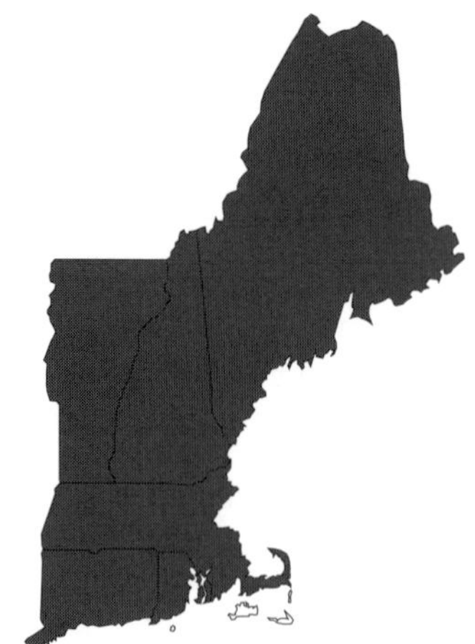

Status in New England: Uncommon to common.

Habitat: Originally thought to be a grassland and open-country species (Young and Jackson 1951). Now known to use a wide variety of forest and field habitats, particularly edges of second-growth forests, open brushy fields, fallow agricultural land, forest openings created by fire or logging, and even urbanized areas (Voigt and Berg 1987:344). Often concentrates in low-lying areas with softwood thermal cover and abundant populations of snowshoe hare and deer in winter (Ozoga and Harger 1966).

Special Habitat Requirements: Open or semiopen country for hunting, sunny well-drained secluded den sites formerly used by foxes and porcupines, also in hollow logs, rocky caves, or in excavated burrows (Bekoff 1982, Harrison and Gilbert 1985, Parker 1995).

Reproduction: Age at sexual maturity: 1 to 2 yr. Breeding period: Late January to February in northern part of range (Parker 1995). Gestation period: 60 to 63 days (Voigt and Berg 1987:344, Parker 1995). Young born: Late March to May (Parker 1995). Litter size: 5 to 9 (summarized in Parker 1995). Litters per year: 1.

Home Range: Ranges in the Northeast tend to be much larger than those in the West due to the differential distribution and availability of food (Harrison 1992). Ranges averaged 52 km^2 and 48 km^2 for males and females in Maine (Caturano 1983). Seasonal range changes occur during pair bonding and breeding, pregnancy, nursing, and winter foraging activities. Winter home ranges averaged 43 km^2 in w. Maine (Major and Sherburne 1987). Minimum dispersal distances of juveniles averaged 94 km (58.4 miles) and 113 km (70.2 miles) for females and males, respectively (Harrison 1992). Eighty-six percent of pups dispersed in their first year; 100 percent of pups dispersed by 1.5 yr (Harrison 1992). Range size may exceed 5 miles (8 km) in diameter depending on food supply and time of year (Godin 1977:199). Range sizes of radio-tracked individuals were greater for males (average 26.3 square miles, 68 km^2) than females (6.3 square miles, 16 km^2) in Minnesota (Berg and Chesness 1978). Pack animals defend well-defined territories, pairs and solitary individuals do not (Bekoff and Wells 1980).

Sample Densities: Eastern coyote winter densities vary from 0.13 to 1.5 coyotes per square mile (0.05 to 0.57 coyotes per km^2) (Voigt and Berg 1987:344, Parker 1995). The winter density of coyotes on an island in Lake Michigan was estimated at 1 animal per 2 square miles (5.2 km^2) (Ozoga and Harger 1966).

Food Habits: Coyotes are opportunistic, consuming snowshoe hare, deer fawns, beaver, muskrat, small mammals (red squirrel and other sciurids, mice, and voles), vegetation (particularly berries, cherries, shadberry, and apple) when available, and invertebrates (Stebler 1951, Hilton 1976, Harrison and Harrison 1984, Parker 1995, Samson and Crete 1997). Winter food in the Northeast is mainly snowshoe hare and carrion of deer (Dibello et al. 1990). Northeastern coyotes tend to be larger and heavier than those in the Midwest and West (Silver and Silver 1969, Schmitz and Lavigne 1987, Thurber and Peterson 1991, Parker 1995). Prey size and availability of food items are major influences in determining the social system (pack versus mated pair) in particular ecosystems (Harrison 1992).

Comments: Coyotes first appeared in New Hampshire (Silver and Silver 1969) and Maine (Richens and Hugie 1974) in the 1930s and in Vermont in the 1940s (Willey 1974). They first occurred in s. New England in 1957 (Pringle 1960b). Eastern coyotes have interbred with wolves and are larger than their western counterparts. They commonly exhibit pack behavior. Coyotes are active mainly in early morning, rather than late afternoon and evening (Parker 1995). Predispersal movements of pups are not sex-specific and increase with age (Harrison et al. 1991). Groups may rarely engage in surplus killing of deer when winter weather conditions are severe (Patterson 1994). Red foxes avoid coyote territories, even when suitable habitat exists (Harrison et al. 1989). Richens and Hugie (1974) argue that eastern coyotes have not replaced wolves ecologically, and have developed a unique niche that depends on food resources that are distinct from those of wolves.

Gray Wolf

(*Canis lupus*)

Range: Generally the higher latitudes of North America, from Alaska e. to Newfoundland, the n. Great Lakes States, and n. Plains (Carbyn 1987:358). Formerly throughout much of the lower 48 states and Mexico outside of the se. United States (Mech 1974, Paradiso and Nowak 1982). Reoccupation of the Yellowstone basin and n. Rockies from dispersing Canadian populations is occurring, supplemented with reintroductions.

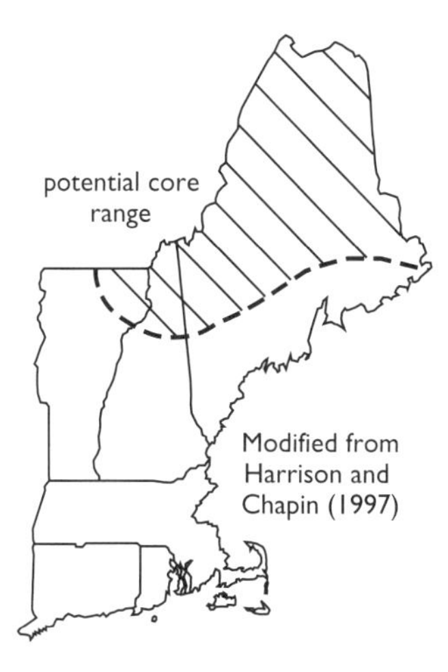

Distribution in New England: Assessment of potential core habitat in New England indicates large contiguous areas exist in n. Maine, n. New Hampshire, and the Gaspé Peninsula and potential dispersal habitat from currently occupied wolf range in southern Ontario and Quebec (Harrison and Chapin 1998). Potential habitats were also identified in the Adirondack Mountains of New York and in the Green Mountains of Vermont, but they have low populations of prey—moose, white-tailed deer, and beaver—and are isolated from other potential core habitat (Harrison and Chapin 1998). Several transients recently documented in Maine may have been dispersing from the Laurentides region of Quebec (Harrison and Chapin 1998, Matula 1997).

Status in New England: Extirpated. No known breeding population at this time. See above.

Habitat: Dependent on prey species habitat availability (Carbyn 1987:358) and ungulate biomass (Fuller 1989) rather than any particular forest cover type or vegetation structure. Human habitation and activity factors (e.g., densities of roads passable by two-wheel-drive vehicles and residences) at the landscape and bioregional scales influence the current use of available habitat and the dispersal to and reoccupation of potential habitat (Thiel 1985, Mech et al. 1988, Fuller et al. 1992, Mladenoff et al. 1995, Harrison and Chapin 1998). Harrison and Chapin (1998) describe potential core habitat in northeastern forested areas with fewer than 4 humans/km², and potential dispersal habitat as forested or forested/agricultural types with fewer than 10 humans/km², and road densities less than 0.7 km roads/km², based on roading characteristics and residence densities in wolf reoccupation areas of the Upper Great Lakes region (Thiel 1985, Mech et al. 1988, Fuller et al. 1992, Mladenoff et al. 1995). Mladenoff and Sickley (1998) model potential wolf habitat in the ne. United States as forest land and wetlands with road densities less than 0.45 km roads/km² and human densities fewer than 1.52 individuals/km².

Special Habitat Requirements: Adequate prey base, lack of persecution, and den locations in sites with good drainage (Carbyn 1987:358).

Reproduction: Age at sexual maturity: 22 to 36 months in the wild (Carbyn 1987:365), as young as 10 months in captivity (Medjo and Mech 1976). Breeding period: January to March depending on latitude (Carbyn 1987:358). Gestation period: 63 days (Carbyn 1987:365). Young born: March to June. Litter size: Average 6, ranging from 1 to 11 (Mech 1974). Litters per year: 1 per pack (Packard and Mech 1980); occasionally more than 1 per pack (Van Ballenberghe 1983).

Home Range: Range from 125 to 344 km² per pack in areas with dispersed prey populations of white-tailed deer/moose (Carbyn 1987:367, Fuller 1989). Ranges commonly exceed 500 km² per pack in areas with moose or caribou as the primary prey species (Carbyn 1987:367).

Sample Densities: Range from 0.7 to 8.4 wolves per 100 km² (Ballard et al. 1987).

Food Habits: Opportunistic predators of ungulates (e.g., moose and deer) and beaver (Carbyn 1987:368), though smaller mammal species may be taken (Mech 1974). Consumption rates when food is plentiful range from 0.1 to 0.21 lb prey/lb wolf/day (0.1 to 0.21 kg prey/kg wolf/day) (Fuller and Keith 1980, Carbyn 1987:368).

Comments: Coyotes first appeared in New England in the 1940s, about a century after wolf extirpation. Wolves are highly social animals organized in packs of related individuals (alpha pair and offspring). Hunting success depends on this social unit, from which reduced competition for mates and deferred reproductive behaviors are also observed (Mech 1974, Carbyn 1987:369). Humans and their activities (e.g., roadkills, shooting, or trapping) are major causes of wolf mortality (Carbyn 1987:366, Fuller 1989), along with other natural factors such as starvation, disease, injury, and interspecific conflicts (Mech 1977, Carbyn 1987). Capable of dispersing long distances, but natural reoccupation of Maine may be unlikely (Wydeven et al. 1998).

Red Fox

(Vulpes vulpes)

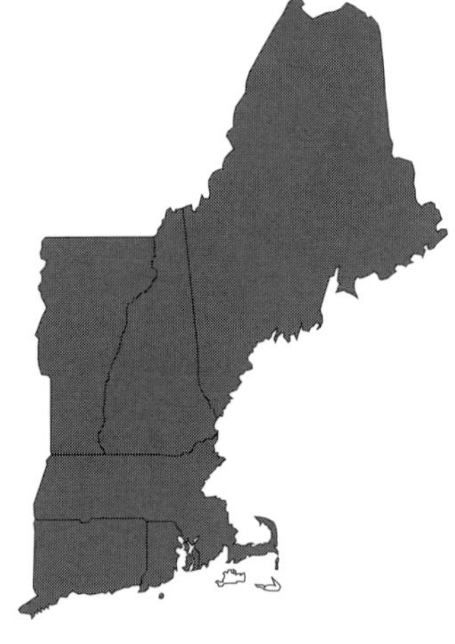

Range: North America from Baffin Island, s. to central Texas; from the East Coast w. through the Rocky Mountains, and throughout the Cascade Range in the Pacific Northwest and n. California, excluding the Great Basin (Samuel and Nelson 1982:475, Voigt 1987:378, Larivière and Pasitschniak-Arts 1996).

Distribution in New England: Widespread throughout the region.

Status in New England: Common to uncommon.

Habitat: Found in a variety of habitats. Preferred habitat is a mosaic of forest, cropland, and pasturage (Voigt 1987:378). Unbroken fields and dense forests less used; edges used heavily (Ables 1975). Red foxes avoid coyotes but coexist in the same area (Voigt and Earle 1983). Intensively use riparian (lakeshore, stream, and river) habitat between coyote territories (Harrison et al. 1989, Voigt and Earle 1983, Major 1983).

Special Habitat Requirements: Suitable den sites, often in woodlands or thick cover (Voigt and Broadfoot 1983).

Reproduction: Age at sexual maturity: Winter after birth (Voigt and Macdonald 1984). Breeding period: Mid January to late February, sometimes extending into March. Peak: Late January. Gestation period: 51 to 56 days, average 53 days. Young born: March or April. Litter size: 1 to 10, average 4 or 5. Litters per year: 1. Female fecundity is influenced by fox density, food supply (abundance of *Microtus* and *Clethrionomys* populations), and mortality (Voigt 1987:382).

Home Range: Red fox home ranges are usually contiguous, nonoverlapping areas, exclusive to a family unit composed of a male-female pair and (seasonally) their pups (Scott 1943, Sargeant 1972, Voigt and Macdonald 1984). Mean annual home range size was reported in Maine at 14.7 to 19.9 km² (Harrison et al. 1989, Major and Sherburne 1987). Territory sizes in Ontario farmland averaged 9 km² (2,224 acres), ranging from 5 to 20 km² (1,235 to 4,940 acres) (Voigt and Tinline 1980), and are comparable with other studies in e. North America (Ables 1969a, Sargeant 1972, Scott 1943, Storm 1965). Phillips et al. (1972) found that 70 percent of the juvenile males on study areas in Iowa and Illinois and 30 percent of the females moved more than 5 miles (8 km) from their natal ranges during their first year. Distances of 15 to 20 miles (24 to 32 km) were common. Seven foxes collared in Wisconsin had home ranges from 0.57 to 1.62 km² (142 to 400 acres) (Ables 1969b).

Sample Densities: Density estimates in s. Ontario range from 0.1 foxes/km² (0.3 foxes per square mile) to 1 fox/km² (2.6 foxes per square mile), with up to 3/km² (7.8 per square mile) on the best habitat (Voigt 1987:383).

Food Habits: Red foxes are omniverous feeders, consuming animals ranging from insects to small mammals and hare (Henry 1986, Major and Sherburne 1987). Commonly take birds, turtles, frogs, snakes, and their eggs. Berries and fruits are eaten when available (Dibello et al. 1990). Surplus food may be buried or cached under leaf litter or snow and marked with urine (Voigt and Earle 1983). In e. and w. Maine, snowshoe hare was the most abundant winter food item in a diet that also included deer carrion (Halpin and Bissonette 1983, Major and Shelburne 1987, Dibello et al. 1990). Dabbling ducks are a major food source during denning season in the northern prairie states (Sargeant et al. 1984).

Comments: Although there were red foxes in North America n. of 40 to 45° N latitude, European red foxes were introduced to the eastern seaboard (Godin 1977:201) about 1750. They are now considered to have been one species (Churcher 1959). May dig dens but prefers to use existing burrows for rearing young and escaping predators. Red fox dens may have an underground tunnel system 25 ft (8 m) long or more (Godin 1977:203). Competition with other wild canids, especially coyotes, and availability of year-round food sources influences fox survival and numbers (Voigt 1987:378). Red foxes avoid suitable lakeshore and riparian habitats within coyote territories (Harrison et al. 1989). Several instances of arboreal foraging activity observed during winter by Sklepkovych (1994).

Gray Fox

(*Urocyon cinereoargenteus*)

Range: Throughout the United States and s. Ontario and Quebec, excluding portions of the mountainous nw. United States and Great Plains (Fritzell and Haroldson 1982, Fritzell 1987:409).

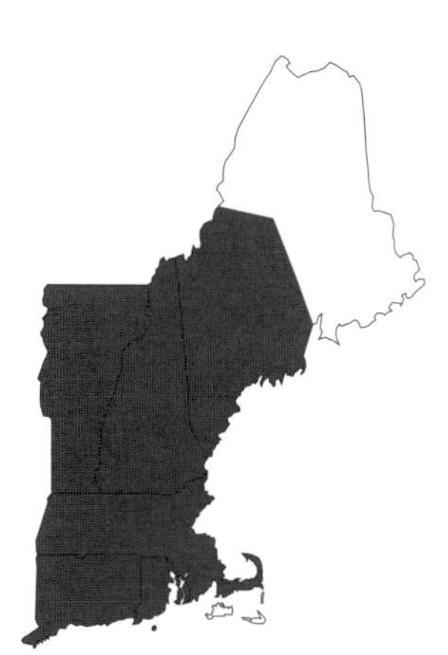

Distribution in New England: Throughout the region except for n. and e. Maine (Fritzell and Haroldson 1982).

Status in New England: Common to uncommon.

Habitat: Dense northern hardwood or mixed forests. May inhabit thickets and swamps. Prefers a mosaic of old fields and hardwood forest types (Wood 1958, Yearsley and Samuel 1980, Haroldson and Fritzell 1984).

Special Habitat Requirements: Den sites are inconspicuously located on warm e.- to s.-facing slopes, in hollow logs, tree cavities, rock crevices, cavities beneath deserted buildings, and abandoned woodchuck burrows (Fritzell 1987:412).

Reproduction: Age at sexual maturity: First year after birth. Breeding period: Mid January to May (Sheldon 1949, Sullivan 1956, Fritzell 1987:410). Peak: Early March (latitude-dependent). Gestation period: 59 days in captivity (Fritzell 1987). Young born: March or April. Litter size: 1 to 7 pups, average 4 (summarized in Fritzell 1987).

Home Range: Varies with food supply, disturbances, denning, and season. Males and females travel throughout large areas in fall and winter. Lactating female home ranges are 20 ha (50 acres), smaller than those of non-lactating females (Nicholson et al. 1985, Haroldson and Fritzell 1984). Range varies from a mile (1.6 km) wide during denning to 5 miles (8 km) in the fall (Godin 1977:206). Yearsley and Samuel (1980) found home ranges from 75 to 185 ha (185 to 457 acres) in West Virginia. Home range varied from 200 to 500 ha (494 to 1,235 acres) in Alabama (Nicholson 1982) and Missouri (Haroldson and Fritzell 1984).

Sample Densities: Fritzell (1987:412) summarized gray fox density estimates of 1.2 to 2.1/km^2 (3.1 to 5.4 per square mile).

Food Habits: Crepuscular and nocturnal. Seasonally omnivorous; small mammals, particularly cottontails and various rodents, comprise the largest dietary fraction in winter diets (MacGregor 1942, Fritzell 1987:413). Birds, reptiles, and amphibians and their eggs, and insects (especially crickets and grasshoppers) are important summer foods. Acorns, apples, grapes, and corn are consumed in the fall (Fritzell 1987); fresh carrion is also eaten.

Comments: The gray fox is native to New England. Though rare in n. parts of the region earlier in this century, gray foxes have reoccupied much of their former range n. of Connecticut since 1930 (Palmer 1956). The northward spread of gray foxes corresponded to that of the eastern cottontail rabbit (Godin 1977:206). Gray foxes are accomplished tree climbers when hunting prey or escaping enemies (Terres 1939, Gunderson 1961, Yeager 1938).

Black Bear

(*Ursus americanus*)

Range: Throughout Canada s. of the barren lands of the n. coast, except in the s. Prairie provinces, extreme s. Ontario, and Prince Edward Island. In the United States, black bears occur throughout the Pacific Northwest, Sierras, and Rockies into Mexico; in the n. Great Lakes region; New England, New York, and through the Appalachians to n. Georgia; Piedmont region, Florida, Gulf Coast, and Ozarks (Kolenosky and Strathearn 1987:443).

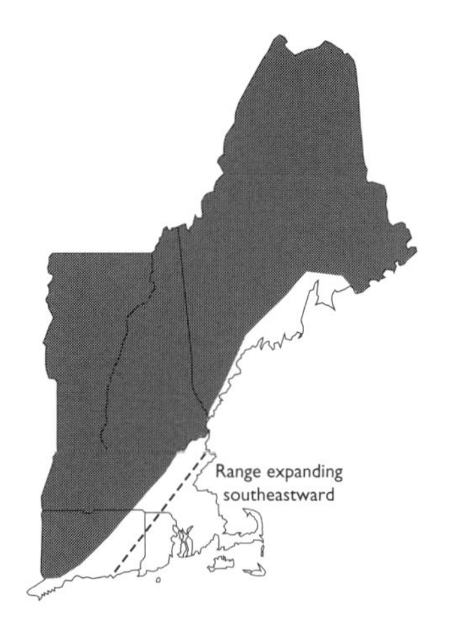

Distribution in New England: Throughout the region except along the coast from Penobscot Bay s. to Long Island Sound (Spencer 1961, Cardoza 1976, McLaughlin et al. 1991, Cardoza and Field 1991).

Status in New England: Common in north to uncommon farther south. Increasing throughout their range in the Northeast.

Habitat: Forest-dominated landscapes (deciduous, coniferous, and mixed) interspersed with a variety of openings, wetlands, and regenerating stands (Schooley 1990, Meddleton 1989, Samson and Huot 1998). Elevational differences in seasonal habitat use (lower elevations in spring/summer; higher elevations in fall) reflect the availability of foods (herbaceous vegetation and insects in spring; various berries, pin cherry, and feral apples in summer/early fall, and beechnuts and mountain-ash berries in fall) (Meddleton and Litvaitis 1990). Regenerating stands, seeded, gated roadbeds, and power-line and railroad right-of-ways are important sources of abundant summer foods in the White Mountains (Meddleton and Litvaitis 1990). Utilizes abundant sources of hard and soft mast within its habitat (Pelton 1982). Females use softwood riparian areas in fall when hard mast crops are minimal (Schooley 1990). Also use tributary buffer zones to move between forested watersheds (Vander Haegen and DeGraaf 1996). Seeps and riparian drainages that green up early in spring provide important herbaceous forage (e.g., skunk cabbage, various grasses and sedges, and tubers) for emerging bears (Elowe 1984).

Special Habitat Requirements: Secure den sites located under fallen trees, in hollow standing trees and logs on the ground, in tree cavities at/above ground level, rocky cavities, rock ledges, slash piles, or other protected areas (Schooley 1990, Alt and Gruttadauria 1984). A low rate of den reuse (4.8 percent) was documented in ne. Pennsylvania (Alt and Gruttadauria 1984). Dens are excavated in the ground where heat conservation is critical (Tietje and Ruff 1980, Pelton et al. 1980). In mild winters bears sometimes "den" in the open or under a laurel bush or thicket. Dens are sometimes in residential areas.

Reproduction: Age at sexual maturity: Females: 3.5 yr to 5 yr (Sauer 1975, Hugie 1982, Willey 1978, Alt 1981, Yodzis and Kolenosky 1986, Elowe and Dodge 1989). Breeding period: Early June through mid July (Erickson et al. 1964). Peak: Mid June. Gestation period: 7 to 8 months (Kurta 1995:217). Delayed implantation of fertilized eggs occurs in the fall (Wimsatt 1963) if the female reaches a minimum body weight of 68 kg (150 lb) in Minnesota (Rogers 1976). Young born: Late December to February in the ne. United States (Erickson et al. 1964, Alt 1983). Litter size: 1 to 5, average 2 to 3, varies with year, locality, early winter maternal condition, and overwinter loss of maternal mass for litters of at least 3 (Alt 1981, Kolenosky and Strathearn 1987:444, Samson and Huot 1995). Litters per year: Females breed once every 2 yr.

Home Range: In mountainous terrain, the elevational gradient provides a variety of vegetative conditions so that home ranges tend to be smaller than those home ranges found on flatter terrain (Rogers 1977, Rogers 1987, Alt et al. 1980, Kolenosky and Strathearn 1987:446, Meddleton and Litvaitis 1990). Adult females maintain well-defined home ranges over their lifetimes, passing their territories on to their daughters (Kolenosky and Strathearn 1987:446). Adult female ranges vary from 15 to 50 km² (6 to 19 square miles); summarized from numerous studies by Kolenosky and Strathearn (1987:446). Subadult female ranges were 10 to 20 km² (4 to 8 square miles). Adult males range over several times the area of adult females.

Sample Densities: Kolenosky and Strathearn (1987:445) summarized 11 studies in North America and described density estimates of 0.1 to 1.3 bears/km² (0.3 to 3.4 bears per square mile). Elowe (1984) estimated densities of 1 bear per 7 km² in Massachusetts. Density estimates were greatest in very diverse forests at a relatively early stage of development.

Food Habits: Black bears are seasonally omnivorous; their diets are largely determined by the seasonal availability of food. Newly emerging grasses, sedges, and forbs, grubs and ants in rotting logs and stumps, *Populus* sp. catkins and shoots, carrion, newborn deer fawns and moose calves, and occasionally adult moose are early spring foods (Pelton 1982, Ozoga and Verme 1982, Franzmann et al. 1980, Schwartz and Franzmann 1991, Austin et al. 1994). Bears first begin to gain weight in early summer when the succession of berries and other fruit appears (e.g., strawberries, pin cherry, raspberries, blueberries, blackberries, sarsaparilla, hawthorn, and apples). Carbohydrate- and fat-rich beechnuts, acorns, and other nuts like hazelnuts are important fall foods that allow the body weight gain necessary for sow and cub survival (Elowe and Dodge 1989). Bears will travel 40 to 80 km (25 to 50 miles) out of their normal ranges in search of these fall foods when they are limited (Kolenosky and Strathearn 1987:446). When summer berry and hard mast crops fail, bears resort to foraging on corn and oats, at garbage dumps, and in dumpsters, and tend to range much more widely than normally expected (Harger 1967, Kolenosky and Strathearn 1987:447). Can become nuisance animals at bird feeders in suburban and rural environments.

Comments: Daily activity cycles influenced by human activity (Kolenosky and Strathearn 1987:447). Trails are used repeatedly, and prominent trees are often marked by either sex by clawing and ripping off bark. Several individuals may mark the same tree. Strongest bonds are between females and offspring (Kolenosky and Strathearn 1987:447). Cubs are born during the winter while the female is denning. Cubs weigh 0.2 to 0.3 kg (0.4 to 0.7 lb) at birth and develop rapidly. Bears experience winter torpor (e.g., lower body temperatures and depressed heart rate) or "carnivorean lethargy" (Kurta 1995:215) during the coldest months and can readily waken if disturbed (Kolenosky and Strathearn 1987:443).

Raccoon

(*Procyon lotor*)

Range: Throughout most of s. Canada and the United States except for the deserts of the Southwest and higher elevations of the Rocky Mountains. Also extends s. from Mexico to Panama (Sanderson 1987:487).

Distribution in New England: Throughout the region.

Status in New England: Common.

Habitat: Wooded areas interspersed with fields and water courses. Not usually found in dense forests, commonly found in wetlands near human habitation (Kaufman 1982:571). Remnant wooded areas within extensive agricultural landscapes with extensive edge habitats (Pedlar et al. 1997).

Special Habitat Requirements: Ground dens, usually abandoned woodchuck burrows or culverts in areas lacking in tree dens (Lehman 1984). Prefers hollow trees (Gysel 1961). Dens are usually located in trees 10 ft (3 m) or more above ground (Banfield 1974:314, Sanderson 1987:490) and commonly located near water.

Reproduction: Age at sexual maturity: 38 to 77 percent of females breed their first fall (Junge and Sanderson 1982, Fritzell et al. 1985). Breeding period: Late February to June in n. latitudes (Fritzell 1978a). Peak: February. Gestation period: Approximately 63 days. Young born: Late April to early May; if the female is not pregnant, a second breeding cycle may begin 2 to 4 months later (Whitney and Underwood 1952). Litter size: 2 to 7 cubs, average 3 to 5 (Sanderson 1987:488, Clark et al. 1989). Litters per year: 1.

Home Range: Size varies with the individual, food availability, and weather. Calculated home range extremes varied from less than 5 ha (12.4 acres) in Ohio suburbs to nearly 5,000 ha (12,355 acres) in North Dakota (Sanderson 1987). Most other home range estimations vary in size from 40 to 100 ha (99 to 247 acres) (Sanderson 1987:491). Juvenile dispersal in n. populations occurs in spring or summer following birth; otherwise juvenile dispersal occurs in the fall and early winter (Lotze and Anderson 1979). Raccoons have traveled up to 165 miles (264 km) in 164 days (Lynch 1967).

Sample Densities: Densities in New Jersey ranged from 1 raccoon per 1.8 ha (4.4 acres) in woodlands near suburban areas, to 1 raccoon per 18.9 ha (47 acres) in mixed forest and agricultural land (Slate et al. 1982). Densities in Quebec farmland ranged from 1 raccoon per 4.6 to 27.7 ha (11.4 to 68.4 acres) (Rivest and Bergeron 1981).

Food Habits: Omnivorous and opportunistic (Hamilton 1936, Stuewer 1943). Animal matter is the major food in spring and early summer. Fruits, grains, and seeds are eaten in summer, fall, and winter. Crayfish, worms, insects, carrion, mollusks, turtle and bird eggs, tender buds and shoots, grass, and garbage are typical foods (Sanderson 1987, Ratnaswamy et al. 1997). During winter dormancy from late November to March, they live off fat accumulated in late summer (Godin 1977:213).

Comments: Primarily nocturnal, may be seen in daylight. Raccoons often defecate and urinate in common latrine sites (Sanderson 1987:492). Dormant through the winter, remaining in dens but not hibernating. Females and young move together as a family group; during cold weather, multiple animals will den together, presumably to conserve energy (Sanderson 1987:492). Raccoons are alert, intelligent animals with a well-developed sense of touch. A significant predator of waterfowl nests (Fritzell 1978b), sea turtle eggs (Ratnaswamy et al. 1997), and gull and seabird colonies (Kadlec 1971, Hartman et al. 1997). Highly susceptible to canine distemper and mid-Atlantic strain of rabies (Kaufmann 1982, Krebs et al. 1994).

American Marten

(*Martes americana*)

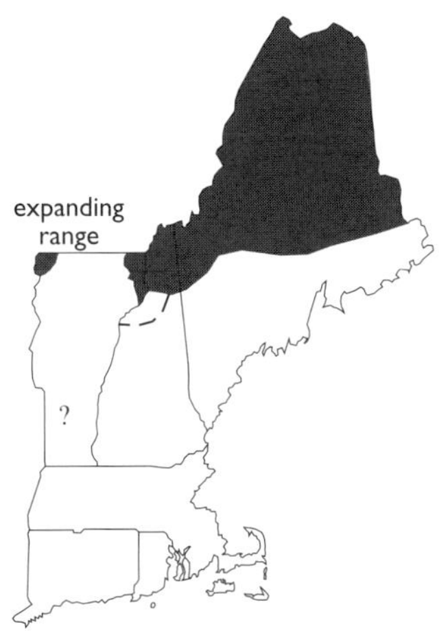

Range: Boreal forests of Canada to Alaska, s. in the Cascade-Sierra Nevada ranges, and the Rockies into New Mexico; extreme n. Minnesota, Wisconsin, and the Upper Peninsula of Michigan; and n. New England and New York (Strickland and Douglas 1987:532, Gibilisco 1994:62).

Distribution in New England: Northern Maine (Krohn et al. 1995) and higher elevations in n. New Hampshire (W. Staats pers. commun.); reintroduced from 1989 to 1991 into s. part of the Green Mountain National Forest in Vermont; establishment uncertain, although martens were detected by remotely triggered cameras at two sites in s. Vermont in winter 1994 (Brooks 1996).

Status in New England: Harvested in Maine; state listed as an endangered species in Vermont and state listed as threatened in New Hampshire (Appendix A).

Habitat: Inhabits a variety of forested habitats including coniferous forests of fir, spruce, and hemlock, cedar swamps, dense mixed hardwood-conifer forests, and forests with substantial unharvested spruce-budworm mortality (Soutiere 1978, Strickland and Douglas 1987:535, Clark et al. 1987, Thompson 1994, Chapin et al. 1997a). Marten used residual forest patches (median size 27 ha) within industrial forest landscapes (Chapin 1995). In the Adirondacks, martens are found in mixed stands as young as 30 yr, as well as in pole and mature hardwood stands, at elevations of 530 m to 1,463 m (1,740 to 4,800 ft) (Brown 1980).

Special Habitat Requirements: Maternal den sites in large [>15 in diameter breast height (40 cm)] hollow trees or logs and subterranean dens; summer resting sites in tree canopies, especially balsam fir witches' brooms (Wynne and Sherburne 1984, Chapin et al. 1997b:166). Females will move kits to different den sites several times prior to independence in late summer (Wynne and Sherburne 1984). Coarse woody debris on the forest floor and dense clusters of small-diameter live conifer stems provide subnivean access points to prey and winter resting places, including subterranean sites (Steventon and Major 1982, Corn and Raphael 1992, Sherburne and Bissonette 1994, Chapin et al. 1997a, Chapin et al. 1997b:166, Gilbert et al. 1997:135).

Reproduction: Age at sexual maturity: Females 2 to 3 yr. Males 1 yr. Breeding period: Late June to early September (Strickland and Douglas 1987:532). Peak: July. Gestation period: After 7 to 8 months of delayed implantation, active gestation of 27 days (Strickland and Douglas 1987:533). Young born: Mid March to late April (Strickland and Douglas 1987:534). Litter size: 1 to 5, typically 3 to 4. Litters per year: 1.

Home Range: Summer/fall home ranges in a harvested population in Maine averaged 5.2 km^2 for males and 2.8 km^2 for females, respectively (Wynne and Sherburne 1984, Katnik et al. 1994). Lactating females had smaller mean ranges (2.0 km^2) than nonreproductive females (3.3 km^2) (Katnik et al. 1994). Mean home-range estimates for an unharvested population in Newfoundland averaged 6.64 km^2 for females and 9.19 km^2 for males (Bissonette et al. 1991:116). Winter home-range estimates in Michigan's Upper Peninsula varied from 2.1 to 3.2 km^2 (Thomasma 1996). Same-sex resident adult martens maintain a spatial or temporal separation of range; immatures and members of the opposite sex are tolerated (Strickland and Douglas 1987:538). A seasonal altitudinal migration may occur in the mountains (Banfield 1974:316). Phillips et al. (1998) found that home-range areas of both sexes did not differ between summer and winter in a Maine forest preserve closed to trapping. Fidelity of males to home-range areas was higher than that of females.

Sample Densities: Unharvested resident adult densities in Algonquin Park were 0.6 to 1.2 adult martens/km^2 (1.6 to 3.1 per square mile) in late winter (Francis and Stephenson 1972). Harvested populations in Maine ranged from 0.1 to 1.2 adult martens/km^2 (Soutiere 1979, Katnik et al. 1994, Phillips 1994).

Food Habits: Marten are opportunistic foragers of meadow and red-backed voles, *Peromyscus* sp., masked shrew and snowshoe hare, passerine birds and ruffed grouse, and vegetation (fruit, berries, nuts) (summarized in Martin 1994:22, Thompson and Colgan 1987). Ungulate and beaver carrion, red squirrels, chipmunks, bird eggs, insects, and some amphibians and reptiles are also eaten (Strickland and Douglas 1987:537). Unusual prey items include a marten-killed goshawk (Paragi and Wholecheese 1994). Active night and day during all seasons. Known as a subnivean (under snow) hunter (Strickland and Douglas 1987:539).

Comments: Commonly called the pine marten. Martens are easily trapped, and populations are susceptible to overharvesting (Strickland and Douglas 1987:535), especially in areas with road accessibility (Soukkala 1983, Hodgman et al. 1994). Median dispersal distances were 14.3 km and 12.0 km for males and females, respectively (Phillips 1994). Similar home-range size, degree of spatial overlap, lactation rates, and age structure were observed in an untrapped forest preserve and adjacent trapped industrial forests (Phillips 1994), even though density was lower in the industrial forest site. Marten productivity may be maintained in patchy logged forest landscapes when home ranges are predominantly located within residual forest patches (Chapin et al. 1997a). On-site delimbing practices, cull tree retention, slash management, and other silvicultural practices that develop structurally complex forest stands, road access limitations, and untrapped preserves are some examples of methods that may conserve forest habitats for martens (Chapin et al. 1997a, Hodgman et al. 1994, Hodgman et al. 1997:86). Higher fisher densities may negatively influence the potential reoccupation of former coniferous habitats by martens (Krohn et al. 1995). Additional mammalian predation by coyote and fisher was documented by Hodgman et al. (1997).

Fisher

(Martes pennanti)

Range: Southeastern Labrador, w. to se. Alaska, s. to the Sierra Nevada of California, the Rocky Mountains to Wyoming, n. Minnesota, the Adirondacks of New York, and in New England s. to Connecticut (Douglas and Strickland 1987:511, Gibilisco 1994:62). Formerly occurred as far s. as w. North Carolina and e. Tennessee; s. distribution is much reduced.

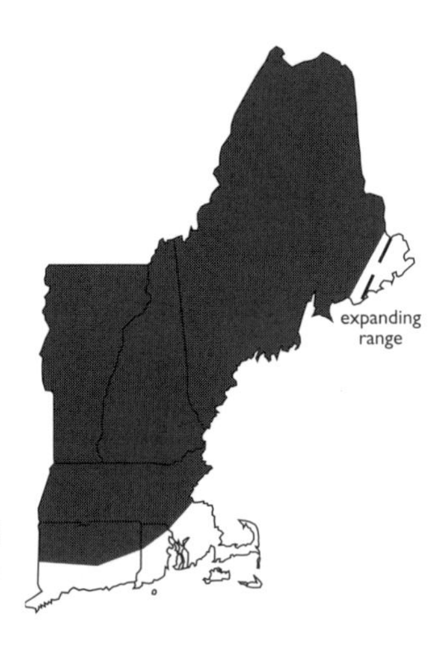

Distribution in New England: Throughout n. New England; spreading s. into se. Massachusetts and central Connecticut (Gibilisco 1994:62).

Status in New England: Common to uncommon.

Habitat: Extensive coniferous or mixed forests with continuous canopy (Powell 1993:88, Buskirk and Powell 1994, Thomasma et al. 1994:316, Thomasma 1996, Proulx et al. 1997) and availability of preferred prey (e.g., snowshoe hare). Favors wetlands (alder) and mixed softwood-hardwood forest types in n. New England in winter (Kelly 1977). A variety of age-classes of coniferous stands used throughout the year (Arthur et al. 1989b). Found less frequently in more open stands or burned areas in winter. Higher energetic costs for fishers to travel and forage in soft deep snow may account for winter use of coniferous types where snow is shallower or more packed (Raine 1983). Most commonly trapped in areas with less than 48 cm monthly snowfall (Krohn et al. 1995).

Special Habitat Requirements: Dens in hollow trees, logs, ground holes under large boulders, or vacant porcupine dens. Rarely digs burrow. Dens may be lined with leaves and are often used as temporary shelters during winter storms; the fisher does not hibernate. Uses tree cavities and nests as resting sites throughout the year; woodchuck burrows are commonly used in the winter (Arthur et al. 1989b). One to five large [>15.7 in (40 cm) diameter breast height] hardwood or softwood cavity trees used per adult female as natal den sites (Paragi et al. 1996, Powell et al. 1997a). Hunts dense patches of regenerating lowland conifer types for snowshoe hare (Powell 1993:117).

Reproduction: Age at sexual maturity: Both sexes become mature before their 12th month of age. Females produce first litter when 2 yr old (Wright and Coulter 1967). Breeding period: Late February to April. Peak: March. Gestation period: After delayed implantation of 11 months, active gestation of 30 days (Paragi et al. 1994). Young born: March to early April (Paragi et al. 1996, Powell et al. 1997a:265). Litter size: 1 to 6 kits, average 2.7 (Strickland et al. 1982:590, Powell 1993, Frost and Krohn 1997:100). Litters per year: 1.

Home Range: Yearly ranges averaged 7.4 square miles (19.2 km^2) in n. New Hampshire (Kelly 1977). Home ranges averaged 16.3 and 30.9 km^2 for females and males, respectively, in Maine but were not significantly different (Arthur et al. 1989a). Males range farther than females, especially during April and May when males are seeking mates and females with kits restrict their movements (Douglas and Strickland 1987:519). Fishers commonly travel along ridges and readily swim streams to reach the next ridge (Coulter 1959).

Sample Densities: Density in summer ranges from 1 animal per 2.8 to 10.5 km^2; in winter from 1 animal per 8.3 to 20 km^2 in s.-central Maine (Arthur et al. 1989a). York (1996) reported average densities of 2.1 fishers per 10 km^2 in n.-central Massachusetts and sw. New Hampshire.

Food Habits: Fishers travel widely in search of prey. Important food items include snowshoe hare, porcupine, passerine birds, deer, ungulate carrion, and vegetation (e.g., apples, nuts, and berries) (Martin 1994:297). Moles, shrews, mice, voles, and squirrels (red, gray, and flying) are also common foods in New England (Kelly 1977, Giuliano et al. 1989, Arthur et al. 1989b). Fishers preyed heavily on raccoons in central Massachusetts where rabies was prevalent there (Powell et al. 1997b:279).

Comments: Fishers are twice as large as martens, and are as agile in trees as they are on the ground. Adults most active before sunrise and after sunset, but some activity occurs throughout the day and night (Arthur and Krohn 1991). Seasonally more active in summer than in winter (Arthur and Krohn 1991). High fisher densities may now limit the reoccupation of former coniferous habitats by marten (Krohn et al. 1995). Fishers are one of the more easily trapped species with historically high pelt prices (Powell 1993:201). Human-related causes (e.g., trapped, shot, hit by vehicle or other) accounted for 94 percent of the mortality observed in a study of radio-collared fishers in s.-central Maine (Krohn et al. 1994:137). Adult females were less susceptible to trapping than either adult males or juveniles (Krohn et al. 1994:137).

Ermine

(*Mustela erminea*)

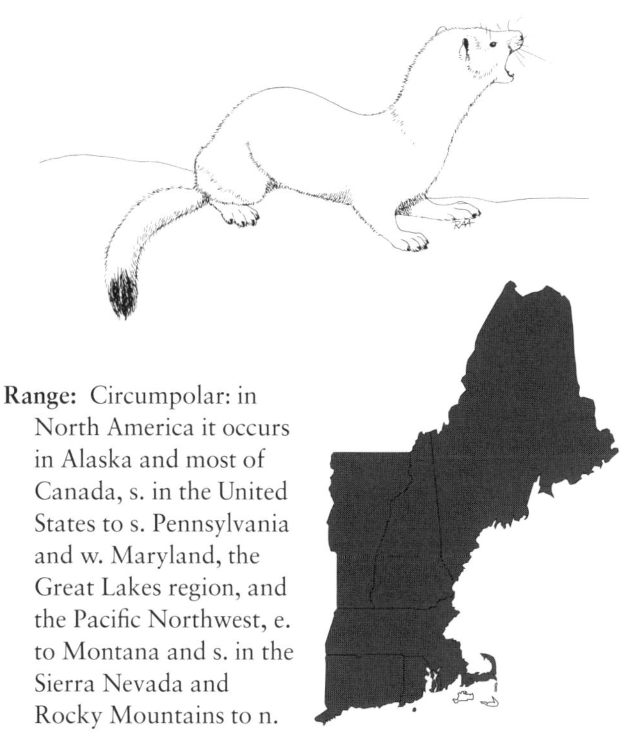

Range: Circumpolar: in North America it occurs in Alaska and most of Canada, s. in the United States to s. Pennsylvania and w. Maryland, the Great Lakes region, and the Pacific Northwest, e. to Montana and s. in the Sierra Nevada and Rocky Mountains to n. New Mexico (Fagerstone 1987:559, King 1990:11).

Distribution in New England: Throughout the region.

Status in New England: Common to uncommon.

Habitat: Successional woodlands and brushland, forest-edge habitats, alpine meadows, marshes, riparian woodlands, and hedgerows (Simms 1979a, King 1983, Fagerstone 1987:560).

Special Habitat Requirements: Plentiful small mammal prey and dense brushy cover.

Reproduction: Age at sexual maturity: Males probably 11 months (Fagerstone 1987:559). Females 4 to 6 weeks (King 1990:124). Breeding period: May to August (King 1990:124). Gestation period: After delayed implantation of 9 to 10 months, 28 days completes the gestation period, totaling 280 days (ranging from 220 to 380 days) (King 1983). Young born: Mid April to early May or 22 to 25 days after the first brown hairs begin to appear on the female snout in Siberia (Ternovsky 1983). Litter size: 4 to 9, typically 6 or 7 (Hamilton 1933b). Litters per year: 1.

Home Range: Males and females exclude members of the same sex from their home ranges (Erlinge 1977), although male and female home ranges overlap somewhat (Fagerstone 1987:560). Simms (1979b) estimated female home ranges between 10 to 15 ha (25 to 37 acres), and that of males between 20 to 25 ha (42 to 62 acres) in mixed habitats in Ontario. In periods of high prey abundance, home ranges can be less than 4 ha (10 acres) (Fagerstone 1987:560). Six hundred meters (1,968 ft) was the longest daily movement observed by Simms (1979b).

Sample Densities: 6 individuals/km² (16 per square mile) in grazed pasture and forest; and 11 individuals/km² (28 per square mile) in overgrown pasture and shrub habitats in Ontario (Simms 1979b).

Food Habits: Mice and voles (staple), chipmunks, moles and shrews, occasionally birds and insects, even rabbits, and rarely snakes, frogs, or fish (King 1983, 1990:63).

Comments: Ermines molt twice annually, in spring and in autumn. The autumn molt from brown to white pelage usually begins in October and is usually complete by late November. The spring molt begins about mid March (Godin 1977:222). Den is usually below ground under fallen tree or stump but may also be in abandoned building, stone wall, hollow log, or almost anywhere there is a small dry enclosure. Nocturnal and active throughout the year. Hunts subnivean (under-snow) runways of small mammals in winter as well as on frozen or compacted snow; burrows easily in soft snow (Simms 1979a, King 1983). Surplus prey cached and defended from other ermines (King 1983). Resting ermines in low temperatures need additional thermal protection and often occupy the nests of prey (Chappell 1980), sometimes improving the thermal characteristics by lining them with vole fur (Erlinge 1979, Fitzgerald 1981). Various raptors including rough-legged hawk, goshawk, and great horned owl, domestic cat, coyote, red and gray fox, and long-tailed weasel are predators of ermine (Fagerstone 1987:559, King 1990:181). Formerly called the short-tailed weasel.

Long-tailed Weasel

(*Mustela frenata*)

Range: Southern Canada to Peru. Not found in the sw. deserts in the United States or nw. Mexico except Baja, n. California (Fagerstone 1987:555, King 1990:11, Sheffield and Thomas 1997).

Distribution in New England: Throughout the region.

Status in New England: Common to uncommon.

Habitat: Open woods and woodland edges, fencerows, river bottomlands. Found at elevations from sea level to the alpine tundra zone. Prefers to be near water (Hall 1951).

Special Habitat Requirements: Uses or enlarges previously excavated small burrows (e.g., chipmunk) or natural holes or crevices for dens; areas of abundant prey (Fagerstone 1987:557).

Reproduction: Age at sexual maturity: Females 3 to 4 months (Svendsen 1982). Males 15 months (Fagerstone 1987:555). Breeding period: July to August. Gestation period: After 7.5 months of delayed implantation, active gestation lasts 23 to 27 days (Wright 1942, Wright 1947). Young born: April to May. Litter size: 1 to 12, average 6 to 9 (P. L. Wright 1948, Heidt 1970). Litters per year: 1.

Home Range: Size and location vary with food availability, cover type, and season. DeVan (1982) estimated home ranges between 10 and 24 ha (25 to 59 acres) in Kentucky when prey was abundant. Summer ranges were larger [16 to 24 ha (39 to 59 acres)] than winter ranges [10 to 18 ha (25 to 44 acres)]. Quick (1944) estimated home ranges between 32 and 160 ha (79 to 395 acres) in southern Michigan mixed agricultural-wooded-marsh habitat; and 80 to 120 ha (198 to 297 acres) in Colorado (Quick 1951). Weasels in open woods traveled farther than weasels in dense brush (Fagerstone 1987:557). Nightly winter distances traveled in new snow by males averaged 704 ft and 346 ft for females (Glover 1943), but normally move farther, 2 miles for males (Quick 1944) and at least a half mile for 1 female (Criddle and Criddle 1925).

Sample Densities: 6.5 to 9 individuals per 100 ha (16.8 to 23 per square mile) in scrub-oak/pitch-pine forest in Pennsylvania (Glover 1943); 0.4 to 3 individuals per 100 ha (1.0 to 7.8 per square mile) in Michigan farmland (Craighead and Craighead 1956, D. L. Allen 1938); 20 to 30 individuals per 100 ha (51.8 to 77.7 individuals per square mile) in an Ontario cattail marsh (Wobeser 1966).

Food Habits: Primarily small mammals including shrews, voles, mice, rabbits, snowshoe hare, squirrels; some ground-nesting birds; insects, carrion (Gamble 1981, Gamble and Riewe 1982), and occasional snakes (King 1990), and bats (Mumford 1969). Vole-sized prey usually almost entirely eaten except for the skull, skin, and vertebral column (Fagerstone 1987:558). Excess prey is cached in burrows for later consumption (Fagerstone 1987:558). May climb trees to catch squirrels or escape predators (Nams and Beare 1982).

Comments: Active year-round; commonly thought to be mainly nocturnal but often seen active during daylight hours. Northern weasel pelage turns white in winter; southern winter pelage is brown (Hall 1951). Molting occurs from mid October to mid November and mid February to mid April.

Mink

(*Mustela vison*)

Range: Throughout Canada s. of the tundra region and the United States except the arid Southwest (Linscombe et al. 1982:629, Eagle and Whitman 1987:615).

Distribution in New England: Throughout the region.

Status in New England: Common to uncommon.

Habitat: A variety of wetland habitats, including lakeshores, streams, and rivers; freshwater and saltwater marshes; and coastal zones (Eagle and Whitman 1987). Favors log-strewn forested wetlands with abundant thicket cover and a dependable supply of food (Eagle and Whitman 1987:616).

Special Habitat Requirements: Den sites inside hollow logs, natural cavities under tree roots, beaver lodges, or in muskrat bank burrows along stream, marsh, or lake edges (Errington 1961, Eagle and Sargeant 1985).

Reproduction: Age at sexual maturity: 10 months. Breeding period: Late February to early April. Peak: March (Mitchell 1961). Gestation period: 40 to 75 days, average 51 days, 28 to 30 days post blastocyst implantation (Enders 1952, Mead 1981). Young born: April or May. Litter size: 1 to 8 kits, average 4 to 5. Litters per year: 1.

Home Range: The average range is 2 to 3 miles (3.2 to 4.8 km) in diameter for males along river in Montana. Two females had home ranges of 19.3 and 50.4 acres (7.8 and 20.4 ha) in similar river habitat. Long-distance travel is common along waterways, and in winter mink may swim under the ice. Home ranges often overlap between juveniles and adults (Mitchell 1961).

Sample Densities: Range from 8.5 to 21.9 mink per square mile (3.3 to 8.5 per km^2) in Montana (Mitchell 1961). Linear densities of 2.6 to 5.8 mink per km (1.6 to 3.6 mink per mile) of shoreline reported on Vancouver Island (Ritcey and Edwards 1956).

Food Habits: Aquatic and terrestrial prey. Importance of prey items varies with season and habitat (Linscombe et al. 1982). Muskrats, shrews, voles, mice, rabbits, ducks, shorebirds, rabbits, fish, frogs, salamanders, crayfish, clams, and insects are eaten (Proulx et al. 1987). Trails prey by scent and often caches food.

Comments: Minks are restless, curious, bold, and mainly nocturnal; they are active year-round. Mink serve as bioindicators of mercury pollution in aquatic habitats (Larivière 1999, Wren et al. 1986). Ranched mink bred in many color phases; wild mink are dark brown with white on the chin (Godin 1977:226).

Striped Skunk

(*Mephitis mephitis*)

Range: Occurs throughout s. Canada, except coastal British Columbia, and throughout the United States except the Mohave Desert region in s. California and Nevada (Godin 1982:674, Wade-Smith and Verts 1982, Rosatte 1987:599).

Distribution in New England: Throughout the region.

Status in New England: Common.

Habitat: Semiopen country, woods and meadows, agricultural lands, suburban areas, and trash dumps. Occurs from sea level to timberline.

Special Habitat Requirements: Den sites may be under houses and abandoned buildings, stumps, in stone walls, rock cavities, culverts, or abandoned burrows, especially those of woodchucks (Rosatte 1987:601).

Reproduction: Age at sexual maturity: Spring following birth (Verts 1967). Breeding period: February to mid April. Peak: Mid March. Gestation period: 59 to 77 days, suggestive of a delayed implantation of the egg and mating date (Wade-Smith et al. 1980). Young born: Late April to early June. Litter size: 2 to 10 kits, typically 6 or 7. Litters per year: 1, though loss of litter may stimulate a second breeding period in May (Wade-Smith and Richmond 1978).

Home Range: Nightly movements cover 0.25 to 0.50 square mile (0.6 to 1.35 km^2), increasing to 4 or 5 square miles (10.4 to 13 km^2) at night during breeding season (Schwartz and Schwartz 1981). Home range size varies from 0.25 square mile (0.6 km^2) in urban habitats to 0.8 to 1.9 square miles (2.2 to 4.9 km^2) in rural areas (Rosatte 1987:602). Adult male seasonal home range sizes (11.6 km^2) are larger than those of adult females (3.7 km^2) in the prairie pothole region of s.-central Saskatchewan (Larivière and Messier 1998).

Sample Densities: Rosatte (1987) averages numerous estimates from studies across a variety of parkland, farmland, and urban habitats at 1.3 to 67 skunks per square mile (0.5 to 26/km^2).

Food Habits: Omnivorous; diet includes insects, snails, small rodents, birds' eggs (especially waterfowl eggs), fruits, grains, nuts, corn, grasses, buds, berries, garbage, and carrion (Wade-Smith and Verts 1982, Pasitschniak-Arts and Messier 1995). In summer the diet may be as much as 43 percent insects (Banfield 1974:339).

Comments: The amount and pattern of white in the skunk's pelage vary greatly; the less white, the more valued by the fur industry. Most people are familiar with skunk musk, composed of butylmercaptan, a malodorous sulfur-alcohol compound secreted by the anal glands. Skunks emit a mist or droplet stream as a defense while facing the threat, arching the back with the tail plumed out high over the back, stamping the forefeet rapidly and shuffling backward. Daily range movements vary from 1.1 to 1.4 km^2 (271.8 to 345.9 acres) (Larivière and Messier 1998) in s. Saskatchewan. Semihibernates during the winter months; young may remain in den with mother. Crepuscular or nocturnal, sometimes active during daylight hours. Routine prey item for great horned owl; also preyed upon by coyote, foxes, and bobcats, usually when in a starved condition (Wade-Smith and Verts 1982). Many skunks are killed by automobiles, especially in spring. Some taxonomists suggest an alternative systematic placement of skunks in the Mephitidae rather than the Mustelidae (Dragoo and Honeycutt 1997, Jones et al. 1997).

River Otter

(*Lontra canadensis*)

Range: Throughout all but the northernmost portions of Canada and Prince Edward Island; Alaska and most of the lower 48 states. Never abundant, otters have been extirpated in many parts of their range since settlement (Godin 1977:234, Toweill and Tabor 1982, Melquist and Dronkert 1987:627, Larivière and Walton 1998) except in deserts or treeless regions.

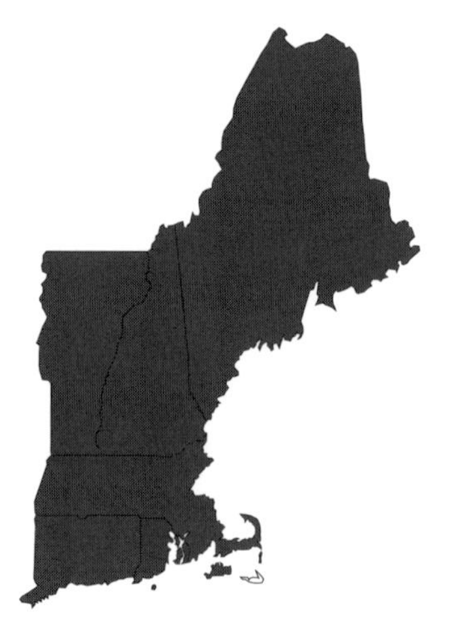

Distribution in New England: Throughout the region.

Status in New England: Uncommon, but probably more common than sighting and trapping would indicate.

Habitat: Riparian borders of streams, lakes or other wetlands in forested areas; watersheds with a high proportion of beaver-influenced drainages in Maine (Dubuc et al. 1990). Also in brackish water.

Special Habitat Requirements: Water body, river, or stream containing fish, suitable den sites, complex riparian vegetation structure (fallen trees, submerged trees, logjams, dense shrub thickets, tall grass patches). Dens may be in crevices, rocky ledges, under fallen trees, in abandoned beaver lodges, in muskrat houses, or in dense thickets bordering water. Shoreline features such as points of land, isthmuses, and mouths of permanent streams were used in higher proportions relative to their availability as spraints (latrine sites) (Newman and Griffin 1994).

Reproduction: Age at sexual maturity: Generally 2 yr (Hamilton and Eadie 1964, Tabor and Wight 1977). Breeding period: March or April. Gestation period: Following delayed implantation of less than 8 months, a true gestation period of 61 to 63 days (Hamilton and Eadie 1964, Larivière and Walton 1998). Young born: February and April. Litter size: 1 to 5, average 2 or 3 (Chilelli et al. 1996). Mean number of corpora lutea in adult females varies from 1.49 in Maine and 2.1 in New Hampshire, to 3.02 in Oregon; and with the sample age structure (Tabor and Wight 1977, Docktor et al. 1987, Hazzard et al. 1993). Litters per year: 1.

Home Range: River otters are highly mobile (Larivière and Walton 1998). Territories maintained within home range (Erlinge 1968). Home range estimates vary from 1.8 to 57 km^2 (0.7 to 22 square miles) as summarized by Melquist and Dronkert (1987:632). Linear distance estimates range from 1 to 78 km (0.6 to 48.5 miles) of shoreline or waterway (Melquist and Dronkert 1987:632). Male home ranges are generally larger than those of females, and sexes exhibit overlapping ranges in w. North America (Bowyer et al. 1995, Reid et al. 1994a).

Sample Densities: Density estimates vary from 1 otter per 1.9 to 2.1 km (1.2 to 1.3 miles) to 1 otter per 10 to 17 km (6.2 to 10.6 miles) of shoreline or waterway, as summarized by Melquist and Dronkert (1987:632).

Food Habits: River otters take fish, frogs, crayfish, salamanders, and turtles, also snakes, small birds, mammals, earthworms, and insects (Wren et al. 1986, Reid et al. 1994b, Larivière and Walton 1998).

Comments: Otters are the most aquatic members of the weasel family and social; wrestling and chasing play is an important socializing otter activity (Liers 1951, Beckel 1991). Otters frequently slide down steep grass-, mud-, or snow-covered slopes or along snow-covered flat terrain. May be active at any time; dawn to midmorning and evening hours are the periods of most activity (Melquist and Hornocker 1979). Active throughout the year. Formerly *Lutra canadensis* (Schreber) (Larivière and Walton 1998).

Mountain Lion

(Puma concolor)

Range: Southern Canada s. in the w. mountains of the United States to Tierra del Fuego, with remnant populations in Florida and perhaps w. Arkansas and e. Oklahoma (Young and Goldman 1946, Dixon 1982:711, Currier 1983, Lindzey 1987:657). Recent sightings documented in Ontario, New Brunswick, and Nova Scotia (Gerson 1988, Cumberland and Dempsey 1994, Stocek 1995). The mountain lion was the most widely distributed mammal in the New World.

Distribution in New England: Unknown. At this time little if any evidence exists of a breeding population in the ne. United States (Tischendorf and Ropski 1996). The last New England specimen was taken in Somerset County, Me., in 1938 (Wright 1961).

Status in New England: Considered extirpated. Unconfirmed sightings are reported each year in New England and the Northeast.

Habitat: Historically used a variety of habitats. If stragglers are present today, they probably inhabit remote mountain forests, swamps, and wooded watercourses.

Special Habitat Requirements: Requires abundant deer-sized prey. Some isolation from humans (Van Dyke et al. 1986a).

Reproduction: Age at sexual maturity: Probably 2 to 3 yr (Currier 1983), though females may breed as early as 20 months (Lindzey 1987:658). Breeding period: Throughout the year at intervals of 18 to 24 months (Lindzey 1987:658). Gestation period: 82 to 96 days (Eaton and Verlander 1977). Young born: In Northern Hemisphere most occur between April and September (Eaton and Verlander 1977) (spring-born cubs have highest survival rates in the North). Litter size: 1 to 6, typically 2 or 3 (Currier 1983, Spreadbury et al. 1996). Litters per year: <1.

Home Range: Western home ranges vary greatly from 12 to 398 square miles (30 to 1032 km²) as summarized by Lindzey (1987:660). Maximum 1-day movements of resident lions range from 3.8 to 8 miles (6.1 to 12.9 km) (Lindzey 1987:660). Adult female mean annual home range size was 55 km² (21.2 square miles); and 151 km² (58.3 square miles) for adult males in se. British Columbia (Spreadbury et al. 1996). Beier's (1993) simulation model suggests 2,200 km² as a minimum habitat area for cougars in California, in which habitat corridors are an important element of the simulation. Home ranges in the Northeast are not known. Summer and winter home ranges shift from higher to lower elevations in mountainous regions following seasonal patterns of ungulate movements (Dixon 1982, Lindzey 1987:660).

Sample Densities: Estimates for areas studied at least 12 months or 2 consecutive winters ranged from 0.13 to 0.013 per square mile (0.5 to 5.0 per 100 km²) (Lindzey 1987:659, Spreadbury et al. 1996). Road track counts may provide useful indices of resident mountain lion density (Van Dyke et al. 1986b).

Food Habits: Nocturnal. Feeds on deer, fox, beaver, porcupine, raccoon, skunk, rabbit, and smaller mammals. Caches large prey.

Comments: Solitary and mobile except for female-kitten groups. Juveniles capable of dispersal distances ranging from 9 to 274 km (5.6 to 170 miles) (Logan et al. 1986). Estimated consumption of 1.8 to 5.5 kg per day, summarized from several western studies (Lindzey 1987:661). There is a small population in Florida (Dixon 1982, Florida Panther Interagency Committee 1987). Many unconfirmed sightings in New England and e. Canada (Wright 1948, 1953, 1959, 1973), and one confirmed track (R. Downing pers. commun.). Some sightings may be captive animals released into the wild or escaped animals (Stocek 1995). Underweight yearlings are implicated in very infrequent attacks on humans in s. California, w. Texas, and e. Colorado (Beier 1991). Also called the cougar or puma; some taxonomists use *Felis concolor* (Linnaeus) (Nowell and Jackson 1996). A mountain lion was killed in Maine near Andover in 1891 (Goodwin 1936); in Vermont in 1881 in Barnard (Titcomb 1901); and about 1885 in the White Mountains (Jackson 1922). Mearns (1900) listed the presence of mountain lions in Rhode Island during the Colonial period. Mountain lions may yet recolonize the more remote parts of New England.

Lynx

(*Lynx canadensis*)

Range: Newfoundland w. to Yukon Territory and Alaska, s. in the United States through the n. Rockies, n. Great Lakes States, and n. New England (Quinn and Parker 1987:684, Tumlison 1987, Koehler and Aubry 1994). Extirpated in much of its southern distribution (Litvaitis et al. 1991b).

Distribution in New England: Lynx are currently detected during winter track surveys in nw. Maine (K. Elowe pers. commun.). Lynx kittens found by radio-tracking an adult female in nw. Maine in spring 1999 (C. McLaughlin pers. commun.). One marked lynx from a recent Adirondacks reintroduction project (Brocke et al. 1990) was shot in central New Hampshire, and another non-Adirondacks released lynx was found dead on I-89 (K. Gustafson pers. commun.).

Status in New England: Uncommon to rare. Recently listed as federally threatened for the contiguous U.S. distinct population segment (Federal Register 24 March 2000: 50 CFR Part 17).

Habitat: Northern forests and other diverse forest landscapes with a significant composition of early successional habitat from either logging, fire, or insect outbreak (Quinn and Thompson 1987). Found at higher densities in boreal mixed wood forests (27 percent early successional habitat; clear-cut sizes average 160 ha [395 acres]) than in boreal coniferous forests (17 percent early successional habitat; clear-cut sizes average 1,384 acres) in ne. Ontario (Quinn and Thompson 1987). Selected successional habitat on Cape Breton Island in both summer and winter (Parker et al. 1983). Favors swamps, bogs, or rocky areas. Deep winter snow cover favors large-pawed lynx over smaller pawed and shorter legged bobcat and may limit n. expansion of bobcat (Parker et al. 1983, Rolley 1987).

Special Habitat Requirements: Successional habitats with high densities of snowshoe hare (Parker et al. 1983). Dense coniferous and deciduous understory cover beneath partially open overstory canopies (Orr and Dodds 1982, Litvaitis et al. 1985b). Rears young in den, which may be among rocks, under fallen trees, in hollow logs, or other sheltered places.

Reproduction: Age at sexual maturity: Females as yearlings and may vary with prey abundance (Brand and Keith 1979, McCord and Cardoza 1982). Breeding period: Mid March to early April (Quinn and Parker 1987:684). Gestation period: About 63 days. Young born: May to early June. Litter size: Mean size of 3.4 to 5.3 depending on hare abundance (Brand and Keith 1979, Slough and Mowat 1996). Litters per year: 1.

Home Range: Mean annual home-range size varied from 16.6 to 62.5 km² (6.4 to 24.1 square miles) in the Northwest Territories, depending on the stage in the hare cycle (Poole 1994). Slough and Mowat (1996) found mean annual home range size varied from 20.7 to 266.2 km² (8.0 to 102.8 square miles) for adult males and 11.2 to 507.1 km² (4.3 to 195.8 square miles) for adult females in sw. Yukon Territory. Three radio-collared adults' home ranges varied from 138 to 221 km² (53.3 to 85.3 square miles) in sw. Manitoba (Carbyn and Patriquin 1983). 6 to 8 square miles (16 to 21 km²) with 2.6-mile (4.2-km) daily cruising radius (Saunders 1963b, Banfield 1974:350). Seasonal home ranges for two radio-collared adults were larger in summer, 25 to 32 km² (9.6 to 12.3 square miles) than in winter, 12 to 18 km² (4.6 to 6.9 square miles) on Cape Breton Island (Parker et al. 1983). Seasonal daily cruising distances for three radio-collared lynx ranged from 7.3 to 10.1 km (4.5 to 6.3 miles) in summer and 6.5 to 8.8 km (4.0 to 5.5 miles) in winter (Parker et al. 1983).

Sample Densities: 2 per 100 km² to 44.9 per 100 km² (0.05 to 1.2 per square mile) in unharvested populations in the Northwest and Yukon Territories (Poole 1994, Slough and Mowat 1996). 18 to 20 per 100 km² (0.47 to 0.52 per square mile) on Cape Breton Island over a 1-yr period (Parker et al. 1983). 2.3 to 10 per 100 km² (0.06 to 0.26 per square mile) over an 8-yr study in Alberta (Brand et al. 1976). Population density in ne. Ontario boreal mixed wood forests appears to be higher than in boreal coniferous forests (Quinn and Thompson 1987).

Food Habits: Mainly snowshoe hare (staple), but also voles, sciurids, and birds (Saunders 1963a). Occasionally carrion of deer, caribou, moose, and beaver. Lynx populations fluctuate with snowshoe hare abundance, reaching peak numbers about once every 10 yr (Saunders 1963a, van Zyll de Jong 1966, Parker et al. 1983).

Comments: Solitary and elusive, lynx are mainly nocturnal and active throughout the year. Capable of dispersing 17 to 930 km (10.6 to 577.9 miles) when local snowshoe hare densities rapidly declined to 0.6 to 1.3 hares/ha (1.5 to 3.2 hares/acre) in Northwest Territories (Poole 1997); and 100 to 1100 km (62.1 to 683.5 miles) in the Yukon (Slough and Mowat 1996). Lynx are easily trapped; mortality from trapping in an increasing lynx population in ne. Ontario ranged from 44 to 86 percent (Quinn and Thompson 1987). Breitenmoser et al. (1993) hypothesize resident lynx form a core population, occupying large stable home ranges throughout the cyclic hare cycle. Where trapped, lynx need a network of untrapped refugia and careful seasonal control and temporary closures, especially during low-density periods in the snowshoe hare cycle when lynx recruitment is suppressed (Parker et al. 1983, Quinn and Gardner 1984, Slough and Mowat 1996, Nowell and Jackson 1996). The northward range extension of bobcat into lynx habitat may be limiting lynx populations along the s. edge of their distribution. Bobcats out-compete lynx except in deep-snow conditions (Parker et al. 1983, Quinn and Parker 1987:684). Whether lynx normally occurred in central or s. New England is uncertain. Sightings and tracks up to the 1960s may be largely due to detections by trappers of southward-wandering individuals in low hare years farther north. Longevity record of 14 yr documented in Labrador (Chubbs and Phillips 1993b).

Bobcat

(Lynx rufus)

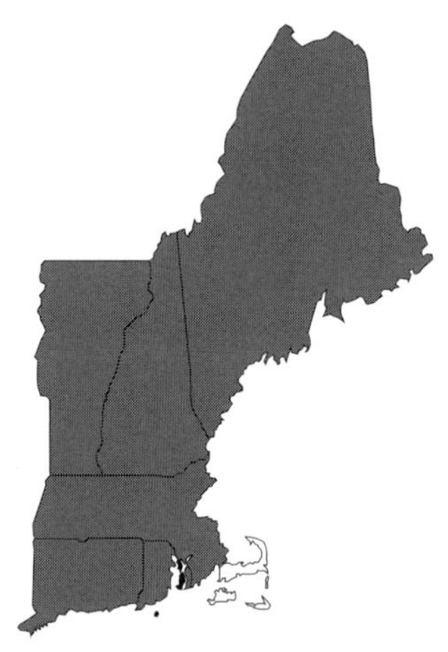

Range: Southern Canada, s. throughout the w. half of the United States; n. Great Lakes states; New England and e. New York; and s. through the Appalachians and associated uplands to Florida and the Gulf Coast (Rolley 1987:671, Larivière and Walton 1997). Reintroduced to n. New Jersey in 1978 (Lund 1980).

Distribution in New England: Throughout the region.

Status in New England: Common to uncommon.

Habitat: Mixed deciduous-coniferous and hardwood forests with high prey densities (Rolley 1987:673). Also brushy, rocky woodlands interspersed with old roads and fields, pipelines, and densely regenerating stands (Rolley 1987:673). Frequently found in lowland conifer stands of white cedar and black spruce in northern parts of their range (Fuller et al. 1985). Softwood cover preferred in winter (May 1982).

Special Habitat Requirements: Dense (>12,000 stem cover units/ha) hardwood and softwood understories with high hare densities and slopes <5 percent (Litvaitis et al. 1986b). Deep snow conditions appear to limit bobcat mobility (McCord 1974, Parker et al. 1983). Rocky ledges critical in Massachusetts (McCord and Cardoza 1982:742). Prefers to den in rock crevices, under windfalls, brush piles, or in hollow logs.

Reproduction: Age at sexual maturity: Females mature within 9 to 12 months after birth (Fritts and Sealander 1978, Parker and Smith 1983). Males mature during second year (Crowe 1975). Breeding period: Late February to March, sometimes extending into June. Gestation period: About 62 days (McCord and Cardoza 1982:733). Young born: Late April to mid May. Litter size: range from 1 to 6; average 2.5 to 3.9 (McCord and Cardoza 1982:733, Parker and Smith 1983). Litters per year: Usually 1. Late summer births in Arkansas attributed to first-time breeders (Fritts and Sealander 1978). Very rarely 2 litters per year (Winegarner and Winegarner 1982).

Home Range: 2 to 5 linear miles (3.2 to 8 km) for nightly travel in Massachusetts (Pollack 1951). McCord (1977) estimated home range to be 26 to 31 acres (10.5 to 12.5 ha) in Massachusetts. In Minnesota, bobcats traveled 3 to 7 miles (4.8 to 11.2 km) while hunting (Rollings 1945). Home-range size varies from 0.6 to 201 km^2 (0.2 to 78 square miles) range-wide (McCord and Cardoza 1982:739). Litvaitis et al. (1986b) found average male bobcat home ranges of 95.7 km^2 (36.9 square miles) and average female ranges of 31.2 km^2 (12.0 square miles) in Maine. Male bobcat home ranges are 2 to 5 times larger than female home ranges (Litvaitis et al. 1986b, Rolley 1987:674). Metabolic home-range size (not influenced by body weight) was inversely correlated with stem cover unit density and hare density (Litvaitis et al. 1986b). Resident bobcats show strong site fidelity; transient bobcats rapidly occupy ranges of removed residents (Litvaitis et al. 1987).

Sample Densities: McCord and Cardoza (1982) summarize reported densities ranging from 0.04/km^2 (0.1 per square mile) to 2.7/km^2 (7.0 per square mile).

Food Habits: Snowshoe hare, cottontails, squirrels, muskrats, small mammals, birds and their eggs (Mills 1984, Major and Shelburne 1987, Litvaitis et al. 1986b, Dibello et al. 1990). Deer is a principal winter diet component in New York and w. Maine (Fox and Brocke 1983, Major and Shelburne 1987, Litvaitis et al. 1986b). Carrion (untainted), snakes, fish, crustaceans, insects, and some vegetation are also eaten. Most prey is taken by stalking. Large bobcats include higher proportions of deer in their diets than smaller bobcats (Litvaitis et al. 1984, Litvaitis et al. 1986a, Matlack and Evans 1992). Deer provided higher levels of dry matter digestibility (95.7 percent) than gray squirrel (76.6 percent), small mammals (72.6 percent), or snowshoe hare (68.3 percent); deer diets were higher in metabolizable energy compared to squirrel, small mammals, or snowshoe hare (Powers et al. 1989).

Comments: Bobcat and coyote home ranges overlap, and both species use the same winter food resources (Major and Shelburne 1987, Litvaitis and Harrison 1989). Coyotes in e. Maine have been implicated in reductions of bobcat carrying capacity by decreasing prey availability (Litvaitis and Harrison 1989). Bobcat and red fox are not competitors for food resources (Major and Shelburne 1987). In e. Maine, radio-collared bobcats used areas with lower densities of USGS class 1 and 2 roads than the road densities generally found at the larger landscape level (Nobel 1993). Favors established routes and uses scent posts. Solitary and elusive, mainly nocturnal but in winter is active during daylight. Northern bobcats regularly seek the sun on exposed, wind-protected slopes and ledges to moderate stressful ambient temperatures in winter (Mautz and Pekins 1989). Avoids crossing bodies of water (generally) but can swim well (Godin 1977:241). Annual survival rates from radio-collared adult bobcats generally vary from 0.56 to 0.67 unless high harvest rates or poaching situations prevail (Fuller et al. 1995, Litvaitis et al. 1987).

White-tailed Deer

(*Odocoileus virginianus*)

Range: Across s. Canada to central British Columbia and throughout the United States, except for Utah, and most of Nevada and California (Smith 1991, Halls 1984). Range extends into South America.

Distribution in New England: Widespread throughout the region.

Status in New England: Common.

Habitat: Forest edges, swamp borders, areas interspersed with fields and woodland openings. During winter months when snow depths exceeds 16 in (40.6 cm), deer will yard in stands of conifers, forming a central resting area with trails packed through the snow.

Special Habitat Requirements: Dense cover for winter shelter, adequate browse.

Reproduction: Age at sexual maturity: Some females can breed at 6 to 7 months, but most breed as yearlings; males are mature at 18 months. Breeding period: Late October to mid December. Peak: November. Gestation period: Roughly 200 days (Verme and Ullrey 1984:97). Young born: May and June with an extreme spread from March to September. Young produced: 1 to 3 fawns, depending on nutritional and genetic factors (Verme and Ullrey 1984:97).

Home Range: Average annual home ranges in nonmigratory deer vary from 146 to 1,285 acres (59 to 520 ha), and seasonally used areas rarely exceed 1.6 km in radius (Marchinton and Hirth 1984:129, Smith 1991). Mean winter range size varied from 730 to 1,859 ha (1,803 to 4,593 acres), and mean summer home range size varied from 1,255 to 3,037 ha (3,101 to 7,504 acres) for migratory deer in Michigan's Upper Peninsula (Van Deelen et al. 1998).

Sample Densities: Maximum densities can easily exceed 30 deer per square mile (11.6 deer/km^2) on agricultural-woodland and suburban landscapes; on heavily forested landscapes densities can range up to 15 deer per square mile (<5.8/km^2) (Baker 1984:12).

Food Habits: Mainly crepuscular. Deer seasonally browse a variety of deciduous and coniferous growth; feeding on twigs, stripping young bark, and eating dried red maple leaves, and hemlock and white cedar twig litter in winter (Crawford 1982, Ludewig and Bowyer 1985, Ditchkoff and Servello 1998). Spring foods include various forbs, grasses, sedges, and ferns. Summer foods shift to various hardwood leaves, grasses, and sedges. Gilled mushrooms, beechnuts, and acorns are important fall foods in addition to the variety of hardwood leaves still available. Deer and moose food habits overlap in fall and winter, causing concern over potential competition for browse in deer wintering areas (Pruss and Pekins 1992).

Comments: Northern limits of deer habitat relate to the depth, quality, and duration of snow cover (Blouch 1984:391, Mattfeld 1984:305). Wolf and mountain lion were historic predators of deer; now domestic dogs are significant predators, and black bear and coyote are significant fawn predators (Ozoga and Verme 1982, Parker 1995). Gregarious, usually forming small groups. Family groups consisting of a doe with her fawns and yearlings are sometimes common in the late fall. Overabundant deer densities preclude tree regeneration and over time alter tree species composition (Tilghman 1989). White-tailed deer are considered a "keystone" herbivore, capable of affecting the distribution and abundance of many other wildlife species and both plant species and communities (Waller and Alverson 1997).

Moose

(Alces alces)

Range: Alaska, the s. half of Canada, n. New England, n. Minnesota and Upper Peninsula of Michigan, and the n. Rockies into Utah (Murie 1934, Franzmann 1981, Franzmann and Schwartz 1997).

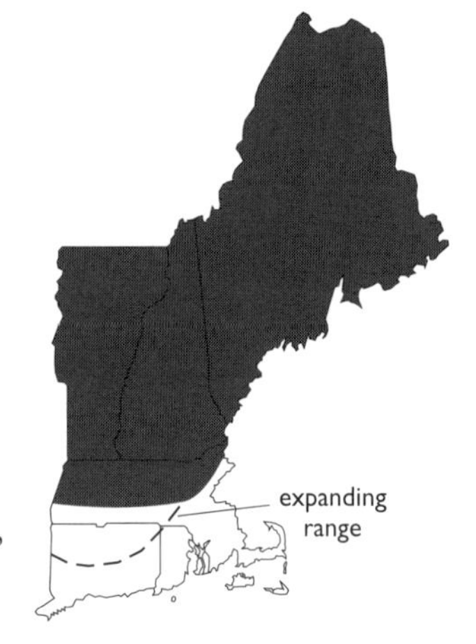

Distribution in New England: Throughout Maine, New Hampshire, Vermont, central and w. Massachusetts (Alexander 1993, Bontaites and Gustafson 1993, Morris and Elowe 1993, Vecellio et al. 1993). Range expanding southward in s. Massachusetts and Connecticut (K. Bontaites pers. commun.).

Status in New England: Locally common to uncommon.

Habitat: Second-growth boreal forests interspersed with semiopen areas and swamps or lakes that offer cover and aquatic plants for food. Mature stands of balsam fir and white birch, and young aspen stands are preferred habitat. Summers are spent near water, winters in drier mixed hardwood-conifer forests. Females use lowland softwood and cedar cover types and regenerating clear-cuts, especially close to wetlands, in summer more than do males; males use hardwood and mixed-wood cover types in summer more than do females (Leptich and Gilbert 1989, Miller and Litvaitis 1992a). Regenerating clear-cuts are important fall and winter foraging habitats (Monthey 1984, Pruss and Pekins 1992, Santillo 1994, Eschholz et al. 1996, Raymond et al. 1996). Hemlock is an important winter cover type in the Algonquin transition zone (Forbes and Theberge 1993). Higher moose densities occur in landscapes subject to significant periodic disturbance from logging, fires, or earthquakes than larger areas less impacted by these types of disturbance (Peek et al. 1976, Forbes and Theberge 1993, MacCracken et al. 1997).

Special Habitat Requirements: Wetlands preferred in summer for relief from mosquitoes and flies and for aquatic plant food items (DeVos 1958, Crossley and Gilbert 1983, Crossley 1985).

Reproduction: Age at sexual maturity: Females mature at 16 to 18 months and produce young in their second year (Peterson 1955, Adams and Pekins 1995). Males probably mature at 1.5 years but are unable to breed until 5 or 6 yr old due to competition from older bulls. Breeding period: Early September to late October (Franzmann 1981). Peak: Late September and early October. Gestation period: 240 to 246 days. Young born: Late May to early June. Young produced: 1, rarely 2.

Home Range: Mean summer home-range size approximately 25 km^2 (range 2 to 60 km^2) in n. Maine (Leptich and Gilbert 1989, Crossley and Gilbert 1983). Miller (1989) found composite home-range size for bulls was 93 km^2, and 153 km^2 for females in n. New Hampshire. Earlier work estimated seasonal home ranges of 5 to 10 km^2 (2 to 4 square miles) throughout North America (LeResche 1974). Mean summer home-range size varied from 15.9 km^2 to 55.7 km^2 elsewhere in North America (Phillips et al. 1973, Taylor and Ballard 1979, Hague and Keith 1981). Mean annual home-range size was 59 ± 5 km^2 on the Copper River Delta, Alaska (MacCracken et al. 1997). Bulls will range farther during breeding season.

Sample Densities: Winter density where moose were abundant in n. New Hampshire ranged from 0.7 to 1.6 moose per km^2 (Adams et al. 1997) using aerial infrared surveys. In e. North America, the average density is 1 moose per 5 square miles (13 km^2) over much of its range; 2 or more moose per square mile (0.8/km^2) approaches carrying capacity (Peterson 1955).

Food Habits: During summer moose prefer to feed in or near clearings, regenerating clear-cuts, burns, or shoreline areas, where they browse on tender leaves, twigs, and bark of deciduous trees, and semiaquatic and aquatic vegetation. They also graze on grasses, lichens, mosses, mushrooms, and herbaceous plants. Winter diet is restricted to conifer (especially balsam fir) and aspen, mountain maple, pin cherry, nannyberry, and beech browse (Ludewig and Bowyer 1985, Pruss and Pekins 1992). Adult daily summer forage intake estimated at 2.6 to 3.5 percent of body weight per day in dry weight; winter intake at 0.5 to 1.3 percent of body weight per day in dry weight (Coady 1982:910,

Schwartz et al. 1984, Renecker and Hudson 1985). Necessary sodium is acquired from roadside salt licks and submergent aquatic plants (Belovsky 1981, Belovsky and Jordan 1981, Bontaites and Gustafson 1993, Miller and Litvaitis 1992b).

Comments: Populations continue to increase in Massachusetts and Connecticut. Timber management operations increase foraging habitat as well as road density and hunter access (Rempel et al. 1997). Glyphosate treatment of naturally regenerating conifer clear-cuts initially reduces hardwood browse production (Santillo 1994, Eschholz et al 1996, Raymond et al. 1996) with increasing foraging activity in clear-cuts 7 to 11 yr post-treatment, especially where dense conifer cover 2.0 to 2.9 m tall is available. Fall foraging moose can reduce browse production available to deer in recent clear-cuts adjacent to deer yards (Pruss and Pekins 1992). Major predators of moose in ne. North America include humans, black bear, and gray wolf (Van Ballenberghe and Ballard 1994). Moose in n. New England tend to winter at higher elevations than random points (Miller and Litvaitis 1992a), though Edwards and Ritcey (1956) report moose in mountainous regions generally seek lower elevations in fall. They may gather together in yards during winter and congregate in lily ponds during summer months but are mainly solitary. Moose are most active at dawn and dusk.

Literature Cited

Ables, E. D. 1969a. Activity studies of red foxes in southern Wisconsin. *J. Wildl. Manage.* 33:145–153.

Ables, E. D. 1969b. Home range studies of red foxes *Vulpes vulpes*. *J. Mammal.* 50:108–120.

Ables, E. D. 1975. Ecology of the red fox in North America. Pages 216–236. In: Fox, M. W. (editor). *The wild canids: Their systematics, behavioral ecology and evolution.* New York: Van Nostrand Reinhold.

Adams, K. P.; Pekins, P. J. 1995. Growth patterns of New England moose: Yearlings as indicators of population status. *Alces* 31:53–59.

Adams, K. P.; Pekins, P. J.; Gustafson, K. A.; Bontaites, K. 1997. Evaluation of infrared technology for aerial moose surveys in New Hampshire. *Alces* 33:129–139.

Adams, L. 1959. An analysis of a population of snowshoe hares in northwestern Montana. *Ecol. Monogr.* 29:141–170.

Aldous, C. M. 1937. Notes on the life history of the snowshoe hare. *J. Mammal.* 18:46–57.

Alexander, C. E. 1993. The status and management of moose in Vermont. *Alces* 29:187–195.

Allen, A. W. 1984. *Habitat suitability index models: Eastern cottontail.* FWS/OBS-82/10.66. Washington, D.C.: U.S. Department of the Interior, Fish and Wildlife Service. 23 pp.

Allen, D. L. 1938. Ecological studies on the vertebrate fauna of 500-acre farm in Kalamazoo County, Michigan. *Ecol. Monogr.* 8:347–436.

Allen, E. G. 1938. The habits and life history of the eastern chipmunk (*Tamias striatus lysteri*). N. Y. State Mus. Bull. 314:1–122.

Allen, J. M. 1954. Gray and fox squirrel management in Indiana. *Pittman-Robertson Bull.* (Indiana Department of Conservation) 1:1–112.

Alt, G. L. 1981. Reproductive biology of black bears in northeastern Pennsylvania. *Trans. Northeast Sect. Wildl. Soc.* 38:88–89.

Alt, G. L. 1983. Timing of parturition of black bears (*Ursus americanus*) in northeastern Pennsylvania. *J. Mammal.* 64:305–307.

Alt, G. L.; Gruttadauria, J. M. 1984. Reuse of black bear dens in northeastern Pennsylvania. *J. Wildl. Manage.* 48:236–239.

Alt, G. L.; Matula, G. J.; Alt, F. W.; Lindzey, J. S. 1980. Dynamics of home range and movements of adult black bears in northeastern Pennsylvania. *Int. Conf. Bear Res. Manage.* 4:131–136.

Althoff, D. P.; Storm, G. L. 1989. Daytime spatial characteristics of cottontail rabbits in central Pennsylvania. *J. Mammal.* 70:820–824.

Anderson, S. 1962. A new northern record of *Synaptomys borealis* in Ungava. *J. Mammal.* 43:421–422.

Anderson, T. J. 1977. Population biology of the masked shrew, *Sorex cinereus*, in hardwood forest areas of the McCormick Experimental Forest, Marquette County, Michigan. M.S. thesis, Northern Michigan University, Marquette. 75 pp.

Anthony, E. L.; Kunz, T. H. 1977. Feeding strategies of the little brown bat, *Myotis lucifugus*, in southern New Hampshire. *Ecology* 58:755–786.

Anthony, E. L. P.; Stack, M. H.; Kunz, T. H. 1981. Night roosting and the nocturnal time budget of the little brown bat, *Myotis lucifugus*: Effects of reproductive status, prey density, and environmental conditions. *Oecologia* 51:151–156.

Anthony, R. G.; Simpson, D. A.; Kelly, G. M. 1986. Dynamics of pine vole populations in two Pennsylvania orchards. *Am. Midl. Nat.* 116:108–117.

Arlton, A. V. 1936. An ecological study of the common mole. *J. Mammal.* 17:349–371.

Arthur, S. M.; Krohn, W. B. 1991. Activity patterns, movements, and reproductive ecology of fishers in south-central Maine. *J. Mammal.* 72:379–385.

Arthur, S. M.; Krohn, W. B.; Gilbert, J. R. 1989a. Home range characteristics of adult fishers. *J. Wildl. Manage.* 53:674–679.

Arthur, S. M.; Krohn, W. B.; Gilbert, J. R. 1989b. Habitat use and diet of fishers. *J. Wildl. Manage.* 53:680–688.

Austin, M. A.; Obbard, M. E.; Kolenosky, G. B. 1994. Evidence for a black bear, *Ursus americanus*, killing an adult moose, *Alces alces*. *Can. Field-Nat.* 108:236–238.

Bailey, V. 1924. Breeding, feeding and other life habits of meadow mice *Microtus*. *J. Agric. Res.* 27:523–536.

Baker, R. H. 1983. *Michigan mammals.* East Lansing: Michigan State University Press. 642 pp.

Baker, R. H. 1984. Origin, classification and distribution. Pages 1–18. In: Halls, L. K. (editor). *White-tailed deer: Ecology and management.* Harrisburg, Pa.: Stackpole Books and Wildlife Management Institute. 870 pp.

Ballard, W. B.; Whitman, J. S.; Gardner, C. L. 1987. Ecology of an exploited wolf population in south-central Alaska. *Wildl. Monogr.* 98:1–54.

Banfield, A. W. F. 1974. *The mammals of Canada.* Toronto, Ontario: University of Toronto Press. 438 pp.

Barbour, M. S.; Litvaitis, J. A. 1993. Niche dimensions of New England cottontails in relation to habitat patch size. *Oecologia* 95: 1–327.

Barbour, R. W.; Davis, W. H. 1969. *Bats of America.* Lexington: University of Kentucky Press. 286 pp.

Barclay, R. M. R. 1982. Night roosting behavior of the little brown bat, *Myotis lucifugus. J. Mammal.* 63:464–474.

Barclay, R. M. R. 1984. Observations on the migration, ecology and behaviour of bats at Delta Marsh, Manitoba. *Can. Field-Nat.* 98:331–336.

Barclay, R. M. R. 1985. Long- versus short-range foraging strategies of hoary (*Lasiurus cinereus*) and silver-haired (*Lasionycteris noctivagans*) bats and the consequences of prey selection. *Can. J. Zool.* 63:2507–2515.

Barclay, R. M. R. 1986. The echolocation calls of hoary (*Lasiurus cinereus*) and silver-haired (*Lasionycteris noctivagans*) bats as adaptations for long- versus short-range foraging strategies and the consequences for prey selection. *Can. J. Zool.* 64:2700–2705.

Barclay, R. M. R. 1991. Population structure of temperate zone insectivorous bats in relation to foraging behavior and energy demand. *J. Anim. Ecol.* 60:165–178.

Barclay, R. M. R.; Faure, P. A.; Farr, D. R. 1988. Roosting behavior and roost selection by migrating silver-haired bats (*Lasionycteris noctivagans*). *J. Mammal.* 69:821–825.

Barclay, R. M. R.; Thomson, C. E.; Phelan, F. J. S. 1982. Screech owl, *Otus asio,* attempting to capture little brown bats, *Myotis lucifugus,* at a colony. *Can. Field-Nat.* 96:205–206.

Barnes, D. M.; Mallik, A. U. 1996. Use of woody plants in construction of beaver dams in northern Ontario. *Can. J. Zool.* 74:1781–1786.

Barnes, D. M.; Mallik, A. U. 1997. Habitat factors influencing beaver dam establishment in a northern Ontario watershed. *J. Wildl. Manage.* 61:1371–1377.

Basey, J. M.; Jenkins, S. H. 1995. Influences of predation risk and energy maximization on food selection by beavers (*Castor canadensis*). *Can. J. Zool.* 73:2197–2208.

Beasley, L. 1978. Demography of southern bog lemmings (*Synaptomys cooperi*) and prairie voles (*Microtus ochrogaster*) in southern Illinois. Ph.D. dissertation, University of Illinois, Champaign. 95 pp.

Beckel, A. L. 1991. Wrestling play in adult river otters, *Lutra canadensis. J. Mammal.* 72:386–390.

Beier, P. 1991. Cougar attacks on humans in the United States and Canada. *Wildl. Soc. Bull.* 11:403–412.

Beier, P. 1993. Determining minimum habitat areas and habitat corridors for cougars. *Conserv. Biol.* 7:94–108.

Bekoff, M. 1977. *Canis latrans. Mammalian Species* 79:1–9. Shippensburg, Pa.: American Society of Mammalogists.

Bekoff, M. (editor). 1978. *Coyotes: Biology, behavior and management.* New York: Academic Press. 384 pp.

Bekoff, M. 1982. Coyote. Pages 447–459. In: Chapman, J. A.; Feldhamer, G. A. (editors). *Wild mammals of North America—Biology, management, and economics.* Baltimore, Md.: Johns Hopkins University Press.

Bekoff, M.; Wells, M. C. 1980. Social ecology and behavior of coyotes. *Sci. Am.* 242:130–148.

Bellocq, M. I.; Bendell, J. F.; Innes, D. G. I. 1994. Diet of *Sorex cinereus,* the masked shrew, in relation to the abundance of lepidoptera larvae in northern Ontario. *Am. Midl. Nat.* 132:68–73.

Belovsky, G. E. 1981. A possible population response of moose to sodium availability. *J. Mammal.* 62:631–633.

Belovsky, G. E. 1984. Summer diet optimization by beaver. *Am. Midl. Nat.* 111:209–222.

Belovsky, G. E.; Jordan, P. A. 1981. Sodium dynamics and adaptations of a moose population. *J. Mammal.* 62:613–621.

Bendel, P. R.; Gates, J. E. 1987. Home range and microhabitat partitioning of the southern flying squirrel (*Glaucomys volans*). *J. Mammal.* 68:243–255.

Beneski, J. T., Jr.; Stinson, D. W. 1987. *Sorex palustris. Mammalian Species* 296:1–6. Shippensburg, Pa.: American Society of Mammalogists.

Benton, A. H. 1955. Observations on the life history of the northern pine mouse. *J. Mammal.* 36:52–62.

Berg, W. E.; Chesness, R. A. 1978. Ecology of coyotes in northern Minnesota. Pages 229–247. In: Bekoff, M. (editor). *Coyotes: Biology, behavior and management.* New York: Academic Press.

Best, T. L. 1996. *Lepus californicus. Mammalian Species* 530:1–10. Shippensburg, Pa.: American Society of Mammalogists.

Best, T. L.; Jennings, J. B. 1997. *Myotis leibii. Mammalian Species* 547:1–6. Shippensburg, Pa.: American Society of Mammalogists.

Bider, J. R. 1961. An ecological study of the hare (*Lepus americanus*). *Can. J. Zool.* 39:81–103.

Bishop, S. C. 1947. Curious behavior of a hoary bat. *J. Mammal.* 28:293–294.

Bissonette, J. A.; Fredrickson, R. J.; Tucker, B. J. 1991. American marten: A case for landscape-level management. Pages 116–134. In: Rodiek, J. E.; Bolen, E. G. (editors). *Wildlife and habitats in managed landscapes.* Washington, D.C.: Island Press. 220 pp.

Bittner, S. L.; Rongstad, O. J. 1982. Snowshoe hare and allies. Pages 146–163. In: Chapman, J. A.; Feldhamer, G. A. (editors). *Wild mammals of North America—Biology, management, and economics.* Baltimore, Md.: Johns Hopkins University Press.

Blair, W. F. 1940a. Notes on home ranges and populations of the short-tailed shrew. *Ecology* 21:284–288.

Blair, W. F. 1940b. Home ranges and populations of the meadow vole in southern Michigan. *J. Wildl. Manage.* 4:149–161.

Blair, W. F. 1940c. Home ranges and populations of the jumping mouse. *Am. Midl. Nat.* 23:244–250.

Blair, W. F. 1941. Some data on the home ranges and general life history of the short-tailed shrew, red-backed mouse and woodland jumping mouse in northern Michigan. *Am. Midl. Nat.* 25:681–685.

Blair, W. F. 1942. Size of home range and notes on the life history of the woodland deer mouse and eastern chipmunk in northern Michigan. *J. Mammal.* 23:27–36.

Blair, W. F. 1948. Population density, life span, and mortality of small mammals in the blue-grass meadow and blue-grass field associations of southern Michigan. *Am. Midl. Nat.* 40:395–419.

Blouch, R. I. 1984. Northern Great Lakes states and Ontario forests. Pages 391–410. In: Halls, L. K. (editor). *White-tailed deer: Ecology and management.* Harrisburg, Pa.: Stackpole Books and Wildlife Management Institute. 870 pp.

Bogan, M. A. 1972. Observations on parturition and development on the hoary bat (*Lasiurus cinereus*). *J. Mammal.* 53:611–614.

Bontaites, K. M.; Gustafson, K. 1993. The history and status of moose and moose management in New Hampshire. *Alces* 29:163–167.

Bookhout, T. A. 1959. Reingestion by the snowshoe rabbit. *J. Mammal.* 40:250.

Boutin, S. 1984. Effect of late winter food addition on numbers and movements of snowshoe hares. *Oecologia* 62:393–400.

Boutin, S.; Birkenholz, D. E. 1987. Muskrat and round-tailed muskrat. Pages 314–325. In: Novak, M.; Baker, J. A.; Obbard, M. E.; Malloch, B. (editors). *Wild furbearer management and conservation in North America.* Toronto: Ontario Ministry of Natural Resources and Ontario Trappers Association.

Bowers, M. A. 1995. Use of space and habitats by the eastern chipmunk, *Tamias striatus. J. Mammal.* 76:12–21.

Bowers, M. A.; Welch, D. N.; Carr, T. G. 1990. Home range size adjustments by the eastern chipmunk, *Tamias striatus,* in response to natural and manipulated water availability. *Can. J. Zool.* 68:2016–2020.

Bowyer, R. T.; Testa, J. W.; Faro, J. B. 1995. Habitat selection and home ranges of river otters in a marine environment: Effects of the Exxon Valdez oil spill. *J. Mammal.* 76:1–11.

Boyce, M. S. 1977. Life histories in variable environments: Applications to geographic variation in the muskrat (*Ondatra zibethicus*). Ph.D. dissertation, Yale University, New Haven, Conn. 146 pp.

Brack, V., Jr.; LaVal, R. K. 1985. Food habits of the Indiana bat in Missouri. *J. Mammal.* 66:308–315.

Brady, J. T.; LaVal, R. K.; Kunz, T. H.; Tuttle, M. D.; Wilson, D. E.; Clawson, R. L. 1983. *Recovery plan for the Indiana bat.* Washington, D.C.: U.S. Department of the Interior Fish and Wildlife Service. 23 pp. + 6 appendices.

Brand, C. J.; Keith, L. B. 1979. Lynx demography during a snowshoe hare decline in Alberta. *J. Wildl. Manage.* 43:827–849.

Brand, C. J.; Keith, L. B.; Fischer, C. A. 1976. Lynx responses to changing snowshoe hare densities in central Alberta. *J. Wildl. Manage.* 40:416–428.

Breitenmoser, U.; Slough, B. G.; Breitenmoser-Würsten, C. 1993. Predators of cyclic prey: Is the Canada lynx victim or profiteer of the snowshoe hare cycle? *Oikos* 66:551–554.

Brigham, R. M. 1991. Flexibility in foraging and roosting behaviour by the big brown bat (*Eptesicus fuscus*). *Can. J. Zool.* 69:117–121.

Brigham, R. M.; Fenton, M. B. 1986. The influence of roost closure on the roosting and foraging behavior of *Eptesicus fuscus* (Chiroptera: Vespertilionidae). *Can. J. Zool.* 64:1128–1133.

Brigham, R. M.; Grindal, S. D.; Firman, M. C.; Morissette, J. L. 1997. The influence of structural clutter on activity patterns of insectivorous bats. *Can. J. Zool.* 75:131–136.

Brocke, R. H. 1970a. Ecological inferences from oxygen consumption data of the opossum. *Ecol. Soc. Am. Bull.* 51:29.

Brocke, R. H. 1970b. The winter ecology and bioenergetics of the opossum, *Didelphis marsupialis,* as distributional factors in Michigan. Unpublished Ph.D. dissertation, Michigan State University, East Lansing. 215 pp.

Brocke, R. H. 1975. Preliminary guidelines for managing snowshoe hare habitat in the Adirondacks. *Trans. Northeast Sect. Wildl. Soc. Fish Wildl. Conf.* 32:46–66.

Brocke, R. H.; Gustafson, K. A.; Major, A. R. 1990. Restoration of lynx in New York: Biopolitical lessons. *Trans. North Am. Wildl. Nat. Resources Conf.* 55:590–598.

Bronson, F. H. 1963. Some correlates of interaction rates in natural populations of woodchucks. *Ecology* 44:637–644.

Bronson, F. H. 1964. Agonistic behavior of woodchucks. *Anim. Behav.* 12:470–478.

Brooks, R. T. 1996. Assessment of two camera-based systems for monitoring arboreal wildlife. *Wildl. Soc. Bull.* 24:298–300.

Brooks, R. T.; Smith, H. R.; Healy, W. M. 1998. Small-mammal abundance at three elevations on a mountain in central Vermont, USA: a sixteen-year record. *Forest Ecol. Manage.* 110:181–193.

Broschart, M. R.; Johnston, C. A.; Naiman, R. J. 1989. Predicting beaver colony density in boreal landscapes. *J. Wildl. Manage.* 53:929–934.

Brower, J. E.; Cade, T. J. 1966. Ecology and physiology of *Napaeozapus insignis* Miller and other woodland mice. *Ecology* 47:46–63.

Brown, A. L.; Litvaitis, J. A. 1995. Habitat features associated with predation of New England cottontails: What scale is appropriate? *Can. J. Zool.* 73:1005–1011.

Brown, M. K. 1980. The status of the pine marten in New York. *Trans. Northeast Sect. Wildl. Soc.* 37:217–226.

Buckner, C. H. 1957. Home range of *Synaptomys cooperi. J. Mammal.* 38:132.

Buckner, C. H. 1964. Metabolism, food capacity and feeding behavior in four species of shrews. *Can. J. Zool.* 42:259–279.

Buckner, C. H. 1966. Populations and ecological relationships of shrews in tamarack bogs of southeastern Manitoba. *J. Mammal.* 47:181–194.

Buckner, C. H.; Ray, D. G. H. 1968. Notes on the water shrew in bog habitats of southeastern Manitoba. *Blue Jay* 26:95–96.

Buech, R. R. 1985. Beaver in water impoundments: understanding a problem of water-level management. Pages 95–105. In: Knighton, M. D. (editor). *Water impoundments for wildlife: A habitat management workshop.* Gen. Tech. Rep. NC-100. St. Paul, Minn.: U.S. Department of Agriculture, North Central Forest Experiment Station.

Buresch, K. 1999. Seasonal pattern of abundance and habitat use by bats on Martha's Vineyard, Massachusetts. M.S. thesis, University of New Hampshire, Durham. 69 pp.

Burt, W. H. 1940. Territorial behavior and populations of some small mammals in southern Michigan. *Univ. Mich. Misc. Publ. Mus. Zool.* 45:1–58.

Burt, W. H. 1957. *Mammals of the Great Lakes region.* Ann Arbor: University of Michigan Press. 246 pp.

Burt, W. H.; Grossenheider, R. P. 1976. *A field guide to the mammals.* 3rd ed. Boston: Houghton Mifflin. 289 pp.

Buskirk, S. W.; Powell, R. A. 1994. Habitat ecology of fishers and American martens. Pages 283–296. In: Buskirk, S. W.; Harestad, A. S.; Raphael, M. G.; Powell, R. A. (editors). *Martens, sables, and fishers: Biology and conservation.* Ithaca, N.Y.: Cornell University Press. 484 pp.

Cagle, F. R.; Cockrum, L. 1943. Notes on a summer colony of *Myotis lucifugus lucifugus. J. Mammal.* 24:474–492.

Calder, W. A. 1969. Temperature relations and underwater endurance of the smallest homeothermic diver, the water shrew. *Comp. Biochem. Physiol.* 30:1075–1082.

Caley, M. J. 1986. Kinship-mediated spacing patterns in muskrats. M.S. thesis, University of Guelph, Guelph, Ontario. 67 pp.

Calhoun, J. B. 1962. The ecology and sociology of the Norway rat. *Publ. Health Serv. Publ.* 1008:1–288.

Cameron, A. W. 1958. Mammals of the islands in the Gulf of St. Lawrence. *Nat. Mus. Can. Bull.* 154:1–165.

Carbyn, L. N. 1987. Gray wolf and red wolf. Pages 358–376. In: Novak, M.; Baker, J. A.; Obbard, M. E.; Malloch, B. (editors). *Wild furbearer management and conservation in North America.* Toronto: Ontario Ministry of Natural Resources and Ontario Trappers Association.

Carbyn, L. N.; Patriquin, D. 1983. Observations on home range sizes, movements and social organization of lynx, *Lynx canadensis*, in Riding Mountain National Park, Manitoba. *Can. Field-Nat.* 97:262–267.

Cardoza, J. E. 1976. The history and status of the black bear in Massachusetts and adjacent New England states. *Mass. Div. Fish Wildl. Res. Bull.* 18:1–113.

Cardoza, J. E.; Field, R. 1991. Massachusetts status report. Pages 38–39. In: Clark, J. D.; Smith, K. G. (editors). *Proc. Tenth Eastern Workshop on Black Bear Research and Management,* Bismarck, Arkansas, April 2–5, 1990. Fayetteville: University of Arkansas Press. 206 pp.

Cardoza, J. E.; Jones, G. S.; French, T. W. 1999. Massachusetts Wildlife's State Mammal List. Massachusetts Division of Fisheries and Wildlife website document at: <http://www.magnet.state.ma.us/dfwele/dfw/dfwmam.htm>.

Carreker, R. G. 1985. Habitat suitability index models: snowshoe hare. Biological Report 82(10. 101). Washington, D.C.: U.S. Department of the Interior, Fish and Wildlife Service. 21 pp.

Caturano, S. L. 1983. Habitat and home range use of coyotes in eastern Maine. M.S. thesis, University of Maine, Orono. 28 pp.

Chamberlain, J. L. 1951. The life history and management of the muskrat on Great Meadows Refuge. M.S. thesis, University of Massachusetts, Amherst. 68 pp.

Chapin, T. G. 1995. Influence of landscape pattern and forest type on use of habitat by marten in Maine. M.S. thesis, University of Maine, Orono. 100 pp.

Chapin, T. G.; Harrison, D. J.; Phillips, D. M. 1997a. Seasonal habitat selection by marten in an untrapped forest preserve. *J. Wildl. Manage.* 61:707–717.

Chapin, T. G.; Phillips, D. M.; Harrison, D. J.; York, E. C. 1997b. Seasonal selection of habitats by resting marten in Maine. Pages 166–181. In: Proulx, G.; Bryant, H. N.; Woodard, P. (editors). *Martens: Taxonomy, ecology, techniques, and management.* Edmonton: Provincial Museum of Alberta. 474 pp.

Chapman, J. A. 1975. *Sylvilagus transitionalis. Mammalian Species* 55:1–4. Shippensburg, Pa.: American Society of Mammalogists.

Chapman, J. A; Hockman, J. G.; Edwards, W. R. 1982. Cottontails (*Sylvilagus floridanus* and allies). Pages 83–123. In: Chapman, J. A.; Feldhamer, G. A. (editors). *Wild mammals of North America—Biology, management, and economics.* Baltimore, Md.: Johns Hopkins University Press.

Chapman, J. A.; Hockman, J. G.; Ojeda C., M. M. 1980. *Sylvilagus floridanus. Mammalian Species* 136:1–8. Shippensburg, Pa.: American Society of Mammalogists.

Chappell, M. A. 1980. Thermal energetics and thermoregulatory costs of small Arctic mammals. *J. Mammal.* 61:278–291.

Chilelli, M.; Griffith, B.; Harrison, D. J. 1996. Interstate comparisons of river otter harvest data. *Wildl. Soc. Bull.* 24:238–246.

Choate, J. R. 1973. Identification and recent distribution of white-footed mice (*Peromyscus*) in New England. *J. Mammal.* 54:41–49.

Christian, D. P.; Daniels, J. M. 1985. Distributional records of rock voles, *Microtus chrotorrhinus,* in northeastern Minnesota. *Can. Field-Nat.* 99:356–359.

Christian, J. J. 1956. The natural history of a summer aggregation of the big brown bat, *Eptesicus fuscus. Am. Midl. Nat.* 55:66–95.

Chubbs, T. E.; Phillips, F. R. 1993a. Unusually high number of embryos in a muskrat, *Ondatra zibethicus,* from central Labrador. *Can. Field-Nat.* 107:363.

Chubbs, T. E.; Phillips, F. R. 1993b. An apparent longevity record for Canada lynx, *Lynx canadensis,* in Labrador. *Can. Field-Nat.* 107:367–368.

Churcher, C. S. 1959. The specific status of the New World red fox. *J. Mammal.* 40:513–520.

Clark, T. W.; Anderson, E.; Douglas, C.; Strickland, M. 1987. *Martes americana. Mammalian Species* 289:1–8. Shippensburg, Pa.: American Society of Mammalogists.

Clark, W. R.; Hasbrouck, J. J.; Kienzler, J. M.; Glueck, T. F. 1989. Vital statistics and harvest of an Iowa raccoon population. *J. Wildl. Manage.* 53:982–990.

Clawson, R. L.; LaVal, R. K.; LaVal, M. L.; Caire, W. 1980. Clustering behavior of hibernating *Myotis sodalis* in Missouri. *J. Mammal.* 61:245–253.

Clough, G. C. 1987a. Relations of small mammals to forest management in northern Maine. *Can. Field-Nat.* 101:40–48.

Clough, G. C. 1987b. Ecology of island muskrats, *Ondatra zibethicus,* adapted to upland habitat. *Can. Field-Nat.* 101:63–69.

Clough, G. C.; Albright, J. J. 1987. Occurrence of the northern bog lemming, *Synaptomys borealis,* in the northeastern United States. *Can. Field-Nat.* 101:611–613.

Coady, J. 1982. Moose. Pages 902–922. In: Chapman, J. A.; Feldhamer, G. A. (editors). *Wild mammals of North America—Biology, management, and economics.* Baltimore, Md.: Johns Hopkins University Press.

Cockrum, E. L. 1952. Mammals of Kansas. *Univ. Kans. Publ. Mus. Nat. Hist.* 7:1–303.

Conaway, C. H. 1952. Life history of the water shrew (*Sorex palustris navigator*). *Am. Midl. Nat.* 48:219–248.

Conaway, C. H. 1958. Maintenance, reproduction, and growth of the least shrew in captivity. *J. Mammal.* 39:507–512.

Conaway, C. H. 1959. The reproductive cycle of the eastern mole. *J. Mammal.* 40:180–194.

Conaway, C. H.; Pfitzer, D. W. 1952. *Sorex palustris* and *Sorex dispar* from the Great Smoky Mountains National Park. *J. Mammal.* 33:106–108.

Connor, P. F. 1959. The bog lemming *Synaptomys cooperi* in southern New Jersey. *Mus. Mich. State Univ. Biol. Ser.* 1:161–248.

Connor, P. F. 1960. The small mammals of Otsego and Schoharie Counties, New York. *Bull. N.Y. State Mus. Ser.* 382:1–84.

Connor, P. F. 1971. The mammals of Long Island, New York. *Bull. N.Y. State Mus. Ser.* 416:1–78.

Constantine, D. G. 1966. Ecological observations on Lasiurine bats in Iowa. *J. Mammal.* 47(1):34–41.

Corn, J. G.; Raphael, M. G. 1992. Habitat characteristics at marten subnivean access sites. *J. Wildl. Manage.* 56:442–448.

Coulter, M. W. 1959. Some recent records of martens in Maine. *Maine Field Nat.* 15:50–53.

Coulter, M. W. 1966. Ecology and management of fishers in Maine. Ph.D. dissertation, Syracuse University, Syracuse, N.Y. 183 pp.

Cowan, I. M. 1936. Nesting habits of the flying squirrel (*Glaucomys sabrinus*). *J. Mammal.* 17:58–60.
Cowan, I. M.; Guiguet, C. J. 1965. The mammals of British Columbia. *British Columbia Prov. Mus. Publ.* 11:1–141.
Craighead, J. J.; Craighead, F. C., Jr. 1956. *Hawks, owls and wildlife.* Harrisburg, Pa.: Stackpole. 443 pp.
Crawford, H. S. 1982. Seasonal food selection and digestibility by tame white-tailed deer in central Maine. *J. Wildl. Manage.* 46:974–982.
Criddle, N.; Criddle S. 1925. The weasels of southern Manitoba. *Can. Field-Nat.* 39:142–148.
Criddle, S. 1932. The red-backed vole (*Clethrionomys gapperi loringi* Bailey) in southern Manitoba. *Can. Field-Nat.* 46:178–181.
Criddle, S. 1973. The granivorous habits of shrews. *Can. Field-Nat.* 87:69–70.
Cronan, J. M.; Brooks, A. 1968. *The mammals of Rhode Island.* Wildlife Pamphlet No. 6. Providence: Rhode Island Department of Agriculture and Conservation, Division of Fish and Game.
Crossley, A. 1985. Summer pond use by moose in northern Maine. M.S. thesis, University of Maine, Orono. 39 pp.
Crossley, A.; Gilbert, J. R. 1983. Home range and habitat use of female moose in northern Maine. *Trans. Northeast Sect. Wildl. Soc.* 40:67–75.
Crowe, D. M. 1975. Aspects of aging, growth and reproduction of bobcats in Wyoming. *J. Mammal.* 56:177–198.
Cumberland, R. E.; Dempsey, J. A. 1994. Recent confirmation of a cougar, *Felis concolor,* in New Brunswick. *Can. Field-Nat.* 108:224–226.
Currier, M. J. P. 1983. *Felis concolor. Mammalian Species* 200:1–7. Shippensburg, Pa.: American Society of Mammalogists.
Curtis, J. D.; Kozicky, E. L. 1944. Observations on the eastern porcupine. *J. Mammal.* 25:137–146.
Dalke, P. D. 1937. A preliminary report of the New England cottontail studies. *Trans. North Am. Wildl. Conf.* 2:542–548.
Dalke, P. D. 1942. The cottontail rabbits in Connecticut. *Conn. Geol. Nat. Hist. Surv. Bull.* 65:1–97.
Dalke, P. D.; Sime, P. R. 1938. Home and seasonal ranges of the eastern cottontail in Connecticut. *Trans. North Am. Wildl. Conf.* 3:659–669.
Dalke, P. D.; Sime, P. R. 1941. Food habits of the eastern and New England cottontails. *J. Wildl. Manage.* 5:216–228.
D'Anieri, P.; Leslie, D. M., Jr.; McCormack, M. L., Jr. 1987. Small mammals in glyphosate-treated clearcuts in northern Maine. *Can. Field-Nat.* 101:547–550.
Danell, K. 1978. Population dynamics of the muskrat in a shallow Swedish lake. *J. Anim. Ecol.* 47:697–709.
Davis, D. E.; Peek, F. 1970. Litter size of the star-nosed mole (*Condylura cristata*). *J. Mammal.* 51:156.
Davis, D. W. 1953. The characteristics of rat populations. *Q. Rev. Biol.* 28:373–401.
Davis, W. B.; Joeris, L. 1945. Notes on the life history of the little short-tailed shrew. *J. Mammal.* 26:370–403.
Davis, W. H.; Hitchcock, H. B. 1965. Biology and migration of the bat, *Myotis lucifugus,* in New England. *J. Mammal.* 46:296–313.
Davis, W. H.; Lidicker, W. Z., Jr. 1956. Winter range of the red bat, *Lasiurus borealis. J. Mammal.* 37:280–281.
Davis, W. H.; Mumford, R. E. 1962. Ecological notes on the bat (*Pipistrellus subflavus*). *Am. Midl. Nat.* 68:394–398.
Davis, W. H.; Barbour, R. W.; Hassell, M. D. 1968. Colonial behavior of *Eptesicus fuscus. J. Mammal.* 49:44–50.
Dean, P. B.; DeVos, A. 1965. The spread and present status of the European hare (*Lepus europaeus hybridus*) in North America. *Can. Field-Nat.* 79:38–48.
DeGraaf, R. M.; Maier, T. J. 1996. Effect of egg size on predation by white-footed mice. *Wilson Bull.* 108:535–539.
DeGraaf, R. M.; Snyder, D. P.; Hill, B. J. 1991. Small mammal habitat associations in poletimber and sawtimber stands of four forest cover types. *Forest Ecol. Manage.* 46:227–242.
DeVan, R. 1982. The ecology and life history of the long-tailed weasel (*Mustela frenata*). Ph.D. dissertation, University of Cincinnati, Cincinnati, Ohio. 300 pp.
DeVos, A. 1958. Summer observations on moose behavior in Ontario. *J. Mammal.* 39:128–139.
Dibello, F. J.; Arthur, S. M.; Krohn, W. B. 1990. Food habits of sympatric coyotes, *Canis latrans,* red foxes, *Vulpes vulpes,* and bobcats, *Lynx rufus,* in Maine. *Can. Field-Nat.* 104:403–408.
Dice, L. R.; Sherman, H. B. 1922. Notes on the mammals of Gogebic and Ontonagon counties, Michigan, 1920. *Univ. Mich. Mus. Zool. Occas. Papers* 109. 40 pp.
Ditchkoff, S. S.; Servello, F. A. 1998. Litterfall: An overlooked food source for wintering white-tailed deer. *J. Wildl. Manage.* 62:250–255.
Dixon, K. R. 1982. Mountain lion. Pages 711–727. In: Chapman, J. A.; Feldhamer, G. A. (editors). *Wild mammals of North America—Biology, management, and economics.* Baltimore, Md.: Johns Hopkins University Press.
Docktor, C. M.; Bowyer, R. T.; Clark, A. G. 1987. Number of corpora lutea as related to age and distribution of river otter in Maine. *J. Mammal.* 68:182–184.
Dodds, D. G. 1965. Reproduction and productivity of snowshoe hares in Newfoundland. *J. Wildl. Manage.* 29:303–315.
Dodge, W. E. 1967. The biology and life history of the porcupine, *Erethizon dorsatum,* in western Massachusetts. Ph.D. dissertation, University of Massachusetts, Amherst.
Dodge, W. E. 1982. Porcupine. Pages 355–366. In: Chapman, J. A.; Feldhamer, G. A. (editors). *Wild mammals of North America—Biology, management, and economics.* Baltimore, Md.: Johns Hopkins University Press.
Dolan, P. G.; Carter, D.C. 1977. *Glaucomys volans. Mammalian Species* 78:1–6. Shippensburg, Pa.: American Society of Mammalogists.
Doucet, C. M.; Ball, J. P. 1994. Analysis of digestion data: apparent and true digestibilities of foods eaten by beaver. *Am. Midl. Nat.* 132:239–247.
Doucet, G. J.; Bider, J. R. 1974. The effects of weather on the activity of the masked shrew. *J. Mammal.* 55:348–363.
Douglas, C. W.; Strickland, M. A. 1987. Fisher. Pages 510–529. In: Novak, M.; Baker, J. A.; Obbard, M. E.; Malloch, B. (editors). *Wild furbearer management and conservation in North America.* Toronto: Ontario Ministry of Natural Resources and Ontario Trappers Association.
Dozier, H. L. 1950. Muskrat trapping on the Montezuma National Wildlife Refuge, New York 1943–1948. *J. Wildl. Manage.* 14:403–412.
Dozier, H. L.; Markley, M. H.; Llewellyn, L. M. 1948. Estimating muskrat populations by house counts. *J. Wildl. Manage.* 12:177–190.
Dragoo, J.; Honeycutt, R. L. 1997. Systematics of mustelid-like carnivores. *J. Mammal.* 78:426–443.

Drickamer, L. C. 1978. Annual reproduction patterns in populations of two sympatric species of *Peromyscus*. *Behav. Biol.* 23:405–408.

Drickamer, L. C. 1987. Influence of time of day on captures of two species of *Peromyscus* in a New England deciduous forest. *J. Mammal.* 68:702–703.

Drueker, J. D. 1972. Aspects of reproduction in *Myotis volans, Lasionycteris noctivagans* and *Lasiurus cinereus*. *Diss. Abstr.* 33B(10): 5065.

Dubuc, L. J.; Krohn, W. B.; Owen, R. B. 1990. Predicted occurrence of river otters by habitat on Mount Desert Island, Maine. *J. Wildl. Manage.* 54:594–599.

Dunn, J .P.; Chapman, J. A.; Marsh, R. E. 1982. Jackrabbits. Pages 124–145. In: Chapman, J. A.; Feldhamer, G. A. (editors). *Wild mammals of North America—Biology, management, and economics.* Baltimore, Md.: Johns Hopkins University Press.

Dutcher, B. H. 1903. Mammals of Mt. Katahdin, Maine. *Proc. Biol. Soc. Wash.* 14:63–72.

Eabry, H. S. 1968. An ecological study of *Sylvilagus transitionalis* and *S. floridanus* of northeastern Connecticut. Storrs: University of Connecticut, Agricultural Experiment Station. 27 pp.

Eabry, H. S. 1970. A feasibility study to investigate and evaluate the possible directions of European hare management in New York. Federal Aid Pittman-Robertson Project W-84-R17. Delmar: New York Department of Environmental Conservation.

Eadie, W. R. 1939. A contribution to the biology of *Parascalops breweri*. *J. Mammal.* 20:150–173.

Eadie, W. R. 1944. The short-tailed shrew and field mouse predation. *J. Mammal.* 25:359–364.

Eadie, W. R. 1948. Shrew-mouse predation during low mouse abundance. *J. Mammal.* 29:35–37.

Eadie, W. R. 1949. Predation on *Sorex* by *Blarina*. *J. Mammal.* 30:308–309.

Eadie, W. R. 1952. Shrew predation and vole populations on a localized area. *J. Mammal.* 33:185–189.

Eadie, W. R.; Hamilton, W. J., Jr. 1956. Notes on reproduction in the star-nosed mole. *J. Mammal.* 37:223–231.

Eagle, T. C.; Sargeant, A. B. 1985. Use of den excavations, decoys, and barrier tunnels to capture mink. *J. Wildl. Manage.* 49:40–42.

Eagle, T. C.; Whitman, J. S. 1987. Mink. Pages 614–624. In: Novak, M.; Baker, J. A.; Obbard, M. E.; Malloch, B. (editors). *Wild furbearer management and conservation in North America.* Toronto: Ontario Ministry of Natural Resources and Ontario Trappers Association.

Eaton, R. L.; Verlander, K. A. 1977. Reproduction in the puma: Biology, behavior and ontogeny. Pages 45–70. In: Eaton, R. L. (editor). *The world's cats.* Volume 3(3). Seattle: University of Washington, Carnivore Research Institute, Burke Museum. 144 pp.

Edwards, R. Y.; Ritcey, R. W. 1956. The migrations of a moose herd. *J. Mammal.* 37:486–494.

Elliott, L. 1978. Social behavior and foraging ecology of the eastern chipmunk *(Tamias striatus)* in the Adirondack Mountains. *Smithsonian Contrib. Zool.* 265:1–107.

Elowe, K. D. 1984. Home range, movements, and habitat preferences of black bears (*Ursus americanus*) in western Massachusetts. M.S. thesis, University of Massachusetts, Amherst. 112 pp.

Elowe, K. D.; Dodge, W. E. 1989. Factors affecting black bear reproductive success and cub survival. *J. Wildl. Manage.* 53:962–968.

Elton, C.; Nicholson, M. 1942. Fluctuations in numbers of muskrat (*Ondatra zibethica*) in Canada. *J. Anim. Ecol.* 11:96–126.

Enders, R. K. 1952. Reproduction in the mink. *Proc. Am. Philos. Soc.* 96:691–755.

Erickson, A. W.; Nellor, J. E.; Petrides, G. A. 1964. *The black bear in Michigan.* Research Bulletin 4. East Lansing: Michigan State University Agricultural Experimental Station. 102 pp.

Erlinge, S. 1968. Territoriality of the otter, *Lutra lutra* I. *Oikos* 19:81–98.

Erlinge, S. 1977. Spacing strategy in stoat *Mustela erminea. Oikos* 28:32–42.

Erlinge, S. 1979. Movements and daily activity pattern of radio-tracked male stoats, *Mustela erminea.* Pages 703–710. In: Amlaner, C. J., Jr.; McDonald, D. W. (editors). *A handbook on biotelemetry and radiotracking.* Oxford: Pergamon Press. 804 pp.

Errington, P. L. 1939. Reactions of muskrat populations to drought. *Ecology* 20(2):168–186.

Errington, P. L. 1961. *Muskrats and marsh management.* Lincoln: Wildlife Management Institute, University of Nebraska Press. 183 pp.

Errington, P. L. 1963. Muskrat populations. Ames: Iowa State University Press. 655 pp.

Errington, P. L.; Errington, C. S. 1937. Experimental tagging of young muskrats for purposes of study. *J. Wildl. Manage.* 1:49–61.

Eschholz, W. E.; Servello, F. A.; Griffith, B.; Raymond, K. S.; Krohn, W. B. 1996. Winter use of glyphosate-treated clearcuts by moose in Maine. *J. Wildl. Manage.* 60:764–769.

Fagerstone, K. A. 1987. Black-footed ferret, long-tailed weasel, short-tailed weasel, and least weasel. Pages 548–573. In: Novak, M.; Baker, J. A.; Obbard, M. E.; Malloch, B. (editors). *Wild furbearer management and conservation in North America.* Toronto: Ontario Ministry of Natural Resources and Ontario Trappers Association.

Fall, M. W. 1971. Seasonal variations in the food consumption of woodchucks, *Marmota monax*. *J. Mammal.* 52:370–375.

Faulkner, C. E.; Dodge, W. E. 1962. Control of the porcupine in New England. *N.H. Conserv. Mag.* 72:9–10.

Faure, P. A.; Fullard, J. H.; Dawson, J. W. 1993. The gleaning attacks of the northern long-eared bat, *Myotis septentrionalis,* are relatively inaudible to moths. *J. Exp. Biol.* 178:173–189.

Fay, F. H.; Chandler, E. H. 1955. The geographical and ecological distribution of cottontail rabbits in Massachusetts. *J. Mammal.* 36:415–424.

Fenton, M. B. 1972. Distribution and overwintering of *Myotis leibii* and *Eptesicus fuscus* (Chiroptera: Vespertilionidae) in Ontario. Royal Ontario Museum Life Science Occasional Paper 21. 8 pp.

Fenton, M. B. 1990. The foraging behaviour and ecology of animal-eating bats. *Can. J. Zool.* 68:411–422.

Fenton, M. B.; Barclay, R. M. R. 1980. *Myotis lucifugus. Mammalian Species* 142:1–8. Shippensburg, Pa.: American Society of Mammalogists.

Ferron, J.; Ouellet, J. P. 1989. Temporal and intersexual variation in the use of space with regard to social organization in the woodchuck (*Marmota monax*). *Can. J. Zool.* 67:1642–1649.

Fitch, H. S.; Sandidge, L. L. 1953. Ecology of the opossum on a natural area in northeastern Kansas. *Univ. Kans. Publ. Mus. Nat. Hist.* 7:305–338.

Fitch, H. S.; Shirer, H. W. 1970. A radiotelemetric study of spatial relationships in the opossum. *Am. Midl. Nat.* 84:170–186.
Fitch, J. H.; Shump, K. A., Jr. 1979. *Myotis keeni. Mammalian Species* 121:1–3. Shippensburg, Pa.: American Society of Mammalogists.
Fitzgerald, B. M. 1981. Predatory birds and mammals. Pages 485–506. In: Bliss, L. C.; Cragg, J. B.; Heal, D. W.; Moore, J. J. (editors). *Tundra ecosystems: A comparative analysis.* Cambridge, U.K.: Cambridge University Press. 813 pp.
Florida Panther Interagency Committee. 1987. Florida panther (*Felis concolor coryi*) recovery plan. Atlanta, Ga.: U.S. Fish and Wildlife Service. 75 pp.
Flyger, V. F. 1969. The 1968 squirrel "migration" in the eastern United States. *Trans. Northeast Sect. Wildl. Soc.* 26:69–79.
Flyger, V. F.; Gates, J. E. 1982a. Fox and gray squirrels. Pages 209–229. In: Chapman, J. A.; Feldhamer, G. A. (editors). *Wild mammals of North America—Biology, management, and economics.* Baltimore, Md.: Johns Hopkins University Press.
Flyger, V. F.; Gates, J. E. 1982b. Pine squirrels (*Tamiasciurus hudsonicus* and *T. douglasii*). Pages 230–238. In: Chapman, J. A.; Feldhamer, G. A. (editors). *Wild mammals of North America—Biology, management, and economics.* Baltimore, Md.: Johns Hopkins University Press.
Forbes, R. B. 1966. Studies of the biology of Minnesotan chipmunks. *Am. Midl. Nat.* 76:290–308.
Forbes, G. J.; Theberge, J. B. 1993. Multiple landscape scales and winter distribution of moose, *Alces alces* in a forest ecotone. *Can. Field-Nat.* 107:201–207.
Forsyth, D. J. 1976. A field study of growth and development of nestling masked shrews (*Sorex cinereus*). *J. Mammal.* 57:708–721.
Fox, L. B.; Brocke, R. H. 1983. Ecology and demography of a northern winter stressed bobcat population. *Trans. Northeast Sect. Wildl. Soc.* 40:98.
Francis, G. R.; Stephenson, A. B. 1972, Marten ranges and food habits in Algonquin Provincial Park, Ontario. Research Report Wildlife 91. Toronto: Ontario Ministry of Natural Resources. 53 pp.
Francq, E. N. 1969. Behavioral aspects of feigned death in the opossum (*Didelphis marsupialis*). *Am. Midl. Nat.* 81:556–568.
Franzmann, A. W. 1981. *Alces alces. Mammalian Species* 154:1–7. Shippensburg, Pa.: American Society of Mammalogists.
Franzmann, A. W.; Schwartz, C. C. (editors). 1997. *Ecology and management of the North American moose.* Washington, D.C.: Smithsonian Institution Press and Wildlife Management Institute. 733 pp.
Franzmann, A. W.; Schwartz, C. C.; Peterson, R. O. 1980. Moose calf mortality in summer on the Kenai Peninsula, Alaska. *J. Wildl. Manage.* 44:764–768.
Fridell, R. A.; Litvaitis, J. A. 1991. Influence of resource distribution and abundance on home-range characteristics of southern flying squirrels. *Can. J. Zool.* 69:2589–2593.
Fritts, S. H.; Sealander, J. A. 1978. Reproductive biology and population characteristics of bobcats (*Lynx rufus*) in Arkansas. *J. Mammal.* 59:347–353.
Fritzell, E. K. 1978a. Aspects of raccoon (*Procyon lotor*) social organization. *Can. J. Zool.* 56:260–271.
Fritzell, E. K. 1978b. Habitat use by prairie raccoons during the waterfowl nesting season. *J. Wildl. Manage.* 49:901–905.
Fritzell, E. K. 1987. Gray fox and island gray fox. Pages 408–420. In: Novak, M.; Baker, J. A.; Obbard, M. E.; Malloch, B. (editors). *Wild furbearer management and conservation in North America.* Toronto: Ontario Ministry of Natural Resources and Ontario Trappers Association.
Fritzell, E. K.; Haroldson, K. J. 1982. *Urocyon cinereoargenteus. Mammalian Species* 189:1–8. Shippensburg, Pa.: American Society of Mammalogists.
Fritzell, E. K.; Hubert, G. F., Jr.; Meyen, B. E.; Sanderson, G. C. 1985. Age-specific reproduction in Illinois and Missouri raccoons. *J. Wildl. Manage.* 49:901–905.
Frost, H. C.; Krohn, W. B. 1997. Factors affecting the reproductive success of captive female fishers. Pages 100–109. In: Proulx, G.; Bryant, H. N.; Woodard, P. (editors). *Martens: Taxonomy, ecology, techniques, and management.* Edmonton: Provincial Museum of Alberta. 474 pp.
Fujita, M. S.; Kunz, T. H. 1984. *Pipistrellus subflavus. Mammalian Species* 228:1–6. Shippensburg, Pa.: American Society of Mammalogists.
Fuller, T. K. 1989. Population dynamics of wolves in north-central Minnesota. *Wildl. Monogr.* 105:1–41.
Fuller, T. K.; Berg, W. E.; Kuehn, D. W. 1985. Bobcat home range size and daytime cover-type use in northcentral Minnesota. *J. Mammal.* 66:568–571.
Fuller, T. K.; Berg, W. E.; Radde, G. L.; Lenarz, M. S.; Joselyn, G. B. 1992. A history and current estimate of wolf distribution and numbers in Minnesota. *Wildl. Soc. Bull.* 20:42–55.
Fuller, T. K.; Berendzen, S. L.; Decker, T. A.; Cardoza, J. E. 1995. Survival and cause-specific mortality rates of adult bobcats (*Lynx rufus*). *Am. Midl. Nat.* 134:404–408.
Fuller, T. K.; Keith, L. B. 1980. Wolf population dynamics and prey relationships in northeastern Alberta. *J. Wildl. Manage.* 44:583–602.
Gaines, M. S.; Rose, R. K.; McClenaghan, L. R. 1977. The demography of *Synaptomys cooperi* in eastern Kansas. *Can. J. Zool.* 55(10):1584–1594.
Gaines, M. S.; Vivas, A. M.; Baker, C. L. 1979. An experimental analysis of dispersal in fluctuating vole populations: Demographic parameters. *Ecology* 60:814–828.
Gamble, A. L. 1981. Distribution in Manitoba of *Mustela frenata longicauda* Bonaparte, the long-tailed weasel, and the interrelation of distribution and habitat selection in Manitoba, Saskatchewan and Alberta. *Can. J. Zool.* 59:1036–1039.
Gamble, R. L.; Riewe, R. R. 1982. Infestations of the nematode *Skrjabingylus nasicola* (Leukart 1892) in *Mustela frenata* (Lichtenstein) and *M. erminea* (L.) and some evidence of a paratenic host in the life cycle of this nematode. *Can. J. Zool.* 60:45–52.
Gardner, A. L. 1982. Virginia opossum (*Didelphis virginiana*). Pages 3–36. In: Chapman, J. A; Feldhamer, G. A. (editors). *Wild mammals of North America—Biology, management, and economics.* Baltimore, Md.: Johns Hopkins University Press.
Gardner, J. E.; Garner, J. D.; Hofmann, J. E. 1991. Summer Roost Selection and Roosting Behavior of *Myotis sodalis* (Indiana Bat) in Illinois. Unpublished report, Illinois Natural History Survey, Champaign.
Garman, S. L.; O'Connell, A. F., Jr.; Connery, J. H. 1994. Habitat use and distribution of the mice *Peromyscus leucopus* and *P. maniculatus* on Mount Desert Island, Maine. *Can. Field-Nat.* 108:67–71.
George, S. B.; Choate, J. R.; Genoways, H. H. 1986. *Blarina brevicauda. Mammalian Species* 261:1–9. Shippensburg, Pa.: American Society of Mammalogists.

Gerson, H. B. 1988. Cougar, *Felis concolor,* sightings in Ontario. *Can. Field-Nat.* 102:419–424.
Getz, L. L. 1960. Home ranges of the bog lemming. *J. Mammal.* 41:404–405.
Getz, L. L. 1961a. Factors influencing the local distribution of shrews. *Am. Midl. Nat.* 65:67–88.
Getz, L. L. 1961b. Home ranges, territoriality and movement of the meadow vole. *J. Mammal.* 42:24–36.
Getz, L. L. 1961c. Factors influencing the local distribution of *Microtus* and *Synaptomys* in southern Michigan. *Ecology* 42:110–119.
Gibilisco, C. J. 1994. Distributional dynamics of modern *Martes* in North America. Pages 59–71. In: Buskirk, S. W.; Harestad, A. S.; Raphael, M. G.; Powell, R. A. (editors). *Martens, sables, and fishers: Biology and conservation.* Ithaca, N.Y.: Cornell University Press. 484 pp.
Gilbert, J. H.; Wright, J. L.; Lauten, D. J.; Probst, J. R. 1997. Den and rest-site characteristics of American marten and fisher in northern Wisconsin. Pages 135–145. In: Proulx, G.; Bryant, H. N.; Woodard, P. (editors). *Martens: Taxonomy, ecology, techniques, and management.* Edmonton: Provincial Museum of Alberta. 474 pp.
Gillette, L. N. 1980. Movement patterns of radio-tagged opossums in Wisconsin. *Am. Midl. Nat.* 104:1–12.
Giuliano, W. M.; Litvaitis, J. A.; Stevens, C. L. 1989. Prey selection in relation to sexual dimorphism of fishers (*Martes pennanti*) in New Hampshire. *J. Mammal.* 70:639–641.
Glover, F. A. 1943. A study of the winter activities of the New York weasel. *Penn. Game News* 14:8–9.
Godin, A. J. 1977. *Wild mammals of New England.* Baltimore, Md.: Johns Hopkins University Press. 304 pp.
Godin, A. J. 1982. Striped and hooded skunks. Pages 674–687. In: Chapman, J. A.; Feldhamer, G. A. (editors). *Wild mammals of North America—Biology, management, and economics.* Baltimore, Md.: Johns Hopkins University Press.
Goehring, H. H. 1972. Twenty-year study of *Eptesicus fuscus* in Minnesota. *J. Mammal.* 53:201–207.
Goertz, J. W. 1971. An ecological study of *Microtus pinetorum* in Oklahoma. *Am. Midl. Nat.* 86:1–12.
Goodwin, G. G. 1932. New records and some observations on Connecticut mammals. *J. Mammal.* 13:36–40.
Goodwin, G. G. 1936. Big game animals in the northeastern United States. *J. Mammal.* 17:48–50.
Goodwin, G. G. 1942. *Cryptotis parva* in Connecticut. *J. Mammal.* 23:336.
Gorman, M.; Stone, R. D. 1990. The natural history of moles. Ithaca, N.Y.: Cornell University Press. 138 pp.
Gould, E.; McShea, W.; Grand, T. 1993. Function of the star in the star-nosed mole, *Condylura cristata. J. Mammal.* 74:108–116.
Grant, P. R. 1976. An 11-year study of small mammal populations at Mont St. Hilaire, Quebec. *Can. J. Zool.* 54:2156–2173.
Green, M. M. 1936. The black rat in Vermont. *J. Mammal.* 17:173.
Griesemer, S. J.; DeGraaf, R. M.; Fuller, T. K. 1994. Effects of excluding porcupines from established winter feeding trees in central Massachusetts. *Northeast Wildl.* 51:29–33.
Griesemer, S. J.; Fuller, T. K.; DeGraaf, R. M. 1996. Denning patterns of porcupines, *Erethizon dorsatum. Can. Field-Nat.* 110:634–637.
Griffin, D. R. 1940a. Notes on the life histories of New England cave bats. *J. Mammal.* 21(2):181–187.
Griffin, D. R. 1940b. Migrations of New England bats. *Harvard Univ. Mus. Comp. Zool. Bull.* 86:217–246.
Griffin, D. R. 1945. Travels of banded cave bats. *J. Mammal.* 26:15–23.
Griffith, L. A.; Gates, J. E. 1985. Food habits of cave-dwelling bats in the central Appalachians. *J. Mammal.* 66:451–460.
Grizzell, R. A., Jr. 1955. A study of the southern woodchuck, *Marmota monax monax. Am. Midl. Nat.* 53:257–293.
Gunderson, H. L. 1961. A self-trapped gray fox. *J. Mammal.* 42:270.
Gysel, L. W. 1961. An ecological study of tree cavities and ground burrows in forest stands. *J. Wildl. Manage.* 25:12–20.
Hague, T. M.; Keith, L. B. 1981. Dynamics of moose populations in northeastern Alberta. *J. Wildl. Manage.* 45:573–597.
Hale, M. B.; Fuller, T. K. 1996. Porcupine (*Erethizon dorsatum*) demography in central Massachusetts. *Can. J. Zool.* 74:480–484.
Hall, E. R. 1951. American weasels. *Univ. Kans. Publ. Mus. Nat. Hist.* 4:1–466.
Hall, E. R.; Kelson, K. R. 1959. *The mammals of North America.* 2 volumes. New York: Ronald Press.
Hall, J. S. 1956. Life history studies of the eastern pipistrelle bat (*Pipistrellus subflavus*) in Massachusetts. M.A. thesis, University of Massachusetts, Amherst. 74 pp.
Hall, J. S. 1960. A life history and taxonomic study of Indiana bat, *Myotis sodalis.* Ph.D. dissertation, University of Illinois, Urbana. 135 pp.
Hall, J. S. 1962. A life history and taxonomic study of the Indiana bat, *Myotis sodalis.* Sci. Publ. No. 12. Reading, Pa.: Public Museum and Art Gallery. 68 pp.
Halls, L. K. (editor). 1984. *White-tailed deer: Ecology and management.* Harrisburg, Pa.: Stackpole Books and Wildlife Management Institute. 870 pp.
Hallett, J. G. 1978. *Parascalops breweri. Mammalian Species* 98:1–4. Shippensburg, Pa.: American Society of Mammalogists.
Halpin, M.; Bissonette, J. 1983. Winter resource use by red fox in eastern Maine. *Trans. Northeast Sect. Wildl. Soc.* 40:158.
Hamilton, W. J., Jr. 1930. The food of the Soricidae. *J. Mammal.* 11:26–39.
Hamilton, W. J., Jr. 1931a. Habits of the short-tailed shrew (*Blarina brevicauda* Say). *Ohio J. Sci.* 31:97–106.
Hamilton, W. J., Jr. 1931b. Habits of the star-nosed mole (*Condylura cristata*). *J. Mammal.* 12:345–355.
Hamilton, W. J., Jr. 1933a. The insect food of the big brown bat. *J. Mammal.* 14:155–156.
Hamilton, W. J., Jr. 1933b. The weasels of New York: Their natural history and economic status. *Am. Midl. Nat.* 14:289–344.
Hamilton, W. J., Jr. 1935. Habits of jumping mice. *Am. Midl. Nat.* 16:187–200.
Hamilton, W. J., Jr. 1936. The food and breeding habits of the raccoon. *Ohio J. Sci.* 36:131–140.
Hamilton, W. J., Jr. 1937. Activity and home range of the field mouse, *Microtus pennsylvanicus pennsylvanicus* Ord. *Ecology* 18:255–263.
Hamilton, W. J., Jr. 1938. Life history notes on the northern pine mouse. *J. Mammal.* 19:163–170.
Hamilton, W. J., Jr. 1939a. Activity of Brewer's mole (*Parascalops breweri*). *J. Mammal.* 20:307–310.

Hamilton, W. J., Jr. 1939b. Observations on the life history of the red squirrel in New York. *Am. Midl. Nat.* 22:732–745.
Hamilton, W. J., Jr. 1940. The biology of the smoky shrew (*Sorex fumeus fumeus* Miller). *Zoologica* 25:473–492.
Hamilton, W. J., Jr. 1941. Reproduction of the field mouse (*Microtus pennsylvanicus* Ord). *Cornell Univ. Agric. Exp. Sta. Mem.* 237:1–23.
Hamilton, W. J., Jr. 1944. The biology of the little short-tailed shrew (*Cryptotis parva*). *J. Mammal.* 25:1–7.
Hamilton, W. J., Jr. 1951. The food of the opossum in New York State. *J. Wildl. Manage.* 15:258–264.
Hamilton, W. J., Jr. 1958. Life history and economic relations of the opossum (*Didelphis marsupialis virginiana*) in New York State. *Cornell Univ. Agri. Exp. Sta. Mem.* 354:1–48.
Hamilton, W. J., Jr.; Eadie, W. R. 1964. Reproduction in the otter (*Lutra canadensis*). *J. Mammal.* 45:242–252.
Hamilton, W. J., Jr.; Whitaker, J. O., Jr. 1979. *Mammals of the eastern United States.* Ithaca, N.Y.: Cornell University Press. 346 pp.
Harestad, A. S.; Bunnell, J. L. 1979. Home range and body weight—A reevaluation. *Ecology* 60:389–402.
Harger, E. M. 1967. Homing behavior of black bears. Research Division Report 118. E. Lansing: Michigan Department of Conservation. 13 pp.
Haroldson, K. J.; Fritzell, E. K. 1984. Home ranges, activity, and habitat use by gray foxes in an oak-hickory forest. *J. Wildl. Manage.* 48:222–227.
Harper, F. 1961. Land and fresh-water mammals of the Ungava Peninsula. *Misc. Publ. Univ. Kans. Mus. Nat. Hist.* 27:57–60.
Harrison, D. J. 1992. Social ecology of coyotes in northeastern North America: relationships to dispersal, food resources, and human exploitation. Pages 53–72. In: Boer, A. H. (editor). *Ecology and management of the eastern coyote.* Fredericton: University of New Brunswick Press.
Harrison, D. J.; Bissonette, J. A.; Sherburne, J. A. 1989. Spatial relationships between coyotes and red foxes in eastern Maine. *J. Wildl. Manage.* 53:181–185.
Harrison, D. J.; Chapin, T. G. 1998. Extent and connectivity of habitat for wolves in eastern North America. *Wildl. Soc. Bull.* 26:767–775.
Harrison, D. J.; Gilbert, J. R. 1985. Denning ecology and movements of coyotes in Maine during pup rearing. *J. Mammal.* 66:712–719.
Harrison, D. J.; Harrison, J. A. 1984. Foods of adult Maine coyotes and their known-aged pups. *J. Wildl. Manage.* 48:922–926.
Harrison, D. J.; Harrison, J. A.; O'Donoghue, M. 1991. Predispersal movements of coyote (*Canis latrans*) pups in eastern Maine. *J. Mammal.* 72:756–763.
Hart, J. A.; Kirkland, G. L., Jr.; Grossman, S. C. 1993. Relative abundance and habitat use by tree bats, *Lasiurus* spp., in southcentral Pennsylvania. *Can. Field-Nat.* 107:208–212.
Hartman, C. G. 1923. Breeding habits, development and birth of the opossum. *Smithsonian Report* (for 1921) 347–363. Washington, D.C.
Hartman, C. G. 1952. *Possums.* Austin: University of Texas Press. 174 pp.
Hartman, L. H.; Gaston, A. J.; Eastman, D. S. 1997. Raccoon predation on ancient murrelets on East Limestone Island, British Columbia. *J. Wildl. Manage.* 61:377–388.
Harvey, M. J. 1976. Home range, movements and diet activity of the eastern mole (*Scalopus aquaticus*). *Am. Midl. Nat.* 95:436–445.
Hatt, R. T. 1929. The red squirrel: Its life history and habits, with special reference to the Adirondacks of New York and the Harvard Forest. *N.Y. State Coll. Forestry Syracuse, N.Y.: Roosevelt Wild Life Ann.* 2:1–146.
Haugen, A. O. 1942. Life history studies of the cottontail rabbit in southwestern Michigan. *Am. Midl. Nat.* 28:204–244.
Haverman, J. R. 1973. A study of population densities, habitats and foods of four sympatric species of shrews. M.S. thesis, Northern Michigan University, Marquette. 70 pp.
Hazard, E. B. 1982. *Mammals of Minnesota.* St. Paul: University of Minnesota Press. 286 pp.
Hazzard, A. M.; Griffin, C. R.; Orff, E. P.; Jones, G. S. 1993. Assessment of river otter harvests in New Hampshire. *Northeast Wildl.* 50:77–84.
Heidt, G. A. 1970. The least weasel, *Mustela nivalis* L. Developmental biology in comparison with other North American *Mustela. Mich. State Univ. Publ. Mus.* (*Biol. Ser.*). 4:227–282.
Henry, J. D. 1986. *Red fox: The catlike canine.* Washington, D.C.: Smithsonian Institution Press. 174 pp.
Henshaw, R. E.; Folk, G. E., Jr. 1966. Relation of thermoregulation to seasonally changing microclimate in two species of bats (*Myotis lucifugus* and *M. sodalis*). *Physiol. Zool.* 39:223–236.
Hickey, M. B. C.; Fenton, M. B. 1990. Foraging by red bats (*Lasiurus borealis*): Do intraspecific chases mean territoriality? *Can. J. Zool.* 68:2477–2488.
Hickey, M. B. C.; Neilson, A. L. 1995. Relative activity and occurrence of bats in southwestern Ontario as determined by monitoring with bat detectors. *Can. Field-Nat.* 109:413–417.
Hilton, H. 1976. The physical characteristics, taxonomic status and food habits of the eastern coyote in Maine. M.S. thesis, University of Maine, Orono. 66 pp.
Hilton, H. 1978. Systematics and ecology of the eastern coyote. Pages 209–228. In: Bekoff, M. (ed.). *Coyotes: Biology, behavior and management.* New York: Academic Press.
Hisaw, F. L. 1923a. Observations on the burrowing habits of moles. *J. Mammal.* 4:79–88.
Hisaw, F. L. 1923b. Feeding habits of moles. *J. Mammal.* 4:9–20.
Hitchcock, H. B. 1949. Hibernation of bats in southeastern Ontario and adjacent Quebec. *Can. Field-Nat.* 63:47–59.
Hitchcock, H. B. 1955. A summer colony of the least bat, *Myotis subulatus leibii* (Audubon and Bachman). *Can. Field-Nat.* 69:31.
Hitchcock, H. B.; Keen, R.; Kurta, A. 1984. Survival rates of *Myotis leibii* and *Eptesicus fuscus* in southeastern Ontario. *J. Mammal.* 65:126–130.
Hodgdon, H. E.; Larson, J. S. 1973. Some sexual differences in behavior within a study of marked beavers (*Castor canadensis*). *Anim. Behav.* 21:147–152.
Hodgdon, K. W.; Hunt, J. H. 1953. *Beaver management in Maine.* Game Div. Bull. 3. Augusta: Maine Department of Inland Fisheries and Game. 102 pp.
Hodgman, T. P.; Harrison, D. J.; Katnik, D. D.; Elowe, K. D. 1994. Survival in an intensively trapped marten population in Maine. *J. Wildl. Manage.* 58:593–600.
Hodgman, T. P.; Harrison, D. J.; Phillips, D. M.; Elowe, K. D. 1997. Survival of American marten in an untrapped forest preserve in Maine. Pages 86–99. In: Proulx, G.; Bryant, H. N.; Woodard, P. (editors). *Martens: Taxonomy, ecology, tech-*

niques, and management. Edmonton: Provincial Museum of Alberta. 474 pp.
Hoff, J. G. 1987. Status and distribution of two species of cottontail rabbits, *Sylvilagus transitionalis* and *S. floridanus*, in southeastern Massachusetts. *Can. Field-Nat.* 101:88–89.
Holmes, A. C. V.; Sanderson, G. C. 1965. Populations and movements of opossums in east-central Illinois. *J. Wildl. Manage.* 29:287–295.
Holmes, D. J. 1991. Social behavior in captive Virginia opossums, *Didelphis virginiana*. *J. Mammal.* 72:402–410.
Hossler, R. J.; McAninch, J. B.; Harder, J. D. 1994. Maternal denning behavior and survival of juveniles in opossums in southeastern New York. *J. Mammal.* 75:60–70.
Howard, R. J.; Larson, J. S. 1985. A stream habitat classification system for beaver. *J. Wildl. Manage.* 49:19–25.
Howell, J. C. 1954. Populations and home ranges of small mammals on an overgrown field. *J. Mammal.* 35:177–186.
Hubert, G. F., Jr.; Hungerford, L. L.; Proulx, G.; Bluett, R. D.; Bowman, L. 1996. Evaluation of two restraining traps to capture raccoons. *Wildl. Soc. Bull.* 24:699–708.
Huey, W. C. 1956. New Mexico beaver management. *New Mex. Dept. Game Fish Bull.* 4:1–49.
Hugie, R. D. 1982. Black bear ecology and management in the northern conifer-deciduous forests of Maine. Ph.D. dissertation, University of Montana, Missoula. 203 pp.
Humphrey, S. R. 1978. Status, winter habitat and management of the endangered Indiana bat, *Myotis sodalis*. *Florida Sci.* 41:65–76.
Humphrey, S. R. 1982. Bats. Pages 52–70. In: Chapman, J. A.; Feldhamer, G. A. (editors). *Wild mammals of North America—Biology, management, and economics.* Baltimore, Md.: Johns Hopkins University Press.
Humphrey, S. R.; Richta, A. R.; Cope, J. B. 1977. Summer habitat and ecology of the endangered Indiana bat, *Myotis sodalis*. *J. Mammal.* 58:334–346.
Ickes, R. A. 1974. Agonistic behavior and the use of space in the eastern chipmunk, *Tamias striatus*. Ph.D. dissertation, University of Pittsburgh, Pittsburgh, Pa.
Innes, D. G. 1978. A reexamination of litter size in some North American microtines. *Can. J. Zool.* 56:1488–1496.
Jackson, C. F. 1922. Notes on New Hampshire mammals. *J. Mammal.* 3:13–15.
Jackson, H. H. T. 1961. *Mammals of Wisconsin.* Madison: University of Wisconsin Press. 504 pp.
Jackson, S. N. 1973. Distribution of cottontail rabbits *Sylvilagus* spp. in northern New England. M.S. thesis, University of Connecticut, Storrs. 48 pp.
Jackson, W. B. 1982. Norway rat and allies. Pages 1077–1088. In: Chapman, J. A.; Feldhamer, G. A. (editors). *Wild mammals of North America—Biology, management, and economics.* Baltimore, Md.: Johns Hopkins University Press.
Jarrell, G. H. 1965. A correction on the range of *Cryptotis parva* in New England. *J. Mammal.* 46:671.
Jenkins, S. H.; Busher, P. E. 1979. *Castor canadensis*. *Mammalian Species* 120:1–8. Shippensburg, Pa.: American Society of Mammalogists.
Jensen, I. M. 1986. Foraging strategies of the mole (*Parascalops breweri* Bachman, 1842). I. The distribution of prey. *Can. J. Zool.* 64:1727–1733.
Johnson, C. E. 1925. The muskrat in New York: Its natural history and economics. *Roosevelt Wild Life Bull.* 3:205–320.
Johnston, C. A.; Naiman, R. J. 1990a. Aquatic patch creation in relation to beaver population trends. *Ecology* 71:1617–1621.
Johnston, C. A.; Naiman, R. J. 1990b. Browse selection by beaver: Effects on riparian forest composition. *Can. J. Forest Res.* 20:1036–1043.
Johnston, C. A.; Naiman, R. J. 1990c. The use of a geographic information system to analyze long-term landscape alteration by beaver. *Landscape Ecol.* 4:5–19.
Johnston, J. E. 1972. Identification and distribution of cottontail rabbits in southern New England. M.S. thesis, University of Connecticut, Storrs.
Jones, C.; Hoffmann, R. S.; Rice, D. W.; Engstrom, M. D.; Bradley, R. D.; Schmidley, D. J.; Jones, C. A.; Baker, R. J. 1997. Revised checklist of North American mammals north of Mexico, 1997. *Occas. Pap. Mus. Tex. Tech. Univ.* 173:1–19.
Jones, J. K., Jr.; Birney, E. C. 1988. *Handbook of mammals of the north-central states.* St. Paul: University of Minnesota Press. 346 pp.
Jordan, J. S. 1948. A midsummer study of the southern flying squirrel. *J. Mammal.* 29:44–48.
Junge, R. E.; Sanderson, G. C. 1982. Age related reproductive success of female raccoons. *J. Wildl. Manage.* 46:527–529.
Kadlec, J. A. 1971. Effects of introducing foxes and raccoons on herring gull colonies. *J. Wildl. Manage.* 35:626–636.
Katnik, D. D.; Harrison, D. J.; Hodgman, T. P. 1994. Spatial relations in a harvested population of marten in Maine. *J. Wildl. Manage.* 58:600–607.
Kaufmann, J. H. 1982. Raccoon and allies. Pages 567–585. In: Chapman, J. A.; Feldhamer, G. A. (editors). *Wild mammals of North America—Biology, management, and economics.* Baltimore, Md.: Johns Hopkins University Press.
Keen, R.; Hitchcock, H. B. 1980. Survival and longevity of the little brown bat (*Myotis lucifugus*) in southeastern Ontario. *J. Mammal.* 61:1–7.
Keith, L. B.; Bloomer, S. E. M. 1993. Differential mortality of sympatric snowshoe hares and cottontail rabbits in central Wisconsin. *Can. J. Zool.* 71:1694–1697.
Keith, L. B.; Windberg, L. A. 1978. A demographic analysis of the snowshoe hare cycle. *Wildl. Monogr.* 58. 70 pp.
Keith, L. B.; Cary, J. R.; Rongstad, O. J.; Brittingham, M. C. 1984. Demography and ecology of a declining snowshoe hare population. *Wildl. Monogr.* 90. 43 pp.
Kelly, G. M. 1977. Fisher (*Martes pennanti*) biology in the White Mountain National Forest and adjacent areas. Ph.D. dissertation, University of Massachusetts, Amherst. 178 pp.
King, C. M. 1983. *Mustela erminea. Mammalian Species* 195:1–8. Shippensburg, Pa.: American Society of Mammalogists.
King, C. M. 1990. *The natural history of weasels and stoats.* Ithaca, N.Y.: Cornell University Press. 253 pp.
King, D. I.; Griffin, C. R.; DeGraaf, R. M. 1998. Nest predator distribution among clearcut forest, forest edge and forest interior in an extensively forested landscape. *Forest Ecol. Manage.* 104:151–156.
King, J. A. (editor). 1968. *Biology of* Peromyscus (*Rodentia*). Special Publication No. 2. Shippensburg, Pa.: American Society of Mammalogists. 593 pp.
Kirkland, G. L., Jr. 1977a. The rock vole, *Microtus chrotorrhinus* Miller (Mammalia: Rodentia), in West Virginia. *Ann. Carnegie Mus.* 46:45–53.

Kirkland, G. L., Jr. 1977b. Responses of small mammals to the clearcutting of northern Appalachian forests. *J. Mammal.* 58:600–609.
Kirkland, G. L., Jr. 1978. Initial responses of small mammals to clearcutting in Pennsylvania hardwood forests. *Proc. Penn. Acad. Sci.* 52:21–23.
Kirkland, G. L., Jr. 1981. *Sorex dispar* and *Sorex gaspensis. Mammalian Species* 155:1–5. Shippensburg, Pa.: American Society of Mammalogists.
Kirkland, G. L., Jr.; Jannett, F. J., Jr. 1982. *Microtus chrotorrhinus. Mammalian Species* 180:1–8. Shippensburg, Pa.: American Society of Mammalogists.
Kirkland, G. L., Jr.; Knipe, C. M. 1979. The rock vole (*Microtus chrotorrhinus*) as a transition zone species. *Can. Field-Nat.* 93:319–321.
Kirkland, G. L., Jr.; Schloyer, C. R.; Hull, D. K. 1976. A novel habitat record for the long-tailed shrew, *Sorex dispar* Batchelder. *Proc. W. Va. Acad. Sci.* 48:77–79.
Kirkland, G. L., Jr.; Schmidt, D. F. 1982. Abundance, habitat, reproduction and morphology of forest-dwelling small mammals of Nova Scotia and southeastern New Brunswick. *Can. Field-Nat.* 96:156–162.
Kirkland, G. L., Jr.; Schmidt, D. F.; Kirkland, C. J. 1979. First record of the long-tailed shrew (*Sorex dispar*) in New Brunswick. *Can. Field-Nat.* 93:195–197.
Kirkland, G. L., Jr.; Van Deusen, H. M. 1979. The shrews of the *Sorex dispar* group: *Sorex dispar* Batchelder and *Sorex gaspensis* Anthony and Goodwin. *Am. Mus. Nov.* 2675:1–21.
Kirkland, G. L., Jr.; Wilkinson, A. M.; Planz, J. V.; Maldonado, J. E. 1987. *Sorex (Microsorex) hoyi* in Pennsylvania. *J. Mammal.* 68:384–387.
Kivett, V. K.; Mock, O. B. 1980. Reproductive behavior in the least shrew (*Cryptotis parva*) with special reference to the aural glandular region of the female. *Am. Midl. Nat.* 103:339–345.
Klein, H. G. 1960. Ecological relationships of *Peromyscus leucopus noveboracensis* and *P. maniculatus gracilis* in central New York. *Ecol. Monogr.* 30:387–407.
Klugh, A. B. 1927. Ecology of the red squirrel. *J. Mammal.* 8:1–32.
Knowles, B. 1992. Bat hibernacula on Lake Superior's north shore, Minnesota. *Can. Field-Nat.* 106:252–254.
Koehler, G. M.; Aubry, K. B. 1994. Lynx. Pages 74–98. In: Ruggiero, L. F.; Aubry, K. B.; Buskirk, S. W.; Lyon, L. J.; Zielinski, W. J. (technical editors). *The scientific basis for conserving forest carnivores: American marten, fisher, lynx, and wolverine in the western United States.* Gen. Tech. Rep. RM-254. Fort Collins, Colo.: U.S. Department of Agriculture, Forest Service, Rocky Mountain Forest and Range Experiment Station. 184 pp.
Kolenosky, G. B.; Strathearn, S. M. 1987. Black bear. Pages 442–454. In: Novak, M.; Baker, J. A.; Obbard, M. E.; Malloch, B. (editors). *Wild furbearer management and conservation in North America.* Toronto: Ontario Ministry of Natural Resources and Ontario Trappers Association.
Koprowski, J. L. 1994. *Sciurus carolinensis. Mammalian Species* 480:1–9. Shippensburg, Pa.: American Society of Mammalogists.
Krebs, C. J.; Gilbert, B. S.; Boutin, S.; Boonstra, R. 1987. Estimation of snowshoe hare population density from turd transects. *Can. J. Zool.* 65:565–567.
Krebs, J. W.; Strine, T. W.; Smith, J. S.; Rupprecht, C. E.; Childs, J. E. 1994. Rabies surveillance in the United States during 1993. *J. Am. Vet. Med. Assoc.* 205:1695–1709.
Krohn, W. B.; Arthur, S. M.; Paragi, T. F. 1994. Mortality and vulnerability of a heavily trapped fisher population. Pages 137–145. In: Buskirk, S. W.; Harestad, A. S.; Raphael, M. G.; Powell, R. A. (editors). *Martens, sables, and fishers: Biology and conservation.* Ithaca, N.Y.: Cornell University Press. 484 pp.
Krohn, W. B.; Elowe, K. D.; Boone, R. B. 1995. Relations between fishers, snowfall, and martens. *Forestry Chron.* 71:97–105.
Krusic, R. A. 1995. Habitat use and identification of bats in the White Mountain National Forest. M.S. thesis, University of New Hampshire, Durham. 86 pp.
Krusic, R. A.; Neefus, C. D. 1996. Habitat associations of bat species in the White Mountain National Forest. Pages 185–198. In: Barclay, R. M. R.; Brigham, R. M. (editors). *Bats and forests symposium,* October 19–21, 1995, Victoria, British Columbia, Canada. Research Branch, British Columbia Ministry of Forestry, Victoria, B. C. Work. Pap. 23/1996.
Krusic, R. A.; Yamasaki, M.; Neefus, C. D.; Pekins, P. J. 1996. Bat habitat use in White Mountain National Forest. *J. Wildl. Manage.* 60:625–631.
Kunz, T. H. 1971. Reproduction of some vespertilionid bats in central Iowa. *Am. Midl. Nat.* 86:477–486.
Kunz, T. H. 1973. Resource utilization: Temporal and spatial components of bat activity in central Iowa. *J. Mammal.* 54:14–32.
Kunz, T. H. 1982. *Lasionycteris noctivagans. Mammalian Species* 172:1–5. Shippensburg, Pa.: American Society of Mammalogists.
Kurta, A. 1985. External insulation available to a non-nesting mammal, the little brown bat (*Myotis lucifugus*). *Comp. Biochem. Physiol.* 82A:413–420.
Kurta, A. 1995. *Mammals of the Great Lakes region.* Ann Arbor: University of Michigan Press. 246 pp.
Kurta, A.; Baker, R. H. 1990. *Eptesicus fuscus. Mammalian Species* 356:1–10. Shippensburg, Pa.: American Society of Mammalogists.
Kurta, A.; Bell, G. P.; Nagy, K. A.; Kunz, T. H. 1989. Energetics of pregnancy and lactation in free-ranging little brown bats (*Myotis lucifugus*). *Physiol. Zool.* 62:804–818.
Kurta, A.; Johnson, K. A.; Kunz, T. H. 1987. Oxygen consumption and body temperature of female little brown bats (*Myotis lucifugus*) under simulated roost conditions. *Physiol. Zool.* 60:386–397.
Kurta, A.; Kath, J.; Smith, E. L.; Foster, R.; Orick, M.; Ross, R. 1993b. A maternity roost of the endangered Indiana bat (*Myotis sodalis*) in an unshaded, hollow, sycamore tree (*Platanus occidentalis*). *Am. Midl. Nat.* 130:405–407.
Kurta, A.; King, D.; Teramino, J. A.; Stribley, J. M.; Williams, K. J. 1993a. Summer roosts of the endangered Indiana bat (*Myotis sodalis*) on the northern edge of its range. *Am. Midl. Nat.* 129:132–138.
Kurta, A.; Kunz, T. H. 1988. Roosting metabolic rate and body temperature of male little brown bats (*Myotis lucifugus*) in summer. *J. Mammal.* 69:645–651.
Kurta, A.; Stewart, M. E. 1990. Parturition in the silver-haired bat, *Lasionycteris noctivagans,* with a description of the neonates. *Can. Field-Nat.* 104:598–600.
Kurta, A.; Teramino, J. A. 1994. A novel hibernaculum and

noteworthy records of the Indiana bat and Eastern pipestrelle (Chiroptera: Vespertilionidae). *Am. Midl. Nat.* 132:410–413.
Kurta, A.; Williams, K. J.; Mies, R. 1996. Ecological, behavioural, and thermal observations of a peripheral population of Indiana bats (*Myotis sodalis*). Pages 102–117. In: Barclay, R. M. R.; Brigham, R. M. (editors). *Bats and forests symposium,* October 19–21, 1995, Victoria, British Columbia, Canada. Research Branch, British Columbia Ministry of Forestry, Victoria, B.C. Work. Pap. 23/1996.
Kwiecinski, G. G. 1998. *Marmota monax. Mammalian Species* 591:1–8. Shippensburg, Pa.: American Society of Mammalogists.
Lackey, J. A.; Huckaby, D. G.; Ormiston, B. G. 1985. *Peromyscus leucopus. Mammalian Species* 247:1–10. Shippensburg, Pa.: American Society of Mammalogists.
Ladine, T. A.; Kissell, R. E., Jr. 1994. Escape behavior of Virginia opossum. *Am. Midl. Nat.* 132:234–238.
Lair, H. 1985a. Length of gestation in the red squirrel, *Tamiasciurus hudsonicus. J. Mammal.* 66:809–810.
Lair, H. 1985b. Mating seasons and fertility of red squirrels in southern Quebec. *Can. J. Zool.* 63:2323–2327.
Lampman, B. H. 1947. A note on the predaceous habit of the water shrew. *J. Mammal.* 28:181.
Larivière, S. 1999. *Mustela vision: Mammalian Species* 608:1–9. Shippensburg, Pa.: American Society of Mammalogists
Larivière, S.; Messier, F. 1998. Spatial organization of a prairie striped skunk population during the waterfowl nesting season. *J. Wildl. Manage.* 62:199–204.
Larivière, S.; Pasitschniak-Arts, M. 1996. *Vulpes vulpes. Mammalian Species* 537:1–11. Shippensburg, Pa.: American Society of Mammalogists.
Larivière, S.; Walton, L. R. 1997. *Lynx rufus. Mammalian Species* 563:1–8. Shippensburg, Pa.: American Society of Mammalogists.
Larivière, S.; Walton, L. R. 1998. *Lontra canadensis. Mammalian Species* 587:1–8. Shippensburg, Pa.: American Society of Mammalogists.
Larson, J. S. 1962. Notes on a recent squirrel emigration in New England. *J. Mammal.* 43:272–273.
Larson, J. S. 1967. Age structure and sexual maturity within a western Maryland beaver (*Castor canadensis*) population. *J. Mammal.* 48:408–413.
LaVal, R. K.; LaVal, M. L. 1979. Notes on the reproduction, behavior, and abundance of the red bat, *Lasiurus borealis. J. Mammal.* 60:209–212.
LaVal, R. K.; LaVal, M. L. 1980. Ecological studies and management of Missouri bats, with emphasis on cave dwelling species. *Terr. Ser.* 8:1–53. Jefferson City: Missouri Department of Conservation.
Lawrence, B. 1946. Brief comparison of the short-tailed shrew and reptile poisons. *J. Mammal.* 26:393–396.
Lawrence, B.; Bossert, W. H. 1969. The cranial evidence for hybridization in New England *Canis. Breviora* 330:1–13.
Lay, D. W. 1942. Ecology of the opossum in eastern Texas. *J. Mammal.* 23:147–159.
Lay, D. W. 1945. Muskrat investigations in Texas. *J. Wildl. Manage.* 9:56–76.
Layne, J. N. 1954. The biology of the red squirrel, *Tamiasciurus hudsonicus* (Bangs) in central New York. *Ecol. Monogr.* 24:227–267.
Layne, J. N.; Hamilton, W. J., Jr. 1954. The young of the woodland jumping mouse, *Napaeozapus insignis insignis* Miller. *Am. Midl. Nat.* 52:242–247.
Lee, D. S.; Funderburg, J. B. 1982. Marmots and allies. Pages 176–191. In: Chapman, J. A.; Feldhamer, G. A. (editors). *Wild mammals of North America—Biology, management, and economics.* Baltimore, Md.: Johns Hopkins University Press.
Leftwich, B. H. 1972. Population dynamics and behavior of the eastern mole, *Scalopus aquaticus machrinoides.* Ph.D. dissertation, University of Missouri, Columbia. 103 pp.
Lehman, L. E. 1984. Raccoon density, home range, and habitat use on south-central Indiana farmland. *Ind. Dept. Nat. Resources Div. Fish Wildl. Pittman-Robertson Bull.* 15. 66 pp.
Leptich, D. J.; Gilbert, J. R. 1989. Summer home range and habitat use by moose in northern Maine. *J. Wildl. Manage.* 53:880–885.
LeResche, R. E. 1974. Moose migrations in North America. *Nat. Can.* 101:393–415.
Lidicker, W. Z. 1966. Ecological observations on a feral house mouse population declining to extinction. *Ecol. Monogr.* 36:27–50.
Liers, E. E. 1951. Notes on the river otter (*Lutra canadensis*). *J. Mammal.* 32:1–9.
Lincoln, A., Jr. 1935. *Sorex dispar* in New Hampshire. *J. Mammal.* 16:223.
Linduska, J. P. 1950. *Ecology and land-use relationships of small mammals on a Michigan farm.* Lansing: Michigan Department of Conservation, Game Division. 144 pp.
Lindzey, F. 1987. Mountain lion. Pages 656–668. In: Novak, M.; Baker, J. A.; Obbard, M. E.; Malloch, B. (editors). *Wild furbearer management and conservation in North America.* Toronto: Ontario Ministry of Natural Resources and Ontario Trappers Association.
Linscombe, G.; Kinler, N.; Aulerich, R. J. 1982. Mink. Pages 629–643. In: Chapman, J. A.; Feldhamer, G. A. (editors). *Wild mammals of North America—Biology, management, and economics.* Baltimore, Md.: Johns Hopkins University Press.
Linzey, A. V. 1983. *Synaptomys cooperi. Mammalian Species* 210: 1–5. Shippensburg, Pa.: American Society of Mammalogists.
Linzey, A. V.; Cranford, J. A. 1984. Habitat selection in the southern bog lemming, *Synaptomys cooperi*, and the meadow vole, *Microtus pennsylvanicus,* in Virginia. *Can. Field-Nat.* 98:463–469.
Linzey, D. W.; Linzey, A. V. 1973. Notes on food of small mammals from Great Smoky Mountains National Park, Tennessee-North Carolina. *J. Elisha Mitchell Soc.* 89:6–14.
Litvaitis, J. A. 1990. Differential habitat use by sexes of snowshoe hares (*Lepus americanus*). *J. Mammal.* 71:520–523.
Litvaitis, J. A. 1991. Habitat use by snowshoe hares, *Lepus americanus,* in relation to pelage color. *Can. Field-Nat.* 105:275–277.
Litvaitis, J. A. 1993. Response of early successional vertebrates to historic changes in land use. *Conserv. Biol.* 7:866–873.
Litvaitis, J. A.; Clark, A. G.; Hunt, J. H. 1986a. Prey selection and fat deposits of bobcats (*Felis rufus*) during autumn and winter in Maine. *J. Mammal.* 67:389–392.
Litvaitis, J. A.; Harrison, D. J. 1989. Bobcat-coyote niche relationships during a period of coyote population increase. *Can. J. Zool.* 67:1180–1188.
Litvaitis, J. A.; Kingman, D.; Lanier, J.; Orff, E. 1991b. Status of lynx in New Hampshire. *Trans. Northeast Sect. Wildl. Soc.* 48:70–75.

Litvaitis, J. A.; Major, J. T.; Sherburne, J. A. 1987. Influence of season and human-induced mortality on spatial organization of bobcats (*Felis rufus*) in Maine. *J. Mammal.* 68:100–106.
Litvaitis, J. A.; Sherburne, J. A.; Bissonette, J. A. 1985a. A comparison of methods used to examine snowshoe hare habitat use. *J. Wildl. Manage.* 49:693–695.
Litvaitis, J. A.; Sherburne, J. A.; Bissonette, J. A. 1985b. Influence of understory characteristics on snowshoe hare habitat use and density. *J. Wildl. Manage.* 49:866–873.
Litvaitis, J. A.; Sherburne, J. A.; Bissonette, J. A. 1986b. Bobcat habitat use and home range size in relation to prey density. *J. Wildl. Manage.* 50:110–117.
Litvaitis, J. A.; Stevens, C. L.; Mautz, W. W. 1984. Age, sex, and weight of bobcats in relation to winter diet. *J. Wildl. Manage.* 48:632–635.
Litvaitis, J. A.; Verbyla, D. L.; Litvaitis, M. K. 1991a. A field method to differentiate New England and eastern cottontails. *Trans. Northeast Sect. Wildl. Soc.* 48:11–14.
Litvaitis, J. A.; Villafuerte, R. 1996. Factors affecting the persistence of New England cottontail metapopulations: The role of habitat management. *Wildl. Soc. Bull.* 24:686–693.
Litvaitis, M. K.; Litvaitis, J. A. 1996. Using mitochondrial DNA to inventory the distribution of remnant populations of New England cottontails. *Wildl. Soc. Bull.* 24:725–730.
Litvaitis, M. K.; Litvaitis, J. A.; Lee, W. J.; Kocher, T. D. 1997. Variation in the mitochondrial DNA of the *Sylvilagus* complex occupying the northeastern United States. *Can. J. Zool.* 75: 95–605.
Llewellyn, L. M.; Dale, F. H. 1964. Notes on the ecology of the opossum in Maryland. *J. Mammal.* 45:113–122.
Logan, K.; Irwin, L.; Skinner, R. 1986. Characteristics of a hunted mountain lion population in Wyoming. *J. Wildl. Manage.* 50:648–654.
Long, C. A. 1972. Notes on habitat preference and reproduction in pygmy shrews (*Microsorex*). *Can. Field-Nat.* 86:155–160.
Long, C. A. 1974. *Microsorex hoyi* and *Microsorex thompsoni. Mammalian Species* 33:1–4. Shippensburg, Pa.: American Society of Mammalogists.
Lotze, J. H.; Anderson, S. 1979. *Procyon lotor. Mammalian Species* 119:1–8. Shippensburg, Pa.: American Society of Mammalogists.
Lovejoy, D. A. 1973. Ecology of the woodland jumping mouse (*Napaeozapus insignis*) in New Hampshire. *Can. Field-Nat.* 87:145–149.
Lovejoy, D. A. 1975. The effect of logging on small mammal populations in New England northern hardwoods. *Univ. Conn. Occas. Pap. Biol. Sci. Ser.* 2:269–291.
Ludewig, H. A.; Bowyer, R. T. 1985. Overlap in winter diets of sympatric moose and white-tailed deer in Maine. *J. Mammal.* 66:390–392.
Lund, R. C. 1980. Return of the tiny tiger. *N.J. Outdoors* September/October:17–19.
Luttich, S.; Rusch, D. H.; Meslow, E. C.; Keith, L. B. 1970. Ecology of red-tailed hawk predation in Alberta. *Ecology* 51:190–203.
Lynch, G. M. 1967. Long-range movement of a raccoon in Manitoba. *J. Mammal.* 48:659–660.
MacArthur, R. A. 1978. Winter movements and home range of the muskrat. *Can. Field-Nat.* 92:345–349.
MacArthur, R. A. 1984. Aquatic thermoregulation in the muskrat (*Ondatra zibethicus*): Energy demands of swimming and diving. *Can. J. Zool.* 62:241–248.
MacCracken, J. G.; Van Ballenberghe, V.; Peek, J. M. 1997. Habitat relationships of moose on the Copper River Delta in coastal south-central Alaska. *Wildl. Monogr.* 136:1–52.
MacDonald, K.; Matsui, E.; Stevens, R.; Fenton, M. B. 1994. Echolocation calls and field identification of the eastern pipistrelle (*Pipistrellus subflavus*: Chiroptera: Vespertilionidae), using ultrasonic bat detectors. *J. Mammal.* 75:462–465.
MacGregor, A. E. 1942. Late fall and winter food of foxes in central Massachusetts. *J. Wildl. Manage.* 6:221–224.
Madden, J. R. 1974. Female territoriality in a Suffolk County, Long Island population of *Glaucomys volans. J. Mammal.* 55:647–652.
Major, J. T. 1983. Ecology and interspecific relationships of coyotes, bobcats, and red foxes in Maine. Ph.D. dissertation, University of Maine, Orono. 64 pp.
Major, J. T.; Sherburne, J. A. 1987. Interspecific relationships of coyotes, bobcats, and red foxes in western Maine. *J. Wildl. Manage.* 51:606–616.
Maly, M. S.; Knuth, B. A.; Barrett, G. W. 1985. Effects of resource partitioning on dispersal behavior of feral house mice. *J. Mammal.* 66:148–153.
Manville, R. H. 1949. A study of small mammal populations in northern Michigan. *Univ. Mich. Misc. Publ. Mus. Zool.* 73:1–83.
Marchinton, R. L.; Hirth, D. H. 1984. Behavior. Pages 129–168. In: Halls, L. K. (editor). *White-tailed deer: Ecology and management.* Harrisburg, Pa.: Stackpole Books and Wildlife Management Institute. 870 pp.
Mares, M. A.; Watson, M. D.; Lacher, T. E., Jr. 1976. Home range perturbations in *Tamias striatus*: Food supply as a determinant of home range and density. *Oecologia* 25:1–12.
Marshall, W. H.; Gullion, G. W.; Schwab, R. G. 1962. Early summer activities of porcupines as determined by radio-positioning techniques. *J. Wildl. Manage.* 26:75–79.
Martell, A. M. 1983. Changes in small mammal communities after logging in north-central Ontario. *Can. J. Zool.* 61:970–980.
Martin, I. G. 1981. Venom of the short-tailed shrew (*Blarina brevicauda*) as an immobilizing agent. *J. Mammal.* 62:189–192.
Martin, R. L. 1966. Redescription of the type locality of *Sorex dispar. J. Mammal.* 47:130–131.
Martin, R. L. 1971. The natural history and taxonomy of the rock vole, *Microtus chrotorrhinus.* Ph.D. dissertation, University of Connecticut, Storrs. 164 pp.
Martin, S. K. 1994. Feeding ecology of American marten and fishers. Pages 297–315. In: Buskirk, S. W.; Harestad, A. S.; Raphael, M. G.; Powell, R. A. (editors). *Martens, sables, and fishers: Biology and conservation.* Ithaca, N.Y.: Cornell University Press. 484 pp.
Mather, D. W. 1933. Gray long-tailed shrew from New Hampshire. *J. Mammal.* 14:70.
Matlack, C. R.; Evans, A. J. 1992. Diet and condition of bobcats, *Lynx rufus,* in Nova Scotia during autumn and winter. *Can. J. Zool.* 70:1114–1119.
Mattfeld, G. F. 1984. Northeastern hardwood and spruce/fir forests. Pages 305–330. In: Halls, L. K. (editor). *White-tailed deer: Ecology and management.* Harrisburg, Pa.: Stackpole Books and Wildlife Management Institute. 870 pp.
Matula, G. J., Jr. (editor). 1997. Wildlife Division Research and Management Report. Augusta: Maine Department of Inland Fisheries and Wildlife. 102 pp.

Mautz, W. W.; Pekins, P. J. 1989. Metabolic rate of bobcats as influenced by seasonal temperatures. *J. Wildl. Manage.* 53:202–205.
May, D. W. 1982. Habitat utilization by bobcats in eastern Maine. *Trans. Northeast Sect. Wildl. Soc.* 39:22.
McCarley, W. H. 1959. An unusually large nest of *Cryptotis parva. J. Mammal.* 40:243.
McClure, H. E. 1942. Summer activities of bats (genus *Lasiurus*) in Iowa. *J. Mammal.* 23:430–434.
McCord, C. M. 1974. Selection of winter habitat by bobcats (*Lynx rufus*) on the Quabbin Reservation, Massachusetts. *J. Mammal.* 55:428–437.
McCord, C. M. 1977. The bobcat in Massachusetts. *Mass. Wildl.* 28:2–8.
McCord, C. M.; Cardoza, J. E. 1982. Bobcat and lynx. Pages 728–766. In: Chapman, J. A.; Feldhamer, G. A. (editors). *Wild mammals of North America—Biology, management, and economics.* Baltimore, Md.: Johns Hopkins University Press.
McDonough, J. J. 1960. *The cottontail in Massachusetts.* Boston: Massachusetts Division of Fisheries and Game.
McKeever, S. 1952. The survey of West Virginia mammals. Pittman-Robertson Project 22-TC. Charleston, W.V.: Conservation Commission West Virginia. Mimeo. 126 pp.
McLaughlin, C. R.; Elowe, K. D.; Caron, M. A. 1991. Maine status report. Pages 30–34. In: Clark, J. D.; Smith, K. G. (editors). *Proc. Tenth Eastern Workshop on Black Bear Research and Management,* Bismarck, Arkansas, April 2–5, 1990. Fayetteville: University of Arkansas Press. 206 pp.
McManus, J. L. 1971. Activity of captive *Didelphis marsupialis. J. Mammal.* 52:846–848.
McManus, J. L. 1974. *Didelphis virginiana. Mammalian Species* 40:1–6. Shippensburg, Pa.: American Society of Mammalogists.
McShea, W. J.; Madison, D. M. 1984. Communal nesting between reproductively active females in a spring population of *Microtus pennsylvanicus. Can. J. Zool.* 62:344–346.
Mead, R. A. 1981. Delayed implantation in mustelids, with special emphasis on the spotted skunk. *J. Reprod. Fertil. Supp.* 29:11–24.
Mearns, E. A. 1900. The native mammals of Rhode Island. *Newport Nat. Hist. Soc. Circ.* 1:1–4.
Mech, L. D. 1974. *Canis lupus. Mammalian Species* 37:1–6. Shippensburg, Pa.: American Society of Mammalogists.
Mech, L. D. 1977. Productivity, mortality, and population trends of wolves in north-eastern Minnesota. *J. Mammal.* 58:559–574.
Mech, L. D.; Fritts, S. H.; Radde, G. L.; Paul, W. J. 1988. Wolf distribution and road density in Minnesota. *Wildl. Soc. Bull.* 16:85–87.
Meddleton, K. M. 1989. Movement patterns and habitat use of black bears in northern New Hampshire. M.S. thesis, University of New Hampshire, Durham. 71 pp.
Meddleton, K. M.; Litvaitis, J. A. 1990. Movement patterns and habitat use of adult female and subadult black bears in northern New Hampshire. *Trans. Northeast Sect. Wildl. Soc.* 47:1–9.
Medjo, D. C.; Mech, L. D. 1976. Reproductive activity in nine- and ten-month-old wolves. *J. Mammal.* 57:406–408.
Meier, P. T. 1991. Response of adult woodchucks (*Marmota monax*) to oral-gland scents. *J. Mammal.* 72(3):622–624.
Melquist, W. E.; Dronkert, A. E. 1987. River otter. Pages 626–641. In: Novak, M.; Baker, J. A.; Obbard, M. E.; Malloch, B. (editors). *Wild furbearer management and conservation in North America.* Toronto: Ontario Ministry of Natural Resources and Ontario Trappers Association.
Melquist, W. E.; Hornocker, M. G. 1979. Methods and techniques for studying and censusing river otter populations. Forest Wildl. Range Exp. Stat. Tech. Rep. 8. Moscow: University of Idaho. 17 pp.
Meng, H. 1959. Food habits of nesting Cooper's hawks and goshawks in New York and Pennsylvania. *Wilson Bull.* 71:169–174.
Merritt, J. F. 1981. *Clethrionomys gapperi. Mammalian Species* 146:1–9. Shippensburg, Pa.: American Society of Mammalogists.
Merritt, J. F.; Merritt, J. M. 1978. Seasonal home ranges and activity of small mammals of a Colorado subalpine forest. *Acta Theriol.* 23:195–202.
Mikesic, D. G.; Drickamer, L. C. 1992. Factors affecting home-range size in house mice (*Mus musculus domesticus*) living in outdoor enclosures. *Am. Midl. Nat.* 127:31–40.
Millar, J. S. 1989. Reproduction and development. Pages 169–232. In: Kirkland, G. L., Jr.; Layne, J. N. (editors). *Advances in the study of* Peromyscus (*Rodentia*). Lubbock: Texas Tech University Press. 366 pp.
Miller, B. K. 1989. Seasonal movement patterns and habitat use of moose in northern New Hampshire. M.S. thesis, University of New Hampshire, Durham. 65 pp.
Miller, B. K.; Litvaitis, J. A. 1992a. Habitat segregation by moose in a boreal forest ecotone. *Acta Theriol.* 37:41–50.
Miller, B. K.; Litvaitis, J. A. 1992b. Use of roadside salt licks by moose, *Alces alces,* in northern New Hampshire. *Can. Field-Nat.* 106:112–117.
Miller, D. H. 1964. Northern records of the pine mouse in Vermont. *J. Mammal.* 45:627–628.
Miller, D. H.; Getz, L. L. 1969. Life history notes on *Microtus pinetorum* in central Connecticut. *J. Mammal.* 50:777–784.
Miller, D. H.; Getz, L. L. 1972. Factors influencing the local distribution of the redback vole, *Clethrionomys gapperi,* in New England. *Univ. Conn. Occas. Pap. Biol. Sci. Ser.* 2:115–138.
Miller, D. H.; Getz, L. L. 1973. Factors influencing the local distribution of the redback vole, *Clethrionomys gapperi,* in New England. II. Vegetation cover, soil moisture and debris cover. *Univ. Conn. Occas. Pap. Biol. Sci. Ser.* 2:159–180.
Miller, D. H.; Getz, L. L. 1977. Factors influencing local distribution and species diversity of forest small mammals in New England. *Can. J. Zool.* 55:806–814.
Miller, G. S., Jr. 1895. On a collection of small mammals from the New Hampshire mountains. *Proc. Boston Soc. Nat. Hist.* 26:177–197.
Mills, J. K. 1984. Food habits of bobcats, *Lynx rufus,* in Nova Scotia. *Can. Field-Nat.* 98:50–51.
Mills, R. S.; Barrett, G. W.; Farrell, M. P. 1975. Population dynamics of the big brown bat (*Eptesicus fuscus*) in southwestern Ohio. *J. Mammal.* 56:591–604.
Mitchell, J. L. 1961. Mink movements and populations on a Montana river. *J. Wildl. Manage.* 25:49–54.
Mladenoff, D. J.; Sickley, T. A. 1998. Assessing potential gray wolf restoration in the northeastern United States: A spatial prediction of favorable habitat and potential population levels. *J. Wildl. Manage.* 62:1–10.
Mladenoff, D. J.; Sickley, T. A.; Haight, R. G.; Wydeven, A. P.

1995. A regional landscape analysis and prediction of favorable gray wolf habitat in the northern Great Lakes region. *Conserv. Biol.* 9:279–294.
Mock, O. B. 1970. Reproduction of the least shrew (*Cryptotis parva*) in captivity. Ph.D. dissertation, University of Missouri, Columbia.
Mohr, C. E. 1936. Notes on the least bats (*Myotis subulatus leibii*). *Proc. Penn. Acad. Sci.* 10:62–65.
Monthey, R. W. 1984. Effects of timber harvesting on ungulates in northern Maine. *J. Wildl. Manage.* 48:279–285.
Monthey, R. W. 1986. Responses of snowshoe hares, *Lepus americanus*, to timber harvesting in northern Maine. *Can. Field-Nat.* 100:568–570.
Monthey, R. W.; Soutiere, E. C. 1985. Responses of small mammals to forest harvesting in northern Maine. *Can. Field-Nat.* 99:13–18.
Morris, K.; Elowe, K. 1993. The status of moose and their management in Maine. *Alces* 29:91–97.
Mumford, R. E. 1969. Long-tailed weasel preys on big brown bats. *J. Mammal.* 50:360.
Mumford, R. E. 1973. Natural history of the red bat (*Lasiurus borealis*) in Indiana. *Period. Biol.* 75:155–158.
Mumford, R. E.; Whitaker, J. O., Jr. 1982. *Mammals of Indiana.* Bloomington: Indiana University Press. 537 pp.
Murie, A. 1934. The moose of Isle Royale. *Univ. Mich. Misc. Publ. Mus. Zool.* 25:7–44.
Muul, I. 1968. Behavioral and physiological influences on the distribution of the flying squirrel, *Glaucomys volans. Univ. Mich. Misc. Publ. Mus. Zool.* 134:1–66.
Naiman, R. J.; Melillo, J. M.; Hobbie, J. E. 1986. Ecosystem alteration of boreal forest streams by beaver (*Castor canadensis*). *Ecology* 67:1254–1269.
Naiman, R. J.; Pinay, G.; Johnston, C. A.; Pastor, J. 1994. Beaver influences on the long-term biogeochemical characteristics of boreal forest drainage networks. *Ecology* 75:905–921.
Nams, V.; Beare, S. S. 1982. Use of trees by ermine, *Mustela erminea. Can. Field-Nat.* 96:89–90.
Neilson, A. L.; Fenton, M. B. 1994. Responses of little brown myotis to exclusion and to bat houses. *Wildl. Soc. Bull.* 22:8–14.
Newman, D. G.; Griffin, C. R. 1994. Wetland use by river otters in Massachusetts. *J. Wildl. Manage.* 58:18–23.
Nicholson, W. S. 1982. An ecological study of the gray fox in east central Alabama. M.S. thesis, Auburn University, Auburn. 93 pp.
Nicholson, W. S.; Hill, E. P.; Briggs, D. 1985. Denning, pup-rearing, and dispersal in the gray fox in east-central Alabama. *J. Wildl. Manage.* 48:33–37.
Nixon, C. M.; Havera, S. P.; Hansen, L. P. 1980. Initial responses of squirrels to forest changes associated with selection cutting. *Wildl. Soc. Bull.* 8:298–306.
Noble, S. M. 1993. Evaluating predator distributions in Maine forest riparian zones using a geographic information system. M.S. thesis, University of Maine, Orono. 54 pp.
Northcott, T. H. 1971. Feeding habits of beaver in Newfoundland. *Oikos* 22:407–410.
Novak, M. 1987. Beaver. Pages 283–312. In: Novak, M.; Baker, J. A.; Obbard, M. E.; Malloch, B. (editors). *Wild furbearer management and conservation in North America.* Toronto: Ontario Ministry of Natural Resources and Ontario Trappers Association.
Nowell, K.; Jackson, P. (compilers). 1996. *Wild cats. Status survey and conservation action plan.* Gland, Switzerland: IUCN/SSC Cat Specialist Group, International Union for Conservation of Nature and Natural Resources. 382 pp.
Obbard, M. E. 1987. Red squirrel. Pages 265–281. In: Novak, M.; Baker, J. A.; Obbard, M. E.; Malloch, B. (editors). *Wild furbearer management and conservation in North America.* Toronto: Ontario Ministry of Natural Resources and Ontario Trappers Association.
O'Donoghue, M. 1983. Seasonal habitat selection by snowshoe hare in eastern Maine. *Trans. Northeast Sect. Wildl. Soc.* 40:100–107.
Oehler, J. D.; Litvaitis, J. A. 1996. The role of spatial scale in understanding responses of medium-sized carnivores to forest fragmentation. *Can. J. Zool.* 74:2070–2079.
Olson, R.; Hubert, W. A. 1994. Beaver: Water resources and riparian habitat manager. Laramie: University of Wyoming. 48 pp.
Orr, C. D.; Dodds, D. G. 1982. Snowshoe hare habitat preferences in Nova Scotia spruce-fir forests. *Wildl. Soc. Bull.* 10:147–150.
Orr, R. T. 1950. Unusual behavior and occurrence of a hoary bat. *J. Mammal.* 31:456–457.
Osgood, F. L. 1935. Four Vermont records of the big-tailed shrew. *J. Mammal.* 16:146.
Ouellet, J. P.; Ferron, J. 1988. Scent-marking behavior by woodchucks (*Marmota monax*). *J. Mammal.* 69:365–368.
Owen, J. G. 1984. *Sorex fumeus. Mammalian Species* 215:1–8. Shippensburg, Pa.: American Society of Mammalogists.
Ozoga, J.; Gaertner, R. 1963. Noteworthy locality records for some Michigan mammals. *Jack-Pine Warbler* 41:89–90.
Ozoga, J.; Harger, E. M. 1966. Winter activities and feeding habits of northern Michigan coyotes. *J. Wildl. Manage.* 30:809–818.
Ozoga, J.; Verme, L. J. 1982. Predation by black bears on newborn white-tailed deer. *J. Mammal.* 63:695–696.
Pack, J.; Mosby, H.; Siegal, P. 1967. Influence of social hierarchy on gray squirrel behavior. *J. Wildl. Manage.* 31:720–728.
Packard, J. M.; Mech, L. D. 1980. Population regulation in wolves. Pages 135–150. In: Cohen, M. N.; Malpass, R. S.; Klein, H. G. (editors). *Biosocial mechanisms of population regulation.* New Haven, Conn.: Yale University Press.
Palmer, R. S. 1956. Gray fox in the Northeast. *Maine Field-Nat.* 12:62–70.
Paradiso, J. L. 1969. Mammals of Maryland. *North Am. Fauna Ser.* 66:1–193.
Paradiso, J. L.; Nowak, R. M. 1982. Wolves (*Canis lupus* and allies). Pages 460–474. In: Chapman, J. A.; Feldhamer, G. A. (editors). *Wild mammals of North America—Biology, management, and economics.* Baltimore, Md.: Johns Hopkins University Press.
Paragi, T. F.; Arthur, S. M.; Krohn, W. B. 1994. Seasonal and circadian activity patterns of female fishers, *Martes pennanti,* with kits. *Can. Field-Nat.* 108:52–57.
Paragi, T. F.; Arthur, S. M.; Krohn, W. B. 1996. Importance of tree cavities as natal dens for fishers. *Northern J. Appl. Forestry* 13:79–83.
Paragi, T. F.; Wholecheese, G. M. 1994. Marten, *Martes americana,* predation on a northern goshawk, *Accipiter gentilis. Can. Field-Nat.* 108:81–82.
Parker, G. R. 1989. Effects of reforestation upon small mammal

communities in New Brunswick. *Can. Field-Nat.* 103:509–519.
Parker, G. R. 1995. *Eastern coyote—The story of its success.* Halifax, Nova Scotia: Nimbus Publishing. 254 pp.
Parker, G. R.; Smith, G. E. J. 1983. Sex- and age-specific reproductive and physical parameters of the bobcat (*Lynx rufus*) on Cape Breton Island, Nova Scotia. *Can. J. Zool.* 61:1771–1782.
Parker, G. R.; Maxwell, J. W.; Morton, L. D. Smith, G. E. J. 1983. The ecology of the lynx (*Lynx canadensis*) on Cape Breton Island. *Can. J. Zool.* 61:770–786.
Parren, S. G.; Capen, D. E. 1985. Local distribution and coexistence of two species of *Peromyscus* in Vermont. *J. Mammal.* 66:36–44.
Pasitschniak-Arts, M.; Messier, F. 1995. Risk of predation on waterfowl nests in the Canadian prairies: Effects of habitat edges and agricultural practices. *Oikos* 73:347–355.
Patric, E. F. 1962. Reproductive characteristics of red-backed mouse during years of differing population densities. *J. Mammal.* 43:200–205.
Patterson, B. R. 1994. Surplus killing of white-tailed deer, *Odocoileus virginianus*, by coyotes, *Canis latrans*, in Nova Scotia. *Can. Field-Nat.* 108:484–487.
Pauls, R. W. 1978. Behavioural strategies relevant to the energy economy of the red squirrel (*Tamiasciurus hudsonicus*). *Can. J. Zool.* 56:1519–1525.
Pedlar, J. H.; Fahrig, L.; Merriam, H. G. 1997. Raccoon habitat use at 2 spatial scales. *J. Wildl. Manage.* 61:102–112.
Peek, J. M.; Urich, D. L.; Mackie, R. J. 1976. Moose habitat selection and relationships to forest management in northeastern Minnesota. *Wildl. Monogr.* 48:1–65.
Pelton, M. R. 1982. Black bear. Pages 504–514. In: Chapman, J. A.; Feldhamer, G. A. (editors). *Wild mammals of North America—Biology, management, and economics.* Baltimore, Md.: Johns Hopkins University Press.
Pelton, M. R.; Beeman, L. E., Eagar, D. C. 1980. Den selection by black bears in the Great Smoky Mountains National Park. *Int. Conf. Bear Res. Manage.* 4:149–151.
Perry, H. R., Jr. 1982. Muskrats (*Ondatra zibethicus* and *Neofiber alleni*). Pages 282–325. In: Chapman, J. A.; Feldhamer, G. A. (editors). *Wild mammals of North America—Biology, management, and economics.* Baltimore, Md.: Johns Hopkins University Press.
Petersen, K. E.; Yates, T. L. 1980. *Condylura cristata. Mammalian Species* 129:1–4. Shippensburg, Pa.: American Society of Mammalogists.
Peterson, R. L. 1955. *North American moose.* Toronto, Ont.: University of Toronto Press. 280 pp.
Peterson, R. L. 1966. *The mammals of eastern Canada.* Toronto, Ont.: Oxford University Press. 465 pp.
Phillips, D. M. 1994. Social and spatial characteristics, and dispersal of marten in a forest preserve and industrial forest. M.S. thesis, University of Maine, Orono. 95 pp.
Phillips, D. M.; Harrison, D. J.; Payer, D. C. 1998. Seasonal changes in home-range area and fidelity of martens. *J. Mammal.* 79:180–190.
Phillips, G. L. 1966. Ecology of the big brown bat (Chiroptera: Vespertilionidae) in northeastern Kansas. *Am. Midl. Nat.* 75:168–198.
Phillips, R. L.; Andrews, R. D.; Storm, G. L.; Bishop, R. A. 1972. Dispersal and mortality of red foxes. *J. Wildl. Manage.* 36:237–248.
Phillips, R. L.; Berg, W. E.; Siniff, D. B. 1973. Movement patterns and range use of moose in northwestern Minnesota. *J. Wildl. Manage.* 37:266–278.
Planz, J. V.; Kirkland, G. L., Jr. 1992. Use of woody ground litter as a substrate for travel by the white-footed mouse, *Peromyscus leucopus. Can. Field-Nat.* 106:118–121.
Pollack, E. M. 1951. Observations on New England bobcats. *J. Mammal.* 32:356–358.
Poole, K. G. 1994. Characteristics of an unharvested lynx population during a snowshoe hare decline. *J. Wildl. Manage.* 58:608–618.
Poole, K. G. 1997. Dispersal patterns of lynx in the Northwest Territories. *J. Wildl. Manage.* 61:497–505.
Powell, R. A. 1981. Hunting behavior and food requirements of the fisher (*Martes pennanti*). Pages 883–917. In: Chapman, J. A.; Pursley, D. (editors). *Proc. Worldwide Furbearer Conf.* Frostburg, Md.
Powell, R. A. 1993. *The fisher: Life history, ecology, and behavior.* 2nd ed., Minneapolis: University of Minnesota Press. 237 pp.
Powell, S. M.; York, E. C.; Scanlon, J. J.; Fuller, T. K. 1997a. Fisher maternal den sites in central New England. Pages 265–278. In: Proulx, G.; Bryant, H. N.; Woodard, P. (editors). *Martens: Taxonomy, ecology, techniques, and management.* Edmonton: Provincial Museum of Alberta. 474 pp.
Powell, S. M.; York, E. C.; Fuller, T. K. 1997b. Seasonal food habits of fishers in central New England. Pages 279–305. In: Proulx, G.; Bryant, H. N.; Woodard, P. (editors). *Martens: Taxonomy, ecology, techniques, and management.* Edmonton: Provincial Museum of Alberta. 474 pp.
Powers, J. G.; Mautz, W. M.; Pekins, P. J. 1989. Nutrient and energy assimilation of prey by bobcats. *J. Wildl. Manage.* 53:1003 1008.
Preble, E. A. 1899. Description of a new lemming mouse from the White Mountains, New Hampshire. *Proc. Biol. Soc. Wash.* 13:43–45.
Preble, N. A. 1956. Notes on the life history of *Napaeozapus. J. Mammal.* 37(2):197–200.
Prince, L. A. 1940. Notes on the habits of the pygmy shrew (*Microsorex hoyi*) in captivity. *Can. Field-Nat.* 54:97–100.
Prince, L. A. 1941. Water traps capture the pygmy shrew (*Microsorex hoyi*) in abundance. *Can. Field-Nat.* 55(5):72.
Pringle, L. P. 1960a. A study of the biology and ecology of the New England cottontail (*Sylvilagus transitionalis*) in Massachusetts. M.S. thesis, University of Massachusetts, Amherst.
Pringle, L. P. 1960b. Notes on coyotes in southern New England. *J. Mammal.* 41:278.
Probert, B. L.; Litvaitis, J. A. 1996. Behavioral interactions between invading and endemic lagomorphs: Implications for conserving a declining species. *Biol. Conserv.* 76:289–295.
Proulx, G.; Bryant, H. N.; Woodard, P. (editors). 1997. *Martens: Taxonomy, ecology, techniques, and management.* Edmonton: Provincial Museum of Alberta. 474 pp.
Proulx, G.; Gilbert, F. F. 1983. The ecology of muskrat, *Ondatra zibethicus* at Luther Marsh, Ontario. *Can. Field-Nat.* 97:377–390.
Proulx, G.; Gilbert, F. F. 1984. Estimating muskrat population trends by house counts. *J. Wildl. Manage.* 48:917–922.
Proulx, G.; McDonnell, J. A.; Gilbert, F. F. 1987. The effect of water level fluctuations on muskrat, *Ondatra zibethicus,* predation by mink, *Mustela vison. Can. Field-Nat.* 101:89–92.
Pruitt, W. O., Jr. 1959. Microclimates and local distribution of

small mammals on the George Reserve, Michigan. *Univ. Mich. Misc. Publ. Mus. Zool.* 109:1–27.

Pruss, M. T.; Pekins, P. J. 1992. Effects of moose foraging on browse availability in New Hampshire deer yards. *Alces* 28:123–136.

Quadagno, D. M. 1968. Home range in feral house mice. *J. Mammal.* 49(1):149–151.

Quay, W. B. 1948. Notes on some bats from Nebraska and Wyoming. *J. Mammal.* 29(2):181–182.

Quick, H. F. 1944. Habits and economics of the New York weasel in Michigan. *J. Wildl. Manage.* 8:71–78.

Quick, H. F. 1951. Notes on the ecology of weasels in Gunnison County, Colorado. *J. Mammal.* 32:281–290.

Quimby, D.C. 1951. The life history and ecology of the jumping mouse, *Zapus hudsonius. Ecol. Monogr.* 21:61–95.

Quinn, N. W. S.; Gardner, J. F. 1984. Relationships of age and sex to lynx pelt characteristics. *J. Wildl. Manage.* 48:953- 956.

Quinn, N. W. S.; Parker, G. 1987. Lynx. Pages 682–694. In: Novak, M.; Baker, J. A.; Obbard, M. E.; Malloch, B. (editors). *Wild furbearer management and conservation in North America.* Toronto: Ontario Ministry of Natural Resources and Ontario Trappers Association.

Quinn, N. W. S.; Thompson, J. E. 1987. Dynamics of an exploited Canada lynx population in Ontario. *J. Wildl. Manage.* 51:297–305.

Rabe, M. L. 1981. New locations for pygmy (*Sorex hoyi*) and water (*Sorex palustris*) shrews in Michigan. *Jack-Pine Warbler* 59:16–17.

Raine, R. M. 1983. Winter habitat use and responses to snow cover of fisher (*Martes pennanti*) and marten (*Martes americana*) in southeastern Manitoba. *Can. J. Zool.* 61:25–34.

Ratnaswamy, M. J.; Warren, R. J.; Kramer, M. T.; Adam, M. D. 1997. Comparisons of lethal and nonlethal techniques to reduce raccoon depredation of sea turtle nests. *J. Wildl. Manage.* 61:368–376.

Raymond, K. S.; Servello, F. A.; Griffith, B.; Eschholz, W. E. 1996. Winter foraging ecology of moose on glyphosate-treated clearcuts in Maine. *J. Wildl. Manage.* 60:753–763.

Reich, L. M. 1981. *Microtus pennsylvanicus. Mammalian Species* 159:1–8. Shippensburg, Pa.: American Society of Mammalogists.

Reid, D. G.; Code, T. E.; Reid, C. H.; Herrero, S. M. 1994a. Spacing, movements, and habitat selection of the river otter in boreal Alberta. *Can. J. Zool.* 72:1314–1324.

Reid, D. G.; Code, T. E.; Reid, C. H.; Herrero, S. M. 1994b. Food habits of the river otter in a boreal ecosystem. *Can. J. Zool.* 72:1306–1313.

Reith, C. C. 1980. Shifts in time of activity by *Lasionycteris noctivagans. J. Mammal.* 61:104–108.

Rempel, R. S.; Elkie, P. C.; Rodgers, A. R.; Gluck, M. J. 1997. Timber-management and natural-disturbance effects on moose habitat: Landscape evaluation. *J. Wildl. Manage.* 61:517–524.

Renecker, L. A.; Hudson, R. J. 1985. Estimation of dry matter intake of free-ranging moose. *J. Wildl. Manage.* 49:785–792.

Richens, V. B. 1974. Numbers and habitat affinities of small mammals in northwestern Maine. *Can. Field-Nat.* 88:191–196.

Richens, V. B.; Hugie, R. D. 1974. Distribution, taxonomic status, and characteristics of coyotes in Maine. *J. Wildl. Manage.* 38:447–454.

Richmond, N. D.; Grimm, W. C. 1950. Ecology and distribution of the shrew, *Sorex dispar,* in Pennsylvania. *Ecology* 31:279–282.

Richter, A. R.; Humphrey, S. R.; Cope, J. B.; Brack, V., Jr. 1993. Modified cave entrances: thermal effect on body mass and resulting decline of endangered Indiana bats (*Myotis sodalis*). *Conserv. Biol.* 7:407–415.

Ritcey, R. W.; Edwards, R. Y. 1956. Live trapping mink in British Columbia. *J. Mammal.* 37:114–116.

Rivest, P.; Bergeron, J. M. 1981. Density, food habits, and economic importance of raccoons (*Procyon lotor*) in Quebec agrosystems. *Can. J. Zool.* 59:1755–1762.

Robinson, W. L. 1975. Birds and mammals. Pages 1–17. In: Robinson, W. L.; Werner, J. K. *Vertebrate animal populations of the McCormick Forest.* Research Paper NC-118. St. Paul, Minn.: U.S. Department of Agriculture Forest Service, North Central Forest Experiment Station. 25 pp.

Rogers, L. L. 1976. Effects of mast and berry crop failures on survival, growth, and reproductive success of black bears. *Trans. North Am. Wildl. Nat. Resources Conf.* 41:431–438.

Rogers, L. L. 1977. Social relationships, movements, and population dynamics of black bears in northeastern Minnesota. Ph.D. dissertation, University of Minnesota, Minneapolis. 194 pp.

Rogers, L. L. 1987. Effects of food supply and kinship on social behavior, movements, and population growth of black bears in northeastern Minnesota. *Wildl. Monogr.* 97:1–72.

Rolley, R. E. 1987. Bobcat. Pages 670–681. In: Novak, M.; Baker, J. A.; Obbard, M. E.; Malloch, B. (editors). *Wild furbearer management and conservation in North America.* Toronto: Ontario Ministry of Natural Resources and Ontario Trappers Association.

Rollings, C. T. 1945. Habits, foods and parasites of the bobcat in Minnesota. *J. Wildl. Manage.* 9:131–145.

Rongstad, O. J.; Tester, J. R. 1971. Behavior and maternal relations of young snowshoe hares. *J. Wildl. Manage.* 35(2):338–346.

Rosatte, R. C. 1987. Striped, spotted, hooded, and hog-nosed skunk. Pages 598–613. In: Novak, M.; Baker, J. A.; Obbard, M. E.; Malloch, B. (editors). *Wild furbearer management and conservation in North America.* Toronto: Ontario Ministry of Natural Resources and Ontario Trappers Association.

Rose, R. K.; Spevak, A. M. 1978. Aggressive behavior in two sympatric microtine rodents. *J. Mammal.* 59:213–216.

Rosen, R. C. 1978. Metabolic ecology of *Microtus pennsylvanicus* and *Microtus chrotorrhinus.* Ph.D. dissertation, University of Vermont, Burlington. 105 pp.

Ross, A. 1967. Ecological aspects of the food habits of insectivorous bats. *Proc. Western Foundation Vertebrate Zool.* 1(4):205–263.

Roze, U. 1989. *The North American porcupine.* Washington, D.C.: Smithsonian Institution Press. 224 pp.

Rusch, D. A.; Reeder, W. G. 1978. Population ecology of Alberta red squirrels. *Ecology* 59:400–420.

Rusch, D. H.; Meslow, E. C.; Doerr, P. D.; Keith, L. B. 1972. Response of great horned owl populations to changing prey densities. *J. Wildl. Manage.* 36:282–296.

Ryan, J. M. 1986. Dietary overlap in sympatric populations of pygmy shrews, *Sorex hoyi,* and masked shrews, *Sorex cinereus,* in Michigan. *Can. Field-Nat.* 100:225–228.

Samson, C.; Crete, M. 1997. Summer food habits and population density of coyotes. *Can. Field-Nat.* 111:227–233.

Samson, C.; Huot, J. 1995. Reproductive biology of female black bears in relation to body mass in early winter. *J. Mammal.* 76:68–77.

Samson, C.; Huot, J. 1998. Movements of female black bears in relation to landscape vegetation type in southern Québec. *J. Wildl. Manage.* 62:718–727.

Samuel, D. E.; Nelson, B. B. 1982. Foxes (*Vulpes vulpes* and allies). Pages 475–490. In: Chapman, J. A.; Feldhamer, G. A. (editors). *Wild mammals of North America—Biology, management, and economics.* Baltimore, Md.: Johns Hopkins University Press.

Sanderson, G. C. 1961. Estimating opossum populations by marking young. *J. Wildl. Manage.* 25:20–27.

Sanderson, G. C. 1987. Raccoon. Pages 486–499. In: Novak, M.; Baker, J. A.; Obbard, M. E.; Malloch, B. (editors). *Wild furbearer management and conservation in North America.* Toronto: Ontario Ministry of Natural Resources and Ontario Trappers Association.

Santillo, D. J. 1994. Observations on moose, *Alces alces,* habitat and use on herbicide-treated clearcuts in Maine. *Can. Field-Nat.* 108:22–25.

Sargeant, A. B. 1972. Red fox spatial characteristics in relation to waterfowl predation. *J. Wildl. Manage.* 36:225–236.

Sargeant, A. B.; Allen, S. H.; Eberhardt, R. T. 1984. Red fox predation on breeding ducks in midcontinent North America. *Wildl. Monogr.* 89:1–41.

Sasse, D. B. 1995. Summer roosting ecology of cavity-dwelling bats in the White Mountain National Forest. M.S. thesis, University of New Hampshire, Durham. 65 pp.

Sasse, D. B.; Pekins, P. J. 1996. Summer roosting ecology of northern long-eared bats (*Myotis septentrionalis*) in the White Mountain National Forest. Pages 91–101. In: Barclay, R. M. R.; Brigham, R. M. (editors). *Bats and forests symposium,* October 19–21, 1995, Victoria, British Columbia, Canada. Research Branch, British Columbia Ministry of Forestry, Victoria, B.C. Work. Pap. 23/1996.

Sauer, P. R. 1975. Relationship of growth characteristics to sex and age for black bears from the Adirondacks region of New York. *N.Y. Fish Game J.* 22:81–113.

Saunders, D. A. 1988. *Adirondack mammals.* Syracuse: State University of New York, College of Environmental Science and Forestry. 216 pp.

Saunders, J. K., Jr. 1963a. Food habits of the lynx in Newfoundland. *J. Wildl. Manage.* 27:384–390.

Saunders, J. K., Jr. 1963b. Movements and activities of the lynx in Newfoundland. *J. Wildl. Manage.* 27:399–400.

Schmitz, O. J.; Lavigne, D. M. 1987. Factors affecting body size in sympatric Ontario *Canis. J. Mammal.* 68:92–99.

Schneider, E. 1990. Hares and rabbits. Pages 254–313. In: Parker, S. B. (editor). *Grizimek's encyclopedia of mammals.* New York: McGraw-Hill.

Schooley, R. L. 1990. Habitat use, fall movements, and denning ecology of female black bears in Maine. M.S. thesis, University of Maine, Orono. 115 pp.

Schowalter, D. B.; Harder, L. D.; Treichel, B. H. 1978. Age composition of some vespertilionid bats as determined by dental annuli. *Can. J. Zool.* 56:355–358.

Schowalter, D. B.; Gunson, J. R.; Harder, L. D. 1979. Life history characteristics of little brown bats (*Myotis lucifugus*) in Alberta. *Can. Field-Nat.* 93:243–251.

Schwartz, C. C.; Franzmann, A. W. 1991. Interrelationship of black bears to moose and forest succession in the northern coniferous forest. *Wildl. Monogr.* 113. 58 pp.

Schwartz, C. C.; Regelin, W. L.; Franzmann, A. W. 1984. Seasonal dynamics of food intake in moose. *Alces* 20:223–244.

Schwartz, C. W.; Schwartz, E. R. 1981. *The wild mammals of Missouri.* Columbia: University of Missouri Press and Missouri Dept. of Conservation. 356 pp.

Scott, D. P.; Yahner, R. H. 1989. Winter habitat and browse use by snowshoe hares, *Lepus americanus,* in a marginal habitat in Pennsylvania. *Can. Field-Nat.* 103:560–563.

Scott, F. W. 1987. First record of the long-tailed shrew, *Sorex dispar,* for Nova Scotia. *Can. Field-Nat.* 101:404–407.

Scott, T. G. 1943. Some food coactions of the northern plains red fox. *Ecol. Monogr.* 13:427–479.

Seidensticker, J.; O'Connell, M. A.; Johnsingh, A. J. T. 1987. Virginia opossum. Pages 249–263. In: Novak, M.; Baker, J. A.; Obbard, M. E.; Malloch, B. (editors). *Wild furbearer management and conservation in North America.* Toronto: Ontario Ministry of Natural Resources and Ontario Trappers Association.

Shadle, A. R. 1951. Laboratory copulations and gestations of porcupine, *Erethizon dorsatum. J. Mammal.* 32:219–221.

Shanks, C. E.; Arthur, G. C. 1952. Muskrat movements and population dynamics in Missouri ponds and streams. *J. Wildl. Manage.* 16:138–148.

Shapiro, J. 1949. Ecological and life history notes on the porcupine in the Adirondacks. *J. Mammal.* 39:247–257.

Sharp, W. M. 1960. A commentary on the behavior of free-ranging gray squirrels. *Penn. Coop. Wildl. Res. Unit Paper* 101:1–13. Mimeo.

Sheffield, S. R.; Thomas, H. H. 1997. *Mustela frenata. Mammalian Species* 570:1–9. Shippensburg, Pa.: American Society of Mammalogists.

Sheldon, C. 1934. Studies on the life histories of *Zapus* and *Napaeozapus* in Nova Scotia. *J. Mammal.* 15:290–300.

Sheldon, C. 1938. Vermont jumping mice of the genus *Napaeozapus. J. Mammal.* 19:444–453.

Sheldon, W. G. 1949. Reproductive behavior of foxes in New York State. *J. Mammal.* 30:236–246.

Sherburne, S. S.; Bissonette, J. A. 1994. Marten subnivean access point use: Response to subnivean prey levels. *J. Wildl. Manage.* 58:400–405.

Short, H. L. 1961. Fall breeding activity of a young shrew. *J. Mammal.* 42:95.

Shump, K. A., Jr.; Shump, A. U. 1982a. *Lasiurus borealis. Mammalian Species* 183:1–6. Shippensburg, Pa.: American Society of Mammalogists.

Shump, K. A., Jr.; Shump, A. U. 1982b. *Lasiurus cinereus. Mammalian Species* 185:1–5. Shippensburg, Pa.: American Society of Mammalogists.

Sievert, P. R.; Keith, L. B. 1985. Survival of snowshoe hares at a geographic range boundary. *J. Wildl. Manage.* 49(4):854–866.

Silver, H.; Silver, W. T. 1969. Growth and behavior of the coyote-like canid of northern New England with observations on canid hybrids. *Wildl. Monogr.* 17. 41 pp.

Simms, D. A. 1979a. North American weasels: Resource utilization and distribution. *Can. J. Zool.* 57:504–520.

Simms, D. A. 1979b. Studies of an ermine population in southern Ontario. *Can. J. Zool.* 57:824–832.

Sklepkovych, B. 1994. Arboreal foraging by red foxes, *Vulpes*

vulpes, during winter food shortage. *Can. Field-Nat.* 108:479–481.

Slate, D.; Wolgost, L. J.; Lund, R. C. 1982. Density and structure of New Jersey raccoon populations. *Trans. Northeast Sect. Wildl. Soc.* 39:19–20.

Slough, B. G.; Mowat, G. 1996. Lynx population dynamics in an untrapped refugium. *J. Wildl. Manage.* 60:946–961.

Smith, C. C. 1968. The adaptive nature of social organization in the genus of three [sic] squirrels *Tamiasciurus. Ecol. Monogr.* 38:31–63.

Smith, L. C.; Smith, D. A. 1972. Reproductive biology, breeding seasons, and growth of eastern chipmunks, *Tamias striatus* (Rodentia: Sciuridae) in Canada. *Can. J. Zool.* 50:1069–1085.

Smith, N. B.; Barkalow, S. F., Jr. 1967. Precocious breeding in the gray squirrel. *J. Mammal.* 48:328–330.

Smith, W. P. 1991. *Odocoileus virginianus. Mammalian Species* 388:1–13. Shippensburg, Pa.: American Society of Mammalogists.

Smolen, M. J. 1981. *Microtus pinetorum. Mammalian Species* 147:1–7. Shippensburg, Pa.: American Society of Mammalogists.

Snyder, D. P. 1956. Survival rates, longevity and population fluctuations in the white-footed mouse, *Peromyscus leucopus,* in southern Michigan. *Univ. Mich. Misc. Publ. Mus. Zool.* 95:1–33.

Snyder, D. P. 1982. *Tamias striatus. Mammalian Species* 168:1–8. Shippensburg, Pa.: American Society of Mammalogists.

Snyder, R. L.; Christian, J. J. 1960. Reproductive cycle and litter size of the woodchuck. *Ecology* 41:647–656.

Sollberger, D. E. 1940. Notes on the life history of the small eastern flying squirrel. *J. Mammal.* 21:282–293.

Sollberger, D. E. 1943. Notes on the breeding habits of the eastern flying squirrel (*Glaucomys volans volans*). *J. Mammal.* 24:163–173.

Sorensen, M. W. 1962. Some aspects of water shrew behavior. *Am. Midl. Nat.* 68:445–462.

Soukkala, A. M. 1983. The effects of trapping on marten populations in Maine. M.S. thesis, University of Maine, Orono. 41 pp.

Soutiere, E. C. 1978. The effects of timber harvesting on the marten. M.S. thesis, University of Maine, Orono. 62 pp.

Soutiere, E. C. 1979. Effects of timber harvesting on marten in Maine. *J. Wildl. Manage.* 43:850–860.

Spencer, A. W.; Pettus, D. 1966. Habitat preferences of five sympatric species of long-tailed shrews. *Ecology* 47:677–683.

Spencer, H. E., Jr. 1961. The black bear and its status in Maine. *Bulletin* No. 4:1–55. Augusta, Me.: Department of Inland Fisheries and Game.

Spreadbury, B. R.; Musil, K.; Musil, J.; Kaisner, C.; Kovak, J. 1996. Cougar population characteristics in southeastern British Columbia. *J. Wildl. Manage.* 60:962–969.

Stabb, M. A.; Gartshore, M. E.; Aird, P. L. 1989. Interactions of southern flying squirrels, *Glaucomys volans,* and cavity-nesting birds. *Can. Field-Nat.* 103(3):401–403.

Stack, J. W. 1925. Courage shown by flying squirrel *Glaucomys volans. J. Mammal.* 6:128–129.

Starrett, A. 1954. Longtail shrew, *Sorex dispar,* in Maine. *J. Mammal.* 35.83–584.

Stebler, A. M. 1951. The ecology of Michigan coyotes and wolves. Ph.D. dissertation, University of Michigan, Ann Arbor. 198 pp.

Steele, M. A. 1998. *Tamiasciurus hudsonicus. Mammalian Species* 586:1–9. Shippensburg, Pa.: American Society of Mammalogists.

Steventon, J. D.; Major, J. T. 1982. Marten use of habitat in a commercially clear-cut forest. *J. Wildl. Manage.* 46:175–182.

Stewart, D. T.; Herman, T. B.; Teferi, T. 1989. Littoral feeding in a high-density insular population of *Sorex cinereus. Can. J. Zool.* 67:2074–2077.

Stocek, R. F. 1995. The cougar, *Felis concolor,* in the Maritime Provinces. *Can. Field-Nat.* 109:19–22.

Storm, G. L. 1965. Movements and activities of foxes as determined by radio-tracking. *J. Wildl. Manage.* 29:1–13.

Strickland, M. A.; Douglas, C. W. 1987. Marten. Pages 531–546. In: Novak, M.; Baker, J. A.; Obbard, M. E.; Malloch, B. (editors). *Wild furbearer management and conservation in North America.* Toronto: Ontario Ministry of Natural Resources and Ontario Trappers Association.

Strickland, M. A.; Douglas, C. M.; Novak, M.; Hunziger, N. P. 1982. Fisher. Pages 586–598. In: Chapman, J. A.; Feldhamer, G. A. (editors). *Wild mammals of North America—Biology, management, and economics.* Baltimore, Md.: Johns Hopkins University Press.

Stuewer, F. W. 1943. Raccoons: Their habits and management in Michigan. *Ecol. Monogr.* 13:203–257.

Sullivan, E. G. 1956. Gray fox reproduction, denning, range, and weights in Alabama. *J. Mammal.* 37:346–351.

Svendsen, G. E. 1982. Weasels. Pages 613–628. In: Chapman, J. A.; Feldhamer, G. A. (editors). *Wild mammals of North America—Biology, management, and economics.* Baltimore, Md.: Johns Hopkins University Press.

Svihla, A. 1930. Breeding habits and young of the red-backed mouse, *Evotomys. Pap. Mich. Acad. Sci. Arts Lett.* 11:485–490.

Svihla, A. 1932. A comparative life history study of the mice of the genus *Peromyscus. Univ. Mich. Misc. Publ. Mus. Zool.* 24:1–39.

Swihart, R. K. 1990. Common components of orchard ground cover selected as food by captive woodchucks. *J. Wildl. Manage.* 54:412–417.

Swihart, R. K. 1991. Influence of *Marmota monax* on vegetation in hayfields. *J. Mammal.* 72:791–795.

Swihart, R. K.; Picone, P. M. 1991. Effects of woodchuck activity on woody plants near burrows. *J. Mammal.* 72:607–611.

Tabor, J. E.; Wight, H. M. 1977. Population status of river otter in western Oregon. *J. Wildl. Manage.* 41:692–699.

Taube, C. M. 1947. Food habits of Michigan opossums. *J. Wildl. Manage.* 11:97–103.

Taylor, K. P.; Ballard, W. B. 1979. Moose movements and habitat use along the Susitna River near Devil's Canyon. *Proc. North Am. Moose Conf. Workshop* 15:169–186.

Teferi, T.; Herman, T. B. 1995. Epigeal movement by *Sorex cinereus* on Bon Portage Island, Nova Scotia. *J. Mammal.* 76:37–140.

Teferi, T.; Herman, T. B.; Stewart, D. T. 1992. Breeding biology of an insular population of the masked shrew (*Sorex cinereus* Kerr) in Nova Scotia. *Can. J. Zool.* 70:62–66.

Ternovsky, D. V. 1983. The biology of reproduction and development of the stoat *Mustela erminea* [Carnivora, Mustelidae]. *Zool. Zh.* 62:1097–1105.

Terres, J. K. 1939. Tree-climbing technique of a gray fox. *J. Mammal.* 20:256.

Thiel, R. P. 1985. The relationship between road densities and wolf habitat suitability in Wisconsin. *Am. Midl. Nat.* 113:404–407.
Thomas, D. W.; Fenton, M. B.; Barclay, R. M. R. 1979. Social behavior of the little brown bat, *Myotis lucifugus*. I. Mating behavior. *Behav. Ecol. Sociobiol.* 6:129–136.
Thomasma, L. E. 1996. Winter habitat selection and interspecific interactions of American martens (*Martes americana*) and fishers (*Martes pennanti*) in the McCormick wilderness and surrounding area. Ph.D. dissertation, Michigan Technological University, Houghton. 116 pp.
Thomasma, L. E.; Drummer, T. D.; Peterson, R. O. 1994. Modeling habitat selection by fishers. Pages 316–325. In: Buskirk, S. W.; Harestad, A. S.; Raphael, M. G.; Powell, R. A. (editors). *Martens, sables, and fishers: Biology and conservation.* Ithaca, N.Y.: Cornell University Press. 484 pp.
Thompson, I. D. 1994. Marten populations in uncut mature and post-logging boreal forests in Ontario. *J. Wildl. Manage.* 57:272–280.
Thompson, I. D.; Colgan, P. W. 1987. Numerical responses of martens to a food shortage in northcentral Ontario. *J. Wildl. Manage.* 51:824–835.
Thomson, C. E. 1982. *Myotis sodalis. Mammalian Species* 163: Â1–5. Shippensburg, Pa.: American Society of Mammalogists.
Thurber, J. M.; Peterson, R. O. 1991. Changes in body size associated with range expansion in the coyote (*Canis latrans*). *J. Mammal.* 72:750–755.
Tietje, W. D.; Ruff, R. L. 1980. Denning behavior of black bears in boreal forest of Alberta. *J. Wildl. Manage.* 44:858–870.
Tilghman, N. G. 1989. Impacts of white-tailed deer on forest regeneration in northwestern Pennsylvania. *J. Wildl. Manage.* 53:524–532.
Timm, R. M. 1975. Distribution, natural history, and parasites of mammals of Cook County, Minnesota. *Univ. Minn. Bell Mus. Nat. Hist. Occas. Pap.* 14. 56 pp.
Timm, R. M.; Heaney, L. R.; Baird, D. D. 1977. Natural history of rock voles (*Microtus chrotorrhinus*) in Minnesota. *Can. Field-Nat.* 91:177–181.
Tischendorf, J. W.; Ropski, S. J. 1996. *Proceedings of the Eastern Cougar Conference, 1994.* Ft. Collins, Colo.: American Ecological Research Institute. 245 pp.
Titcomb, J. W. 1901. Animal life in Vermont. *Vermonter* 5:2.
Todd, A. W.; Keith, L. B.; Fischer, C. A. 1981. Population ecology of coyotes during a fluctuation of snowshoe hares. *J. Wildl. Manage.* 45:629–640.
Toweill, D. E.; Tabor, J. E. 1982. The northern river otter. Pages 688–703. In: Chapman, J. A.; Feldhamer, G. A. (editors). *Wild mammals of North America—Biology, management, and economics.* Baltimore, Md.: Johns Hopkins University Press.
Trent, T. T.; Rongstad, O. S. 1974. Home range and survival of cottontail rabbits in southwestern Wisconsin. *J. Wildl. Manage.* 38:459–472.
Tryon, C. A.; Snyder, D. P. 1973. Biology of the eastern chipmunk, *Tamias striatus:* Life tables, age distributions, and trends in population numbers. *J. Mammal.* 54:145–168.
Tumlison, R. 1987. *Felis lynx. Mammalian Species* 269:1–8. Shippensburg, Pa.: American Society of Mammalogists.
Tyndale-Biscoe, C. H. 1973. *Life of marsupials.* New York: American Elsevier. 254 pp.
Van Ballenberghe, V. 1983. Two litters raised in one year by a wolf pack. *J. Mammal.* 64:171–172.
Van Ballenberghe, V.; Ballard, W. B. 1994. Limitation and regulation of moose populations: The role of predation. *Can. J. Zool.* 72:2071–2077.
Van Deelen, T. R.; Campa, H., III; Hamady, M.; Haufler, J. B. 1998. Migration and seasonal range dynamics of deer using adjacent deeryards in northern Michigan. *J. Wildl. Manage.* 62:205–213.
Vander Haegen, W. M.; DeGraaf, R. M. 1996. Predation on artificial nests in forested riparian buffer strips. *J. Wildl. Manage.* 60:542–550.
VanDruff, L. W. 1971. The ecology of the raccoon and opossum, with emphasis on their role as waterfowl nest predators. Ph.D dissertation, Cornell University, Ithaca, N.Y. 140 pp.
Van Dyke, F. G.; Brocke, R. H.; Shaw, H. G. 1986a. The use of road track counts as indices of mountain lion presence. *J. Wildl. Manage.* 50:102–109.
Van Dyke, F. G.; Brocke, R. H.; Shaw, H. G.; Ackerman, B. B.; Hemker, T. P.; Lindzey, F. G. 1986b. Reactions of mountain lions to logging and human activity. *J. Wildl. Manage.* 50:95–102.
van Zyll de Jong, C. G. 1966. Food habits of the lynx in Alberta and the Mackenzie District, N. W. T. *Can. Field-Nat.* 80:18–23.
van Zyll de Jong, C. G. 1983. *Handbook of Canadian mammals—Marsupials and insectivores.* Ottawa, Canada: National Museum of Natural Sciences. 210 pp.
van Zyll de Jong, C. G. 1985. *Handbook of Canadian mammals—Bats.* Ottawa, Canada: National Museum of Natural Sciences. 212 pp.
Vecillio, G. M.; Deldinger, R. D.; Cardoza, J. E. 1993. Status and management of moose in Massachusetts. *Alces* 29:1–7.
Vera, C. J.; Servello, F. A. 1994. Effects of paper mill sludge in spruce-fir forests on wildlife in Maine. *J. Wildl. Manage.* 58:719–727.
Verme, L. J.; Ozoga, J. J. 1981. Changes in small mammal populations following clear-cutting in upper Michigan conifer swamps. *Can. Field-Nat.* 95:253–256.
Verme, L. J.; Ullrey, D. E. 1984. Physiology and nutrition. Pages 91–118. In: Halls, L. K. (editor). *White-tailed deer: Ecology and management.* Harrisburg, Pa.: Stackpole Books and Wildlife Management Institute. 870 pp.
Verts, B. J. 1963. Movements and populations of opossums in a cultivated area. *J. Wildl. Manage.* 27:127–129.
Verts, B. J. 1967. *The biology of the striped skunk.* Urbana: University of Illinois Press. 218 pp.
Vickery, W. L.; Bider, J. R. 1978. The effect of weather on *Sorex cinereus* activity. *Can. J. Zool.* 56:291–297.
Villafuerte, R.; Litvaitis, J. A.; Smith, D. F. 1997. Physiological responses by lagomorphs to resource limitations imposed by habitat fragmentation: Implications for condition-sensitive predation. *Can. J. Zool.* 75:148–151.
Voigt, D. R. 1987. Red fox. Pages 378–392. In: Novak, M.; Baker, J. A.; Obbard, M. E.; Malloch, B. (editors). *Wild furbearer management and conservation in North America.* Toronto: Ontario Ministry of Natural Resources and Ontario Trappers Association.
Voigt, D. R.; Berg, W. E. 1987. Coyote. Pages 344–357. In: Novak, M.; Baker, J. A.; Obbard, M. E.; Malloch, B. (editors). *Wild furbearer management and conservation in North America.* Toronto: Ontario Ministry of Natural Resources and Ontario Trappers Association.

Voigt, D. R.; Broadfoot, J. 1983. Locating pup-rearing dens of red foxes with radio-equipped woodchucks. *J. Wildl. Manage.* 47:858–859.
Voigt, D. R.; Earle, B. D. 1983. Avoidance of coyotes by red fox families. *J. Wildl. Manage.* 47:852–857.
Voigt, D. R.; Macdonald, D. W. 1984. Variation in the spatial and social behavior of the red fox, *Vulpes vulpes. Acta Zool. Fenn.* 171:261–265.
Voigt, D. R.; Tinline, R. L. 1980. Strategies for analyzing radio tracking data. Pages 387–404. In: Amlaner, C. J., Jr.; Macdonald, D. W. (editors). *A handbook on biotelemetry and radio tracking.* Oxford, U.K.: Pergamon Press.
Vonhof, M. J. 1996. Roost-site preferences of big brown bats (*Eptesicus fuscus*) and silver-haired bats (*Lasionycteris noctivagans*) in the Pend d'Oreille Valley in southern British Columbia. Pages 62–79. In: Barclay, R. M. R.; Brigham, R. M. (editors). *Bats and forests symposium,* October 19–21, 1995, Victoria, British Columbia, Canada. Research Branch, British Columbia Ministry of Forestry, Victoria, B. C. Work. Pap. 23/1996.
Wade-Smith, J.; Richmond, M. E. 1978. Reproduction in captive striped skunks (*Mephitis mephitis*). *Am. Midl. Nat.* 100:452–455.
Wade-Smith, J.; Richmond, M. E.; Mead, R. A.; Taylor, H. 1980. Hormonal and gestational evidence for delayed implantation in the striped skunk, *Mephitis mephitis. Gen. Comp. Endocrinol.* 42:509–515.
Wade-Smith, J.; Verts, B. J. 1982. *Mephitis mephitis. Mammalian Species* 173:1–7. Shippensburg, Pa.: American Society of Mammalogists.
Wai-Ping, V.; Fenton, M. B. 1988. Nonselective mating in little brown bats (*Myotis lucifugus*). *J. Mammal.* 69:641–645.
Walker, E. P.; Warnick, F.; Lange, K. I.; Vible, H. E.; Hamlet, S. E.; Davis, M. A.; Wright, P. F. 1975. *Mammals of the world.* 2 vols. Baltimore, Md.: Johns Hopkins University Press. 1,500 pp.
Waller, D. M.; Alverson, W. S. 1997. The white-tailed deer: A keystone herbivore. *Wildl. Soc. Bull.* 25:217–226.
Weigl, P. D.; Osgood, D. W. 1974. Study of the northern flying squirrel, *Glaucomys sabrinus,* by temperature telemetry. *Am. Midl. Nat.* 92:482–486.
Weller, M. W. 1978. Management of freshwater marshes for wildlife. Pages 267–284. In: Good, R. E.; Whigham, D. F.; Simpson, R. L. (editors). *Ecological processes and management potential.* New York: Academic Press.
Wells-Gosling, N. 1985. *Flying squirrels: Gliders in the dark.* Washington, D.C.: Smithsonian Institution Press. 128 pp.
Wells-Gosling, N.; Heaney, L. R. 1984. *Glaucomys sabrinus. Mammalian Species* 220:1–8. Shippensburg, Pa.: American Society of Mammalogists.
Wetzel, R. M.; Gunderson, H. L. 1949. The lemming vole, *Synaptomys borealis,* in northern Minnesota. *J. Mammal.* 30:437.
Whitaker, J. O., Jr. 1963a. A study of the meadow jumping mouse, *Zapus hudsonius* Zimmerman, in central New York. *Ecol. Monogr.* 33:215–254.
Whitaker, J. O., Jr. 1963b. Food, habitat and parasites of the woodland jumping mouse in central New York. *J. Mammal.* 44:316–321.
Whitaker, J. O., Jr. 1967. Hoary bat apparently hibernating in Indiana. *J. Mammal.* 48:663.
Whitaker, J. O., Jr. 1972a. Food habits of bats from Indiana. *Can. J. Zool.* 50:877–883.
Whitaker, J. O., Jr. 1972b. *Zapus hudsonius. Mammalian Species* 11:1–7. Shippensburg, Pa.: American Society of Mammalogists.
Whitaker, J. O., Jr. 1974. *Cryptotis parva. Mammalian Species* 43:1–8. Shippensburg, Pa.: American Society of Mammalogists.
Whitaker, J. O., Jr. 1995. Food of the big brown bat *Eptesicus fuscus* from maternity colonies in Indiana and Illinois. *Am. Midl. Nat.* 134:346–360.
Whitaker, J. O., Jr. 1998. Life history and roost switching in six summer colonies of eastern pipistrelles in buildings. *J. Mammal.* 79:651–659.
Whitaker, J. O., Jr.; French, T. W. 1984. Foods of six species of sympatric shrews from New Brunswick. *Can. J. Zool.* 62:622–626.
Whitaker, J. O., Jr.; Gummer, S. L. 1992. Hibernation of the big brown bat in buildings. *J. Mammal.* 73:312–316.
Whitaker, J. O., Jr.; Hamilton, W. J., Jr. 1998. *Mammals of the eastern United States.* 3rd ed. Ithaca, N.Y.: Cornell University Press. 583 pp.
Whitaker, J. O., Jr.; Mumford, R. E. 1972. Notes on occurrence and reproduction of bats in Indiana. *Proc. Indiana Acad. Sci.* 81:376–383.
Whitaker, J. O., Jr.; Schmeltz, L. L. 1973. Food and external parasites of *Sorex palustris* and food of *Sorex cinereus* from St. Louis County, Minnesota. *J. Mammal.* 54:283–285.
Whitaker, J. O., Jr.; Wrigley, R. E. 1972. *Napaeozapus insignis. Mammalian Species* 14:1–6. Shippensburg, Pa.: American Society of Mammalogists.
Whitney, L. F.; Underwood, A. B. 1952. *The raccoon.* Orange, Conn.: Practical Science.
Willey, C. H. 1978. *The Vermont black bear.* Vermont Fish and Game Department. Federal Aid Wildlife Restoration Project W-8-R. 73 pp.
Willner, G. R.; Feldhamer, G. A.; Zucker, E. E.; Chapman, J. A. 1980. *Ondatra zibethicus. Mammalian Species* 141:1–8. Shippensburg, Pa.: American Society of Mammalogists.
Wilson, D. E.; Reeder, D. M. (editors). 1993. *Mammal species of the world.* Washington, D.C.: Smithsonian Institution Press. 1,206 pp.
Wimsatt, W. A. 1945. Notes on breeding behavior, pregnancy and parturition in some Vespertilionid bats of the eastern United States. *J. Mammal.* 26:23–33.
Wimsatt, W. A. 1963. Delayed implantation in the Ursidae, with particular reference to the black bear (*Ursus americanus* Pallus). Pages 49–76. In Enders, A. C. (editor). *Delayed implantation.* Chicago, Ill.: University of Chicago Press.
Winchell, J. M.; Kunz, T. H. 1996. Day-roosting activity budgets of the eastern pipistrelle bat, *Pipistrellus subflavus* (Chiroptera: Vespertiliondae). *Can. J. Zool.* 74:431–441.
Winegarner, C. E.; Winegarner, M. S. 1982. Reproductive history of a bobcat. *J. Mammal.* 63:680–682.
Wiseman, G. L.; Hendrickson, G. O. 1950. Notes on the life history and ecology of the opossum in southeast Iowa. *J. Mammal.* 31:331–337.
Wobeser, G. A. 1966. Ecology of the long-tailed weasel (*Mustela frenata novaboracensis* [Emmons] in Rondeau Park, Ontario. M.S. thesis, University of Guelph, Guelph, Ontario. 101 pp.
Wolff, J. O. 1980. The role of habitat patchiness in the population dynamics of snowshoe hares. *Ecol. Monogr.* 50(1):111–130.

Wolff, J. O. 1985. The effects of density, food, and interspecific interference on home range size in *Peromyscus leucopus* and *Peromyscus maniculatus*. *Can. J. Zool.* 63:2657–2662.

Wood, J. E. 1958. Age structure and productivity of a gray fox population. *J. Mammal.* 39:74–86.

Woods, C. A. 1973. *Erethizon dorsatum. Mammalian Species* 29:1–6. Shippensburg, Pa.: American Society of Mammalogists.

Wren, C. D.; Stokes, P. M.; Fischer, K. L. 1986. Mercury levels in Ontario mink and otter relative to food levels and environmental acidification. *Can. J. Zool.* 64:2854–2859.

Wright, B. 1973. The cougar is alive and well in Massachusetts. *Mass. Wildl.* 24(3):2–8.

Wright, B. S. 1948. Survival of the northeastern panther (*Felis concolor*) in New Brunswick. *J. Mammal.* 29:235–246.

Wright, B. S. 1953. Further notes on the panther in the northeast. *Can. Field-Nat.* 67:12–28.

Wright, B. S. 1959. *The ghost of North America—The story of the eastern panther.* New York: Vantage Press. 140 pp.

Wright, B. S. 1961. The latest specimen of the eastern puma. *J. Mammal.* 42:144–148.

Wright, P. L. 1942. Delayed implantation in the long-tailed weasel (*Mustela frenata*), the short-tailed weasel (*Mustela cicognani*), and the marten (*Martes americana*). *Anat. Rec.* 83:341–353.

Wright, P. L. 1947. The sexual cycle of the male long-tailed weasel (*Mustela frenata*). *J. Mammal.* 28:343–352.

Wright, P. L. 1948. Breeding habits of captive long-tailed weasels (*Mustela frenata*). *Am. Midl. Nat.* 39:338–344.

Wright, P. L. 1950. *Synaptomys borealis* from Glacier National Park, Montana. *J. Mammal.* 31:460.

Wright, P. L.; Coulter, M. W. 1967. Reproduction and growth in Maine fishers. *J. Wildl. Manage.* 31:70–86.

Wrigley, R. E. 1969. Ecological notes on the mammals of southern Quebec. *Can. Field-Nat.* 83:201–211.

Wrigley, R. E. 1972. Systematics and biology of the woodland jumping mouse, *Napaeozapus insignis*. *Ill. Biol. Monogr.* 47. Urbana: University of Illinois. 117 pp.

Wrigley, R. E.; Dubois, J. E.; Copland, H. W. R. 1979. Habitat, abundance and distribution of six species of shrews in Manitoba. *J. Mammal.* 60:505–520.

Wydeven, A. P.; Fuller, T. K.; Weber, W.; MacDonald, K. 1998. The potential for wolf recovery in the northeastern United States via dispersal for southeastern Canada. *Wildl. Soc. Bull.* 26:776–784.

Wynne, K. M.; Sherburne, J. A. 1984. Summer home range use by adult marten in northwestern Maine. *Can. J. Zool.* 62:941–943.

Yahner, R. H. 1978. The adaptive nature of the social system and behavior in the eastern chipmunk, *Tamias striatus*. *Behav. Ecol. Sociobiol.* 3:397–427.

Yahner, R. H. 1980. Burrow system use by red squirrels. *Am. Midl. Nat.* 103:409–411.

Yahner, R. H. 1987. Feeding-site use by red squirrels, *Tamiasciurus hudsonicus*, in a marginal habitat in Pennsylvania. *Can. Field-Nat.* 101(4):586–589.

Yamasaki, M. 1997. White Mountain National Forest Small Mammal Identification and Collection Report—1996. Unpublished report on file. U.S. Department of Agriculture, Forest Service, Northeastern Forest Experiment Station, Durham, N.H. 38 pp.

Yates, T. L.; Schmidly, D. J. 1978. *Scalopus aquaticus. Mammalian Species* 105:1–4. Shippensburg, Pa.: American Society of Mammalogists.

Yeager, L. E. 1938. Tree climbing by a gray fox. *J. Mammal.* 19:376.

Yearsley, E. F.; Samuel, D. E. 1980. Use of reclaimed surface mines by foxes in West Virginia. *J. Wildl. Manage.* 44:729–734.

Yerger, R. W. 1953. Home range, territoriality and populations of the chipmunk in central New York. *J. Mammal.* 34:448–458.

Yerger, R. W. 1955. Life history notes on the eastern chipmunk, *Tamias striatus lysteri* Richardson, in central New York. *Am. Midl. Nat.* 53:312–323.

Yodzis, P.; Kolenosky, G. B. 1986. A population dynamics model of black bears in eastcentral Ontario. *J. Wildl. Manage.* 50:602–612.

York, E. C. 1996. Fisher population dynamics in north-central Massachusetts. M.S. thesis, University of Massachusetts, Amherst.

Young, S. P.; Goldman, E. A. 1946. *The puma, mysterious American cat.* Washington, D.C.: American Wildlife Institute. 358 pp.

Young, S. P.; Jackson, H. H. T. 1951. *The clever coyote.* Washington, D.C.: Wildlife Management Institute. 411 pp.

Part II

Species/Habitat Relationships

Using the Matrices

Species/habitat matrices present summary information in a simple, condensed, tabular form (fig. 6, p. 396). These matrices show the typical habitats used by the various species in New England and reveal the importance of management in providing habitat for early-successional forest and non-forest species. The information is based upon the literature reviewed in the species accounts. Two sets of matrices are provided, one for the forest cover types and groups, and another for the nonforest types—terrestrial, wetland and deepwater, and other habitat types.

Forest Cover Types and Groups

Eleven forest cover types are used to describe forest/wildlife habitat associations. These eleven forest cover types are based on descriptions in *Forest Cover Types of the United States and Canada* (Eyre 1980). Similar types are grouped, especially as they reflect similarities in wildlife species composition and responses to silvicultural treatment, into six cover-type groups, as in *New England Wildlife: Management of Forested Habitats* (DeGraaf et al. 1992). These cover-type groups relate well to standard silvicultural prescriptions and are readily useful to foresters.

Forest development is indicated by size class as follows:

S *Regeneration through seedlings:* Live trees and associated vegetation less than 1 in (2.5 cm) dbh (diameter breast height) and at least 1 ft (30 cm) tall.

Sp *Sapling through poletimber:* Saplings are live trees 1.0 to 3.9 in (2.5 to 9.9 cm) dbh; poles are live trees 4.0 to 8.9 in (10.0 to 22.0 cm) dbh for softwoods and 4.0 to 11.9 in (10.0 to 30.0 cm) dbh for hardwoods. The matrix assumes that stands are fully stocked, that is, contain approximately 75 square feet of basal area per acre.

St *Sawtimber:* A stand with at least half of the stocking in sawtimber-size trees—at least 9.0 in (23 cm) dbh for softwoods or 12.0 in (31 cm) for hardwoods.

L *Large sawtimber:* A stand with at least half of the stocking in large sawtimber trees—at least 20 in (51 cm) dbh for softwoods and 24 in (61 cm) dbh for hardwoods.

U *Uneven-aged:* Stands of northern hardwood cover types that contain trees of all size-classes.

These size-classes apply to all forest cover types under even-age management with one exception. Only in the northern hardwoods cover-type group do we list wildlife habitat associations for uneven-aged stands.

Common and scientific names of trees follow the *Checklist of United States Trees* (Little 1979). Names of understory plants follow *Gray's Manual of Botany* (Fernald 1950).

The forest cover groups and their respective types are as follows.

Aspen-Birch

Aspen. This type includes quaking aspen (*Populus tremuloides*) and bigtooth aspen (*P. grandidentata*), but in New England, quaking aspen is more likely to occur in

Aspen. Twin Mountain, New Hampshire, August 1985.
Photo: R. M. DeGraaf.

Paper birch. Gorham, New Hampshire, August 1985.
Photo: R. M. DeGraaf.

pure stands. Common associates are paper birch (*Betula papyrifera*) and pin cherry (*Prunus pensylvanica*), which, when occurring in admixture, die out early. These species occur on a variety of sites and soil types. The aspen type occurs on most soils except very dry sands or very wet swamps. Aspen is unique in that almost all stands originate as suckers arising from existing root systems. It will sometimes reproduce from seed on burns, clear-cuts, and other scarified sites.

Aspen is a relatively short-lived pioneer type—it does not reproduce under its own shade. On dry sites aspen is replaced by red pine (*Pinus resinosa*), red maple (*Acer rubrum*), or oaks (*Quercus* sp.); on mesic sites by white pine (*P. strobus*); on fertile sites by northern hardwoods, and on fertile wet sites by balsam fir (*Abies balsamea*) (Brinkman and Roe 1975).

Paper birch. Paper birch is pure or dominant. Associated species include quaking and bigtooth aspen, balsam fir, red spruce (*Picea rubens*), white pine, yellow birch (*B. alleghaniensis*), and in southern New England, eastern hemlock (*Tsuga canadensis*). The type pioneers on burned areas and clear-cuts, and grows best on deep fertile, well-drained sites. Raspberries and blackberries (*Rubus* sp.) make up a high proportion of the ground cover at the time of establishment of paper birch stands. These are shaded out in about 10 years, but pin cherry can persist for 30 or more years. Paper birch is succeeded by spruce-fir in northern parts of its range, and to the south by northern hardwoods and hemlock on fertile, well-drained sites (Safford 1980).

Northern Hardwoods

Northern hardwoods. This category includes sugar maple/ash, sugar maple/beech/yellow birch, and beech/red maple cover types. True northern hardwoods are dominated by sugar maple (*Acer saccharum*), beech (*Fagus grandifolia*), and yellow birch and occur widely as a pure type in northern New England. They grade into a mixed hardwood or transition type in southern New England; associated species throughout the region include basswood (*Tilia americana*), red maple, hemlock, white ash (*Fraxinus americana*), white pine, balsam fir (*Abies balsamea*), black cherry (*P. serotina*), paper birch, sweet birch (*B. lenta*), and red spruce. Northern hardwood is the basic hardwood type in northern New England, and occurs to an elevation of 2,500 ft (760 m). It prefers fertile loamy soils and good moisture conditions. Striped maple (*A. pensylvanicum*), witch-hazel (*Hamamelis virginiana*), and hobblebush (*Viburnum alnifolium*) are common in the understory throughout the region. Best development of the type occurs on moist, fertile, well-drained loamy soils. On drier sites, beech becomes more prominent. On wetter sites, the type blends into a red maple/yellow birch/hemlock or a red spruce mixture. The type tends to be climax. From New England to Pennsylvania, the beech *Nectria* complex has gradually reduced the proportion of beech in many stands (Berglund 1980).

Northern hardwoods (sugar maple, beech, yellow birch). Berlin, New Hampshire, July 1984. Photo: R. M. DeGraaf.

Swamp Hardwoods

Red maple. Red maple is pure or dominant. In New England, red maple and associated species are common on wet sites; the type is essentially pure in southern New England. Associates are yellow birch, balsam fir, and sugar maple in northern New England, and black gum (*Nyssa sylvatica*), sycamore (*Platanus occidentalis*), and silver maple (*A. saccharinum*) in southern New England. In New England and the Upper Peninsula of Michigan, it occupies moist to wet muck or peat soils in swamps, depressions, or along sluggish streams, often found as an inclusion in northern hardwoods on wetter sites (Powell and Erdmann 1980). It can be readily differentiated from northern hardwoods by the absence of beech and the increased proportion of yellow birch and red spruce.

Spruce-Fir

Balsam fir. Balsam fir is characteristically pure or predominant. There are many associates, mostly on moist or wet-site soils in northern New England; these include paper birch, quaking and bigtooth aspen, red spruce, and in swamps, northern white cedar (*Thuja occidentalis*). In southern New England, hemlock and red maple are common associates. The type is common in northern New England, occurring on upland sites, on low-lying moist flats, and in swamps. Pure stands result (usually) from heavy cutting, blowdown, or following infestation of spruce budworm. This type is common in northern New England, and may be climax in the zone below timberline. Only black spruce (*P. mariana*) grows above it (Westveld 1953).

Red maple. Amherst, Massachusetts, July 1985.
Photo: R. M. DeGraaf.

Balsam fir. West Milan, New Hampshire, August 1985.
Photo: R. M. DeGraaf.

The type occurs extensively in Quebec, where five distinct subtypes are recognized. In the United States, the type is not as complex; however, balsam fir is an important component in the following types in northern New England: red spruce/balsam fir, black spruce, aspen, and paper birch. Common understory species include speckled alder (*Alnus rugosa*), mountain maple (*Acer spicatum*), and pin cherry, among large shrubs and small trees. Low understory plants include Canada yew (*Taxus canadensis*), red raspberry, blueberries (*Vaccinium* sp.), and hobblebush (Frank et al. 1980).

Red spruce/Balsam fir. The type may consist of red spruce and balsam fir or may predominate in a mixture of associated trees—the composition varies by site and disturbance history. We include here the northern white-cedar type and associates, which are commonly associated in northern New England. This is a northern New England type, occupying moderately to poorly drained flats, but not swamps. Associates are red maple, paper birch and yellow birch, and aspens primarily, but also white pine, hemlock, and occasionally black spruce and tamarack (*Larix laricina*).

The type occurs near sea level in eastern Maine, from an elevation of 2,400 to 4,500 ft (730 to 1,370 m) in the White Mountains of New Hampshire, from an elevation of 2,500 to 3,800 ft (760 to 1,160 m) in the Green Mountains of Vermont, and occurs on the tops of some of the higher Berkshire Hills in western Massachusetts.

Red spruce–balsam fir. West Milan, New Hampshire, August 1985. Photo: R. M. DeGraaf.

Red spruce. Mt. Tabor, Vermont, June 1948. U.S. Forest Service Photo.

The type occurs on two kinds of sites in New England: (1) poorly drained flats and ridges or benches at lakeshores, streams, and swamps and bogs; and (2) well-drained to dry, shallow soils on steep, rocky, upper mountain slopes. Stands are usually very dense; the ground may be essentially devoid of plants except for mosses and a few seedlings of red spruce and balsam fir. Regenerating stands, however, produce a thick growth of blueberry, creeping snowberry (*Symphoricarpos mollis*), mountain holly (*Nemopanthus mucronata*), raspberry, and downy serviceberry (*Amelanchier arborea*), among others (Griffin 1980).

Red spruce. Red spruce is pure or accounts for a majority of the stocking; common associates in northern New England are balsam fir, paper birch, and yellow birch; others include sugar maple, red maple, mountain-ash (*Sorbus americana*), eastern white pine, and eastern hemlock. Red spruce occurs near sea level in eastern Maine and from an elevation of 1,500 to 4,500 ft (450 to 1,370 m) inland throughout northern New England on moderately well-drained to poorly drained flats (but not true swamps), and on well-drained slopes, including thin soils on upper slopes. Red spruce pioneers on abandoned fields and pastures in northern New England, and on these fairly well-drained sites it is usually replaced by shade-tolerant hardwoods, especially sugar maple and beech. Red spruce is long-lived and, barring major disturbance, is very stable. Older stands develop an uneven-aged character even though of even-aged origin. The understory is frequently sparse, or even absent; the ground beneath red spruce stands is covered with tree litter and patches of short-lived red spruce seedlings. Old-field red spruce contains a ground cover of bunchberry (*Cornus canadensis*) on wet sites and hobblebush on well-drained sites. Regenerating stands usually produce raspberries in abundance (Blum 1980).

Hemlock

Eastern hemlock. Eastern hemlock is pure or predominant over any associate, but associates are numerous; these commonly include beech, sugar maple, yellow birch, red

Eastern hemlock (virgin stand). Petersham, Massachusetts, August 1985. Photo: R. M. DeGraaf.

maple, black cherry, white pine, northern red oak (*Quercus rubra*), white oak (*Q. alba*), sweet birch, and in northern New England, paper birch, balsam fir, and red spruce. In southern New England, the type prefers cool locations such as moist ravines and north-facing slopes; in the northern parts of its New England distribution, warmer, drier sites are tolerated. It occurs from sea level to elevations of 3,000 ft (915 m) in New England.

Eastern hemlock is very shade tolerant. Its long life span and ability to respond to release after two centuries of suppression have allowed the type to persist; early logging, and the fires that followed, greatly reduced the occurrence of this shallow-rooted climax species. Under mature stands, understory development is sparse; openings to admit light commonly produce striped maple, hobblebush, and mapleleaf viburnum (*Viburnum acerifolium*), among others. False lily-of-the-valley (*Maianthemum canadense*) is probably the most common herb (Wiant 1980).

White pine–northern red oak–red maple. Belchertown, Massachusetts, August 1985. Photo: R. M. DeGraaf.

Oak-Pine

Northern red oak. Northern red oak accounts for a majority of the stocking. Associates vary according to site and locale, and include black oak (*Q. velutina*), scarlet oak (*Q. coccinea*), and chestnut oak (*Q. prinus*), hickories (*Carya* sp.), and red maple. In New England, the type has a spotty distribution, occupying ridge crests and upper north slopes. On better sites, associates are black cherry, sugar maple, white ash, and beech. The type is rare in northern New England and reaches best development in New England in western Massachusetts and northern Connecticut on loam and silt-loam soils. The type is subclimax; shade-tolerant species such as beech and sugar maple increase in proportion over time (Trimble 1980).

Northern red oak. Ware, Massachusetts, August 1985. Photo: R. M. DeGraaf.

White pine/Northern red oak/Red maple. Northern red oak, eastern white pine, and red maple predominate; white ash is the most common associate, but others include paper, yellow, and sweet birches, sugar maple, beech, hemlock, and black cherry. This type ocurs across southern and central New England to elevations of 1,500 ft (450 m), generally on deep, well-drained fertile soils. This type is common in the transition between northern hardwoods and spruce-fir types in northern New England, and between northern hardwoods and oak types—characteristic of central types—in southern New England. The type often follows "old field" white pine in New England, where hardwood seedlings and saplings form the understory (Baldwin and Ward 1980). Common understory shrubs include witch-hazel, alternate-leaf dogwood (*Cornus alternifolia*), mapleleaf viburnum, and mountain-laurel (*Kalmia latifolia*).

Eastern white pine. Eastern white pine is pure or usually predominant. We include red pine, which has a spotty distribution throughout New England on sandy, gravelly, or sandy loam soils, and white pine/hemlock, a common subtype in central and southern New England, where it occupies a range of soil types in cool locations such as ravines and north slopes (in the southern parts of its range). These

Eastern white pine. Sunderland, Massachusetts, July 1985.
Photo: R. M. DeGraaf.

Pitch pine. Montague, Massachusetts, August 1985.
Photo: R. M. DeGraaf.

other pine types are included primarily because they support similar wildlife communities.

Eastern white pine frequently occurs in pure stands; common New England associates on light soils are pitch pine (*P. rigida*), gray birch (*B. populifolia*), quaking aspen, bigtooth aspen, red maple, and white oak. On heavier soils, paper, yellow, and sweet birches, white ash, black cherry, northern red oak, sugar maple, hemlock, red spruce, and northern white cedar are associated in New England, but none are characteristic. The type is widespread in central New England from sea level to elevations of 2,500 ft (760 m). This type occurs over a wide range of conditions and sites; establishment is often easier on poor sites because hardwood competition is less. Once established on better sites, white pine will usually grow faster than hardwoods.

White pine is a common pioneer on abandoned agricultural land in New England. The type seldom succeeds itself, but on dry, sandy soils it may persist a long time and even approach permanence. On heavier soils, white pine is usually succeeded by northern hardwoods, white pine/hemlock, or white oak.

Eastern white pine is a major component of two other New England forest cover types—white pine/northern red oak/red maple, and white pine/hemlock—and occurs in various proportions in other types throughout the region.

In pure or almost pure white pine stands, the understory is composed primarily of ericaceous shrubs such as blueberries, huckleberries (*Gaylussacia* sp.), azaleas (*Azalea* sp.), and mountain-laurel. In New England, common ladyslipper (*Cypripedium* sp.) is common on light soils and highbush blueberry (*V. corymbosum*) on wetter sites (Wendel 1980).

Terrestrial, Wetland and Deepwater, and Other Nonforest Habitat Types

Twenty-seven nonforest cover types are used to present forest/wildlife habitat associations. Terrestrial nonforested cover types are based on descriptions in *Fisheries &*

Hayfield. Guildhall, Vermont, August 1987. Photo: R. M. DeGraaf.

Pasture. Amherst, Massachusetts, 1985. Photo: R. M. DeGraaf.

Orchard. Amherst, Massachusetts, May 1990. Photo: R. M. DeGraaf.

Wildlife Habitat Management Handbook (U.S. Dept. of Agriculture, Forest Service, Eastern Region, 1975, Milwaukee, Wis.) and *Land Resources in New Hampshire* (U.S. Department of Agriculture, Soil Conservation Service, 1982, Durham, N.H., 43 pp.). Wetland and deepwater types are based on descriptions in Cowardin et al. (1979). Similar types are grouped, especially when they reflect similarities in wildlife species composition and responses to treatments that comprise the nonforested habitat matrix. Other nonforest types are based on smaller scale habitats that are not usually typed on maps, such as stable banks, cliffs, caves, and structures, but offer distinct habitat elements for a variety of wildlife species.

The nonforested habitat types are as follows:

Terrestrial

Upland fields.

Cultivated—tilled agricultural cropland.

Grass—hayfields, etc.

Forb—broadleaved herbaceous cover [e.g., goldenrod (*Solidago*), sensitive fern (*Onoclea*)].

Shrub/old fields—abandoned agricultural fields reverting to forest, characterized by grasses, shrubs, and small trees.

Pastures—grazing lands, usually too wet or rocky for cultivation.

Savanna. Grasslands with shrubs and widely and irregularly scattered trees, which result from either soil-moisture regimes or disturbances such as fire or grazing.

Orchard. Old or contemporary plantings of fruit trees, usually with grassy ground cover.

Krummholz zone. The transition zone from subalpine forest to alpine tundra, characterized by dwarfed, deformed, and wind-sheared trees.

Alpine zone. The elevated slopes above the timberline, characterized by low, shrubby, slow-growing woody plants and ground cover of boreal lichens, sedges, and grasses.

Krummholz. Kilkenny, New Hampshire, June 1989. Photo: R. M. DeGraaf.

Alpine habitat. White Mountains, New Hampshire, August 1985. Photo: R. M. DeGraaf.

Wetlands/Deepwater

In general, wetlands are lands where saturation with water largely determines the nature of soil development and the types of plant and animal communities living in the soil and on its surface (Cowardin et al. 1979). The dominant plants are hydrophytes. The single feature that most wetlands share is soil or substrate that is at least periodically saturated or covered by water. Wetlands are transition zones between terrestrial and aquatic systems, where the water table is usually at or near the ground surface, or where the land is covered by shallow water.

Deepwater habitats are permanently flooded lands lying below the deepwater boundary of wetlands (Cowardin et al. 1979). Deepwater habitats include environments where surface water is permanent and often deep, so that water, rather than air, is the principal medium within which the dominant organisms live, whether or not they are attached to the substrate. As in wetlands, the dominant plants are hydrophytes; however the water is generally too deep to support emergent vegetation.

Palustrine. These are nontidal wetlands dominated by emergent mosses, lichens, persistent emergents, shrubs, or trees.

Sedge meadow—dominated by sedges (*Carex*), cattails (*Typha*), etc.; surface water depths to 6 in (15 cm) in winter and early spring; soil surface exposed but saturated in summer. Typed as fen or northern sedge meadow by Curtis (1959) and persistent emergent wetland in Cowardin et al. (1979).

Shallow marsh—characterized by persistent emergent herbaceous hydrophytes and water depths to 1.5 ft (45 cm); tend to maintain the same appearance as the years pass. Typed as inland shallow fresh marsh by Golet and Larson (1974) and persistent emergent wetland by Cowardin et al. (1979).

Deep marsh—characterized by emergent and floating-leaved plants and water depths to 6 ft (1.8 m). Typed as inland deep fresh marsh by Golet and Larson (1974) and emergent wetland and aquatic bed by Cowardin et al. (1979).

Shrub swamp—dominated by woody vegetation less than 20 ft (6 m) tall; soil seasonally or permanently flooded to a depth of 1 ft (30 cm). Typical woody species include alders (*Alnus* sp.), buttonbush (*Cephalanthus occidentalis*), and red osier dogwood (*Cornus stolonifera*). Typed as shrub swamp by Golet and Larson (1974) and scrub-shrub wetland by Cowardin et al. (1979).

Bog—characterized by peat accumulation due to cold, acidic conditions; sometimes a floating mat of *Sphagnum,* which tends to be dominant; sundew (*Drosera* sp.) and pitcher plant (*Sarracenia* sp.) are generally common. Woody species often include Labrador tea (*Ledum groenlandicum*), bog rosemary (*Andromeda glaucophylla*), bog laurel (*Kalmia polifolia*), and leatherleaf (*Chamaedaphne calyculata*), among others. Typed as bog by Dansereau and Segadas-vianna (1952) and scrub-shrub wetland, forested wetland, and moss-lichen wetland by Cowardin et al. (1979).

Pond—permanent palustrine water body, characterized by emergent and/or floating-leaved plants, up to 20 acres (8 ha) in size. Typed as open water by Golet and Larson (1974) and aquatic bed, unconsolidated bottom, and emergent wetland by Cowardin et al. (1979).

Palustrine wetland–shrub swamp. Berlin, New Hampshire, July 1983. Photo: R. M. DeGraaf.

Palustrine wetland–beaver pond. New Salem, Massachusetts, June 1990. Photo: R. M. DeGraaf.

Lacustrine. Deepwater habitats with all of the following characteristics: (1) situated in a topographic depression or a dammed river channel; (2) lacking trees, shrubs, persistent emergents, emergent mosses or lichens with greater than 30 percent areal coverage; and (3) total area exceeds 20 acres (8 ha).

Lake—characterized by a littoral zone to depths of 6.5 ft (2 m) and a deeper limnetic zone. Typed as open water by Golet and Larson (1974) and aquatic bed, unconsolidated bottom, and emergent wetland by Cowardin et al. (1979).

Lake. Berlin, New Hampshire, July 1984. Photo: R. M. DeGraaf.

Riverine. Wetland and deepwater habitats that are contained within a channel through which water flows.

Stream—characterized as intermittent or permanent, up to 30 cubic feet per second at high flow. Typed as intermittent and upper perennial systems with unconsolidated bottom and shore, aquatic bed, and emergent, scrub-shrub and forested wetlands by Cowardin et al. (1979).

River—characterized by low- and high-gradient nontidal flows of at least 30 cubic feet per second at low flow. Typed as lower and upper perennial systems by Cowardin et al. (1979).

Riparian zone—stream and river floodplains and associated woody, shrubby, and herbaceous vegetation.

Estuarine. Deepwater tidal habitats and adjacent tidal wetlands that are usually semienclosed by land but have open, partly obstructed, or sporadic access to the open ocean, and in which ocean water is at least occasionally diluted by freshwater runoff from land.

Estuary/salt marsh—characterized as an intertidal, low-energy system in terms of wave action (Cowardin et al. 1979).

Marine. Marine habitats described in Cowardin et al. (1979) are condensed in this publication into two general types. These are:

Coastal beach/rocks—unconsolidated and rocky shore subject to tidal and wave action (Cowardin et al. 1979).

Bay, ocean—open ocean overlying the continental shelf and the exposed coastline.

Other Nonforest Habitat Types

Stable banks. Excavated sand or gravel banks or naturally cut stream banks topped by an overhanging grassy top.

Ledge/cliff. Exposed bedrock, cliffs, talus accumulation.

Cave, mine. Natural caves and old mine shafts characterize this specialized habitat type.

Structure, building. Bridge foundations, trestles, cribwork, etc.

Derelict building, debris. Abandoned buildings and piles of scrap material characterize this specialized habitat type.

Bank swallow colony. Bellows Falls, Vermont, August 1990.
Photo: R. M. DeGraaf.

Special Habitat Features

Special habitat features are listed for many species. These features are considered to be essential for that species to occur regularly or to reproduce. Many species are generally associated with a given forest-type group, but the presence of a specific habitat feature is critical to their occurrence. For example, cavity-nesting waterfowl use a variety of forest cover types but the special habitat feature(s), in this particular case water, must also be present. Thus, the species/habitat associations must be viewed as a complex of within-stand or special habitat requirements occurring in species' overall or general habitat. Some special habitat features can be provided through forest management—the aforementioned cavities, for example, either by delayed rotation or streamside buffer strips where timber harvest is minimal—but the stream or pond cannot. Those special habitat features that can be provided through forest management are noted with an asterisk in the following list of classification criteria:

Forest Components

Canopy Closure:

*<15%—very open canopy
*15 to 30%—open canopy
*31 to 70%—intermediate canopy
*>70%—closed canopy

Perch Types:

*High exposed—supracanopy nesting and exposed hunting sites
*Low exposed—exposed hawking sites low to the ground

Overstory Inclusions:

*Deciduous—one tree or group of deciduous trees in a coniferous stand
*Coniferous—one tree or group of coniferous trees in a deciduous stand

Tree Boles:

*Dead ≥6 in dbh—adjacent to water
*Live ≥12 in dbh—adjacent to water
*Live ≥18 in dbh—adjacent to water
*Dead and soft, <6 in dbh—general forest
*Dead and hard, 6 to 12 in dbh—general forest
*Dead and hard, 12 to 18 in dbh—general forest
*Live, columnar decay, 8 to 12 in dbh—general forest
*Live, broken top, 12 to 18 in dbh—general forest
*Live, broken top or large limb >18 in dbh—general forest
*Live, hollow >20 to 24 in dbh—general forest

Midstory Layer:

*Woody vegetation 10 to 30 ft in height

Shrub Layer:

*Deciduous seedlings, saplings, shrubs 2 to 10 ft in height
*Coniferous seedlings, saplings, shrubs 2 to 10 ft in height
*Mixed deciduous and coniferous vegetation 2 to 10 ft in height
*Ericaceous shrubs 2 to 10 ft in height
*Wetland shrubs

Ground Cover Vegetation:

*<30% Upland herbaceous ground cover 0 to 2 ft—sparse
*30 to 75% Upland herbaceous ground cover 0 to 2 ft—intermediate
*>75% Upland herbaceous ground cover 0 to 2 ft—abundant
*Wetland vegetation

Duff and Ground Layer:

*Forest litter and moss—leaves, twigs, moss
*Exposed soil
Rocky forest floor
*Dead and down woody debris—trees, larger limbs and branches
*Waterside decaying logs—basking sites adjacent to water

Subterranean Habitats:

Boulder fields—rapid permeability
Cobbles—rapid soil permeability

Sand and gravel—rapid soil permeability
Loams—moderate soil permeability
Silts—slow soil permeability
Clay—slow soil permeability

Mast and Fruit:

*Hard mast—nut-bearing trees
*Soft mast—fleshy fruit-producing trees and shrubs

Miscellaneous Features:

*Seeps
* Vernal/autumnal temporary pools
*Woods roads—unpaved roads and tracks
*Slash piles
*Gravel pits and exposed soil sites
*Log landings

Upland Nonforest Components

Opening Type:

Lawns, golf courses, etc.
Cultivated crop land
Fallow field
Pasture
Blueberry fields
*Gravel pit
*Log landing

Wetland and Aquatic Components

System:

Palustrine
Lacustrine
Riverine
Estuarine
Marine

Water Depth:

Open water—limnetic zone >6.5 ft (2 m)
Aquatic bed—littoral zone <6.5 ft (2 m)—*Ceratophyllum, Nuphar,* and *Nymphaea* present
Emergent wetland—littoral zone <6.5 ft (2 m)—*Typha* or *Scirpus* present
Scrub-shrub wetland—littoral zone <1.5 ft (0.5 m)
Seasonally wet/flooded
Intermittent drainage

Bottom Composition:

Bedrock
Boulder—Cobble
Gravel—Sand
Silt—Organic

pH Level:

Low pH <5.6
Moderately low pH 6.9 to 5.6
Neutral pH 7.0
Moderately high pH 7.1 to 8.4
High pH > 8.4

Water Temperature:

32°F to 50°F (0° to 10°C)
51°F to 70°F (11°C to 21°C)
71° F to 80° F (22° C to 27° C)
>81° F (>27° C)

Adjacent Riparian Vegetation:

Aquatic bed
Unconsolidated shore
Emergent wetland
Moss-lichen wetland
Scrub-shrub wetland
*Forested wetland
*Upland nonforest

Species Activities/Seasonal Use

Habitat utilization by species is rated separately for life history activities and seasons as follows for birds and mammals:

B—Breeding season, shelter (for mammals, refers to the period when young are born and are being nurtured)
BF—Breeding season, feeding
W—Winter shelter
WF—Winter feeding

For amphibians and reptiles, habitat use is shown for breeding (B) and nonbreeding (NB) activity periods only, because with few exceptions, they are inactive during winter, and overwinter underground or in bottom sediments. Consult the previous species accounts for the time periods of these activities.

Habitat Suitability

The suitability (quality) of each community type for a given species was based on ratings by the experts acknowledged, and on our field experience. Although they are subjective, they represent the best estimates currently available. On the matrix, the light shading indicates utilized habitat, and the dark shading indicates preferred habitat.

Literature Cited

Baldwin, H. I.; Ward, W. W. 1980. White pine-northern red oak-red maple. Pages 27–28. In: Eyre, F. H. (editor). *Forest cover types of the United States and Canada.* Washington, D.C.: Society of American Foresters.
Berglund, J. V. 1980. Sugar maple-beech-yellow birch. Page 31. In:

Eyre, F. H. (editor). *Forest cover types of the United States and Canada.* Washington, D.C.: Society of American Foresters.

Blum, B. M. 1980. Red spruce. Page 19. In: Eyre, F. H. (editor). *Forest cover types of the United States and Canada.* Washington, D.C.: Society of American Foresters.

Brinkman, K. A.; Roe, E. I. 1975. *Quaking aspen: Silvics and management in the Lake States.* Agriculture Handbook 486. Washington, D.C.: U.S. Department of Agriculture. 52 pp.

Cowardin, L. M.; Carter, V.; Golef, F. C.; LaRoe, E. T. 1979. *Classification of wetlands and deepwater habitats of the United States.* FWS/OBS-79/31. Washington, D.C.: U.S. Department of the Interior, Fish and Wildlife Service. 103 pp.

Curtis, J. T. 1959. *The vegetation of Wisconsin.* Madison: University of Wisconsin Press. 657 pp.

Dansereau, P.; Segadas-vianna, F. 1952. Ecological study of the peat bogs of eastern North America. I. Structure and evolution of vegetation. *Can. J. Bot.* 30:490–520.

DeGraaf, R. M.; Yamasaki, M.; Leak, W. B.; Lanier, J. W. 1992. *New England wildlife: Management of forested habitats.* Gen. Tech. Rep. NE-144. Radnor, Pa: U.S. Department of Agriculture, Forest Service, Northeastern Forest Experiment Station. 271 pp.

Eyre, F. H. (editor). 1980. *Forest cover types of the United States and Canada.* Washington, D.C.: Society of American Foresters. 148 pp.

Fernald, M. L. 1950. *Gray's manual of botany.* 8th ed. New York: D. Van Nostrand. 1,632 pp.

Frank, R. M.; Majcen, Z.; Gagnon, G. 1980. Balsam fir. Pages 10–11. In: Eyre, F. H. (editor). *Forest cover types of the United States and Canada.* Washington, D.C.: Society of American Foresters.

Golet, F. C.; Larson, J. S. 1974. *Classification of freshwater wetlands in the glaciated Northeast.* Resource Publication 116. Washington, D.C.: U.S. Fish and Wildlife Service. 56 pp.

Griffin, R. H. 1980. Red spruce-balsam fir. Pages 19–20. In: Eyre, F. H. (editor). *Forest cover types of the United States and Canada.* Washington, D.C.: Society of American Foresters.

Little, E. L., Jr. 1979. *Checklist of United States trees (native and naturalized).* Agriculture Handbook 541. Washington, D.C.: U.S. Department of Agriculture. 375 pp.

Powell, D. S.; Erdmann, G. G. 1980. Red maple. Pages 34–35. In: Eyre, F. H. (editor). *Forest cover types of the United States and Canada.* Washington, D.C.: Society of American Foresters.

Safford, L. O. 1980. Paper birch. Page 18. In: Eyre, F. H. (editor). *Forest cover types of the United States and Canada.* Washington, D.C.: Society of American Foresters.

Trimble, G. R., Jr. 1980. Northern red oak. Pages 43–44. In: Eyre, F. H. (editor). *Forest cover types of the United States and Canada.* Washington, D.C.: Society of American Foresters.

U.S. Department of Agriculture. 1975. *Fisheries and wildlife habitat management handbook.* Milwaukee, Wis.: Forest Service, Eastern Region.

U.S. Department of Agriculture. 1982. *Land Resources in New Hampshire.* Durham, N.H.: Soil Conservation Service.

Wendel, G. W. 1980. Eastern white pine. Pages 25–26. In: Eyre, F. H. (editor). *Forest cover types of the United States and Canada.* Washington, D.C.: Society of American Foresters.

Westveld, M. 1953. Ecology and silviculture of spruce-fir forests of eastern North America. *J. Forestry* 51:422–430.

Wiant, H. V. 1980. Eastern hemlock. Page 27. In: Eyre, F. H. (editor). *Forest cover types of the United States and Canada.* Washington, D.C.: Society of American Foresters.

The Matrices

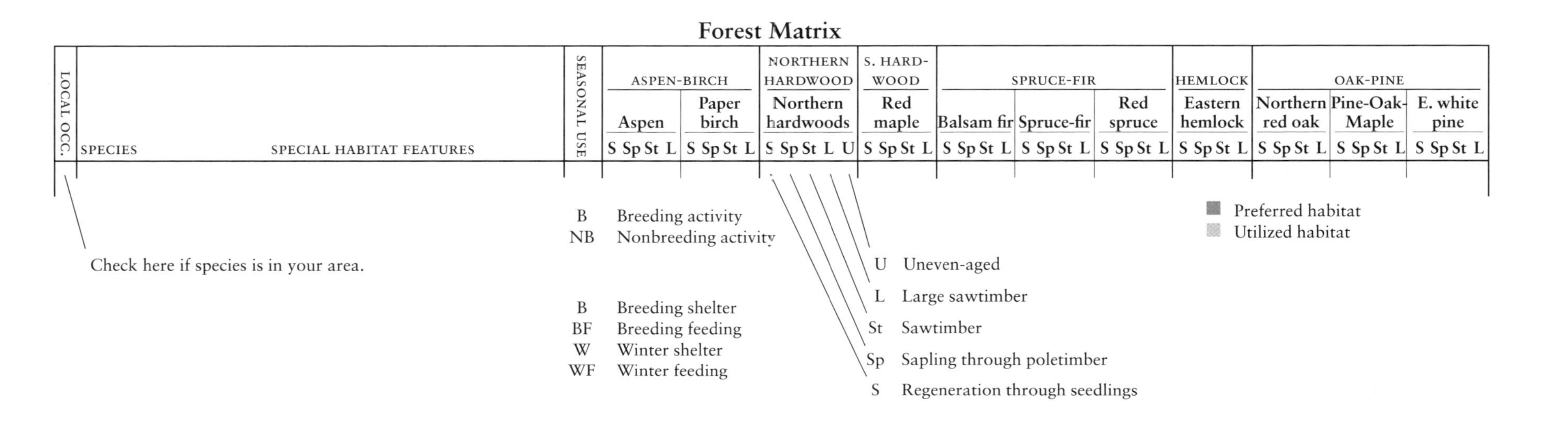

Nonforested Matrix

LOCAL OCCURRENCE | SPECIES | SPECIAL HABITAT FEATURES | SEASONAL USE

TERRESTRIAL
- Upland Field: Cultivated, Grass, Forb, Shrub/old field, Pasture
- **Savanna**, **Orchard**, **Krummholz**, **Alpine**

WETLAND/DEEP WATER
- Palustrine: Sedge meadow, Shallow marsh, Deep marsh, Shrub swamp, Bog, Pond
- **Lake**
- Riverine: Stream, River, Riparian
- **Estuary/salt marsh**
- Marine: Coastal beach/rocks, Bay/ocean

OTHER
- **Stable bank**, **Ledge, cliff**, **Cave, mine**, **Structure**, **Derelict bldg., debris**

Check here if species is in your area.

B Breeding activity
NB Nonbreeding activity

B Breeding shelter
BF Breeding feeding
W Winter shelter
WF Winter feeding

Preferred habitat
Utilized habitat

Fig. 6. The key to elements in the species/habitat matrices.

Forest Matrix

LOCAL OCC.	SPECIES	SPECIAL HABITAT FEATURES	SEASONAL USE	ASPEN-BIRCH		NORTHERN HARDWOOD	S. HARD-WOOD	SPRUCE-FIR			HEMLOCK	OAK-PINE		
				Aspen	Paper birch	Northern hardwoods	Red maple	Balsam fir	Spruce-fir	Red spruce	Eastern hemlock	Northern red oak	Pine-Oak-Maple	E. white pine
				S Sp St L	S Sp St L	S Sp St L U	S Sp St L	S Sp St L	S Sp St L	S Sp St L	S Sp St L	S Sp St L	S Sp St L	S Sp St L
	Marbled Salamander *Ambystoma opacum*	Temporary (autumnal) pools or swamps in upland woods for breeding.	B NB											
	Jefferson Salamander *Ambystoma jeffersonianum*	Vernal pools or semipermanent water adjacent to upland habitat for breeding.	B NB											
	Blue-spotted Salamander *Ambystoma laterale*	Vernal pools or semipermanent water for breeding; adjacent upland habitat.	B NB											
	Spotted Salamander *Ambystoma maculatum*	Mesic woods, vernal pools or semipermanent water for breeding.	B NB											
	Red-spotted Newt *Notophthalmus v. viridescens*	Water bodies with aquatic vegetation for adult newts; juveniles (efts) are terrestrial.	B NB											
	Northern Dusky Salamander *Desmognathus fuscus*	Permanent forested seeps, springs, headwater streams.	B NB											
	Northern Redback Salamander *Plethodon cinereus*	Logs, stumps, rocks, etc.	B NB											
	Northern Slimy Salamander *Plethodon glutinosus*	Rock outcroppings, logs in wooded areas.	B NB											
	Four-toed Salamander *Hemidactylium scutatum*	Wet woodlands containing sphagnum moss.	B NB											
	Northern Spring Salamander *Gyrinophilus p. porphyriticus*	Woodland headwater streams; high gradient seeps or springs.	B NB											
	Northern Two-lined Salamander *Eurycea bislineata*	Well-shaded streams for breeding.	B NB											

Forest Matrix (*continued*)

LOCAL OCC.	SPECIES	SPECIAL HABITAT FEATURES	SEASONAL USE	ASPEN-BIRCH: Aspen (S Sp St L)	ASPEN-BIRCH: Paper birch (S Sp St L)	NORTHERN HARDWOOD: Northern hardwoods (S Sp St L U)	S. HARDWOOD: Red maple (S Sp St L)	SPRUCE-FIR: Balsam fir (S Sp St L)	SPRUCE-FIR: Spruce-fir (S Sp St L)	SPRUCE-FIR: Red spruce (S Sp St L)	HEMLOCK: Eastern hemlock (S Sp St L)	OAK-PINE: Northern red oak (S Sp St L)	OAK-PINE: Pine-Oak-Maple (S Sp St L)	OAK-PINE: E. white pine (S Sp St L)
	Eastern Spadefoot *Scaphiopus holbrookii*	Sandy soils; temporary, seasonally or periodically flooded areas (5 to 30 cm deep).	B NB											
	Eastern American Toad *Bufo a. americanus*	Shallow, still water for breeding.	B NB											
	Fowler's Toad *Bufo fowleri*	Well-drained sandy soils; shallow water for breeding.	B NB											
	Northern Spring Peeper *Pseudacris c. crucifer*	Pools for breeding adjacent to forested or shrub habitat.	B NB											
	Gray Treefrog *Hyla versicolor*	Seeps or shallow water adjacent to upland forests for breeding; hibernates under leaves, logs.	B NB											
	Green Frog *Rana clamitans melanota*	Margins of riverine or lacustrine habitats for breeding.	B NB											
	Mink Frog *Rana septentrionalis*	Margins of lacustrine habitats with abundant lily pads; adjacent wetlands.	B NB											
	Wood Frog *Rana sylvatica*	Vernal pools in forest habitat.	B NB											
	Northern Leopard Frog *Rana pipiens*	Shallow pools of standing water adjacent to wet meadows.	B NB											
	Pickerel Frog *Rana palustris*	Shallow, clear water of bogs, woodland streams, and lake margins.	B NB											
	Spotted Turtle *Clemmys guttata*	Wetland complex with vernal pools, adjacent upland forest and well-drained openings for nesting.	B NB											
	Wood Turtle *Clemmys insculpta*	Forested rivers, streams within 200 m of well-drained sandy or gravelly substrates for nesting.	B NB											
	Eastern Box Turtle *Terrapene c. carolina*	Old fields, clearings, and ecotones with sandy soils.	B NB											

Species	Habitat	
Blanding's Turtle *Emydoidea blandingii*	Wetland complex with shallow water (<0.5 m); adjacent upland fields or forest for nesting.	B NB
Five-lined Skink *Eumeces fasciatus*	Steep, rocky open woods with logs and slash piles.	B NB
Northern Brown Snake *Storeria d. dekayi*		B NB
Northern Redbelly Snake *Storeria o. occipitomaculata*	Woodland debris—bark and rotting wood.	B NB
Common Garter Snake *Thamnophis sirtalis*		B NB
Ribbon Snake *Thamnophis sauritus*	Permanent shallow water in open grassy habitat.	B NB
Eastern Hognose Snake *Heterodon platyrhinos*	Sandy soils, open woodlands.	B NB
Northern Ringneck Snake *Diadophis punctatus edwardsii*	Mesic areas with abundant cover.	B NB
Eastern Worm Snake *Carphophis a. amoenus*	Well-drained, loose soils for burrowing or covering.	B NB
Northern Black Racer *Coluber c. constrictor*	Upland areas that are periodically cleared or mowed; adjacent forest and wetlands.	B NB
Eastern Smooth Green Snake *Liochlorophis vernalis*	Upland grassy openings.	B NB
Black Rat Snake *Elaphe o. obsoleta*		B NB
Eastern Milk Snake *Lampropeltis t. triangulum*	Slash, wood piles, debris or loose soils for egg laying.	B NB
Northern Copperhead *Agkistrodon contortrix mokasen*	Rocky hillsides, talus slopes.	B NB
Timber Rattlesnake *Crotalus horridus*	Rock ledges on forested hillsides.	B NB

Forest Matrix (*continued*)

LOCAL OCC.	SPECIES	SPECIAL HABITAT FEATURES	SEASONAL USE	ASPEN-BIRCH: Aspen (S Sp St L)	ASPEN-BIRCH: Paper birch (S Sp St L)	NORTHERN HARDWOOD: Northern hardwoods (S Sp St L U)	S. HARDWOOD: Red maple (S Sp St L)	SPRUCE-FIR: Balsam fir (S Sp St L)	SPRUCE-FIR: Spruce-fir (S Sp St L)	SPRUCE-FIR: Red spruce (S Sp St L)	HEMLOCK: Eastern hemlock (S Sp St L)	OAK-PINE: Northern red oak (S Sp St L)	OAK-PINE: Pine-Oak-Maple (S Sp St L)	OAK-PINE: E. white pine (S Sp St L)
	Great Blue Heron *Ardea herodias*	Tall trees for nesting.	B BF W WF											
	Green Heron *Butorides virescens*	Forested wetlands or open water habitats.	B BF W WF											
	Turkey Vulture *Cathartes aura*	Forest clearings and fields; large branch stubs of dead trees for roosting.	B BF W WF											
	Wood Duck *Aix sponsa*	Trees ≥16 in (40.6 cm) dbh with large cavities and 4 in (10.2 cm) diameter entrance holes.	B BF W WF											
	American Black Duck *Anas rubripes*	Wooded wetlands, streambanks (inland).	B BF W WF											
	Common Goldeneye *Bucephala clangula*	Cavity trees ≥20 in (51 cm) dbh; clear cold, shallow water.	B BF W WF											
	Hooded Merganser *Lophodytes cucullatus*	Undisturbed wooded areas with cavity trees ≥25 in (38 cm) dbh; clear fresh water.	B BF W WF											
	Common Merganser *Mergus merganser*	Large cavity trees at water's edge.	B BF W WF											
	Bald Eagle *Haliaeetus leucocephalus*	Large undisturbed water bodies containing fish; large, live supracanopy nest trees within 0.25 mile of shore.	B BF W WF											

Species	Habitat	Season
Sharp-shinned Hawk *Accipiter striatus*	Extensive undisturbed, open mixed woodlands.	B BF W WF
Cooper's Hawk *Accipiter cooperii*	Mature deciduous or coniferous woodlands.	B BF W WF
Northern Goshawk *Accipiter gentilis*	Extensive mature mixed woodlands.	B BF W WF
Red-shouldered Hawk *Buteo lineatus*	Cool, moist mature forest.	B BF W WF
Broad-winged Hawk *Buteo platypterus*	Extensive woodlands with roads or clearings.	B BF W WF
Red-tailed Hawk *Buteo jamaicensis*	Mature forests for nesting; nonforest openings for foraging.	B BF W WF
Golden Eagle *Aquila chrysaetos*	Cliffs for nesting; wide expanses of open land.	B BF W WF
American Kestrel *Falco sparverius*	Cavity trees ≥12 in (30.5 cm); open flat terrain.	B BF W WF
Merlin *Falco columbarius*	Old corvid nests, natural cavities, or old woodpecker holes in open woodlands.	B BF W WF
Peregrine Falcon *Falco peregrinus*	High cliffs or tall urban buildings; abundant avian prey.	B BF W WF
Ruffed Grouse *Bonasa umbellus*	Drumming log sites amid dense saplings; herbaceous cover on old roads, small openings.	B BF W WF

Forest Matrix (*continued*)

LOCAL OCC.	SPECIES	SPECIAL HABITAT FEATURES	SEASONAL USE
	Spruce Grouse *Falcipennis canadensis*	Coniferous forest.	B BF W WF
	Wild Turkey *Meleagris gallopavo*	Open, mast-producing woodlands; large conifers for roosting; herbaceous cover in woodland clearings, small openings.	B BF W WF
	Northern Bobwhite *Colinus virginianus*	Brushy field edges; well-drained sandy or loamy soils.	B BF W WF
	American Woodcock *Scolopax minor*	Fertile moist soils containing earthworms; small clearings and dense swales.	B BF W WF
	Mourning Dove *Zenaida macroura*	Open land with bare ground.	B BF W WF
	Black-billed Cuckoo *Coccyzus erythropthalmus*	Low, dense thickets.	B BF W WF
	Yellow-billed Cuckoo *Coccyzus americanus*	Low, dense thickets.	B BF W WF
	Eastern Screech-Owl *Otus asio*	Cavity trees ≥12 in (30.5 cm) dbh.	B BF W WF
	Great Horned Owl *Bubo virginianus*	Large abandoned hawk nests; large tree cavities.	B BF W WF

Column groups: ASPEN-BIRCH (Aspen: S Sp St L; Paper birch: S Sp St L); NORTHERN HARDWOOD (Northern hardwoods: S Sp St L U); S. HARDWOOD (Red maple: S Sp St L); SPRUCE-FIR (Balsam fir: S Sp St L; Spruce-fir: S Sp St L; Red spruce: S Sp St L); HEMLOCK (Eastern hemlock: S Sp St L); OAK-PINE (Northern red oak: S Sp St L; Pine-Oak-Maple: S Sp St L; E. white pine: S Sp St L).

Species	Habitat	
Northern Hawk Owl *Surnia ulula*	Coniferous forests.	B BF W WF
Barred Owl *Strix varia*	Cool, damp lowlands; cavity trees ≥20 in (50.8 cm) dbh.	B BF W WF
Great Gray Owl *Strix nebulosa*	Meadows, swamps near woodlands.	B BF W WF
Long-eared Owl *Asio otus*	Dense conifer thickets in open country; coniferous forested wetlands.	B BF W WF
Boreal Owl *Aegolius funereus*	Cavity trees, often with large holes; mixedwood or coniferous forest and coniferous forested wetlands.	B BF W WF
Northern Saw-whet Owl *Aegolius acadicus*	Cavity trees ≥12 in (30.1 cm) dbh near forest clearings.	B BF W WF
Common Nighthawk *Chordeiles minor*	Barren areas with rocky substrate; graveled rooftops.	B BF W WF
Whip-poor-will *Caprimulgus vociferus*	Ungrazed woodlands with openings.	B BF W WF
Ruby-throated Hummingbird *Archilochus colubris*	Tubular nectar-bearing flowers, preferably red.	B BF W WF
Red-headed Woodpecker *Melanerpes erythrocephalus*	Cavity trees in savanna or open country.	B BF W WF
Red-bellied Woodpecker *Melanerpes carolinus*	Extensive mature woodlands with dead trees or trees with large dead limbs.	B BF W WF

Forest Matrix (*continued*)

LOCAL OCC.	SPECIES	SPECIAL HABITAT FEATURES	SEASONAL USE	ASPEN-BIRCH: Aspen (S Sp St L)	ASPEN-BIRCH: Paper birch (S Sp St L)	NORTHERN HARDWOOD: Northern hardwoods (S Sp St L U)	S. HARDWOOD: Red maple (S Sp St L)	SPRUCE-FIR: Balsam fir (S Sp St L)	SPRUCE-FIR: Spruce-fir (S Sp St L)	SPRUCE-FIR: Red spruce (S Sp St L)	HEMLOCK: Eastern hemlock (S Sp St L)	OAK-PINE: Northern red oak (S Sp St L)	OAK-PINE: Pine-Oak-Maple (S Sp St L)	OAK-PINE: E. white pine (S Sp St L)
	Yellow-bellied Sapsucker *Sphyrapicus varius*	Trees ≥10 in (26 cm) dbh; particularly aspens containing decayed wood.	B BF W WF											
	Downy Woodpecker *Picoides pubescens*	Decay columns in trees ≥6 in (15 cm) dbh or larger limbs.	B BF W WF											
	Hairy Woodpecker *Picoides villosus*	Decayed wood column in trees ≥10 in (25 cm) dbh or larger limbs.	B BF W WF											
	Three-toed Woodpecker *Picoides tridactylus*	Spruce and fir trees ≥12 in (31 cm) dbh with column of decayed wood; dead trees with loose bark.	B BF W WF											
	Black-backed Woodpecker *Picoides arcticus*	Spruce and fir trees ≥10 in (25 cm) dbh with column of decayed wood; dead trees with loose bark; often near water.	B BF W WF											
	Northern Flicker *Colaptes auratus*	Open areas, forest edges; trees ≥12 in (31 cm) dbh with column of decayed wood.	B BF W WF											
	Pileated Woodpecker *Dryocopus pileatus*	Mature forest; trees ≥20 in (51 cm) dbh with column of decayed wood.	B BF W WF											
	Olive-sided Flycatcher *Contopus cooperi*	Tall perch adjacent to low, wet coniferous thicket, bog.	B BF W WF											
	Eastern Wood-Pewee *Contopus virens*	Forest edge or open woods.	B BF W WF											

Species	Habitat	
Yellow-bellied Flycatcher *Empidonax flaviventris*	Low, wet coniferous forest.	B BF W WF
Acadian Flycatcher *Empidonax virescens*	Extensive mature, deciduous woodlands; closed canopy; open understory for foraging.	B BF W WF
Alder Flycatcher *Empidonax alnorum*	Dense woody regeneration, low shrubs; thickets and clearings.	B BF W WF
Willow Flycatcher *Empidonax traillii*	Low deciduous trees and shrubs in clearings; thick hardwood regeneration.	B BF W WF
Least Flycatcher *Empidonax minimus*	Open deciduous or mixed forest; forest edges.	B BF W WF
Eastern Phoebe *Sayornis phoebe*	Exposed perches, cliffs or ledges in streamside clearings; buildings or structures in openings for nesting.	B BF W WF
Great Crested Flycatcher *Myiarchus crinitus*	Mature deciduous forest; cavity trees on forest edges.	B BF W WF
Eastern Kingbird *Tyrannus tyrannus*	Clearings, fields, orchards; perches for aerial foraging.	B BF W WF
Loggerhead Shrike *Lanius ludovicianus*	Open country with short grasses, scattered trees, shrubs.	B BF W WF
Northern Shrike *Lanius excubitor*	Scattered trees or shrubs in open country.	B BF W WF
White-eyed Vireo *Vireo griseus*	Low shrubs, thickets.	B BF W WF

Forest Matrix (*continued*)

LOCAL OCC.	SPECIES	SPECIAL HABITAT FEATURES	SEASONAL USE	ASPEN-BIRCH		NORTHERN HARDWOOD	S. HARD-WOOD	SPRUCE-FIR			HEMLOCK	OAK-PINE		
				Aspen	Paper birch	Northern hardwoods	Red maple	Balsam fir	Spruce-fir	Red spruce	Eastern hemlock	Northern red oak	Pine-Oak-Maple	E. white pine
				S Sp St L	S Sp St L	S Sp St L U	S Sp St L	S Sp St L	S Sp St L	S Sp St L	S Sp St L	S Sp St L	S Sp St L	S Sp St L
	Yellow-throated Vireo *Vireo flavifrons*	Mature deciduous forest.	B BF W WF											
	Blue-headed Vireo *Vireo solitarius*	Mixedwood or coniferous forest.	B BF W WF											
	Warbling Vireo *Vireo gilvus*	Scattered deciduous trees in open areas; often adjacent to water.	B BF W WF											
	Philadelphia Vireo *Vireo philadelphicus*	Hardwood forest edges; early successional forests.	B BF W WF											
	Red-eyed Vireo *Vireo olivaceus*	Closed deciduous forest canopy.	B BF W WF											
	Gray Jay *Perisoreus canadensis*	Coniferous forests.	B BF W WF											
	Blue Jay *Cyanocitta cristata*		B BF W WF											
	American Crow *Corvus brachyrhynchos*	Open areas for foraging.	B BF W WF											
	Common Raven *Corvus corax*	Cliffs, abandoned raptor nests.	B BF W WF											

Species	Habitat	Season											
Purple Martin *Progne subis*	Martin houses, tree cavities, abandoned woodpecker cavities near ponds, lakes, openings.	B BF W WF											
Tree Swallow *Tachycineta bicolor*	Cavity trees ≥10 in (25 cm) dbh; old woodpecker cavities; near open areas.	B BF W WF											
Northern Rough-winged Swallow *Stelgidopteryx serripennis*	Sandy or clay banks; stabilized by grassy overhang near, but up to 1 km from water.	B BF W WF											
Bank Swallow *Riparia riparia*	Stabilized steep banks of sand, gravel, or clay; near open areas, preferably flowing water.	B BF W WF											
Cliff Swallow *Petrochelidon pyrrhonata*	Open areas, water; vertical wall with an overhang; mud for nest construction.	B BF W WF											
Barn Swallow *Hirundo rustica*	Abandoned or little-used buildings for nesting.	B BF W WF											
Black-capped Chickadee *Poecile atricapillus*	Cavity trees or stubs in small woodlands; clearings or open woodlands.	B BF W WF											
Boreal Chickadee *Poecile hudsonicus*	Softwood snags, stubs.	B BF W WF											
Tufted Titmouse *Baeolophus bicolor*	Cavity trees ≥8 in (20 cm) dbh.	B BF W WF											
Red-breasted Nuthatch *Sitta canadensis*	Cavity trees ≥12 in (31 cm) dbh in mixed or coniferous woods.	B BF W WF											
White-breasted Nuthatch *Sitta carolinensis*	Cavity trees ≥12 in (31 cm) dbh in hardwoods and mixed woods.	B BF W WF											

Forest Matrix *(continued)*

LOCAL OCC.	SPECIES	SPECIAL HABITAT FEATURES	SEASONAL USE	ASPEN-BIRCH		NORTHERN HARDWOOD	S. HARD-WOOD	SPRUCE-FIR			HEMLOCK	OAK-PINE		
				Aspen	Paper birch	Northern hardwoods	Red maple	Balsam fir	Spruce-fir	Red spruce	Eastern hemlock	Northern red oak	Pine-Oak-Maple	E. white pine
				S Sp St L	S Sp St L	S Sp St L U	S Sp St L	S Sp St L	S Sp St L	S Sp St L	S Sp St L	S Sp St L	S Sp St L	S Sp St L
	Brown Creeper *Certhia americana*	Woodland trees with sloughing or loose bark.	B BF W WF											
	Carolina Wren *Thryothorus ludovicianus*	Cavity trees amidst brushy vegetation, thickets, swamps.	B BF W WF											
	House Wren *Troglodytes aedon*	Cavity trees, shrubs.	B BF W WF											
	Winter Wren *Troglodytes troglodytes*	Moist, mixed or coniferous woodlands with down logs; low woody vegetation.	B BF W WF											
	Golden-crowned Kinglet *Regulus satrapa*	Dense conifer thickets or stands, especially spruce.	B BF W WF											
	Ruby-crowned Kinglet *Regulus calendula*	Conifer stands, or mixed stands with a predominance of softwoods, especially spruce or fir.	B BF W WF											
	Blue-gray Gnatcatcher *Polioptila caerulea*	Open, deciduous woodlands.	B BF W WF											
	Eastern Bluebird *Sialia sialis*	Low cavity trees or nest boxes; open country.	B BF W WF											
	Veery *Catharus fuscescens*	Moist woodlands with understory of low trees and shrubs.	B BF W WF											

Species	Habitat	
Bicknell's Thrush *Catharus bicknelli*	Stunted coniferous forest at high elevations.	B BF W WF
Swainson's Thrush *Catharus ustulatus*	Coniferous or mixed forest.	B BF W WF
Hermit Thrush *Catharus guttatus*	Coniferous or mixed woodlands with dense undergrowth.	B BF W WF
Wood Thrush *Hylocichla mustelina*	Cool, moist mature deciduous or mixed forest.	B BF W WF
American Robin *Turdus migratorius*		B BF W WF
Gray Catbird *Dumetella carolinensis*	Low, dense shrubby vegetation in open country or forest understory.	B BF W WF
Northern Mockingbird *Mimus polyglottus*	Low thickets; high exposed perches; persistent fruit.	B BF W WF
Brown Thrasher *Toxostoma rufum*	Hardwood forest-field ecotone.	B BF W WF
European Starling *Sturnus vulgaris*	Cavity trees ≥10 in (25 cm) dbh.	B BF W WF
Bohemian Waxwing *Bombycilla garrulus*	Soft mast (fruit, berries).	B BF W WF
Cedar Waxwing *Bombycilla cedrorum*	Soft mast (fruit, berries).	B BF W WF

Forest Matrix (*continued*)

LOCAL OCC.	SPECIES	SPECIAL HABITAT FEATURES	SEASONAL USE	ASPEN-BIRCH: Aspen (S Sp St L)	ASPEN-BIRCH: Paper birch (S Sp St L)	NORTHERN HARDWOOD: Northern hardwoods (S Sp St L U)	S. HARD-WOOD: Red maple (S Sp St L)	SPRUCE-FIR: Balsam fir (S Sp St L)	SPRUCE-FIR: Spruce-fir (S Sp St L)	SPRUCE-FIR: Red spruce (S Sp St L)	HEMLOCK: Eastern hemlock (S Sp St L)	OAK-PINE: Northern red oak (S Sp St L)	OAK-PINE: Pine-Oak-Maple (S Sp St L)	OAK-PINE: E. white pine (S Sp St L)
	Blue-winged Warbler *Vermivora pinus*	Old fields with scattered shrubs and small trees.	B BF W WF											
	Golden-winged Warbler *Vermivora chrysoptera*	Brushy edges; openings with saplings, forbs or grasses.	B BF W WF											
	Tennessee Warbler *Vermivora peregrina*	Brushy, semiopen country.	B BF W WF											
	Nashville Warbler *Vermivora ruficapilla*	Young hardwood/mixedwood stands; open mixedwood stands interspersed with brush, thickets.	B BF W WF											
	Northern Parula *Parula americana*	Presence of bearded lichen (*Usnea*).	B BF W WF											
	Yellow Warbler *Dendroica petechia*	Small scattered trees or dense shrubs; commonly near water.	B BF W WF											
	Chestnut-sided Warbler *Dendroica pensylvanica*	Early successional deciduous stands; dense hardwood regeneration 3 to 10 ft (1 to 3 m) tall.	B BF W WF											
	Magnolia Warbler *Dendroica magnolia*	Young stands of spruce or fir, sometimes of hemlock.	B BF W WF											
	Cape May Warbler *Dendroica tigrina*	Spruce-fir forests.	B BF W WF											

Species	Habitat	Status
Black-throated Blue Warbler *Dendroica caerulescens*	Hardwood/mixedwood stands well-developed understory; often in with mountain laurel.	B BF W WF
Yellow-rumped Warbler *Dendroica coronata*	Coniferous trees (summer); bayberry thickets (winter).	B BF W WF
Black-throated Green Warbler *Dendroica virens*	Mixedwood/coniferous stands.	B BF W WF
Blackburnian Warbler *Dendroica fusca*	Coniferous/mixedwood stands.	B BF W WF
Pine Warbler *Dendroica pinus*	Pine stands.	B BF W WF
Prairie Warbler *Dendroica discolor*	Coniferous cover in old fields; hardwood saplings, slash.	B BF W WF
Palm Warbler *Dendroica palmarum*	Young open coniferous stands, bogs; thick shrub growth.	B BF W WF
Bay-breasted Warbler *Dendroica castanea*	Dense coniferous stands, especially of spruce and fir.	B BF W WF
Blackpoll Warbler *Dendroica striata*	Stunted spruce, especially at higher elevations.	B BF W WF
Cerulean Warbler *Dendroica cerulea*	Tall deciduous trees; often near water.	B BF W WF
Black-and-white Warbler *Mniotilta varia*	Deciduous/mixedwood stands.	B BF W WF

Forest Matrix (*continued*)

LOCAL OCC.	SPECIES	SPECIAL HABITAT FEATURES	SEASONAL USE	ASPEN-BIRCH: Aspen (S Sp St L)	ASPEN-BIRCH: Paper birch (S Sp St L)	NORTHERN HARDWOOD: Northern hardwoods (S Sp St L U)	S. HARDWOOD: Red maple (S Sp St L)	SPRUCE-FIR: Balsam fir (S Sp St L)	SPRUCE-FIR: Spruce-fir (S Sp St L)	SPRUCE-FIR: Red spruce (S Sp St L)	HEMLOCK: Eastern hemlock (S Sp St L)	OAK-PINE: Northern red oak (S Sp St L)	OAK-PINE: Pine-Oak-Maple (S Sp St L)	OAK-PINE: E. white pine (S Sp St L)
	American Redstart *Setophaga ruticilla*	Open deciduous/mixedwood stands with a dense shrub/ midstory.	B BF W WF											
	Worm-eating Warbler *Helmitheros vermivorus*	Well-developed understory.	B BF W WF											
	Ovenbird *Seiurus aurocapillus*	Open, mature mesic or dry forest.	B BF W WF											
	Northern Waterthrush *Seiurus noveboracensis*	Cool, shaded, wet ground with shallow pools.	B BF W WF											
	Louisiana Waterthrush *Seiurus motacilla*	Woodlands with flowing water.	B BF W WF											
	Mourning Warbler *Oporornis philadelphia*	Larger openings of hardwood regeneration; dense shrubs.	B BF W WF											
	Common Yellowthroat *Geothlypis trichas*	Dense hardwood regeneration, shrub layer.	B BF W WF											
	Hooded Warbler *Wilsonia citrina*	Dense understory.	B BF W WF											
	Wilson's Warbler *Wilsonia pusilla*	Cold shrub swamps, bogs; thickets.	B BF W WF											

Species	Habitat	Season
Canada Warbler *Wilsonia canadensis*	Dense deciduous/ericaceous understory; especially along streams, bogs.	B BF W WF
Yellow-breasted Chat *Icteria virens*	Dense thickets and vines with scattered young trees, often near water.	B BF W WF
Scarlet Tanager *Piranga olivacea*	Hardwood/mixedwood forest canopy.	B BF W WF
Eastern Towhee *Pipilo erythrophthalmus*	Dense, brushy understory; well-drained soils.	B BF W WF
American Tree Sparrow *Spizella arborea*	Open country; brushy cover; weedy fields (winter).	B BF W WF
Chipping Sparrow *Spizella passerina*	Clearings with bare ground; coniferous or thorny shrubs.	B BF W WF
Field Sparrow *Spizella pusilla*	Old fields.	B BF W WF
Fox Sparrow *Passerella iliaca*	Dense herbaceous vegetation; field-forest ecotones.	B BF W WF
Song Sparrow *Melospiza melodia*	Brushy cover, hardwood regeneration; conspicuous song perches.	B BF W WF
Lincoln's Sparrow *Melospiza lincolnii*	Brushy thickets along field edges or drainways; wet grassy or sedge meadows.	B BF W WF
Swamp Sparrow *Melospiza georgiana*	Brushy wetlands with emergent vegetation.	B BF W WF

Forest Matrix (*continued*)

Local Occ.	Species	Special Habitat Features	Seasonal Use	Aspen-Birch: Aspen	Aspen-Birch: Paper birch	Northern Hardwood: Northern hardwoods	S. Hardwood: Red maple	Spruce-Fir: Balsam fir	Spruce-Fir: Spruce-fir	Spruce-Fir: Red spruce	Hemlock: Eastern hemlock	Oak-Pine: Northern red oak	Oak-Pine: Pine-Oak-Maple	Oak-Pine: E. white pine
				S Sp St L	S Sp St L	S Sp St L U	S Sp St L	S Sp St L	S Sp St L	S Sp St L	S Sp St L	S Sp St L	S Sp St L	S Sp St L
	White-throated Sparrow *Zonotrichia albicollis*		B BF W WF											
	Dark-eyed Junco *Junco hyemalis*	Woods roads, cut banks; uprooted trees.	B BF W WF											
	Northern Cardinal *Cardinalis cardinalis*	Thickets, vines.	B BF W WF											
	Rose-breasted Grosbeak *Pheucticus ludovicianus*	Forest edges, dense hardwood thickets and sapling stands; brushy fields.	B BF W WF											
	Indigo Bunting *Passerina cyanea*	Scattered tall trees in grassy openings; forest-field ecotones.	B BF W WF											
	Red-winged Blackbird *Agelaius phoeniceus*	Emergent vegetation adjacent to open fields.	B BF W WF											
	Rusty Blackbird *Euphagus carolinus*	Northern forested wetlands; alder swales.	B BF W WF											
	Common Grackle *Quiscalus quiscula*	Wet, open country, shrub swamps, pond edges.	B BF W WF											
	Brown-headed Cowbird *Molothrus ater*	Open grassy habitat.	B BF W WF											

Species	Habitat	
Orchard Oriole *Icterus spurius*	Open country with scattered deciduous trees.	B BF W WF
Baltimore Oriole *Icterus galbula*	Tall, scattered deciduous trees (preferably elm).	B BF W WF
Pine Grosbeak *Pinicola enucleator*	Northern conifer forest.	B BF W WF
Purple Finch *Carpodacus purpureus*	Coniferous trees.	B BF W WF
House Finch *Carpodacus mexicanus*	Open ground with low seed-producing plants.	B BF W WF
Red Crossbill *Loxia curvirostra*	Northern conifer forest.	B BF W WF
White-winged Crossbill *Loxia leucoptera*	Northern conifer forest.	B BF W WF
Common Redpoll *Carduelis flammea*	Open country (winter).	B BF W WF
Hoary Redpoll *Carduelis hornemanni*	Open country (winter).	B BF W WF
Pine Siskin *Carduelis pinus*	Coniferous trees.	B BF W WF
American Goldfinch *Carduelis tristis*	Open, weedy fields and marshes with scattered small trees.	B BF W WF

Forest Matrix (*continued*)

LOCAL OCC.	SPECIES	SPECIAL HABITAT FEATURES	SEASONAL USE	ASPEN-BIRCH		NORTHERN HARDWOOD	S. HARD-WOOD	SPRUCE-FIR			HEMLOCK	OAK-PINE		
				Aspen	Paper birch	Northern hardwoods	Red maple	Balsam fir	Spruce-fir	Red spruce	Eastern hemlock	Northern red oak	Pine-Oak-Maple	E. white pine
				S Sp St L	S Sp St L	S Sp St L U	S Sp St L	S Sp St L	S Sp St L	S Sp St L	S Sp St L	S Sp St L	S Sp St L	S Sp St L
	Evening Grosbeak *Coccothraustes vespertinus*	Spruce and fir forest (breeding season).	B BF W WF											
	Virginia Opossum *Didelphis virginiana*	Hollow log or tree cavity.	B BF W WF											
	Masked Shrew *Sorex cinereus*	Damp woodlands, ground cover.	B BF W WF											
	Water Shrew *Sorex palustris*	Herbaceous cover, cold-water wetlands.	B BF W WF											
	Smoky Shrew *Sorex fumeus*	Loose, damp leaf litter.	B BF W WF											
	Long-tailed Shrew *Sorex dispar*	Rocky, wooded sites.	B BF W WF											
	Pygmy Shrew *Sorex hoyi*	Moist leaf mold near water.	B BF W WF											
	Northern Short-tailed Shrew *Blarina brevicauda*	Low vegetation, damp, loose leaf litter.	B BF W WF											
	Least Shrew *Cryptotis parva*	Loose soil.	B BF W WF											

Species	Habitat	
Hairy-tailed Mole *Parascalops breweri*	Loose, moist, well-drained soil.	B BF W WF
Eastern Mole *Scalopus aquaticus*	Soft, moist soils containing earthworms.	B BF W WF
Star-nosed Mole *Condylura cristata*	Wet muck, humus.	B BF W WF
Little Brown Myotis *Myotis lucifugus*	Females: dark, warm sites for maternity colonies. Forest openings for feeding.	B BF W WF
Northern Long-eared Bat *Myotis septentrionalis*	Large cavity trees for roosting. Caves, mine shafts for hibernation near 40°F (2–7°C); high humidity and calm air.	B BF W WF
Indiana Myotis *Myotis sodalis*	Tree cavities and loose bark for roosting. Cool stable temperatures in caves (hibernacula).	B BF W WF
Eastern Small-footed Myotis *Myotis leibii*	Caves and mine shafts for hibernacula.	B BF W WF
Silver-haired Bat *Lasionycteris noctivagans*	Dead trees with loose bark or cavities; wetlands nearby for foraging.	B BF W WF
Eastern Pipistrelle *Pipistrellus subflavus*	Warm, draft-free damp hibernacula; open woods for foraging.	B BF W WF
Big Brown Bat *Eptesicus fuscus*	Cold, dry caves or buildings for hibernacula.	B BF W WF
Red Bat *Lasiurus borealis*	Deciduous trees on forest edges for roosting.	B BF W WF

Forest Matrix *(continued)*

LOCAL OCC.	SPECIES	SPECIAL HABITAT FEATURES	SEASONAL USE	ASPEN-BIRCH: Aspen (S Sp St L)	ASPEN-BIRCH: Paper birch (S Sp St L)	NORTHERN HARDWOOD: Northern hardwoods (S Sp St L U)	S. HARDWOOD: Red maple (S Sp St L)	SPRUCE-FIR: Balsam fir (S Sp St L)	SPRUCE-FIR: Spruce-fir (S Sp St L)	SPRUCE-FIR: Red spruce (S Sp St L)	HEMLOCK: Eastern hemlock (S Sp St L)	OAK-PINE: Northern red oak (S Sp St L)	OAK-PINE: Pine-Oak-Maple (S Sp St L)	OAK-PINE: E. white pine (S Sp St L)
	Hoary Bat *Lasiurus cinereus*	Roosts in coniferous and deciduous foliage.	B BF W WF											
	Eastern Cottontail *Sylvilagus floridanus*	Brush piles, stone walls, dens or burrows; herbaceous and shrubby cover.	B BF W WF											
	New England Cottontail *Sylvilagus transitionalis*	Closely spaced patches of young dense woody cover.	B BF W WF											
	Snowshoe Hare *Lepus americanus*	Dense brushy or softwood cover.	B BF W WF											
	European Hare *Lepus europaeus*	Fields, meadows.	B BF W WF											
	Eastern Chipmunk *Tamias striatus*	Tree or shrub cover; elevated perches, decaying stumps/logs, stone walls.	B BF W WF											
	Woodchuck *Marmota monax*	Open land with well-drained soils in which to burrow.	B BF W WF											
	Gray Squirrel *Sciurus carolinensis*	Mast-producing trees; tall trees for dens and leaf nests.	B BF W WF											
	Red Squirrel *Tamiasciurus hudsonicus*	Woodlands with mature trees; conifers preferred.	B BF W WF											

Species	Habitat	Season
Southern Flying Squirrel *Glaucomys volans*	Mature woodland with cavity trees; favors cavities with entrance diameters of 1.6–2 in (4–5 cm).	B BF W WF
Northern Flying Squirrel *Glaucomys sabrinus*	Mature trees, cavities for winter dens; arboreal lichens.	B BF W WF
Beaver *Castor canadensis*	Low-gradient woodland streams with adjacent young hardwoods.	B BF W WF
Deer Mouse *Peromyscus maniculatus*	Down logs, rotting stumps.	B BF W WF
White-footed Mouse *Peromyscus leucopus*	Down logs, rotting stumps, cavities.	B BF W WF
Southern Red-backed Vole *Clethrionomys gapperi*	Springs, brooks, seeps, bogs; debris or slash cover.	B BF W WF
Meadow Vole *Microtus pennsylvanicus*	Herbaceous vegetation; loose soils.	B BF W WF
Rock Vole *Microtus chrotorrhinus*	Cool moist, rocky woodlands with herbaceous ground cover and flowing water.	B BF W WF
Woodland Vole *Microtus pinetorum*	Moist, well-drained soils; variable depths of leaf litter or grass cover.	B BF W WF
Southern Bog Lemming *Synaptomys cooperi*	Moist soils; grass-sedge ground cover.	B BF W WF
Northern Bog Lemming *Synaptomys borealis*	Moist to wet loose soils; decaying leaf litter or herbaceous ground cover.	B BF W WF

Forest Matrix (*continued*)

LOCAL OCC.	SPECIES	SPECIAL HABITAT FEATURES	SEASONAL USE	ASPEN-BIRCH: Aspen	ASPEN-BIRCH: Paper birch	NORTHERN HARDWOOD: Northern hardwoods	S. HARD-WOOD: Red maple	SPRUCE-FIR: Balsam fir	SPRUCE-FIR: Spruce-fir	SPRUCE-FIR: Red spruce	HEMLOCK: Eastern hemlock	OAK-PINE: Northern red oak	OAK-PINE: Pine-Oak-Maple	OAK-PINE: E. white pine
				S Sp St L	S Sp St L	S Sp St L U	S Sp St L	S Sp St L	S Sp St L	S Sp St L	S Sp St L	S Sp St L	S Sp St L	S Sp St L
	Meadow Jumping Mouse *Zapus hudsonius*	Open wooded stands; herbaceous ground cover; loose soils. Hibernates in winter.	B BF W WF											
	Woodland Jumping Mouse *Napaeozapus insignis*	Moist cool woodlands; loose soils, herbaceous ground cover. Hibernates in winter.	B BF W WF											
	Porcupine *Erethizon dorsatum*	Rock ledges or den trees.	B BF W WF											
	Coyote *Canis latrans*	Well-drained secluded den sites.	B BF W WF											
	Gray Wolf *Canis lupus*	Seclusion; abundant prey base. Potential habitat in north.	B BF W WF											
	Red Fox *Vulpes vulpes*	Well-drained den sites. Tends to hunt more open or semiopen habitats.	B BF W WF											
	Gray Fox *Urocyon cinereoargenteus*	Hollow logs, tree cavities, rock crevices. Tends to hunt forest edges.	B BF W WF											
	Black Bear *Ursus americanus*	Dens in any semiprotected areas (fallen trees, talus slopes, slash piles); seeps and wet areas in early spring; mast.	B BF W WF											
	Raccoon *Procyon lotor*	Hollow trees, dens usually >10 ft (3 m) above ground.	B BF W WF											

Species	Habitat features	Regions
American Marten *Martes americana*	Hollow trees, coarse woody debris.	B BF W WF
Fisher *Martes pennanti*	Hollow trees, logs; dense regenerating softwoods.	B BF W WF
Ermine *Mustela erminea*	Dense brushy cover, slash. Areas of abundant prey.	B BF W WF
Long-tailed Weasel *Mustela frenata*	Areas of abundant prey. Previously excavated den sites. Wooded edges.	B BF W WF
Mink *Mustela vison*	Hollow logs, natural cavities, under tree roots. Forest-wetland edges.	B BF W WF
Striped Skunk *Mephitis mephitis*	Well-drained soils for burrows/den sites. Open uplands; around human habitation.	B BF W WF
River Otter *Lontra canadensis*	Water (e.g., stream, pond, lake or river); bank dens.	B BF W WF
Mountain Lion *Puma concolor*	Seclusion; abundant prey base (e.g., deer).	B BF W WF
Lynx *Lynx canadensis*	Abundant prey base (e.g., snowshoe hare). Dense softwood regeneration; secluded den sites.	B BF W WF
Bobcat *Lynx rufus*	Dense woody understory; rock ledges and talus; successional habitats with abundant prey base (e.g., lagomorphs).	B BF W WF
White-tailed Deer *Odocoileus virginianus*	Softwood yarding cover in north; adequate winter browse; summer herbaceous forage and mast.	B BF W WF

Forest Matrix *(continued)*

LOCAL OCC.	SPECIES	SPECIAL HABITAT FEATURES	SEASONAL USE	ASPEN-BIRCH: Aspen	ASPEN-BIRCH: Paper birch	NORTHERN HARDWOOD: Northern hardwoods	S. HARDWOOD: Red maple	SPRUCE-FIR: Balsam fir	SPRUCE-FIR: Spruce-fir	SPRUCE-FIR: Red spruce	HEMLOCK: Eastern hemlock	OAK-PINE: Northern red oak	OAK-PINE: Pine-Oak-Maple	OAK-PINE: E. white pine
				S Sp St L	S Sp St L	S Sp St L U	S Sp St L	S Sp St L	S Sp St L	S Sp St L	S Sp St L	S Sp St L	S Sp St L	S Sp St L
	Moose *Alces alces*	Wetlands (in summer).	B											
			BF											
			W											
			WF											

Nonforested Matrix

LOCAL OCCURRENCE	SPECIES	SPECIAL HABITAT FEATURES	SEASONAL USE	TERRESTRIAL									WETLAND/DEEP WATER													OTHER				
				Upland Field									Palustrine							Riverine				Marine						
				Cultivated	Grass	Forb	Shrub/old field	Pasture	Savanna	Orchard	Krummholz	Alpine	Sedge meadow	Shallow marsh	Deep marsh	Shrub swamp	Bog	Pond	Lake	Stream	River	Riparian	Estuary/salt marsh	Coastal beach/rocks	Bay/ocean	Stable bank	Ledge, cliff	Cave, mine	Structure	Derelict bldg., debris
	Common Mudpuppy *Necturus maculosus*	Flowing water ≥1 m deep, hiding cover for breeding.	B												□	□		□	■	■	■									
			NB																■	■	■									
	Marbled Salamander *Ambystoma opacum*	Temporary (autumnal) pools or swamps in upland woods for breeding.	B													□						□								
			NB																			□								
	Jefferson Salamander *Ambystoma jeffersonianum*	Vernal pools or semipermanent water adjacent to upland habitat for breeding.	B										□	□	□	□	□	■				□								
			NB																											□
	Blue-spotted Salamander *Ambystoma laterale*	Vernal pools or semipermanent water for breeding; adjacent upland habitat.	B										■	■	□	■	□	■				□								
			NB																											□
	Spotted Salamander *Ambystoma maculatum*	Mesic woods, vernal pools or semi-permanent water for breeding.	B										□	□	□	□	□	■				□								
			NB																											
	Red-spotted Newt *Notophthalmus v. viridescens*	Water bodies with aquatic vegetation for adult newts; juveniles (efts) are terrestrial.	B										□	□	■	□	□	■	■	□										
			NB										□	□	□	□	□	□	□	□		□								
	Northern Dusky Salamander *Desmognathus fuscus*	Permanent forested seeps, springs, headwater streams.	B															□		■		■								
			NB																	□		□								
	Northern Redback Salamander *Plethodon cinereus*	Logs, stumps, rocks, etc.	B																											□
			NB																											□
	Northern Slimy Salamander *Plethodon glutinosus*	Rock outcroppings, logs in wooded areas.	B																								□	□		□
			NB																								□	□		□
	Four-toed Salamander *Hemidactylium scutatum*	Wet woodlands containing sphagnum moss.	B										□	□		□	■			□										
			NB													□	■													□
	Northern Spring Salamander *Gyrinophilus p. porphyriticus*	Woodland headwater streams; high gradient seeps or springs.	B																	■		□								
			NB																	■		□								

Nonforested Matrix *(continued)*

LOCAL OCCURRENCE	SPECIES	SPECIAL HABITAT FEATURES	SEASONAL USE
	Northern Two-lined Salamander *Eurycea bislineata*	Well-shaded streams for breeding.	B NB
	Eastern Spadefoot *Scaphiopus holbrookii*	Sandy soils; temporary, seasonally or periodically flooded areas (5 to 30 cm deep).	B NB
	Eastern American Toad *Bufo a. americanus*	Shallow, still water for breeding.	B NB
	Fowler's Toad *Bufo fowleri*	Well-drained sandy soils; shallow water for breeding.	B NB
	Northern Spring Peeper *Pseudacris c. crucifer*	Pools for breeding adjacent to forested or shrub habitat.	B NB
	Gray Treefrog *Hyla versicolor*	Seeps or shallow water adjacent to upland forests for breeding; hibernates under leaves, logs.	B NB
	Bullfrog *Rana catesbeiana*	Deep permanent water with floating or emergent vegetation.	B NB
	Green Frog *Rana clamitans melanota*	Margins of riverine or lacustrine habitats for breeding.	B NB
	Mink Frog *Rana septentrionalis*	Margins of lacustrine habitats with abundant lily pads; adjacent wetlands.	B NB
	Wood Frog *Rana sylvatica*	Vernal pools in forest habitat.	B NB
	Northern Leopard Frog *Rana pipiens*	Shallow pools of standing water adjacent to wet meadows.	B NB
	Pickerel Frog *Rana palustris*	Shallow, clear water of bogs, woodland streams, and lake margins.	B NB

Habitat columns: TERRESTRIAL — Upland Field: Cultivated, Grass, Forb, Shrub/old field, Pasture; Savanna, Orchard, Krummholz, Alpine. WETLAND/DEEP WATER — Palustrine: Sedge meadow, Shallow marsh, Deep marsh, Shrub swamp, Bog, Pond; Lake; Riverine: Stream, River, Riparian; Estuary/salt marsh; Marine: Coastal beach/rocks, Bay/ocean. OTHER — Stable bank, Ledge, cliff, Cave, mine, Structure, Derelict bldg., debris.

Common Snapping Turtle *Chelydra s. serpentina*	Wetlands within 10 km of well-drained sandy, gravelly, or loamy areas for nesting.	B NB
Spotted Turtle *Clemmys guttata*	Wetland complex with vernal pools, adjacent upland forest and well-drained openings for nesting.	B NB
Bog Turtle *Clemmys muhlenbergii*	Shallow, spring-fed wetlands.	B NB
Wood Turtle *Clemmys insculpta*	Forested rivers, streams within 200 m of well-drained sandy or gravelly substrates for nesting.	B NB
Eastern Box Turtle *Terrapene c. carolina*	Old fields, clearings, and ecotones with sandy soils.	B NB
Map Turtle *Graptemys geographica*	Shallow water bodies adjacent to open areas; loamy, sandy, gravelly substrates for nesting.	B NB
Plymouth Redbelly Turtle *Pseudemys rubriventris bangsi*	Muddy-bottom shallows with thick aquatic vegetation; near open uplands for nesting.	B NB
Painted Turtle *Chrysemys picta*	Wetlands with basking sites (floating logs); areas of open water (15 m^2 minimum)	B NB
Blanding's Turtle *Emydoidea blandingii*	Wetland complex with shallow water (≤0.5 m); adjacent upland fields or forest for nesting.	B NB
Common Musk Turtle *Sternotherus odoratus*	Permanent water bodies 0.3–3.0 m deep within 15 m of well-drained areas for nesting.	B NB
Eastern Spiny Softshell *Apalone s. spinifera*	Shallow water with muddy bottoms for burrowing; aquatic vegetation.	B NB
Five-lined Skink *Eumeces fasciatus*	Steep, rocky open woods with logs and slash piles.	B NB
Northern Water Snake *Nerodia s. sipedon*	Branches, logs overhanging water, or boulders of dams and causeways in reservoirs.	B NB
Northern Brown Snake *Storeria d. dekayi*		B NB
Northern Redbelly Snake *Storeria o. occipito-maculata*	Woodland debris—bark and rotting wood.	B NB

Nonforested Matrix (*continued*)

LOCAL OCCURRENCE	SPECIES	SPECIAL HABITAT FEATURES	SEASONAL USE	TERRESTRIAL									WETLAND/DEEP WATER													OTHER				
				Upland Field									Palustrine							Riverine				Marine						
				Cultivated	Grass	Forb	Shrub/old field	Pasture	**Savanna**	**Orchard**	**Krummholz**	**Alpine**	Sedge meadow	Shallow marsh	Deep marsh	Shrub swamp	Bog	Pond	**Lake**	Stream	River	Riparian	**Estuary/salt marsh**	Coastal beach/rocks	Bay/ocean	**Stable bank**	**Ledge, cliff**	**Cave, mine**	**Structure**	**Derelict bldg., debris**
	Common Garter Snake *Thamnophis sirtalis*		B		□	□	□	□	□	□	□	□																		
			NB		□	□	□	□	□	□	□	□	□	□	□	□	□	□		□		□							□	□
	Ribbon Snake *Thamnophis sauritus*	Permanent shallow water in open grassy habitat.	B										■	□	□	□	□	□		■		■								
			NB										□	□	□	□	□	□		■		■								
	Eastern Hognose Snake *Heterodon platyrhinos*	Sandy soils, open woodlands.	B	□	□	□	□	■	□	□												□							□	
			NB	□	□	□	□	□	□	□				□								□							□	□
	Northern Ringneck Snake *Diadophis punctatus edwardsii*	Mesic to xeric areas with abundant cover.	B					□																						■
			NB					□																						■
	Eastern Worm Snake *Carphophis a. amoenus*	Well-drained, loose soils for burrowing or covering.	B	□		□	□	□																						
			NB	□		□	□	□																						
	Northern Black Racer *Coluber c. constrictor*	Upland areas that are periodically cleared or mowed; adjacent forest and wetlands.	B		■	□	□	□	□	□				□		□						□							□	■
			NB		□	□	□	□	□	□				□		□						□					□		□	■
	Eastern Smooth Green Snake *Liochlorophis vernalis*	Upland grassy openings.	B		■	□	□	■	□																					
			NB		□	□	□	□	□				□	□		□	□													□
	Black Rat Snake *Elaphe o. obsoleta*		B		□	□	■	□	□	□																			□	■
			NB		□	□	■	□	□	□																	□		□	□
	Eastern Milk Snake *Lampropeltis t. triangulum*	Slash, wood piles, debris or loose soils for egg laying.	B		□	□	□	□	□	□																			■	■
			NB		□	□	□	□	□	□																	□		■	■
	Northern Copperhead *Agkistrodon contortrix mokasen*	Rocky hillsides, talus slopes. Near water in late summer.	B																								■			
			NB																								■			
	Timber Rattlesnake *Crotalus horridus*	Rock ledges on forested hillsides. Near streams in late summer.	B																								■			
			NB																								■			

Species	Habitat	Season
Common Loon *Gavia immer*	Large undisturbed water bodies with stable water levels and islands.	B BF W WF
Pied-billed Grebe *Podilymbus podiceps*	Marshes with emergent vegetation >5 ha (12.5 acres).	B BF W WF
Double-crested Cormorant *Phalacrocorax auritus*	Undisturbed nesting sites.	B BF W WF
American Bittern *Botaurus lentiginosus*	Undisturbed tall marsh vegetation.	B BF W WF
Least Bittern *Ixobrychus exilis*	Deep marshes with clumps of emergent vegetation.	B BF W WF
Little Blue Heron *Egretta caerulea*	Open water or wetland habitat.	B BF W WF
Great Blue Heron *Ardea herodias*	Tall trees for nesting.	B BF W WF
Great Egret *Ardea alba*	Open water or wetland habitats near woods or forested wetlands.	B BF W WF
Cattle Egret *Bubulcus ibis*	Wetlands for nesting; upland openings for feeding.	B BF W WF
Snowy Egret *Egretta thula*	Wetlands.	B BF W WF
Green Heron *Butorides virescens*	Forested wetlands or open water habitats.	B BF W WF

Nonforested Matrix (*continued*)

TERRESTRIAL — Upland Field: Cultivated, Grass, Forb, Shrub/old field, Pasture; Savanna, Orchard, Krummholz, Alpine

WETLAND/DEEP WATER — Palustrine: Sedge meadow, Shallow marsh, Deep marsh, Shrub swamp, Bog, Pond; Lake; Riverine: Stream, River, Riparian; Estuary/salt marsh; Marine: Coastal beach/rocks, Bay/ocean

OTHER — Stable bank, Ledge, cliff, Cave, mine, Structure, Derelict bldg., debris

LOCAL OCCURRENCE	SPECIES	SPECIAL HABITAT FEATURES	SEASONAL USE
	Black-crowned Night-Heron *Nycticorax nycticorax*	Open water or forested wetlands.	B BF W WF
	Yellow-crowned Night-Heron *Nyctanassa violacea*	Forested wetlands.	B BF W WF
	Glossy Ibis *Plegadis falcinellus*	Wetlands.	B BF W WF
	Turkey Vulture *Cathartes aura*	Forest clearings and fields; large branch stubs of dead trees.	B BF W WF
	Canada Goose *Branta canadensis*	Open water; herbaceous ground cover. Elevated nest sites.	B BF W WF
	Mute Swan *Cygnus olor*	Shallow ponds, rivers with abundant aquatic vegetation.	B BF W WF
	Wood Duck *Aix sponsa*	Trees ≥16 in (40.6 cm) dbh with large cavities and 4 in (10.2 cm) diameter entrance holes.	B BF W WF
	Gadwall *Anas strepera*	Moderate to large bodies of water with submerged aquatic plants.	B BF W WF

Species	Habitat	Season
American Wigeon *Anas americana*		B BF W WF
American Black Duck *Anas rubripes*	Wooded wetlands, streambanks (inland).	B BF W WF
Mallard *Anas platyrhynchos*		B BF W WF
Blue-winged Teal *Anas discors*		B BF W WF
Northern Shoveler *Anas clypeata*	Shallow water with muddy bottoms, surrounded by dry grassy areas.	B BF W WF
Northern Pintail *Anas acuta*	Drakes need mudbanks or exposed shorelines.	B BF W WF
Green-winged Teal *Anas crecca*		B BF W WF
Canvasback *Aythya valisineria*	Stretches of open water and emergent vegetation.	B BF W WF
Ring-necked Duck *Aythya collaris*	Expanses of open water.	B BF W WF
Bufflehead *Bucephala albeola*	Winters on coastal bays and on open water inland.	B BF W WF
Common Goldeneye *Bucephala clangula*	Cavity trees ≥20 in (51 cm) dbh; clear cold, shallow water.	B BF W WF

Nonforested Matrix (*continued*)

LOCAL OCCURRENCE	SPECIES	SPECIAL HABITAT FEATURES	SEASONAL USE
	Hooded Merganser *Lophodytes cucullatus*	Undisturbed wooded areas with cavity trees ≥15 in (38 cm) dbh; clear fresh water.	B BF W WF
	Common Merganser *Mergus merganser*	Large cavity trees at water's edge.	B BF W WF
	Red-breasted Merganser *Mergus serrator*	Clear water for feeding.	B BF W WF
	Osprey *Pandion haliaetus*	Clear lakes, rivers containing fish; supracanopy or exposed nest sites.	B BF W WF
	Bald Eagle *Haliaeetus leucocephalus*	Large undisturbed water bodies containing fish; large, live supracanopy nest trees within 0.25 mile of shore.	B BF W WF
	Northern Harrier *Circus cyaneus*	Marshes or open country with low vegetation.	B BF W WF
	Sharp-shinned Hawk *Accipiter striatus*	Extensive undisturbed, open mixed woodlands.	B BF W WF
	Cooper's Hawk *Accipiter cooperii*	Mature deciduous or coniferous woodlands in open or semi-open country.	B BF W WF

Habitat columns: TERRESTRIAL — Upland Field: Cultivated, Grass, Forb, Shrub/old field, Pasture; Savanna, Orchard, Krummholz, Alpine. WETLAND/DEEP WATER — Palustrine: Sedge meadow, Shallow marsh, Deep marsh, Shrub swamp, Bog, Pond; Lake; Riverine: Stream, River, Riparian; Estuary/salt marsh; Marine: Coastal beach/rocks, Bay/ocean. OTHER — Stable bank, Ledge, cliff, Cave, mine, Structure, Derelict bldg., debris

Species	Habitat	Season
Northern Goshawk *Accipiter gentilis*	Extensive undisturbed mature mixed woodlands.	B BF W WF
Red-shouldered Hawk *Buteo lineatus*	Cool, moist mature forest, riparian woodlands.	B BF W WF
Broad-winged Hawk *Buteo platypterus*	Extensive woodlands with roads or clearings.	B BF W WF
Red-tailed Hawk *Buteo jamaicensis*	Mature forests for nesting; nonforest openings for foraging.	B BF W WF
Rough-legged Hawk *Buteo lagopus*	Open country.	B BF W WF
Golden Eagle *Aquila chrysaetos*	Cliffs for nesting; wide expanses of open land.	B BF W WF
American Kestrel *Falco sparverius*	Cavity trees ≥12 in (30.5 cm); open flat terrain.	B BF W WF
Merlin *Falco columbarius*	Natural cavities or old woodpecker holes in open woodlands.	B BF W WF
Peregrine Falcon *Falco peregrinus*	High cliffs; abundant avian prey base.	B BF W WF
Gray Partridge *Perdix perdix*	Grasslands, grain crops.	B BF W WF
Ring-necked Pheasant *Phasianus colchicus*	Cultivated fields, farmland.	B BF W WF

Nonforested Matrix (*continued*)

(■ = dark square; □ = light square)

LOCAL OCCURRENCE	SPECIES	SPECIAL HABITAT FEATURES	SEASONAL USE	Cultivated	Grass	Forb	Shrub/old field	Pasture	Savanna	Orchard	Krummholz	Alpine	Sedge meadow	Shallow marsh	Deep marsh	Shrub swamp	Bog	Pond	Lake	Stream	River	Riparian	Estuary/salt marsh	Coastal beach/rocks	Bay/ocean	Stable bank	Ledge, cliff	Cave, mine	Structure	Derelict bldg., debris
				TERRESTRIAL									WETLAND/DEEP WATER													OTHER				
				Upland Field									Palustrine							Riverine				Marine						
	Ruffed Grouse *Bonasa umbellus*	Drumming log sites amid dense saplings; herbaceous cover on old roads, small openings.	B				□			□																				
			BF																											
			W																											
			WF				□			□																				
	Spruce Grouse *Falcipennis canadensis*	Coniferous forest.	B								□						□													
			BF								□						□													
			W								□						□													
			WF								□						□													
	Wild Turkey *Meleagris gallopavo*	Open, mast-producing woodlands; large conifers for roosting; herbaceous cover in woodland clearings, small openings.	B																											
			BF	□	□	□	□	□	□	□																				
			W																											
			WF	□	□	□	□	□	□	□																				
	Northern Bobwhite *Colinus virginianus*	Brushy field edges; well-drained sandy or loamy soils.	B	□	■	□	□	□	□	□																				
			BF	□	■	□	□	□	□	□																				
			W	□	■	□	□	□	□	□																				
			WF	□	■	□	□	□	□	□																				
	King Rail *Rallus elegans*	Stable water levels.	B										□	■	□								□							
			BF										□	■	□								□							
			W																				□							
			WF																				□							
	Virginia Rail *Rallus limicola*	Wetlands with sedges and cattails.	B											■	□								□							
			BF											■	□								□							
			W																											
			WF																											
	Sora *Porzana carolina*	Emergent vegetation.	B										□	■	□	□		□					□							
			BF										□	■	□	□		□					□							
			W																											
			WF																											
	Common Moorhen *Gallinula chloropus*	Emergent vegetation in water 1–3 ft (0.3–1 m) deep.	B											■	□			□	□				□							
			BF											■	□			□	□				□							
			W																											
			WF																											

American Coot *Fulica americana*	Emergent vegetation in water 1–4 ft (0.3–1.2 m) deep.	B BF W WF
Killdeer *Charadrius vociferus*	Bare ground, sparse ground vegetation.	B BF W WF
Willet *Catoptrophorus semi-palmatus*	Coastal marshes and nearby grassy areas.	B BF W WF
Spotted Sandpiper *Actitis macularia*	Margins of freshwater bodies.	B BF W WF
Upland Sandpiper *Bartramia longicauda*	Large open grasslands.	B BF W WF
Common Snipe *Gallinago gallinago*	Moist organic soils; large open spaces, low scattered vegetation, bogs, swamps.	B BF W WF
American Woodcock *Scolopax minor*	Fertile moist soils containing earthworms; small clearings and dense swales.	B BF W WF
Ring-billed Gull *Larus delawarensis*		B BF W WF
Herring Gull *Larus argentatus*		B BF W WF
Great Black-backed Gull *Larus marinus*		B BF W WF
Common Tern *Sterna hirundo*	Gravelly sandy beaches; grassy uplands along marine shore.	B BF W WF

Nonforested Matrix *(continued)*

TERRESTRIAL — Upland Field: Cultivated, Grass, Forb, Shrub/old field, Pasture; **Savanna**, **Orchard**, **Krummholz**, **Alpine**

WETLAND/DEEP WATER — Palustrine: Sedge meadow, Shallow marsh, Deep marsh, Shrub swamp, Bog, Pond; **Lake**; Riverine: Stream, River, Riparian; **Estuary/salt marsh**; Marine: Coastal beach/rocks, Bay/ocean

OTHER — **Stable bank**, **Ledge, cliff**, **Cave, mine**, **Structure**, **Derelict bldg., debris**

LOCAL OCCURRENCE	SPECIES	SPECIAL HABITAT FEATURES	SEASONAL USE
	Black Tern *Chlidonias niger*	Aquatic habitats with extensive stands of emergent vegetation and areas of shallow open water.	B BF W WF
	Rock Dove *Columba livia*		B BF W WF
	Mourning Dove *Zenaida macroura*	Open land with bare ground.	B BF W WF
	Black-billed Cuckoo *Coccyzus erythropthalmus*	Low, dense thickets.	B BF W WF
	Yellow-billed Cuckoo *Coccyzus americanus*	Low, dense thickets.	B BF W WF
	Barn Owl *Tyto alba*	Barns, silos, abandoned buildings, tree cavities.	B BF W WF
	Eastern Screech-Owl *Otus asio*	Cavity trees ≥12 in (30.5 cm) dbh.	B BF W WF
	Great Horned Owl *Bubo virginianus*	Large abandoned hawk nests; large tree cavities.	B BF W WF

Species	Habitat	Season
Snowy Owl *Nyctea scandiaca*	Open country.	B BF W WF
Northern Hawk Owl *Surnia ulula*	Coniferous forests.	B BF W WF
Barred Owl *Strix varia*	Cool, damp lowlands; cavity trees ≥20 in (50.8 cm) dbh.	B BF W WF
Great Gray Owl *Strix nebulosa*	Meadows, swamps near woodlands.	B BF W WF
Long-eared Owl *Asio otus*	Dense conifer thickets in open country; coniferous forested wetlands.	B BF W WF
Short-eared Owl *Asio flammeus*	Extensive open grasslands, dunes.	B BF W WF
Boreal Owl *Aegolius funereus*	Cavity trees, often with large holes; mixedwood or coniferous forest and coniferous forested wetlands.	B BF W WF
Northern Saw-whet Owl *Aegolius acadicus*	Cavity trees ≥ 12 in (30.1 cm) dbh near forest clearings.	B BF W WF
Common Nighthawk *Chordeiles minor*	Barren areas with rocky substrate; graveled rooftops.	B BF W WF
Whip-poor-will *Caprimulgus vociferus*	Ungrazed woodlands with openings.	B BF W WF
Chimney Swift *Chaetura pelagica*	Chimneys; large hollow dead trees in openings and wetlands.	B BF W WF

Nonforested Matrix (*continued*)

LOCAL OCCURRENCE	SPECIES	SPECIAL HABITAT FEATURES	SEASONAL USE	TERRESTRIAL									WETLAND/DEEP WATER													OTHER				
				Upland Field									Palustrine							Riverine				Marine						
				Cultivated	Grass	Forb	Shrub/old field	Pasture	**Savanna**	**Orchard**	**Krummholz**	**Alpine**	Sedge meadow	Shallow marsh	Deep marsh	Shrub swamp	Bog	Pond	**Lake**	Stream	River	Riparian	**Estuary/salt marsh**	Coastal beach/rocks	Bay/ocean	**Stable bank**	**Ledge, cliff**	**Cave, mine**	**Structure**	**Derelict bldg., debris**
	Ruby-throated Hummingbird *Archilochus colubris*	Tubular nectar-bearing flowers, preferably red.	B				□			□																				
			BF			□	□			□																				
			W																											
			WF																											
	Belted Kingfisher *Ceryle alcyon*	Perches over streams, ponds; banks for nest sites.	B															□	□	□	□	□		□		■				
			BF															□	□	□	□	■	□							
			W																			■								
			WF															□	□	□	□	■	□							
	Red-headed Woodpecker *Melanerpes erythro-cephalus*	Cavity trees in savanna or open country.	B						□	□												□								
			BF					□	□	□												□								
			W							□												□								
			WF							□												□								
	Red-bellied Woodpecker *Melanerpes carolinus*	Extensive mature woodlands with dead trees or trees with large dead limbs.	B																			■								
			BF																			■								
			W							□												■								
			WF							□												■								
	Yellow-bellied Sapsucker *Sphyrapicus varius*	Trees ≥10 in (26 cm) dbh; particularly aspens containing decayed wood.	B																			□								
			BF																			□								
			W							■												□								
			WF							■												□								
	Downy Woodpecker *Picoides pubescens*	Decay columns in trees ≥6 in (15 cm) dbh or larger limbs.	B																			■								
			BF																			■								
			W							□												■								
			WF							□												■								
	Hairy Woodpecker *Picoides villosus*	Decayed wood column in trees ≥10 in (25 cm) dbh or larger limbs.	B																			■								
			BF																			■								
			W							□												■								
			WF							□												■								
	Three-toed Woodpecker *Picoides tridactylus*	Spruce and fir trees ≥12 in (31 cm) dbh with column of decayed wood; dead trees with loose bark.	B														□													
			BF														□													
			W														□													
			WF														□													

Species	Habitat	
Black-backed Woodpecker *Picoides arcticus*	Spruce and fir trees ≥10 in (25 cm) dbh with column of decayed wood; dead trees with loose bark; often near water.	B BF W WF
Northern Flicker *Colaptes auratus*	Open areas, forest edges; trees ≥12 in (31 cm) dbh with column of decayed wood.	B BF W WF
Pileated Woodpecker *Dryocopus pileatus*	Mature forest; trees ≥ 20 in (51 cm) dbh with column of decayed wood.	B BF W WF
Olive-sided Flycatcher *Contopus cooperi*	Tall perch adjacent to low, wet coniferous thicket, bog.	B BF W WF
Eastern Wood-Pewee *Contopus virens*	Forest edge or open woods.	B BF W WF
Yellow-bellied Flycatcher *Empidonax flaviventris*	Low, wet coniferous forest.	B BF W WF
Acadian Flycatcher *Empidonax virescens*	Extensive mature, deciduous woodlands; closed canopy; open understory for foraging.	B BF W WF
Alder Flycatcher *Empidonax alnorum*	Dense woody regeneration, low shrubs; thickets and clearings.	B BF W WF
Willow Flycatcher *Empidonax traillii*	Low deciduous trees and shrubs in clearings; thick hardwood regeneration, streamside thickets.	B BF W WF
Least Flycatcher *Empidonax minimus*	Open deciduous or mixed forest; forest edges.	B BF W WF
Eastern Phoebe *Sayornis phoebe*	Exposed perches, cliffs or ledges in streamside clearings; buildings or structures in openings for nesting.	B BF W WF

Nonforested Matrix (*continued*)

				TERRESTRIAL									WETLAND/DEEP WATER													OTHER				
				Upland Field									Palustrine							Riverine				Marine						
LOCAL OCCURRENCE	SPECIES	SPECIAL HABITAT FEATURES	SEASONAL USE	Cultivated	Grass	Forb	Shrub/old field	Pasture	Savanna	Orchard	Krummholz	Alpine	Sedge meadow	Shallow marsh	Deep marsh	Shrub swamp	Bog	Pond	Lake	Stream	River	Riparian	Estuary/salt marsh	Coastal beach/rocks	Bay/ocean	Stable bank	Ledge, cliff	Cave, mine	Structure	Derelict bldg., debris
	Great Crested Flycatcher *Myiarchus crinitus*	Mature deciduous forest; cavity trees on forest edges.	B BF W WF																											
	Eastern Kingbird *Tyrannus tyrannus*	Clearings, fields, orchards; perches for aerial foraging.	B BF W WF																											
	Loggerhead Shrike *Lanius ludovicianus*	Open country with short grasses, scattered trees, shrubs.	B BF W WF																											
	Northern Shrike *Lanius excubitor*	Scattered trees or shrubs in open country.	B BF W WF																											
	White-eyed Vireo *Vireo griseus*	Low shrubs, thickets.	B BF W WF																											
	Yellow-throated Vireo *Vireo flavifrons*	Mature moist deciduous forest.	B BF W WF																											
	Warbling Vireo *Vireo gilvus*	Scattered deciduous trees in open areas; often adjacent to water.	B BF W WF																											
	Philadelphia Vireo *Vireo philadelphicus*	Hardwood forest edges; early successional forests.	B BF W WF																											

Species	Habitat	
Red-eyed Vireo *Vireo olivaceus*	Closed deciduous forest canopy.	B BF W WF
Gray Jay *Perisoreus canadensis*	Coniferous forests.	B BF W WF
Blue Jay *Cyanocitta cristata*		B BF W WF
American Crow *Corvus brachyrhynchos*		B BF W WF
Fish Crow *Corvus ossifragus*		B BF W WF
Common Raven *Corvus corax*	Cliffs, abandoned raptor nests.	B BF W WF
Horned Lark *Eremophila alpestris*	Bare, exposed soil.	B BF W WF
Purple Martin *Progne subis*	Martin houses near ponds, lakes, openings.	B BF W WF
Tree Swallow *Tachycineta bicolor*	Cavity trees ≥10 in (25 cm) dbh; old woodpecker cavities; near open areas.	B BF W WF
Northern Rough-winged Swallow *Stelgidopteryx serripennis*	Sandy or clay banks; stabilized by grassy overhang near, but up to 1 km from water.	B BF W WF
Bank Swallow *Riparia riparia*	Stabilized steep banks of sand, gravel, or clay; near open areas, preferably flowing water.	B BF W WF

Nonforested Matrix (*continued*)

LOCAL OCCURRENCE	SPECIES	SPECIAL HABITAT FEATURES	SEASONAL USE	TERRESTRIAL: Upland Field					TERRESTRIAL				WETLAND/DEEP WATER: Palustrine						WETLAND/DEEP WATER	WETLAND/DEEP WATER: Riverine			WETLAND/DEEP WATER	WETLAND/DEEP WATER: Marine		OTHER				
				Cultivated	Grass	Forb	Shrub/old field	Pasture	**Savanna**	**Orchard**	**Krummholz**	**Alpine**	Sedge meadow	Shallow marsh	Deep marsh	Shrub swamp	Bog	Pond	**Lake**	Stream	River	Riparian	**Estuary/salt marsh**	Coastal beach/rocks	Bay/ocean	**Stable bank**	**Ledge, cliff**	**Cave, mine**	**Structure**	**Derelict bldg., debris**
	Cliff Swallow *Petrochelidon pyrrhonata*	Open areas, water; vertical wall with an overhang; mud for nest construction.	B																					□			□		■	■
			BF		■	□	□	□	□	□			□	□	□			□												
			W																											
			WF																											
	Barn Swallow *Hirundo rustica*	Abandoned or little-used buildings for nesting.	B																										■	■
			BF		■	□	□	□	□	□			□	□	□			□		□	□	□	□	□						
			W																											
			WF																											
	Black-capped Chickadee *Poecile atricapillus*	Cavity trees or stubs in small woodlands; clearings or open woodlands.	B							□	□											□								
			BF							□	□											□								
			W																			□								
			WF				□		□	□												□								
	Boreal Chickadee *Poecile hudsonicus*	Softwood snags, stubs.	B														□													
			BF														□													
			W														□													
			WF														□													
	Tufted Titmouse *Baeolophus bicolor*	Cavity trees ≥8 in (20 cm) dbh.	B																			■								
			BF							□												■								
			W																			■								
			WF							□												■								
	Red-breasted Nuthatch *Sitta canadensis*	Cavity trees ≥12 in (31 cm) dbh in mixed or coniferous woods.	B								□						□													
			BF								□						□													
			W														□													
			WF							□							□													
	White-breasted Nuthatch *Sitta carolinensis*	Cavity trees ≥12 in (31 cm) dbh in hardwoods and mixed woods.	B							□												□								
			BF							□												□								
			W							■												□								
			WF						□	■												□								
	Brown Creeper *Certhia americana*	Woodland trees with sloughing or loose bark.	B														□					□								
			BF														□					□								
			W														□					□								
			WF							□							□					□								

Species	Habitat	Season
Carolina Wren *Thryothorus ludovicianus*	Cavity trees amidst brushy vegetation, thickets, swamps.	B BF W WF
House Wren *Troglodytes aedon*	Cavity trees, shrubs.	B BF W WF
Winter Wren *Troglodytes troglodytes*	Moist, mixed or coniferous woodlands with down logs; low woody vegetation.	B BF W WF
Sedge Wren *Cistothorus platensis*	Sedge meadows or drier edges of marshes.	B BF W WF
Marsh Wren *Cistothorus palustris*	Marshes.	B BF W WF
Golden-crowned Kinglet *Regulus satrapa*	Dense conifer thickets or stands, especially spruce.	B BF W WF
Ruby-crowned Kinglet *Regulus calendula*	Conifer stands, or mixed stands with a predominance of softwoods, especially spruce or fir.	B BF W WF
Blue-gray Gnatcatcher *Polioptila caerulea*	Open, deciduous woodlands.	B BF W WF
Eastern Bluebird *Sialia sialis*	Low cavity trees or nest boxes; open country.	B BF W WF
Veery *Catharus fuscescens*	Moist woodlands with understory of low trees and shrubs.	B BF W WF
Bicknell's Thrush *Catharus bicknelli*	Stunted coniferous forest at high elevations.	B BF W WF

Nonforested Matrix (*continued*)

LOCAL OCCURRENCE	SPECIES	SPECIAL HABITAT FEATURES	SEASONAL USE	TERRESTRIAL: Upland Field: Cultivated	Grass	Forb	Shrub/old field	Pasture	Savanna	Orchard	Krummholz	Alpine	WETLAND/DEEP WATER: Palustrine: Sedge meadow	Shallow marsh	Deep marsh	Shrub swamp	Bog	Pond	Lake	Riverine: Stream	River	Riparian	Estuary/salt marsh	Marine: Coastal beach/rocks	Bay/ocean	OTHER: Stable bank	Ledge, cliff	Cave, mine	Structure	Derelict bldg., debris
	Hermit Thrush *Catharus guttatus*	Coniferous or mixed woodlands with dense undergrowth.	B				□										□													
			BF				□										□													
			W													□														
			WF													□														
	Wood Thrush *Hylocichla mustelina*	Cool, moist mature deciduous or mixed forest.	B																			□								
			BF																			□								
			W																											
			WF																											
	American Robin *Turdus migratorius*		B					□		□	□					□						□								□
			BF	■	□			□		□	□					□						□								
			W							□						□	□					□								
			WF							□						□	□					□								
	Gray Catbird *Dumetella carolinensis*	Low, dense shrubby vegetation in open country or forest understory.	B				□	□		□						□	□					■								
			BF				□	□		□						□	□					■								
			W																											
			WF																											
	Northern Mockingbird *Mimus polyglottus*	Low thickets; high exposed perches; persistent fruit.	B				■	□		□																				
			BF				□	□		□																				
			W				□	□		□																				
			WF				□	□		□																				
	Brown Thrasher *Toxostoma rufum*	Hardwood forest-field ecotone, multiflora rose in pastures.	B				□	■														□								
			BF				□	■														□								
			W				□	□																						
			WF				□	□																						
	European Starling *Sturnus vulgaris*	Cavity trees ≥10 in (25 cm) dbh.	B																			□							□	■
			BF	□				□		□												□	□	□						
			W																			□							□	■
			WF	■	□	□		□	□	□												□	□	□						
	American Pipit *Anthus rubescens*	Alpine or arctic tundra; rocky, hummocky terrain.	B									■																		
			BF									■																		
			W	□																			□	□						
			WF	□																			□	□						

Species	Habitat	Season
Bohemian Waxwing *Bombycilla garrulus*	Soft mast (fruit, berries).	B BF W WF
Cedar Waxwing *Bombycilla cedrorum*	Soft mast (fruit, berries).	B BF W WF
Blue-winged Warbler *Vermivora pinus*	Old fields with scattered shrubs and small trees.	B BF W WF
Golden-winged Warbler *Vermivora chrysoptera*	Brushy edges; openings with saplings, forbs or grasses.	B BF W WF
Tennessee Warbler *Vermivora peregrina*	Brushy, semiopen country.	B BF W WF
Nashville Warbler *Vermivora ruficapilla*	Young hardwood/mixedwood stands; open mixedwood stands interspersed with brush, thickets.	B BF W WF
Northern Parula *Parula americana*	Presence of bearded lichen (*Usnea*).	B BF W WF
Yellow Warbler *Dendroica petechia*	Small scattered trees or dense shrubs; commonly near water.	B BF W WF
Chestnut-sided Warbler *Dendroica pensylvanica*	Early succesional deciduous stands; dense hardwood regeneration 3 to 10 ft (1 to 3 m) tall.	B BF W WF
Yellow-rumped Warbler *Dendroica coronata*	Coniferous trees (summer); bayberry thickets (winter).	B BF W WF
Prairie Warbler *Dendroica discolor*	Coniferous cover in old fields; hardwood saplings, slash.	B BF W WF

Nonforested Matrix (*continued*)

				TERRESTRIAL									WETLAND/DEEP WATER													OTHER				
				Upland Field									Palustrine							Riverine				Marine						
LOCAL OCCURRENCE	SPECIES	SPECIAL HABITAT FEATURES	SEASONAL USE	Cultivated	Grass	Forb	Shrub/old field	Pasture	**Savanna**	**Orchard**	**Krummholz**	**Alpine**	Sedge meadow	Shallow marsh	Deep marsh	Shrub swamp	Bog	Pond	**Lake**	Stream	River	Riparian	**Estuary/salt marsh**	Coastal beach/rocks	Bay/ocean	**Stable bank**	**Ledge, cliff**	**Cave, mine**	**Structure**	**Derelict bldg., debris**
	Palm Warbler *Dendroica palmarum*	Young open coniferous stands, bogs; thick shrub growth.	B BF W WF													□ □	■ ■					□ □								
	Bay-breasted Warbler *Dendroica castanea*	Dense coniferous stands, especially of spruce and fir.	B BF W WF				□ □															□ □								
	Blackpoll Warbler *Dendroica striata*	Stunted spruce, especially at higher elevations.	B BF W WF								■ ■																			
	Cerulean Warbler *Dendroica cerulea*	Tall deciduous trees; often near water.	B BF W WF																			■ ■								
	Black-and-white Warbler *Mniotilta varia*	Deciduous/mixedwood stands.	B BF W WF						□ □	□ □							□ □					□ □								
	American Redstart *Setophaga ruticilla*	Open deciduous/mixedwood stands with a dense shrub/midstory understory.	B BF W WF					□ □	□ □	□ □												□ □								
	Northern Waterthrush *Seiurus noveboracensis*	Cool, shaded, wet ground with shallow pools.	B BF W WF													□ □	□ □													
	Louisiana Waterthrush *Seiurus motacilla*	Woodlands with flowing water.	B BF W WF														□ □			□ □		■ ■								

Species	Habitat	Season
Mourning Warbler *Oporornis philadelphia*	Larger openings of hardwood regeneration; dense shrubs.	B BF W WF
Common Yellowthroat *Geothlypis trichas*	Dense hardwood regeneration, shrub layer.	B BF W WF
Hooded Warbler *Wilsonia citrina*	Dense understory.	B BF W WF
Wilson's Warbler *Wilsonia pusilla*	Cold shrub swamps, bogs; thickets.	B BF W WF
Canada Warbler *Wilsonia canadensis*	Dense deciduous/ericaceous understory; especially along streams, bogs.	B BF W WF
Yellow-breasted Chat *Icteria virens*	Dense thickets and vines with scattered young trees, often near water.	B BF W WF
Eastern Towhee *Pipilo erythrophthalmus*	Dense, brushy understory; well-drained soils.	B BF W WF
American Tree Sparrow *Spizella arborea*	Open country; brushy cover; weedy fields (winter).	B BF W WF
Chipping Sparrow *Spizella passerina*	Clearings with bare ground; coniferous or thorny shrubs.	B BF W WF
Field Sparrow *Spizella pusilla*	Old fields.	B BF W WF
Vesper Sparrow *Pooecetes gramineus*	Open country with short, herbaceous vegetation; conspicuous perches.	B BF W WF

Nonforested Matrix (*continued*)

TERRESTRIAL — Upland Field: Cultivated, Grass, Forb, Shrub/old field, Pasture; **Savanna**, **Orchard**, **Krummholz**, **Alpine**

WETLAND/DEEP WATER — Palustrine: Sedge meadow, Shallow marsh, Deep marsh, Shrub swamp, Bog, Pond; **Lake**; Riverine: Stream, River, Riparian; **Estuary/salt marsh**; Marine: Coastal beach/rocks, Bay/ocean

OTHER — **Stable bank**, **Ledge, cliff**, **Cave, mine**, **Structure**, **Derelict bldg., debris**

LOCAL OCCURRENCE	SPECIES	SPECIAL HABITAT FEATURES	SEASONAL USE
	Savannah Sparrow *Passerculus sandwichensis*	Open country with herbaceous cover of moderate height.	B BF W WF
	Grasshopper Sparrow *Ammodramus savannarum*	Larger openings; continuous, tall herbaceous cover; conspicuous perches.	B BF W WF
	Fox Sparrow *Passerella iliaca*	Dense herbaceous vegetation; field-forest ecotones.	B BF W WF
	Song Sparrow *Melospiza melodia*	Brushy cover, hardwood regeneration; conspicuous song perches.	B BF W WF
	Lincoln's Sparrow *Melospiza lincolnii*	Brushy thickets along field edges or drainways; wet grassy or sedge meadows.	B BF W WF
	Swamp Sparrow *Melospiza georgiana*	Brushy wetlands with emergent vegetation.	B BF W WF
	White-throated Sparrow *Zonotrichia albicollis*		B BF W WF
	Dark-eyed Junco *Junco hyemalis*	Woods roads, cut banks; uprooted trees.	B BF W WF

	Species	Habitat	Season
	Lapland Longspur *Calcarius lapponicus*	Open areas with sparse vegetation (winter).	B BF W WF
	Snow Bunting *Plectrophenax nivalis*	Open areas with sparse vegetation (winter).	B BF W WF
	Northern Cardinal *Cardinalis cardinalis*	Coniferous thickets, vines.	B BF W WF
	Rose-breasted Grosbeak *Pheucticus ludovicianus*	Forest edges, dense hardwood thickets and sapling stands; brushy fields.	B BF W WF
	Indigo Bunting *Passerina cyanea*	Scattered trees in grassy openings; forest-field ecotones.	B BF W WF
	Bobolink *Dolichonyx oryzivorus*	Large expanses of grassland or old hayfields with little or no alfalfa.	B BF W WF
	Red-winged Blackbird *Agelaius phoeniceus*	Emergent vegetation adjacent to open fields.	B BF W WF
	Eastern Meadowlark *Sturnella magna*	Grasslands; elevated song perches.	B BF W WF
	Rusty Blackbird *Euphagus carolinus*	Northern forested wetlands; alder swales.	B BF W WF
	Common Grackle *Quiscalus quiscula*	Wet, open country, shrub swamps, pond edges.	B BF W WF
	Brown-headed Cowbird *Molothrus ater*	Open grassy habitat.	B BF W WF

Nonforested Matrix (*continued*)

LOCAL OCCURRENCE	SPECIES	SPECIAL HABITAT FEATURES	SEASONAL USE
	Orchard Oriole *Icterus spurius*	Open country with scattered deciduous trees.	B BF W WF
	Baltimore Oriole *Icterus galbula*	Tall, scattered deciduous trees (preferably elm).	B BF W WF
	Pine Grosbeak *Pinicola enucleator*	Northern conifer forest.	B BF W WF
	Purple Finch *Carpodacus purpureus*	Coniferous trees, old fields, n. pastures invaded by red spruce, Christmas tree plantations.	B BF W WF
	House Finch *Carpodacus mexicanus*	Open ground with low seed-producing plants, residential areas, farms.	B BF W WF
	Red Crossbill *Loxia curvirostra*	Northern conifer forest.	B BF W WF
	White-winged Crossbill *Loxia leucoptera*	Northern conifer forest.	B BF W WF
	Common Redpoll *Carduelis flammea*	Open country (winter).	B BF W WF

TERRESTRIAL — Upland Field: Cultivated, Grass, Forb, Shrub/old field, Pasture; Savanna, Orchard, Krummholz, Alpine

WETLAND/DEEP WATER — Palustrine: Sedge meadow, Shallow marsh, Deep marsh, Shrub swamp, Bog, Pond; Lake; Riverine: Stream, River, Riparian; Estuary/salt marsh; Marine: Coastal beach/rocks, Bay/ocean

OTHER — Stable bank, Ledge, cliff, Cave, mine, Structure, Derelict bldg., debris

Species	Habitat	Season
Hoary Redpoll *Carduelis hornemanni*	Open country (winter).	B BF W WF
Pine Siskin *Carduelis pinus*	Coniferous trees.	B BF W WF
American Goldfinch *Carduelis tristis*	Open, weedy fields and marshes with scattered small trees.	B BF W WF
Evening Grosbeak *Coccothraustes vespertinus*	Spruce and fir forest (breeding season).	B BF W WF
House Sparrow *Passer domesticus*	Open areas around human habitation. Avoids heavily forested areas.	B BF W WF
Virginia Opossum *Didelphis virginiana*	Hollow log or tree cavity.	B BF W WF
Masked Shrew *Sorex cinereus*	Damp woodlands, ground cover.	B BF W WF
Water Shrew *Sorex palustris*	Herbaceous cover, cold-water wetlands.	B BF W WF
Smoky Shrew *Sorex fumeus*	Loose, damp leaf litter.	B BF W WF
Long-tailed Shrew *Sorex dispar*	Rocky, wooded sites.	B BF W WF
Pygmy Shrew *Sorex hoyi*	Moist leaf mold near water.	B BF W WF

Nonforested Matrix (*continued*)

LOCAL OCCURRENCE	SPECIES	SPECIAL HABITAT FEATURES	SEASONAL USE
	Northern Short-tailed Shrew *Blarina brevicauda*	Low vegetation, damp, loose leaf litter.	B BF W WF
	Least Shrew *Cryptotis parva*	Loose soil.	B BF W WF
	Hairy-tailed Mole *Parascalops breweri*	Loose, moist, well-drained soil.	B BF W WF
	Eastern Mole *Scalopus aquaticus*	Soft, moist soils containing earthworms.	B BF W WF
	Star-nosed Mole *Condylura cristata*	Wet muck, humus.	B BF W WF
	Little Brown Myotis *Myotis lucifugus*	Females: dark, warm sites for maternity colonies. Forest openings for feeding.	B BF W WF
	Northern Long-eared Bat *Myotis septentrionalis*	Large cavity trees for roosting. Caves, mine shafts for hibernation near 40° F (2–7° C); high humidity and calm air.	B BF W WF
	Indiana Myotis *Myotis sodalis*	Tree cavities and loose bark for roosting. Cool stable temperatures in caves (hibernacula).	B BF W WF

Habitat columns — TERRESTRIAL: Upland Field (Cultivated, Grass, Forb, Shrub/old field, Pasture), Savanna, Orchard, Krummholz, Alpine; WETLAND/DEEP WATER: Palustrine (Sedge meadow, Shallow marsh, Deep marsh, Shrub swamp, Bog, Pond), Lake, Riverine (Stream, River, Riparian), Estuary/salt marsh, Marine (Coastal beach/rocks, Bay/ocean); OTHER: Stable bank, Ledge, cliff, Cave, mine, Structure, Derelict bldg., debris

	Species	Habitat	Season
	Eastern Small-footed Myotis *Myotis leibii*	Caves and mine shafts for hibernacula.	B BF W WF
	Silver-haired Bat *Lasionycteris noctivagans*	Dead trees with loose bark or cavities; wetlands nearby for foraging.	B BF W WF
	Eastern Pipistrelle *Pipistrellus subflavus*	Warm, draft-free damp hibernacula; open woods for foraging.	B BF W WF
	Big Brown Bat *Eptesicus fuscus*	Cold, dry caves or buildings for hibernacula.	B BF W WF
	Red Bat *Lasiurus borealis*	Deciduous trees on forest edges for roosting.	B BF W WF
	Hoary Bat *Lasiurus cinereus*	Roosts in coniferous and deciduous foliage.	B BF W WF
	Eastern Cottontail *Sylvilagus floridanus*	Brush piles, stone walls, dens or burrows; herbaceous and shrubby cover.	B BF W WF
	New England Cottontail *Sylvilagus transitionalis*	Closely spaced patches of young dense woody cover.	B BF W WF
	Snowshoe Hare *Lepus americanus*	Dense brushy or softwood cover.	B BF W WF
	European Hare *Lepus europaeus*	Fields, meadows.	B BF W WF
	Black-tailed Jackrabbit *Lepus californicus*	Sand dunes; adjacent upland openings.	B BF W WF

Nonforested Matrix *(continued)*

LOCAL OCCURRENCE	SPECIES	SPECIAL HABITAT FEATURES	SEASONAL USE	TERRESTRIAL									WETLAND / DEEP WATER													OTHER				
				Upland Field									Palustrine							Riverine				Marine						
				Cultivated	Grass	Forb	Shrub/old field	Pasture	**Savanna**	**Orchard**	**Krummholz**	**Alpine**	Sedge meadow	Shallow marsh	Deep marsh	Shrub swamp	Bog	Pond	**Lake**	Stream	River	Riparian	**Estuary/salt marsh**	Coastal beach/rocks	Bay/ocean	**Stable bank**	**Ledge, cliff**	**Cave, mine**	**Structure**	**Derelict bldg., debris**
	Eastern Chipmunk *Tamias striatus*	Tree or shrub cover; elevated perches, decaying stumps/logs, stone walls.	B		□		□				□																			
			BF		□		□				□																			
			W		□		□				□																			
			WF																											
	Woodchuck *Marmota monax*	Open land with well-drained soils in which to burrow.	B	□	■	■	□	■	■	■																■				■
			BF	□	■	■	□	■	■	■																				
			W	□	■	■	□	■	■	■																■				■
			WF																											
	Gray Squirrel *Sciurus carolinensis*	Mast-producing trees; tall trees for dens and leaf nests.	B																			□								
			BF																			□								
			W																			□								
			WF																			□								
	Red Squirrel *Tamiasciurus hudsonicus*	Woodlands with mature trees; conifers preferred.	B								□																			
			BF								□																			
			W								□																			
			WF								□																			
	Southern Flying Squirrel *Glaucomys volans*	Mature woodland with cavity trees; favors cavities with entrance diameters of 1.5–2 in (4–5 cm).	B							□																				□
			BF							□																				
			W							□																				□
			WF							□																				
	Northern Flying Squirrel *Glaucomys sabrinus*	Mature trees, cavities for winter dens; arboreal lichens.	B																											□
			BF																											
			W																											□
			WF																											
	Beaver *Castor canadensis*	Low-gradient woodland streams with adjacent young hardwoods.	B															■	■	■	■	■								
			BF											□	□	□	□	■	■	■	■	■								
			W															■	■	■	■	■								
			WF											□	□	□	□	■	■	■	■	■								
	Deer Mouse *Peromyscus maniculatus*	Down logs, rotting stumps.	B				□				□	□																	■	■
			BF				□				□	□																	■	■
			W				□				□	□																	■	■
			WF				□				□	□																	■	■

White-footed Mouse *Peromyscus leucopus*	Down logs, rotting stumps, cavities.	B BF W WF
Southern Red-backed Vole *Clethrionomys gapperi*	Springs, brooks, seeps, bogs; debris or slash cover.	B BF W WF
Meadow Vole *Microtus pennsylvanicus*	Herbaceous vegetation; loose soils.	B BF W WF
Rock Vole *Microtus chrotorrhinus*	Cool moist, rocky woodlands with herbaceous ground cover and flowing water.	B BF W WF
Woodland Vole *Microtus pinetorum*	Moist, well-drained soils; variable depths of leaf litter or grass cover.	B BF W WF
Muskrat *Ondatra zibethicus*	Wetlands with dense emergent vegetation; stable water level.	B BF W WF
Southern Bog Lemming *Synaptomys cooperi*	Moist soils; grass-sedge ground cover.	B BF W WF
Northern Bog Lemming *Synaptomys borealis*	Moist to wet loose soils; decaying leaf litter or herbaceous ground cover.	B BF W WF
Norway Rat *Rattus norvegicus*	Buildings, dumps or loose soil for digging burrows amidst human habitation.	B BF W WF
House Mouse *Mus musculus*	Buildings in winter; around human habitation.	B BF W WF
Meadow Jumping Mouse *Zapus hudsonius*	Open wooded stands; herbaceous ground cover; loose soils. Hibernates in winter.	B BF W WF

Nonforested Matrix (*continued*)

LOCAL OCCURRENCE	SPECIES	SPECIAL HABITAT FEATURES	SEASONAL USE
	Woodland Jumping Mouse *Napaeozapus insignis*	Moist cool woodlands; loose soils, herbaceous ground cover. Hibernates in winter.	B BF W WF
	Porcupine *Erethizon dorsatum*	Rock ledges or den trees.	B BF W WF
	Coyote *Canis latrans*	Well-drained secluded den sites.	B BF W WF
	Gray Wolf *Canis lupus*	Seclusion; abundant prey base. Potential habitat in north.	B BF W WF
	Red Fox *Vulpes vulpes*	Well-drained den sites. Tends to hunt more open or semiopen habitats.	B BF W WF
	Gray Fox *Urocyon cinereoargenteus*	Hollow logs, tree cavities, rock crevices. Tends to hunt forest edges.	B BF W WF
	Black Bear *Ursus americanus*	Dens in any semiprotected areas (fallen trees, talus slopes, slash piles); seeps and wet areas in early spring; mast.	B BF W WF
	Raccoon *Procyon lotor*	Hollow trees, dens usually ≥10 ft (3 m) above ground.	B BF W WF

TERRESTRIAL — Upland Field: Cultivated, Grass, Forb, Shrub/old field, Pasture; **Savanna**, **Orchard**, **Krummholz**, **Alpine**

WETLAND/DEEP WATER — Palustrine: Sedge meadow, Shallow marsh, Deep marsh, Shrub swamp, Bog, Pond; **Lake**; Riverine: Stream, River, Riparian; **Estuary/salt marsh**; Marine: Coastal beach/rocks, Bay/ocean

OTHER — **Stable bank**, **Ledge, cliff**, **Cave, mine**, **Structure**, **Derelict bldg., debris**

Species	Habitat	Region
American Marten *Martes americana*	Hollow trees, logs; coarse woody debris, abandoned burrows.	B BF W WF
Fisher *Martes pennanti*	Hollow trees, logs; dense regenerating softwoods, abandoned burrows.	B BF W WF
Ermine *Mustela erminea*	Dense brushy cover, slash. Areas of abundant prey.	B BF W WF
Long-tailed Weasel *Mustela frenata*	Areas of abundant prey. Previously excavated den sites. Wooded edges.	B BF W WF
Mink *Mustela vison*	Hollow logs, natural cavities, under tree roots. Forest-wetland edges.	B BF W WF
Striped Skunk *Mephitis mephitis*	Well-drained soils for burrows/den sites. Open uplands; around human habitation.	B BF W WF
River Otter *Lontra canadensis*	Water (e.g., stream, pond, lake or river); bank dens.	B BF W WF
Mountain Lion *Puma concolor*	Seclusion; abundant prey base (e.g., deer).	B BF W WF
Lynx *Lynx canadensis*	Abundant prey base (e.g., snowshoe hare). Dense softwood regeneration; secluded den sites.	B BF W WF
Bobcat *Lynx rufus*	Dense woody understory; rock ledges and talus; successional habitats with abundant prey base (e.g. lagomorphs).	B BF W WF
White-tailed Deer *Odocoileus virginianus*	Softwood yarding cover in north; adequate winter browse; summer herbaceous forage and mast.	B BF W WF

Nonforested Matrix (*continued*)

LOCAL OCCURRENCE	SPECIES	SPECIAL HABITAT FEATURES	SEASONAL USE	TERRESTRIAL: Upland Field: Cultivated	Grass	Forb	Shrub/old field	Pasture	TERRESTRIAL: Savanna	Orchard	Krummholz	Alpine	WETLAND/DEEP WATER: Palustrine: Sedge meadow	Shallow marsh	Deep marsh	Shrub swamp	Bog	Pond	Lake	Riverine: Stream	River	Riparian	Estuary/salt marsh	Marine: Coastal beach/rocks	Bay/ocean	OTHER: Stable bank	Ledge, cliff	Cave, mine	Structure	Derelict bldg., debris
	Moose *Alces alces*	Wetlands (in summer).	B													░														
			BF				░						░	░	▓	░	░	▓	▓	▓	▓	▓								
			W				░				▓					░														
			WF				░				▓					░														

Appendices

Appendix A. Special Status Designations

Many wildlife species or subspecies in New England are protected by state or federal legislation, hunting regulations, or have been noted by conservation groups and agencies as species of special concern due to their relative scarcity.

The following is a regional list of vertebrates in New England that are currently designated as either state or federally threatened or endangered. We have included species of special concern that are not formally listed but that have been proposed by various conservation organizations. These listings are based on current federal, state, and natural heritage database postings and numerous agency responses, and include both upland and coastal/marine species.

Vertebrate species receive special status designations for a number of reasons: (a) habitats in which they occur are naturally limited or are at the range limit; (b) habitats in which they occur are in the process of changing, either through succession (e.g., abandoned farmland → forest land) or through increasing human habitation (e.g., abandoned farmland → suburban development); (c) increasing development pressures on shoreline and wetland habitats continue; and (d) species that are infrequently observed or rare in their occurrence and so need additional monitoring to assess their abundance.

If interested in further information regarding a species' status in a particular state, the user should contact the state agency with jurisdiction over wildlife, the Natural Heritage Program of The Nature Conservancy in that state, or a local Audubon Society chapter.

The following are the relevant contacts:

Maine Department of Inland Fisheries & Wildlife
Endangered Species Group
650 State Street
Bangor, ME 04401–5654

New Hampshire Fish and Game Department
Nongame Wildlife Program
2 Hazen Dr.
Concord, NH 03301

Vermont Agency of Natural Resources
Dept. of Fish and Wildlife
103 South Main St.
Waterbury, VT 05671–0501

Massachusetts Division of Fisheries and Wildlife
Route 135
Westborough, MA 01581

Rhode Island Division of Fish, Wildlife and Estuarine Resources
PO Box 218
Kingston, RI 02892–0218

Connecticut Department of Environmental Protection, Wildlife Division
Session Woods Wildlife Management Area
PO Box 1550
Burlington, CT 06013

U.S. Fish and Wildlife Service, Region 5
300 Westgate Center Drive
Hadley, MA 01035–9589

Species regulated by state hunting laws are not included in this list.

Special Status Designations

Species	Last updated:	Maine 6/10/97, 9/25/96	N.H. 7/94	VT. 6/13/96	Mass. undated	Conn. undated	R.I. undated
Amphibians							
Common mudpuppy *Necturus maculosus*				*			
Marbled salamander *Ambystoma opacum*			*	*	ST		
Jefferson salamander *Ambystoma jeffersonianum*			*	SC	SC	SC (complex)	
Blue-spotted salamander *Ambystoma laterale*				SC	SC	SC (complex) ST (diploid)	
Allegheny dusky salamander *Desmognathus ochrophaeus*				SC			
Northern slimy salamander *Plethodon glutinosus*			*			ST	
Four-toed salamander *Hemidactylium scutatum*		SC		SC	SC		
Northern spring salamander *Gyrinophilus p. porphyriticus*		SC			SC	ST	*
Eastern spadefoot *Scaphiopus holbrookii*					ST	SE	ST
Fowler's toad *Bufo fowleri*			*	SC			
Northern leopard frog *Rana pipiens*		SC					*
Western chorus frog *Pseudacris triseriata*				SE			
Reptiles							
Common musk turtle *Sternotherus odoratus*		SC		SC			
Spotted turtle *Clemmys guttata*		ST	*	ST	SC		*
Bog turtle *Clemmys muhlenbergii*					FT/SE	FT/SE	
Wood turtle *Clemmys insculpta*		SC	*	SC	SC	SC	*
Eastern box turtle *Terrapene c. carolina*		SE	*			SC	*
Northern diamondback terrapin *Malaclemys t. terrapin*					ST		SE
Map turtle *Graptemys geographica*				SC			
Plymouth redbelly turtle *Pseudemys rubriventris bangsi*					FE/SE		
Blanding's turtle *Emydoidea blandingii*		SE	*		ST		
Eastern spiny softshell *Apalone spinifera*				ST			
Loggerhead *Caretta caretta*		FT/ST	FT		FT/ST	FT/ST	FT

Codes:

SE: State endangered.
ST: State threatened.
FE: Federally endangered as of 5/22/00.
FT: Federally threatened as of 5/22/00.
SC: State special concern.
SH: State historical.
* Not formally listed but tracked by states and Heritage programs.

Special Status Designations

Species	Last updated:	Maine 6/10/97, 9/25/96	N.H. 7/94	VT. 6/13/96	Mass. undated	Conn. undated	R.I. undated
Green turtle *Chelonia mydas*					FT/ST	FT/ST	FT
Hawksbill *Eretmochelys i. imbricata*					FE/SE		
Atlantic ridley *Lepidochelys kempi*		FE	FE		FE/SE	FE/SE	FE
Leatherback *Dermochelys coriacea*		FE	FE		FE/SE	FE/SE	FE
Five-lined skink *Eumeces fasciatus*				SE		ST	
Northern brown snake *Storeria d. dekayi*		SC					
Eastern ribbon snake *Thamnophis s. sauritus*		SC		*		SC	SC
Eastern hognose snake *Heterodon platirhinos*			*			SC	SC
Eastern worm snake *Carphophis a. amoenus*					ST		*
Northern black racer *Coluber c. constrictor*		SE		SC			
Black rat snake *Elaphe o. obsoleta*				SC	SE		*
Northern copperhead *Agkistrodon contortrix mokeson*					SE		
Timber rattlesnake *Crotalus horridus*			SE	SE	SE	SE	SH
Birds							
Common loon *Gavia immer*			ST	SE	SC	SC	
Pied-billed grebe *Podilymbus podiceps*			SE	SC	SE	SE	SE
Leach's storm-petrel *Oceamodroma leucorhoa*		SC			SE		
Great cormorant *Phalacrocorax carbo*		SC					
Double-crested cormorant *Phalacrocorax auritus*			*				
American bittern *Botaurus lentiginosus*			*	*	SE	SE	SE
Least bittern *Ixobrychus exilis*		SC	*	SC	SE	ST	SC
Great blue heron *Ardea herodias*			*	*			*
Great egret *Ardea albus*						ST	*
Snowy egret *Egretta thula*				*		ST	*
Little blue heron *Egretta caerulea*						SC	*
Cattle egret *Bubulcus ibis*				*		SC	*
Black-crowned night-heron *Nyctanassa nycticorax*		SC	*	*			*
Yellow-crowned night-heron *Nyctanassa violacea*						SC	*

Special Status Designations

Species Last updated:	Maine 6/10/97, 9/25/96	N.H. 7/94	VT. 6/13/96	Mass. undated	Conn. undated	R.I. undated
Glossy ibis *Plegadis falcinellus*					SC	*
Gadwall *Anas strepera*					SC	*
Green-winged teal *Anas crecca*		*				SC
Blue-winged teal *Anas discors*		*			ST	*
Ring-necked duck *Aythya collaris*		*				
Common eider *Somateria mollissima*		*				
Harlequin duck *Histrionicus histrionicus*	ST					
Common goldeneye *Bucephala clangula*		*				
Barrow's goldeneye *Bucephala islandica*	SC					
Hooded merganser *Lophodytes cucullatus*						SC
Turkey vulture *Cathartes aura*			*			
Osprey *Pandion haliaetus*		ST	SE			
Bald eagle *Haliaeetus leucocephalus*	FT/ST	FT/SE	FT/SE	FT/SE	FT/SE	FT
Northern harrier *Circus cyaneus*		ST	SC	ST	SE	SE
Sharp-shinned hawk *Accipiter striatus*				SC	SE	SH
Cooper's hawk *Accipiter cooperii*	SC	ST	SC	SC	ST	*
Northern goshawk *Accipiter gentilis*	SC	*				*
Red-shouldered hawk *Buteo lineatus*		*			SC	
Golden eagle *Aquila chrysaetos*	SE	SE				
American kestrel *Falco sparverius*					SC	
Merlin *Falco columbarius*		*				
Peregrine falcon *Falco peregrinus*	SE	SE	SE	SE	SE	
Spruce grouse *Falcipennis canadensis*		*	SE			
Northern bobwhite *Colinus virginianus*		*				
Yellow rail *Coturnicops noveboracensis*	SC					
Black rail *Laterallus jamaicensis*					SE	
Clapper rail *Rallus longirostris*						*
King rail *Rallus elegans*		*		ST	SE	*

Special Status Designations

Species	Last updated:	Maine 6/10/97, 9/25/96	N.H. 7/94	VT. 6/13/96	Mass. undated	Conn. undated	R.I. undated
Virginia rail *Rallus limicola*			*				
Sora *Porzana carolina*			*	SC			*
Common moorhen *Gallinula chloropus*		SC	*	*	SC	SE	*
American coot *Fulica americana*		SC					
Piping plover *Charadrius melodus*		FT/SE	FT/SE		FT/ST	FT/ST	FT
American oystercatcher *Haematopus palliatus*						SC	*
Willet *Catoptrophorus semipalmatus*			*			SC	SC
Upland sandpiper *Bartramia longicauda*		ST	SE	ST	SE	SE	SE
Eskimo curlew *Numenius borealis*		FE			FE/SE	FE/SC	
Whimbrel *Numenius phaeopus*		SC					
Red-necked phalarope *Phalaropus lobatus*		SC					
Laughing gull *Larus atricilla*		SC					
Ring-billed gull *Larus delawarensis*			*				
Great black-backed gull *Larus marinus*			*	*			
Roseate tern *Sterna dougallii*		FE/SE	FE/ST		FE/SE	FE/SE	FE/SH
Common tern *Sterna hirundo*		SC	SE	SE	SC	SC	
Arctic tern *Sterna paradisaea*		ST	ST		SC		
Least tern *Sterna antillarum*		SE	ST		SC	ST	ST
Black tern *Chlidonias niger*		SE		ST			
Razorbill *Alca torda*		ST					
Black guillemot *Cepphus grylle*			*				
Atlantic puffin *Fratercula arctica*		ST					
Yellow-billed cuckoo *Coccyzus americanus*			*				
Common barn-owl *Tyto alba*				SC	SC	SE	SE
Eastern screech-owl *Otus asio*		SC	*				
Long-eared owl *Asio otus*			*	SC	SC	SE	SC
Short-eared owl *Asio flammeus*		SC		SC	SE	ST	
Northern saw-whet owl *Aegolius acadicus*						SC	SC

Special Status Designations

Species	Last updated:	Maine 6/10/97, 9/25/96	N.H. 7/94	VT. 6/13/96	Mass. undated	Conn. undated	R.I. undated
Common nighthawk *Chordeiles minor*			ST	SC		ST	SC
Whip-poor-will *Caprimulgus vociferus*			*	SC		SC	
Red-headed woodpecker *Melanerpes erythrocephalus*			*	SC		SE	SC
Three-toed woodpecker *Picoides tridactylus*		SC	*	SC			
Black-backed woodpecker *Picoides arcticus*			*	SC			
Pileated woodpecker *Dryocopus pileatus*							*
Olive-sided flycatcher *Contopus cooperi*		SC				SC	
Acadian flycatcher *Empidonax virescens*			*				*
Alder flycatcher *Empidonax alnorum*						SC	
Horned lark *Eremophila alpestris*			*			ST	*
Cliff swallow *Hirundo pyrrhonota*							SH
Purple martin *Progne subis*			ST			SC	
Gray jay *Perisoreus canadensis*			*	SC			
Fish crow *Corvus ossifragus*			*				SC
Common raven *Corvus corax*						SC	
Carolina wren *Thryothorus ludovicianus*				*			
Winter wren *Troglodytes troglodytes*							SC
Sedge wren *Cistothorus platensis*		SE	SE	ST	SE	SE	SC
Marsh wren *Cistothorus palustris*			*				
Blue-gray gnatcatcher *Polioptila caerulea*			*	*			
Golden-crowned kinglet *Regulus satrapa*							SC
Bicknell's thrush *Catharus bicknelli*		SC		SC			
Brown thrasher *Toxostoma rufum*						SC	
American pipit *Anthus rubescens*		SE	*				
Loggerhead shrike *Lanius ludovicianus migrans*		SC	SE	SE	SE		
Philadelphia vireo *Vireo philadelphicus*				*			
Blue-winged warbler *Vermivora pinus*			*	*			
Golden-winged warbler *Vermivora chrysoptera*			*	*	SE	ST	SH

Special Status Designations

Species	Last updated:	Maine 6/10/97, 9/25/96	N.H. 7/94	VT. 6/13/96	Mass. undated	Conn. undated	R.I. undated
Tennessee warbler *Vermivora peregrina*			*	*			
Northern parula *Parula americana*					ST	SC	ST
Cape May warbler *Dendroica tigrina*			*	*			
Black-throated blue warbler *Dendroica caerulescens*							SE
Blackburnian warbler *Dendroica fusca*							ST
Pine warbler *Dendroica pinus*				*			
Prairie warbler *Dendroica discolor*				*			
Palm warbler *Dendroica palmarum*			*				
Bay-breasted warbler *Dendroica castanea*				*			
Blackpoll warbler *Dendroica striata*					SC		
Cerulean warbler *Dendroica cerulea*			*	SC			ST
Prothonotary warbler *Protonotaria citrea*							*
Worm-eating warbler *Helmitheros vermivorus*							SC
Connecticut warbler *Oporornis agilis*			*				
Mourning warbler *Oporornis philadelphia*					SC		
Wilson's warbler *Wilsonia pusilla*			*	SC			
Yellow-breasted chat *Icteria virens*						SE	SH
Vesper sparrow *Pooecetes gramineus*		SC	*	SC	ST	SE	SH
Savannah sparrow *Passerculus sandwichensis*			*			SC	
Grasshopper sparrow *Ammodramus savannarum*		SE	*	SC	ST	SE	ST
Henslow's sparrow *Ammodramus henslowii*			SE	SE	SE	SC	SH
Sharp-tailed sparrow *Ammodramus caudacutus*			*			SC	
Seaside Sparrow *Ammodramus maritimus*			*			SC	SC
White-throated sparrow *Zonotrichia albicollis*							SC
Dark-eyed junco *Junco hyemalis*							SC
Eastern meadowlark *Sturnella magna*		SC	*			SC	
Rusty blackbird *Euphagus carolinus*		SC		SC			
Orchard oriole *Icterus spurius*		SC	*	*			SC

Special Status Designations

Species	Last updated:	Maine 6/10/97, 9/25/96	N.H. 7/94	VT. 6/13/96	Mass. undated	Conn. undated	R.I. undated
Red crossbill *Loxia curvirostra*			*				
White-winged crossbill *Loxia leucoptera*			*				
Mammals							
Water shrew *Sorex palustris*				*	SC		*
Smoky shrew *Sorex fumeus*							SC
Long-tailed shrew *Sorex dispar*		SC		SC	SC		
Pygmy shrew *Sorex hoyi*				*			
Least shrew *Cryptotis parva*						SE	
Little brown myotis *Myotis lucifugus*		SC					
Northern long-eared myotis *Myotis septentrionalis*		SC		*			
Indiana myotis *Myotis sodalis*			FE/*	FE/SE	FE/SE	FE/SE	
Eastern small-footed bat *Myotis leibii*		SC	SE	ST	SC	SC	
Silver-haired bat *Lasionycteris noctivagans*		SC		*		SC	
Eastern pipistrelle *Pipistrellus subflavus*		SC	*	*			
Big brown bat *Eptesicus fuscus*		SC					
Red bat *Lasiurus borealis*		SC				SC	
Hoary bat *Lasiurus cinereus*		SC				SC	
New England cottontail *Sylvilagus transitionalis*		SC	*	SC			SC
Southern flying squirrel *Glaucomys volans*		SC					
Eastern woodrat *Neotoma floridana*						SC	
Penobscot meadow vole *Microtus pennsylvanicus shattucki*		SC					
Block Island meadow vole *Microtus pennsylvanicus provectus*							SC
Rock vole *Microtus chrotorrhinus*		SC		SC			
Woodland vole *Microtus pinetorum*				*			
Southern bog lemming *Synaptomys cooperi*				*	SC	SC	SC
Northern bog lemming *Synaptomys borealis*		ST	*				
Gray seal *Halichoerus grypus*					SC	SC	
Harbor seal *Phoca vitulina*			*				

Special Status Designations

Species	Last updated:	Maine 6/10/97, 9/25/96	N.H. 7/94	VT. 6/13/96	Mass. undated	Conn. undated	R.I. undated
Finback whale *Balaenoptera physalus*		FE	FE		FE/SE	FE	FE
Sei whale *Balaenoptera borealis*		FE	FE		FE/SE	FE	FE
Blue whale *Balaenoptera musculus*					FE/SE	FE	FE
Humpback whale *Megaptera novaeangliae*		FE	FE		SE	FE	FE
Northern right whale *Balaena glacialis*		FE	FE		SE	FE	FE
Harbor porpoise *Phocoena phocoena*						SC	
Sperm whale *Physeter catodon*		FE	FE		FE/SE	FE	FE
American marten *Martes americana*			ST	SE			
Fisher *Martes pennanti*							*
Gray wolf *Canis lupus*		FE	FE	FE	FE	FE	FE
Lynx *Lynx canadensis*		FT/SC	FT/SE	FT/SE			
Bobcat *Lynx rufus*							ST
Mountain lion *Puma concolor couguar*		FE	FE/*	FE/SE	FE	FE/SH	FE

Appendix B. New England Rare Terrestrial, Coastal, and Migratory Species

Species	Habitat	Season/comment
Amphibians		
Allegheny dusky salamander *Desmognathus ochrophaeus*	Reported from one location in Vt.	Resident/combined
Western chorus frog *Pseudacris triseriata*	Endangered status in Vt.	Resident/not heard in last 10 yrs
Reptiles		
Northern diamondback terrapin *Malaclemys t. terrapin*	Coastal marshes	Resident
Red-eared slider *Trachemys scripta elegans*	Ponds, drainage ditches, and shallow edges of lakes	Introduced resident/does not breed successfully
Birds		
Red-throated loon *Gavia stellata*	Coastal bays	Winter
Horned grebe *Podiceps auritus*	Coastal bays	Winter
Red-necked grebe *Podiceps grisegena*	Coastal bays	Winter
Leach's storm-petrel *Oceanodroma leucorhoa*	Coastal rocks	Spring, summer
Great cormorant *Phalacrocorax carbo*	Coastal rocks	Permanent resident
Tricolored heron *Egretta tricolor*	Coastal marsh	Spring, summer
Snow goose *Chen caerulescens*	Coastal bays, estuaries	Winter
Brant *Branta bernicla*	Coastal bays	Winter
Redhead *Aythya americana*	Coastal bays (inland on rare occasions)	Winter
Greater scaup *Aythya marila*	Coastal bays, estuaries	Winter
Lesser scaup *Aythya affinis*	Coastal bays, estuaries	Winter
King eider *Somateria spectabilis*	Coastal bays, ocean	Winter
Common eider *Somateria mollissima*	Coastal bays, ocean	Breeding, winter
Harlequin duck *Histrionicus histrionicus*	Coastal bays	Winter
Surf scoter *Melanitta perspicillata*	Coastal bays, ocean	Winter

New England Rare Terrestrial, Coastal, and Migratory Species (*continued*)

Species	Habitat	Season/comment
White-winged scoter *Melanitta fusca*	Coastal bays, ocean	Winter
Black scoter *Melanitta nigra*	Coastal bays, ocean	Winter
Oldsquaw *Clangula hyemalis*	Coastal bays, ocean	Winter
Barrow's goldeneye *Bucephala islandica*	Coastal bays, ocean	Winter
Ruddy duck *Oxyura jamaicensis*	Coastal marshes, bays	Rare breeder and in winter, common spring migrant
Clapper rail *Rallus longirostris*	Coastal marsh	Rare breeder and in winter
Black-bellied plover *Pluvialis squatarola*	Coastal mud flat	Summer nonbreeder, rare in winter
American oystercatcher *Haematopus palliatus*	Coastal beach, rocks	Summer
Greater yellowlegs *Tringa melanoleuca*	Fresh and coastal marshes	Spring and fall migrant, rare in winter
Lesser yellowlegs *Tringa flavipes*	Fresh and coastal marshes	Spring and fall migrant
Solitary sandpiper *Tringa solitaria*	Fresh marshes	Spring and fall migrant
Whimbrel *Numenius phaeopus*	Coastal mud flat	Uncommon spring, locally common fall migrant
Hudsonian godwit *Limosa haemastica*	Coastal mud flat	Rare spring, locally common fall migrant
Marbled godwit *Limosa fedoa*	Coastal mud flat	Rare spring, uncommon fall migrant
Ruddy turnstone *Arenaria interpres*	Coastal sand and pebble beach	Uncommon in winter
Red knot *Calidris canutus*	Coastal mud flat	Common fall migrant, occasional winter resident
Sanderling *Calidris alba*	Ocean beaches	Common migrant and winter resident
Semipalmated sandpiper *Calidris pusilla*	Coastal mud flat	Common and fall migrant
Least sandpiper *Calidris minutilla*	Marshes, mud flats	Common spring and fall migrant, inland and along the coast
White-rumped sandpiper *Calidris fuscicollis*	Coastal marsh, mud flat	Uncommon spring, common fall migrant
Baird's sandpiper *Calidris bairdii*	Coastal marsh, mud flat	Uncommon fall migrant
Pectoral sandpiper *Calidris melanotos*	Coastal marsh, mud flat	Uncommon spring, common fall migrant
Purple sandpiper *Calidris maritima*	Coastal rocks	Locally common winter resident
Dunlin *Calidris alpina*	Coastal beach	Abundant spring and fall migrant, local winter resident
Stilt sandpiper *Calidris himantopus*	Coastal marsh, mud flat	Rare spring, uncommon fall migrant
Short-billed dowitcher *Limnodromus griseus*	Coastal marsh, mud flat	Common spring, abundant fall migrant
Red-necked phalarope *Phalaropus lobatus*	Coastal marsh, estuary, ocean	Common offshore migrant
Red phalarope *Phalaropus fulicaria*	Coastal marsh, estuary, ocean	Abundant offshore migrant
Laughing gull *Larus atricilla*	Coast	Local breeding, abundant fall migrant

New England Rare Terrestrial, Coastal, and Migratory Species (*continued*)

Species	Habitat	Season / comment
Bonaparte's gull *Larus philadelphia*	Coast	Common to abundant spring and fall migrant and winter resident
Iceland gull *Larus glaucoides*	Coast	Uncommon winter resident
Glaucous gull *Larus hyperboreus*	Coast	Uncommon winter resident
Roseate tern *Sterna dougallii*	Coastal beach	Locally common breeder, abundant migrant
Arctic tern *Sterna paradisaea*	Coastal beach	Local breeder, uncommon migrant
Least tern *Sterna antillarum*	Coastal beach	Locally common breeder
Black skimmer *Rynchops niger*	Coastal beach	Rare breeder and spring/summer resident
Razorbill *Alca torda*	Coastal rocks	Breeder/winter resident
Black guillemot *Cepphus grylle*	Coastal rocks	Breeder/winter resident
Atlantic puffin *Fratercula arctica*	Coastal rocks	Breeder/winter resident
Gray-cheeked thrush *Catharus minimus*	Coniferous forests	Very uncommon spring and fall migrant
Orange-crowned warbler *Vermivora celata*	Woodlands	Rare spring, uncommon fall migrant
Prothonotary warbler *Protonotaria citrea*	Wooded swamps in s. New England	Exceedingly rare breeder
Saltmarsh sharp-tailed sparrow *Ammodramus caudacutus*	Coastal marsh	Locally common breeder, rare in winter resident
Seaside sparrow *Ammodramus maritimus*	Coastal marsh	Very local breeder, uncommon migrant, rare in winter
White-crowned sparrow *Zonotrichia leucophrys*	Field edges	Uncommon spring, common fall migrant, rare in winter
Dickcissel *Spiza americana*	Coastal marsh	Rare spring, uncommon fall migrant, rare in winter resident
Mammals		
European rabbit *Oryctolagus cuniculus*	Boston Harbor islands	Introduced resident
Beach vole *Microtus breweri*	Moist grassy beach sites on Muskeget Island only	Resident
Gray seal *Halichoerus grypus*	Secluded rocky ledges	Resident
Harbor seal *Phoca vitulina*	Coastal bays, estuaries, rocky shores	Resident
Harp seal *Pagophillus groenlandicus*	Coastal waters	Migrant
Hooded seal *Cystophora cristata*	Coastal waters	Migrant

Sources

American Ornithologists' Union. 1998. *Check-list of North American birds.* 7th ed. Washington, D.C.: American Ornithologists' Union. 829 pp.

Collins, J. T. 1997. *Standard common and current scientific names for North American amphibians and reptiles.* 4th ed. Society for the Study of Amphibians and Reptiles Herpetological Circular 25.

Jones, C.; Hoffmann, R. S.; Rice, D. W.; Engstrom, M. D.; Bradley, R. D.; Schmidly, D. J.; Jones, C. A.; Baker, R. J. 1997. Revised checklist of North American mammals north of Mexico, 1997. *Occas. Pap. Mus. Tex. Tech Univ* 173:1–19.

Appendix C. Marine Species Not Covered in Text, Excluding Accidentals

Species	Habitat	Season/comment
Reptiles		
Loggerhead *Caretta caretta*	Pelagic and coastal waters	Late summer/fall
Green turtle *Chelonia mydas*	Pelagic and coastal waters	Late summer/fall
Hawksbill *Eretmochelys i. imbricata*	Pelagic and coastal waters	Late summer/fall
Atlantic ridley *Lepidochelys kempi*	Pelagic and coastal waters	Late summer/fall
Leatherback *Dermochelys coriacea*	Pelagic and coastal waters	Late summer/fall
Birds		
Northern fulmar *Fulmarus glacialis*	Pelagic	—
Cory's shearwater *Calonectris diomedea*	Pelagic	Summer/fall
Greater shearwater *Puffinus gravis*	Pelagic	Summer/fall
Sooty shearwater *Puffinus griseus*	Pelagic	Summer/fall
Manx shearwater *Puffinus puffinus*	Pelagic	Summer/fall
Wilson's storm-petrel *Oceanites oceanicus*	Pelagic	Summer/fall
Northern gannet *Morus bassanus*	Pelagic	Summer/fall migrant
Great skua *Catharacta skua*	Pelagic	Summer
Pomarine jaeger *Stercorarius pomarinus*	Pelagic	Summer/fall migrant
Parasitic jaeger *Stercorarius parasiticus*	Pelagic	Summer/fall migrant
Long-tailed jaeger *Stercorarius longicaudus*	Pelagic	Summer/fall migrant
Black-legged kittiwake *Rissa tridactyla*	Pelagic and coastal waters	Winter
Dovekie *Alle alle*	Pelagic and coastal waters	Winter
Common murre *Uria aalge*	Pelagic and coastal waters	Winter
Thick-billed murre *Uria lomvia*	Pelagic and coastal waters	Winter

Marine Species Not Covered in Text, Excluding Accidentals (*continued*)

Species	Habitat	Season/comment
Mammals		
Finback whale *Balaenoptera physalus*	Pelagic and coastal waters	Migrant
Sei whale *Balaenoptera borealis*	Pelagic and coastal waters	Migrant
Minke whale *Balaenoptera acutorostrata*	Pelagic and coastal waters	Migrant
Blue whale *Balaenoptera musculus*	Pelagic	Migrant
Humpback whale *Megaptera novaeangliae*	Pelagic	Migrant
Northern right whale *Balaena glacialis*	Pelagic and coastal waters	Migrant
Beluga or White whale *Delphinapterus leucas*	Pelagic and coastal waters	Migrant
Bottle-nosed dolphin *Tursiops truncatus*	Coastal waters	—
Striped dolphin *Stenella coeruleoalba*	Pelagic	—
Common or Saddle-backed dolphin *Delphinus delphis*	Coastal waters	—
Atlantic white-sided dolphin *Lagenorhynchus acutus*	Coastal waters	—
White-beaked dolphin *Lagenorhynchus albirostris*	Coastal waters	—
Grampus or Risso's dolphin *Grampus griseus*	Pelagic	—
Long-finned pilot whale *Globicephala melas*	Pelagic and coastal waters	Migrant
Killer whale *Orcinus orca*	Pelagic and coastal waters	Migrant
Harbor porpoise *Phocoena phocoena*	Coastal bays and inlets	Resident
Goose-beaked whale *Ziphius cavirostris*	Pelagic	—
Northern bottle-nosed whale *Hyperoodon ampullatus*	Pelagic	—
North Atlantic beaked whale *Mesoplodon bidens*	Pelagic and coastal waters	—
Dense-beaked whale *Mesoplodon densirostris*	Pelagic and coastal waters	—
True's beaked whale *Mesoplodon mirus*	Pelagic	—
Pygmy sperm whale *Kogia breviceps*	Pelagic	—
Sperm whale *Physeter macrocephalus*	Pelagic	—

Sources

American Ornithologists' Union. 1998. *Check-list of North American birds.* 7th ed. American Ornithologists' Union, Washington, D.C.: 829 pp.

Collins, J. T. 1997. *Standard common and current scientific names for North American amphibians and reptiles.* 4th ed. Society for the Study of Amphibians and Reptiles Herpetological Circular 25.

Jones, C.; Hoffmann, R. S.; Rice, D. W.; Engstrom, M. D.; Bradley, R. D.; Schmidly, D. J.; Jones, C. A.; Baker, R. J. 1997. Revised checklist of North American mammals north of Mexico, 1997. *Occas. Pap. Mus. Tex. Tech Univ.* 173:1–19.

Tufts, R. W. 1986. *Birds of Nova Scotia.* 3rd ed. Nimbus Publishing Limited and the Nova Scotia Museum, Halifax, Nova Scotia. 478 pp.

Index

Note: Italic page numbers indicate figures; boldface type identifies the main entry for the species.